CHARLES DARWIN'S NATURAL SELECTION

CHARLES DARWIN'S
NATURAL SELECTION

BEING THE SECOND PART OF
HIS BIG SPECIES BOOK
WRITTEN FROM 1856 TO 1858

—————

EDITED FROM MANUSCRIPT BY

R. C. STAUFFER

University of Wisconsin, Madison

CAMBRIDGE UNIVERSITY PRESS

Published by the Syndics of the Cambridge University Press
Bentley House, 200 Euston Road, London NW1 2DB
American Branch: 32 East 57th Street, New York, N.Y.10022

Library of Congress Catalogue Card Number: 72-95406

ISBN: 0 521 20163 2

First published 1975

Printed in Great Britain
at the University Printing House, Cambridge
(Euan Phillips, University Printer)

CONTENTS

A section of manuscript facsimiles is to be found between pages 16 and 17.

To Nora Barlow
On behalf of all who study her grandfather, Charles Darwin

COMMONLY USED SYMBOLS AND
ABBREVIATION CODE

L & L *The Life and Letters of Charles Darwin*, edited by Francis Darwin, 3 vols. (London, 1887)

NY New York edition of *Life and Letters*, 2 vols. (1888)

ML *More Letters of Charles Darwin*, edited by Francis Darwin and A. C. Seward, 2 vols. (London, 1903)

SYMBOLS RELATING TO THE MANUSCRIPT

Reference numbers are those written on the individual folios and other pieces of paper, either given by Darwin or added later to unnumbered pieces

FC Refers to folios of fair copies of parts of the manuscript, where Darwin occasionally made changes from his earlier holographs

( ) Darwin's parentheses

⟨ ⟩ Darwin's cancellations

[ ] Editor's additions to Darwin's text or notes

[?] Uncertain reading

/ Normal end of manuscript folio, note slip, etc.

// End of existing piece of manuscript where Darwin sheared off a portion to be transferred for use elsewhere

.... Unrestorable gap in text

s Seite, Darwin's usage for page citations in German sources

ACKNOWLEDGEMENTS

Acknowledgements are inherently endless and incomplete, and I ask indulgence from the many friendly, helpful institutions and people whom I do not name here.

The dedication to Nora, Lady Barlow represents the great appreciation of many scholars to the whole Darwin family, especially Sir Robin Darwin and to the late Sir Alan Barlow for their preservation of the Darwin papers as an invaluable intact collection generously made available through their gifts to the Cambridge University Library, as well as deep personal gratitude for her friendly help.

Had this work appeared in two volumes, the second would have been dedicated to Dr Sydney Smith of St Catharine's College, Lecturer in Zoology in the University of Cambridge, for his persistent diplomatic and generous assistance to Darwinists and for his personal friendship. Peter Gautrey of the University Library, Cambridge, has been a warm helpful friend in many ways besides his invaluable contribution of careful copy reading of my final edited typescript against the original manuscript.

I am happy to record my indebtedness to my wife, Velma Mekeel Stauffer, specifically for inspired aid in deciphering some of the most illegible words in Darwin's handwriting and for many other hours of partnership in working together on the manuscript.

My colleagues and students of the Department of the History of Science of the Madison University of Wisconsin have supported my work with friendly encouragement and loyal patience. Helpful fellow scholars and friends include Loren Eiseley, John Brooks, William Stearn, Bert J. Loewenberg, Thompson Webb, Jr, Thomas R. Buckman, the late Sir Gavin de Beer, Sten Lindroth, Mr O'Grady of the Linnean Society of London, and Mr Robinson of Down House.

Friendly and indispensable fellow workers on the manuscript include as research assistants Elizabeth Nash, who made the original typed transcript of chapter 11, and Alice Guimond who transcribed all the other chapters except five (which I did myself). M. J. S. Hodge and Albert A. Baker contributed valuable locations of Darwin's exact source and source editions. Graham Pawelec was inspired and inspiring in his editing of the bibliography. The Index and Concordance were complied under the supervision of Sydney Smith (for details see p. 630).

Edna Dahl, in typing the complete final manuscript, was never ruffled by the countless vagaries of my typing and handwriting, nor by the necessary editorial conventions unpredicted in standard secretarial training.

My indebtedness to institutions ranges widely indeed. My understanding of Darwin's field work basic for his research career, is founded on my own field experience supported for two summers by the Minnesota State Geological Survey under my father, Clinton R. Stauffer, and for three summers by the Woods Hole Oceanographic Institution under Alfred C. Redfield.

Background research was aided by library hospitality *savante*, *gemütlich* and *kosmopolit* in the *Bibliothèque Nationale* and the *Muséum National d'Histoire Naturelle* in Paris and the University Libraries in Vienna and Uppsala.

West of the Channel my indebtedness for generous library assistance includes Widener, the Memorial Library of the Madison University of Wisconsin, the British Museum Reading Room and Manuscript Room, the British Museum (Natural History), the London Library, the Linnean Society of London, University College, London, and the Wellcome Institute of the History of Medicine. Then in Cambridge, England, Christ's College, the Balfour Library, the Botany School Library, and of course, happily indispensable, The University Library where I enjoyed borrowing privileges and much other aid thanks to H. R. Creswick, E. B. Ceadel, A. Tillotson, J. Claydon, and most of all P. Gautrey in the Anderson Room.

Financial support of the necessary research and editorial work came from the Wisconsin Alumni Research Foundation through the University Research Committee, travel funds from the American Philosophical Society, the essential fifteen months research grant NSF G-13032 from the National Science Foundation, and an appreciated advance from the University Press, Cambridge. Without additional private support from departed Stauffer and Webb family this work would not have been completed.

Indispensable encouragement and cosmopolitan hospitality came from P. G. Burbidge, A. Winter and the Syndics of the Cambridge University Press and from the University Combination Room and the Senior Common Room of St Catharine's College, Cambridge.

To all, my warm and grateful thanks; and complete absolution for any errors still persisting despite their assistance. Any sins of omission or commission are my own.

ROBERT C. STAUFFER
July 10, 1973

GENERAL INTRODUCTION

On The Origin of Species was literally only an abstract of the manuscript Darwin had originally intended to complete and publish as the formal presentation of his views on evolution. Compared with the *Origin*, his original long manuscript work on Natural Selection, which is presented here, has more abundant examples in illustration of Darwin's argument plus an extensive citation of sources. It had reached a length of over one quarter of a million words and was well over half completed when Darwin's writing was dramatically interrupted by the celebrated letter from the other end of the world outlining Alfred Russel Wallace's astonishingly parallel but independently conceived theory of natural selection. Darwin felt obliged to change his plans for initial publication; and, after the brief preliminary announcement was presented jointly with Wallace's paper at the Linnean Society of London, he rapidly wrote out in eight months the new abstract of his views which appeared as the *Origin of Species* in 1859. But, he still planned to publish a more extensive account of his views on evolution, and he did not abandon his long manuscript, nor write on the unused backs of the sheets for drafting other new publications as he so often did with other manuscripts.[1] As we shall see, the first two chapters of the manuscript became the two volumes of his *Variation of Animals and Plants under Domestication* (1868). The following eight and a half chapters are published here under the title, *Natural Selection*, which Darwin gave to this work in the 1857 letter to Asa Gray published in the preliminary announcement of 1858.[2]

Judging from my own experience in tracing and checking the references for the present work I believe any attempt on Darwin's part to recheck a normal number of references in order to include sources in footnotes for the *Origin* could have added a number of months to the time he needed to prepare his *Origin* manuscript for the copyist and then for the press. Because of the pressure to publish as quickly as possible a fuller statement of Darwin's evolutionary views once the preliminary announcement had been made at the July 1, 1858 meeting of the Linnean Society, the *Origin of Species* was, I believe, unique among Darwin's published

[1] L & L, I, 121; NY, I, 99; e.g. Darwin MSS. 17 (i).
[2] L & L, II, 123; NY, I, 480.

1

books and formal scientific papers in appearing without a single footnote. Therefore the *Origin* tantalizes us with questions as to what were the immediate sources of Darwin's facts and ideas, and Darwin was ahead of his critics in mentioning the desirability of publishing the references for these sources. Already on the second page of the first edition of the *Origin* he stated: 'This Abstract, which I now publish, must necessarily be imperfect. I cannot here give references and authorities for my several statements...No one can feel more sensible than I do of the necessity of hereafter publishing in detail all the facts, with references, on which my conclusions have been grounded; and I hope in a future work to do this.' The first part of this moral obligation to publish his sources Darwin satisfied in 1868 when he published his two volumes on *Variation under Domestication*. For his own selection of his other sources we must examine the present work on Natural Selection. One already published illustration of the use that can be made of this work is the specific confirmation of earlier speculation about the derivation of Darwin's ecological concept of the economy of nature from the Linnean dissertations.[1]

The manuscript proves to support Darwin's comment about it: 'I fear my M.S. for the bigger book...would be illegible,'[2]... partly because of his handwriting and partly because his drastic revisions often obscure the continuity of the text. Transcription of a considerable portion of the manuscript seemed the best way to start studying its content, and I have completed the transcription and editing of the manuscript in order to make the work generally accessible and with the particular hope that it will promote informed analysis of other aspects of Darwin's work.

The first desideratum in introducing the manuscript would seem to be to supply the reader with the background information most useful for understanding the work. Certain historical material seems important here, and I have tried to present a reasonably full account of the immediate history of the manuscript as well as of the editorial procedure followed. Some important details relate specifically to individual chapters, and these will be presented separately before the parts of the text immediately concerned.

My editorial aim has been to take time not only to clarify the lengthy text to make it as readable as possible but also to clarify Darwin's source citations. His abbreviated references, in a form

---

[1] R. C. Stauffer, 'Haeckel, Darwin, and Ecology', *Quart. Rev. Biol.*, 32 (1957), 138–144; '*On the Origin of Species*: an unpublished Version', *Science*, 130 (1959), 1449–52, 'Ecology in the Long Manuscript Version of Darwin's *Origin of Species* and Linnaeus' *Oeconomy of Nature*', *Amer. Phil. Soc., Proc.*, 104 (1960), 235–41.

[2] L & L, II, 281; NY, II, 75.

natural to an unfinished draft, are often somewhat cryptic. Where necessary I have added to Darwin's notes just enough clues to key them clearly to the cumulative bibliography where fuller titles, editions, and dates have been included; and blank citations and important details have been filled in as far as possible from the clues Darwin left in his papers and his notebooks listing titles and dates for the books he had read.

The citation of almost 750 books and articles makes the bibliography of the long manuscript an extensive guide to the sources Darwin selected out of his very comprehensive reading as most valuable for his own purposes. It is valuable as well for the modern student of the evolution of biological thought, whether as scientist or historian; for in giving us his own selective reading list for the preceding century of natural science Darwin has pointed out a pathway offering a representative view of a scientific literature formidably vast for exhaustive examination by any single scholar.

DARWIN'S PAPERS AND LIBRARY AS WORKING MATERIALS

The Natural Selection manuscript not only has a prominent place in a considerable sequence of Darwin papers touching on evolution, but it is related to many more of the notebooks, papers, letters and annotated books, journals, and pamphlets which, together with his memories of his extensive field experience particularly in South America, and his continuing observations and experiments, constituted Darwin's working materials for his writing. The Darwin family, Down House authorities, the British Museum (Natural History), and the University officials at Cambridge have done everything feasible to make these available to scholars. *Nature*[1] published something of the contents of these papers, but scholars must proceed to the invaluable *Handlist of Darwin Papers at the University Library Cambridge* (Cambridge, 1960) for an indispensable survey, which gives a preliminary account and listing of the more than 150 major parts or groups of items in that part of the collection then already at Cambridge. The manuscript material here published from this part of the collection will be identified by the reference numbers published in this Handlist. Since this Handlist was published, important new portions of Darwin's papers have been located and made available in the University

---

[1] 'Darwin Manuscripts and Letters, Gifts to Cambridge and Down House', *Nature*, 150 (1942), 535.

Library Cambridge by Sir Robin Darwin.[1] These latter items will be designated by the reference numbers for the sections of the collection in which they occur which were assigned in the hand-written 'Catalogue of the MSS, papers, letters, and printed books of Charles Darwin now at Gorringes, Downe, Kent, July, 1932', made by Mrs Catherine Ritchie Martineau.

In addition to these notebooks, note sheets and slips, scientific journals, and manuscript drafts included in these papers, we must note the extensive marginalia and note slips in Darwin's scientific library of books, journals, reprints and pamphlets. Those books with significant annotations are now in the University Library Cambridge together with Darwin's reprint collection, on loan from the Botany School by courtesy of the Professor of Botany.[2]

Darwin's papers and library constituted an interrelated set of working materials for Darwin, and when studied together they help reveal the development of his thought as Sydney Smith has elegantly shown.[3] Many of these papers, valuable as background for understanding the present work on natural selection have now been published. Four of the pocket note books filled with evolutionary evidence and ideas from July 1837 to July 1839 have been published by Sir Gavin de Beer.[4] The two early drafts of Darwin's evolutionary argument written in 1842 and 1844 were published together in 1909 by Francis Darwin.[5]

For general background, the *Life and Letters of Charles Darwin* is of prime importance, particularly the chapter, entitled 'The Unfinished Book', which is devoted to the Natural Selection manuscript. Then besides the *More Letters* also published by Francis Darwin, and the complete *Autobiography* published by

[1] Sir Gavin de Beer, M. J. Rowlands, and B. M. Skramovsky, 'Darwin's Notebooks on Transmutation of Species, Part VI', *Br. Mus. nat. Hist. Bull.* (Hist. ser.) 3 (1967), 131–2.

[2] For titles see the mimeographed 'Darwin Library: List of books received in the University Library Cambridge, March–May, 1961'; see also: Cambridge University. Botany School., *Catalogue of the Library of Charles Darwin...*comp. Henry William Rutherford (Cambridge, 1908). Peter Vorzimmer notes in *Isis* 54 (1963), 374, n. 11 that there are also about a quarter of a million words of Darwin's marginalia in his 2500 reprints.

[3] 'The Origin of "The Origin" as discerned from Charles Darwin's Notebooks and his Annotations in the Books he read between 1837 and 1842', *Advmt. Sci.*, London, 16 (1960), 391–401.

[4] *B. M. Bull.* 2 (1960) 25–183; 2 (1961) 185–200; 3 (1967) 129–76; see also 'A Transcription of Darwin's First Notebook on "Transmutation of Species"', ed. Paul H. Barrett, *Mus. Comp. Zool. Harvard, Bull.*, 122 (1960) 245–96.

[5] First published as *The Foundations of the Origin of Species: Two Essays Written in 1842 and 1844* by Charles Darwin. Ed. Francis Darwin, (Cambridge, 1909); republished in *Evolution By Natural Selection*. With a foreword by Sir Gavin de Beer (Cambridge, 1958).

Nora Barlow, we should note the valuable chronological details in the Pocket Diary,[1] kept by Darwin from 1838 to 1881, of which de Beer has published an old copy made by an amanuensis.[2] More recently the long sought original diary has been found, so I have been able to rely upon that.

'MY BIG BOOK': THE NATURAL SELECTION MANUSCRIPT

The more immediate background of what Darwin came to call 'my big book'[3] starts before the middle of the 1850s. In 1853 Darwin's first major scientific honour came to him in the Royal Society's award of the Royal Medal in recognition of his books on the *Geology of the Voyage of the Beagle* and his comprehensive taxonomy of the barnacles.[4] The latter work, which confirmed his position as a professionally qualified biologist,[5] was then near enough to completion so that he could mention to Hooker his expectation to be at work on his 'species book' in a year or two.[6]

The next year he was ready to pack up his barnacle specimens, arrange for distributing copies of his publication; and, with the decks thus clear, he recorded in his Pocket Diary, for September 9, 1854: 'Began sorting notes for Species theory.' In March of 1855 he wrote to his second cousin and close college friend, William Darwin Fox, that 'I am hard at work at my notes collecting and comparing them, in order in some two or three years to write a book with all the facts and arguments, which I can collect, *for and versus* the immutability of species.'[7]

DARWIN'S WORKING NOTES AND PAPERS

These notes on his thoughts and on his extensive reading in the search for relevant facts formed an important part of Darwin's working materials along with his current observations and experi-

---

[1] So designated in: Cambridge. University. *Order of the Proceedings at the Darwin Celebration held at Cambridge June 22–June 24, 1909, with a Sketch of Darwin's Life.* (Cambridge, 1909), p. [13] n. Francis Darwin referred to it as 'Diary' or 'Pocket-Book', e.g. L & L, I, p. iv; ML, I, p. xvii: 'Outline of Charles Darwin's Life based on his Diary, dated August, 1838', and *Foundations*, xiv, xvii.

[2] 'Darwin's Journal', *B. M. Bull.*, 2 (1959), 1–21. Darwin's original diary is with the C.U.L. C.D. MSS. item D 5.

[3] L & L, II, 85; NY, I, 443.

[4] *Roy. Soc. London, Proc.*, 6 (1854), 355–6.

[5] Cf. opinions of Hooker and Huxley, L & L, I, 346–8; NY, I, 314–16.

[6] L & L, II, 41; NY, I, 402.

[7] L & L, II, 46; NY, I, 406; Christ's College (Cambridge) Library, Darwin–Fox letter no. 87.

ments and his memories of his fundamentally important field experience in South America.

Charles Darwin and his son Francis have both described his procedure in regard to his working papers;[1] and examination of the extant manuscripts allows us to understand in some significant detail how his system actually worked. Initially and at least until 1839, Darwin jotted his notes and thoughts on his readings in small bound notebooks, such as those on evolution published by Sir Gavin de Beer. Later, he changed to the opinion of Alphonse de Candolle: 'The essential is to be able to compare, classify, and rearrange the materials up until the definitive writing, without being obliged to tear apart a notebook or to copy and recopy what one has written.'[2] In his *Autobiography*, Darwin explained: 'I keep from thirty to forty large portfolios, in cabinets with labelled shelves, into which I can at once put a detached reference or memorandum' (p. 137). Thus for example his assembled notes and correspondence containing useful facts on the struggle for existence are still together in volume 46(i) of the Darwin Papers at Cambridge. Notes for other chapters of his evolution book are similarly grouped together. Finally he resolved even to select, separate, and sort out the many pages of his early evolution notebooks which had material he might use in his species book. Inside the front cover of Notebook B, the first of them, he wrote: 'All useful pages cut out. Dec. 7/1856.' The other three notebooks also have numerous pages cut out and have Darwin entries such as that inside the front cover of the second: 'All good References selected Dec 13 1856.'[3] Of these selected reference notes, pp. 253–4 from the second notebook were attached to the verso of folio 14 of chapter VII of the Natural Selection manuscript along with other note slips including one stating: 'In Portfolio "Instinct" some excellent facts from Bachman on change of ranges in N. American Birds...' (See Appendix for chapter VII.) Similarly, page five selected and cut out of the third notebook, is also explicitly related to the Natural Selection manuscript by Darwin's pencilled classification: 'Ch IX Mongrels & Hybrids.'[4] As we shall see, Darwin's scientific papers can provide clear identification of some of the references in his manuscript which he abbreviated too drastically to be self-explanatory.

Besides sorting his notes and selected letters into classified

[1] L & L, I, 100, 151–2; NY, I, 80, 127–9; *Autobiography*, 137–8.
[2] *La phytographie*...(Paris, 1880), p. 37; cf. L & L, III, 333; NY, II, 505–6.
[3] de Beer, *B. M. Bull.*, 2 (1960), 41, 82, 128, 160.
[4] de Beer ed., *B. M. Bull.*, 3 (1967), 157.

portfolios, Darwin also compiled useful surveys and an index of his reading in the form of notebooks listing, in roughly dated sequence, short titles of the books of both scientific and non-scientific which he had read.[1] His papers include a long series of abstracts of books, pamphlets, and articles from scientific journals.[2]

WRITING

Would the facts noted from thousands of pages of reading really support the theory of evolution by natural selection Darwin had sketched out in 1842 and developed in the 1844 essay? This question seems to have been in his mind when he wrote Hooker on March 26, 1854: 'How awfully flat I shall feel, if, when I get my notes together on species, &c., &c., the whole thing explodes like an empty puff-ball', and again in 1855: 'I should have less scruple in troubling you if I had any confidence what my work would turn out. Sometimes I think it will be good; at other times I really feel as much ashamed of myself as the author of the *Vestiges* ought to be of himself.'[3]

Robert Chambers' still anonymous *Vestiges of Creation* had indeed stirred up wide discussion of evolution among the reading public, but it had not persuaded experienced scientists that they might need to re-examine their adverse verdict against the mutaability of species.[4] From the point of view of desiring favourable consideration of the strictly scientific value of an evolutionary theory, Darwin could well write 'Lamarck...has done the subject harm, as has Mr. Vestiges.'[5] Darwin could maintain that 'without speculation there is no good and original observation',[6] but he was not interested in abstract theorizing for its own sake. In his mature period, works such as his grandfather's *Zoonomia* could only leave him 'much disappointed, the proportion of speculation being so large to the facts given'.[7]

In 1855 Darwin's tactical problem was clear. Previous discussion of evolution as presented in the works of Lamarck and in the still anonymous *Vestiges of Creation* had only led to its being rejected or ignored by the vast majority of scientists. To win much unprejudiced consideration of his views, Darwin had to succeed where others had failed.

[1] Darwin MSS., items 119, 120, 128.   [2] Darwin MSS., Vols. 71–5, item 116.
[3] L & L, II, 44; NY, I, 404; ML no. 43.
[4] Milton Millhauser, *Just before Darwin: Robert Chambers and Vestiges* (Middletown, Conn., 1959), Chs. 5, 6, esp. pp. 125, 148–9.
[5] L & L, II, 39; NY, I, 399.
[6] L & L, II, 108; NY, I, 399.   [7] *Autobiog.*, p. 49.

Friends such as Charles Lyell and Lyell's brother-in-law Charles J. F. Bunbury, the squire of Mildenhall, certainly deserve credit for encouraging Darwin at this period. In a letter dated April 16, 1856, Bunbury wrote to Darwin: 'I am exceedingly interested by all you tell me about your researches & speculations on species & variation & distribution, & am delighted that you are going on working at the subject. I trust that you will not on any account give up the idea of publishing your views upon it; tho' neither you nor any one else may be able to unravel the whole mystery, or to command the universal assent of naturalists, still the research of one who has studied the whole question so long, & with such extensive knowledge & in so philosophical a spirit, cannot fail to be of very great advantage to science. The whole subject,— I mean every thing connected with the geography of plants & animals, including all the questions of distribution & variation, is to me particularly interesting & delightful; but how much we have yet to learn upon it! The difficulties which appear to attend upon each & every one of the theories,—of specific centres, of multiple creation, & of transmutation,—are so many, that what is most clear to me is the necessity of caution & candour, of avoiding dogmatism, & of giving a fair consideration to every fact & argument on any side. I say this, because the theory to which you lean is the most remote from that to which *I* incline, & yet I am quite ready to admit that your notion *may* be the right one.'[1]

Early in May Darwin was corresponding with Lyell and with Hooker about the former's urgent recommendation that Darwin publish a preliminary sketch of his views on evolution,[2] and in his Pocket Diary Darwin recorded for May 14, 1856: 'Began by Lyell's advice writing Species Sketch.' But the initial doubts Darwin expressed to Hooker about publishing a preliminary announcement of his views without giving supporting evidence grew stronger. Meanwhile his letters began to dwell on problems concerning the geographical distribution of plants and animals. On July 8, 1856, he wrote to Lyell: 'I have just been quoting you in my essay on ice carrying seeds in the S. Hemisphere...

---

[1] Darwin MSS. c. 40. c. The end of this letter with the signature is missing, but the reference to 'my Cape book' and Mildenhall as the place of writing both suggest Bunbury as the writer, and this is confirmed by Darwin's reply, given in the introduction to chapter XI, p. 528f. Bunbury had heard Darwin talk about his evolutionary views as early as Nov. 23, 1845, see *The Life of Sir Charles J. F. Bunbury, Bart.* Edited by his sister-in-law Mrs. Henry Lyell. (London, 1906), I, pp. 213–14. About Bunbury, see also the obituaries by J. D. Hooker, *Roy. Soc. London, Proc.* 46 (1889), xiii–xiv, and by J. W. Judd, *Geol. Soc. London, Proc.* Session 1886–7, pp. 39–40.

[2] L & L, II, 67–71; NY, I, 426–30.

[See *Natural Selection*, chapter XI, folio originally numbered 39 (now 47).]

Hooker, with whom I have formerly discussed the notion of the world or great belts of it having been cooler...I think is much inclined to adopt the idea.—With modification of specific forms it explains some wondrous odd facts in distribution.

But I shall never stop if I get on this subject, on which I have been at work, sometimes in triumph, sometimes in despair, for the last month.'[1]

By mid-July he had so far enlarged his proposed scale of writing as to mention (apparently as already written) forty pages just on the influence of the glacial period on distribution.[2] And this was the scale and scope of the first draft of chapter XI of the present manuscript.

As he soon explained to Lyell, 'I have found it quite impossible to publish any preliminary essay or sketch; but I am doing my work as completely as my present materials allow without waiting to perfect them. And this much acceleration I owe to you.'[3]

Thus Darwin was under way on actually writing his species book. The first chapter on stock breeding and on variation under domestication he left in an imperfect state,[4] but he was sufficiently satisfied with the second chapter to record its completion on October 13, 1856 in his Pocket Diary. In November he wrote to Lyell, 'I am working very steadily at my big book',[5] and as he finished each succeeding chapter or major section he continued to record his progress in his Pocket Diary by noting the dates, which appear at the start of the chapters in this edition.

Darwin not only wrote first drafts of his chapters, but he also revised them, rewrote, reorganized, expanded, and supplemented. On special points he consulted many authorities such as Hooker and Huxley by letter, and even had fair copies made of sections such as those on variations in large and small genera and on geographical distribution to send them to Hooker for his general opinion. Such details of the history of the manuscript will be covered in the appropriate chapter introductions.

By the spring of 1858 Darwin had completed his tenth chapter and had recently finished for chapter IV a major supplement on divergence, when, on June 18, his writing was interrupted by the arrival of Wallace's letter with its sketch of evolutionary processes in terms so surprisingly close to Darwin's own. Darwin's agreement,

---

[1] C.U.L. C.D. MSS.; 146, Lyell letter no. 54.
[2] ML, no. 49 to Hooker, July 13, 1856.
[3] L & L, II, 85: cf. 71, 84; NY, I, 443, cf. 430, 442.
[4] ML no. 84.   [5] L & L, II, 85; NY, I, 443.

following the strong urging by Lyell and Hooker, to present along with Wallace's letter brief selections of his own writings which had been read in previous years by Hooker and Asa Gray is well known.

After the harrowing interval with both scarlet fever and diphtheria spreading from the village of Downe into his own house, with nurses sick as well as children, culminating in the death of his youngest child three days before his paper was presented at the Linnean Society meeting, Darwin started to write a formal article on his views for the Linnean Society. This article by March 1859 had grown into a complete book, well characterized by his proposed title: 'An Abstract of an Essay on the Origin of Species and Varieties through Natural Selection.' In regard to this title, he wrote Lyell: 'I am sorry about Murray [publisher of the *Origin*] objecting to the term Abstract, as I look at it as the only possible apology for *not* giving references and facts in full, but I will defer to him and you.'[1]

When Wallace's letter interrupted Darwin's writing program on June 18, 1858, the long manuscript had covered about two thirds of the topics later presented in the *Origin of Species*. If we estimate the length of the surviving eight and a half chapters of Natural Selection at 225,000 words, and project to the fourteen chapters as in the *Origin*, this would indicate a length of about 375,000 words if the work had been completed. This would have made a book perhaps slightly longer than Murchison's *Silurian System* but certainly shorter than Lyell's *Principles of Geology*, and the scale does not seem inordinate considering the standards of the days of double-decker and triple-decker novels. Of the fourteen chapters of the *Origin*, nine had been preceded by extensive treatment in *Natural Selection*. The table below not only shows the correlation between the two works but also suggests some of the reorganization of the argument in the later work. In comparing the two works we can agree with Darwin's remark to Hooker that writing the *Origin* as an Abstract of his long manuscript 'has clarified my brains very much, by making me weigh the relative importance of the several elements'.[2] Yet in view of the great amount of writing on Natural Selection actually completed and the more than 1,800 pages which Darwin published just in the decade after 1858, the assertion that without the pressure arising from Wallace's 1858 letter Darwin would never have finished his Species Book seems unpersuasive.

In 1859 Darwin presented the first edition of the *Origin* as a preliminary announcement, simply an abstract of his work, stating

---

[1] L & L, II, 153; NY, I, 508.   [2] L & L, II, 138; NY, I, 494.

TABLE. *Comparing Natural Selection and Origin of Species*

Chapter	MSS.	Chapter	1859
I	Variation under Domestication	I	Variation under Domestication
II	Variation under Domestication (cont.)		
III	On possibility of all organisms crossing: on susceptibility to change		Partly in IV
IV	Variation in Nature	II	Variation under Nature
V	Struggle for Existence	III	Struggle for existence
VI	Natural Selection	IV	Natural Selection
VII	Laws of Variation	V	Laws of Variation
VIII	Difficulties in Transitions	VI	Difficulties on Theory
IX	Hybridism & Mongrelism	VIII	Hybridism
X	Instinct	VII	Instinct
	Section on Geographical Distribution	XI	Geographical Distribution

that 'No one can feel more sensible that I do of the necessity of hereafter publishing in detail all the facts, with references, on which my conclusions have been grounded; and I hope in a future work to do this.' and 'My work is now nearly finished; but...it will take me two or three more years to complete it...'[1]

As we have seen, even the Natural Selection manuscript had been for Darwin a condensed form of the presentation he preferred for his material, and he recorded in his Pocket Diary that in January 1860, he 'Began looking over MS for work on Variation.' As he wrote to Asa Gray, this was to be 'the first part forming a separate volume, with index etc. of the three volumes which will make my bigger work'.[2] By June he recorded the completion of the second chapter of the work eventually published in 1868 as *The Variation of Animals and Plants under Domestication*, and he continued to record his writing progress in his Pocket Diary until 1867 when in March he received the first proof. Thus instead of completing the *Natural Selection* manuscript he expanded the scale of his treatment, so that the two volumes on *Variation* represent the first two chapters of *Natural Selection*. He also published material from other parts of *Natural Selection* in *Variation*. There are now folios missing from the surviving Natural Selection manuscript and other folios with part of the text cut away. These gaps can often be related to topics which were treated in both works[3] and it seems evident that he simply incorporated passages

---

[1] *Origin*, pp. 2, [1].   [2] L & L, II, 270; cf. II, 318; NY, II, 64, cf. II, 111.
[3] E.g. *Natural Selection*, ch. 9, fols. 36 v and 40.

from the older manuscript into the new one by transferring what he had already written to save himself recopying.[1] A further such transfer and incorporation of materials on variation from the first two chapters of *Natural Selection* would easily account for the fact that of those initial chapters only one folio (here published in the appendix) has been preserved with the remainder of the manuscript.[2] It also could account for the fact that some few of the pages selected and cut out of the transmutation notebooks seem to be lost permanently. Unfortunately aside from the preliminary draft on Pangenesis practically none of the manuscript of *Variation under Domestication* seems to have survived. Apparently the rest of the two initial chapters of *Natural* Selection were thus used up and discarded.[3]

In 1867, when he finished writing his *Variation under Domestication*, he still considered this as the first part of his big Species Book, which was to be completed with two more works, and he still expected to publish the material covered by the Natural Selection manuscript, which he had so carefully saved and to which he then returned to write addenda.[4] In the introduction after describing the scope of the two volumes to be published in 1868, he announced that the 'problem of the conversion of varieties into species...will form the main subject of my second work'.[5] Here, 'after treating of the Variation of organisms in a state of nature, of the Struggle for Existence and the principle of Natural Selection, I shall discuss the difficulties which are opposed to the theory. These difficulties may be classed under the following heads: the apparent impossibility in some cases of a very simple organ graduating by small steps into a highly perfect organ; the marvellous facts of Instinct; the whole question of Hybridity; and, lastly, the absence, at the present time and in our geological formations, of innumerable links connecting all allied species.'[6]

This prospectus of the 'second work' fits the present manuscript, except that the latter does not include a discussion of missing fossil links. Instead it includes a section on the effects of the ice age as the only completed part of Darwin's fuller discussion of geographical distribution.

This section is the only portion of the manuscript which

---

[1] See *Natural Selection* MS. ch. 9, fol. 136 v where Darwin wrote in regard to a missing note, presumably on a separate slip of paper: 'Note used in Domestic Animals Chapter 15, Crossing.'

[2] Darwin MSS. vol. 51, see Robert C. Olby, 'Charles Darwin's Manuscript of "Pangenesis"', *Brit. J. Hist. Sci.*, 1 (1963), 251–63.

[3] Cf. L & L, I, 121; NY, I, 99. [4] See addendum dated 1867, on fol. 67 of ch. 4.

[5] *Variation*, I, 5. [6] *Variation*, I, 8.

seems to fit best with Darwin's prospectus for the concluding part of his full-scale Species Book: 'In a third work I shall try the principle of natural selection by seeing how far it will give a fair explanation of...several large and independent classes of facts; such as the geological succession of organic beings, their distribution in past and present times, and their mutual affinities and homologies.'[1]

This program, which Darwin outlined in the introduction to *Variation under Domestication*, he never completed. His letter of July 6, 1868, to Alphonse de Candolle explains: 'You ask me when I shall publish on the "Variation of Species in a State of Nature." I have had the MS. for another volume almost ready during several years, but I was so much fatigued by my last book that I determined to amuse myself by publishing a short essay on the "Descent of Man"...Now this essay has branched out into some collateral subjects, and I suppose will take me more than a year to complete. I shall then begin on "Species", but my health makes me a very slow workman.'[2]

For the *Descent of Man* (1st ed. 1871), Darwin again evidently quarried in his Natural Selection manuscript. On folio 13 of the manuscript for chapter x on instinct he scrawled in the margin: "Used Man Book", and the textual gaps created when he sheared off portions of folios 11 and 12 of that chapter can be filled from the corresponding passages in the *Descent*. As Dr Alice Guimond discovered when she was my research assistant, Darwin published more material from the manuscript in 1868 in an article on specific differences in Primula, later incorporated in his book on *The Different Forms of Flowers in Plants of the same Species* (1877) and other material in his book on *The Effects of Cross and Self Fertilisation in the Vegetable Kingdom* (1876).[3]

Besides Darwin's own use of the materials in the Natural Selection manuscript, its history also includes loans of sections to scientist friends, and some authorized posthumous publication. In November, 1859, Huxley had begun to consult Darwin in preparation for the lecture 'On Species and Races, and their Origin', which he gave at the Royal Institution on February 10, 1860,[4] and Darwin soon loaned him the manuscript of chapter ix on hybridism and of his

---

[1] *Variation*, I, 9.

[2] L & L, III, 100; NY, II, 280.

[3] For example, compare his discussion on the primrose and the cowslip on folios 68–79 of ch. IV with his article in *Linn. Soc. J.* (Botany) 10 (1868) 437–54 and with ch. II of *The Different Forms of Flowers*...(London, 1877) and cf. fols. 27, 27 v, and 30 of ch. 3 with *Cross Fertilisation*, pp. 378–9, 395.

[4] 'On Species and Races, and their Origin', *Roy. Inst. G. B., Proc.*, 3 (1860), 195–200.

TABLE. *Stages of Darwin's Organized Writing
on the Origin of Species*

Version	Short Title	Dates of Writing (from Pocket Diary)	Estimated Length
I	1842 Sketch	May, June, 1842	15,000 words
II	1844 Essay	Finished July 5, 1844	52,000 words
III	Natural Selection	July 1856–June 1858	225,000 words extant
IV	Origin of Species	July 1858–March 1859	155,000 words (c. 80,000 in parts corresponding to present text of Natural Selection)
		(6th ed., June–Oct. 1871)	
V	Variation under Domestication	March, 1860– January, 1867	315,000 words

discussion of pigeons, presumably from chapter II.[1] The manuscript materials on instinct which Darwin loaned G. J. Romanes, who published portions of them, will be discussed in the editorial introduction to chapter X. After Darwin's death, his son Francis loaned some of the manuscript to Wallace, and allowed him to publish excerpts, particularly about variation among wild species, in his book on *Darwinism*.[2]

In reviewing the history of Darwin's organized writing on evolution we can see that the Natural Selection manuscript forms part of a sequence of versions which can be summarized in the table above.

[1] L & L, II, 251, 281; NY, 46, 75; ML, nos. 84, 85.
[2] 1st ed. (London, 1889), pp. viii, 46, 69, 79–80. These quotations are from *Natural Selection* ch. IV, fols. 25 to 33.

EDITORIAL CONSIDERATIONS

Since Darwin painstakingly wrote and revised his manuscript with publication in view, the first aim of this edition is to print the book Darwin had in mind.

The text is so long that I believe readability should take precedence over the inclusion of minor details of the manuscript such as insignificant cancellations. For such details, the original manuscript is available in the Anderson Room of the University Library, Cambridge, and a microfilm is available in the library of the University of Wisconsin at Madison. Examination of the accompanying facsimiles of manuscript passages and comparison with the printed text will reveal some of the problems and illustrate the editorial procedure followed.

The first edition of *Variation of Animals and Plants under Domestication* offers a model of format, including the setting of subordinate material in reduced type in the text. Today, however, long footnotes even covering more than a full page of text (e.g. *Variation*, II, pp. 375–6), which did not discourage thousands of Victorian book buyers, now do seem extreme; and in the present work, where long notes could be smoothly incorporated into the main text, this has been done (e.g. chapter III, fol. 64 v).

Occasional gaps occur in the manuscript where Darwin later apparently used passages in preparing his published books. In such cases the continuity has been supplied by quotations on the same subject matter from his other works.[1] For example on folio 105 of chapter VII, shown in the accompanying facsimile, the content of the surviving top and bottom portion of the cut-up manuscript sheet corresponds closely to the text on page 163 of the first edition of the *Origin*, which thus supplies 'a double shoulder stripe' as the missing subject of the incomplete sentence fragment at the top of 105 A. The information in the cancelled passage on the common donkey Darwin repeated in *Variation under Domestication* (I, 63) where he added a source reference to Martin's *The Horse*. Perhaps Darwin cut off the missing middle third of this folio in order to attach this reference to the expanded manuscript he was making up for chapter II of *Variation*, while reserving the information on Hemionus at the foot of the sheet (now 105 A) for

---

[1] E.g. ch. IX, fol. 36 v, where the surviving MS. note corresponds to note 12, *Variation* II, p. 105.

use much farther on in chapter XIII, note 36 (*Variation*, II, 43). This probably explains Darwin's pencil scrawl: 'All used' on 105 A. In chapter IX, the present folio 21 (which was renumbered, since it was folio 6 of the earlier draft) has been cut up, and part is gone. (See the accompanying facsimiles.) The missing portion of the quotation from Herbert's *Amaryllidaceae* can be restored from the text quoted in *Variation under Domestication* as well as from Herbert's original text. Again in chapter IX the note surviving on the lower part of a sheared-off folio now numbered 36 v corresponds to note 12, p. 105 of the second volume of *Variation* and the text for the missing upper part of the folio can be restored from the published text.

Besides the gaps left by these selective excisions, Darwin left occasional blank spaces to be filled in when he might later find the appropriate names, numbers, or citations, and these have been filled in from the sources Darwin used, where this is feasible.

DARWIN'S PROCEDURE IN WRITING AND REVISING

Before the text reached its present form, Darwin had worked it over in ways which leave many traces in the manuscript. He customarily wrote in ink on one side of folios of paper measuring about $8\frac{1}{4}$ by $12\frac{3}{4}$ inches. In revising, he cancelled by making horizontal lines through words or lines or by vertical lines through longer passages so that the earlier wording is usually readable. (See the facsimile of folio 9 of chapter V.) Some revising he did immediately by cancelling an incomplete sentence and starting anew, as in the middle of folio 94, chapter IX, shown in the facsimile. Similarly in chapter III, in the long note following folio 32, Darwin made three false starts: '⟨It is almost superfluous, but I may state⟩ that ⟨Yet somewhere [?] I have observed instances quite off [?] Although I have seen quite enough to convince me that this claim is quite fanciful, yet⟩ Nevertheless some facts could be given ⟨to⟩ in favour of ⟨it⟩ such a view:' Some cancellations merely show that Darwin alternated in his mind between equivalent wordings so that a complete reproduction of the manuscript would read: '⟨no doubt in all probability⟩ no doubt' (ch. IV, fol. 7), '⟨identical absolutely similar⟩ identical' (ch. VII, fol. 66), 'makes ⟨us one⟩ us feel' (ch. IX, fol. 71), and 'instinctive actions, wondrous though they ⟨are be are⟩ be' (ch. X, fol. 3). For the sake of readability such minor variants and cancelled passages which were rephrased with essentially the same content have been omitted

16

Chapter VII, folios 105 and 105A exemplify a discontinuity in the text resulting from cutting up the manuscript sheet. Presumably the missing middle third was used up in putting together the manuscript for *Variation under Domestication.*

Sheared off portions of present folio 21 of chapter IX. Originally this was folio 6 of the earlier draft of this chapter. The missing portion of the quotation from Herbert's *Amaryllidaceae* can be restored from the text quoted in *Variation under Domestication* as well as from the original. See p. 399 n 4.

Cancelled upper portion of folio 9, chapter v, showing Darwin's original beginning discussion of the struggle for existence. For transcription of this rejected passage see p. 569. For revised text, see chapter v, folio 9.

Folio 94, chapter IX. In this sample folio of the manuscript, a sentence in the middle was cancelled before completion, and was reworded immediately thereafter.

from the printed text. In other clearly important instances, such as alternatives for the phrase 'struggle for existence' the worked-over original text has been printed in the appendix. Other cancelled words, phrases, and passages which seem to amplify or clarify Darwin's thought have been printed within angle brackets in the regular text wherever feasible and otherwise have been placed in the appendix.

Besides cancellations, Darwin's revision led to additions of new material. Sometimes he wrote words or phrases in between the lines. Sometimes he wrote additions on the blank versos of his folios. He wrote some additions on separate slips of paper pinned or pasted on to the manuscript. (See the facsimile of folio 94, chapter IX, where he signalled an interpolation by adding 'a text'.) With rare exceptions I have found no useful clues such as watermarks to help date these additions. On a few occasions differences in the colour of the ink reveal a lapse of time between the writing of text and of revisions, but the length of the interval is uncertain. For a very few addenda or notes, Darwin supplied dates; these range from October 10, 1856 (ch. XI, fol. 6 v), through June, 1858 (ch. VI, fol. 53A) to 1867 (ch. IV, fol. 67). Addenda longer than a line or two written on the backs of folios have been designated by the folio number followed by 'v' for verso and this same sign designates many of the additions made on slips apparently later pinned to the backs of manuscript sheets.

If we can generalize from two instances where Darwin turned over a sheet after cancelling a false start of a few lines on a folio he had already numbered, he numbered his folios as he wrote. Interpolations he often designated by lettering; for example in chapter VI he interpolated a sequence running from folio 26a to 26nn, discarded folio 27, and designated the following sheet '27 & 28'. Some chapters he reorganized drastically, cancelling the original folio numbers and supplying new ones. In chapter IX, the folios originally numbered 5 and 6 were renumbered 20 and 21 (see facsimile); folios 13 to 16 were numbered 30 to 33, and the folio originally numbered 20 became 38.

In the printed text the end of each piece of paper is marked by a slant sign and the numbering and lettering of the new manuscript folios and slips is given followed by another slant sign, and thus the reader can recognize Darwin's additions where they amount to more than a line or two of writing. Where Darwin destroyed the continuity of the text by shearing off parts of folios, the location of the cut is signalled by double slant signs.

Besides cancellations and additions applying to the text as it was to be printed, Darwin wrote an occasional instruction to the copyist or to the printer such as 'Lead', (ch. vi, fol. 28), 'Small Type – notes run into text', (ch. x, fol. 69), and 'Large type again', (ch. x, fol. 78). (See top of facsimile of folio 105, chapter vii.) These have been taken account of without special editorial comment.

On the manuscript Darwin sometimes scrawled pencil memoranda to himself such as: 'Get Huxley to read over for this.' (ch. iii, fol. 16). Part of these have since been rubbed out, but where they are intelligible they have been printed, usually as footnotes.

The Darwin papers also contain some reading notes and letters directly related to the manuscript. These have been printed in connection with the associated portions of the text. They have been cited according to the volume or item numbers in the *Handlist of Darwin Papers* or in Mrs Martineau's catalogue.

Finally, on the manuscript there are some signs of its later use. In the margin of folio 13 of chapter x, Darwin pencilled 'Used [　] Man Book', and Lonsdale's anecdote about snails was published in 1871 on page 325 of volume one of *The Descent of Man*. Similarly, there are other jottings whose meaning is more or less obvious. On the verso of folio 136 of chapter ix, where one would expect the pinned-on slips with the reference previously cited, instead one finds that Darwin wrote in ink: 'Note used in Domestic Animals Chapter 15, Crossing.' Since note 9 of chapter 15 (*Variation*, ii, p. 88) fits the context of the Natural Selection manuscript, this jotting is crystal clear. Elsewhere in chapters iii, vi, vii, and ix we find the jottings 'all used', 'used' or an encircled U alone or with light vertical cancel lines down the page.[1] (See the facsimile of folio 105 of chapter vii and of folio 21 of chapter ix.) Many of these jottings are easily connected with passages in *Variation* where Darwin used material from the Natural Selection manuscript. In chapter iv, 27 and elsewhere we find Francis Darwin's initials. With a few special exceptions, vertical cancellings and jottings of this sort have been ignored in my editing.

DARWIN'S HANDWRITING, SPELLING AND PUNCTUATION

As Darwin described it 'My handwriting, I know, is dreadfully bad.'[2] This often forces the reader to guess at words, and even his family had difficulty in reading it.[3] In the fair copies made of

---

[1] In regard to a similar use of vertical cancel lines, see Francis Darwin's introduction to the *Foundations* p. xxi.

[2] ML no. 636.

[3] H. E. Litchfield, ed., *Emma Darwin, Wife of Charles Darwin. A Century of Family Letters.* Cambridge, 1904. i, p. 436, and *ML* no. 2.

portions of this work, the copyist misread enough words which Darwin did not correct so that we must go back to Darwin's holograph for the basic text.

Darwin himself sometimes misread his own writing. For example, in chapter IX folio 19 he correctly quoted Herbert's 'in cases of natural impregnation', but later when he reviewed this passage he could not read the final 's' in 'cases' and so he added a 'the' above the line to make the phrase read 'in the case of natural impregnation'. In chapter X, folio 112 he quoted Kirby and Spence's 'utmost activity' then later misread his 'utmost' as 'almost' and added 'incessant' to restore meaning by saying 'almost incessant activity'.[1]

Such handwriting makes it practically impossible to reproduce every detail of the text exactly as Darwin intended it to be printed. I have given the best reading I could, but only in the cases where my best interpretation seems to make doubtful sense or where several different words – often proper nouns – would fit the handwriting equally well and I have found no clues as to which Darwin probably meant, have I specifically warned the reader of a particular uncertainty in the manuscript by adding a question mark within square brackets.

The reader should be generally warned about certain specific difficulties in the handwriting. The following pairs of words are frequently indistinguishable: 'to' and 'the', 'when' and 'where', 'could' and 'would', 'than' and 'then', 'man' and 'men'. Considerable uncertainty often arises in the choice of alternative readings between 'that' and 'the', and, unfortunately, between 'probable' and 'possible'. As the last example suggests, Darwin's long 's' is a particularly obscure letter. The final 's' in plurals must usually be determined from the context. The pair of letters 'r' and 'o' are sometimes indistinguishable as in the case of 'grow' (ch. v, fol. 42 v). In an unusual proper name such as Gouan this can be troublesome. Of course other commonly indistinguishable letter pairs such as 'e' and 'i' also occur, so that a choice between possible proper names such as Marten's, Martens', Martin's, and Martins' must depend upon the context.

In some words, letters instead of being merely uncertain seem to be entirely missing, so that a reading such as 'Gret Britain' seems clear. Such omission seems specially frequent for letters before 'y': 'may' for 'many', 'thy' for 'they', and 'vey' for 'very'. Particularly for the cases of proper names, students of Darwin manuscripts should remember the possibility that the

---

[1] Cf. L & L I, 119; NY, I, 97.

correct form may be other than that which Darwin seems to have written; in chapter IV, folio 55, the reading seems to be Magillvray instead of Macgillivray, for example. When writing words such as Gaertner, he apparently intended to ligature the 'a' and the 'e', but the 'e' is usually undetectable so that the word seems to be written as Gartner. Such apparent lapses of the pen are rarely quite clear-cut, however, and the normal spelling has ordinarily been used in the text without special editorial comment.

DARWIN'S SPELLING

A reading such as 'chesnut' might appear to represent either a lapse of the pen or an error in spelling, but it is one of a group of un-expected spellings including 'plaister', 'owzel', and 'Feroe Islands' for which Darwin had had reasonable precedents in his sources.[1]

He was inconsistent in spelling as is illustrated in chapter VII where on folio 37 he wrote both 'connection' and 'connexion' and where on folio 113 he clearly wrote 'organization' and on folio 118 'organisation'. He also made clear-cut errors such as 'thoroughily'.[2] Where the handwriting is clear, the spelling of the manuscript is followed without any particular editorial comment. Where Darwin's spelling is uncertain, the normal English form is used.

Some quite clearly written words still puzzle me. In chapter V, 40, he mentions Inverorum, and this spelling also appears on page 295 of volume one of the *Life and Letters*, but I have not found exactly this place name in any of the numerous gazetteers I could consult. Perhaps Inveroran is the closest. In the same chapter on folio 29, the reading Colinsay seems quite certain. This exact place name I have not located either, and Colonsay does not seem to fit the context. Particularly in cases such as this, it seems best not to change Darwin's apparent spelling lest an essential clue to Darwin's meaning be discarded.

PUNCTUATION AND CAPITALIZATION

Just as about his spelling so about his punctuation Darwin's handwriting leaves many uncertainties. Clearly he often used colons where we would use semicolons. This suggests a system of

---

[1] Or an unreasonable precedent. See the introduction to the *Catalogue of Charles Darwin's Library* (Cambridge, 1908), p. x, where Francis Darwin comments on his father's copying the spelling 'ciliae' from Robert Grant.

[2] Ch. VII, fol. 117; cf. Nora Barlow's discussion of his spelling in her preface to *Charles Darwin's Diary of the Voyage of H.M.S. 'Beagle'.* (Cambridge, 1933), p. xix.

punctuation similar to that set forth in Lindley Murray's *English Grammar* which was so widely used that the book averaged an edition a year during the first half of the nineteenth century. In presenting this manuscript, Darwin's punctuation is retained in so far as it is clear. The many doubtful points such as distinctions between colons and semicolons have been interpreted to conform with present-day usage.

I have retained the parentheses Darwin used, but I have discarded his square brackets. Most often these simply set off material to be printed as a footnote. Occasionally Darwin used square brackets to mark the beginning or the end of a paragraph. This is most clear on folio 78 v of chapter VIII and on folio 64 of chapter X, where he added an ordinary paragraph sign as well. (See also facsimile of folio 94 of chapter IX.) In Darwin's text as here printed and in the related footnotes, square brackets will have the customary function of indicating material added by the editor.

Similarly in regard to a more frequent use of capital letters, Darwin's practice seems to have been different from ours; but, here again the handwriting often leaves his intentions uncertain. Such uncertainties have been resolved in favour of present-day practice. Where Darwin in revising changed the beginning or ending of sentences without completing a corresponding change in capitalization, I have changed this without special comment, e.g. chapter X folio 137 v.

I have also silently expanded Darwin's contractions for words such as should, island, and reverend, and have spelled out 'Natural Selection' where Darwin used the abbreviation 'Nat. Sel.' in his pencilled addition to folio 51 in chapter VI. I have omitted words which were unintentionally written twice.

The preceding editorial discussion applies to the work as a whole. Comments about points concerning single chapters and their history will appear in the separate introduction immediately preceding the individual chapters in a form similar to the following comment on Darwin's own tabulation of the contents of his manuscript.

COMMENT ON DARWIN'S TABLE OF CONTENTS

The following extensive table of contents, which Darwin himself wrote out, merits some special consideration for what it tells us about the detailed history of the Natural Selection manuscript. First of all it supplies a full outline for the two initial chapters, now missing except for one single stray survivor, folio 40 from chapter one. Presumably Darwin's manuscript for these chapters was incorporated and used up in the course of writing the two later volumes on *The Variations of Animals and Plants under Domestication*. The first folio of the manuscript for the Table of Contents has been cancelled by a single diagonal pencil line in the same manner as Darwin employed to mark passages farther on in the manuscript which he had used in later publications. The contents for these two missing chapters can be compared with those for the two volumes on *Variation*.

The fact that the very first entry for the table is for folio 16 of the first chapter raises the question, what about the preceding fifteen folios? I believe these formed Darwin's preface, which we know that he wrote because he referred to it in the postscript of a letter to Baden Powell on January 18, 1860: 'I have just bethought me of a Preface which I wrote to my larger work, before I broke down & was persuaded to write the now published abstract. In this Preface I find the following passage, which on my honour I had completely forgotten as if I had never written it. 'The "Philosophy of Creation" has lately been treated in an admirable manner by the Rev. Baden Powell in his Essay &c 1855. Nothing can be more striking than the manner in which he shows that the introduction of new species is a "regular not a casual phenomenon", Or as Sir John Herschel expresses it "a natural in contradistinction to a miraculous process"'.[1] To my particular regret I have as yet been unable to find any further trace of this Preface in the surviving Darwin manuscripts.

Darwin seems to have written his table of contents chapter by chapter and not very long after each part or chapter was finished. In the case of chapter VI, for example, Darwin completed his original table at the end of folio 4 of his table of contents section and continued straight on with the contents for chapter VII on folio 5 of his table. Then a year later in the spring of 1858 he returned to chapter VI to revise and expand it, particularly by interpolating a new discussion of divergence some forty folios long. Thereupon he cancelled the original table of contents for the latter part of the chapter, which had been notably transformed, and wrote out for this second draft a new table on a folio which he had to number '4 bis' to fit it into its place in the table as a whole. Here in my edition, in the following table only the new version is included; the older cancelled portion is to be found later, in my introduction for chapter VI, which immediately

---

[1] Gavin de Beer, ed., 'Some unpublished Letters of Charles Darwin', Roy. Soc. London, *Notes and Records*, 14 (1959), 54.

precedes Darwin's text and discusses the specific history of that chapter. Similarly for chapter IV, about a year after he finished his original draft he wrote a long additional section. For this he wrote out a table of contents on a folio he had to number '3 bis' to fit it into its proper place. As is evident in the case of chapter IX and therefore probably for the other chapters he did wait until he had finished the chapter before he wrote out the table of contents for it. In the case of this chapter after finishing his original draft he revised it drastically. The table of contents for this chapter starts on folio 6 of his manuscript table immediately after the end of the table for chapter VIII, yet it fits the revised form of chapter IX and has no cancelled references relating to the earlier draft, so that it could only have been written out after the second version of chapter IX had been completed. In the case of the two portions of the manuscript of which Darwin had fair copies made, namely the addition to chapter IV and the discussion of geographical distribution which I have called chapter XI, he waited at least until the copyist had finished because the folio reference numbers in his tables fit only the fair copies and not the drafts he himself wrote out.

Most exceptionally, considering the Natural Selection manuscript as a whole, watermarks in five out of the eleven foolscap sheets used for the table of contents supply dates; these are compatible with the previous assumptions about the different times of writing of the different parts of the table of contents. The sheet for chapter XI, written first of all, bears the date 1856, as do folios 1, 3, and 5 of the table. Folio 3 bis giving the table for the addition written in 1858 for chapter IV bears the watermark date 1858. The complete manuscript for the table now consists of folios 1 to 3, 3 bis, 4, 4 bis, and 5 to 8. These are mounted at the very beginning of volume 8 of the Darwin Papers. The unnumbered folio with the contents of chapter XI is now item 58 in volume 72.

On the manuscript Darwin pencilled certain notes as memoranda or agenda which should be recorded for consideration. After the entry for folio 3 of chapter II, he scrawled: 'N.B. Why not animals domesticated *We do not want them* Goats & Asses.' He bracketed the next two entries, for 'The Cabbage' and 'Dog' adding: 'Sports in Plants? isolation' [?] On the verso of folio 1 opposite the last lines of the table for chapter II on folio 2 he wrote: 'I think here all naturalised animals—P. Santo Rabbits—Mice in different countries—Naturalisation Cardoon—Naturalised plants in N. America.' At the beginning of chapter IV: 'Antiquity of variation under nature of shells in *Madeira must* be included in this chapter', and following the entry 'conspicuous and useful plants not culti-vated' for folio 44 of chapter IV, Darwin scribbled: 'ʃʃFeral animals & plants not domesticated.' For chapter VII, Darwin seems to have cancelled the entry for folio 41 on Brullé's law, presumably after he had received from Huxley the adverse com-ment on Brullé which is given in the introduction to chapter VII.

For Darwin's table of contents, there has been no attempt to reproduce the apparently insignificant exact details of his holo-graph. His abbreviations, such as 'Var. under domest.' for the title of chapter I, have been expanded without the use of square brackets. Similarly for folio 25 of chapter I, Darwin's 'avitism' has been corrected to 'atavism'. The numbers in the table, which precede each topic, are folio references to Darwin's holograph manuscript, largely as Darwin himself supplied them. Thus any topic indicated in the table of contents can be located in the printed text by referring to the folio numbers given between slant signs in the text. For the addition to chapter IV and for chapter XI Darwin's reference numbers correlated with the some-times inaccurate contemporary fair copy, and in these cases I have replaced them by new numbers (not signalled by square brackets) which correlated with the folio numbers of the holograph manu-scripts used as the basis for the published text.

DARWIN'S TABLE OF CONTENTS FOR HIS
MANUSCRIPT ON NATURAL SELECTION

CHAPTER III ON POSSIBILITY OF ALL ORGANISMS CROSSING: ON SUSCEPTIBILITY OF REPRODUCTION TO CHANGE

CHAPTER VI NATURAL SELECTION

CHAPTER X INSTINCT.

CHAPTER XI GEOGRAPHICAL DISTRIBUTION.

ON THE POSSIBILITY OF ALL ORGANIC BEINGS OCCASIONALLY CROSSING, & ON THE REMARKABLE SUSCEPTIBILITY OF THE REPRODUCTIVE SYSTEM TO EXTERNAL AGENCIES

INTRODUCTION

On October 3, 1856, Darwin wrote to his second cousin, W. D. Fox, that 'I...am now drawing up my work as perfect as my materials of nineteen years' collecting suffice, but do not intend to stop to perfect any line of investigation beyond current work...I find to my sorrow it will run to quite a big book.'[1] Ten days later, according to his Pocket Diary, he finished his second chapter, and presumably he then proceeded to this chapter three and wrote rather fluently, for the manuscript is less laboured than for many of the later chapters. After several mentions of the theme of crossing in letters to Hooker[2] he wrote to him on December 10. 'It is a most tiresome drawback to my satisfaction in writing, that though I leave out a good deal & try to condense, every chapter runs to such an inordinate length: my present chapter on the causes of fertility & sterility & on natural crossing has actually run out to 100 pages M.S., & yet I do not think I have put in anything superfluous.—'[3] The completion date for this third chapter was December 16, 1856, according to the Pocket Diary.

Although for this chapter Darwin made very few revisions, one involving terminology is worth comment. On folio 20 of this chapter Darwin states: 'All the vertebrata are bisexual', here clearly meaning not hermaphrodite but having two separate and distinct sexes. This same usage occurs in the *1842 Sketch*[4] where he wrote 'All bisexual animals must cross, hermaphrodite plants do cross, it seems very possible that hermaphrodite animals do cross.' Similarly in the *Monograph on the Fossil Lepadidae*, published in 1851, he wrote: 'Ibla cumingii...is bisexual; one or two males being parasitic near the bottom of the sack of the female...hence *Ibla cumingii* is exactly analogous to *Scalpellum ornatum*. On the other hand, the closely allied Australian *Ibla Cuvierii*, like *Scalpellum vulgare*, is hermaphrodite...'[5] In print later in this same year of 1851 he changed his usage completely around in his monograph

---

[1] L & L, II, 84; NY, I, 442.
[2] Letters of Nov. 15, ML no. 334; Dec. 1, C.D. MSS. vol. 114, no. 185.
[3] C.D. MSS., vol. 114, no. 186; ML no. 337.
[4] *Foundations*, p. 2–3, and also in 1844 Essay, p. 92.
[5] (London, 1851), p. 16, n. 1. Regarding this work Darwin wrote to Hooker in June, 1849: 'I am going to press....' L & L, II, 37; NY, I, 397.

on recent Lepadidae to equate bisexual with hermaphrodite when he stated that: 'Ibla, though externally very different in appearance from Scalpellum, is more nearly related to that genus than to any other; in both genera some species have the sexes separate, the imperfect males being parasitic on the female, and other species are bisexual or hermaphrodite,' (p. 182). Discussing the affinities of the species which I. E. Gray had named *Ibla Cuvierana* and which Darwin named *Ibla quadrivalvis*, he wrote: 'Considering these so slight differences, it is highly remarkable that this species should be hermaphrodite, whilst *I. cumingii* is unisexual' (p. 207), and farther on in another general discussion of a genus stated that '*Scalpellum ornatum* and perhaps *S. rutilum*, are unisexual; the other species are hermaphrodite', (p. 221). This same complete reversal of intended denotation regarding the term 'bisexual' appears in the manuscript of this chapter. Darwin clearly made the change well before he had finished the original version although he had some difficulty in keeping this change of usage in mind. Evidence of the standard usage occurs as early as folio 21, but even on folio 68 he started to write 'bisexual' to denote the opposite of hermaphrodite but caught himself before he had finished the word, drew a line through his error, and continued writing on the same line to make the phrase now read: 'in closely allied groups of hermaphrodite & ⟨bisex⟩ unisexual plants.' Study of folios 21 to 26 reveals a confusing oscillation of usage in Darwin's original drafting of this portion of the manuscript. (Since my editorial system does not signal interlinear interpolations, I should assure the critical reader that, for these folios, the cancellations in every case occur only on the original lines of writing, not in interpolations added between the original lines at some later but uncertain time.) In revising these passages Darwin cancelled and changed all but the first occurrence of his earlier usage of 'bisexual'; and he often tried to make his terminology clearer by substituting 'hermaphrodite' for 'bisexual'.

DARWIN'S LATER USE OF MANUSCRIPT

Evidences of Darwin's later uses of the text and material of this chapter abound in the manuscript. Dr Alice Guimond, while working as my research assistant, discovered direct quotations and close paraphrases from the manuscript in the published text of Darwin's *Variation of Animals and Plants under Domestication* (London, 1868), and also that use of information to be found in the manuscript was published in his *Effects of Cross and Self-Fertilisation in the Vegetable Kingdom* (London, 1876). At the top of folio 72 Darwin wrote: 'All used to p. 102', that is to the end of the manuscript for this chapter, and in his *Variation under Domestication* volume II, pp. 148–72, starting with the section headed 'Sterility from changed Conditions of Life', the use is clearly evident. This folio 72 was also cancelled with a vertical line to indicate use and so were the following sheets to the top of folio 97. The lower part of folio 97 was marked with an encircled 'U' as were folios 98–102. One or both of these symbols, or occasionally a criss-cross cancel, also makes folios 1 to 44, 27, 37 to 40, 52 to 54, and 56 and 57 as used.

More drastic signs of use also occur. A quotation from William Herbert's *Amaryllidaceae* at the foot of folio 5 has been sheared off completely and is now missing, although the end of the quotation appears at the top of the next sheet of the manuscript. The missing text can be restored from p. 127 of volume II of *Variation under Domestication* which gives the full quotation.

Similarly the bottom fourth of folio 74 and the top fourth of folio 75 have been removed. The remaining manuscript text for the second paragraph of folio 74 is closely paralleled by the opening wording of the paragraph beginning towards the top of page 150 of volume II of *Variation*. After the gap the manuscript proceeds: 'Four wild species of the Horse genus have been bred in Europe', which the printed text in this same paragraph repeats as 'Four wild species of the horse genus have bred in Europe', and then paraphrases the rest of the manuscript remaining for this folio. It seems almost certain that while writing the manuscript for *Variation under Domestication* Darwin cut off these parts of the older Natural Selection manuscript and simply pinned or pasted them on to his new manuscript in the same way that certain existing sheets of the *Natural Selection* manuscript are pieced together from passages selected and sheared from earlier drafts (e.g ch. 9, fol. 84). The full list of cut off folios for this chapter is: 5, 6, 9, 13, 36, 39, 74, 75, 78 and 79. Similarly certain note slips are now missing from the Natural Selection manuscript, and were probably transferred to the later manuscript.

Much, both of the short passages of missing text and the missing notes can be restored with confidence from the first edition of *Variation under Domestication*.

ON THE POSSIBILITY OF ALL ORGANIC BEINGS OCCASIONALLY CROSSING, & ON THE REMARKABLE SUSCEPTIBILITY OF THE REPRODUCTIVE SYSTEM TO EXTERNAL AGENCIES

[Completed Dec. 16, 1856]

1/The subject of the present chapter is related to some points discussed in the previous chapters as to breeds being kept constant by the blending of slight & individual differences & to several questions which follow, & may therefore be as well intercalated here as elsewhere.[1]

On the ill effects of close breeding in & in. That evil arises from this process carried to an extreme has been a general opinion in various countries & times, is universally known.[2] That general beliefs of

---

[1] [As Dr Alice Guimond, my research assistant, discovered, Darwin used material from this chapter in his *Variation of Animals and Plants Under Domestication* (London, 1868) and in his *Cross and Self-Fertilisation* (London, 1876). The first fourteen MS. folios he cancelled with a vertical line, marked with a large 'U' or both, presumably to indicate he had used the material written there. Moreover, he cut up his manuscript, and apparently removed some reference slips which I have not found, so that to restore the gist of these missing parts of the MS. I have borrowed from his published text on the same topics in the *Variation*.]

[2] Sir G. Grey in his most interesting Journal of Expeditions into Australia Vol 2, p. 243 says that anything approaching to the crime of incest is held in abhorrence

this nature have often no foundation is very true; but in this case it may perhaps [be] more readily trusted as the breeder is often most unwilling to act on his beliefs, as it must seriously interfere with his process of continued selection of some peculiarity in his own stock./1v/Independently of the undoubted evil of matching animals having the same infirmity, which must always tend to be the case when relations unite/1/the general belief seems to be that decrease in size, & of general vigour is the first result of close interbreeding, & then lessened fertility.

I have never met a Pigeon Fancier who did not believe in the evil of close interbreeding; & he has the best/2/opportunity of judging, from pigeons being paired for life, & many generations raised in a short period: when size is an object as in the Pouter, it is asserted[1] that the ill effects are very soon perceived, not so when small birds are wanted as in the Almond Tumbler; but in such cases many of the birds become shy breeders.—The high price of many fancy dogs, which have long been closely selected & interbred, I have been assured is, due more to the difficulty in getting them to breed freely, than in their throwing inferior animals; I have known the female requiring to be held, exactly as in the production of some Hybrids[2] & indeed if no such difficulty existed the high price of such dogs would be quite inexplicable. The particulars have been given me of one gentleman who long had kept a small family of blood-hounds, & from being very unwilling to cross his breed, he almost lost them, so infertile had they become, until he was obliged to resort to a cross when his breed became fertile./

2 bis/The evidence of an acute observer like Sir John Sebright, who bred all sorts of animals during his whole life, & who boasted that he could produce any feather in [three] years & any form in [six] years;[3] & who always worked by crossing & thereby closely interbreeding, is very good; & he was a most firm believer[4] in the ill effects of this process carried on too long. I was assured by

by the Australians. So it is with the aborigines of N. America, & Dobrizhoffer makes the same in regard to the Abipones of S. America. [I, 71] It is singular that this feeling does not appear to have been felt by the Kingly class of the Polynesians; but Ellis [Tour through Hawaii (London, 1826), pp. 414–15.] does not doubt that evil followed from their incestuous marriages. Prescott [See William H. Prescott, Conquest of Peru, book 1, ch. 3, passage associated with note 47.]—Ohio [cf. Variation, ch. 17, n. 21.] Chinese of same name [?].

[1] A Treatise on Fancy Pigeons by J. M. Eaton, p. 56 [citation clearly to 1st. ed.]
[2] Hunter's Animal Economy in regard to a she wolf too [1837 ed., p. 323.]
[3] [Darwin left blank spaces for these two numbers. They are attributed to Sebright without source reference in Eaton's Fancy Pigeons, 1852, p. iv.]
[4] [The Art of Improving the Breeds of Domestic Animals (London, 1809), pp. 8, 11–12.]

Mr Yarrell that Sir John had for so long interbred his Owl-Pigeons, that he nearly lost his whole stock by their extreme infertility: I have seen some silver Bantams, bred from Sir John Sebrights', which were nearly as sterile as Hybrids for they had laid in that season two full nests of eggs, not one of which produced a ⟨single⟩-chicken. The cock, also, seemed to have lost its secondary male characters, for it had not saddle-hackles, & was scarcely more brilliant ⟨plumage⟩ than the hen./

3/On the other hand some competent judges have doubted the ill effects of interbreeding. The case of Bakewells cattle has often been quoted, & it shows that a man with a large flock may continue the process for a considerable time; but Youatt[1] speaking of the subsequent deterioration of this breed says 'it had acquired a delicacy of constitution inconsistent with common management' & 'many of them had been bred to that degree of refinement that the propagation of the species was not always certain.'—In most of the cases of closely selected cattle & sheep there has been much mystery, & crosses have been suspected. The English Race horse & Mr. Meynell's hounds[2] have also been advanced as instances of pretty close interbreeding without any ill effect. In these cases it may be suspected from what we shall presently see, that individuals being taken to different parts of the country & differently treated, & then occasionally brought together & matched, would lessen the ill effects of interbreeding. Again the case of the half-wild cattle in Chillingham/4/which have gone on interbreeding for the last 400 or 500 years[3] seems a strong case; but Lord Tankerville, the owner, expressly states that 'they are bad breeders'.[4] Those in the Duke of Hamilton's Park, are believed to have degenerated in size; I am informed by Mr. D. Gairdner that the stock kept, in the park of 200 acres, varies from 65 to 80, & that only about 8 or 10 are yearly killed, which seems to show no great fertility.

In the closely analogous case of Fallow Deer in parks, I find that the owners go to the trouble of occasionally obtaining bucks from other parks to cross the breed. In the case of the aurochs of Lithuania, which have a much wider range than the cattle of the British park, some authors believe that they have become considerably reduced in size. So it certainly is[5] with the Red Deer of Scotland; but in the latter & indeed in the other cases it seems

[1] Cattle p. 199.
[2] Karkeek, Veterinary Journal Vol. 4. p. 4 & Mr. Appleby [actually Apperley] in Encyclopedia of Rural Sports p. 280
[3] Culley on Live Stock [Introd. pp. x–xi.]
[4] British Association Zoolog Sect. 1838
[5] Scrope [*Deerstalking*, cf. pp. 10–11, 170 of 1839 ed.]

impossible to decide how much of the decrease of size to attribute to less varied food, & in the case of the Red Deer to sportsmen having picked out for many generations the finest Bucks; the less fine having been thus allowed to propagate their kind./

5/*Good effects of crossing.* However difficult it may be to obtain quite satisfactory evidence of the ill effects of close interbreeding, the converse of the proposition, namely that good arises as far as increased size, vigour & fertility comes from crossing distinct families & breeds, I think admits of no doubt. I have never met any breeder of animals who doubted it; & it seems useless to adduce authorities or facts. But in regard to plants [,] as *varieties* have been much more seldom crossed than in animals, I will go into some details to show that the same rule holds with them.

Gaertner, whose accuracy & caution seem most trustworthy, believes in the good effect of taking the pollen from another individual of the same species; he states[1] that he observed this many times, especially in exotic genera, as in Passiflora, Lobelia or Fuchsia.

⟨Herbert[2] says⟩//[‘I am inclined to think that I have derived advantage from impregnating the flower from which I wished to obtain seed with pollen from another individual of the]/6/same variety, or at least from another flower, rather than with its own.’

In these cases we have referred to crossing individuals of the same variety; we now come to crosses of distinct varieties.//

6A/Andrew Knight[3] found that the offspring of crossed varieties of Peas were *remarkably* tall & vigorous; & that crossed wheat resisted blight better than the pure kinds./

7/We have seen in crossing varieties, that the offspring gains in size, vigour & fertility; in crossing distinct species it would appear that size & vigour is gained in an equal or apparently greater degree, but fertility is greatly impaired or very often wholly lost. Every traveller has been struck with the vigour & health of the common mule & this holds good with the hybrid Yak in the Himalayas; in the almost quite sterile hybrid from the fowl & pheasant, marked increase of size has been often noticed. In plants every single experimentiser Kolreuter, Gaertner, Sageret, Lecoq Herbert &c have been struck with the wonderful height,

---

[1] Beiträge zur Kenntniss der Befruchtung 1844, S. 366

[2] Amaryllidaceae p. 371 [The bottom of MS. fol. 5 is sheared off; the missing part of the quotation ending at the top of fol. 6 is supplied from *Variation*, ii, 127, ch. 17 n. 41.]

[3] Philosoph. Transact. 1799 p. 200 [At end of this sentence Darwin pencilled a memo for addendum: ‘Loudon's Gardener's Magazine for grapes & other cases.’]

size vigour, tenacity of life, precocity number of flowers, power of resisting cold &c of most of their Hybrid productions. Kolreuter[1] is astonished at the portentous size of some of his hybrids & gives numerous precise measurements in comparison with both parents. Gaertner[2] sums up his conviction on this subject in the strongest manner. Kölreuter attributed these facts to the sterility of hybrids, owing, I presume, a sort of compensation, in the same manner that capons, emasculated cats, some breeds of oxen are larger than unmutilated males. But Gaertner (p. 394 & 526) has shown that there is much difficulty/8/in admitting this explanation to its full extent; for there is no parallelism between the degree of sterility & the increase of size or luxuriance of growth; indeed the most striking cases have been observed in not very sterile hybrids. It deserves notice that the mass [?] ⟨luxuriance⟩ & enormous size of the roots in a crossed Mirabilis of unusual fertility for a hybrid[3] was found to be inherited. It seems probable that the result is due both to nutriment which ought to have gone to the sexual function being applied to general growth, & secondly to that same general law which as we have seen gives to mongrels, animals & plants not only increased fertility but greater constitutional vigour & size. It is not a little remarkable thus to see under such opposite contingencies as increased & decreased fertility, an accession of size & vigour.

It is well ascertained[4] that hybrids will always breed more easily with one of their parents, & indeed not rarely with a third distinct species, than when self-fertilised or crossed inter se.—Herbert would [have] explained even this fact by the advantage of a fresh cross, but Gaertner far more justly accounts for it, by the pollen of the hybrid plant, being in itself in some degree vitiated, whereas the pollen/9/of either parent species or of a third distinct species is sound. Nevertheless there are some facts on record, which seem to show that even in hybrids a fresh cross does do some good in respect to their fertility. Herbert states[5] that having in flower at same time nine hybrid[6]//

---

1 ["Fortsetzung", 1763, s. 29; "Dritte Fortsetzung", S. 44, 96; "Act. Acad. St. Petersburg", 1782, part ii, p. 251; "Nova Acta", 1793, pp. 391, 394; "Nova Acta", 1795, pp. 316, 323.' as cited in *Variation*, ii, 130, ch. 17 n. 53.]

2 ["'Bastarderzeugung", s. 259, 518, 526 *et seq*.' as cited in *Variation*, ii, 130, ch. 17 n. 52.]

3 [Kölreuter] Nova Acta 1795. p. 316.

4 Gaertner Bastarderzeugung p. 430 [cf. Herbert Amaryllidaceae, p. 352.]

5 Amaryllidaceae 1837 p. 371. The statement is confirmed after experiments tried during several years in Horticultural Soc. Journal. Vol. 2. p. 19

6 [Fol. 9 is sheared off at this point and the rest of the paragraph is quoted from *Variation*, ii, 138–9, ch. 17.]

[Hippeastrums, of complicated origin, descended from several species, he found that "almost every flower touched with pollen from another cross produced seed abundantly, and those which were touched with their own pollen either failed entirely, or formed slowly a pod of inferior size, with fewer seeds." In the 'Horticultural Journal' he adds that, "the admission of the pollen of another cross-bred Hippeastrum (however complicated the cross) to any *one* flower of the number, is almost sure to check the fructification of the others." In a letter written to me in 1839, Dr. Herbert says that he had already tried these experiments during five consecutive years, and he subsequently repeated them, with the same invariable result. He was thus led to make an analogous trial on a pure species, namely, on the *Hippeastrum aulicum*, which he had lately imported from Brazil: this bulb produced four flowers, three of which were fertilised by their own pollen, and the fourth by the pollen of a triple cross between *H. bulbulosum, reginae,* and *vittatum*; the result was, that "the ovaries of the three first flowers soon ceased to grow, and after a few days perished entirely: whereas the pod impregnated by the hybrid made vigorous and rapid progress to maturity, and bore good seed, which vegetated freely." This is, indeed, as Herbert remarks, "a strange truth," but not so strange as it then appeared.]

9A/Now considering how many crossed Hippeastrums were experimentised on, & that they were crossed in all sorts of ways, & that the pollen in each case applied to the stigma of one plant was from some other hybrid, & therefore not sound, I can understand the strong &/10/overpowering ⟨marked⟩ good effect of its application, only on the abstract good from crossing, as seen in crossing varieties. Moreover this case of the hybrid Hippeastrums is confirmed as we shall hereafter see in the chapter on Hybridism in some degree by some extraordinary cases, well ascertained by Gaertner, Kolreuter, & Herbert, in which pure species of Lobelia, Passiflora, Hippeastrum, Verbascum, had both pollen & germ in proper condition as shown by their fertilising, & being fertilised by, other species, but yet were incapable of, self-fertilisation, when their own pollen was placed on their own stigmas. These facts seem to show that in hybrids from distinct species, independently of the greater vigour & luxuriance often acquired, that even in regard to fertility, which is undoubtedly almost universally diminished or quite annihilated, there is some slight counterbalancing good in the act of crossing which occasionally appears in the intercrossing of hybrid with hybrid./

11/*Good from slight changed conditions*—I think some little light can be thrown on the good resulting from crossing the breed, from considering the effects on the individual of slightly changed conditions. It has been a very general belief from ancient times to the present day, in many countries that ⟨decided⟩ good results from taking the seed, tuber or bulb of a plant grown in one kind of soil or situation & planting it in another; the most opposite kinds of soil being chosen, seeds, tubers &c being often interchanged between residents thus situated. I should have thought less of this belief, if it had been confined to cottagers or common farmers, but I find on enquiring from some [who] attend especially to raising seed-corn, & whose success is testified by their obtaining the highest prices in the market, that they find it indispensable to change their seed every few years.—One eminent gentleman in this line has two farms at different heights & on very different soils, so that he is able to exchange his own seed, but even with this advantage, he yet finds it advantageous to purchase occasionally fresh seed grown on other land.[1] Mr. Robson, a practical gardener,[2] /12/positively states that he has seen himself decided advantage in obtaining bulbs of the onion, tubers of potatoes & seed from different soils, & from distant parts of England. Oberlin[3] attributed in great part the surprising good he effected amongst the poor of the Vosges in the cultivation of the potato, (the yield having been reduced in between 50 & 60 years from 120–150 to only 30 or 40 bushels in the year 1767) to changing the sets. In the cases of good resulting from the exchanging of seeds, I should think it could not be explained on the same chemical principles as in the rotation ⟨of crops⟩ of different species, namely by the seed obtaining some ⟨chemical⟩ element in one soil good for use with say for wheat not found in sufficient abundance in another soil also good for wheat for how small a difference in a single grain could the excess be, & this one grain has to influence the whole yield of the plant. Such a chemical view has more probability, & yet not much I think, when applied to the exchange of tubers of potatoes; but even in this case the slice planted bears but a small proportion to the yield of tubers./

13/As animals are less fixed to one spot & the same conditions it is less easy to get evidence of the good of change. But with invalids,/14/no medical man doubts of such good being most

---

[1] The Rev. D. Walker in his Prize Essay of Highland Agricult. Soc. Vol. 2. p. 200, expresses a strong opinion on this subject. See also Marshall's 'Minutes of Agriculture' Nov. 1775. ⟨Mr. Loiseleur Deslongchamps in his 'Considerations sur les Céreales 1843. p. 200' gives numerous references on this subject.⟩

[2] Cottage Gardener 1856 p. 186   [3] Memoirs of p. 73

evident. Small farmers again find their cattle prosper best when they can occasionally change their pasture. It seems very doubtful whether in these cases the good can simply be accounted for by some fresh element in their food, which was before wanting. It would rather appear as if the marvellous & complicated play of affinities & constant change by which life is kept up, was somehow stimulated by almost any sort of slight change in the conditions to which the individual is exposed. Judging from plants, as both those which are useful from the number & quality of their seeds, & those which are useful from their organs of vegetation seem to be benefitted by a change, we may infer, that as in the case of crossing, both general luxuriance & fertility are increased.

If the facts here just given can be trusted, I think we can in some degree understand the good of crossing, for the individual with a blended constitution, derived from the union of the male & female from two varieties, differing in/15/structure or constitution or even two individuals of different families will be exposed during its life whatever the conditions of its existence may be, to a somewhat different relation with external things to what either of its simple parents can have been;—⟨for I presume it will be admitted that every part of the structure is related either to the external conditions or to other portions of its own structure.⟩

Considering the various cases now discussed,—obscure as many of the facts are & doubtful the evidence—namely the apparent ill effects of close interbreeding, the good from crossing individuals of distinct families or varieties, & even of species in this latter case with the great exception of fertility, considering what little light is thrown on the subject from the good of changed conditions to the individual, I should be strongly tempted to believe with Mr. Andrew Knight,[1] that it was an essential part of the great laws of propagation that *occasionally* there should be the concourse of two separate individuals in the act of reproduction. But instantly it will occur to everyone that there are very many hermaphrodite organisms, with/16/the two sexes united in one individual. How it may be asked can in such cases two individuals occasionally cross? If an organism can from the day of its creation go on most strictly interbreeding, that is self-fertilising itself from the day of its creation to its extinction, one may well doubt all the foregoing

---

[1] Philosophical Transactions 1799. p 202. Mr. Knight argues "that nature intended a sexual intercourse should take place between neighbouring plants of the same species."—Kolreuter in Mem. de l'Acad. St. Petersbourg Vol. 3. p. 197 makes striking similar remarks: I think in the *Portfolio* on "Dichogamy".—[Darwin's notes on dichogamy are now in vol. 49 of his papers in the Cambridge University Library. His notes on this article of Kölreuter's are on fol. 163 of vol. 49.]

facts & put them all down to popular prejudices. I can hardly believe this. The subject has sufficient importance for us, in relation to crossing of *slight* varieties being a powerful means of keeping a breed or species true,—in relation to some points in geographical distribution,—perhaps to the extinction of species when become very rare,—& to some other points, that I must discuss it at some little length.[1]

First for some general considerations, which seem to me to have considerable weight. In land animals, after attending to the subject for several years, I have not been able to find any one case, in which the concourse of two individuals is not requisite;[2] yet there are a good many hermaphrodite animals/17/as land-shells, certain annelids, as earthworms, land-leeches & planariae, but these all unite in pairs for propagation. In aquatic animals there are numerous cases of hermaphrodite ⟨bisexual⟩ animals which can certainly propagate by self-fertilisation; but in these forms the fluid medium in which they live, & from the fluid nature of the liquor seminis there is a *possibility* of an occasional cross, & we shall presently see that this is favoured by their structure.

In land animals, on the other hand from the nature of the liquor seminis it is obvious there never could be a cross between two individuals, without their close contact or union; & this, as far as I can find out, is the universal rule in land bisexual animals. This fact is the more striking, when we contrast land animals & land plants; in these latter hermaphroditism ⟨bisexuality⟩ & self fertilisation is the rule &/18/unisexuality (monoecious & dioecious plants) the exception; but in plants the fertilising element or pollen is not liquid & can easily, as is well known, be carried through the air from individual to individual by insects & the wind.

Secondly, in plants it is known[3] that damp winds & rain are very injurious to their fertilisation; yet the general rule is that flowers are open & fertilisation takes place ⟨sub jove⟩ under the open sky. Such cases as the snap-dragon & papilionaceous flowers cannot be considered as exceptions, but rather as confirming the remark, for though they protect the stigma & anthers from rain, as do drooping tubular & bell-shaped flowers, yet they are not sealed up, but frequently opened & visited by insects. The few cases in which fertilisation appears to take place in really closed flowers will be presently discussed. I am far from pretending that

---

[1] [Here Darwin pencilled the following memorandum: 'Get Huxley to read over for this.']

[2] The Acarus mentioned by *Owen* See the authority [?].

[3] Gaertner Bastarderzeugung p. 11.

43

there may not be some other additional & quite different explanation of the generality of the fact of the fertilisation of plants taking place, exposed to the injurious effects of climate & to an enormous loss of pollen/19/by the consumption of insects, but yet if an occasional cross with another individual is a law of nature, we have an explanation of these facts.

Thirdly, in animals & plants there are many instances of hermaphrodite ⟨bisexual⟩ & unisexual species in the same group & even frequently in the same genus; that is, we have the two sexes united in the same individual, or in two separate individuals in organisms, in all other respects very closely allied. Now if there be no such thing in nature, as an hermaphrodite fertilising itself throughout its whole existence;—if the only difference be in degree, the hermaphrodite *occasionally* crossing with another individual, the unisexual at every act of propagation, then the concurrence of bisexual & unisexual organisms in the same groups is less surprising & Nature in this case, as in other cases, has not moved per saltum./

20/Now for some details showing that in all animals the *occasional* crossing of two individuals seems to be possible: if it could be demonstrated that the structure of any animal was at all times such that the access of the liquor seminis from another individual was impossible, then the conclusion towards which I am tending that an occasional cross is a law of nature would be proved to be erroneous.—I shall pass over those low animals, the protozoa, barely distinguishable from plants, for I believe true sexual generation has not been observed in them; but the steady progress of knowledge of late years should make us very cautious in assuming that they have not sexes. In the lower plants as mosses & lichens there are many cases of species for long periods & in certain districts which have here at most rarely been seen to fructify, being propagated by generation but which are known in other districts & at other times to follow the ordinary law;[1] & so it may be with some of the lower animals.

All the vertebrata are bisexual, except as it would appear some fish of the genus Serranus[2] but from what we know of the habits of fish, an/21/occasional cross seems far from improbable. In the enormous Kingdom of true articulata (excluding annelids) all are ⟨bisexual⟩ unisexual, except the acarus previously alluded to, & the order of Cirripedia. In Cirripedes I have shown that a very

---

[1] [Durieu in] Silliman's Journal vol. 21. p. 171. Several instances are here given taken from the Transactions of the Linnean Soc. of Bordeaux.

[2] Quatrefages. Revue [des deux mondes, 1856, tome 4, p. 80, n. 2.] I have not seen the memoir of M. [Dufossé?].

few are ⟨bisexual⟩ unisexual; & that the fertilisation of some other very few which are hermaphrodite are aided by what I have called complemental males, which are distinct individuals; these few species, therefore, can be crossed. But by a piece of good fortune I met with some monstrous specimens[1] of Balanus bala- noides, a truly hermaphrodite form, in which the male organs were rudimentary, & the channel absolutely imperforate, never- theless three of these specimens included developed larvae; proving (without we admit lucina sine concubitu) that the liquor seminis from other individuals had gained access to the open sack of these monstrous individuals.

In the other ⟨two⟩ great animal Kingdoms, there are many ⟨bisexual⟩ hermaphrodite forms; but it deserves notice that during the last 20 or 30 ⟨twenty⟩ years a surprising number of these lower animals, which were formerly thought to be ⟨bisexual⟩ hermaphrodite are now known to be unisexual.—Of the ⟨bisexual⟩ hermaphrodite animals/22/many, as[2] the gasteropod univalve shells, & marine worms or annelids require the concourse of two individuals. Until lately all acephalous mollusca, or bivalve-shells, were thought to be hermaphrodite, but now many as the common mussel & cockle[3] are known to be ⟨bisexual⟩ unisexual, & their fertilisation is probably ⟨must be⟩ effected by the spermatozoa being drawn in by the same ciliary currents by which food is obtained; & this same method could facilitate an occasional cross in the hermaphrodite bivalves. I long thought from the description which I had read that the common oyster was a case of perpetual self-fertilisation, but it now seems as I am informed by Prof. Huxley, from the observation of M. Devaine that the male & female products are matured at different periods & therefore that the oyster though in structure an hermaphrodite, in function would appear to be ⟨bisexual⟩ unisexual. /22 v/This likewise, according to Prof. Huxley's own observations is the case with the ⟨bisexual⟩ hermaphrodite ascidians./22/From the analogy of plants, I should expect that this maturity at different periods would prove to be of frequent occurrence with/23/animals. In parasitic worms or Entozoa, many are ⟨bisexual⟩ unisexual, but some which are hermaphrodite[4] mutually unite; & Dr. Creplin remarks that in

[1] Monograph on the Cirripedia, published by the Ray Soc. 1854. p. 102.
[2] [Here Darwin pencilled 'all?' before 'the gasteropod', a line under and a question mark after 'gasteropod' and two question marks after 'annelids'. In the margin he pencilled: 'V. Owen & Huxley.']
[3] Von Siebold in Wiegmann's Archiv fur Naturgesch 1837. p. 51
[4] Dr. Creplin in appendix to Steenstrup's Untersuchungen über das Vorkommen des Hermaphroditismus tr. Hornschuch 1846. I have seen a translation of this owing to the kindness of Mr. Busk

those in which Von Siebold discovered an internal passage from the male to the female organs, apparently insuring perpetual self-fertilisation, the so-called cirrus exists, which would lead from analogy to the conclusion that there must be at least occasional mutual fertilisation.

Distrusting my own knowledge I applied to Professor Huxley, whose knowledge of the invertebrate animals is well known to be profound, whether he knew of any animals whose structure was such that an occasional cross was physically impossible. He informs me that some of the jelly-fish (Beroidae) seem to offer the greatest difficulty, but even in them it is not positively known whether or not the eggs are discharged fertilised;/24/& that as these animals derive their food from indrawn currents of water, which bathe the ovaria, it is certainly quite possible that the spermatozoa of other individuals might come into action.[1] Again Prof. Huxley informs me that he should have thought that the hermaphrodite Bryozoa or Polyzoa (certain corallines) would have offered insuperable difficulties to an occasional cross, had it not been for Mr. Hinck's observations, who saw in some species the spermatozoa pouring out from pores between the tentacula; & as Prof. Huxley remarks what could this be for, except to fertilise some other individual. Moreover there are some ⟨unisexual Polyzoa⟩ Bryozoa, with the sexes distinct,[2] which proves that fertilisation can be effected between the separated, & yet fixed polyps.[3]

In all these cases of aquatic animals it is well [to] remember Spallanzani's curious experiment,[4] namely that three grains of the liquor seminis of a frog thoroughly diffused in a /25/pound & a half of water retained its full power, & that the same quantity when diffused in 22 pounds sufficed to vivify some of the eggs. The weight of the liquor seminis serving to fertilise a single egg was calculated to be only $\frac{1}{199,468,7500}$ of a grain! Finally as far as I can discover, under our present state of knowledge, no animal is known, the structure of which would prevent an occasional cross; & this fact, considering the astounding diversity of nature, seems to me an improbable coincidence, without the capacity of such occasional crossing be one of the laws of reproduction ⟨propagation⟩.[5]

---

[1] [Pencilled memorandum:] Nordmann & Owen on sexes separate in Flustra. [Nordmann] *L'Institut* 1839 p. 95—on sexes in coralline allied to Flustra & on zoospermatic animalcules!

[2] Hincks Brit. Assoc. 1852 Proc. of Sect. p. 75.

[3] [Pencilled memorandum:] I might put in notes other examples of Pollen.

[4] Dissertation II relative to Natural History. English Translat. [See vol. II, dissertation II, § CXLIII pp. 149–50.]

[5] In Bee v. Siebold shows semen keeps power for 4 or 5 years. [*Parthenogenesis in Moths and Bees*, p. 14].

26/*Crossing of Plants.* To show that all plants are capable of being occasionally crossed by another individual of the same species, is more difficult than with animals. Hermaphroditism ⟨Bisexuality⟩ with self-fertilisation is here the rule, & the separation of the sexes the exception. The mere proximity of the male & female organs in the same flower,—the apparent frequency of the pollen & stigma being ready at the same time,—the explosion of the anthers close to the stigma & the lightness of the pollen,—the movement of the stamens to the pistil & of the pistil towards the stamens, would all at first lead to the conclusion that self-fertilisation would be almost invariable. But I think we shall see that such a conclusion would be hasty. Besides the comparatively few monoicous & dioicous ⟨mono- and dioecious⟩ plants, C. C. Sprengel[1] has shown that many hermaphrodite plants are what he calls dichogamous,—namely that either the pollen is mature & has been shed in one flower before its stigma is ready to receive it, or on the contrary (which is a less frequent case) the stigma is ready before the anthers have burst;[2] hence in these cases, the plants are essentially ⟨bisexual⟩ unisexual, being fertilised by the pollen of an older or younger flower ⟨or at least an occasional cross is greatly facilitated; I cannot doubt from the observations of others, & even from my own that these cases are frequent.⟩ I may state that I have tested during several years many of Sprengel's observations, in those cases in which I could judge by the clefts of the stigma, opening &c, & am convinced of his general accuracy.[3]/

26 bis/It would be useless to give examples of dichogamy from Sprengel; they are so numerous, for instance in many Scrophulariaceae, & in all or in most Umbelliferae;[4] in many Onagraceae as I have myself observed in genera not noticed by Sprengel. So again Kölreuter observed similar facts long ago[5] in many Mal-

---

[1] The curious work containing these observations is entitled "Das Entdeckte Geheimniss der Natur 1793." The greatest living Botanist, Robert Brown, thinks highly of Sprengel's power of observation, as I ⟨know from conversation⟩ have heard from him. I mention this because Gaertner in his two admirable works does not seem to think highly of Sprengel; He gives however, (Bastarderzeugung p. 65) one strong case of inevitable dichogamy & admits (p. 659) that in *many* plants, as in whole families, that the pollen and stigma do not come to maturity at the same time, but in many of these cases, probably in most, the pollen is retained close at hand so that it may easily fertilise the pistil in the same flower. A. F. Wiegmann (Uber die Bastardzeugung 1828. S. x) says after careful observation he is convinced that most of the statements of Sprengel are correct, & that he could write a commentary on his work.

[2] [Pencilled memorandum] Mr Schubert [?]

[3]   "   add half-dichogamous

[4] Das Entdeckte &c p. 28, 154, 50, 322 &c.

[5] Mem. de l'Acad. St. Petersburg. Tom 3. p. 197

vaceae. This seems to be the case from Cassini's observations[1] "nearly throughout the Compositae."; & the pollen in this great Family was observed by Kolreuter to be aculeate, & specially adapted to adhere to insects. In Lobelia, judging from my own examination of a few species, the pollen is swept clean out of the united anthers, in the same manner as in the Compositae, by the fringe on the style, some time before the stigma is ready for its reception. That the growth of the pistil in these cases is really adapted to sweep the pollen out of the anthers, before the stigma is mature, I must think from having observed the same process effected by very different means in the Crucianella stylosa [;] here the mouth of the corolla is much contracted, so that the anthers, which open whilst the flower is in bud, instead of being united together as in Lobelia & the Compositae, are pressed close round the pistil. The style is of remarkable length, & lies zig-zag in the bud; as soon as the flower opens it is rather quickly & sometimes suddenly protruded by its elasticity; & in this movement owing to the largely knobbed & rugose stigma, it pushes out the pollen; & not till some time afterward does the stigma open & becoming humid is apparently ready for fertilisation./

27/It is known that many cultivated varieties of plants, not only are capable of occasionally crossing, but without great care are actually crossed very frequently. The Cruciferae are particularly apt to be adulterated, a single cabbage*ª[2] plant sufficing to contaminate whole beds of other varieties. I had a radish plant which flowered in the same bed with several other varieties: I saved a few seeds from one plant, & out of the 22 plants which I raised only 12 came true to their kind. /27 v/So, again, with turnips [.][3]

---

[1] quoted in Linn. Transact Vol XIII. p. 595

[2] [Scholars should be warned that for fols. 27 & 30 Darwin's note slips seem disarranged or interchanged, and their proper insertion points seem uncertain. Darwin used reference marks such as '*ª' to identify notes and to indicate their insertion points. These will be given along with the present folio reference numbers of the note slips, so that the reader may rearrange them if he does not accept the order adopted here.]

*ª [30 v] Gaertner (Bastardzeugung p. 566) gives an experiment on 4 plants of Matthiola annua, the flowers of which he castrated, & kept two in his room unfertilised & they produced no seed; the other two he placed in his garden, *100 yards* distant from some other plants; both of them produced some poor pods, containing 68 apparently good seed, from which, however, only 20 seedlings were raised. See p. 573 for an analogous fact with Nicotiana. *ª [30 v] ⟨Wiegmann's experiments (Über die Bastardzeugung s. 32, 33) on cabbages show the extraordinary degree in which they cross without artificial aid. A most intelligent foreman in a nursery garden assured me that he had known seed of a plot of Brussels-sprouts spoiled by a bed of Drum-head cabbages about 200 yards off. Mr. Masters of Canterbury has known a single Red Cabbage spoil the seed of Savoys, Cabbage & Broccoli in neighbouring gardens.—Many other instances could be given.—⟩   [3] Gardener's Chronicle 1856 p. 729

In the Cruciferae according to Gaertner,[1] the pollen & stigma are not ready at the same time; but I doubt whether this alone will account for the extent to which they blend, & I suspect that the pollen of another variety must have a prepotent effect over the pollen of the stigma's own flower, in the same way as it is known that the pollen of one *species* is prepotent over & obliterates the effect of the pollen of another species, previously placed on a stigma.—/27/Gallesio in his treatise ⟨on oranges⟩ does not doubt that oranges very commonly cross.[2] It is impossible to prevent the different varieties of Rhubarb (as I have known myself) from crossing, if grown near each other./27 v/The various species of Crinum sent by W. Herbert[3] to Calcutta cross so freely in the garden, that true seed cannot be saved./27/And many other instances in Rhododendron, Berberis, Poppies (in which latter I know of case in which not one seedling came true) could be given./27 v/The mere circumstance of great beds of one variety being cultivated in any one place is alone a considerable protection that seeds shall not be adulterated; & hence certain villages[4] have become famous for pure seed of certain varieties, owing to the masses of the same variety there cultivated & to the exclusion of other kinds./

27/But by far the strongest proof, as it seems to me, of the extent to which the pollen from one flower is carried to other flowers of the same species is incidentally offered by hybridisers. Without a single exception, all these naturalists, several of whom have devoted their lives to the subject, insist in the strongest manner on the *absolute necessity of perfect isolation* of the castrated flower,*[b5] so as to preclude the possibility of access of its own pollen. Herbert[6] positively states that it is not always sufficient to enclose/28/a flower in gauze; so subtle are the means by which pollen can be introduced: from some experiments in hybridising which I have made myself, I suspect that it is the minute Thrips which in these cases brings pollen, as I have found that it crawls into flowers protected by gauze & I have often found this insect

---

[1] Bastardzeugung s. 659   [2] [Traité du Citrus, pp. 40–1.]

[3] Amaryllidaceae p. 32. [cf. Darwin's *Effects of Cross and Self Fertilisation* (London, 1876) ch. 10, note on p. 395. Another note slip marked 27 v, having no indication of insertion point, reads:] Kolreuter Dritte Fortsetzung. s. 56. Hybrid Pinks often arise naturally in gardens.

[4] Lindley's Horticulture p. 319.

*[b) 5] [30 v] See Gaertner's, the most admirable of all observers on this subject, strong expression on this subject in his Bastardzeugung s. 670. Experiments made in the open air, he says, must be absolutely rejected. (Beitrage zur Kenntniss s. 510, 573), See also Lecoq De la Fecondation &c. 1845. p. 27. [cf. Darwin's *Cross Fertilisation*, ch. 10, note on pp. 378–9.]

[6] Amaryllidaceae p. 349

dusted with pollen. In the first season of Gaertners grand series of observations he crossed after castration 20 distinct genera, & obtained, as he thought hybrids from ⟨nearly⟩ all; but Herbert, who had been in the field before, at once published his entire disbelief of these experiments, & asserted that the isolation had not been sufficient; which was subsequently acknowledged[1] with perfect candour by Gaertner. Prof. Henschel's experiments, worthless in all other respect[2] are interesting as showing the extent to which crossing goes on without they be completely isolated; he castrated flowers of 37 species (belonging to about 22 genera,) & either put on no pollen or pollen of other genera &c, & yet obtained seedlings from all.—Other parallel cases of experiments made by Dr. Mauz might have been given. No doubt in many of these cases the fertilisation has been effected by pollen carelessly left in the castrated flowers. But a most curious table published by Gaertner[3]/29/shows I think conclusively to what a wonderful extent pollen is carried from flower to flower. In 1825 he castrated 520 flowers & placed in them pollen of other species & genera; & as he says he thought it laug[h]able to suppose[4] that pollen could be brought to his castrated flowers from other flowers of the same species growing between 500 & 600 yards distant, he did not isolate the plants more perfectly. The result was[5]

Flowers

19.   Which produced seed that did not germinate, & therefore have no bearing whatever on the result, & may be eliminated.

29.   which produced true hybrids, & therefore the pollen, intentionally placed on them produced its effect.—

270   which remained unimpregnated, & therefore on which the foreign pollen had no effect, & on which the pollen of its own kind had not been brought by any agency.

202.   produced seed, which yielded pure plants, & therefore on which the foreign pollen had produced no effect, but pollen of its own kind had somehow been introduced.

520   total number of flowers experimentised on in 1825.

Now one's first impression is that in the 202 castrated flowers, attempted to be impregnated with other pollen, but which produced their own kind, is that their own pollen must have been carelessly left in; but Gaertners tables[6] shows that this explanation is not

---

[1] Bastardzeugung. s. 128   [2] given by Gaertner in his Kentniss s. 574

[3] Kenntniss s 550   [4] Kenntniss. s. 539, & 575

[5] Kenntniss s. 576   [6] Kenntniss &c s. 555 & 576

sufficient, for during the 18 subsequent years he castrated no less than 8042/30/flowers, & always kept them in a closed room, so that they could not possibly get pollen from other individuals of the same species, & then during these many years & out of the 8042 flowers he had only 70 cases of seed producing pure plants, showing, that the castration had been imperfect; whereas in 1825 when he experimentised in the open air we have seen that out of 520 flowers 202 produced pure seedlings! Yet plants of the same species in several of the cases did not grow within 500 or 600 yards distance!*[1] It should, however, be not overlooked that when a flower is castrated, the stigma retains its capacity for fertilisation[2] for a considerable time, so that a castrated flower would have a much better chance of being fertilised by pollen from another individual, than would a plant having pollen of its own, the action of which would fertilise the pistil.

Now considering that there are some monoecious & dioicious plants,—that there are many dichogamous plants of C. C. Sprengel, which are in fact monoecious—considering the many cases of the intercrossing of varieties in our gardens—& especially considering the astonishing care which all hybridisers have found absolutely essential to prevent the pollen of/31/its own kind being brought to the castrated flowers—& considering the hybridisers have indiscriminately from varied motives worked on nearly all kinds of flowers,/31 v/—lastly considering the many cases, in our gardens & in a state of nature, of hybrids spontaneously being formed—/31/ I must conclude that the transmission of the pollen from individual to individual is not only very generally possible, but that it actually is so transmitted. Some further facts will be presently given.

Before considering the many grave cases of difficulty opposed to the foregoing conclusion being made universal, it may be interesting briefly to consider the means of transmission. It is known that in many plants with the sexes in separate flowers the pollen is carried by the wind, & hence has to be produced in such astonishing quantities, that many buckets full of the pollen of various fir-trees have been swept off the decks of ships on the shores of N. America.[3] In some associated hermaphrodite plants, as in Gramineae, in which the stigma is large, branched & at the period of fertilisation fully exposed, in which the pollen is but/32/

---

[1]*[27 v] But even in the largest nurseries, it is surprising the trouble which the owners are compelled to take to keep their seed crops unadulterated; thus Messrs. Sharp "have land engaged in the growth of seed in no less than eight parishes." (Gardeners' Chronicle 1856. p. 823). [Cf. *Cross Fertilisation*, ch. 10, p. 395.]

[2] Kenntnis S. 145   [3] [Anon.] Silliman's Journal Jan. 1842.

little coherent, & the long slender filaments seem formed to scatter the pollen, I do not doubt that crosses must be often effected by the wind./31/But in most hermaphrodite ⟨bisexual⟩ flowers, owing to their structure, or to the small quantity of the pollen, or to its coherence or to the small size of the stigma, I think it may safely be concluded that the wind can but seldom bring sufficient pollen (for several grains are almost always required for the act of fertilisation) from one flower to the other so as to effect a cross between two individuals./32/Insects of various orders, more especially Bees,[1] are the great agents. Many flowers cannot be fertilised without their agency, as is admitted, though very unwillingly, by Gaertner:[2] it is impossible to read C. C. Sprengels details & then examine many flowers, as most Irideae, Passiflora, Viola, most Orchideae &c &c & doubt this: in regard to all the Ascelpiadeae which have been carefully examined, Robert Brown, says the 'absolute necessity' of the assistance of insects is manifest,[3] & Sprengel ⟨Gaertner⟩ believes that their agency is rendered more effectual by their extraordinary activity, due to the intoxicating effect of the nectar.—It would be tedious to give other examples. All those who have personally attended to the subject have become strongly impressed with the efficiency of insect agency in the fertilisation of flowers./

32 n/There can be little doubt that C. C. Sprengel has pushed his views to a quite fanciful degree; as for instance, when he accounts for all the streaks of colour on the petals, as serving ⟨formed⟩ to guide insects to the nectary. Nevertheless some facts could be given in favour of such a view: Thus in a patch of the little blue Lobelia, which was incessantly visited by Hive Bees, I found that the flowers from which the corolla, or the lower streaked petal alone had been cut off, were no longer visited./ ⟨Whether the Bees were then led to think that these flowers were withered, or whether the absence of this convenient alighting place on the lower petal was the cause, I know not. But I feel sure that⟩ Bees seem to work against each other with excessive competition ⟨industry⟩, so that they grudge the least loss of time: thus when visiting flowers with several nectaries if one be dry, they do not try the others; again when visiting flowers which have been bored, if one has accidentally not been bored I have seen Bee after Bee pass over it & not stop to bite a hole nor will they enter the open tubular flower, though having to crawl over it, but will dash on to another bored flower. By the way, if proof were wanted how little Bees require any guide to the nectary, their habit of biting holes in the lower part of the corolla or through the calyx, so as to reach the nectary without the loss of time of crawling in at the mouth of the flower would prove it. These holes when once formed are known to

[1] Bees are found in all parts of the world; even in the extreme Arctic regions they have been seen sucking the flowers. But I must add that on the little coral islets, called the Keeling Islands, in the Indian Ocean, I found no Bees; but there were other insects.

[2] Beiträge zur Kenntniss. s. 335   [3] Linnean Transact. vol. 16, p. 731

& used by Bees/32 n. 1/of various species & genera: when as in Kidney beans, the hole has been bored on the lower side of the calyx, bee after bee flies to the under side with unerring precision. Bees, as far as I can judge are guided by various senses to flowers, & more especially by knowledge of the position of each tuft of flowers in a garden. It is well known that the same Bee keeps as much as it can to the same species, when getting nectar; & I have repeatedly seen them flying in a ⟨direct line⟩ clearly determined course from plant to plant of the same species, when round a corner & so out of sight.—They are good Botanists, & know well that plants of the same species may have brilliantly different colours; but they know that they are only varieties & visit them indiscriminately. I think they recognise a plant by its general habit; I have seen Humble bees after visiting a tall blue Larkspur fly to another plant, of which the buds were so little open, that they were hardly tinged with blue. They seem often to be aware if another Bee has *almost instantly before* visited a flower, & will then not try; but I have seen one blunder & itself visit the same flower twice./

33/It is, I think, impossible to doubt that the structure of very many flowers has been formed in direct relation to the part which insects play in their fertilisation. What can be a more beautiful adaptation than that shown by R. Brown to exist in the Asclepiadeae & Orchideae, between the stickiness of the gland of the pollen masses, of their separate grains one to another & to the surface of the stigma, by which it follows that the instant an insect touches the gland it draws out the whole pollen mass out of its case, & then the sticky stigmas of the several flowers, as the insect crawls from one to the other, each take a few grains of pollen from the coherent mass. It is worth anyones while to watch a Bee visiting a Salvia, or to push some thin body like the Bees head down the tube of this flower, & notice how the anthers & stigma are protruded & rubbed on the Bee's back; then let him cut open the flower & see the cause of this is two projections near the base of the stamens, closing the passage, & the movements of which by the/34/Bees proboscis, causes the protrusion of the anthers & stigma from beneath their hood: I can no more doubt the final cause of this structure than I can of a certain mouse-trap. I have seen a Bee enter the flower of a Mimulus & in doing this the two-lipped stigma fairly licked the back of the insect which was thickly dusted with the pollen from another flower, & then the two lips of the stigma slowly closed on the pollen which it had thus obtained. It is pretty to compare in those species of Fumaria, in which either one or both nectaries secrete honey, the different movements of the parts of the flower as a Bee enters. Even in trifling details as in the position of the stamens ⟨anthers⟩ & pistil, in relation to the nectary, I believe that there is very generally a distinct relation to the action of insects: thus in the

Dictamnus Fraxinella, I noticed during several days that the stamens & pistil were placed so that a Bee visiting the nectary would not touch them; but then came a hot day & the anthers all burst & stigma was humid, & I found their positions all changed & their tips now stood in the direct gangway to the nectary, & were/ 35/brushed by every Bee which entered. I could fill pages full of other instances from C. C. Sprengel & from my own observations.

I will only allude to the case of the Berberis/35 v/in which the stamens move to the pistil, & in which consequently, it might have been thought there would be seldom any chance of a cross with another individual: but Kolreuter has shown[1] that they never move till touched by some insect; so that insects are necessary to their fertilisation; & their flying from flower to flower could hardly fail to bring pollen from individual to individual. Indeed the extent to which the American evergreen Barberries (Mahonia) have been hybridised together, so that it is almost difficult in our nursery gardens, as I have found, to get a pure plant, shows that this has occurred not only with the individuals of the same species, but of different species. Similar remarks are applicable to some other plants, of which either the stamen or pistils move on being touched./35 v/Thus in regard to the pistil of a Goldfussia, it is scarcely possible to doubt from Ch. Morrens[2] remarks & curious observations that the movement of the stigma when touched towards the lower side of the corolla, where the fallen pollen is collected, stands in direct relation to the action of insects. Again in Stylidium, as described by Ch. Morren[3] I can see no difficulty, from the proportions in the parts, of a Bee carrying pollen from flowers, when by sucking at the nectary it causes the sudden & remarkable movement of the column; though Ch. Morren may be quite correct that this movement, also, aids the fertilisation of the flower by its own pollen.—/35 v/In Parnassia palustris the stamens slowly move one after the other over the pistil; but Sprengel[4] positively asserts that the pistil at this period is not fit for fertilisation, & ⟨therefore that the plant is strictly dichogamous⟩ he supposes that it is fertilised by pollen from a younger flower brought by some nocturnal insect. Allium[5] is in nearly the same case./

35/Bees & other insects visit flowers both for the pollen & nectar. The nectar cannot be supposed to be formed, any more than the pollen, for the sole purpose of attracting insects; for nectar is sometimes secreted outside flowers, as by the bracts of various Legumi-

---

[1] Nova Acta Petrop. 1788 p. 214
[2] Nouv. Mém. de l'Acad. Roy. de Bruxelles. Tome xii 1839.
[3] Ib. Tome xi. 1838   [4] Geheimniss &c 167   [5] ib. p. 186

nosae,[1] but nature has utilised this secretion for the very distinct purpose of facilitating fertilisation, & as I believe occasional crossing.

When Kolreuter first discovered[2] that the Malvaceae, owing to the adhesive pollen & stigma not being ready at the same time in the same flower, can be fertilised only by the agency of insects, he says he was astonished that so important a function should have been left, as he then thought to accident,—to a mere happy chance; but he adds that further observation convinced him that the wise Creator has thus used the most/36/sure means. Hardly any means, I am convinced, could be surer; & in regard to our present discussion, it should be borne in mind that in every case in which insect agency is essential to fertilisation, & indeed in every case in which insects habitually visit flowers during this period, it is hardly possible to doubt that pollen is often brought from flower to flower of distinct individuals, & thus a cross between ⟨separate individuals of the same species⟩ effected. I have repeatedly seen many minute beetles, dusted with pollen, fly from flower to flower: some flowers, which are very rarely visited by bees as the Phloxes (which I have never seen visited except during one year) are frequently visited by butterflies:[3]//36 v/I have remarked this particularly with the Rhingia rostrata on the Lychnis dioica, on an Ajuga & on many others, I have seen the same thing with Volucella plumosa on a Myosotis,—I may add that I have never seen a Bee visiting a Daisy but I have seen the Rhingia, Scaeva iris (?) & Hilara globulipes all thickly dusted with the pollen of this plant/37/flight, dusted like millers with pollen. I have seen several times the same thing with Thrips, an insect hardly larger than a bit of chopped bristle: one day I watched with a lens, one in the flower of a convovulus having four grains of pollen on its head, & these I saw left on the stigma, as it crawled over it. The crossing of the great flowers of foreign lands, may well be aided by Humming & other birds: I remember shooting in S. America, a mocking thrush, which had its head of so bright an orange from the pollen of a Cassia that I at first thought it was a new species.—

But Bees are the most important of all insects for this end. Until I watched I was not at all aware how quickly they work. In exactly one minute I saw one Humble-Bee visit 24 of the closed flowers of a toad-flax (Linaria cymbalaria); another 22 flowers of the Snow-berry tree (Symphoricarpos ⟨Chiococca⟩ racemosa); another 17 flowers of a Larkspur on two separate plants &c.—

---

[1] I called attention to this fact in the Gardeners Chronicle, 1855 July 21; & had at the time quite forgotten that it had been previously noticed by Sprengel
[2] Vorlaufige Nachricht 1761 p. 22
[3] [The rest of fol. 36 is sheared off.]

The top flower of an Oenothera was visited eight times by Humble Bees in 15 minutes/37 v/& I noticed that one Bee visited in the course of a few minutes every single plant of Oenothera in a large flower garden; passing over, without regard, other plants having large yellow flowers, like Escholtzia:/37/in 19 minutes each flower of a tuft of Nemophila insignis was visited twice: in a large plant of Dictamnus Fraxinella with 280 flowers, from the rate at which Bees visited it, as observed during/38/several days, each flower at *lowest* computation must have been visited daily 30 times. It is no wonder that the beauty of many flowers, as I have noticed in some Mimuli & Lathyrus grandiflora, is greatly destroyed by the scratching of the hooked tarsi of the bees. Some flowers seem never visited by Bees; but with the exception of the Gramineae in all other cases of *indigenous* plants to which I have attended I have found that they were visited by other insects. Night-blooming flowers, which are often sweet-scented & of a white colour, I have reason to believe are visited by moths. One must be very cautious before assuming that any flower is not visited by Bees: in the first summer of my observations on this subject I watched many times daily for 14 days the Linaria cymbalaria & never saw a Bee look at it, when suddenly after a hot day Bees were most industriously at work. So again for a fortnight I saw Bees visiting White & Red clovers, but never looking at the little yellow Trifolium minus; & as the flowers were so minute I doubted whether they would ever visit them; when suddenly I one day found innumerable bees hard at work at this species over the whole country, & neglecting the other kinds. In ⟨all these⟩ most cases I believe that the secretion of the nectar, which determines the visits of the Bees is coincident/39/with the flowers being ready for fertilisation. The secretion of the nectar seems in close relation to temperature: I have observed in a little blue Lobelia, that if the sun went behind a cloud for even half an hour, the visits of the Bees immediately slackened & soon ceased.[1]/39*/I may give one more instance in regard to the action of insects. In Viola odorata Sprengel has shown that the pollen cannot escape owing to the manner in which the anthers with their scales close round the pistil, till disturbed by the proboscis of an insect: he proved this by covering up some flowers & leaving others uncovered,[2] & finding pollen shed in latter, but never in the protected flowers.

---

[1] [Fol. 39 is sheared off just after an asterisk at this point. The lower part of fol. 39 is now numbered 40 A. In the C.D. MSS., fol. 27 of vol. III is a full-sized sheet of gray foolscap marked: '* to p. 39', and is inserted here where it seems to belong.]

[2] Das Entdeckte &c p. 390.

Now in 1841 I watched almost daily & many times a day several patches of the V. tricolor or Heartsease for seven weeks & never saw an insect of any kind visit them; when suddenly on two successive days I saw several small Humble-bees visiting all the flowers. In the next year after a fortnight watching in vain I again saw two or three species of Bees (& a fly dusted with pollen) visiting most of the flowers & I found pollen profusely shed on the lower petals, all around the stigma; & I noticed the same fact on the same day with some plants of wild V. tricolor. Now in both these years, I noticed a few days after the visit of the Bees & of the Fly (I marked the flowers visited by the Fly) a great number of the flowers on the several clumps suddenly withered as if the germens had been set. Hence I cannot in the least doubt that I saw in these Humble Bees, the priests who celebrated the marriage ceremony of the Heartsease./

40 A./In a flower garden containing some plants of Oenothera, the pollen of which can easily be recognised from its great size & shape, I found not only single grains, but whole masses within many flowers, of Mimulus, Digitalis, Antirrhinum, & Linaria. Other kinds of pollen were likewise distributed in the same flowers. A large part of the stigmas of a plant of Thyme in which the anthers were completely [?] aborted were likewise examined & their stigmas, though scarcely larger than a split needle, were covered not only with the pollen of Thyme brought by the bees from other plants, but with several other kinds of pollen. ⟨but I was not/40/surprised at this, seeing how much Thyme is frequented by Bees & flies⟩/40 v/Those who have not attended to the subject of Hybridism; may feel inclined to exclaim that if pollen is carried from distinct species to species, so freely as these facts show in the cases, an endless number of hybrids would be formed. But nature has provided a most efficient check to this, namely in the prepotent effect of each species own pollen; so that all effect from the pollen of another species is obliterated by the previous or subsequent action of its own.—/40/I found a hybrid Rhododendron which [was] quite destitute of pollen, & which was so seldom visited by Bees, that after long watching the branch [?] for many days I never saw but four Bees visit it: yet on one morning I found from 50 to 100 grains of pollen of Azalea or Rhododendron on the stigmas of these flowers: another day I examined the stigmas of 19 flowers & on 13 of them there was the same ⟨I found some⟩ pollen. Kölreuter relates[1] a curious experiment bearing on this subject: in an Hibiscus, which is necessarily fertilised by insects,

[1] Fortsetzung &c 1763 p. 69

because its pollen is shed before the stigmas are ready, he marked 310 flowers & daily put pollen on their stigmas & left the same number of other flowers to the agency of insects which did not work during some days as the weather was cold with continued rain. He then counted the seeds of both lots; the flowers which he fertilised with such astonishing care produced 11,237 seeds & those left to the insects 10,886—that is only 351 fewer seeds./

41/From the facts now given, at too great length, though I could have given many more, I think it can hardly be doubted that insects play a very important part in the fertilisation of flowers; & furthermore that in those cases in which their agency may be not at all necessary, yet that they can hardly fail occasionally to bring pollen from one individual to another. Nor must the action of the wind be quite overlooked, which probably is highly efficient for an occasional cross in some hermaphrodite flowers, as it undoubtedly is in some mono-oecious & dioecious plants for their ordinary fertilisation. I should, indeed, have been inclined boldly to affirm the proposition that all plants are not only capable, but do actually receive an occasional cross, had it not been for the following cases of serious difficulty.

Facts opposed to the doctrine that in plants an occasional cross is necessary. Very many statements may be found in the works of Botanists not only that the pollen is often matured & the anthers burst before the bud is opened, which admits of no doubt, but that in certain plants the stigma is regularly fertilised in the unopened flowers.[1] which would render an occasional cross a physical impossibility. But there are many difficulties in the way of/42/ascertaining this; & observations made on only a few flowers during one season cannot avail much; for Gaertner has shown[2] that the bursting of the anthers & relative maturity of the stigmas depends much on the weather & varies in the same species; & there seems to be no doubt that a plant may be occasionally fertilised in the unopened bud, of which the pollen is ordinarily ready only when the flower is fully expanded. Again Gaertner has shown[3] that an abnormal precocity not rarely affects many flowers & that in this abnormal state it can be fertilised in the bud. But Gaertner was a firm believer that in many plants, even in whole Families,[4] as the Leguminosae, Cruciferae, Onagraceae, Campanu-

---

[1] Thus Aug. de Saint. Hilaire in his admirable Lecons de Botanique 1841. p. 572 says, without entering in details, "Chez une foule de plantes c'est dans le bouton ....que la fecondation s'opère."

[2] Beiträge &c. s. 104   [3] Beiträge zur Kenntniss &c. s. 571

[4] Bastardzeugung s. 655

laceae &c, fertilisation takes place, not only some hours, but even from one to two days before the corolla opens. Now I am quite unable to reconcile this statement with others: of the Leguminosae I shall speak afterwards: in regard to the Campanulaceae there has been much discussion on this very point, & notwithstanding Gaertner's statement[1] that the stigma can be fertilised before the clefts are fully marked, I can hardly doubt that Sprengel formerly & Wilson lately[2] are correct/43/in believing that the fertilisation takes place after the flower is fully opened; if Gaertner is correct that the fertilisation takes place in the bud there is an inconceivable waste of pollen on the curiously organised, & retractile collecting hairs of the pistil; & the manner in which Bees, as I have often watched, frequent the flowers is admirably adapted to bring the pollen from the collecting hairs of one flower on to the stigma of another./43 v/In Phyteuma, one of the Campanulaceae, Sprengel found[3] plenty of the coloured pollen on the open stigma; but if a branch, with unopened flowers was put into a glass of water in a room where there were no insects, not a grain could be discovered on the stigmas./43/So again in regard to the Onagraceae, I must think the weightiest evidence would be required to overthrow Sprengels statements[4] in regard to Epilobium & Oenothera (which as far I can judge from repeated observation seem strictly true) that far from being fertilised in the bud, they are dichogamous, & invariably fertilised by the pollen of younger, which he saw effected by Humble Bees: Gaertner himself, elsewhere[5] admits that in some Fuchsias, the pollen is not shed for some days after the flower is fully expanded, as is well known to Hybridisers.[6] Lastly with respect to the Cruciferae, Gaertner's statement that they are fertilised in the bud seems to me quite extraordinary, considering the everyday experience of gardeners with cabbages, turnips, Radishes, &c. & to my mind throws doubt on/44/his other statements in regard to habitual fertilisation in the bud.

M. Loiseleur-Deslongchamps[7] believes, though confessedly on imperfect observations & in opposition to some other authors that Wheat is fertilised within the closed flowers. This surprises me much, for I have repeatedly seen the florets widely open, with the feathery stigma protruded on one side, with the dangling anthers not fully discharged, & with the grains of pollen sticking over all parts of the florets: in most grasses, all the florets open at the

---

[1] Beitrage s. 338
[2] Hookers. [*Lond. J. Bot.* 7 (1848) 92–7.]
[3] Das Entdeckte. s. 117
[4] Das Entdeckte &c. s. 225
[5] Beiträge &c. s. 104
[6] Lecoq de la Fecondation &c. p. 129
[7] Consid. sur les Céréales 1842. p. 80

same time & with the protruded stigma, the plant for the time, as
every one must have observed has a very different appearance: in
wheat each floret opens separately & keeps open for only 3 or
4 hours, leaving the empty anthers dangling outside so that the
whole phenomenon is far less conspicuous than in most grasses; &
if the Chinese are at all to be trusted some varieties as Huc states
flower in the night.[1] The structure is such that I can hardly
understand how an occasional cross from another individual can
be avoided. A. Knight[2] asserts that by sowing different varieties
together,/45/"I obtained as many varieties as I wished." Col Le
Couteur whose great experience makes his opinion valuable, though
he gives no precise facts believes[3] that wheats cross. Puvis[4] asserts
that nearly all the varieties which were grown near each other in
an Agricultural Garden under his charge were each year modified;
but his evidence seems to me of little value, as he attributes to
the action of the pollen on the grain itself that kind of change
which it is known results from climate & culture.—Opposed to
these statements we have a much more precise one from M. Loiseleur Deslongchamps.[5] Namely that during eight years he cultivated
from 100 to 200 varieties very near each other, & that he never
saw a hybrid appear. Making some allowance for different varieties,
as noted in this very respect by Col. Le Couteur, flowering at
different times, & even from the positions of the beds with respect
to the wind, this statement is very remarkable, & at first seems
almost conclusive against occasional crosses. But I do not think
the experiment has been fairly tried, until the different varieties
are sown close together, as Knight sowed/46/them; for wheat is
not, as far as I can observe visited by insects & a cross could
take place only by the wind, & as the pollen though pretty
plentiful bears no sort of comparison, to the quantity in those
dioecious plants, in which the wind is the fertilising agent, crosses
could very rarely, as several grains of pollen are probably required,
take place without the two individual grew quite close together.
This remark is probably applicable to most Graminea; but the
social habits of most of the species in the Family makes the
difficulty of an occasional cross less than it would be in less social
plants./46 v/Water plants are very apt, I think, in proportion to their
numbers, to have their sexes in separate flowers: these, also, are
very social plants, as remarked by M. Alph. De Candolle. According
to all analogy, the division of labour, or in this instance separation

---

[1] [Huc, *Chinese Empire* (London, 1855), II, 312–13.]
[2] Philosophical Transactions 1799 p. 200   [3] On the Varieties of Wheat. p. 66 &c.
[4] De la Dégénération 1837. p. 77   [5] Ib. p. 81

of sexes, is advantageous to all living beings, & therefore it may
be that water plants can safely partake of this advantage, because
they grow nearer each other, & therefore can be more easily
fertilised by pollen brought by the wind or insects.—/

46/It may, perhaps, be objected, that large trees with thousands
or tens of thousands of flowers, (like a large bed of the same
variety of a plant in a garden with respect to another *variety*)
could hardly ever be crossed with the pollen of another distinct
individual;—that crossing between the several flowers on the same
tree at best would be like the crossing of near relations in animals,—
& this, I think is a valid objection. But on the other hand it is
a curious fact, ⟨which I have heard remarked on by Botanists, &
which will strike⟩ that if any one will turn over a Synopsis of the
Vegetable Kingdom/47/on the Linnean system, he will find that
the Monoecious, Dioecious, & Polygamous classes include a surprising number of trees,—that is that trees are apt to have their
sexes separated. Now it is obvious that in flowers, which can be
fertilised only by the pollen from another flower, there will be
a better chance, (whether the pollen be habitually brought by
wind or insects) that pollen should be brought from a quite distinct
individual, than in the case of a hermaphrodite flower having its
own pollen close at hand.—/47 v/⟨Let any one run over in his
mind the trees even in our own small island, & he will find many
in this predicament; & even some that are hermaphrodite, I have
reason to believe are according to Sprengel dichogamous./47/More-
over trees are very apt to grow together or to be social as may be
inferred from the much greater frequency of forest-clad-land, than
of single scattered trees: This relation of sociability may not be
so fanciful as it at first seems:—single trees would interbreed &
would produce seedlings not so well able to struggle with surrounding
vegetation, as the crossed offspring of the same species, & therefore
the species might be able to take root & grow only where several
individuals existed. I am aware that there are very numerous
exceptions to the above remark that trees have their sexes in separate flowers; but yet the above coincidence of trees being so often
mono or dioicous under our present point of view seems worth
notice./

47a/To test the foregoing remark a little further, I find that
in Great Britain there are 32 indigenous trees[1] of these 19 or more
than half (5.93) have their sexes separated,—an enormous proportion compared with the remainder of the British Flora: nor

---

[1] I have taken the 4th Edit. of the London Catalogue as my guide for the indigenous
trees, & Loudon's Encyclop. to distinguish trees from bushes.

is this wholly owing to a chance coincidence in some one Family having many trees & having a tendency to separated sexes: for the 32 trees belong to nine Families, & the trees with separate sexes belong to five Families. This result, as far as the number of species of trees with separated sexes would have been greater had I included all the tall dioicous willows, but I have counted only half-a-dozen willows in the thirty-two.[1] Remembering that Dr. Hooker[2] had observed that the very peculiar Flora of New Zealand was characterised by the number of its trees, & by the number of the plants with more or less separated sexes; I thought the foregoing relation might here be thus well tested: hence I applied to Dr. Hooker, who, not remembering his former results, & as this/47 b/subject is open to doubt under several points of view, has gone over his materials & thinks the following a fair result. There are about 756 phanerogamous plants; & of them no less than 108 are trees. Of the 108, fifty-two or very nearly half, have the sexes separated: of shrubs, there are 149, & of these 61 or considerably less than a third have the sexes separated: of herbaceous plants there are 500, & of these only 121, or not one-fourth have sexes separated. So that we have here the same relation as in Great Britain, with Shrubs shown to be in an intermediate condition. In this case, also, the trees are not confined to some one or two Families, which chanced to have their sexes separated, for these 108 trees belong to no less than 38 Families, & the 52 trees with sexes separated belong to 18 Families, or exactly half. Whether or not, in the above record the trees which have not their sexes separated may be dichogamous in C. C. Sprengel's sense, I do not here consider./

47 bis/Some water plants seem to flower always under water with their corolla perfectly closed: if this could be shown to be invariably the case in any species, it would demonstrate that a cross with another individual could *never* take place. All British Botanists describe the rare Subularia aquatica as flowering under water with the corolla perfectly closed: Prof. Dickie is the only Botanist, whom I know to have examined it often, & he informs me that he has invariably found it near Aberdeen submerged, with the corolla closed, with fully developed anthers & plenty of

---

[1] [Here Darwin pencilled: 'Dr Asa Gray', presumably in connection with the statistics on separation of sexes in trees in a letter by Asa Gray dated 1 June 1857, part of which is mounted as ULC vol. 8, fol. 47 bA. This letter amplifies statistics given in: Asa Gray 'Statistics of the Flora of the Northern United States', Amer. Journ. Sci. and Arts, vol. XXIII, no. 69, ART XXXVII, p. 400, May 1857. Darwin's copy is ULC, vol. 135(3).]

[2] Introduction to the Flora of New Zealand. p. xxviii

seed in Autumn: but in Germany Koch[1] expressly states that 'sub aqua clandestine floret, extra aquam flores parvi albi explicantur.'—The same thing happens with several other marsh plants; thus Limosella aquatica which in this country generally flowers in the open air, was seen by Dr. Hooker in Kerguelen land flowering with closed corolla under the ice.—The Menyanthes trifoliátà is hardly a parallel case, for it is not said to flower under water, but on account of the very humid situations in which it grows, it has been asserted[2]/47 tres/to shed in Russia its pollen & be fertilised with the flower closed; but in Staffordshire I found that this was by no means the case. A more curious instance is offered by Podostemon, some species of which Dr. Hooker[3] informs me flower under water with their corollas closed, carpeting the rocky beds of the torrents of the Khasia mountains in Bengal. The species referred to are annual, & appear only in the rainy season when the torrents are swollen, & Dr. Hooker has never seen them flowering in the open air; but he will not assert that this may not sometimes occur, when the torrents sink. Some Podostemaceous species raise their caulescent stems above water, when they flower; & some few species are monoicous or dioicous, & it is not known whether the pollen in these latter species is carried under water from flower to flower, or whether they are fertilised above water. So that until the natural history of the Family is more thoroughly worked out, this case is not quite so fatal to the views here advocated as it at first appears. There are several other water plants, belonging to the Naiadaceae & allied Families,/47(4)/which seem to offer much difficulty to an occasional cross; but in most of them, the manner of fertilisation is imperfectly known, & several of them are monoicous or dioicous, & therefore it would seem that there must be some means of conveying the pollen under water, from flower to flower.[4]/

48/The following appears a strong case against my doctrine: M. Auguste Saint-Hilaire[5] states that in Goodenia the pollen is shed in the bud, & then becomes enclosed in a cup surrounding

---

[1] Synopsis Florae Germ. Edit. 2. p. 73. I am indebted for this reference to Mr Babington & to Mr H. C. Watson

[2] M. Gillibert. Act. Acad. St. Petersburg. 1777. p. 45. [on verso of fol. 47 bis, Darwin scribbled (& later cancelled) the following: 'Subularia: one of the strongest cases. See J. E. Smith & others; I long thought a case of clear self impregnation.
Babington   like Limosella— Menyanthes   Podostemon.
—   Zostera. Lindley Veg. Kingdom.']

[3] Himalayan Journal vol. 2. p. 314

[4] This is the conclusion of P. Cavolini in regard to Zostera oceanica, & of Willdenow in regard to Najas etc., see Annals of Botany vol. 2. p. 43. 1806.—It seems to be now made out that Ruppia maritima rises to the surface to flower.—

[5] Lecons de Botanique p. 572

the stigma & is then hermetically sealed; so that here a cross would appear physically impossible. But I observe that R. Brown[1] speaks of the cup enclosing the pollen till the stigma is ready. & Ch. Morren[2] speaks of the cup as being excitable, & 'qui se ferme après avoir recu quelques grains de pollen;' therefore I infer it may open itself again—As the cup seems to be analogous with the collecting hairs in the Campanulaceae & Lobeliaceae (to which Families the Goodenia is allied) one must doubt whether the cup would act in so opposite a manner as the collecting hairs./

49/The following is a somewhat different case: Fabricius & Sprengel[3] have shown that Flies are necessary for the fertilisation of Aristolochia clematitis; but they believe, that when a Fly once enters the tubular flower, it is imprisoned for life by the thick set hairs on the inside of the corolla: if this be so a cross with another individual could never be effected. But having been myself deceived in a somewhat parallel case I am sceptical on this subject: in the common Arum maculatum, I found in some flowers from 30 to 60 midges & minute Diptera of three species, & as many were lying dead at the bottom, & as the filaments on the spadix above the anthers seemed to offer some difficulty to their escape, I concluded that after once entering a flower they probably never left it. To try this I quietly tied gauze over a flower & came back in an hour's time, when I found that several had crawled out of the spathe & were in the gauze: I then gathered a flower & breathed hard into it several times, soon several very minute Flies crawled out/[4]50/dusted all over, even to their wings, with pollen, & flew away; three of them I distinctly saw fly to another arum about a yard off; they alighted on the spathe & then suddenly flew down into the flower. I opened this flower & found that not a single anther had burst, but at the bottom of the spathe, near to but not on the stigmas, I found a few grains of pollen, which must unquestionably have been brought by the above or other midges from another individual arum. I may mention that in some other arums which had their anthers burst I saw these midges crawling over the stigmas & leave pollen on them.

⟨I have given *all* the facts, which I have been able to collect, which seem to be opposed to the doctrine of occasional crossing

---

[1] Appendix to Flinders Voyage p. 560. [Brown's statement refers to the order Goodenoviae, to which Goodenia belongs; see p. 561.]

[2] Nouveaux Mem. de l'Acad. Roy. de Bruxelles Tom xi. 1838 p. 4

[3] Das Entdeckte &c. p. 418 [See cols. 424–5.]

[4] [Darwin revised the top of fol. 50 extensively, but the resulting text is fragmentary and obscure. The original unrevised version is given here. For the cancellations and scribbled additions of his revision see appendix.]

in more detail, than those which seem to favour it. And there still remain, three cases, viz Hollyocks, certain Orchideae & the Leguminosae.⟩

Hollyocks (Alcea). Loudon, Herbert & others have stated that the several differently coloured varieties come true from seed. As from the observation of Kölreuter & Sprengel there can be no doubt that the stigmata/51/are fertilised by the coherent pollen of younger flowers, by the agency of Bees, which I have actually witnessed myself in a carefully castrated flower; so this asserted trueness of the many varieties seemed to me very surprising. Hence I brought 18 packets of the best German seed, & raised 18 little beds of plants; but though generally very true, there were seven beds with one or more plants false; altogether out [of] 111 plants 85 came up quite true & 26 not true to their colour. Now if the seed-beds were, as is probable, large, from which it would follow that generally each flower would have pollen brought to it from the same variety, there is nothing in this proportion (even if we attribute, as we ought, some of the false plants to variability,) to cast doubt on the crossing of Hollyocks./

52/*Orchidaceae*: that in very many genera of this Family, the agency of insects is necessary for their fertilisation cannot be doubted, & therefore an occasional cross from another individual is probable.—Mr. R. Brown believes in this necessity, but adds that all the capsules of a dense spike not infrequently producing seeds, seems hardly reconcileable with impregnation by insects[1] I will therefore give a few facts to show how efficient insects are in the Family. It is known that in Orchis, Gymnadenia, ⟨Habenaria⟩ & Listera the pollen-masses cannot be shaken out of their pouches, & can be drawn out only by something touching the sticky gland; yet in a plant of Orchis maculata with 44 flowers open, twelve beneath the buds had neither pollen-masses removed, but everyone of the 32 lower flowers had one or generally both removed: in a stem of Listera ovata, every one of the 17 lowest flowers had pollen on the stigmatic surface: in Gymnadenia conopsea with 54 open flowers, 52 flowers had their pollen masses removed; in another plant with 45 open flowers, 41 had been visited by insects; in another individual I found three pollen masses on *one* stigma. Four small plants of Orchis Morio grew in my orchard; I covered one with bell-glass;/53/the other three plants had 23 quite or partially opened flowers & day after day I found some of the pollen masses disappearing till all were gone with the exception of one single flower which withered with the pollen-masses in their

[1] Linnean Transacts. vol. 16, p. 704

pouch: but one or two terminal flowers, in each plant not included in the 23, & which opened subsequently never had their pollen masses removed. I then looked at the plants under the bell-glass & found not one single pollen-mass removed; & though then left uncovered every flower withered in the course of six days with *all* pollen-masses in their pouches & the germens did not swell. From this fact, I infer that whatever nocturnal insect (for I never saw an insect visit the plants by day) haunts this orchis had ceased its visits, as indeed might be inferred from the extreme terminal flowers of the three plants which had never been covered, retaining their pollen-masses.

I have repeatedly seen in Listera ovata, Gymnadenia conopsea, Habenaria bifolia & Orchis morio, plenty of pollen on the stigmatic surface, but with pollen-masses of the same flower in their pouches; & still oftener the reverse case, namely the pollen-masses removed, but no pollen on the stigmas,—which clearly shows that each flower in these species is very generally fertilised not by its own pollen, but by that of another flower or individual. After having/ 54/attended to this subject at intervals during several years I have seen no insect visit an Orchid, except once a Butterfly sucking an Orchis pyramidalis & once a Gymnadenia conopsea; but Sprengel[1] has been more fortunate for several times he saw a Hymenopterous insect visiting Listera ovata, & he saw the pollen-masses removed & the pollen left on the stigma by these insects: on Epipactis latifolia, also, he saw a Fly with the pollen adhering to its back. I do not doubt that usually moths are the agents for fertilisation; & I must think in the Butterfly orchis[2] (Habenaria) the white-coloured flower, the sweet smell at night, the abundant nectar contained in a nectary with which only a tube as fine as needle can be inserted all stand in direct & beautiful relation to the visits of nocturnal Lepidoptera.—[3]

It is well known that in certain exotic Orchidaceous plants, parts of the flower have the power of movement, when irritated. In Mormodes the pollen-masses are jerked out with such force as sometimes to hit a person's face; & I was told by Mr. Loddiges that he thought not one in a hundred would miss hitting the stigmatic surface: but I am not able to say what the result would be on the chances of two individuals crossing in this case, & in that of those Australian genera[4]/55/in which the labellum when

[1] Das Entdeckte &c. s. 409, 415
[2] Das Entdeckte s. 405
[3] [Scribbled addition:] Entomologists have been often puzzled by finding their glass sticking to   & flower feeding Beetles.
[4] Lindley, Vegetable Kingdom. 1853, p. 179

touched by an insect suddenly turns round & shutting up a box-like cavity, imprisons the insect./

56/We now come to a case in which it appears, though the flower is open, that there is a direct mechanical provision for perpetual self-fertilisation: in certain species of Ophrys, R. Brown[1] has shown that the pollen-masses readily fall out of their pouches, but being retained by their glands, & the stalk being of the proper length, they swing downwards, strike on & adhere to the stigmatic surface: hence insects as Mr. Brown remarks are not at all necessary for their fertilisation: to test this I covered up under a case of gauze some plants of Ophrys apifera, so that no insect could visit them or the wind agitate them, yet in every flower I found the pollen masses fallen on the stigmas. Again during three years I have examined many plants, one day looking at every flower in 18 plants, of this Ophrys, & I have never found the pollen-masses removed or pollen on the stigma of a flower excepting its own proper pollen. Hence I should have concluded that this was/57/ certainly a case of perpetual self-fertilisation; had it not been, firstly, that the sticky glands are here present, & if insects did ever visit this flower[2] a cross might readily be effected, & if they never do why are the glands sticky? Secondly in Ophrys muscifera, the pollen-masses cannot be shaken out, as I have repeatedly tried; & therefore the agency of insects is required as in the other Orchidaceous genera for their removal & apposition of the stigma; but upon examining 102 fully expanded flowers, during different years, I found in this number that only in 13 flowers had one or both the pollen masses been removed; in the other 89 flowers (most of them withered) the pollen-masses were still in their pouches. Hence we see that in Ophrys muscifera in the district in which I live[3] the agency for the ordinary fertilisation of the plant is far less effective than in other orchids & it may be that in Ophrys apifera the less important agency for an occasional cross is here likewise highly defective;—consequently that both species are here living under conditions unfavourable in one respect, but so favourable in some other, that they are able to survive;— nor need we be surprised at this, as there are many cases of plants

[1] Linnean Transact. vol. 16, p. 739
[2] Mr. Brown suspects that the flowers of Ophrys resemble insects in order to deter other insects visiting them. But I cannot avoid feeling very sceptical on this head. As we shall immediately see in Ophrys muscifera the agency of insects seems requisite.
[3] In Spandow in Germany, Sprengel found that in Orchis militaris, of 138 flowers, only 31 had seed-capsules; & he attributes this to deficient fertilisation, & contrasts it with the case of Gymnadenia conopsea in which nearly all the germens had set.

living in a country, in which they seldom or never are known to seed. But as seeding is the normal condition with these very species in other countries or times; so may an occasional cross possibly be the normal condition with Ophrys apifera./

58/*Leguminosae.* We now come to our last & ⟨perhaps⟩ most difficult case. The stamens & pistil are here beautifully enclosed within the keel shut up as in a bivalve shell; & as the pollen is shed in profusion at an early period, I am not surprised that Pallas & some other authors have advanced this great order as an instance in which a hybrid, could never be naturally formed. Yet if I trusted only to Sprengel's observations on the action of insects & to my own after having attended especially to these flowers during several years, I should have inferred that they could not have escaped frequent crosses, between individuals of the same or another variety. The flowers in this Family are especially frequented by Bees, & I have seen on them certain flies, butterflies & the minute winged Thrips, all covered by pollen. It is really beautiful to see what takes place, when a large bee alights on the wing-petals of we will say a common bean; how its weight depresses the wing-petals & with them the keel, by which the rectangularly bent pistil & already shed pollen are forced out & rubbed against the hairy body of the Bee, as it visits flower after flower. In many Leguminosae the hairs beneath the stigma act in the prettiest manner to brush out the pollen in/59/masses against the bee. Even such very minute flowers as those of the yellow[1] clover (Trifolium minus) are visited by Bees & the keel in them is generally split open: in Coronella after a hot day, I have seen the keel open of itself. But before anyone comes to a conclusion on the part which insects play in the fertilisation of the Leguminosae, long observation is required: for weeks together a Bee will not be seen even to look at a certain species, & then that species will suddenly be visited by thousands: Bees can suck the nectar as I have seen in the common Pea, without moving in the least the stamen & pistil; but then again I have seen at another time a Bee whilst sucking this flower force out the pollen in profusion & get its under surface well dusted against which the stigma was rubbed. Other Bees will visit the already fertilised flower & collect the old pollen. Other Bees frequently bite holes at the bottom of the calyx & corolla & so get the nectar, without aiding in any way its fertilisation ⟨performing what I believe is their proper function⟩ whilst Humble Bees are thus robbing the flower of the nectar hive-bees may be

---

[1] [Here between the lines on the MS. there occurs the pencilled comment: 'No'.]

collecting its pollen./60/But the case which convinces me that there is a direct relation between the structure of papilionaceous flowers & the agency of insects, is that of the Kidney Bean, (Phaseolus) which it is worth any one's while to notice: the tubular keel, with the included pistil & stamens, is here curled like a french-horn & has its little open end directed to the right side: when a Humble-bee alights on the wing-petals, the tubular keel is so acted on that the pistil is protruded & the hairs on it brush out quantities of pollen, & the pollen & stigma are rubbed against the bee's side. Now I have noticed (which was overlooked by Sprengel) that the nectary is so placed as to induce both humble & hive bees invariably to alight on that side towards which the pistil is protruded. And that this is not a mere chance relation may be inferred from the structure of Lathyrus grandiflorus,[1] in which the keel, though not actually spiral is distorted towards one side, & again it is on this side that bees are induced invariably to alight, & in so alighting they cause the pollen to be protruded against them.

But now let us see what direct evidence we have of the crossing of our many cultivated leguminous varieties. A. F. Wiegmann[2] asserts that by merely planting together varieties of Phaseolus/61/Vicia, Pisum & Ervum, he procured various hybrids, & that the seeds in the pure ⟨female⟩ parent were affected by the pollen of the other varieties; as some of these crosses were bigeneric, I should not have even alluded to these statements; had not the accurate Gaertner, a most hostile witness,[3] after most *careful* experiments in artificially crossing the varieties of Peas, come unwillingly to the conclusion that the pollen of one variety does sometime affect the seed of the castrated female plant, in the same way as happened with Wiegmann's plant, when left spontaneously to cross with each other. The only possible error in Gaertners experiments, which I can see is that it might have been the act of castration & not the pollen of the other varieties which affected the colour of the peas in the artificially fertilised pods. Certainly in some varieties, as I have witnessed (but Gaertner selected the most constant) the colour of the pea is extremely apt

---

[1] A writer in Loudon's Gardener's Magazine (vol. 8. 1832 p. 50) [Letter signed G.C.] says that having observed that this plant never set its pods, by moving the keel & so causing the stigma & anther to protrude, he found that the greater number of the flowers, thus treated, formed pods. But this does not always succeed ([Godsall] Ib. p. 733), as may perhaps be accounted for by our climate being unfavourable to this plant. I should mention that Bees seem to visit this exotic plant only during certain seasons.

[2] Ueber die Bastarderzeugung 1828

[3] Bastarderzeugung s. 89

to vary.—I was led by their statements to apply to Mr. Masters of Canterbury, a great raiser of pea-seed & the author of an article on this subject, & he/62/answered me that undoubtedly some varieties of Peas & Beans occasionally become crossed with other varieties, but that he had never known a whole crop deteriorated./ 62 v/Again in Mr. Sharps great seed nurseries[1] it is said that Peas are grown very extensively & as they are considered liable to be adulterated, 'considerable precautions are employed to secure separation.'—/62/But in these cases, I must remark that it must always be very difficult to distinguish in close varieties between the effects of a cross & of simple variation. Lastly it is incidentally asserted in the Memoirs of the Board of Agriculture of New York[2] that the varieties of the Kidney-bean easily cross with each other when grown together.

But now let us look to the evidence on the other side. A. Knight[3] castrated several pea-flowers; on some he put pollen of other varieties, & some he left without any; & these latter did not set, showing that no pollen was brought to them by bees. Secondly I applied to Messrs. [   ] great raisers of seed-peas & they do not believe that their varieties cross, & they take no especial precautions to prevent it; & this seems to be the general practice of gardeners. Thirdly a friend had planted during two generations, three varieties of Peas & three of Beans in rows close together all in flower at the same time, & I saw their produce or third/63/ generation & they seemed to run *all* true; but most of these varieties were closely allied, & between some of them a cross would not easily have been detected. Lastly, (& this case has struck me most) Mr. Cattell of Westerhaven regularly has beds of five varieties of the Sweet Pea, (Lathyrus odoratus) for seed grown close together; these varieties differ in no respect whatever except in colour, they flower at the same time & are frequented by Bees; yet each variety comes up, as I know from experience, true. Here certainly there can be hardly any crossing; probably none whatever; but it would be rash to conclude positively that there was none, for I have noticed sometimes a plant of one variety growing amongst the others, which I have attributed to a stray pea having got into the wrong packet; but possibly such might be the result of a cross[4];

---

[1] [Anon.] Gardener's Chronicle 1856 p. 823.   [2] [J. Armstrong] vol. 2. p. 100
[3] Philosoph. Transact. 1799 p. 196
[4] I have failed in my endeavours to test this, for all the flowers which I castrated, both those on which I put pollen of other varieties & those which I left without any pollen [,] fell off unimpregnated. This difficulty in manipulation is well known to hybridisers, & I presume explains the reason so few hybrids have been formed in this Family; I have heard of only three ⟨two⟩ viz one in the genera Erythrina & two in Cytisus.—

for it is known that in very close varieties differing only in colour, the offspring sometimes are not intermediate but take after either parent: thus Kölreuter[1] crossed red Hollyock with the pollen of yellow & the two seedlings were yellow; I crossed a dull purple Hollyock with the pollen of a bright yellow & the seedlings was red. Kolreuter crossed a white one with pollen of red, & the several offspring were red, with one purple./

64/With respect, then, to the Leguminosae, bearing in mind the facts given on their structure in relation to insects; bearing in mind Wiegmann & Mr. Masters & Messrs. Sharp's statements; & on the other hand the opposed facts just given, more especially the case of the Sweet Peas, it is difficult to come to any sure conclusion. But, I think, we may conclude that crosses between individual & individual, if such do occur, can take place but rarely in the Leguminosae; & the facts here given seem to me more strongly opposed to the law, which I am attempting to establish, than any others, at present known to me.—/64 v/We have seen in a former part of this discussion, that forest-trees, when hermaphrodite, offer a difficulty to my notion of general crossing from the simple occurrence of very numerous flowers on the same individual close together. Therefore as the papilionaceous structure alone offers a difficulty, this is much aggravated in forest-trees belonging to the papilionaceous division of the Leguminosae; of which, as I am informed by Mr. Bentham, there [are] a good many in Tropical countries of gigantic size; & of which the Robinia, pseudacacia offers a well-known example./

64/I will now sum up the discussion in this chapter, on the question whether it be a subordinate law in the mysterious act of reproduction that *occasionally* the concurrence of two distinct individuals is necessary. First for plants, the numerous cases of varieties which are known to cross freely if grown near each other; —the extraordinary precautions which hybridisers unanimously agree are necessary to prevent a castrated plant receiving pollen from another individual, thus obliterating the action of the foreign pollen;—the many cases of dichogamous/65/plants, or those in which the pollen is shed when the stigma is mature ⟨at different times⟩;—the many cases in which insects are necessary for the fertilisation of plants; & the other cases in which they are not necessary, but in which they are frequently visited by insects, & in which there seems an obvious relation in their structure to the visits of insects,—all tend to show that crosses between

[1] Acta. Acad. Petropol. 1782 p 256

individual & individual must at least, be frequent. A camel-hair brush which may be aptly compared with the hairy body of an insect is found useful by hybridisers to bring pollen from flower to flower; but ask any one, if he were to remove the pollen out of one flower with a brush, & use the same brush to bring foreign pollen, whether he could thus make a hybrid, & he will tell you that there would not be slightest chance of success.—

As it is known that protection from rain & damp is favourable to the fertilisation of flowers, it is remarkable how extremely general it is that the act takes place fully exposed. The reported cases of habitual fertilisation within ⟨the bud or any closed chamber⟩ the closed corolla are comparatively very few; & as has been shown are mostly open to some doubt [.] I cannot but suspect that such cases as that of/66/Subularia, Podostemon, Goodenia[1] of Ophrys apifera, & even of the Sweet Pea & of other papilionaceous flowers will be modified & explained with the progress of knowledge. How comes it, with the almost infinite modifications of structure in the vegetable kingdom, that no case, as far as I can find out, is known of the anthers bursting actually on the stigma: in Stylidium ⟨Goldfussia⟩ there is a near approach to it, but here there is a wonderful contrivance of self movement & of collecting hairs, of nectariferous organs which I can hardly doubt would favour by the agency of insects an occasional cross: in several ⟨many⟩ of those cases in which the anthers move to the stigma or the stigma to the anthers, insects are requisite to excite the movement, & not only would favour a cross, but, in Mahonias ⟨barberries⟩ at least, crosses do frequently take place. What again is the meaning of the superfluity of pollen in many hermaphrodite flowers? Kölreuter has shown that in Hibiscus[2] sixty/67/grains of pollen are sufficient to fertilise all the seeds in a flower, the anthers of which he calculated had 4863 grains of pollen; but Hibiscus though hermaphrodite is a dichogamous plant, & therefore might require a very great excess: in Geum urbanum the pollen is only ten times in excess[3]: Gaertner thinks[4]

---

[1] [In the MS., a question mark within parentheses is pencilled after Goodenia.]

[2] Vorläufige Nachricht 1761. s 9..& the statement is confirmed by Gaertner in his Beitrage zur Kenntnis. s. 346

[3] Gaertner Beitrage &c s. 346. [After this note came the words 'in the' followed by a blank space practically large enough for the following slip now numbered fol. 67 a reading:] Gardeners' Chron. Nov. 21. 1845. Article on there being 7000 pollen grains to every *ovule* or seed in Glycine—I mention *because* ⟨good as⟩ Papilionaceae, as argument for cross impregnation. We recognise use of numerous pollen grains in Zea, why not here? ['Wrong reference' is scribbled in the margin of this slip; the citation should be 1846 vol., p. 771.]
Beitrage zur Kennt. s. 440

that this superfluity of pollen is simply for ensuring the fertilisation of the plant: but on this view it must be admitted that generally flowers have been formed, without any object which we can see, with a structure rendering self-fertilisation so far difficult, that this difficulty is compensated by a great superfluity of so highly wrought an organic product as pollen! On the other hand we can understand the act of fertilisation taking place so generally in open flowers,—the maturity of the pollen & stigma being at different times—the many & very curious relations of structure to the visits of insects—the superfluity of pollen—/68/the presence in closely allied groups of hermaphrodite & ⟨bisex⟩ unisexual plants,—if the occasional concourse of two individuals be a law of nature. From the well-known elective power between various kinds of pollen specifically different & the stigmatic surface, it seems to me not improbable that the pollen of a distinct individual or slight variety may be prepotent over the flowers own pollen; & from the facts given in regard to the greater vigour of the crossed offspring of varieties, I believe that such crosses would have a better chance of surviving in the severe struggle for existence to which all living beings are subjected, than the offspring of self-fertilisation.—Although I believe good results from crossing & that probably the occasional concourse of two individuals is even a law of Nature, yet I come very far from supposing that such is the sole good of the separation of the sexes, (which necessitates a cross each time); for analogy leads to the belief that division of labour, to use Milne Edwards expression, tends to the perfection of every function.[1]/

69/Turning to animals, although many are hermaphrodite, we have the remarkable fact that not one single land animal, in which a fortuitous cross is obviously impossible by the same agencies, viz insects & wind that are so efficient with plants, is hermaphrodite in the strict sense of the word or self sufficient. Again amongst aquatic animals, not one case is positively known, in which a perfectly enclosed structure would render a fortuitous cross impossible; & that in aquatic animals fortuitous crosses are not improbable we may infer from the fact that many fixed aquatic animals have separate sexes. These facts, & others identical with those just referred to under plants, as to maturity of the ova & spermatozoa at different times, we can understand if the occasional concourse of two individuals be a necessity; & I think it would be difficult to offer any other explanation.—/

[1] [Milne-Edwards, *Introduction à la zoologie générale*, Paris, 1851, introd., pp. 35, 56–7.]

70/If it be asked why the occasional concourse of two individuals should be a law, I think the facts given showing that the crossed offspring of two varieties, & even of two individuals in hermaphrodite plants, have their vigour & fertility increased, afford a sufficient answer. Even hybrids from between distinct species gain in stature & vigour compared with their pure parents; & in some strange cases their fertility which is always deteriorated seems somewhat improved by further crossing./70 v/On the other hand close interbreeding, even in animals with separated sexes in which a cross, between two individuals, is a necessary accompaniment, seems injurious./70/It would appear as if the good from crossing was like that felt by the individual from some slight change in the conditions of its existence. But if it be further asked, why changed conditions should do good to the individual, & why a slight cross should add to the vigour & fertility of the offspring, no answer can be given, or can be expected seeing how utterly ignorant we are in regard to Life & its Reproduction.—

Finally weighing all the evidence as well as I can, I certainly think that it will hereafter/71/be found, that the occasional concourse of two individuals, & these individuals not very closely related, is a subordinate Law in Reproduction.—I have stated in full all the facts opposed to this view, which are known to me, but have not given all those in favour of it. The difficulties many of which as we have seen are grave enough, I must leave to the judgment of the reader./

72/*On changes of condition causing lessened fertility or complete sterility.*—[1]As we have in this chapter so largely discussed the good apparently derived from crossing varieties & individuals, & from slight changes in the conditions of existence, it will be convenient here, also, to discuss the effects of those changes which lessen or quite destroy the fertility of organic beings; though the subject is, I think it will be seen, more intimately related to hybridism than to the points hitherto treated of—

There is a wide difference, as strongly insisted on by Isidore Geoffroy St. Hilaire[2] between taming an animal, & getting it to breed in captivity, which alone can be called domesticating it. The one is very easy, but domestication, as the experience of all ages shows, is very difficult. One's first impulse is to attribute

---

[1] [Here in the MS. is pencilled the note: 'All used to p. 102' (i.e. to the end of this chapter.). See *Variation*, II, 148–72, ch. 18, section headed 'Sterility from changed Conditions of Life' to the end of the chapter.]

[Space for citation left blank. In *Variation* ch. 18, note 9, Darwin cited: 'Essais de Zoologie Générale, 1841, p. 256.']

the whole difficulty to the sexual instinct being affected, as has often been the explanation with respect to the Elephant in India; & in the case of birds in some instances to a proper place or materials for nidification. This in some instances may be a sufficient explanation,/73/but in very many cases, animals couple but very rarely or even never conceive; & here it cannot be an instinct which fails: moreover we shall find in plants a large parallel series of facts.—

Why many animals taken young, perfectly tamed, quite healthy & living long, should not breed, it is impossible to explain. One must attribute it to some change in the conditions of its existence. Sometimes one may infer that it is not owing to any change of climate, as when captive animals will not breed in their native country; in other cases it would appear not to be caused by want of exercise; in others not by change of food. Perhaps it may be due to these several slight changes combined. Some orders are far more affected than others, without any assignable reason; but it often happens that certain species in the orders usually least affected will not breed; & on the other hand that some species in those orders which are generally most affected will breed. In some cases the animals are never known to couple; in others they do couple but never or most rarely produce young. An apparently very slight change in the condition of existence has sometimes caused an animal to breed, which had never done so before./

74/I will now give some facts. My materials are derived from scattered notices; from an M.S. report from the Zoological Garden, between the years 1838 & 1846 inclusive, of all the animals which were seen to couple & of those which produced young; from subsequently published Reports, & from inquiries which I made from the keeper of the Birds at the Surrey Zoological Gardens, I should premise that I have no doubt that under very slightly different management, in other menageries, the results would be somewhat different; & that in the long course of years individuals of the least fruitful species would be found to produce young under the same treatment which rendered all other individuals sterile.[1]

First for the most notorious case of the Elephant, in its native

---

[1] [The following dubiously legible pencilled comment occurs on the verso of fol. 74:] I lay particular stress on animals not breeding when thoroughly tamed & left considerable liberty in their own country—In menageries very many do not breed, or breed rarely & produce few young.—There it must be at part [?] through not ranging & attributable to ill-health, but some [?] live long & others suddenly double away like sheep as highly fertile—We shall now see that the lessened fertility runs in classes without any apparent rule—Instincts.

country of India, though kept in great numbers in perfect health, has with one or two exceptions, been never known to couple; but if we go[1] [a little eastward to Ava, we hear from Mr. Crawfurd[2] that their "breeding in the domestic state, or at least in the half-domestic state in which the female elephants are generally kept, is of everyday occurrence;" and Mr. Crawfurd informs me that he believes that the difference must be attributed solely to the females being allowed to roam the forests with some degree of freedom. The captive rhinoceros, on the other hand, seems from Bishop Heber's account[3] to breed in India far more readily than the elephant]/75/In captivity. Four wild species of the Horse genus, have been bred in Europe, but generally one species with another already [?] hybrid here; though the conditions of their existence must be very different from those of their native desert home.— Most wild species of the Pig breed readily; & the Peccary [Dicotyles torquatus] has bred in the Zoological Gardens; but this animal, in its // [species, the *D. labiatus*, though rendered so tame as to be half-domesticated, breeds so rarely in its native country of Paraguay, that according to Rengger[4] the fact requires confirmation.]

76/The carnivora generally breed nearly, or quite as freely, as the Ruminants in captivity, but the plantigrade division must be excepted. Bears of several species couple most freely in the Zoological Gardens but ⟨with the exception of the cinnamon bear, have never bred⟩ have bred only thrice. I have heard of the Badgers having bred twice, once in Germany[5] & once in the Zoological Gardens; I suppose it must be very rare in Germany, as the fact was published. The Cuati or Nasua in its native country of Paraguay, though kept in pairs for many years, & perfectly tamed has never been known to breed there, or to show any sexual passion.[6] So according to this same author it has been thus with two other plantigrades, Procyon or Raccoon, & the Gulo: these three genera, have been kept in the Zoological gardens, & the two former have been known to couple, but have never bred.— In the Dog-Family of the Carnivora, it is very different, as most breed, but it has very rarely taken place with Foxes & Jackalls.— In the Cat Family, breeding is likewise very general; but even here they couple far more freely than conceive; in the M.S. return

---

[1] [The MS. is cut up here. The missing portions of text are supplied from *Variation*, II, ch. 150, 18, portion relating to notes 13–15, which are quoted as notes 2–4.]
[2] ['"Embassy to the Court of Ava", vol. i. p. 534.']
[3] ['"Journal", vol. i. p. 213.' (i.e. *Journey through India*.)]
[4] ['"Saügethiere", s. 327.']
[5] [Sicmuszowa-Pietruski] Wiegmann's Archiv. für Naturgesch. 1837. p. 162
[6] Rengger. ib. p. 106

from the Zoological Gardens for eight years the coupling was noticed between various species 73 times, but young were produced only 15 times. It is remarkable that on a change of treatment with the Carnivora at these Gardens, & when they were freely exposed in open cages to a much colder temperature, they were found to breed very much/77/more freely. I have never been able to hear of the Tiger, though known to couple, breeding in India: nor does the hunting Leopard or Chetah; but [in] this latter case pains may have been taken to prevent their breeding, as animals which have hunted in a state of nature are alone worth taming.[1] Every one knows under what unnatural conditions, shut up in a small cage, the Ferret breeds; & even the otter has once bred in the Zoological Gardens; whereas the Herpestes griseus, though many have been kept in the gardens, & some species of Viverra & Paradoxurus have never bred there.

In regard to Rodents, the Rabbit breeds most freely in wretched little hutches, ⟨as does the Guinea Pig,⟩ where the common Hare, though it has many times been tamed, most rarely will breed. Some few Rodents as the Chinchilla, some mice, a porcupine, a Lemming have bred in the Gardens; some have coupled & never bred & some have done neither. To give one example no Squirrel, has ever bred, though the Sciurus cinereus has been known to couple, & as many as fourteen of the S. palmarum have been kept together. Nor have I ever heard of the English squirrel breeding in captivity. What a strange contrast to the free breeding of the rabbit, guinea-pig & white mice!

Lastly in regard to the many species of Monkeys; most couple freely, but during the eight years, of which I had a return, there[2]... [Monkeys, in the nine-year Report from the Zoological Gardens, are stated to unite most freely, but during this period, though many individuals were kept, there were only seven births.]...

78/*Birds*. We have seen that the Carnivora, with the exception of the plantigrades, breed pretty freely in captivity; but the case is very different with Hawks. It is said//[3] [that as many as eighteen species have been used in Europe for hawking, and several others in Persia and India;[4] they have been kept in their native country

---

[1] Sleemans Rambles in India Vol. 2. p. 10

[2] [MS. fol. 78 has been cut up to leave only a narrow remnant. The preceding and following text is pieced out from *Variation*, II, 153, ch. 18, whose notes 28–30 follow as the next three notes.]

[3] ['"Encyclop. of Rural Sports", p. 691.']

[4] ['According to Sir A. Burnes ("Cabool", &c., p. 51), eight species are used for hawking in Scinde.']

in the finest condition, and have been flown during six, eight, or nine years;[1] yet there is no record of their having ever produced young.]...//79/African, American & Australian Ostriches have often bred in confinement: yet what a change in habits, climate & nature of food they must have suffered!—Most Gallinaceous birds brought from all quarters of the world, breed very freely. We see what an astonishing change the Guinea-fowl, from the dry deserts of Africa; & the Peacock from the jungle of India have undergone, & yet breed freely. At Lord Derby's some Ortyges, Grouse, & even Partridges have bred. The Capercailzie has bred in the Regents Park; but in Sweden it has been found[2] [   ] that the [   ] grouse would not breed without the birds were kept in a space, though small one, of enclosed wood. On the other hand it is well known that Partridges will not breed in captivity; but one case is recorded of the red-legged partridge having bred[3] when kept in a large court with other birds./

80/Pigeons, again, breed much more readily than most birds in confinement: in the return from the Regents Park for the eight years, thirteen species bred, & only two were seen to couple with no result.—Both the magnificent crowned Pigeons have bred in the Gardens; but Mr. Crawfurd informs me that nearly fifty birds were kept in a pleasure ground for several years in Prince Edward Island, in a climate one would have thought admirably adapted to them, & that they never bred.

Parrots, of which such numbers are kept & which have often lived to such extraordinary ages, showing that they are healthy, breed so rarely that paragraphs in the newspapers[4] are sometimes inserted when such occurs: in the Regents Park, & in the Surrey Zoological Gardens some few species couple, but I believe the Australian Euphema pulchella is the only species which has ever produced fertile eggs: Sir R. Schomburgk says[5] that Parrots kept tame & loose in Guyana do not breed.—What a singular & inexplicable contrast is thus presented by Parrots with Pigeons.—

Of the small birds or insessores, several as the linnet, Goldfinch, Siskin &c are known freely to breed with the canary bred in confinement; but very many others, as the Bull-finch have with the exception of one or two crosses with the Canary, have never

---

[1] [Hoy, 'Loudon's "Mag. of Nat. Hist.", vol. vi., 1833, p. 110.']
[2] [Darwin left an unfilled space for the reference here.]
[3] [Defay] Journal de Physique Tom. 25. p. 294.
[4] [Denny] Athenaeum 1843. p. 829. Brit. Assoc. Report [for 1843, part 2, p. 71.]
[5] [Darwin left an unfilled space for the citation here. *Variation*, ch. 18 at note 42 indicates this information was given in a personal communication from Schomburgk to Darwin.]

been known/81/to breed. Though Larks, (Alauda) of four species are kept in numbers, & I have known of some which lived in a large aviary for seven years; yet none, as I have been assured by a great Bird Fancier, here in their native country have ever been known to breed. In the 8 year returns from the Zoological Gardens, I have particulars of 24 confined species which have never bred, & of which only four have been known to couple.

Waders or Grallatores, as a class, seem eminently sterile in captivity; but many of them are short-lived in this state, so that the fact is not so remarkable as it would otherwise be.—I have heard only of three breeding: namely a Water-Hen (Gallinula chloropus) in the Regents Park; a Crane (Scops paradisea) at Lord Derby, & Grus antigone at Calcutta.[1]

The great Duck Family, Anatidae, seems the most fertile of all, apparently more so than even the Gallinaceous birds or Pigeons; yet one would have thought that their conditions of existence, considering their aquatic & generally wandering habits & insect food, would have been singularly affected by confinement. Between 20 & 30 species have bred in the Zoological Gardens. On the other hand, Sir R. Schomburgk[2] says/82/that he has never heard of the Dendrocygna viduata, though easily tamed & frequently kept by the Indians of Guyana, breeding. Lastly with respect to Gulls (Larus) ⟨& Pelicans⟩, though kept in numbers in their native country, in the Regents Park & Surrey gardens, are never known to couple or to breed, with the exception of the Herring Gull in the 1850–51, in the Regents Park. But their condition of existence & food, it might have been thought, would have been not more unnatural than with marine Ducks in confinement. Insects seem to suffer in their fertility like the larger animals. [The bottom half of this folio is blank.]

83/I have been informed in the ⟨Regents Park⟩ Zoological Gardens, that even those Mammals & Birds, which do breed in confinement very rarely breed for the first year or two. The secondary male characters seem sometimes to be affected, as in the ⟨case of the crimson breast of the Cock Linnet⟩ loss of the brilliant colours of many cock birds under confinement.[3] The young are apt to be born dead or to die immediately,—of which fact Rengger gives several instances in Paraguay: the flow of milk is often checked;, which all shows disturbance in the repro-

---

[1] [J. E. Gray] The Knowsley Menageries 1846. Pl. xiv.—Mr. Blyth Report Asiatic Soc. of Bengal, May. 1855.
[2] Geograph. Journal vol. XIII. 1844. p. 32
[3] Bronn Ges[ch]ichte der Natur B.2. s. 96: [Barrington] Phil. Transact. 1722. p. 278.

ductive functions. I have fancied that even the strangely perverted maternal instinct, so frequently leading animals in confinement to devour their new-born young, may likewise be connected with the same general disturbance.

Considering all the facts which I have been able to collect, most of which I have given, it seems impossible to come [to] any more definite conclusion, than that captivity has an especially injurious influence on the reproductive system; & more injurious in some orders than in others, but with many exceptions in every case. Generally, the cause can hardly be change of food, for the difference in the effect produced by captivity is vast, when we compare/84/ carnivorous mammals & birds; nor can it be generally want of exercise, when we consider for instance the case of Ostrich tribe, so cooped up in confinement, & ranging so widely in their natural state: nor can it be generally change of climate, when we see captive animals so frequently sterile in their own climate.

The case of domestic animals, perhaps, is hardly appropriate with respect to climate, as it may be said that their constitutions are enured to change; but it is remarkable that those Dogs, as the Bull-Dog, which degenerate in India, yet breed freely there as I am informed by Dr Falconer, as do likewise, according to Dr Daniel dogs imported from Britain into Sierra Leone. From the latter country, I have received owing to the kindness of Dr. Daniell, Poultry & Pigeons, & though brought here in Autumn & so exposed to a great change of climate the males were ready at once to procreate their kind. Rabbits breed pretty well in India. The only instance of the fertility of domestic animals having been affected of which I have heard, that of Geese & Poultry given by Roulin when first imported into Bolivia:[1] Dr. Falconer, also, informs me that the eggs of Turkeys in the hot & dry province of Delhi are extremely apt to be infertile: Geese,/85/as I am informed by Mr. Crawfurd, do not lay at Manilla.[2] Lastly we cannot generally account for the infertility of animals in captivity by the want of health, for many of them live to old age; & in the case Hawks, used for Hawking, must have been in robust health. Moreover the diseases of which animals die in menageries, (& numerous post-mortem examinations of the cases in the Zoological gardens have been published in the Veterinary Journal), are chiefly inflammations of the internal viscera & membranes, & tubercular cases. Such diseases are known in mankind not to affect the

[1] Bronn Ges[ch]ichte B.2. p. 100.
[2] [See *A descriptive dictionary of the Indian islands and adjacent countries*, London, 1856, p. 145.]

reproductive system. Of all domestic animals, the sheep, perhaps, is the most subject to disease, yet it is very fertile. In captive animals, the reproductive organs, do not appear to be diseased; but their proper function is often most gravely interfered with. The case seems quite an especial one: I do not know if there are instances of any other organs, not diseased, yet not performing their function. We can attribute this deficient action only to general constitutional derangement./

86/*Plants* In the vegetable kingdom there is a large class of facts in regard to sterility analogous with those in the animal kingdom. But the subject is here much obscured by several considerations. It is notorious that very many plants in hot-houses & in our gardens, though living in apparently the most perfect health, & often more vigorous than in their native habitat, never produce seed. I do not allude to the cases in which the seed-pod, for want of heat or other causes does not ripen, (though this may be analogous to the frequent births of dead offspring in menageries) but to those cases in which the ovules, as far as we can judge, are not fertilised. Many productions of the temperate region, for instance most of our fruit trees, when grown in tropical countries do not flower; so it sometimes is with plants in our own country when treated with an excess of manure or kept too hot & damp in greenhouses: but it seems very doubtful whether such cases come under our present subject, for here the reproductive individual is not produced, & therefore cannot be classed as sterile./86 v/To check over luxuriance, gardeners in India mutilate in the oddest way European plants which they cultivate./86/But there are many foreign plants in our gardens, which do not seem injured by our climate, in which the pollen seems perfectly good, & in which the pistil seems perfectly formed, which nevertheless never or most rarely set their seeds. These cases seem analogous to those/ 87/of captive animals, in which the reproductive system seems far more sensitive to change than any other part of the organisation.

Linnaeus long ago remarked[1] that alpine plants when cultivated in gardens, though in their natural site loaded with seed, produce very few or entirely abort; but with care, & planted in favourable situations some will produce abundant seed, as in the case of Draba sylvestris, "one of our most thoroughly alpine plants"[2]

---

[1] Swedish Acts. vol. I 1739 p. 3. [See Linné 'Versüch von Pflanzung der Gewächse...' *Abhandlungen aus der Naturlehre* 1 (1749), 13.]—Pallas makes the same remark in his travels vol. I. p. 292.

[2] Cybele Britannica vol I. p. 131 [Watson's statement here is about *D. rupestris*.]

which multiplies itself by seed in Mr. H. C. Watson's garden. Zuccarini has remarked[1] that scarcely any of the genus Oxalis from the Cape of Good Hope will seed in Europe./

87 v/In the genus Syringa, which seems perfectly hardy in our climate, I cannot hear that the Persian or Chinese Lilac ever set their seed; & I find that their pollen in water does not swell like that of common Lilac, which does produce (I do not know whether always) seed, which I have found to germinate: whereas in Germany, Gaertner instances[2] the common Lilac, as never producing seed, though having well-formed seed-capsules, in the same manner as many quite sterile hybrids here./87/Many hardy liliaceous plants are quite sterile in our gardens; as are many Bog plants: Numberless instances could be given: but in some of these cases the subject is much obscured by what Gaertner has called *contabescence*, namely the abortion of the anthers, which in some cases at least seems to be in no way connected with any change of conditions; but I shall have to return to this subject.

There are several cases on record, as in Lobelia, Passiflora, Gladiolus, Lilium candidum &c, of plants having good pollen, as known by its fertilising/87 bis/other plants, but in which the female organ either cannot be fertilised anyway, or only by pollen of another individual or other species:[3] some of these may be special cases, like those of the contabescent anthers, but as they generally occur in exotic genera, they are probably due to something unfavourable in the conditions of ⟨existence⟩ the cultivated plants. Pollen, when once in process of formation does not appear easily injured; a plant may be transplanted or a branch may [be] gathered with flowers in early bud, & ⟨if⟩ placed in water the pollen will be perfectly matured. But the female organs seem much more sensitive, for Gaertner found[4] that generally with dicotyledons, previous transplanting, even if the plant did not flag at all, prevented the act of fertilisation; & this resulted even with plants in pots, if the root had grown out of the hole at the bottom but in some few cases as in Digitalis the transplanting did not prevent fertilisation. According to the testimony of Mauz, Brassica rapa ripened its seed, with the plant pulled up & placed with its roots in water, as have several monocotyledons when cut from their roots. But I do not know whether in these cases the flower had previously been fertilised, for this, judging from W. Herbert

---

[1] [Darwin left an unfilled space for the reference here.]
[2] Beitrage zur Kenntnis. s. 560, 564
[3] Gaertner Bastarderzeugung s. 333, 356, [3]66
[4] Beitrage zur Kenntniss s. 252, 333 [Mauz' work is described here.]

experiment on Crocus makes a great difference; for he found that after the act of fertilisation, neither transplantation or mutilation prevented the seed from being perfected, but that "no application of its own pollen would fertilise the flower after transplantation."[1]/

88/In accordance to the nature of the species acted on, excess of food or manure, & some believe especially ammoniacal manures, will produce sterility. Nothing is easier, as I have tried to produce on some plants, as the common primrose, absolute sterility by manuring it too much. Plenty of perfect flowers are produced, but these produce no seed, or seed which will not grow: Gaertner also[2] alludes to the excessive flowering of some sterile species, & compares the fact to the excessive flowering of sterile hybrids: in other cases too much manure, especially if accompanied by too much heat, as before alluded to, prevents flowering. The effect of much manure depends on the nature of the plant; in some cases it is hardly possible to give too much; & Gaertner enumerates[3] Gramineae, Cruciferae & Leguminosae as standing much manure, whereas succulent, & bulbous-rooted plants &c are thus easily rendered sterile. Hence in some case potting by checking the supply of food increases the fertility of hybrid plants, & in other cases lessens them.[4] The extreme poverty of soil seems to have much less effect than too much richness on causing sterility, although of course the number of seeds is lessened, owing to the lesser size of the plants: but in/89/some plants of Trifolium minus & repens, flowering on an old lawn never manured, not one seed seemed to be produced: some other plants produced very few. I have tried starving kitchen garden plants & very small & few Pods can be produced.[5]

The period of growth during which the plant is watered often seems to affect greatly the fertility of a plant; so also does bottom heat. Many pelargoniums are extremely sterile (many of them no doubt owing to their being hybrids) but seeds have been obtained from some by extremely slight changes in treatment. So Kölreuter[6] after comparing the manner in which some pure species of Mirabilis shed their flowers like hybrids, says that some were rendered more fertile by being kept dryer in pots. Very slight changes in position as on a slight bank, inste[a]d of at its foot, will sometimes make the difference, of a plant which appeared equally healthy in both positions, setting its seed or not producing one.—

[1] Journal of the Horticultural Soc. Vol. 2. 1847. p. 83
[2] Bastarderzeugung s. 370   [3] Beitrage zur Kenntnis s. 333
[4] Gaertner Bastard. s. 378 s. 519: Kölreuter Act. Acad. St. Petersburgh 1781 Part II p. 303.
Herbert in Hort. Journal on Crocus   [6] Nova Acta Petrop 1793. p 391

No doubt temperature has a very important influence on the fertility of plants: but it is surprising what changes, in this respect some species will bear to which they are not naturally subjected. To give one example; Dean Herbert showed me in his garden Zephyranthes candida seeding well after having been just covered by/90/snow; but this plant, he informs me is a native of La Plata, where snow does not fall; & it runs wild & spreads itself in the dry & hot climate of Lima.—

Several cultivated plants, like domesticated animals, will endure the greatest change of climate & yet retain their fertility; & what makes the case far more remarkable, have their natures so far changed that their chemical composition is sensibly modified: thus Dr. Falconer informs me that Hemp seeds well on the plains & on the mountains of India, but its fibre is brittle; Linum does the same, but its seeds contain 25 per cent more oil: the poppy contains on the plains much more narcotin in proportion to morphine; & in wheat there is a similar difference in the proportions of starch & gluten; yet these plants in both situations seed well.—I suspect cultivation allows a plant to undergo change without sterility. I have alluded to the more or less complete abortion of the anthers, called by Gaertner, contabescence: until I read Gaertners able discussion on this subject,[1] I attributed all these causes to sterility from changed conditions. The cases are very numerous: Kölreuter gives many[2] in Dianthus & Verbascum: Herbert[3] adduces the N. America Azaleas,/91/which anyone may compare (as I have often done) with the most sterile hybrids, & the anthers will be found to be in exactly the same aborted condition. Gaertner has shown, that contabescence varies in different plants in intensity;— that it occasionally affects very many species in all classes but is most apt to occur in certain orders, as in Caryophyllaceae, Liliaceae (& Ericaceae may, I think be added);—that when one flower is affected generally all are affected;—⟨that whatever the degree of contabescence may be plants propagated by cuttings, layers etc retain[4] the same degree of contabescence⟩ & that it comes on at a very early period in the bud.

These facts alone, would not have convinced me that contabescence was due to some cause distinct to exposure to unnatural conditions; for in plants, very differently from in animals, we may I think infer that the fertility of the reproductive individual or flower is fully as much affected by the conditions to which the

---

[1] Beitrage zur Kenntniss &c s 117 et seq.
[2] Zweite Fortsetzung p. 10, p. 121—Dritte F. p 57.
[3] Amaryllidaceae p 355   [4] [Beiträge] s. 119.

whole plant, or vegetative individuals have been exposed, as by those to which the reproductive individual itself is exposed; we see this in the effect of previous treatment on the bearing of fruit trees, & this perhaps would account for contabescence coming on very early in life, & for all the flowers on the same plant being affected. But Gaertner further/92/shows that contabescence, when it once comes on, is permanent (with one exception) in degree for life;—that it can be propagated by layers cuttings &c;—that no change in treatment, as potting &c affect the degree;—that it is doubtfully hereditary in hybrids from a contabescent plant;—& lastly that the female organs generally not affected or only rendered precocious, & that in some instances in which after artificial fertilisation the seeds were counted, the full normal number were produced. These facts more especially the last one seem quite incompatible with the view that contabescence can be caused by unnatural conditions of existence; for it seems incredible that the female organs should not be at all affected whilst the male were rendered completely sterile: some degree of inequality of affection would be not at all improbable, from the frequent production of hybrids in those captive animals which very rarely produce pure [?] young in confinement. Moreover many endemic plants are contabescent, which seems equally incompatible with the above view. One potent cause of contabescence probably is a tendency to become dioicous, as indicated by Gaertner in the case of Silene; & that may have nothing to do with external conditions. On the other hand, as exotic plants seem very often affected; & as Kölreuter[1] seems to think that it is most apt to affect indigenous plants, when transplanted into a garden; & as Wiegmann[2]/93/states that the contabescent wild plants of Dianthus & Verbascum which he found, grew on a dry, sunny sterile bank, the affection may in some instances be due to exposure to unnatural conditions.[3]

Double flowers: *seedless fruit.*—Flowers are often made, (as commonly expressed) nearly or even quite infertile by doubling. The male organs are much more often affected than the female, as everyone may see.[4] The tendency to double depends on the nature of the species; for we have some species extremely double, as the Gorze, in classes which very rarely have double flowers. It depends, also, on the structure, as flowers with many stamens & petals are most apt to become double. Luxuriant growth & rich soil no

---

[1] Dritte Fortsetzung s. 57   [2] Über die Bastarderzeugung s. 27
[3] [Here Darwin scribbled in pencil: 'Anthemis nobilis.']
[4] Gaertner Bastardzeugung s. 363 s 569

doubt are highly favourable to doubling; & Prof. Lehmann[1] found several wild plants double near a hot spring: on the other hand I may mentioned that I found many stunted wild plants of Gentiana campestris,[2] growing on a very poor chalky bank very double; I have also noticed a Staphylea & Aesculus pavia, & some other plants growing very poorly under favourable conditions, with a distinct tendency to become double: therefore luxuriant growth & good soil are not absolutely necessary concomitants./

94/Again when the fruit is largely developed seeds are rarely perfected[3]: we see this in our best pears: the Enville pineapple which is a poor one is the only kind having seeds: this is notoriously the case with the Banana & Bread-fruit; it being extremely rare to find even a single good seed, except in some poor varieties. So again it is generally believed that a great development of tubers or roots often (certainly not always as in carrots, turnips &c) causes infertility; as does a great tendency to propagate by runners, & suckers.

These several affections have always been considered as the causes of the lessened or destroyed fertility, owing to an antagonism or compensation in growth. I strongly suspect the effect has been here held for the cause. I do not doubt that if any cause whatever produced a great development, especially if in the proximity to the reproductive organs, this would tend to produce infertility: but we have to consider what so frequently gives in cultivated plants the first tendency to such development often in connection with lessened fertility. There can be no doubt that the first tendency having been given, selection, taking advantage of the hereditary principle has played a most important part in nearly every case, & as we know/95/in the history of several double flowers, in which the work commenced in the seed of a flower having one or two stamens converted into petals. I believe that the first cause is lessened fertility from the plant being exposed to unnatural conditions, more especially to excess of food; & that the doubling of the flowers, the great size & succulence of the fruit, of the roots, & the tendency to form suckers &c is the result of, ⟨or is compensation of⟩ organic matter not being consumed in the formation of seeds, together with generally an excess of food ⟨the process having been perfected by man's selection⟩. I have come to this conclusion, from finding an exactly parallel series of facts, but not perfected & added to by continual selection in a case in

---

[1] Quoted by Gaertner Bastard. s. 567   [2] Gardener's Chronicle.
[3] See Prof. Lindleys excellent remarks on this subject in Theory of Horticulture p 175–179

which lessened fertility or entire sterility has supervened from an entirely independent cause; namely from hybridity. Gaertner has shown[1] that hybrid plants are more inclined to produce double flowers than pure species; & the tendency is hereditary; in hybrids & in double flowers the male organs are first affected; in both there is a strong tendency to yield innumerable flowers. Again Gaertner insists[2]/96/most strongly on the very general tendency of hybrids, even utterly sterile kinds, to produce the perfect receptacles of the seed or fruit: thus, Sabine on Passion Flower.

With respect to the development of roots, Kölreuter expresses his unbounded astonishment at the size of those of hybrid Mirabilis. All hybridisers, also,[3] are unanimous in the strong tendency in hybrids to increase by their roots, & throw up suckers &c.— Considering this strictly parallel series of facts, & that it can hardly be disputed that unnatural conditions have a special action in lessening the fertility of organic beings, it seems to me, that the view here adopted, that the lessened fertility is the first cause aided by excess of food & selection, & that double flowers, fine fruit, large roots, &c is the result. Therefore the enormous class of facts here alluded to, come, I think, fairly under the present discussion, & support the conclusion that considerable changes of condition have an especial action on the reproductive system. I may add that horticulturists have often/97/spoken of infertility as the bane of horticulture; but on the views here advocated they ought to confess that though this may be so, they owe to it, their choicest productions.—

96 v/How far the several known & extraordinary cases of plants never flowering or never seeding in their native country, when they are abundant, come under our present subject, I am doubtful. Certain plants ascend mountains to a height, & in the arctic regions to a latitude in which they do not produce seed. In such cases I presume that there can be no doubt that their infertility is owing to the climate to which they are exposed, but that they have some other advantage over their few competitors in these sterile regions, which allows them to hold their own. We may suppose this to be the case in the curious instance mentioned by Kalm that the coniferous trees which cover in an impenetrable mass the swamps on the shores of N. America, never seed there; but only when growing in the higher country. Certain water-plants in our own country rarely or never seed. Dr Bromfield[4] gives a still more curious

[1] Bastard. s. 565
[2] Bastard. "Fruchtungsvermögen der Bastarde: diese Eigenschaft ist sehr ausgebreitet bei den Bastarden." s. 537
[3] Gärtner Bastard. s. 527   [4] [*Phytologist*, 1848, 3.i, p. 376]

instance, namely in the common ivy which abounds in Russia & over the North of Europe but never flowers.

97/Although we have seen so many animals in captivity & so many plants under cultivation are rendered more or less infertile: yet those animals which do submit to the particular changes of conditions implied by domestication, are far from having their fertility checked; on the contrary the more abundant & regular supplies of food which domestic animals probably receive in comparison with wild ones, appears, as might have been expected, to increase their fertility./97 v/If it be denied that domestic animals which are often fattened & which are protected from famine, do receive more food on average than wild ones then I know not how to test the dictum/97/I have compared the produce of nearly all our domestic animals,[1] with their wild prototypes, when known or with the most nearly allied animals. Of course there is often doubt about the rate of increase of wild animals, but as far as known all domestic animals, without it be the Peacock, bear either a greater number of young at a birth or at shorter intervals, probably at a younger age, than wild. In some domestic animals selection/98/may have increased their fertility, by the most fertile individuals, but in others as in cat, Pigeons &c I do not suppose this point has ever been attended to. In regularly cultivated plants, some as we have seen are nearly sterile; but these are such as can be propagated by cuttings, grafting &c; & in most of these the infertility, in accordance with the views just advocated has been of use, as causing greater development of some useful product, & therefore here infertility has been selected. In many plants, cultivated for their seed, selection probably will have increased their fertility: but there are many other plants propagated by seed, but yet which would never have been selected for this advantage; as the carrot, parsnip, cabbage, asparagus. As in these instances the wild prototype is known, I have taken the finest wild plants which I could find, & ordinarily fine cultivated plants, & I find that the cabbage has about...[2] [Seeds vary so much in number that it is difficult to estimate them; but on comparing beds of carrots saved for seed in a nursery garden with wild plants, the former seemed to produce about twice as much seed. Cultivated cabbages yielded thrice as many pods by measure as wild cabbages from the rocks of South Wales. The excess of

[1] [Memorandum pencilled on verso: 'Bechstein—Ferret, Rabbit—*Wild Pigeon* Poultry'.]

[2] [Here well before the end of the page Darwin stopped in the middle of his sentence. The text is pieced out from the last two paragraphs of ch. 16 in *Variation*, II, 112–13.]

berries, produced by the cultivated Asparagus in comparison with the wild plant is enormous...with plants like carrots, cabbages, and asparagus, which are not valued for their prolificacy, selection can have played only a subordinate part; and their increased fertility must be attributed to the more favourable conditions of life under which they have long existed.][1]

99/I have alluded to this last subject more particularly on account of Mr Doubleday's[2] theory, which is that an abundance of food checks fertility & poverty increases it or "that prolificness is in the ratio of the state of depletion". Independently of mankind, in regard to whom, I should have thought that the Malthusian explanations of restrained or reckless marriages, would have accounted for the asserted facts, the only evidence appears to me the undoubted fact that you can fatten individual animals to such an excess, as to check their fertility;[3] & that in plants the same can be easily done by excess of manure.[4] If indeed it could be proved that the most flourishing wild animals & plants, which exist in the greatest numbers in any country, from this very cause of their flourishing so much, had their fertility checked, it would be a most serious objection to the principles hereafter to be elucidated in the chapter on selection. ⟨But to me, all the facts seem to point in an opposite direction.⟩/

[1] [At the foot of the blank portion of fol. 98, Darwin pencilled the following dubiously legible memorandum: 'In carrot [?] I did not measure but after selecting the finest wild plant compared it with—The wild one grew in cultivated ground & had more than those growing in natural ground.']

[2] [The following note, now to be found in ULC vol. 46.1, fol. 24, appears to belong here:] The True Law of Population. I have read this work, an article by Mr. Hickson in the Westminster & Foreign Quarterly Review Oct. 1849, 'Godwin on Population' & various other Essays written against Malthus' great work, with all the attention of which I am capable, but I cannot say that they have had any weight with me, in opposition to the few facts given in this chapter, & which could have been largely added to. I am bound to add that so eminent an authority as Dr Carpenter (Principles of Comparative Physiology 1854 p 122) seems to admit Mr Doubleday's doctrine; so again that shrewd observer Hugh Miller (Schools and Schoolmasters p. 266) seems of same opinion, & remarks, that "when hardship presses on the life of the individual, so as to threaten its extinction, it is rendered more fruitful."—

[3] [The following note on the verso of fol. 99 seems to belong here:] Gaertner in his Bastardzeugung s. 378, gives references to Henschel & Girou de Buzareingues, that domestic animals produce more in fruitful years, than when food fails. But in plants we have seen that there can be no question that by poverty of soil the number of seeds can be lessened. No one can doubt that few[er] ears of any corn will be produced on very poor land than on rich. In the case of the most wonderful increase on record, namely that of the domestic animals become feral & rapidly spreading over America, can it be believed, that this astonishing increase was owing to lessened fertility, for want of food: if there was any change whatever in fertility, which may be doubted, in all probability this would be increased.

[4] [A note slip seems lost here. Note 76 in ch. 18 of *Variation*, which may well represent the missing reference, cites Gaertner: "Beiträge zur Kenntniss der Befruchtung", 1844, s. 333.]

100/In concluding this Chapter, it must be admitted that the evidence on the several points discussed in it, has been often very dubious & partly rests on the weakest possible grounds [,] general belief. Yet to my mind the evidence does seem to weigh in favour of the following conclusions; that slight changes in the condition of existence are favourable to the life of both animals & plants;— that in both, close interbreeding between the nearest relations is unfavourable to vigour & fertility, & that, on the converse hand, crossing with a distinct individual or variety (& even distinct species in *some* respects) is favourable in all respects; & further that there is some probability, though many of the gravest difficulties at present stand in the way, that it is a fundamental principle in the act of reproduction that there should be, perhaps at very wide intervals, the occasional concourse of two distinct individuals.—On the other hand, I think it must be admitted that greater changes of condition, or more strictly changes of a particular nature with respect to each species, have an special tendency, in both animals & plants, to cause infertility, that the cause seems to us to act most capriciously, affecting/101/one order far more than another; but with numerous exceptions in each order. That as slight changes of condition ⟨& slight crosses⟩ are good to the individual & as the offspring of the crossing of closely allied forms are more vigorous & fertile so we have a parallel series, in greater changes of condition causing more or less sterility in the individual & in the notorious fact of the lessened fertility or utter sterility in the hybrids produced by the crossing of distinct species or unlike forms. Neither in hybrids, or in an individual species placed out of its natural conditions, can we tell, till we try, whether the fertility will be greatly or slightly affected, so ignorant are we of the exact cause. But to the subject of Hybridity we shall hereafter to return.—

Hence I cannot doubt the truth of the propositions that in all living beings the reproductive system is acted on in an especial manner, unlike any other part of the organisation, by the conditions of existence; that both male & female element is acted on, the action appearing to us most capricious either for good or evil. This proposition seems to me important, for it brings into connection all the facts in this Chapter with the variability of organic beings when placed out of their natural conditions under domestication. If the reproductive system is so easily acted on, that/102/ changes of condition, which do not in the least affect the health of the individual, yet seriously affect or entirely stop its function; surely it is not surprising that the product of the reproductive system should be affected; should depart from its normal function of like producing like, but that varieties should arise,[1] also, the conclusion which seemed in the last chapter most probable, that the chief cause of variation did not supervene, in accordance with the common view during pregnancy, or the formation of the seed, or during the act of impregnation; but in the action of the life of the parents, on the separate male & female elements of reproduction.

[1] [There is a reference mark here, but the note apparently on a slip formerly pinned to fol. 102 has not been found.]

VARIATION UNDER NATURE

INTRODUCTION

Darwin wrote the original draft of chapter IV during the period from mid December 1856 to late January 1857 according to his Pocket Diary. A year later he wrote a fifty sheet section on the contrasts in variation in genera with large and small numbers of species, which he intended for insertion towards the end of his original chapter. The history of these two parts of the chapter is best considered separately. In comparison with some chapters, the original section of chapter four shows relatively little revision. On folio 67, pencilled additions by Darwin dated 1861 and 1867 indicate his return to this section of the manuscript. From folio 68 to 72 cancellations signal that Darwin made later use of the material as Dr Guimond discovered. But here in contrast to chapter three Darwin did not take passages directly from the Natural Selection manuscript but in 1868 he only used some of his earlier examples and references incorporated with new material including experiments of his own.[1] At some time after the final revision of the section on variation among Indian elephants, on folios 48 and 49, Darwin had a fair copy made of these two sheets.[2]

The large section written later on the commonness, range, and variation of species in large and small genera has a history rather separate from the rest of the fourth chapter. In an earlier memorandum dated January 4, 1855, Darwin indicated one theme of this section:

it may be concluded, as Mr. Watson remarks (Cybele Brit. vol. I, p. 18) that "those most widely & generally distributed, even in large spaces, being usually also the most common species."... Hence we may rudely conclude, that *wide-ranging species are commonest*: this harmonises with fact that they range far & are numerous, from same cause, viz successfully struggling with the organic & Physical conditions of area.—

The number of individuals must especially depend on struggle with other individuals.[3]

In regard to extensive numerical analyses of catalogues of regional flora, including helpful volumes borrowed from Hooker, all to provide quantitative evidence for his view of varieties as incipient species, Darwin later wrote Hooker:

[1] 'On Specific Differences in Primula', *Linn. Soc. J.* (Botany) 10 (1868), 441–2. *The Different Forms of Flowers on Plants of the Same Species* (London, 1877), ch. II, pp. 60–2.

[2] Fair copy now in C.D. MSS. vol. 45, fols. 18–19.

[3] C.D. MSS. vol. 15.1, fols. 36–7 of 2nd no. sequence. The last sentence was added in pencil along the margin.

I was led to all this work by a remark of Fries, that the species in large genera were more closely related to each other than in small genera; and if this were so, seeing that varieties and species are so hardly distinguishable, I concluded that I should find more varieties in the large genera than in the small...[1]

Fries' statement appears in Darwin's reading notes on the *Botanical Gazette*, where Darwin wrote:

p. 188 "In genera containing many species, the individual species stand much closer together than in poor genera; hence it is well in the former case to collect them around certain types or principal species, about which, as around a centre, the others arrange themselves, as satellites." This very important, it shows that extinction has *not* been at work in the large genera.—But some of the small growing genera ought to have close species.—[2]

In regard to the Fries quotation, Darwin later added in pencil the note that: "Bentham, Hooker & Thompson say Hieracium not large genus only forms. All three *greatly* doubted truth of statement & quoted case of Senecio & others where species very distinct." He also added on a pinned on note slip: "Waterhouse does not in least believe in Fries statement that large genera have closer species.—"

Darwin saved about 300 foolscap sheets of tabulations of genera & species from standard catalogues and calculations of relative proportions of species to genera and varieties to species.[3] At the beginning, tabulations and calculations on Boreau's *Flore du centre de la France* are marked 'wrong & useless'. In July, 1857, John Lubbock pointed out some fundamental error in procedure which Darwin had been making in his calculations, thus vitiating his initial labours on statistics from Boreau and Fürnrohr.[4]

Having appealed to Hooker for a loan of these Floras so that he could rework them Darwin continued his tabulations and calculations of ratios of variation and speciation in parallel with his writing of successive chapters of Natural Selection, and he frequently mentioned this statistical work in his letters to Hooker. On August 22, 1857, he wrote:

I am very glad to hear that you have been tabulating some Floras about varieties. Will you just tell me roughly the result?—Do you not find it takes much time? I am employing a laboriously careful Schoolmaster who does the tabulating & dividing part into two great cohorts more carefully than I can. This being so, I sh^d be

---

[1] L & L, II, 102–3; NY, I, 460.

[2] C.D. MSS. vol. 73, fol. 118, notes regarding Fries article, 'A Monograph of the Hieracia', *Bot. Gaz.* 2 (1850), 85–92, 185–8, 203–19.

[3] C.D. MSS. vols. 15.2, & 16.1, 16.2.

[4] L & L, II, 103–4; NY, I, 461. Fred. Somkin, 'The Contribution of Sir John Lubbuck, F.R.S., to the *Origin of Species*: Some Annotations to Darwin.' Roy. Soc. London, *Notes & Rec* 17 (1962) 185.

very glad some time to have Koch-Webb's Canaries—& Ledebour, & Grisebach....

On September 11: 'The magnificent & awful Box of Books arrived quite safely this morning....I shall not, of course, try to do all, but will invest a handsome sum with our Schoolmaster...." Then on September 30th: "I hope you are not getting impatient for your books back: for I have done only a few of them which I shd like to do; for it is very slow work, & our Schoolmaster has only his evenings to spare.'[1]

The following spring, on March 10, 1858, the day after finishing chapter x, on instinct, Darwin mentioned to Hooker that he was putting notes together on large and small genera, and the next day he warned Hooker he would want him to read his draft when it was finished.[2]

A month later, on April 10, Darwin wrote Hooker:

I have almost finished my discussion; but it will take some little time to have it copied; & as my health has been lately wretched, I start in 9 days for a fortnight of Hydropathy & rest. On my return I will send it, & most grateful I am to you being willing to take the trouble to read it. I enclose a memorandum on way which I want you to consider my M.S. which please keep & read, when I send the M.S.—[3]

DARWIN'S MEMORANDUM:

'Is the whole worth publishing? I do not promise to be guided by your judgment, but it will have *great* weight when in some ⟨year⟩ months time I reconsider subject.

Have I fairly stated the *more important* objections in *abstract*: to have given all in full would have made my now tedious discussion intolerably tedious.

I shd be very glad to hear any criticisms in detail; & you & Watson have done me an enormous service in drawing my attention to & enumerating the numerous objections but what I want you to do now is, in as candid a frame as you can, to balance all the vague probabilities on both sides of question.—

Remember that my book is written for geologists & zoologists, so that on some points I daresay my remarks may appear to you trivial.

I have discussed some *extra* hypothetical points chiefly for sake,

---

[1] C.D. MSS. vol. 114, Ltr. no. 208 (cf. ML, no. 53), no. 211, 210 (sic i.e. order inverted here). For instructions from Darwin to the schoolmaster, Mr Norman, see vol. 16.1, fols. 133A, 136A.

[2] Ltr. 228, L & L, ii, 103; NY, i, 460.   [3] C.D. MSS. vol. 114, Ltr. 231.

here & in other places to show what points ought to be considered in theory of the descent of species, rather than in hopes of throwing light on the many points of present inextricable confusion.—'

[HOOKER'S COMMENT:]

'My pencil ⟨notes⟩ alterations were intended to make passages clear to myself not for corrections or hints to you so do not mind them.'   J. D. Hooker"[1]

In the Pocket Diary, the first two lines entered for 1858: 'March 9th Finished Instinct Chapter April 14 Discussion on Large genera & small' together with the letter of April 10 strongly suggest to me that the April 14 entry was intended to record the completion date of this additional section. Then upon his return from Dr Lang's hydropathic establishment at Moor Park, Farnham, Surrey, on May 6 Darwin sent off the fair copy to Hooker.[2] Early in June Hooker sent Darwin an encouraging note about the manuscript to which Darwin replied most gratefully on June 8.[3] Then came the startling interruption of Wallace's letter on natural selection, and Hooker's joint efforts with Lyell to secure fair recognition for both Wallace and Darwin. Apparently only on July 13, could Hooker complete his examination and send Darwin his verdict of considered approval:

I went deep into your MS. on variable species in big and small genera and tabulated Bentham after a fashion, but not very carefully. After very full deliberation I cordially concur in your view and accept it with all its consequences.[4]

Hooker's immediate comments on Darwin's draft are recorded on the fair copy and these appear as notes in this portion of chapter IV.

VARIATION UNDER NATURE

1/In this Chapter we have to discuss the variability of species in a state of nature. The first & obvious thing to do would be to give a clear & simple definition of what is meant by a species; but this has been found hopelessly difficult by naturalists, if we may judge by scarcely two having given the same.

I will copy the latest & most laboured definition by Alph. De Candolle[5] who has carefully discussed the subject in relation to plants: he says species are "collections d'individus qui se ressemblent assez pour 1º avoir en commun des caractéres nombreux et

[1] C.D. MSS. vol. 15.1, fol. 0.   [2] L & L, ii, 107; NY, i, 465.
[3] ML, no. 64.   [4] Hooker, L & L, i, 458.
[5] Géographie Botanique p. 1072

importants, qui se continuent pendant plusieurs générations, sous l'empire de circonstances variées; 2° s'ils ont des fleurs, se féconder avec facilité les uns les autres et donner des graines presque toujours fertiles; 3° se comporter à l'égard de la température et des autres agents extérieurs d'une manière semblable ou presque semblable; 4° en un mot, se ressembler comme les plantes analogues de structure, que nous savons positivement être sorties d'une souche commune, depuis un nombre considérable de générations." M. De Candolle lays stress on making the element of descent subordinate to that of resemblance, so that the definition may be less hypothetical. But as animals & plants must be here equally considered, I agree with Dr. Carpenter who gave at Glasgow to the British Association an interesting lecture on this subject,[1] that descent does come in as a prominent idea. Although when speaking of the resemblance of two forms, the comparison should of course extend to all ages & sexes, yet as zoologists/2/have often described these stages as specifically distinct, an error instantly corrected when their descent was known, it is very natural that they should bring this idea prominently forward. Thus if the development of ⟨Trichoda lynceus⟩ had not been known, the stages through which it passes, as M. Quatrefages[2] has remarked, would have been considered as forming eight distinct genera: I am convinced that in the cirripede Ibla without knowledge of its descent, the male & female & its two larval stages would have formed four distinct Families in the eyes of most systematic naturalists. Again the most ill-shapen monster is rendered home to its species the instant we know its parents.

Let us test M. de Candolles definition with a plant. Assuming for the moment that it was *demonstrated* (& we shall presently see that the evidence is very strong) that the primrose & cowslip can be produced from the same stock; would they be called by any Botanist distinct species in the ordinary acceptation of the word? Yet/3/the individuals of the cowslip & the individuals of the primrose accord in every single respect with M. Decandolle's definition of two distinct species: for (1st) the individuals in each agree in many important characters, which are constant during many generations under different conditions, for they are found in distinct parts of Europe: (2nd) they do not fertilise each other with facility, as the best experimentiser, who ever lived, Gaertner, found after repeated trials during many years: (3rd) they do not behave in the same way in regard to temperature & soil, for they

---

[1] [See *The Athenaeum* (1855), 1090.]
[2] Revue des deux Mondes [(1856 tome 3), 871.]

have different ranges & inhabit different situations (4th they cannot be said to resemble each other as much as *analogous* plants do, which we positively & habitually know to have descended from a common source. Hence I conclude, that descent is a prominent idea under the word species as commonly accepted.

The idea of descent almost inevitably leads the mind to the first parent, & consequently to its first appearance, or creation. We see this in Morton's pithy definition of "primordial forms",[1] adopted by Agassiz. The same idea is supreme, & resemblance goes for nothing, with those zoologists, who consider two forms, absolutely similar as far as our senses serve, when inhabiting distant countries, or distant geological/4/times, as specifically distinct. Having the idea of the first appearance of a form prominently in their minds, they argue logically that as most of the forms in the two countries or times are distinct, the distinction being in some great, in others less & less, they naturally ask, why forms apparently absolutely identical should not have been separately created, & which they in consequence would call distinct species.—As we have to discuss in this work whether forms called by all naturalists distinct species are not lineal descendants of other forms, this minor question will fall or rise with the greater question; & is here only alluded to in connection with the definition of the word species.—

Some authors, as Kölreuter, take the fertility of the offspring of two forms as the sole ⟨or leading⟩ test of what to consider as species; & however unlike two forms may be, if they produce quite fertile offspring, they consider them as specifically the same. The great importance of this difference in fertility in what are ordinarily called varieties/5/& species, has in my opinion of late years been much undervalued by some authors. In the chapter on Hybridism we shall fully consider this subject & we shall find that there are great difficulties (I do not mean merely practical ones in its application) in taking lessened fertility in the offspring as an unerring guide what forms to call species. I will here only remark, that perfect fertility & utter sterility glide into each other, in so insensible a manner that it is hardly possible to draw any line; hence the two most laborious experimentisers who ever lived, Kölreuter & Gaertner after numerous experiments in regard to certain forms, have come to diametrically opposite conclusions; the one concluding that certain forms are varieties, & the other that they are undoubted species.—

Short as this discussion has been it suffices, I think, to show

---

[1] [See J. C. Nott and G. R. Gliddon, *Types of Mankind*, p. 375.]

how various are the ideas, that enter into the minds of naturalists when speaking of species./5 v/With some, resemblance is the reigning idea & descent goes for little; with others descent is the infallible criterion; with others resemblance goes for almost nothing, & Creation is everything; with others sterility in crossed forms is an unfailing test, whilst with others it is regarded of no value./5/ At the end of this chapter, it will be seen that according to the views, which we have to discuss in this volume, it is no wonder that there should be difficulty in/6/defining the difference between a species & a variety;—there being no essential, only an arbitrary difference. In the following pages I mean by species, those collections of individuals, which have commonly been so designated by naturalists. Everyone loosely understands what is meant when one speaks of the cabbage, Radish & sea-kale as species; or of the Broccoli, & cauliflowers as varieties: between such extremes there is often a wide neutral territory in which the term species & varieties are bandied about according to the state of our knowledge & our ideas of the term species.—

Botanists in discussing the subject of variation have usually included ⟨together⟩ that variation which occurs under domestication & that under natural conditions; & this is probably the best plan, though not for our particular object. They have divided[1] varieties into "variations" in which the varying characters are not fixed even in the individual plant, all the buds produced on the same plant being here considered as one individual. In animals we have very few instances of this class; but as the black colour in cage birds produced by hemp-seed goes off with change of food; & slight changes in the/7/hairy covering of animals when transported into a different climate[2] have been observed. The term 'Variety' is applied to forms often offering considerable differences, & which can be securely propagated by buds, grafts, cuttings, suckers &c, but which are believed not to be inheritable by seed. This class nearly corresponds with "abanderungen" in Bernhardi's classification[3] in which the form is not hereditary or only so in certain soils; & likewise in a lesser degree with his "Spielarten" in which the form tends to go back in one or more generations to the parent type. As we know scarcely anything of the variation of those lower animals which can be propagated by division, the class "Variety" in the above strictest sense is not applicable to the animal Kingdom; though no doubt, in the less strict sense of

---

[1] M. Alp. De Candolle has given a full discussion on this subject. Geograph. Bot. p. 1078
[2] The cat in West Africa   [3] Ueber den Begriff der Pflanzenart 1834. p. 5

being hereditary in only a slight degree, there are very many cases amongst animals, & some even in a state of nature. Lastly we have the class "Race", corresponding with "Abarten" of Bernhardi/ 8/& with subspecies of some authors, in which the form is strictly inherited, often even under changed conditions; of this class we know there are plenty under domestication, some known, & more suspected in a state of nature, as in the geographical races of some Zoologists. But the term subspecies is used by some authors, to define (& corresponds in this sense with "unterart" of Bernhardi) very close species, in which they cannot determine whether to consider them as species or varieties. The existence of these doubtful forms has lately been explicitly admitted by M. Alp. Decandolle in regard to plants, & by implication by Mr. Wollaston[1] in regard to insects: M. Decaisne & Dr. Hooker use the term without expressing more than that the difference between such subspecies is slight, yet permanent. As these authors are of the highest authority, this admission is important as sub-species fill up a gap, between species, admitted by everyone & varieties admitted by everyone. Between varieties & individual differences there seems a gradual passage but to this subject we shall recur. In species we should remember how extremely close some undoubtedly distinct forms are, as many plants, & as in some of the willow wrens, which are so close that the most experienced ornithologists can hardly distinguish them except by their voice, & the materials with which they line their nests; yet as these wrens inhabit the same country [? county] & always exhibit the same/9/ difference, no one can doubt that they are good species. So that between individual differences & undoubted species naturalists have made various short steps.

In the above classification of several varieties the main difference rests on the hereditariness of the characters. Though the classes blend insensibly into each other, this classification is of some use when applied to domestic productions; & no doubt it holds good in varieties in a state of nature, which we are here considering. But it seems to me that we are far too ignorant to apply it to varieties under natural conditions, more especially in regard to animals. We have seen in our first chapter that the same character is inherited in very different degrees by different species, & even in different individuals of the same species; we have reason to suspect that a character becomes more fixed by long continued generation; although on the other hand, a character suddenly appearing is sometimes strongly inherited. Who can tell how much

----

[1] On the Variation of Species. p. 185

the dwarfed character of a plant or the dark colour of an insect on a mountain, or of a shell in brackish water/9 v/or of the improved character of the fur of Beavers Martins &c the further we go north[1]/9/is due to inheritance & how much to the exposure of the individual from its earliest days to the condition in question. Probably in all such cases, the/10/form would change when placed under other circumstances; & some in fewer generations than others; but then it might be argued that this was not a fair test, as many races or strongly hereditary varieties change in some degree under new conditions. ⟨I am inclined to believe that with the rarest exceptions every changed structure is in some degree inheritable.⟩ In animals perfectly black individuals are not very rarely born, even in the same litter with ordinary coloured individuals: & in some places these appear much more frequently than in others, thus I am informed by Mr. Crawfurd that black Leopards are far more commonly produced in Java, than elsewhere; & in such cases I know not whether to attribute this to a strange hereditary principle, or to some unknown conditions acting on the parents. Fish of the same species are well known to present distinguishable differences in different lakes: Sir H. Davy[2] states that red-fleshed dark-banded trout were taken from one Scotch lake & put into another, where the trout were white-fleshed; the young here produced had their flesh less red, & in 20 years the variety was lost. Laying on one side the probability of crosses having taken place, we see here that the red flesh was in some degree inherited; & some would assert that if the red trout in their own lake had/11/transmitted their character for some additional hundred-thousand generations, the character would have kept truer. From these & similar considerations I have thought it advisable to use only the term "variety", & where it is known or almost known to be strictly inherited "race": and I use the term variety loosely, simply in accordance with common acceptation, as I do the term species. ⟨for the same reason in both cases⟩ If the distinction could be drawn between hereditary & temporary variation in a state of nature it would be of great importance for our object; for variations in a state of nature which are not inherited are of little signification, & deserve notice, (perhaps) only as showing the possibility of change in structure.—

Practically the systematic naturalist, without troubling himself more than he can help about descent & creation, considers those forms as one species which he can unite by other intermediate & graduated forms. It is his golden rule. But those who have not

[1] Kalms Travels vol 3. p. 58   [2] Salmonia p 53

themselves worked, can form little idea of the irksome labour required in its application. For example look at the case of Aqilegia vulgaris, as worked out by/12/Dr. Hooker in his Flora Indica [1, 44], who devoted weeks to the examination of specimens from all parts of Asia & Europe, & who ends in uniting about 16 species of other authors into one. I may state, as I know that similar cases have occurred with others, that in Lepas anatifera & Balanus tintinnabulum. I at first wrote out full descriptions of several supposed species; then after getting more specimens from various parts of the world, I thought that I ought to run them all into one, & tore up my separate descriptions: after an interval of some months I looked over my specimens & could not persuade myself to call such different forms one species & rewrote separate descriptions; but lastly having got still more specimens, I had again to tear up those & finally concluded that it was impossible to separate them! When the Naturalist has got the intermediate forms between two supposed species, the work though laborious is *generally* simple; but he is very often obliged to judge by analogy. And here springs up an endless source of doubt. On how widely distinct groups may he draw for his analogies?/13/it is a remark repeated in almost every systematic work, that the very same organ whether or not of physiological importance will be constant in one group & so afford good specific characters, & will be ⟨highly⟩ variable in another. His power of drawing analogies will not only obviously depend on his amount of knowledge, but on the frame of his mind. Is it then surprising that naturalists should differ in the extreme degree in which they do, in determining what forms to call by the various ⟨defined &⟩ recognised term species?

I have remarked that generally when the naturalist has got intermediate stages he unites with confidence two forms distinct in appearance. But here, also, he sometimes has cause to doubt. The intermediate forms may be hybrids; these he may often recognise by their sterility, but by no means always, at least without counting their seeds & comparing them in number with those of both presumed pure parents; but Gaertner thinks that a hybrid should be artificially made for comparison; or he may discover that they are not hybrids by one of the supposed parent forms not growing in the neighbourhood./14/But independently of this source of doubt, which perhaps has been over-rated by some authors, there is another & more important one, namely the probability of one of two forms, or of two forms which deserve in every sense to be called species, both varying greatly & running so closely together that the extreme varieties become undistinguishable. This is the more probable, as we shall afterwards see that, certainly varieties of one form tend to mock the characters of other species in the same group. To give a very few examples: Drs. Torrey & Gray, in speaking of the N. American Asters say "that several species, which we cannot but consider as distinct do frequently present very puzzling intermediate forms; & that an apparent transition is not always real."[1]/14 v/—Such cases more or less striking do not seem to be very rare, for even in the small British Flora, Mr. Hewett C. Watson has marked for me 15 cases, (not including the protean forms in Rubus, Hieracium &c) in which two species & in some cases three species apparently distinct are/15/united, more or less perfectly, by intermediate forms:[2] to give a single example,—Geum urbanum & rivale are universally thought to be distinct; but between them we have the var[iety] G. intermedium (considered a distinct species by some authors) & several intermediate forms, breaking down every character between the two types: in this instance Dr. T. Bell Salter has stated that he produced G. intermedium by crossing the above two species; but from observations in the Flora 1848 p. 42 [Hornschuch], in regard to the absence of the two parents in a place where G. intermedium was found, we perhaps have here two distinct origins of the connecting links, making the confusion doubly confounded.

Mr. Watson, who has paid the closest attention to the subject under discussion,[3] & to whose assistance I am under great obligation, in a letter, ⟨which he has permitted me to publish⟩, has pointed out in a very clear manner the following four categories in our British plants.

Mr. Watson's note./

15 A–1/ ⟨Categories of Species⟩

1. Plants distinguishable from each other by positive characters, & generally received as true Species.
2. Same as No 1; but so closely resembling each other as to be frequently mistaken one for the other, & by botanists even of some experience.

[1] Silliman's American Journal of Science vol. 40. p 280 [actually vol. 41 (1841), p. 281.]
[2] Mr. H. C. Watson has given me a list of examples divided into three groups. (1) of two species actually passing into each other by intermediate varieties. (2) of two species closely approximating to each other by intermediate varieties. (3) & more commonly of one two species varying & its varieties assuming some of the characters of the other species, either positive or negative, but without actually passing into that other species.
[3] An admirable paper ⟨entitled on the Theory of Progressive Development⟩ from which I have largely borrowed views & facts by Mr. H. C. Watson on the relations of species to each other & the varieties is given in the Phytologist 1845 p. 140 & 161.

3. Same as No 1; & not liable to be mistaken in their typical forms; but accompanied by intermediate or transition forms, approximating so much to each or both, as not to be quite satisfactorily assigned to either. ⟨(N.B. The primrose & cowslip would be in this category, but it has been there proved that the intermediate produces both the alleged species from the same year's seed)⟩/

15 A–2/4. Plants deemed true species where their typical & most general forms only are looked at; but the limit of the species is rendered uncertain by the existence of forms closely allied, deemed varieties of the type by some botanists, distinct species by other botanists. As is the case with the intermediates of no. 3, so these varieties or sub-species of No 4 are usually much more rare or local than the type species. They differ from the intermediates of No 3 only as varieties or quasi species clustering around one, instead of uniting together two supposed genuine species./

15 A–3/Altho' four such categories are easily defined on paper, & illustrated by selected examples, they glide together by other examples; & thus, as groups, they are different in degree rather than in kind. To give examples of the four categories,

1. The Apricot, plum, & Cherry are commonly placed under one genus, *Prunus*; & as species these are very readily distinguished by any body. 2. But there are two Cherries spontaneous in England, an arborescent & a fruticose, which by most botanists are deemed two real though very similar species, & between which in a wild state we can hardly point out any connecting links./

15 A–4/3. Many botanists deem the wild sloe of England to be quite a distinct species from the cultivated & probably imported plum-tree of the gardens. Nevertheless, between the plum-tree of the garden & the sloe-bush of the hedges, there exist numerous intermediate forms or links, which render it highly difficult to say, 'here ends the sloe & its varieties, there begins the plum & its varieties! If we hold the intermediate Bullace a good species, this also passes insensibly down to the sloe, & improves almost as insensibly into the plum, so numerous & fine are the steps or links either way.

4. Linné described the fruticose bramble as *a* species, under name of Rubus fruticosus; but various modern botanists make out 50 to 100 supposed species of Bramble which others call varieties of R. fruticosus, & others again group into a small number of species, say half a dozen./

16/To the naturalist who looks at species as not essentially differing from varieties, being only more permanent, with the connecting links extinct, the occasional blending by intermedial forms of two or more apparently distinct species, will not be wonderful; indeed the wonder is to us, with our restricted notions of the lapse of time, that many more cases are not on record.—

Individual differences.—Besides the varieties recorded by naturalists we have individual differences, which are not thought worthy of separate notice, either from being so slight, or from being believed to be so little permanent or forms graduating or blending into each other so that they cannot be divided even into distinct varieties./16 v/Nothing can be looser than this distinction; no doubt a multitude of what perhaps should be called individual variations, with no degree of permanence figure as recognised varieties; moreover it is quite a common practice with naturalists to pick out of a graduated & inextricable mass of forms, a few leading types & designate them as varieties as does Mr. Wollaston[1] when speaking of his "technical" varieties. In other cases, when this has not been done, it might be easily effected, especially if a few of the intermediate forms were to become lost; as remarked to me by Mr. Watson in regard to Polygonum aviculare. But on the other hand if we take the extreme case of well marked & permanent varieties, & the difference, just perceptible though hereditary, between a brother & sister organism, some such distinction does exist, as no one would put these differences into the same class. M. Boreau, who has so carefully studied the Flora of France[2] has called attention to this distinction & says "les varieties proprement dites sont plus tranchees"./16/Individual differences from being generally very slight compared with the difference between species have not I think always been sufficiently noticed by naturalists. When discussing the subject of varieties one is apt, except in very variable forms [,] after a short preliminary study to forget them; but let any one collect specimens in almost any group of beings, about which he is profoundly ignorant, & he/17/will be for a short time, at least I have been, utterly perplexed to tell what are individual & what specific characters. This indeed is tacitly acknowledged by every cautious naturalist, by their dislike to define a new species, without it be some strongly marked form, if he possesses only a single specimen. I have been in the habit during many years of marking in all careful monographs & works, in which measurements have been given of several individuals, with care taken to note sexes & age, & I cannot doubt that individual differences are very often considerable; & no one doubts that this is the case with plants [.] It is impossible to give instances: many cases might be selected from Mr. Waterhouse's excellent work on two great orders of Mammalia, & likewise in Macgillivray's elaborate work on British Birds. I will refer only to one other instance, as it, also, relates to birds, generally considered, & I believe truly, as very fixed in form: Graba[3] who particularly attended to this subject, says that he shot hundreds of seabirds

---

[1] Variation of Species. p. 5.
[2] Flore du centre de la France 1840 p 101
[3] Tagebuch auf eine Reise nach Färo 1830 s. 103: he gives details of measurement of beaks &c of Anthus s. 56 & 67.—of beak & tarsi in Larus s 65 & 80.—& in Colymbus s. 118 etc

at Faroe & that he seldom omitted to measure very one, & the result was that rarely did two individuals of the same species agree throughout in their measurements./

18/These individual differences differ in amount to a surprising degree in various species & in various groups of species, one part or organ being affected in one species or group, & the same part being very constant in another set of species [.] Some forms are extremely constant in their whole organisation others as variable, causing to the naturalist an odious amount of perplexity. Generally the characters which individually vary, are of slight physiological importance, but this is not always the case; & I will immediately give a table of some of the more important & curious cases of variation (the slighter ones not being worth notice) which do not seem to be characteristic of any breed or variety, & therefore are not marked as separate varieties by Naturalists.

But here arises a perplexing question; are these individual differences of the same order & have they the same origin as those other differences, either greater, more permanent, or less closely linked together, which separate recognised varieties. ⟨Many authors seem to consider that each species was created with a certain fixed amount of variability, or to use an expression in a letter of Prof. Dana, with "its system of librations under the influences of nature to which it may be subject", & this would include both recognised varieties & individual variations.⟩ No one will pretend that any clear line of demarcation can be drawn between these two classes of facts; but some authors as Dr Prosper Lucas/19/ think that the production of slight differences is the normal & invariable function of the reproductive system in all organisms, independently of their conditions of existence; & the universality of some slight individual differences countenance this conclusion; but this view I presume no one would extend to marked varieties, & thus even a fundamental difference between individual differences & varieties seem to be indicated. But to me it seems a simpler view to account for all individual differences, which cannot be explained by differences in the parents or more remote ancestors, by the effects of varied ⟨external⟩ conditions acting on the parents & ancestral forms & thus affecting indirectly (as we have seen in the last chapter) the reproductive system & consequently its products. According to this view if we could start with quite similar organisms & bred them for many generations during their whole lives under absolutely similar conditions, the offspring would be absolutely similar; & consequently we should look at all individual differences (independently of those produced by crossing) as having the same nature & origin with those marked by naturalists as varieties.

In favour of this view we have the broad facts that there is much more individual variability as well as distinct varieties in domestic/20/productions, than in those under their natural & unchanged conditions. M. Boreau thinks that it is the very common plants, which vegetate in all places & under all exposures, which offer innumerable slight differences. It is certain that some species which are extremely constant in one area are extremely variable in another: thus the Helix aspersa one of our most constant land-shells in the South of France, as I am informed by Sir C. Lyell is very variable; & many instances might be given. On the other hand the general impression which I have taken, is that a variable species is in all places & all times variable; but I have not met with careful observation on this head. Variable sea-shells seem to be variable everywhere, but these in most cases are attached shells, as Limpets & oysters & cirripedes & they would everywhere be modified by the surfaces of attachment. In Beetles Coccinella seems everywhere variable in its spotted colouring. I applied to Dr. Hooker on this subject & he went through the Tasmanian & New Zealand Flora with this idea, & he found that those genera which were very variable in Europe were there also very variable; but in the Himalaya, the species of Willows, Rubus,/21/Senecio Gnaphalium, which are so eminently variable in Europe & in N. America were there not so./

21a/I have applied, also, to Mr. Davidson, whose vast experience in Brachiopodous shells, makes his opinion of the highest value & I find he has specially attended to this subject & is puzzled by it equally with myself: he says that certainly many fossil shells, as Spirifer rostratus of which he has examined vast numbers of specimens from various places & periods, present everywhere the same quite extraordinary amount of variability: on the other hand some other shells of this same order vary but little either in time or space: innumerable examples could be given of the foregoing cases & this was all that I could learn on this subject from the late Prof. E. Forbes & from Mr. Woodward. Under certain conditions the same species, of which Mr. Davidson has given me examples, will be very variable in one space & constant in another: thus, also, Mr. Searles Wood, who is so intimately acquainted with the Crag fossil shells, informs me that several species, from the Mammaliferous stage are remarkably variable more so than the same shells at the present day, & which he is inclined to attribute to the former estuary conditions of the site:

on the other hand Mr Wood has not found the same degree of variability in the Eocene estuary shells of Hampshire/

21/These facts, & more especially the existence in every great class of organisms of groups of species adapted to varied conditions & growing in different countries eminently variable, as the genera of plants just mentioned & many others & as in the Brachiopods in various geological formations seem to indicate that the variability is here innate & independent of the conditions of existence: or that according to the common view, that they have been created with this tendency, each having its own system of libration to use an expression of Prof. Dana in a letter to me. But this tendency can seldom be predicated of every species in the variable group; thus even in Rubus, the R. [   ] is a very fixed form: in the eminently variable genus of shells, Pleurotomaria M. Eudes-Deslongchamps[1] states that some vary hardly at all, some, so to speak without any limit. How variable are the species of Squirrels, yet Dr Bachman who has so carefully studied the N. American species, informed me that some are very true to their characters. As under cultivation forms are often produced which are characterised by being variable, it/22/is perhaps possible, according to the views we are examining in this work, to account for groups of variable species by their inheriting this tendency from a common parent; but I am not satisfied with this conjecture.

If it could be rendered probable that in the course of time some one or two of the forms of a species individually very variable might become fixed, then with the extinction of the intermediate forms we should see the stages & in some cases better understand the origin of the more permanent varieties. The occurrence of certain constant species in the most variable groups harmonises with such a view.—M. Lecoq, I presume is of opinion that this would happen, & likewise that the more fixed varieties would be converted into & deserve to be called species, for he speaks of such genera as Rubus; as being genera in process of formation. But I must leave the case of these many Protean groups an open question; not doubting, however, that in very many instances there is no real distinction in nature or origin between individual differences & more strongly marked & permanent varieties./

23/I will now give a few ⟨selected⟩ examples of individual variation or differences, not known to characterise a recognised variety; & I shall select them from various motives, some from the physiological importance of the organ affected, or from such part being in the group in question generally constant &c.—

[1] Mem. de la Soc. Linn. de Normandie Tom. 8. 1849. p. 23

Several other cases might have been added, & will be subsequently given, illustrating the variability of rudimentary organs, of greatly developed parts, & of sexual characters &c. One chief object in the following list, is to show that the common remark that organs called important by naturalists *never* vary is not quite correct, but anyone, unacquainted with Natural History, who might infer that because this or that part varies in certain species given as examples, it would likewise vary in other groups, would err greatly.—/

24/In Utricularia nelumbifolia,[1] in the perfect (sexual) flower, especially where only one stamen is antheriferous the anther is commonly found to be one celled. The lobes of the style are variable in number, as are the scales of corolla & calyx.

In Zannichellia palustris[2] "the form of the stigma the length of the style, the number of anther-cells...the fruits more or less stipitate are very variable."

In the common Beech Fagus sylvestris[3] Persoon has described a wild individual with extraordinary large leaves & fruit, & another with the bark & manner of branching so precisely like an oak, that the country people consider it a cross.

Prof. Vaucher says that he has found the kind of gemmation[4] with one exception always the same in the same species of tree, & that it generally is a generic character; but that in the common Lilac, Syringa vulgaris, he has observed two forms, "bourgeon terminal" & "presentant ruptures"./

25/Papaver bracteatum & orientale[5] present indifferently two sepals & four petals or three sepals & six petals, which is sufficiently rare with the other species of the genus.

In the Primulaceae, & in the great class to which this Family belongs[6] the unilocular ovarium is free, but M. Duby has often found individuals in Cyclamen hederaefolium "ou la base de l'ovaire etait soudée jusque à un tiers de la longeur avec la partie inferieure un peu charnue et dilatée du calice."

M. Aug. St. Hilaire[7] speaking of some bushes of the Gomphia oleaefolia, which he at first thought formed a quite distinct species, says, "Voilà donc dans un même individu des loges et un style

---

[1] Dr. Asa Gray, Silliman's American Journal vol. 45. p 215 [where reference is to Oakesia conradii not Utricularia.]

[2] Sir W. J. Hooker & Arnott's British Flora 1855 p 486

[3] Linnaean Transactions vol 5. p. 232

[4] Mem. Soc. Phys. de Geneve Tom. [1 (1822)] p. 300

[5] Decandolle, Mém. Soc. Phys de Genève. Tom 2. Part 2. p 127

[6] Duby Mém Soc. Phys de Genève Tom. x. p 406

[7] Sur le Gynobase, Mem. du Mus. d'Hist. Nat. Tom x. (1823) p 134

qui se rattachent tantôt a un axe vertical, et tantôt a un gynobase; donc celui-ci n'est qu'un axe véritable; mais cet axe est deprimé au lieu d'etre vertical." He adds (p 151) "Tout ce qui precédè[1] n'/26/indiqueroit-il pas que la nature s'est en quelque sorte essayé dans la famille des Rutacees a former d'un seul ovaire multiloculaire, monostylé et symetrique, plusieurs ovaires uniloculaires munis chacun d'un style." And he subsequently shows (p. 364) that in Zanthoxylum monogynum "il arrive souvent que sur le même pied, sur la même panicule [text seems actually 'le même panicle'] on trouve des fleurs à un ou deux ovaires." And that this [is] an important character, from the Rutaceae, to which Zanthoxylum belongs being placed "dans la cohorte (Tom. XI. p. 48) a ovaire solitaire."—The same author (Tom XI. p. 49) referring to this same character differing in the different species of Helianthemum, states that in the H. mutabile "une lame, *plus ou moins large*, s'etend entre le pericarpe et le placenta."/

27/De Candolle has divided the Cruciferae into five sub-orders in accordance with position of radicle & cotyledons, yet M. ⟨Monnard &⟩ J. Gay[2] found in 16 seeds of Petrocallis Pyrenaica the form of the embryo so uncertain that he could not tell whether it ought to be placed in "pleurorhi[z]ée" or "notorhizée": so again (p 400) in Cochlearia saxatilis M. Gay examined 29 embryos & of these 16 were rigorously "pleurorhizées" 9 had character intermediate between pleuro- & Notorhizees & 4 were pure notorhizées: a few other examples are given of variability in a character of great importance in this large Family.

In the Cruciferae it is well known, that Bracteae are generally absent, but these have been observed[3] in certain individuals of Cardamine pratensis, in Erucastrum Pollichii & in (cultivated) Wall-flowers.—In regard to bracts, I may add that W. Herbert[4] says that there are varieties natural & arising from cultivation of Crocus aureus, with & without bracts.[5]/

28/The insertion of petals & stamens is a character of high generality; but M. J. Gay[6] found in Arenaria tetraquetra, that in var. uniflora, which is polygamous, that in the hermaphrodite flowers the insertion was ambiguous neither visibly perigynous or

---

1 [Pencil note at bottom of fol. 25: 'I forget & I am not sure that this has bearing'.]
2 Annales des Scien. Naturelles 1. S. Tom 7. p. 389 [citation should be p. 391.]
3 [Anon.], Henfrey's. Botanical Gazette vol. 3. p. 82. & vol 1 p. 307
4 Journal of Hort. Soc. vol. 2 p. 283
5 [Here at the foot of the sheet Darwin added in pencil: 'Hooker says there are species of crocus with Bracts.' as if this were a memorandum later producing the previous sentence]
6 Ann des Sci. nat. Tom. 3 (1 series) p. 27 [citation should be p. 35.]

hypogynous, whereas in the female individuals, the insertion was perigynous: in *var.* aggregata (thought by some to be a distinct species) the insertion was ambiguous in all the individuals.

M. Raspail asserts[1] that a grass Nastus Borbonicus is so eminently variable in its floral organization, that the varieties might serve to make a Family with sufficiently numerous genera & tribes,— a remark which shows that important organs must be here variable.

In Globularia nudicaulis[2] the upper lip of the corolla varies remarkably, being sometimes entirely wanting, sometimes very small & divided to the base./

29/In some species of Hern[i]aria[3] on the same individual, the divisions of the calyx are regular or irregular with four or five sepals.

In Suaeda, the vertical or horizontal position of the seeds in the pericarp has been thought a character of some importance, but M. A. Moquin[4] found that S. altissima "presente des grain[e]s tantôt droit[e]s, tantôt obliques et quelquefois couchées." With the different position of the seeds the point of attachment of the umbilicus varies./

30/M. Milne Edwards[5] has given a curious table of measurements of 14 specimens of Lacerta, & taking the length of the head of standard, he finds, neck, trunk, tail, front & hind legs, second toes of posterior legs, colour & femoral pores all varying wonderfully; & so it is more or less with eleven other species. So apparently trifling a character, as the scales on the head, affording almost the only constant character.

Mr. Couch[6] has seen the common ling Gadus molva with two cirri on the throat & G. mustela with five barbs.

The eggs of many Birds, especially of the Crow genus, of Shrikes, & Gulls vary in tint of colour, in spotting & size, even sometimes in the same nest.[7]

The Beak of birds, though generally so constant in character that most of the systematic divisions are founded on it, varies sometimes considerably in length; & I was shown in the British Museum by Mr. G. R. Gray three examples of/31/a Nutcracker

[1] Annal. des Sci. Nat. 1 ser. Tom. 5. p. 440
[2] Cambessèdes in Annal. Nat. Scien. 1 ser. Tom. 9. p. 15 [see p. 17.]
[3] Decaisne, in Annal. Sc. Nat. 1 ser. Tom. 22. p. 97
[4] Annal. des Soc. Nat. 1 Ser. Tom. 23 p. 274
[5] Annal. des Scienc. Nat. 1 Series. Tom. 16, p. 50
[6] Linn. Transacts. vol. xiv p. 73
[7] [Sheppard,] Linn. Transact. vol xv Part i. p. 9. See, also, for numerous cases W. C. Hewitsons British Oology where the variations are shown by coloured Plates.

(Nucifraga) shot in some forest, with beaks of remarkably different length: he showed me, also, a Himalayan Nuthatch (Sitta) with beaks similarly varying. I observed the same fact in two S. American birds[1] the Uppucerthia & Opetiorhynchus. The conspicuous character of the tooth on the upper mandible, varies in some Hawks, as in the Jer Falcon.[2] In whole Families of Birds the number of tail feathers is constant; but in some, as in Swans & in some Gallinaceae the number is variable; & this is the case according to [   ] (Isis [   ]) in many short-tailed Birds as the King-fisher: in the N. American coot[3] the number varies from 10 to 16. In some Hawks & Owls, the proportional lengths of the primaries, a character perpetually used to separate species, varies.[4] I have already quoted from Graba instances of variations in length of the/32/tarsi in several sea-birds & so it is with Anser Canadensis.[5]

Is. Geoffroy St. Hilaire[6] has mentioned the case of a Monkey with an extra pair of molar teeth.[7] Such cases, I may remark, are often called monstrosities; but if the teeth are well formed, I hardly see that they should be so called without every deviation from the normal structure be so designated. ⟨Mr. Bellamy exhibited to Brit. Association in 1841, the head of Arvicola agrestis with fangs to its teeth, a character known to separate two groups of Mice.[8]⟩ Dr. J. E. Gray has found considerable variability[9] in the molars of certain seals. The form of the lower jaw seems also to vary[1] considerably in Sloths. So according to M. De Blainville it is with the lower jaws of the Hippopotamus/

33/Dr. Andrew Smith[2] in speaking of the antelope Cephalopus Natalensis, "the females are almost always found without horns, yet individuals are occasionally killed in which they exist; hence it would appear that their presence or absence ought not to be highly considered in establishing the generic characters."

In some species of Shrews (Sorex) & in some field-mice Arvicolae, the Rev[d] L. Jenyns[3] found the proportional length of the intestines to vary considerably. He found the same variability in the number

[1] Zoology of Voyage of Beagle: Birds p. 66, 67
[2] Sir J. Richardson & Swainson, Fauna Boreali-Americana. p. 27
[3] Fulica Americana, in Richardson's Fauna Bor. Americ. p. 404
[4] Ib. Birds. p. 58, 60, 80, 90   [5] Ib. p. 469
[6] [Here Darwin left an unfilled blank space for the citation.] See *Histoire des anomalies*, I, p. 660.]
[7] [Here Darwin later added: 'Owen Ourang Outang'.]
[8] [See Br. Ass. Rep. for 1841 (1842) part 1, p. 68.]
[9] Proceed Zoolog. Soc June 12 1849 [see 'On the variation in the teeth of the crested seal...']
[1] Ib. May 8 1849 [see 'On the genus Bradypus of Linnaeus'].
[2] Illust. Zoolog. of S. Africa. 1849 Pl. 32
[3] Annals of Nat. Hist. vol 7. 1841. p. 267. 272.

of the caudal vertebrae. In three specimens of an Arvicola,[1] he found the Gall-Bladder having a very different degree of development, & there is reason to believe it is sometimes absent. Prof. Owen has shown[2] that this is the case with the gall-bladder of the Giraffe/

34/It has been long known that the presence of nails on the posterior thumbs of the ⟨Borneo⟩ Ourang[3] is variable; & Prof. Owen has shown that with the nail there is an additional joint & bone. Prof. Owen informs me that he has seen a specimen having that muscle of the index-finger, which has been thought characteristic of man; but in another specimen it ran to the second finger as well as to the index.

In Spiders, from six cases recorded by Mr. Blackwall[4] the more or less complete absence of pairs of the eyes, & even the presence of a symmetrical superpernumerary one does not seem to be so rare a variation, as might have been anticipated in so important an organ.

In the sea-urchins (Clypeastroida) the position of the anal orifice is highly variable, being even in the same undoubted species, sometimes above, sometimes below, & sometimes on the border of the shell.[5]/

34 v/In many insects of several widely different classes, the presence of wings, is extremely variable within the limits of the same undoubted species; as in one British beetle Calathus mollis, in some Hymenoptera, & in several aquatic hemiptera.[6] In a rare case described by Mr. Wollaston (p. 96) the connateness of the elytra varied.—/

35/It has been remarked by some authors, that the difficulty in determining what forms are really species, is due simply to want of knowledge. Undoubtedly this is often true, more especially in regard to the different stages of growth & sex of animals. But I suppose the Flora of Great Britain may be considered well-known, & yet how differently is the number of species estimated by different authors! Mr. Hewett C. Watson informs me that after examining the London Catalogue (4th Edit) for this object, he finds that there are about 1800 names which have been considered by some Botanists as Species, but that out of this number, about 450 are

[1] Ib. p. 272
[2] [Darwin left space for citation. See *Zool. Soc. London, Proc.* 6 (1838), 10.]
[3] Sir J. Brooke in Annals of Nat. Hist. vol. 9. 1842 p 58
[4] Annals of Nat. Hist Vol xi. 1843. p 166
[5] Agassiz & Desor in Annal. des Scienc. Nat. 3 series. Tom. 6. p. 318
[6] Westwood Modern Classification of Insects Vol. 2. p. 431. & Wollaston Variation of Species. p. 43. 101

considered by other Botanists as mere varieties: moreover he has given me curious details, showing how opinions have alternated in successive periods two forms having been considered varieties, then species then varieties & lastly species again; these opinions being probably at no time unanimous. In certain Protean British genera the following table, published by Mr. Watson[1] shows at a glance how unfixed is the criterion of a species.—And it particularly deserves notice that most of these genera in our own country/ 36/have been the subject of special monographs, sometimes by successive authors, who have devoted the closest attention to these genera./

36 v/	Salix.	Mentha.	Rosa.	Rubus.	Saxifraga.
Hudson (1791)	18	6	5	5	9
Smith (1824–8)	64	13	22	14	25
Lindley (1835)	29	9	17	21	24
Hooker (1842)	70	13	19	14	16
Babington (1843)	57	8	19	24	20
London Catalogue (1844)	38	8	7	34	16

["The table is intended to show the number of indigenous species in some of these genera, varying according to the author who describes and catalogues them." H. C. Watson, *loc. cit.*]/36 v/ Atriplex is another protean genus. The Rev. Leighton told me that he had some seeds of several species collected in various places in his garden, & that a mass of plants came up, which defied the powers of the two botanists most skilful in this tribe, to classify.—/

36/So again M. Ch. Des Moulins[2] in his discourse on the well-known Flora of central France, says that in 2332 phanerograms, there are still 250 forms under litigation.

I suppose no two land-shells are better known than Helix hortensis & nemoralis. Mr. Bean[3] of Scarborough has collected 152 vars. of H. hortensis 58 of H. pullata of some authors or the white-mouthed var. of this species; 236 vars. of H. nemoralis, & 21 of its variety or supposed species H. notabilis. Notwithstanding all this attention, & notwithstanding the fact, as I am informed by Sir C. Lyell, that H. hortensis ranges further north than H. nemoralis & is alone found in Canada, yet some great conchologists, as Deshayes doubt whether H. hortensis & nemoralis are not the same species./

[1] Phytologist. May 1845. p. 143   [2] Actes de la Soc. Linn. Tom. 16. 1849 p. 56
[3] As quoted in Forbes Report Brit. Assoc. 1839 p. 136

37/To give another example, not so much to show that there is difficulty in deciding what form to call species & what varieties, but that even in a class, generally having such fixed characters as Birds, there is some appreciable amount of variation. In Germany, according to common authors, there are about 282 Birds, but Brehm[1] by dividing species, adds to this number 576 species, making a total of 856 species: thus he divides the tit-lark (Anthus pratensis) into 12 species & the Nightingale into 6 etc.—Now I have never met an ornithologist who thought these species worthy of consideration, & it has been asserted in Germany that many have been formed on single specimens.—On the other hand Brehm was a laborious observer: he collected[2] more than 4000 skins, & he positively asserts that his new species are often found paired together, that they can be found on the same spot in successive years, & that they can often be distinguished by their voices & habits; & lastly that Bird catchers practically make similar distinctions. He grounds his distinctions chiefly on slight differences in the shape of the skull, beak, tail & feet. Though it may be very proper to/38/ignore these fine differences as specific, I can hardly doubt but that they exist. I believe this the more as our great ornithologist Mr. Gould has lately shown me some of our commonest birds from different districts, certainly presenting an appreciable difference.—

Lamarck long since remarked that there was not much difficulty in distinguishing species ⟨from varieties⟩ as long as specimens were brought from a single country,—not that this can be considered, as we have just seen, as always quite correct—but that the real difficulty begins when specimens pour in from every region inhabited by the genus.—Though this may be very true, yet with cautious & sound naturalists, how often do these numerous specimens if collected from continuous regions clear away doubts;[3]

---

[1] Vögel Deutschlands 1831

[2] Ib. Introduct p. xix

[3] I am far from wishing to assert that this always the case: on the contrary I was formerly much struck, when witnessing Mr. Waterhouse (than whom a more ⟨cautious⟩ accurate naturalist can hardly be found) examining the large collection of Mice, which I made in S. America: when the specimens came all from the same exact locality, or from very distant localities, there was seldom much difficulty in distinguishing the species; but when a specimen or two had been collected at a moderate distance from any other locality, then I repeatedly observed there was very great doubt & difficulty. Exactly the same thing was noticeable in the difficult genus of Birds, Synallaxis, of which I collected many specimens.— Probably if I had collected still more numerous specimens, from every intermediate station, there would have been less difficulty, but the difficulty would have been removed only by admitting considerable variations, or by designating every infinetesimal difference as specific.

but the doubts are generally dispelled by admitting considerable variation;—intermediate forms connecting others which might have been classed as specifically distinct. Hence apparently it arises that those who study local floras are apt to admit more forms as species, than those who take a wider field. But the/39/difficulty rises to a climax & indeed seems insuperable where very closely similar forms are compared coming from islands & from countries apparently now quite separated:[1] I was much struck how entirely arbitrary the distinction is between varieties & species, when I witnessed different naturalists comparing the organic productions which I brought home from the islands, off the coast of S. America. In such cases there is no intermediate territory for the existence of intermediate forms; & the naturalist must rely wholly on analogy. North America & Europe offer the most striking example of this difficulty: let it be observed to what different conclusions the best naturalists have come to in regard to many quadrupeds, birds, insects & plants[2] of these two quarters of the world; some

[1] Instances innumerable could be given in regard to the islands of several great archipelagoes; & even from so small a one as the Galapagos group. Mr. G. R. Gray showed me some small pigeons (Peristera Macro-dactylus, Brasiliensis, brevipennis &c) from the W. Indian Islands & mainland, which certainly differed ⟨slightly⟩ sensibly in length of wings, toes & plumage; & yet so cautious a naturalist as Mr. Gray, is strongly inclined to believe that they are only local races. I have quoted this instance, because Mr. Blyth has instanced in a letter to me another genus Treron in this family, as offering in the East the very same cause of doubt; but he leans to considering these slight differences as specific; for he remarks if these be given up, where can we stop; & well may he ask this: the answer in future years, will be, as I believe, no where.—It is known that large rivers in S. America form an impassible barrier to monkeys, & on their opposite sides the monkeys often differ slightly; these forms have been described by many authors as distinct species; but Dr. Natterer, a most careful observer who resided many years in Brazil (Note by Mr. Waterhouse in Annals of Nat. History 1844. vol. 13. p. 48) was convinced that these forms were only races of the same species See Mr. Wollastons works in regard to the insects of Madeira. See Mr. Layards & Blyths remarks in regard to the Birds of Ceylon.—

[2] Dr. Asa Gray has lately published a truly admirable paper on the Statistics of the Flora of the Northern United States (⟨American⟩ Journal of Science 2 ser. vol 23. p 80) & he gives a list of 15 varieties of plants common to Europe "which not only have been, but are not unlikely to be again distinguished as species", & another list of 42 N. American species, "almost all of which are more or less liable to be reduced to geographical varieties", of European plants. Had the United States been worked as carefully by *local* botanists as have the different parts of Europe, there can be no question, that a number of forms, which Dr. Asa Gray considers identical with European plants, would have been cut off by Botanists having less widely extended knowledge, as distinct North American species.

For Birds compare Prince Napoleons list with that by Sir J. Richardson & Swainson & other work. For Coleoptera, compare Mr. Murrays remarks on (Proc. Phys. Soc. Edinburgh in Zoologist. vols. 11 & 12. p. 3894) the differences, which he (as Kirby likewise) considers too slight to be specific, with M. Leconte in Agassiz Lake Superior p. [blank—see pp. 239*–240*] who seems to consider that all? [sic] or nearly all, should be specifically separated.

calling the slight differences which can undoubtedly be observed in nearly all the animal productions from the old & new world, varieties, & some calling them species.

At present a considerable number of naturalists cut the knot by calling all forms from distinct regions, distinct species, even if the differences are excessively slight & even if apparently they are/40/identical. To those who rest on the hypothesis of distinct creation as the criterion of a species, this may be logical; but who can say what regions should be called distinct? Can we say we know all the means of distribution; past & present; as what part was land & what sea, & what was the exact temperature of either, within comparatively recent geological times? In regard to distance, as Mr. S. Haldeman & Wollaston[1] have well remarked where shall we draw the line; if N. America & Europe are so distant from each other, that we may call their most closely allied inhabitants distinct species; are the Azores or Madeira sufficiently distant in regard to Europe to justify the same distinction. Must we extend the same view to Madeira & Porto Santo, within [   ] miles of each other, but with so many shells & insects quite distinct, & so many forms presenting marked varieties? Lastly must we extend it to Ireland & England, with only extremely few species distinct, but with some few, as generally considered, well marked varieties? Practically each naturalist arbitrarily decides the question for himself, in accordance/41/with his hypothetical idea of the term species, in accordance with what he knows of the amount of variation witnessed during the present time, & according to his tendency to trust in analogy.

We have seen that in the best known countries there is much uncertainty in deciding what to call species & what varieties. And further it seems to me that very generally if an animal or plant inhabits different districts or even if very common in one district, if it be conspicuous for any quality, or if it be valuable or in any way attracts man's notice, so as to be thoroughly well studied varieties will have been observed, & the more striking varieties will often have been considered as distinct species. Look to the King of beasts, as popularly called, how naturalists have doubted whether or not the Maneless Lion of Persia[2] is a distinct species: Some few think that of Nubia also distinct; & the great lion-slaughterer Mr. Gordon/42/Cumming is convinced that there is

[1] Boston Journ. of Nat Hist vol 4. p 480. Wollaston Variation of Species p. 38.
[2] Capt. Smee in Zoolog. Transacts. vol III [actually vol. I] p. 165 concludes that the Maneless lion of Guzerat is only a variety; I believe many naturalists now think it distinct. The Hyaena of Persia (Harlan's Researches p. 535) is, also, said to differ from that of Morocco only in wanting a mane.

more than one even in the Cape district.[1] or look to the Elephant
in India, but the variation in this animal is so curious that I shall
presently enter into some little detail on the subject; as I shall
on the well-known & persecuted Fox of Europe. What disputes
there have been in regard to the Bears of Scandinavia, there so
ardently hunted, whether these there be one or more species. How
many moles may a person casually examine without perceiving
the slightest difference, yet being a thoroughly well known, animal,
we hear from Mr. Bell, in his excellent history of British Quadrupeds
(p. 106) that there are several remarkable varieties.[2] The Sportsman[3]
can distinguish the Red Deer (Cervus elaphus)/43/of the different
Scotch forests; "the Braemar deer are allowed to be quite different
from those of Atholl, they stand higher & are in general of greater
weight": those of Corrichebar are again different & have larger
head than those of Atholl: the red deer of the outer Hebrides are
very small[4] So in Germany three varieties of this deer are distinguished & inhabit different localities.[5] Other instances could be
given as with the common Hare. So with Fish, it is certain that
the salmon of many different rivers can be distinguished by fisher-
men; & the Herring which has been so closely studied, is found
to present a vast range of variation.[6] To descend lower in the
scale; Fishmongers can distinguish whence their oysters come, &
so they can on the coast of N. America with the clam, of which
they distinguish five varieties.[7]/

44/In plants most of those useful or much noticed by man are
cultivated, & therefore do not come in here, as their variations
may be all due to cultivation. To begin with a humble example;
varieties of the water-cress (Nasturtium officinale) are hardly
noticed by botanists, but those who cultivate acres of this plant
(not seedlings raised under cultivation) for the London market
distinguish three varieties, which are not caused by any difference
in the quality of the water, for they may be seen growing together;
they differ in hardiness & other qualities; & the large brown-leaved
variety is the only one which will grow well, when the water is

---

1 Lichtenstein in his Travels vol 2. p. 31. says the country people distinguish three
different sorts of Lions at the Cape.

2 The Rev R. Sheppard in Linn. Transact. vol. xiv. p 587, describes a remarkable
variety with a white snout, & white line on the head, belly orange, forming a line
on the chest; tail covered with long white hairs, & with the tip quite white.

3 See W. Scrope's Art of Deer Stalking,—a most interesting work. [c. p. 408.]

4 [Here Darwin left a blank space for a reference.]

5 Bechstein Naturgesch. Deutschlands. 1801. p 458.

6 Wilson's Voyage round Scotland vol 2. p. 206. The Herring Fishery was one of
the points especially attended to in this voyage.

7 Venus mercenaria. Dr. Mitchill in Sillimans Journal of Sciences. vol 10. p. 287.

not very shallow.[1] What is the tree, which ought to be best known in Britain? assuredly the Oak; yet I see that Mr. Babington, Hooker & Arnott with Dr Greville in their last Edition, treat Quercus robur & sessiliflora as varieties, whereas Dr Lindley in the Gardener's Chronicle speaks decisively of them as distinct species, & Sir James Smith seems to entertain no doubt on this subject. Every forester can distinguish the two forms: it is asserted that they come true to seed[2]/45/though this has been denied: the quality of their timber is said to be different[3] & Quercus sessiliflora is hardier & ascends the Scotch mountains higher than Q. robur.[4] On the other hand the existence of a perfect gradation of inter- mediate forms is admitted by everyone & Dr. Bromfield quotes[5] with approval the remarks of another most careful observer Mr. Bree that 'though there are sessile oaks bearing fruit on peduncles & pedunculated oaks bearing almost sessile fruit, there is yet a certain indescribable something about the trees, by means of which I can always distinguish each, without minutely examining either the acorns or the leaf-stalks.' So that according to these two excellent observers the distinction of the two varieties or two species, (& the highest possible authority can be quoted for either term) of our one most conspicuous tree can be best recognised, like a man's face, by 'a certain indescribable something.'

It would be superfluous to give other examples; but parallel ones could be given in regard to our Elms, to our Birches, & most striking ones/46/in regard to the Scotch Fir, in which the varieties or species, call them which you please & you will have high authority for doing so, are adapted to different situations, produce different kinds of timber & are hereditary in their quality.[6] In the Yew, the highest authority Dr. Asa Gray thinks the Canadian form perhaps only a variety; & this seems the general opinion of Botanists in regard to that most remarkable bush, the Irish yew, found growing wild in Ireland with its upright dwarf habit, & large scattered leaves; but I presume, if this plant had been found of both sexes, growing abundantly in some distant region, no

[1] Mr. Bradbery account of the cultivation of the water-cress. [See Ker,] Transact. Hort. Soc. Vol IV. p. 537.

[2] Gardener's Chronicle [1855, p. 776; 1856, 191–2, 405.]

[3] Sir J Smith English Flora. vol IV. p 149 & Gardener's Chronicle [1856, pp. 191–2]

[4] Mr. Farquarson in Hooker's Bot. Misc. vol 3. p 127.

[5] Phytologist Vol. 3. p. 883.

[6] In regard to the Elm, see Dr. Bromfields remarks in Phytologist. vol 3. p. 837. Mr. H. C. Watson exhibited before the Bot. Soc. of London (Annals of Nat History. vol 12. 1843 p 450) specimens showing that Betula alba, pendula, glutinosa, & pubescens are all mere fleeting varieties of the common Birch. For the Pinus sylvestris, see Loudon's Arboretum p. 2189 & 2150 & Gardeners Chronicle [   ].

botanist would have hesitated to name it as a distinct species. The last example which I will give is that of the noble Cedar of Lebanon: it appears in our gardens most/47/distinct from the Deodar, yet when old, Botanists cannot point out any good character between these two forms & the Cedar of the Atlas, & as the seedlings vary hence are inclined to consider them as varieties, a conclusion indignantly repudiated by other Botanists.[1] The question in these several cases, is not whether these forms deserve a name, popular usage has settled that point, but whether they should be designated by the undefined title of Species.—

Incidentally several cases of variation in a state of nature have now been given, & incidentally others will be hereafter given. It would be as easy as useless to quote the almost numberless instances of forms, which have been considered on good authorities as permanent varieties having much of the character of species; & I will conclude this chapter by giving from various motives, a few additional instances of variation, in which the evidence is rather better than in most cases./

48/*Indian Elephants*. Dr Falconer who has had great experience in Elephants, & who has seen as many as 1200 at a fair, informs me that they differ considerably, more than horses of the same breed, in size, general proportions, manner of carrying the head, form of tusks, shape of feet & in the absence of the nail on one toe: Mr Corse has given a nearly similar account[2] & says that the different castes have their proper names. In the Ayeen Akbery, written about the year 1600, four kinds of Elephants are specified. Most of these differences probably come under our class of merely individual differences; but both Dr. Falconer & Mr. Corse believe that some of the breeds inhabit different, adjoining districts; & animals which are thought to be cross-bred, are occasionally caught. As far as size is concerned, climate appears influential; at least, as I am informed by Mr. Crawfurd, elephants northward of a certain latitude are excluded by the government contracts. Dr. Falconer tells me that there are two marked breeds, one thicker in its general proportions, more courageous, & with short tusks directed downwards; in the other breed, the tusks are upturned, & the/49/animal when attacked by a tiger tries to pitch his opponent into the air; whereas the breed with the downward directed tusks, when attacked, falls as if instinctively on its knees, & endeavours to crush & pin the tiger to the ground; this breed is consequently more dangerous to ride, as sometimes even experienced hunters

---

[1] [Darwin here added in pencil: 'Hooker—Gardeners Chronicle'.]

[2] Philosoph. Transacts 1799 p. 206.

are thrown on to the tiger. Now such differences in structure & habits, I think all zoologists, will agree, would in most cases be thought of specific value; but I believe no one has even suspected that there are two species in India. In Ceylon, there is, also, a distinct breed, but this has by some[1] been thought to form another species. Until quite lately the Elephant of Sumatra, was thought to be the same,[2] but now from differences in its skeleton it is thought to be a distinct species./

50/*Foxes*. These are well known to be variable animals & all over the world the species are discriminated with difficulty. British sportsmen speak[3] of three kinds, but it is doubtful whether these are anything but individual indifferences [sic]. In Scotland the accurate Macgillivray[4] describes four kinds, but he uses besides general proportions the tail being tipped with white, which Bechstein[5] has shown is a quite variable point. But the Highland or mountain Fox of Scotland seems certainly to form a distinct race: Mr Colquhoun[6] a very good observer says any one can distinguish this animal even at a distance from the small fox of the low grounds; he stands higher, his head broad, nose not so pointed, his coat more shaggy & mixed with white hairs: he is much more powerful & preys on young sheep, & rears his young, not in holes, but in clefts in the rocks; is less nocturnal in his habits,/51/& altogether, as Mr. St. John remarks, is more like a wolf, than a lowland fox. In Scandinavia it has been a question disputed both by naturalists & hunters, whether the common red, the black & crucigerous Foxes are distinct species or only varieties. So in N. America a parallel series occurs & it has been disputed whether the red Fox, (ranked as a different species from that of Europe) the black & silver & crucigerous (ie with a dorsal stripe & a transverse one on the shoulders) foxes are distinct or not: Sir J. Richardson[7] inclines to consider them all as varieties. So much interest has this question excited in Scandinavia as the differences are said not to be confined to colour alone that a fox colony was established by some gentlemen near Stockholm[8] & in it two crucigerous foxes produced in the course of four years 19 cubs; of these 9 were crucigerous; 8 were black (including those with white tipped tails),

---

[1] Mr. Hodgson in Asi[a]t. Soc. of Bengal vol I (1832) p. 345.
[2] Crawfurd Descriptive Dict. of Indian Islands. 1856 p. 136.
[3] Encyclop. of Rural Sports. p. 448.
[4] Transacts of Wernerian Soc. vol VII. p. 481.
[5] Naturgesch. Deutschlands. B I. s. 627.
[6] The Moor & the Loch. p. 97. Ch. St. John, Wild Sports & Nat. History of the Highlands p 232.
[7] Frauna Boreali-america p 93.
[8] L. Lloyd Scandinavian Adventures vol II. (1854) p. 52.

& 2 red: two of the black cubs, also, produced young & these, six in number, were all black. Mr. Lloyd infers from these experiments that the crucigerous fox is a cross from the black & red, which he seems to consider, with/52/many of the inhabitants, as distinct species, producing, as it thus seems perfectly fertile hybrids, but Prof Nillson's conclusion that they are proved by these experiments to deserve only the name of varieties, seems to me the most probable. It is clear that these variations are in some degree hereditary, & the whole case is interesting as showing how difficult it is to decide what to call species & what varieties. The occurrence, also, of strictly analogous varieties in N. America from the generally received distinct American & Arctic Foxes (C. fulvus & lagopus) is an interesting fact; & the more interesting from these forms, not being produced in Great Britain, though they are, according to Bechstein, in Germany./

53/*Raven.* It has long been known that pied Ravens are found at the little islands of Faroe. This bird is white somewhat symmetrically marked with black, & as the beak "is much larger being not only higher at the base, but more elongated, & in form more attenuated at the end" than that of the Ravens,[1] it has been admitted by Brisson, Vie[i]llot, Wagler, Temminck, & others the most distinguished ornithologists, as a distinct species under the name of Corvus leucophaeus./53 v/As this particular race is known no where else, (though partial albino ravens do occur elsewhere) this fact has been used as an argument that it is a distinct species; but perhaps the argument might be reversed with equal force, as not one other bird or indeed other production is endemic in this small spot./53/When, however the ornithologist Graba visits these islands, & investigates the case he finds that the pied ravens (at first quite white, the black feathers appearing with age), are produced in the same nest with ordinary ravens; & that in one case when black & pied were mated either exclusively black birds or one white bird with the others black were produced.[2] The fact of the black & pied ravens being sometimes/54/mated & producing either black or white young, is not, as we shall immediately see in the case of the Hooded crow, so conclusive as Graba seems to think; but combined with the white appearing in the nests of common ravens, & more especially with the fact of the

---

[1] Magillvray [sic] History of British Birds. vol. 3. p 745.

[2] Tagebuch auf einer Reise nach Fåro 1830. p. 51. Graba's description of the beak nearly agrees with that of Macgill[i]vray. I may add that Landt in 1810 in his Description of Feroe p 220, says that black & speckled ravens are sometimes seen paired & that both kinds are sometimes found in the same nest.

two birds described by Graba, the one by Macgillivray, & that by Temminck, differing very considerably in their colouring, even sometimes on opposite sides of the same individual, I think this can leave no doubt that the C. leucophaeus is only a variety. Graba says that they are not very rare, & he states the interesting fact of which he was a witness that the pied birds are persecuted & driven away by the common ravens (p. 51, 54.); & Macgillvray once saw on the Hebrides a bird of this kind, apparently a wanderer, which he describes as "a neglected & persecuted stranger". Now suppose whatever the cause may be, which gives rise to this variety in Faroe to act with rather more intensity, so that pied ravens alone were to hold possession of these islands, how utterly impossible it would be ever to ascertain whether it was right to call this form a variety or species./55/No doubt any chance wandering black raven would be persecuted & driven away by the pied majority, as these latter now are by the black birds; & crosses being thus prevented, it is probable that the pied colouring & other characters would become in the course of many generations more fixed & constant.

Now let us turn to the Carrion & Hooded crows (Corvus corone & cornix): these birds are so much alike that as Magillvray observes[1] "were the colours the same in both it would be almost impossible to distinguish them". "The extent & tint of the grey-coloured space varies greatly in the Hooded crow ["]² & Bechstein asserts that in Siberia some are quite black, but how these can be distinguished from carrion crows I know not. The eggs of the two species are undistinguishable as are their digestive organs & their general habits are alike. Numerous cases are on record in Germany, England, Scotland & Ireland of these two forms being seen paired, & the young are either/56/quite like one of the parents or intermediate in colour.[3] Hence several respectable ornithologists have looked at these birds as varieties; yet, as their voice is slightly different & as different districts are often inhabited separately by either one or the other form; & as when occurring together they keep separate; & as the carrion crow seems to have a more southern range than the Hooded crow & more especially as ordinary specimens of both can be distinguished with the utmost facility, I must agree with Mr. Macgillvray that, in common parlance, "the two species are perfectly distinct".

[1] History of British Birds vol I. p. 529.   [2] Ib. p. 534.

[3] Bechstein, refers to three cases in his Naturgeschichte B. 2. s. 1170; Mr. Slater informs me that he has known of a case in Hampshire W. Thompson gives cases [See Nat. Hist. Ireland, vol. I p. 309] Macgillivray vol 3 p. 721 gives cases in Yorkshire & Scotland.

Lastly let us consider one other case; we have in Britain one single well-known bird, the red Grouse, (Tetrao Scoticus) which has been almost universally ranked as a distinct species, & is confined to the British islands. On the other hand the Tetrao saliceti of Scandinavia, is a bird which we might have expected to inhabit Great Britain, but is not found here./57/Gloger alone, as I believe, has argued at length[1] that they are certainly only local varieties of the same species.—Mr. Gould after studying T. saliceti in Scandinavia tells me that they agree perfectly in eggs, in the immature plumage, in habits, in voice & in summer plumage, with the exception of the white primary feathers & that he cannot avoid the suspicion that they may possibly be varieties./ 57 v/The Red Grouse is very variable in plumage, & easily runs into sub-local races.[2] Macgillvray says that it differs from T. saliceti in having a lesser beak; but Nilson, as quoted by Gloger says he examined 30 specimens of T. saliceti, & the beak was scarcely alike in two./57/I apprehend if these birds had been found together, & it does not seem improbable that colonies of the one might now be established in the territory of the other; no ornithologist whatever would have thrown a suspicion on their specific distinctness; hence their geographical separation & consequent exposure to a different climate seems to have been the sole cause of their specific diversity having been suspected; & undoubtedly as Britain has no other endemic bird this is an argument of some apparent weight ⟨in favour of the two forms being identical⟩: on the other hand if we had possessed a few more endemic species the argument might have been reversed, notwithstanding it might most truly be said that every gradation exists in the proportional number of endemic forms possessed/58/by a country, & why should not insular Britain possess its single endemic Bird?

I have entered into these three last cases at some little length in order to show how difficult it is to determine what to call a species & what a variety, even with using all sorts of collateral evidence in well-known Birds, which are amongst the least varying animals. The series seems to me an interesting one, from the pied & black ravens which must be considered as varieties, though hitherto esteemed by most ornithologists as distinct & which inhabit the same little island, but with some tendency to keep separate—to the carrion & Hooded crow, considered by a vast majority of ornithologists as distinct, often inhabiting distinct regions, but when mingling, often crossing—to the red & willow grouse,

[1] Das Abandern der Vögel 1833. p. 117.
[2] Macgillivray British Birds. vol I p. 174 & p. 186.

almost universally considered as distinct & inhabiting quite distinct countries, but yet with a taint of suspicion hanging over them./

59/As it is so rare that varieties of Birds, sufficiently distinct to have been esteemed species by first rate naturalists can be proved to be not distinct that I will give one more case. The common & ring eyed Guillemot (Uria troile & U. ringvia or lacrymans) have been by about an equal number of ornithologists esteemed as good or doubtful species or as mere varieties. The ring-eyed form inhabits the northern islands & is generally rare; but Graba found that the Faroe islands[1] were its home, one out of about five existing as this form; at first Graba thought it was specifically distinct, for besides the conspicuous ring of white round the eyes & from the eye backwards, it differed in other respects; but these differences Graba found were not constant, & he subsequently himself twice saw it paired with the common Guillemot; & the inhabitants affirm that sometimes from the two eggs of the common Guillemot, one will be ring-eyed.

In Madeira, there is only one endemic bird, but some of the European birds are slightly smaller, & some are slightly duskier; & the Redpole (Fringilla cannabina) retains/60/its crimson breast throughout the year.[2] The black-cap (Sylvia atricapilla) besides being sometimes duskier, presents a variety, in which the black colour extends from the cap to the shoulders & occasionally even over all the under parts of the body; this has been described by so good an ornithologist as Sir W. Jardine as a distinct species; but as the inhabitants believe that it is produced from the same nest as the common black cap, there cannot be much doubt that Dr. Heineken & Mr. Harcourt are right in esteeming it as a variety.

I will now give a single case in Fish taken from Bronn;[3] the Cyprinus gibelio & carassius have generally been considered distinct species, for they differ in almost every part in proportion, as shown by the table given by Bronn; but Eckstrom narrates that the offspring of the C. carassius removed from a large lake into a small pond, assumed an intermediate form; & on the other hand the offspring from C. gibelio from a small pond turned into a large lake 40–50 years before, had become changed into C. carassius./ 61/I have selected the foregoing instances, from being able to adduce some other evidence besides the mere existence of a graduated series of intermediate forms. But I will now give two instances from Mr. Wollaston's works of Variation deduced from

---

[1] Tagebuch. p. 106, p 150.
[2] E. Vernon Harcourt Annals & Mag. of Nat. History. June 1855 and Sketch of Island of Madeira 1851.   [3] Ges[ch]ichte der Natur. B. 2. s. 106.

intermediate forms observed with sepecial care by this excellent entomologist in the confined locality of Madeira. Harpalus vividus[1] is perhaps the best example; if very many specimens from many sites had not been collected, ⟨clearly showing a perfectly graduating series,⟩ the varieties would have been described as forming several species; those from the lowland & the wooded mountain slopes appearing "altogether distinct". It is an interesting fact, that it attains its maximum of sculpture & minimum of size at about the elevation of 3000 to 4000 feet; both above & below which height, "as it recedes from the upper & lower limits of the sylvan districts, it becomes gradually modified, & almost in a similar manner". It varies greatly in colour/62/shape, in puncturing & in striation & what is even more important in the degree to which the elytra are soldered together: the united elytra are found only rarely in the sylvan districts. This beetle, also offers an instance, of which very many could be cited in the most distinct genera, namely of the individuals inhabiting the rocky islet called the Deserta grande, attaining a larger size than elsewhere. To take another very different genus of beetles, namely Ptinus[2] in which some species of which "do not attain half the bulk on many of the adjacent rocks, that they do in more sheltered districts; & so marvellously is this verified in a particular instance, that I have but little doubt that five or six species, so called, might have been recorded out of one". Ptinus albopictus has a separate radiating form on every islet of the group, but all merge together by innumerable intermediate links. Very many other examples might have been adduced of each islet & even rock of different altitude having its separate variety./

63/   *Plants*

Centaurea nigrescens has been separated by some botanists from C. niga, (the common Knap-weed) by several characters, of which the most conspicuous is that the heads are rayed. The Rev. Prof. Henslow informs me that this form kept true for two generations in his garden, but that in the fourth year it was clearly reduced to C. nigra. I mention this case, because, the var. C. nigrescens, as I am informed by Prof. Henslow occupies nearly the whole of Hampshire to the exclusion of the common forms; and here we have the argument from range, on a small scale, as with the Red Grouse of Britain, which may be used on either side.

[1] Insecta Maderensia p. 54: The Variation of Species. p. 67.
[2] Insecta Maderensia p. 260, 267. For other cases see p. 11, 30, 36, & 78.—

Koch raised the ensuing year from seeds of a dandelion (Taraxi-cum palustri)[1] T. palustri, T. officinale, T erectum, T nigricans, & T corniculatum,—forms which have been admitted by some Botanists as species, & two of which were first named by De Can-dolle. Prof. Henslow on the other hand, though not doubting that T. palustre, is a variety, has found it/64/come up true for three or four generations when self-sown in his garden Koch has, also, raised from seed of one species of Isatis[2] tinctoria, campestris, praecox, dasycarpa,—forms as species by De Candolle, Ledebour & other distinguished Botanists: most of these forms inhabit different parts of Europe & Siberia. From cultivating another cruciferous plant Sisymbrium austriacum, Koch concludes that S. eckart[s]bergense, Willd. & taraxacifolium & acutangulum, both of De Candolle, are only varieties.

Mr. Hewett C. Watson is one of the few British botanists who has experimentally tried to test species by cultivation; thus he has succeeded in raising on plants of Festuca loliacea "stems which a botanist would assuredly have assigned to F. pratensis";[3] & he almost succeeded in running together the common & Italian Rye grass (Lolium perenne & multiflorum).[4] But Mr. Watsons experiments on seeds & living plants which he collected at the Azores, are particularly interesting: thus plants raised from Azorean seed of the Polygonum maritimum "partook much of/65/ the physical characters of P. Raii from the shores of Great Britain". Seeds of the Tolpis crinita from the Azores, produced plants un-distinguishable from T. umbellata; yet these plants differ in the pappus of the fruit, in a manner on which distinct genera have been founded by some authors.[5] Again Mr. Watson has found[6] that cultivation during four generations in England of the forms of Raphanus raphanistrum found in the Azores has partially obliterated a character in the pods which was at first obvious. The rich deep colour of Myosotis Azorica[7] tends to fail in our country; & the seedlings have varied so much that Mr. Watson is unable to say which should be referred to M. Azorica & which to M. maritima; & some approximate to the Canary species, M. sylvatica: yet in their wild state they were as easily distinguished as any other/66/species of the genus.

[1] Annal. des Scienc. Nat. 2 Series. Bot. Tom 2. p. 119.
[2] Annal des Scienc. Nat. Bot. 2 series. Tom. 3. p. 375.
[3] Phytologist, June 1845. p. 166.
[4] I may add to these cases of conversion in Graminea that [of] Bernhardi (Ueber den Begriff der Pflanzenart. 1834. s. 30) that by repeated sowings Panicum ciliare was perfectly changed into P. sanguinale.   [5] [Watson,] Phytologist. 1845. p. 167.
[6] London Journal of Botany 2d ser. vol. 6. p. 385   [7] Ib. p. 388.

The accurate Kölreuter[1] asserts that he has seen the Digitalis thapsi, when cultivated in northern Europe, & when artificially fertilised, so as to preclude any possibility of a cross, after four or five generations assume the characters of D. purpurea, & at last was completely converted into it. The hybrid offspring from the reciprocal crosses of D. thapsi & purpurea were perfectly fertile. Dr Lindley in his Monograph on Digitalis expresses some doubt whether Kölreuter may not have taken a variety of D. purpurea for D. thapsi, but as he speaks of his D. thapsi, as that of Spain, he may probably be trusted. These two forms are considerably unlike in many respects, & have generally been received as good species.[2]

E. von Berg gives a curious account[3] of the extreme variability of the seedlings of cultivated plants of Iris so that Dr. Horns-[ch]uch[4] asserts that he raised twenty reputed species from Iris sambucina or Germanica; I confess that owing to some other recorded experiments of E. von Berg/67/I should have thought that there had been some mistake here, had not his results in the case of the genus Iris been strongly corroborated by quite independent testimony. For M. C. Bouchés[5] by sowing seeds of I. Germanica raised 13 reputed species; & what is important for us, three of these, namely I. florentina, Germanica & pallida are Linnean species & have been found growing in separate districts, & in their own native habitats remain unaltered.

The blue & red pimpernel (Anagallis arvensis & coerulea)[6] have by a good many botanists been considered as distinct species, for besides in the colour of the flower, they differ in some other

[1] Journal de Physique Tom 21. p 291.
[2] In Loudon's Arboretum vol. 3. p 1374 it is stated that Mr. Masters of Canterbury a great raiser of Elms, & therefore one who ought to judge well, is convinced that the Ulmus Americana is identical with the Huntingdon Elms, a variety undoubtedly of English origin. In Bronn's Ges[ch]ichte de[r] Natur B. 2. p. 85, there is a marvellous account of the change of a plant of Lobelia lutea into L. bellidifolia; & by Link of Ziziphora intermedia from Z. dasyantha, & of a great change in Ribes alpinumi. Mr. Gordon of Birnie in his Flora of Moray-shire [p.iv] says the 'Avena pratensis is confined to soils of this description (calcareous), changing its habits, as the proportion of their ingredients differs. Where there is a superabundance of limy matter, the plant often assumes a glaucous-rigid appearance, which has probably originated the A. alpina and causes it still to hold a place as a distinct species.'—
[3] Flora 1833 Beiblatter & 1835. B. 2. s. 564.
[4] Flora 1848 p 55.
[5] Flora 1833, Nachschrift, Horns[ch]uch. s. 44.
[6] [Here Darwin later scribbled in pencil: '1861 a new var. Eugenia [?] I read [?] latterly [?] came red and blue
'In 1867 Red & Blue var[ietie]s of A. grandiflora produced both var[ietie]s & intermediate—See notes on crossing plants.
'1867 Both vars. extra fertile when crossed.']

respects./67 v/It is certain that each kind can be long perpetuated by seed & keep true.[1] On the other hand/67/the Rev^d. Prof. Henslow's experiments,[2] though nearly can hardly be considered absolutely decisive, in showing that one form can be raised from the other: Dr. Bromfield[3] has seen bright blue & flesh-coloured flowers on actually the same plant, when cultivated in a garden. Dr. Asa Gray says that in the United States whither this species has been introduced all the coloured varieties are met with, having flowers of variable size. Bernhardi[4]/68/says that it is almost certain (& I have received corroborative evidence) that the allied Anagallis collina produces blue & red flowered varieties. Considering these several statements the probability seems to me strong that the A. coerulea & arvensis should be considered only as varieties. I have alluded to this case chiefly owing to the remarkable fact, that Gaertner with all his experience failed after repeated & reciprocal trials[5] to raise a single hybrid between these two forms, whence he concludes that they are distinct: Herbert succeeded with Anagallis collina; & if Gaertner had shown that he could artificially fertilise either variety with its own pollen one would then have had more confidence in his result.—

The most interesting case on record is that of the Primrose, ⟨common⟩ oxlip Bardfield oxlip & cowslip (Primula vulgaris elatior & veris). These plants differ, as everyone knows, in their flowers foliage & habit; they all three differ in the forms of the capsule & seed:[6] the primrose & cowslip have a/69/different scent: they flower at somewhat different times: they ordinarily inhabit different stations, the cowslip in open fields & the primrose on banks & in shaded woods, but they are sometimes mingled; they abound in different districts in different proportions;/69 v/& in Switzerland the P. vulgaris & elatior ascend to different heights the primrose being the more tender.[7] They have, also, different geographical ranges; Dr. Bromfield has remarked[8] "that the primrose is absent from all the interior regions of Northern Europe, where the cowslip

---

[1] [Teesdale,] Linnean Transactions vol. 5. p. 44 & [Wiegmann] Flora 1821. B. 1. s 15.

[2] Loudon's Mag. of Nat. Hist. vol. 3. 1830. p. 537, but compare with vol. 5. p. 493.

[3] Phytologist vol. 3. p. 699.   [4] Begriff der Pflanzenart. p 9.

[5] Bastarderzeugung s. 309. [On the verso of this folio Darwin asked: ⟨"could Gaertner have by chance tried only a male plant or with female pollen??"⟩ and later cancelled this query.]

[6] Rev^d. W. Leighton in Annals of Nat. History vol 2. 2 series. 1848. p. 164.

[7] Annals of Nat. History. vol. ix. 1842. [P. J. Brown,] p. 156. & [E. Doubleday,] p. 515. See, also, Boreau Flore du centre de la France 1840. Tom 2. p 376. and Hooker's & Arnotts British Flora 1855, on the rarity of P. veris in Scotland.

[8] [E. Doubleday,] p. 515. Phytologist vol. 3. p. 694.

is indigenous": & Messrs. Bentham & Hooker inform me that in the East, the primrose is found only in the Caucasus; that the oxlip ranges from the Caucasus to about the latitude of Moscow & the Cowslip from the Caucasus to four degrees northwards to the latitude of St. Petersburgh.[1]

Lastly Gaertner laboriously experimentised on these several forms during four years, & actually castrated & crossed no less than 170 flowers, & yet ⟨strange to say⟩ he only twice succeeded in getting any good yet scanty seed[2]/

70/He expressly states that the Primulaceae offer no mechanical difficulties to crossing,[3] but yet it would have been far more satisfactory if he had shown that he could artificially fertilise a Primula with its own pollen. On the supposition which seems to me most probable that this extreme infertility is not real, but only apparent, & caused by some want of skill or knowledge, we have, nevertheless, as good as, indeed far better evidence than is attainable, in most cases, of the infertility of these forms together,— seeing how perseveringly the experiment was tried by the most practised operator who ever lived.—

Considering these several statements, it seems to me difficult to imagine better evidence than in this case that the primrose & cowslip deserve to be called distinct species. But now let us look to the other side: it is universally acknowledged[4] that in England there are so many intermediate forms found wild that it is most difficult to draw any strict line of demarcation between the two extremes of the primrose & cowslip. And what is the result of the many experiments/71/which have been made? Several years ago, the Hon. & Rev[d]. W. Herbert[5] raised from the seed of a highly manured red cowslip, a primrose, cowslip, oxlips of various colours, ⟨a black polyanthus,⟩ a hose-in-hose cowslip, & a natural primrose bearing its flowers on a polyanthus stalk: from the seedling hose-in-hose cowslip, he raised a hose-in-hose primrose. Subsequently the Rev[d]. Prof. Henslow[6] doubting Mr. Herbert's experiment

---

[1] In Britain see Cybele Brit. W. C. Watson Cybele Britannica vol. 2. p. 293. says only on range of Primrose & Cowslip that "P. veris would seem to be an uncommon plant in the W. of Scotland".

[2] Bastarderzeugung. s. 721; & s. 178; but the table is not quite correct for a cross is mentioned at p. 247 not introduced into the table.

[3] Le[c]oq[?]—at Maer gardener assured me he had known whole bed of Polyanthus spoilt by P. veris growing near. [See Lecoq, *De la Fécondation naturelle...*, Paris, 1845.]

[4] See Phytologist vol. 3. p 43. for some excellent observations on the intermediate states by Mr. Watson.

[5] Transactions of the Horticult. Soc. vol. IV, p 19.

[6] Loudon's Mag. of Nat. Hist. vol. 3. 1830. p. 409.

took the seed of some cowslips growing in a shady part of his garden, & raised seedlings which varied considerably, approaching more or less closely to certain wild oxlips ⟨which Prof Henslow had observed⟩; "& one was a perfect primrose". These experiments were not thought sufficient;[1] & that most critical observer Mr. H. C. Watson raised at several periods many seedlings, from the cowslip, (P. veris), from a Claygate oxlip, & from an oxlip, truly intermediate in most points, but with/72/the primrose predominating, & the conclusion at which he arrives[2] is "that seeds of a cowslip can produce cowslips & oxlips; & that seeds of an oxlip can produce cowslips, oxlips & primroses."

The experiments of Mr. Sidebotham[3] are, perhaps, the most important of all, for the plants from which he procured seed, were covered by bell-glasses & so crossing was prevented. He performed all the operations with his own hands. Moreover he experimentised on the Bardfield oxlip (P. Jacquinii or P. elatior of Jacq.), which has very generally been received as a third distinct species; though in this case, as with the common oxlip, Mr. Watson & Dr. Bromfield[4] have "seen exceptional instances to all the characters, taken singly, by which this plant is distinguished from P. vulgaris & P. veris"; but Dr. Bromfield admits that it certainly has much the air of a distinct species./73/Mr. Sidebotham's experiments were as follows, & they are the more important as he was a hostile witness, & confesses that the experiments "disappointed me greatly & interfered very materially with my previous idea of specific identity".

These experiments bring out clearly the hereditary tendency in all five forms. Both here & in Mr. Watson's experiments there is no direct passage from a true cowslip to a primrose or reversely; but Mr. Herbert experimentised on a cultivated red cowslip, highly manured, & from it he raised "a natural primrose on a polyanthus

---

[1] ⟨Some nurserymen (as I have been myself informed) are convinced that such changes take place in their seed beds, others have strongly denied them, as in Gardener's Magazine vol. VII p. 123, 247⟩.

[2] Phytologist Vol. 3. p 43, and vol 2. p. 217. p. 852.

[3] Phytologist vol. 3. p. 703 [Darwin later rejected Sidebotham's claims. Note sheet 69 v on H. C. Watson's *Cybele Britannica* states: 'Vol 3. p. 488—doubts Mr. Sidebotham experiments, so I had better not speak so enthusiastically—doubts them from want of general accuracy—& from his want of Botanical knowledge,—relates more especially to P. elatior from P. veris; & to P. vulgaris from P. elatior Allows they support P. vulgaris P. veris coming from an intermediate form, & reverse case.—Thinks he did not take sufficient precautions, what I know not—Express a doubt about P. elatior.—'. See also *The Different Forms of Flowers on Plants of the Same Species.* 1st ed., London, 1877. note on p. 60, where Darwin states that these experiments 'may be passed over as valueless.']

[4] Phytologist vol 1. p. 1001 & vol. 3. p. 695.

Names of seedlings produced	Seed from P. veris produced (cowslip)	Seed from P. veris, var. major produced (oxlip)	Seed from P. vulgaris an intermedia produced (Claygate oxlip)[1]	Seed from P. vulgaris produced (primrose)	Seed from P. Jacquinii = P. elatior of Jacq. produced (Bardfield oxlip)
P. veris (common cowslip)	412	9			
P. veris var. major of Lond. Cat. (oxlip) (3 being hose-in-hose)	30	21			
P. veris approaching Poly-anthus (oxlip) (some hose-in-hose) some dark-coloured	13	20			
P. vulgaris var. intermedia (Claygate oxlip)		7	19		
—var. caulescens		3	3	1	1
P. vulgaris (primrose)		2	1	15	1
P. Jacquinii (Bardfield oxlip)		1			24
Plants producing no flowers.	18	2	4	2	6
Total number of seedlings raised	473	65	27	18	32

[Table compiled from lists in Sidebotham]

stalk" & again on the succeeding year from his seedling hose-in-hose (calycantha) cowslip he raised a hose-in-hose/74/primrose. The Rev^d. Prof. Henslow's cowslip, whence he raised "a perfect primrose", was a garden plant & grew in a shady place. It goes for nothing that some authors have planted seeds, especially if gathered from wild plants,[2] & have found that all the seedlings, have come true to their kind; it only shows how true the kind is, when not disturbed by cultivation.—

No one, I believe, has disputed the accuracy of the statements of these four Botanists, Messrs. Herbert, Henslow, Watson & Sidebotham; three of whom, I may add, commenced their experiments in a sceptical frame of mind. But the results have been attempted to be explained away by the supposition of the inter-crossing of the several forms. Now laying on one side Gaertner's laborious & careful experiments, ⟨which nearly all failed,⟩ &

---

[1] From seeds of this form Mr. Watson (Phytologist vol. 2. p 218. & Dr. Bromfield remarks vol 3. 69 [695?]) raised 88 seedlings of which 63 were intermediate, 5 were genuine cowslips & 20 true primroses.

[2] Phytologist vol. 3 p 180.

assuming that insects could effect, that which he could not; do the results agree with this view of crossing? It seems to me most decidedly not. Mr. Sidebotham expressly/75/states that he protected his flowers by glasses; & this having been done, it seems quite incredible that there should have been so much crossing in all his five cases. ⟨indeed apparently as much variations in the offspring in most of the experiments.⟩ Moreover on the mountains of Switzerland the P. elatior or supposed hybrid between P. veris & vulgaris grows "by thousands in places within many leagues of which the P. vulgaris is absolutely unknown".[1] so it must be with the oxlip from its Northern range in Russia; so with the oxlip (or P. Jacquinii) of Bardfield, round which place "the primrose does not occur for some miles".[2] Lastly, & I may venture to say that I speak after a careful study of all well ascertained facts on Hybridism, there is no known instance of one species fertilised by the pollen of another species producing pure forms of both or either parent as must have occurred on this view with Herbert's & Henslow's cowslips & with Mr. Sidebotham's P. Jacquinii, if they had been fertilised by the pollen of the primrose. Moreover the common oxlip, or supposed Hybrid between the primrose & cowslip, yielded, as we have seen in Mr. Watson's & Sidebotham's/76/ experiments, various oxlips & *pure* primroses & *pure* cowslips; whether we choose to imagine these hybrids were self-fertilised, or were fertilised by either pure supposed parent, so sudden & absolute a reversion to either or both parent-forms is ⟨in the case of *species*⟩ without any known analogy[3] in carefully recorded experiments on the crossing of *species*. From these several & combined reasons I think we are justified in absolutely rejecting the view that all the forms produced in the foregoing several recorded experiments, & likewise existing in nature, can be accounted for by the crossing of two or three aboriginally distinct species; their origin I think must be attributed to variation, but I am far from wishing to assert that some or many of the graduated intermediate forms may not likewise be in large part due to their having at some time crossed, which no doubt would increase their variability & probably aid in their tendency to reversion to either one or both of the parent varieties./

77/In all the experiments, the common oxlip seems the most

---

[1] [P. J. Brown,] Annals of Nat. History vol IX. 1842. p 156.

[2] [Doubleday,] Annals of Nat. Hist. vol. IX. p. 515.

[3] The well-known & marvellous case of Cytisus adami would be analogous in the individual, (though not in seedlings) if it could be shown that this tree was really a Hybrid: some competent judges firmly believe that it was produced by the union of two buds of the two species.—

variable form; though the cowslip is sometimes little less so, for in Prof. Henslow's seedlings "not one had the decided characters of the common cowslip" [p. 409]. Unfortunately no one, except Mr. Sidebotham seems to have tried the seed of the pure primrose; & it would be very rash to draw any conclusions from the apparent greater trueness of the primrose; but if this one experiment were confirmed, the primrose probably should be looked at as the primordial form, whence has been derived through intermediate oxlip-forms the cowslip, & the Bardfield oxlip. It is, perhaps, the most probable view that the common oxlips are varieties of the cowslip, easily reverting back towards the primrose; some of the forms having been complicated by crosses with either the primrose or cowslip. I have entered into this case with great detail because, considering the structure habitat, range in height & latitude, & *apparent* infertility of the two forms, & the many careful experiments made on them, this seems the most interesting case on record. An able Botanist has remarked[1] that if the primrose & cowslip are proved to be specifically identical, "we may question 20,000 other/78/presumed species." If common descent is to enter into the definition of a species, as is almost universally admitted, then I think it is impossible to doubt that the primrose & cowslip are one species. But if, in accordance to the views which we are examining in this work, all the species of the same genus have a common descent; this case differs from ordinary cases, only in as much as the intermediate forms still exist in a state of nature, & that we are enabled to prove experimentally the common descent. ⟨Hence common practice & common language is right in giving to the primrose & cowslip distinct names.⟩

I will end this long discussion by recalling attention to another statement by Mr. Herbert in regard to the species of Primula, which, though it may seem incredible I think ought not to be lightly rejected, as Mr. Herberts observations on the common cowslip & on various other subjects/79/have stood the test of subsequent observation. Mr. Herbert affirms[2] that he raised a powdered Auricula (P. auricula) from P. nivalis; & that he likewise raised P. Helvetica (described as a species by Don, but treated as a variety of viscosa in Steudel) from P. nivalis; & that thirdly he raised P. Helvetica likewise from P. viscosa. Hence Mr. Herbert concludes that these Swiss Primulas are only local varieties.[3]

[1] Phytologist vol. 2. p 875   [2] Transact. Hort. Soc. vol. IV. p. 19.
[3] [On the blank lower third of this folio, Darwin pencilled: 'Here discussion on large genera.']

A 1/*Wide ranging, common and much diffused species tend most to vary*:—The elder De Candolle, & several other Botanists[1] have insisted that it is the widely ranging, the common & vigorous plants which vary most./A 1 v/Alph. De Candolle[2] gives a list of 117 species which range over at least a third of the terrestrial surface, & he states that the greater part of these offer varieties. I have attempted to test this proposition conversely; that is by taking the species which present varieties, & seeing whether a large proportion of them are common & widely diffused in their own country./A 1/Ledebour divides the enormous territory, included in his Flora Rossica into 16 Provinces; & to each species he appends the number of Provinces which it inhabits. There are 999 phanerogamic species which present varieties, marked by Greek letters, & these on an average range over 4.94 Provinces; whereas there are 5347 species which have no varieties, & these range over only 2.43 provinces; so that the varying species range over rather more than twice as large an area as the other species. The rule holds very nearly the same when each of the four volumes is tried separately. But we shall presently see & have to discuss the many difficulties which arise in considering the value of the varieties appended by Botanists to their species/

A 2/In the London Catalogue of British Plants the number of the 18 provinces, in which each species has been found, is added from Mr H. C. Watson's Cybele Britannica. The number of varieties given in this Catalogue is not great, but Mr Watson has added for me in M.S. some others; the principle on which he has acted in doing this, & the reasons for omitting some varieties & some few whole genera, are given in the Supplement to this Chapter; but I may add that all the varieties here included have been ranked as species by some one or more botanists. Now there are 1053 species which have no such varieties appended to them, & these on an average range over 10.76 of the Provinces; whereas there are 169 species which have such varieties, & these range over an average of 14.55 provinces. I have, also, tried these species in another way, not by taking an average, but by seeing how many species range over all 18 provinces; & I find that of the 1053 non-varying species, 216 occur in the whole 18 provinces, or in the proportion of $\frac{205}{1000}$; whereas of the 169 species which present varieties, there are 70 which range over the 18 provinces, that is the proportion of $\frac{414}{1000}$; so that proportionally twice as many of the varying species range throughout the eighteen provinces, as of the non-varying species./

---

[1] Boreau. Flore du Centre de la France. Tom. 1. p. 101.

[2] Geographie Botanique ⟨1855⟩ p. 586. [Actually 564–81.]

A 3/With respect to '*commonness*', it is evident that a species might, as indeed is the case with many aquatic plants, range over an enormous territory, & yet not be common or individually numerous anywhere. In a small area, like Britain, where a plant is found in every province, diffusion & commonness almost blend together. Boreau in his Flora of the Central part of France (See supplement to this chapter, for particulars on this & other works quoted) has marked by C. C the very common species; & I find he has 1280 species not presenting ⟨any marked⟩ variety, of which 240 are very common,—that is in the proportion of $\frac{187}{1000}$; there are other 193 species with varieties recorded, & of these 78 are very common, or in the proportion of $\frac{404}{1000}$; so that proportionally more than twice as many of the varying species are very common in comparison with the non-varying. ⟨I may here remark that Boreau draws a distinction between the polymorphic species, which vary almost indefinitely & are not included in the above number, & those species which present varieties sufficiently distinct to be marked by Greek letters.⟩/

A 4/ Miquel in his list of the plants of Holland, marks a very few species having varieties & marks all the very common species; but the recorded varieties are so few, & no particulars specified in regard to them, that the list is not satisfactory: there are 1133 non-varying species, of which 201 are common or in proportion of $\frac{177}{1000}$; & on [the] other hand there are 46 varying species of which 27 are common, or in proportion of $\frac{586}{1000}$; hence more than thrice as many of the varying species are common than of the non-varying species, but the proportion is probably here exaggerated.

Again Prof. Asa Gray in his Flora of the N. United States, appends the word common to many species, & I find that of the 1851 non-varying species, 439 are marked as common, i.e. $\frac{237}{1000}$; whereas there are other 202 species which present varieties (either marked in small or large type, see supplement to this chapter), of which 82 are marked as common,—i.e.$\frac{405}{1000}$, here then, not far from proportionally twice as many varying species are common as of the non-varying.[1]

From the foregoing cases, we see, that such numerical evidence as can be obtained, subjected as it is [to] doubts on the value of

---

[1] In Mr Wollaston's Insecta Maderensia (Introduct. p. xiii) 12 Coleoptera are mentioned as the most abundant in individuals in this group of islets, to which may be added, as I am informed by Mr Wollaston, a Ptinus and Oxytelus. Hence out of the 482 species, about one in 34 of all the species is very common. But of the 61 species, which present varieties, six are very common, i.e. one tenth of the varying species are very common.

the recorded varieties, supports the opinion of those botanists, who believe that the much diffused & common/A 5/species are most liable to vary, or to present varieties, which have been thought sufficiently distinct to be recorded. We can understand why wide-ranging species, which live under various climates, & which come into contact with diverse groups of organic beings (a much more important consideration, as I think will be seen in a future chapter) should vary more than local species. Wide ranging species will also generally from/A 5A/the mere fact of their inhabiting many places, & from the vigour which they show in thus ranging far & coming into successful competition with many organic beings under different climates, will generally be common or individually numerous: indeed Dr. Asa Gray after examining this question says, "so true is it as a general rule that species of wide range in our country are species of frequent occurrence, that I have not noticed any strongly marked exceptions to it".[1] Even in regard to species strictly confined to a moderately sized & uniform locality, which are not exposed to very different conditions, we may, I think, see why such species, when common & much diffused in their own country, should present more varieties than when rare. If we suppose varieties to be mere fleeting productions, like monstrosities, then, if originating in exactly the same proportional numbers in common & rare species say one in a million individuals, they would, within the life-time of Botanists, be far oftener encountered amongst the common than the rare species; & so would be oftener /A 6/recorded in botanical works. But of two species, if one were common & one rare during the whole or greater part of their existence on the earth, then a greater number of such fleeting varieties would, it is probable, actually originate in the common than in the rare species. Now I believe, though we are here fore-stalling what we shall have hereafter to discuss, that by far the most effective origin of well marked varieties and of species, is the natural selection or preservation of those successive, slight, & accidental (as we in our ignorance must call them) variations, which are in any way advantageous to the individuals thus characterized: hence there would be a better chance of varieties & species being thus formed amongst common than amongst rare. I may add, to illustrate what I mean, that a nurseryman who raises seedlings of a plant by the hundreds of thousand far oftener succeeds in his life-time in producing a new & valuable variety, than does a small amateur florist. So it would be with a common,

---

[1] Statistics of the Flora of the N. United States, in American Journal of Science, 2nd. Series, 1857, Vol. 23, p. 393.

in comparison with a rare species, raised by the hand of nature in millions on millions during the incomparably longer period of its existence on the earth.

But botanists do not actually wish (though unintentionally it is often done) to record, & define as varieties, mere fleeting variations or monstrosities./A 7/Boreau, for instance, & others have expressly stated that they record only the more strongly defined varieties: more than one-third of the varieties marked by Asa Gray are considered by him as possibly deserving to be called species: in the London Catalogue, the greater number of the most trifling varieties have been removed for me by Mr. Watson & all those which are left (182 in number) have been ranked by some one botanist as species. Of the degree of permanence of varieties in plants we know hardly anything: but when a variety is the common form throughout any province or even quite small district, we must suppose that it is in some degree permanent. We have seen in the case of certain land-shells of Madeira that some of the varieties are of extremely high antiquity. Now when a variety is in some degree permanent, whether it has originated in a single accidental variation, or by the addition of several such successive variations through natural selection, or through the direct & gradual action of external conditions, as of climate, its first origin is even of less importance to it, than its preservation; for in order to become in any degree permanent, it has to struggle with all other organic beings in its own country; & this shows that it has/ A 8/at least nearly equal, or has perhaps acquired even some greater, constitutional advantages, in comparison with its parent-species. The mere fact of a species being very common or widely extended shows that it is advantageously situated in respect to the inorganic conditions of its life, & in respect to all the other organic beings, animal & vegetable, with which it has to come into competition; & the varieties produced from such common species, from differing little from them, will gradually partake of (or have in excess) their advantages, whatever they may be. Finally then, I suppose that common species present more varieties, when these are in some degree permanent, than do rare species, from partaking of the advantages which make the parent species common; and that varieties (not now considering those wholly due to the direct action of climate &c) originate more frequently amongst common species than amongst rare, owing to more accidental (as we must call them) variations arising during the whole existence of a species which abounds in individuals, than during the existence of a species which has presented much fewer individuals.

The law first enunciated by M. M. d'Archiac & Verneuil & since confirmed by several geologists, that the species which range over a very wide area, are those which have existed for the longest period, seems at first opposed to the/A 9/foregoing conclusion, taken in connexion with my view that closely allied species do not essentially differ from varieties; for it implies that the species which have ranged furthest have longest remained immutable. But if we reverse the proposition, which can be done with equal truth, it is not so discordant;—namely that species which have existed longest, have had, owing to geological & other changes, the best chance of spreading furthest. The majority of such species we may, without contradicting the law, suppose to have become modified either into varieties or into new species, but that a certain number having undergone no change (& it has never been pretended that wide ranging species universally vary) has given rise to the foregoing palaeontological law./

A 10/*Geographical Range of Varieties themselves*:—I have met with scarcely any observations on this head. When two varieties inhabit two distinct countries, as is often the case & as is very generally the case with the higher animals, it is obvious that the two varieties separately have a much narrower range than the parent species. A variety, for instance, inhabiting N. America & another variety of the same species inhabiting Europe will both have a very much more confined range than the parent form; so on a much smaller scale, the many varieties of endemic species, confined to the separate islets of the same small archipelago (for instance in the case of the insects of the small Madeira group described by Mr Wollaston) follow the same rule.* So again the numerous alpine, maritime, shade or moisture-loving varieties of species, which commonly live in other and different habitats, have confined ranges compared with their parent-Types. These considerations alone make it probable that the far greater number of varieties have narrower ranges than the species whence they have sprung. I have looked to many local Floras, & as far as I could judge, the recorded varieties seem usually to have restricted ranges. In the London Catalogue (1857) the range within Britain is given by Mr Watson of some, namely 53 varieties, & I find that on an average they range over 7.7 Provinces; whereas the/A 11/46 species, to which these varieties belong, range over 14.3 of the provinces;—

* All this depends on the arbitrary assumption of which is var. & which species. [J.D.H.] Begin with stating that it is a truism Probably not worth giving so much of a *truism*. [C.D.]

or over nearly twice as wide an area. At my request Mr Watson was so kind as to append remarks on the nature of the habitats & of the ranges of those varieties of British plants with which he was personally acquainted; but as he stated to me, it was not possible to arrive at any definite conclusions from the numerous sources of error; but I may add that from this list it seems that a large number are alpine, maritime, &c forms; sometimes confined to one or to a few localities, but often pretty widely diffused: a good many varieties are, as far as known, strictly local, & some of them have become extinct since having been first noticed: in several cases the varieties, when not strictly confined to any particular locality, or habitat, seem to be rarer than the type-forms:—

The only published observation which I have met with on the range of varieties is by Mr. C. B. Adams,—a competent judge in regard to the terrestrial mollusca on which he treats:[1] he states that the several/A 12/varieties of a species seldom have the same range with it or with each other; 'each variety has its own limits of distribution;' one variety will often have an 'extent of distribution equal to that of two or more other varieties' of the same species. He believes that varieties follow the same laws of geographical distribution with species; and hence he concludes that they have been aboriginally created as varieties. But it follows from his remarks that varieties generally have more confined ranges than their type-species.*

In all cases, this latter remark, is to a large extent a mere truism; for when two forms are so closely similar, that one is called a species and the other a variety, the commoner of the two, is almost sure to be called the species, and the less common one, the variety: for we cannot tell which of the two has branched off from the other.[2]

As by our theory two closely allied species do not differ essentially from a species & its strongly defined variety, I was anxious to ascertain anything about the ranges of such closely allied species but I can advance only one single case, as follows: Mr Watson has marked for me in/A 13/the London Catalogue (4th Edit:), which is a pretty well sifted list & does not include the most doubtful

---

* This is reasoning in a ○ [circle.] The idea of a var[iety] is founded on variety. [J.D.H.]

[1] Contributions to Conchology. No. 10. On the nature & origin of the species of Mollusca in Jamaica. p. 193.

[2] See an excellent discussion on this point in Dr. Hooker's Introductory Essay to the Flora of New Zealand. p. xvii & note.—Dr. Asa Gray, also has remarked to me that mere priority of description has in some cases determined which form has been called the species & which the variety.

species, the forms therein admitted as species, which he considers as most like varieties: he has marked 63, & adds that most of these have been of late years, as it were, cut out of other species: they have all been considered by some few botanists as mere varieties, but by the large majority of local authors have been ranked as good species. Now I find that these 63 species in the London Catalogue range on an average over 6.9 provinces; so that, they have very nearly the same extent of range, with that (7.7) of the 53 printed varieties in this same catalogue.*/

A 14/*On the relation of the commonness and diffusion of species to the size of the orders and genera in which they are included*:—My object in looking at this question regards Variation:—As we have seen that a large proportion of the common and widely diffused species present varieties, if these common species occur most frequently in the numerically large groups, it would be some indication that a greater number of varying species would occur in them—& this latter subject is an important one which we shall presently have to discuss./A 14 v/There is, as it seems to me, some a priori probability that the species in the large groups would be generally common & more widely diffused than in the small groups; for the simple fact of many closely allied species inhabiting any country shows that there is something in its condition, organic or inorganic favourable to them; & this by itself would tend to make the species numerous in individuals & widely diffused within that country beyond the common average.[1]/

A 14/Alph: De Candolle has shown[2] that there is some but very slight evidence that the Orders numerically large in a country, include more common or "vulgar" species than do the smaller Orders; but that the species of such large orders generally have

---

* Very good remark. [J.D.H.]

[1] Alph. De Candolle (Geograph. Bot. p. 562) takes a directly opposite view. ⟨He supposes that when the conditions of life are most favourable to a group, many delicate species could live. "Or, les espèces les plus delicates doivent avoir l'aire la plus restrainte" But we have seen that there is generally a relation between the extent of range and commonness of a species: and how is a species known to be delicate except from being rare and having a narrow range? Is it not saying that certain species are delicate *because* they are rare & have a confined range, & *therefore* they are rare and have a confined range? The rarity and confined range of a species, depends, I believe, in the vast majority of cases, on its not being able to compete with or withstand other organic beings; and by no means on the conditions of its existence being favourable.⟩ [Deleted in fair copy.] M. De Candolle throughout his admirable work seems to me very often to greatly underrate the predominant importance of the struggle for life on all organic beings,—a subject to be discussed in our next chapter.—

[2] Geograph. Bot. p. 465–470. p. 562.

more confined ranges; & he concludes with some doubt that where only a few species of an order exist, these will be the more robust & the widest rangers. It has appeared to me, from reasons not worth giving, that if any such rule did hold good, it would be more likely to appear in smaller groups or genera rather than in orders.[1] But whether in genera or orders/A 15/there are very many causes which would tend to conceal such a result. Namely, our best classifications are considered by many able botanists as still highly artificial. The species in large genera are as remarked to me by Mr H. C. Watson, more difficult to identify, & he believes that many species in such large genera, which are now ranked as, distinct in distant countries, would on close examination often be found to be identical; & consequently such species in the larger genera would really have wider ranges than they appear to have in books; moreover there would sometimes be the greatest difference in the range of a species, according to the value set on its specific characters; for instance a European species having a variety in N. America would have an enormous range, but if that variety were ranked as a species, the range of the European form would be immensely reduced. Aquatic & littoral plants generally have very wide ranges, quite independently of the question whether they form parts of large or small genera. Lowly organised plants as a general rule range further than the more highly organised, and lastly when two areas, separated by the sea or by other/A 16/ barriers, are considered, the capacity for dissemination in the species in common, would probably come into play.

(Some of these multiform causes of error may, I think, be in some degree eliminated by not considering the whole range of the species, but only the degree of diffusion & commonness of the species, described by a single botanist, within one continuous territory, more especially if not of vast size. And for my special object of finding out whether more varieties have originated in any country (or if originating elsewhere, are in this country enabled to subsist) amongst the larger or the smaller genera, it

---

[1] Dr. Asa Gray (in American Journal of Science, 2nd Series, Vol. 23 p. 391) has distributed under their orders 430 species which are the widest rangers in the northern U. States & at the same time the most common species. I have had these orders so arranged that all the species (977) included in the larger orders in the U. States are nearly equal in number to all the species (939) included in the smaller orders. And I find that the number of the wide-ranging & common species are more numerous in the smaller than in the larger orders, in the proportion of 233/1000 to 215/1000. If the species could have been arranged by *genera* instead of by *orders*, namely if all the larger genera had been put on one side & the smaller genera in the other: I hardly doubt from the following Table (Tab. A), that the larger *genera* would have included a larger proportion of these common & widely ranging species.

TABLE A[1]

The numerator gives the number of the much diffused or the common species in each country.

The denominator gives the number of species in the left column in the larger genera & in the right hand column in smaller genera—See Supplement to this chapter for titles of Works etc.

		Larger Genera		Smaller Genera	
Britain:	London Catalogue (1857) H. C. Watson—Larger genera with 5 species and upwards, smaller with 4 species and downwards—The numerator expresses the number of species found in all the 18 Provinces, into which Britain is divided.	$\dfrac{148}{592}$	$=\dfrac{250}{1000}$	$\dfrac{138}{629}$	$=\dfrac{219}{1000}$
Russia:	Ledebour (Dicotyledonae alone). Larger Genera with 10 species and upwards, smaller genera with 9 species and downwards. The numerator expresses the number of species found in at least 8 of his 16 Provinces. The species inhabiting 8 Provinces have about thrice the average range of all the phanerogamic plants:—	$\dfrac{239}{3385}$	$=\dfrac{70}{1000}$	$\dfrac{131}{1937}$	$=\dfrac{67}{1000}$
Centre of France:	Boreau—Larger genera with 5 species and upwards, smaller with 4 and downwards. The numerator expresses the species marked C.C. or very common.	$\dfrac{163}{732}$	$=\dfrac{222}{1000}$	$\dfrac{155}{741}$	$=\dfrac{209}{1000}$
Holland:	Miquel—Larger Genera with 4 species and upwards, smaller with 3 species and downwards. The numerator expresses the number of common species.	$\dfrac{120}{622}$	$=\dfrac{192}{1000}$	$\dfrac{108}{557}$	$=\dfrac{193}{1000}$
Ratisbon:	Furnrohr—Larger genera with 4 species and upwards, smaller with 3 species and downwards. The numerator expresses the number of species marked "sehr gemein".	$\dfrac{102}{533}$	$=\dfrac{191}{1000}$	$\dfrac{79}{526}$	$=\dfrac{150}{1000}$
N. United States:	Asa Gray—Larger genera with 5 species and upwards, smaller with 4 species and downwards. The numerator expresses the number of species marked as "common".	$\dfrac{326}{1136}$	$=\dfrac{286}{1000}$	$\dfrac{195}{917}$	$=\dfrac{212}{1000}$

[1] [Darwin's holograph draft of this table is ULC Darwin MSS. vol. 16.1, fol. 172.]

seems to me quite immaterial whether the same species in other countries have very wide or narrow ranges,—are very common or rare.)

(The following short table (Tab. A.) gives the proportions of the common & of the most widely diffused species, in the larger & in the smaller genera, in six countries.)

We here see a slight preponderance, in the larger genera in all the cases except in Holland, and Miquel's tables differ more or less, in every single respect, as far as I have tried them, from those of other Botanists. The slight preponderance would probably/ A17/be somewhat increased, more especially in such large territories as those included in the Flora Rossica, if some of the many above-specified causes of error could be removed: for instance the influence of peculiar stations on the range, which is independent of the size of the genera./A17 v/I may add, as supporting the table that Dr. Asa Gray finds that 75 per cent of the *widest ranging* species in N. America belong to genera having above the average number of species[1] and in regard to "commonness", we see in the table that a greater number of species marked as "common" are included in the larger genera; ⟨& indeed as already remarked Dr Asa Gray has shown that the common & widely ranging species are almost invariably the same.⟩ Dr. Hooker also finds a similar result by tabulating the species common to Europe & N. America, which have a vast range & these usually belong to large genera. Conversely, in regard to commonness, Dr. Hooker has remarked to me in a letter* that in a general Herbarium, genera with single species are represented by a single specimen far oftener than large† genera, showing that the genera with a single species are usually rarer in individuals./

A17/In regard to the extent of diffusion, the preponderance small as it is in Table A, quite or almost disappears, if an average of the ranges of all the species in the larger & smaller genera be taken, instead of, as in the Table, the proportional numbers of the species having unusually wide ranges. Thus in the Flora Rossica,

---

* I cannot now find your letter on this subject, but I hope I shall & I quote now only vaguely from memory. [C.D.]

† Or more local [J.D.H.]

[1] American Journal of Science, 2nd series. Vol. xxiii. 1857 p. 380, Dr. Gray remarks that the converse of the above proposition does not hold good for out of 33 species which have the narrowest range of all the species, 21 belong to large genera. But it is conformable with my views that many species in the large genera should like varieties be extremely local. The species with a wide but disjointed range (p. 387) seem to make a real exception: but with disjointed species, several interfering causes, as extinction, the action of the Glacial epoch, chance dissemination, may have come into play.

all the species (3955 in number) in the larger genera (for the size of the genera see the table) have an average range of 2.8 provinces: whereas the species (2407 in number) in the smaller genera have a slightly larger average range over 2.88 Provinces. Again in the London Catalogue of British plants (5th edit.), the species in the larger genera range on an average over 11.4 provinces, in the smaller over 11.2 provinces. Nor according to the views, which we are in this work discussing, is this surprising; for we here look at species as first branching off into varieties, & these then becoming modified (by means which it will hereafter be attempted to be explained) into closely allied, & ultimately into quite distinct species: now we have seen that varieties generally have narrow range, as have those closely allied forms which were marked for me by Mr Watson, &/A18/which are admitted in the London Catalogue as true species; & such forms, when a general average is struck, would greatly reduce the range of the widely diffused species,—including those species, of which the varieties had not as yet become converted into local species.

On our theory, however, another cause of doubt and difficulty here comes in. We have no reason to suppose that all forms, even within the same class, undergo modification at the same rate; indeed geology leads to the belief that the more highly organised forms,—as Vertebrata compared with most other animals—brachiopods in comparison with acephala, & these with gasteropoda—are replaced at a quicker rate than the more lowly organised. Hence of two sets of species, having originally exactly equal ranges, one set might become after a given period converted into a greater number of new specific forms having restricted ranges, whilst the other set remained unaltered with their original wide ranges. I suspect that, on our theory, this may be the explanation of the Compositae,/A19/for instance, which are considered by many botanists as very highly organised plants, having species on an average with very narrow ranges.[1] This view may perhaps, also,

---

[1] Mr Gould in his Introduction to the Birds of Australia (1848, p. 122) divides this country into five sections & adds one for a few outlying regions: he gives the range of each species in these six divisions. As Birds are very highly organised beings, & as Mr Gould admits extremely slight modifications of structure to be of specific value, I have thought it worth while to have the species (omitting the Natatores) tabulated in genera having four species & upwards & into genera with three species & downwards. The 300 species in the larger genera range over an average of 1.84 sections; whereas the 228 species in the smaller genera range over an average of 2.24 sections. Here we see that the closely allied species in Mr Gould's larger genera have narrower ranges than those species which have not according to my view, been converted into representative races & species in the several sections of the country.

throw light on the general rule[1] of lowly organised plants having wider ranges than the more highly organised: though probably the greater facility of dissemination in most of the lowest plants has largely influenced the result. On this view, it is not that the more highly organised productions of nature have originally had narrower ranges, but that they soonest become changed into local & distinct species.*/

A 19 A/The undoubted fact that not rarely species in the smallest genera in a country are extremely common & range very widely is not opposed to our view; for a species, before it can have become modified into several distinct species inhabiting distant localities, must have ranged, according to our theory, over the whole area, inhabited by the forms derived from it, either in its original un-altered specific state, or during its successively modified states. On the other hand, some cases are on record of groups, possessing numerous species, all of which are individually very rare & have very confined ranges, & yet with nothing special in the stations inhabited by them to account for this. Dr Hooker has given[2] a most striking instance of this fact in the Coniferae of New Zealand & Tasmania; & whilst examining the fossil Lepadidae of the Chalk period, I was much struck with the number of the species of certain genera in comparison with those now living; & yet all were very scarce in individual specimens. We may, perhaps, hypothetically account for such cases, by supposing that such genera are on the road towards extinction: for E. Forbes & others have remarked that the first step in this road is marked by a reduction of the individuals of the species.†/

A 20/*On species with recorded varieties being more frequent in large than in small genera:—/*[3]

/Fair copy 15 A/From looking at species as only strongly marked & well defined varieties, I was led to anticipate that the species of the larger genera in each country would oftener tend to present varieties, than the species of the smaller genera; for on this view wherever many closely related species, (i.e. species of the same genus)/A 20/have been formed [,] many varieties, or as I look at them incipient species ought, as a general rule, to be now forming.

---

* Good [J.D.H.]
† how can it be otherwise? [J.D.H.]
  by catastrophe it would be otherwise [C.D.]

[1] Alph. De Candolle. Géographie Botanique. p. 499, 519.
[2] Dr. Hooker in [*Flora Novae-Zelandiae*, I, xxix.]
[3] [See Appendix for Darwin's earlier version of the opening for this section.]

Where many large trees grow, we expect to find saplings. But if we look at each species as a special act of creation, there is no apparent reason why more varieties should occur in a group having many species, than in one having few. On the other hand, where many species of a genus have been formed through variation, circumstances have been favourable for variation; & hence we might expect that the circumstances should generally be still favourable to variation & that varieties should occur there at the present day in larger numbers than elsewhere./

A 21/To explain my meaning further by a loose simile,—if a nation consisted of clans of very unequal sizes, & if we ⟨knew that these clans in ancient times had been very different in size, some much larger, some much smaller & some not then existing, & yet imagine ourselves quite ignorant of the cause of the difference of size, whether due to immigration or some other influence; then⟩ if we divided the population into two nearly equal halves, all the large clans on one side, & the many small clans on the other side; we should expect to find, on taking a census at a moderately long interval that the rate of births over deaths was greater in the larger clans than in the smaller; and we should expect to find it so, notwithstanding that we knew that some of the small clans were now rapidly increasing in size & some of the larger clans declining./A 21 v/If we found this to be the case in several nations composed of clans, we should conclude that the greater rate of births over deaths was the cause of the size of the larger clans: & not, for instance, the recent immigration of the large clans./A 21/ What the rate of births over deaths is to our clans, I suppose the production of varieties to be to the number of species in a genus; but unfortunately in looking to the varieties existing at any one time, we are acting as if we took a census of the clans at excessively short intervals. Each child does not grow up to man's estate, nor by any means do I suppose that each variety becomes converted into a species. What death is to the individual & ultimately to the clan, I suppose extinction to be to the varieties, to the species, & ultimately to the genus. I may add that if we found any trace of the breaking up of the larger clans into smaller clans, we should infer that this was the origin of any new clans, which, had arisen since ancient historical times./

A 22/I was strengthened in my expectation of finding more varieties in the larger genera by a remark of Fries,[1] that, "in genera containing many species, the individual species stand much closer together than in poor genera: hence it is well in the former case to collect them around certain types or principal species,

---

[1] Quoted in Henfrey's Bot. Gazette. Vol. I [actually vol. II]. p. 188.

about which, as around a centre, the others arrange themselves as satellites." And according to our theory the closer two or more species stand together, the more nearly do they in so far approach the character of varieties; we should also bear in mind, as has been shown in the earlier parts of this chapter, with how much difficulty naturalists distinguish species from varieties, even in the best known countries, How many debateable forms there are amongst the plants of Great Britain, of France and of the United States, ranked confidently by one eminent botanist as a species, by another as only a variety. In regard to insects, Mr. Westwood has made[1] nearly the same remark with Fries: he says 'in very extensive genera, the distinctions of the species are so minute, that it requires the most practised eye to separate them'. I consulted Dr. Hooker on Fries' remark, & though he at first dissented* he subsequently quite concurred in its substance; & indeed this I find is an extremely general impression with all good observers. I likewise consulted Mr. H. C. Watson, of whose caution & judgment I have the highest opinion: after some deliberation he wrote to me, that although the difficulty/A 23/in distinguishing in a genus of 50 species, each species from 49 others, is obviously much greater than in distinguishing one species from two others in a genus of three species; yet he believes that generally the extremes are more remote in the larger genera than in the smaller, & moreover that the species in the smaller genera are more distinct from each other.

He represented the difference in the following diagram. Larger genus with ten species.—1, 2, 3, 4, 5, 6, 7, 8, 9, 10 Smaller genus with four species, 2, 4, 6, 8.

No one will pretend that the rule is universal; some small genera having very closely related species; & some few large genera having very distinct species. Further, I feel sure that all these naturalists would allow that in very many genera, some few species stand out much more distinctly than the others; & that the remaining closely allied species are not all equally related to each other: this might have been represented by the figures in the above two rows being placed at unequal distances from each other; some being crowded, like satellites, as Fries would have called them, around certain figures.—

I have tried to test numerically this doctrine of large genera including many very closely related species. But numerous dif-

---

* Because Fries does not observe that all? [sic] large genera are made up of two sets of species, one set as distinct inter se as those of small genera—the other all inosculate. [J.D.H.]

[1] Quoted in the Boston Journal of Nat. Hist. Vol. 4,p. 474. [In article by Haldeman.]

ficulties interfere: thus all the genera with a single species have to be entirely removed, as such genera/A 24/could not include two closely related species; but one species is sometimes equally related closely to two or even three other species, & then one does not know what to do for a standard of comparison. Moreover in these very closely related forms, the difference of opinion between botanists, whether or not they have been rightly classed as species, is carried to an extreme. However, I may briefly state that Mr Watson marked for me in the London Catalogue 71 forms therein admitted as true species, but which are very closely related to other species, & have indeed all been ranked by at least some one botanist as only varieties: of these, 57 occur in genera having five species and upwards, & only 14 in genera having 4, 3 or 2 species; so that in proportion to the number of species in these two great bodies of genera, the very closely related species stand as .90 in the larger genera to .35 in the smaller. Dr. Asa Gray has kindly gone through his Flora of the N. United States & has marked for me all the closest-allied forms, which he has classed as & believes to be nearly all, true species, but which he considers as the most likely hereafter to be ranked as varieties: he has marked these in couplets & sometimes in triplets: in the 996 species included in genera having six species & upwards, there are 296 close species: in the 696 species included in genera, having 5, 4, 3 & 2 species, there are 192 close species: so that the close species in the larger genera are as .297 to .275 in the smaller genera. Dr. Hooker also marked for me the closest allied species in his Flora of New Zealand (see supplement for certain omissions & for manner in which the genera are divided) & they occurred in the larger genera, in the proportion of .175 to .166 in the smaller genera./

A 25/To return to our question whether a greater number of varieties occur in the larger genera, which, as we have just seen, appear to include a larger proportion of closely allied forms, distinguishable with difficulty, or indistinguishable with any approach to certainty, from varieties. At first, I thought it would be a simple affair to discover this by dividing all the species in a Flora into two nearly equal masses,—all those in the larger genera on one side, & all those in the smaller on the other side, & then count the number of species presenting varieties./A 25 v/ I chose Floras, because these are much better known than any considerable Faunas, & plants are highly variable. But I have taken two well-worked out insect faunas./A 25/I soon found, however, owing to the kind suggestions of Mr Watson & Dr Hooker

TABLE I[1]

For particulars on the works here tabulated and on the few corrections made, see the Supplement to this Chapter.	The numerators in the columns give the number of species presenting varieties; the denominators the number of species in the larger and smaller genera: these fractions are all reduced to common denominators of a thousand for comparison, and are printed in larger type to catch the eye. The right hand rows of figures in the three columns, with decimals, show the average number of varieties which each varying species has,— thus the number 1.50 shows that each two varying species have on average between them three varieties.

	Larger Genera	Smaller Genera (including those with single species)	Genera with a single species
Great Britain. Bentham			
Great Britain: Babington —Larger Genera with 5 species and upwards, smaller with 4 species and downwards [Pencil note by C.D.: Write this column larger'.]	$\frac{101}{663} = \frac{152}{1000}$ 1.40	$\frac{89}{745} = \frac{119}{1000}$ 1.30 [Pencil note by C.D.: 'Write this larger'.]	$\frac{24}{255} = \frac{94}{1000}$ 1.50
Great Britain, Henslow— Larger Genera with 5 species and upwards, smaller with 4 species and downwards. The Varieties are divided into two groups, the less strongly marked, and those which have been ranked by some eminent Botanists as species. Lesser Vars:	$\frac{69}{560} = \frac{123}{1000}$ 1.55	$\frac{67}{692} = \frac{96}{1000}$ 1.40	
Stronger Vars:	$\frac{33}{560} = \frac{58}{1000}$ 1.33	$\frac{29}{692} = \frac{41}{1000}$ 1.20	
Great Britain—London Catalogue (1853) (see Supplement for nature of Varieties)—Larger Genera with 5 species and up-wards, smaller with 4 species and downwards	$\frac{97}{616} = \frac{157}{1000}$ 1.35	$\frac{85}{642} = \frac{132}{1000}$ 1.27	
Great Britain—London Catalogue—forms ranked as species in this catalogue but which have been thought by some authors to be varieties. In this second line, larger genera with 5 species and up-wards, smaller with 4, 3, and 2 species	$\frac{57}{559} = \frac{101}{1000}$	$\frac{14}{377} = \frac{37}{1000}$	

[1] [Darwin's holograph draft for this table is in ULC vol. 16.1, fol. 167.]

Table I cont.

	Larger Genera			Smaller Genera (including those with single species)			Genera with a single species		
Centre France: Boreau—Larger Genera with 5 species and upwards, smaller with 4 species and downwards.	$\frac{113}{732}$	$= \frac{154}{1000}$	1.38	$\frac{84}{741}$	$= \frac{107}{1000}$	1.47	$\frac{19}{267}$	$= \frac{721}{1000}$	1.47
Holland: Miquel—Larger Genera with 4 species and upwards, smaller with 3 species and downwards.	$\frac{22}{622}$	$= \frac{35}{1000}$		$\frac{25}{557}$	$= \frac{44}{1000}$				
Germany & Switzerland: Koch—Larger Genera with 7 species, and upwards, smaller with 6 species and downwards	$\frac{390}{2093}$	$= \frac{186}{1000}$	1.72	$\frac{162}{1365}$	$= \frac{118}{1000}$	1.79	$\frac{32}{345}$	$= \frac{92}{1000}$	1.50
Dalmatia: Visiani—Larger Genera with 5 species and upwards, smaller with 4 species and downwards.	$\frac{164}{1007}$	$= \frac{162}{1000}$	1.37	$\frac{130}{899}$	$= \frac{144}{1000}$	1.31	$\frac{46}{290}$	$= \frac{158}{1000}$	1.26
Rumelia: Grisebach—Larger Genera with 6 species and upwards, smaller with 5 species and downwards.	$\frac{98}{1136}$	$= \frac{86}{1000}$	1.45	$\frac{54}{1083}$	$= \frac{49}{1000}$	1.14	$\frac{12}{326}$	$= \frac{36}{1000}$	1.16
Russia, Ledebour (All 4 vols together) Larger Genera with 10 species and upwards, smaller with 9 species and downwards	$\frac{692}{3955}$	$= \frac{174}{1000}$	1.48	$\frac{307}{2407}$	$= \frac{127}{1000}$	1.39	$\frac{45}{475}$	$= \frac{94}{1000}$	1.26
Ledebour—Vol: I separately.	$\frac{207}{1237}$	$= \frac{167}{1000}$	1.42	$\frac{62}{576}$	$= \frac{107}{1000}$	1.32			
—— Vol: II ——	$\frac{192}{1243}$	$= \frac{154}{1000}$	1.56	$\frac{94}{767}$	$= \frac{122}{1000}$	1.35			
—— Vol: III ——	$\frac{171}{905}$	$= \frac{188}{1000}$	1.49	$\frac{94}{595}$	$= \frac{157}{1000}$	1.50			
—— Vol: IV ——	$\frac{122}{570}$	$= \frac{214}{1000}$	1.45	$\frac{57}{470}$	$= \frac{121}{1000}$	1.36			
N. United States. A. Gray—Larger Genera with 5 species and upwards, smaller with 4 sp. and downwards. The two kinds of vars. marked in this work are here classed together.	$\frac{112}{1136}$	$= \frac{98}{1000}$	1.40	$\frac{65}{917}$	$= \frac{70}{1000}$	1.36	$\frac{32}{361}$	$= \frac{88}{1000}$	1.37

Table I cont.

	Larger Genera	Smaller Genera (including those with single species)	Genera with a single species
Canary Islands, Webb & Berthelot—Larger Genera with 4 species and upwards, smaller with 3 and downwards.	$\frac{49}{421} = \frac{116}{1000}$	$\frac{42}{551} = \frac{76}{1000}$	
India (part of Flora) Hooker & Thomson—Larger Genera with 7 species and upwards, smaller with 6 species and downwards.	$\frac{21}{258} = \frac{81}{1000}$ 1.61	$\frac{13}{165} = \frac{78}{1000}$ 1.53	
Tierra del Fuego: Hooker—Larger Genera with 3 species and upwards, smaller with 2 species and downwards	$\frac{19}{177} = \frac{107}{1000}$ 1.57	$\frac{16}{163} = \frac{98}{1000}$ 1.37	
New Zealand: Hooker—Larger Genera with 4 species and upwards, smaller with 3 species and downwards	$\frac{52}{361} = \frac{149}{1000}$ 1.82	$\frac{37}{323} = \frac{114}{1000}$ 2.05	$\frac{15}{159} = \frac{94}{1000}$ 2.00
Insecta: Coleoptera Madeira: Wollaston—Larger Genera with 4 species and upwards, smaller with 3 species and downwards	$\frac{35}{225} = \frac{155}{1000}$ 1.71	$\frac{26}{257} = \frac{101}{1000}$ 1.34	
Sweden-Gyllenhal—Larger Genera with 11 species and upwards, smaller with 10 species and downwards	$\frac{512}{1344} = \frac{380}{1000}$ 1.85	$\frac{151}{485} = \frac{311}{1000}$ 1.43	$\frac{11}{43} = \frac{255}{1000}$ 1.54

that there were many great difficulties in the way. The subject is so highly important to us, as we shall see in a future chapter, that these difficulties must be discussed at tedious length; but it will be convenient first to give the tables./

A 26/In Table 1, we have several of the best known local Floras, (some of which were selected for me by Dr. Hooker) with the species divided into two great groups, those in the larger & those in the smaller genera. On the extreme right hand we have the genera with only a single species, but these are likewise included amongst the smaller genera. Some of the smaller Floras have been selected simply from giving remote countries under different climates. I may premise that I have given every single Flora (&

TABLE II[1]

	Larger Genera			Smaller Genera (with the smallest wholly removed)		
Great Britain: Bentham Great Britain: Babington— Larger Genera with 8 species and upwards, smaller with 7–4 species both included.—	$\frac{79}{455}$	$= \frac{173}{1000}$	1.41	$\frac{53}{360}$	$= \frac{147}{1000}$	1.24
Centre of France: Boreau— Larger Genera with 8 species and upwards, smaller with 7–4 species both included	$\frac{86}{505}$	$= \frac{170}{1000}$	1.40	$\frac{41}{343}$	$= \frac{119}{1000}$	1.31
Germany & Switzerland: Koch —Larger Genera with 11 species and upwards, smaller with 10–5 species both included	$\frac{257}{1216}$	$= \frac{211}{1000}$	1.99	$\frac{114}{683}$	$= \frac{166}{1000}$	1.95
Dalmatia: Visiani—Larger Genera with 8 species and upwards, smaller with 7–4 species both included	$\frac{120}{707}$	$= \frac{169}{1000}$	1.39	$\frac{71}{492}$	$= \frac{144}{1000}$	1.36
Rumelia: Grisebach—Larger Genera with 8 species and upwards, smaller with 7–4 species both included	$\frac{78}{917}$	$= \frac{85}{1000}$	1.44	$\frac{33}{513}$	$= \frac{64}{1000}$	1.33
Russia: Ledebour—Larger Genera with 16 species and upwards, smaller with 15–6 species both included	$\frac{573}{3285}$	$= \frac{174}{1000}$	1.48	$\frac{234}{1437}$	$= \frac{162}{1000}$	1.42
N. United States: A Gray— Larger Genera with 9 species and upwards, smaller with 8–5 species both included. (The two kinds of varieties classed together.)	$\frac{76}{710}$	$= \frac{107}{1000}$	1.36	$\frac{34}{426}$	$= \frac{79}{1000}$	1.26

two Entomological Faunas) which I have had tabulated, & have not picked out those which favoured my views. Nor have I divided the genera first in one way & then in another; but before knowing what the result would be, I determined to divide the smaller Floras nearly equally, but in the larger floras to have a greater number of species on the side of the larger genera, & then reduce

[1] [The holograph draft for this table is in ULC vol. 16.1, fol. 170.]

TABLE III.[1] *Decandolle Prodromus, Vols. 2, 10, 11, 12, 13, 14*

Name of Orders The Numerator and Denominator as in the foregoing Tables.	Genera with 11 species and upwards	Genera with 10 species and downwards
Leguminosae	$\frac{223}{2937} = \frac{75}{1000}$ 1.34	$\frac{38}{619} = \frac{61}{1000}$ 1.39
Rosaceae	$\frac{103}{562} = \frac{183}{1000}$ 3.09	$\frac{24}{144} = \frac{166}{1000}$ 2.20
Borragineae	$\frac{59}{480} = \frac{122}{1000}$ 1.38	$\frac{10}{111} = \frac{90}{1000}$ 1.40
Scrophulariaceae	$\frac{118}{1413} = \frac{83}{1000}$ 1.15	$\frac{24}{347} = \frac{68}{1000}$ 1.29
Acanthaceae	$\frac{232}{1088} = \frac{213}{1000}$ 1.43	$\frac{65}{335} = \frac{194}{1000}$ 1.35
Verbenaceae	$\frac{21}{506} = \frac{41}{1000}$ 1.00	$\frac{5}{82} = \frac{60}{1000}$ 1.00
Labiatae	$\frac{207}{1999} = \frac{103}{1000}$ 1.34	$\frac{32}{278} = \frac{115}{1000}$ 1.62
Solanaceae	$\frac{258}{1419} = \frac{181}{1000}$ 1.45	$\frac{11}{139} = \frac{79}{1000}$ 1.72
Proteaceae	$\frac{153}{912} = \frac{167}{1000}$ 1.41	$\frac{6}{72} = \frac{83}{1000}$ 1.16
Polygonaceae	$\frac{130}{614} = \frac{211}{1000}$ 1.63	$\frac{8}{62} = \frac{129}{1000}$ 1.37
Nineteen Small Orders	$\frac{112}{1120} = \frac{100}{1000}$ 1.50	$\frac{48}{406} = \frac{118}{1000}$ 1.14
All the species in the 6 Vols.	$\frac{1616}{13050} = \frac{123}{1000}$ 1.51	$\frac{271}{2595} = \frac{104}{1000}$ 1.43
Urticaceae—Weddell— Tabulated by Dr. Hooker	$\frac{65}{334} = \frac{194}{1000}$ 1.89	$\frac{26}{100} = \frac{260}{1000}$ 1.57*
All six volumes together	Genera with 17 species & upwards	Genera with 16–8 species both included
	$\frac{1510}{12103} = \frac{124}{1000}$ 1.52	$\frac{188}{1472} = \frac{127}{1000}$ 1.48

* Is Weddell's Urticaceae part of Decandolle or a separate work? [C.D.]

[1] [Darwin's draft of this table is in ULC vol. 16.1, fol. 169.]

Table III cont.

All six volumes	Largest Genera (76 in number) including half the species	Smaller Genera (1088 in number) including the other half of the species
	$\dfrac{959}{7815} = \dfrac{122}{1000}$ 1.59	$\dfrac{929}{7830} = \dfrac{118}{1000}$ 1.40
28 Largest Genera, each including on average 134 species, taken out of all the Orders in the Six Vols.	$\dfrac{455}{3772} = \dfrac{120}{1000}$ 1.74	

all to a common denominator: for if the larger Floras had been divided equally, from the great size of many of the genera, but comparatively few would have been included amongst the "Larger Genera": & as we cannot suppose that the larger genera go on varying or increasing in species for ever, it requires a considerable number of genera, as will presently be more fully explained, to strike a fair average. In the very large Flora Rossica, I have given in the table, the result for each volume separately, just to show that the excess of varieties in the larger genera is common to the whole/A 27/Flora: I did the same in some other cases with the same results. I have given Great Britain as worked out by several Botanists,/A 27 v/not as being particularly well-known, but in order to show that personal differences in estimating the value of species & varieties, makes no essential difference in the general result./

A 27/Now if we look to the two columns, under the larger & smaller genera, printed in larger type, in which the number of species, presenting varieties, are reduced to a common denominator, we see that with one single exception, the species in the large genera present decidedly more species having varieties, than do the species in the smaller genera. Moreover the average number of varieties to the varying species, with few exceptions, is larger in the larger than in the smaller genera: this is seen in the right hand columns of decimals,—the figures 1.50 for instance, showing that each two varying species have an average of three varieties. The one exception in the table, just alluded to, is Miquel's list of the plants of Holland: but so extremely few varieties are here marked, & as the results deduced from his list differ in several other respects from those obtained by other botanists, it may, I think, be disregarded.

In Table II, I have selected a few (& given all which I have selected) of the larger local Floras, & have entirely removed the smallest genera: & by looking at the columns printed in the larger type, & at the column with decimals we see the same rule throughout, namely of a greater number of varying species, & a greater average number of varieties, in the larger than in the smaller genera.—

If, then, local floras are to be trusted, & if the varieties recorded by various botanists (& two celebrated Entomologists) are worth anything, & if the varieties have been recorded fairly or nearly equally in the larger & smaller genera.—/A 28/all subjects presently to be discussed—we must conclude that there is a decided preponderance of varieties in the larger in comparison with the smaller genera.—

Table III gives the results of the tabulation of all the species (15,645 in number) in six volumes of De Candolle's Prodromus: selected for me by Dr. Hooker, & done at his suggestion. We here see a very different result from that deduced from the local Floras. In the genera having only 11 species & upwards there are more recorded varieties than in the genera with 10 species, & downwards; this holds good for the summary of the six volumes, & for most of the separate orders, but fails in some orders, especially in the great, natural & most carefully worked out (by Bentham) order of the Labiatae. The rule, however, does not hold good, (see Table) if all the genera with seven species & downwards be wholly excluded; so that all that can be said, is that the smallest genera usually present fewer recorded varieties. It deserves remark, how closely similar the result is when all the genera with 10 [11] species & upwards, with 17 species & upwards, when the 76 largest genera which include half the species, & when the 28 very largest genera are taken:—the proportion of the species having varieties in these several cases varying only from $\frac{120}{1000}$ to $\frac{124}{1000}$. The larger the genera are, however, the average number of varieties to the varying species seems to increase being in the 28 gigantic genera, as much as 1.74: so that each two varying species has on an average more than three varieties./

A 29/Now what is the evidence from these three Tables worth? The first question to consider is, whether it is best to take local Floras, or parts of the whole vegetable kingdom. The latter though having some advantages, has, for my special purpose several most serious sources of error. Geology tells us that in the long course of time, small groups have increased, come to a maximum, then declined, & ultimately disappeared. Hence we may feel pretty sure that some groups of plants, now numerically large, have nearly or quite arrived at their maximum, or are now declining; & that other small groups are now increasing more or less rapidly in numbers./A 29 v/Greatly as genera differ in size, yet there is a limit ⟨in number of species⟩ beyond which they rarely pass; & therefore, on my view of varieties being incipient species, there must always come a period when the largest genera will cease to increase at least as a single genus; though it does not by any means follow that sections or portions of such genera may not go on increasing, & other sections decline & be lost./A 29/It is idle to speculate what would be the precise effect on varieties of the declination, from less favourable conditions of life, of a group of species; but as the individual numbers of most of the species would probably decrease, from the relations lately pointed out, the amount of variation at any one time would probably be less: we do not even at all know, whether commencing extinction would generally first act on the species in the larger or smaller genera: though one may surmise on the latter: the ultimate result, we shall in a future chapter see, would probably be to leave in any group, those forms which are most distinct from each other. Now in a local Flora any genera, still large, which had come to/A 30/vary in a less degree, or a small genus which was varying largely, would, supposing for the moment our rule to be true of the species in large genera varying more than those in small genera, be on an average compensated by the other genera of the same country: so it should be in a Prodromus of the whole vegetable kingdom, if such existed, & there were no other causes of error: but looking to each separate order we might expect, if there be any truth in my view, to find some orders in which the large genera varied little,[1] & some in which the small genera varied greatly.

Secondly it is known[2] that the same order or genus often has

---

[1] I suspect that the Labiatae, viewed as a whole are now undergoing some great change in development. When divided in the three different ways shown in Tab III the smaller genera have a preponderance of varying species: yet there are two gigantic genera containing together no less than 653 species, & these contain fewer varying species (viz. $\frac{99}{1000}$ & only 1.20 varieties to each varying species) than the smaller genera however divided. If the sub-order Satureieae, (including only between $\frac{1}{5}$ and $\frac{1}{6}$ of the Labiatae) be removed, the larger genera have a preponderance of varying species. In the smaller genera of Labiatae the average number of varieties to the varying species is unusually large. Lastly looking to some of the local Floras, I find that in Boreau, Koch & Visiani the smaller genera in this order have more varying species than the larger: on the other hand in Babington & Ledebour, the large genera in this order, as generally throughout all these several Floras, have a preponderance of varieties.

[2] Alph: De Candolle, Geographie Bot. p. 1237–1245. In Hooker's Bot. Miscell: (Vol. 2 p. 257) there is given from Ledebour several curious cases of the great predominance of certain genera in the Altai: for instance there are 62 species of Personatae, & one-third of these belong to the genus Pedicularis: of the 130 Leguminosae, three-fourths belong to Astragalus, Oxytropis & Phaca.—

many more species in one country, than in another, either owing
to differences of climate or other unknown conditions. Where many
species of a genus exist, relatively to the other inhabitants of the
country, we have seen that there is some evidence that, on an
average, a large number of them are common & widely diffused;
and that of such common & diffused species a large number
present varieties. This at least is possible, but it could be hardly
detected except in a local Flora; for when all the species of the
genus were collected in a general Prodromus, the supposed greater
amount of variation where the species were numerous, & the less
amount, where thinly scattered & where the genus did not seem
to flourish, would tend to counterbalance each other, & conceal
the result. Again there are many moderately-sized genera with
all their species confined to one country, & which in that country
would be a large or rich genus, & which, according to my general
theory ought to be largely varying, as they have in that/A 31/
country become modified into many species; but the greater number
of such moderately-sized endemic genera would in a general
Prodromus have to be tabulated amongst the smaller genera, &
would vitiate the result. In fact such genera with absolutely few
species in comparison with genera in the whole vegetable kingdom,
but rich in species in their own country, are exactly those genera
which we might expect would yield the best evidence on our
view. Gigantic genera are often widely distributed over a large
portion of the world; & we must believe (as Sir C. Lyell has remarked
in his Principles in regard to the wide range of the *same* species)
that owing to the slowness of geological changes, of climate, &c.,
this spreading of the species of the same genus (descendants from
common parents according to our theory) must have taken an
enormous length of time: hence, although in a very large widely-
spread genus there must have been, on our view, a great amount
of modification, this modification may have been slow. On the
other hand in local genera, we may believe from the very fact
of their not having ranged widely, that they often are not of
such ancient origin as the widely spread genera; & in taking
a census of such comparatively fleeting objects as varieties, we
ought to look as much as possible to those groups of species, which
are undergoing the most rapid change; & it is just these very
endemic genera/A 32/rich in the species in their own country, which
would be lost, or rather would give a directly false answer when
tabulated in a general prodromus.

To take as a final illustration, the case alluded to in a previous
note of the genera Pedicularis and Astragalus, so extraordinarily

rich in species in the region of the Altai. As so many species have been formed there, we ought to look to these two genera/A 32 v/in that quarter, in order to see the manufactory of species at work: that is, according to my view, we ought there to find in these two genera, a greater than average number of varieties. And if this rule were found generally to hold good in local Floras, namely that the genera which had many species had many varieties, it would throw much light on the origin of species. But what can it signify under this special point of view, whether or not other species of Pedicularis and Astragalus are varying in other quarters of the world?*[1]

A 32/Hence I conclude from the several reasons just assigned, namely that some large genera must have arrived at their maxima and be now declining, & some small genera be rapidly increasing in number of species,—that some genera have been largely developed in certain countries, and elsewhere much more feebly,—that endemic genera probably have in many cases increased at a quicker rate than mundane genera, & yet would be ranked as small genera in a general Prodromus,—from these several reasons, I conclude/A 33/that a fragment of a Prodromus would be of little service, and an entire Prodromus of far less service for our special purpose than local floras. Nor should I have tabulated the six volumes of De Candolle, had it not been for Dr. Hooker's advice, nor should I have published the results, had not honesty compelled me, as they are on the whole unfavourable. Nevertheless I am bound to confess that from the wide diffusion of plants, and from genera largely dominant being generally everwhere numerous, I had expected more favourable results.

The best territories for my special object, would be those with all the species endemic, for all the species will probably have originated in such areas and where many species of the same genus have been formed, there as a general rule we ought now to find most variation in progress. Under this point of view, New Zealand & Madeira are the best areas in Tab. i, but they would have been better, had they included a greater number of species. I can, however, see no valid objection to taking, as a representative of the whole, fragments of one natural area, as (in Tab. i) the several kingdoms of Europe. Another advantage in local floras over a Prodromus, in which latter the orders are worked out by different men, is that there would be generally more uniformity

---

* Hence the smaller the area the better the result? [J.D.H.]

[1] [From here until the middle of fol. A 41, the text of the draft is not in Darwin's handwriting.]

in the value attached to varieties & species; there must be a prodigious difference in the value of the species as given by Dunal in the Solanaceae and by Bentham in the Scrophulariaceae, & though it is quite immaterial for us whether a greater or less amount of difference causes two forms to be called species or varieties, it is of some consequence that there should/A 34/be some approach to uniformity in the relative value of the species & varieties when all are tabulated together.

Now comes the question, what is the value of the varieties recorded in Botanical works? Am I justified in hypothetically looking at them as incipient species? do they differ in the same manner, only less in degree, from their types, as one closely allied species differs from another? I do not doubt that mere monstrosities have been recorded sometimes as varieties, though I do not suppose that any botanist would intentionally do so, & some authors have expressly stated that they have endeavoured to avoid this. Some also have stated, for instance Boreau, Visiani & Wollaston, that they have endeavoured to record as varieties not mere fleeting differences, but those alone with some degree of permanence. So again I do not doubt that a good many varieties are merely nominal, & owe their origin to doubts & confusion; & as such would be more likely to arise in large genera, than in small, this would directly vitiate our tables. That varieties even in the most carefully worked out floras are of very unequal values must be admitted; but it would have been a serious objection to my view of varieties being incipient species in various stages of modification, had they been all equally like or unlike each other and their parental types. I may here repeat that I am far from supposing that all varieties become converted into what are called species; extinction may equally well annihilate varieties, as it has so infinitely many species. That many varieties have in some degree the character of species I cannot doubt, for so many have been ranked as species by one botanist or another. Thus in the small British Flora, we have in Mr. Watson's list (Tab i) 182 varieties, so ranked by the greater number of sound botanists,/A 35/but which have all been considered as species by some one botanical author; & we have in addition 71 other forms called species in the well sifted London Catalogue, but which have been ranked as varieties by some one botanist. So again in Professor Henslow's list there are 62 forms considered by him as varieties, but which have been ranked by such eminent men as the elder De Candolle, Sir J. Smith, Sir W. Hooker & Lindly as true species.

Dr. Hooker objects to my whole manner of treating the present subject because varieties are so ill defined; had he added that species were likewise ill defined, I should have entirely agreed with him; for my belief is that both are liable to this imputation; varieties more than closely allied species, & these more than strongly marked species.

Mr. Watson & Dr. Hooker have also objected that there are many species so highly variable, & with the varieties running so closely into each other, that botanists do not attempt to mark them as distinct; hence in my tables, some of the most variable species do not appear to have any varieties. Boreau & Mr Wollaston also state that such polymorphic forms are not included amongst their recorded varieties. In the former part of this chapter we have seen how difficult it is to decide whether Polymorphism is of the same nature with more defined variation,/A 36/so that I am inclined to think that it is an advantage that such polymorphic species are partly excluded from my tables. That they are not by any means wholly excluded I am aware; for botanists occasionally mark by Greek letters ideal types which cannot really be defined from an inextricable mass of varying forms. So again when only a few specimens have been collected of some *rare* polymorphic species, the varieties would necessarily appear far more defined than they really are, & so would be liable to be recorded as distinct. I do not suppose that polymorphism which is partly excluded from our tables is much commoner in small than in large genera, or conversely; if it were so, it would have seriously vitiated our tables,—that is, if we suppose Polymorphism to be essentially of the same nature with more definite variation. In some of the floras I have excluded the most notorious polymorphic genera, which abound with doubtful species & doubtful varieties; but this has *never* been done except with the larger genera; & the result has invariably been to make the preponderance of varieties in the larger genera, *less* than it would have been, had these genera been admitted.

Mr Watson & Dr Hooker likewise object that* our best classifications are very far from natural; but any great perfection on this head is not material for my purpose: I divide all the species in a country/A 37/into two great bodies; all those in the larger genera on one side, all those in the smaller on the other side; & I presume it will not be disputed that the species in the larger genera taken together present a greater number of forms more closely allied together in little groups, than do the species in the smaller genera. I have however, found in tabulating the British

* remind me [J.D.H.]

Flora that the species of some few genera when split up into smaller genera, had to be placed among the smaller genera, whereas in other British floras they stood on the other side. But the several British floras in Tab. I show that this has not materially affected the result.

I cannot look at any of these causes of error as very important; they would, I think, to a large extent disappear when averages are taken; & the uniform result in Tab I & II bears out this conclusion. But now comes a far more serious cause of doubt, suggested to me by Dr. Hooker after seeking some of my tables; namely that botanists have recorded varieties more fully in the large than in the smaller genera. He believes this to have been the case from several reasons, but more especially from floras serving in part as mere dictionaries; & as it is obviously more difficult to name a species in a large than in a small genus, he thinks botanists have guarded against error by more carefully recording the varieties in the larger genera. I have consulted several other botanists, & though it does not appear that they had previously thought on this point, they generally/A 38/concur in this view. One botanist, however, Dr. A. Gray, whose opinion will be considered by all as of the greatest weight, after deliberation does not believe that he has himself so acted: he at first thought that he might have unfairly recorded a greater number of varieties in the smaller genera, which, from what little systematic work I have myself done, was my impression owing to the greater interest of mono-typic genera. Now if Dr. Hooker & the others who concur with him be right, all the foregoing tables are utterly worthless;* for they do not show nature's work only the imperfect handiwork of botanists. It is presumptious in me to believe that botanists have worked more philosophically than they themselves think they have; but I can hardly avoid this conclusion.

For in the first place it is somewhat remarkable that so many botanists & two Entomologists should all unconsciously & un-intentionally have produced so uniform a result, as may be seen in the first two tables: more especially as the varieties recorded by different authors are of such different values. To test Dr. Hooker's capital objection, I selected some of the principal local floras, & entirely removed the genera of least size; these are all given in Tab. II; here the larger genera (larger than in Tab. I) still show a marked preponderance in the proportional number of varying species over the smaller genera,† here not so small as in

* vitiated though perhaps not overturned [J.D.H.]
† give the case of Rubus [J.D.H.]

Tab. I. Dr. Hooker/A 39/would probably account for this fact by saying that the larger the genera & the more difficult the species were to identify, the greater the number of the recorded varieties would be; but as the difficulty goes on regularly increasing with the size of the genus the excess is not so great or so uniform as might have been expected on this view. The excess in the number of the varieties in the larger genera not regularly increasing with the size of the genera, may be explained on my hypothesis by some of the largest genera having reached their maxima. If we now look to the genera with a single species (right hand column in Tab. I) the difficulty in identifying the species is reduced to a minimum, yet we find that the number of species in these monotypic genera[1] which have varieties, though proportionally less than in the next group of larger genera, is by no means diminished in an extreme degree, as might have been confidently expected on Dr. Hooker's view: in two instances, namely in the U. States & Dalmatia, the number is actually greater than in the next group of larger genera. All this may be seen by comparing the right hand & middle columns in Tab. I.

If we look to the rows of figures with decimals in Tab. I & II, which give the average numbers of varieties which the varying species include, we find a degree of uniformity, especially in Tab. II very remarkable as it seems to me on Dr. Hooker's view. For my own part I look at these rows of figures as shewing, that not only/A 40/more species present varieties, but that the varying species generally present more varieties in the larger than in the smaller genera.

In the monotypic genera (right hand column in Tab. I) where the difficulty in naming species is reduced, as already remarked, to a minimum, we find the average number of varieties to the varying species, in five cases, either equal to, or actually greater, than in the next group of larger genera. ⟨This fact, I think, if the average from the small number of species in the monotypic genera can be trusted, might be explained on my view, but the explanation is not worth giving.*⟩ On Dr. Hooker's view that the species in the larger & smaller genera really have on an average an equal number of varieties; but that the varieties have not been fully

* Small genera being few in individuals do not present so many Herbarium varieties. [J.D.H.]

[1] [   ] says p. 574 that some have thought that monotypic species do not vary. He does not give any authority except [Puvis] (De la Dégénération p. 37) who refers only to varieties raised under [cultivation], and adduces the supposed fact in regard to all variations being due to intercrossing.

recorded by botanists in the smaller genera, we are driven to conclude (as may be seen by comparing the middle & left hand columns in Tab. 1) that although Boreau in France, Koch in Germany, & Hooker in New Zealand, did not† fully & fairly record all the species having varieties in the small genera: yet that in these very genera/A 41/they have recorded a greater than average number of the varieties themselves. This strikes me as improbable, & on the whole it seems to me far more probable that the tables make some approach to a fair representation of the manner in which species vary in nature. Any how I have endeavoured to give an abstract of the more important facts & arguments on both sides, & those few naturalists who are interested in the subject, can form their own judgement.

Finally, then, if we review our whole discussion on local Floras, which alone are well adapted for our purpose, it may I think be concluded, that on an average, a greater number of species in the large genera are common & widely diffused in their own country, than in the smaller genera; but that this greater number is (according to our theory) being slowly & steadily diminished by these species tending to vary, & thus being converted first into local varieties & then into local species. We can understand why a species which ranges widely & thus becomes exposed to somewhat different conditions of life is the most likely to vary; and a species numerous in individuals has a better chance, within any given/A 42/period, of breaking into varieties, which from possessing some advantage might be preserved & so become more or less permanent. Moreover common & widely diffused species must generally be better adapted to the conditions of life, to which they are exposed than the rarer & more local species, as will be more fully discussed in the next chapter when we treat of the severe competition to which every being is exposed; hence varieties from such favoured species will have the best chance of enduring for a long period & of increasing in numbers. It may be added that if a variety has ever increased so largely in individual numbers that it has come to exceed those of its parental type; it assuredly will have been called the species, & the original species the variety.

From these relations, & more especially from the actual facts given in the tables of the local Floras, I believe that the species in the larger genera, which as a general rule are very closely related to each other & in so far themselves approach in character to varieties, oftener present varieties (& a greater number of

† had [not] the means [J.D.H.]

6-2

varieties) than do the species in the smaller genera./A 42 v/It is not that the species of very small genera never vary, or that the species of large genera invariably present a great number of varieties; for if it were so, it would be fatal to my theory, as genera of all sizes have to increase & decline. Nor by any means is it, that all the species of a genus present varieties; for this is a very rare case;—it is only that more species have varieties clustered round them in the larger than in the smaller genera. And in regard to the close affinity of the species to each other in the large genera, it is not that all are equally related to each other; but, that some species are closely clustered round other species; causing the genus to consist of smaller & unequal sub-groups. These/A 43/conclusions as far as they can be trusted, strengthen our general theory, that species do not essentially differ from varieties, & that varieties by further modification may be converted into species. But our tables more especially throw light on the origin of the species of a genus, where very many are endemic in a moderately sized territory, & where we may suspect that they have been formed within comparatively recent times; for it is in local floras alone, that we invariably find more recorded varieties in the large genera, than in the small; & I have given my reasons for putting some faith in the records of so many Botanists, whose works agree in this respect. Furthermore, I believe, that the rule of the species in the larger genera on an average varying more, & therefore as I look at it, increasing in the number of their species at a quicker rate, than the species in the smaller genera, when taken in connexion with a large amount of extinction & with a principle, hereafter to be explained, which may be called that of divergence—taken together throw a clear light on the affinities of all organic beings within the same great classes; for we invariably see organic beings related to each other in groups within groups—or somewhat like the branches of a tree sub-dividing from a central trunk.

Conclusion. From the various facts now given in this chapter, & innumerable others might have been added, I cannot doubt that there is much variability in organic beings in a state of nature/80 v/ The widely-ranging, the much diffused & common, in short the vigorous species are those which are the most apt to vary./80/The variation differs greatly in degree; in some it is scarcely perceptible, in others strongly marked; so that we have a graduated series from the finest shades of individual differences, to well defined races, distinguishable with great difficulty, if really distinguishable at all, from sub-species & closely allied species. In certain protean genera, the variability may in part be of a different nature; but on this point it seems difficult to arrive at any definite conclusion. From what we have seen of the effects of domestication or changed conditions on organisms of all kinds, & which beings, it has been shown in the second chapter, could not have been originally selected from the plasticity of their organisation, & knowing well that the history of the world is emphatically that/81/of Change, it would have been a discordant result if there had been no variability in a state of nature. Judging from the effects of domestication it is indeed surprising that we do not clearly see in nature more organic change, but if such greatly changed organisms do exist, they would be universally called species & not varieties.

According to the views discussed in this work, species do not differ essentially from varieties;—two closely allied species usually differing more from each other than two varieties, & being much more constant in all their characters. This greater constancy may be looked at as partly due to the several causes of variability having acted less energetically on the two species under comparison than on the one species yielding the two or more varieties; and partly to the characters of the two species having been long inherited, & by this very cause having become more/82/fixed. The greater amount of difference between the two species than between the two varieties, may be looked at as simply the result of a greater amount of variation; the intermediate varieties between the two species or between them & a common parent having become extinct. Hence as a general rule, species may be looked at as the result of variation at a former period; & varieties, as the result of contemporaneous variation.

But the forms generally considered as varieties & those considered as species differ in one other most important respect; namely in the perfect fertility of varieties together & the lessened fertility of the offspring of two species. This subject will be discussed in a separate chapter; & I will here only repeat that the infertility of species when crossed graduates away so insensibly/83/that the two most experienced observers who ever lived have come to diametrically opposite results when experimentising on the same forms;—that the infertility does not closely go with the general amount of difference between the two forms, but follows laws of its own;—that it is most powerfully affected by the sex in reciprocal or reversed crosses of the very same two species;—and finally that, as we have seen in the last chapter, the reproductive system is eminently subject to disturbance & that infertility of an analogous kind to that resulting from hybridism supervenes from other & totally distinct causes. Hence, as it will be attempted to be shown in the chapter devoted to this subject, there is no valid reason, why the different "sexual affinity" (to use Gaertner's expression) of different species to each other should be thought a character of overpowering weight, in comparison with the other differences between species when contrasted with the difference between varieties. ⟨with each other; as, for example, in the tendency to adhere when grafted together.⟩ /

84/It seems to me that the term species is one arbit[r]arily given for convenience sake to a set of individuals closely like each other; &, that it is not essentially different from the term variety, which is given to less distinct & more fluctuating forms. The term variety, in comparison with mere individual differences, is applied, also, arbit[r]arily & for convenience. Practically if two forms are tolerably constant in their characters & are not known to be connected by a nearly perfect series of intermediate forms they are called species; & according to the views here given, even should the two distinct forms be thus connected, if the intermediate forms are comparatively rare, so as seldom to cause much difficulty in naming an individual specimen, there seems no good reason why they should not be called species; & in that case science & common language would accord in giving names of equal value, to the primrose & cowslip,—/85/to the deodar & cedar of Lebanon, —to the Durmast and common oak,—as well as to the many fine species distinguished by the naturalists on characters of little physiological importance.

As the only known cause of close similarity in two organic beings, is descent from a common parent, it is natural that the idea of descent should have entered into almost every definition of the term species. A monster may be abnormal in any degree, but the instant we know its parentage, we do not doubt about referring it to its species.—On the views here discussed, the idea of the common descent of all the individuals of the same species equally comes into play; but it is not confined, as in the ordinary definition, to the individuals of the same species, but is extended to the species themselves belonging to the same genus & family, or to whatever higher group our facts will lead us.—/

86/According to these views it is not surprising that naturalists should have found such extreme difficulty in defining to each other's satisfaction the term species ⟨as distinct from variety.⟩ It ceases to be surprising, indeed it is what might have been expected, that there should exist the finest gradation in the differences between organic beings, from individual differences to quite distinct species;—that there should be often the gravest difficulty in knowing what to call species & what varieties in the best known countries, & amongst the most conspicuous & best known organic beings if ranging over a wide territory; & that the difficulty should be hopelessly great in two adjoining but now perfectly, or almost perfectly separated regions./86 v/We can understand why it is that the species in large genera are generally more closely related to each other & related in little clusters like satellites around certain other species, why they are apparently often confined in their distribution, & lastly why they oftener present varieties & a greater number of varieties, than do the species in small genera: for, on our views, where, in any country, many species of a genus have been formed there has been in such genus a greater than average amount of modification within the existing geological period; & hence we might expect that the resultant forms would tend to resemble varieties in closely re- sembling each other & in being grouped around certain species, like varieties around their parents & in being local. We might moreover, expect, on these views that where there has been lately much specific modification, there generally would be now most variation in progress.

The conclusion that there is no/86/essential difference, only one of degree & often in the period of variation, between Species & Varieties, seems to me at least as simple an explanation of the many/87/difficulties by which naturalists are beset, as that each species should have been, independently created with its own system of variability,—the varieties imitating the characters of other species, supposed to have also been independently created, so closely as to defy in many cases the labours of the most experienced Naturalists.

CHAPTER IV, SUPPLEMENT

a/Phanerogamic plants alone have been tabulated out of the following works./av/In the counting the number of varieties them- selves, I have not except in a very few cases which are specified counted those marked *a*: for these seem generally to be the type- forms more fully described: or the type forms in an exaggerated degree. I would, however, here make no important difference for our object whether counted or not, as they would have been counted both for the large & small genera.—/

a/C. C. Babington. *Manual of British Botany. 3 Edit. 1851* The naturalized & doubtful plants, included in brackets and marked by asterisks are all omitted. The genera Rubus, Rosa, Salix & Hieracium are, also, omitted; from the extreme doubts, almost universally entertained, which forms to consider varieties and which species: as these are large genera & have many varieties had they been admitted the proportional number of varieties would have been greater in the larger genera. Mr Babington is generally considered to admit very fine species.

The Rev. Prof. Henslow. A. Catalogue of British Plants. 2nd Edit. 1835. The species certainly not indigenous have been expunged; but those marked (4) as *possibly* introduced by man have been left in. The genera *Rubus, Rosa,* Salix & Hieracium have been excluded for reasons given above; if left in, the result would have been as above stated. The Varieties are marked by Greek letters, but certain varieties (62 in number) are preceded by (-); & this signified that these forms have been considered by one of the four following eminent Botanists, as true species, namely De Candolle in Bot. Gallicum: Hooker in British Flora, Lindley in Synopsis of British Flora, & Sir J. E. Smith in English Flora. These varieties, as so ranked by Prof. Henslow, must have much of the character of species./

b/*Mr H. C. Watson & J. T. Syme, London Catalogue of British Plants,* 4th Edit. 1853. All the species printed in italics, thought to be naturalized, are expunged. Genera Rosa, Rubus, Hieracium & Salix for reasons & with results already assigned have been omitted. In this well sifted list, only few varieties are recorded: but Mr Watson has added for me some which have been ranked by at least one Botanist as a species: he has, also, expunged some few of the most trifling printed varieties, which have not been considered by any one botanist as a species. He has, also, marked for me some of the forms ranked in this catalogue as species, but which have been considered by some Botanists as varieties. If considered, as in the second line of the Tab. I, as varieties, the number of species is diminished both amongst the large & small genera; when considered as species as in another part of our discussion, such could not occur in genera having only a single species, so that these have been also removed in our calculation, though not strictly necessary; & their removal makes the result less striking than it would otherwise have been. For the calculation of the Ranges I have used the 5th Edition 1857. The number of the Pro-

vinces is not appended to some of the species, & these are wholly
excluded: as are those confined to Ireland & the Channel Islands.—/

c/*A. Boreau. Flore du Centre de la France, 1840.* Cultivated plants
omitted. Genera Rubus, Rosa, Salix omitted, for reasons & with
the result before assigned. M. Boreau expressly states (Tom. 1,
p. 101) that he distinguishes "les varieties [sic] proprement dites",
which are "plus tranchées" from the endless variations, which
many common & widely ranging plants display.

*Ant. Miquel: Disquisitio Geographico—Botanica de Plantarum
Regni Batavi. 1837.* This list is unsatisfactory for our purpose so
few varieties being indicated: owing to a mistake in the printed
list, I am doubtful about one variety, but have admitted it. I should
not have given the results from this list, had I not felt bound to
do so from honesty, as the result differed from those in all the
other Floras, in several respects. The certainly naturalized plants
(marked with †) are omitted.

Koch. Synopsis Florae Germanicae et Helveticae Edit. 2. 1843. I
have here made no omissions: in counting the number of the
varieties themselves, I have counted those marked *a* as well as
B &c. I have not counted the subvarieties of varieties. This is
a very large Flora including 3458 species.

Rob. de Visiani: Flora Dalmatica 1842–1852. I have excluded the
cultivated plants. In counting the number of varieties to each
species, I have not counted those marked (*a*). Visiani seems to
have carefully distinguished varieties from variations./

d/*A. Grisebach, Spicilegium Florae Rumelicae et Bithynicae 1843.*
Doubtful species excluded. Monstrosities marked 'lusus' not included:
in counting the number of varieties, those marked (*a*) not included.

C. Ledebour Flora Rossica. 1842. I have made no exclusions, but
I have not taken the trouble to add the species in the Addenda,
I have counted as varieties, only those marked by Greek letters, &
not those species which are merely said to be variable. In counting
the number of varieties themselves, those marked (a) not counted:
nor the sub-varieties of varieties.

*Asa Gray. Manual of the Botany of the Northern United States.
2nd. Edit. 1856.* The naturalized plants are omitted & all in the

	Larger Genera	Smaller Genera
Small-type varieties	$\dfrac{74}{1136} = \dfrac{65}{1000}$ 1.43	$\dfrac{34}{917} = \dfrac{37}{1000}$ 1.36
Large-type varieties	$\dfrac{38}{1136} = \dfrac{33}{1000}$ 1.34	$\dfrac{31}{917} = \dfrac{33}{1000}$ 1.38

large genus Salix, according to Dr. Gray's advice. The varieties are divided into two classes, the ordinary ones which are less strongly marked, & others printed in full-faced type, which have been thought to be species by some Botanists & about which Dr. Gray is doubtful: the two kinds of varieties are classed together in the tables, as the number of the more strongly marked varieties is so small: taken separately we have as follows.

e/*Webb & Berthelot. Hist Nat. des Iles Canaries: Phytographie.* I have not been able to exclude the many naturalised plants. I have here included not only the varieties marked by Greek-letters, but those polymorphic species of which the variations are divided into groups.

Hooker & Thompson. Flora Indica. 1855. This is a mere fragment, including only 428 species; & was taken only because illustrating a tropical country. The species "dubiae" have been excluded. The variations marked "variants" not counted, only those marked by Greek letters.

Hooker. Flora Antartica 1844. I have taken only the portion including Tierra del Fuego, the Falkland Islands & Kerguelen Land. This Flora including only 340 species is too small for our purposes: & was taken only from giving so distant a locality.

Hooker. Flora of New Zealand 1853. This is an interesting Flora for our purpose from containing so many endemic species. The large genera Senecio, Coprosma & Veronica have been omitted, as Dr. Hooker informs me that the species are so variable, that it is difficult to say what are species & what varieties. Had these been included the proportion of species having varieties would have been larger in the larger genera.—In calculating the "close-species",/ev/the above genera, & Carex & Uncinia have been omitted: these latter because Dr. Hooker had not himself described them: the monotypic genera have, also, of course been excluded for this latter purpose:—/

Wollaston—Catalogue of the Coleopterous Insects of Madeira. 1857.
The certainly & probably naturalised species have been omitted.
Several new species have been added since the publication of the
Insecta Maderensia: I tabulated the insects in this latter work
without removing the naturalised species, & the result is for the
large genera $\frac{148}{1000}$ & for the small $\frac{86}{1000}$. The varieties have been most
carefully attended to in these admirable works.—

Gyllenhal. Insecta Suecica 1808–1827. I selected this work on the
advice of Mr. Wollaston. The species given in the addenda have
not been added in.—The numerous variations are mostly of a very
trifling nature, being chiefly confined to colour.—

*Furnrohr: Flora Ratisbonensis 1839 (in Naturhist. Topog von
Regensburg)* This list has been used only for the species marked
"sehr gemein".

Alph: de Candolle Prodromus. Vols. 2, 10, 11, 12, 13 & 14. These
volumes were kindly selected for me by Dr. Hooker for various
reasons, as containing several large & well worked out Orders &
several small Orders. The Proteaceae are remarkable for their
confined range. These six volumes/g/include 15,645 species. Those
"non satis notae" & "dubiae" have been excluded. In counting
the number of varieties, the few cultivated ones have been excluded:
those marked (*a*) have not been enumerated, as being generally
only the typical forms in excess. I have experienced some doubt
about some of the varieties marked by asterisks.

⟨I may here add that in Tables II & III, several of the works
were selected for & kindly lent to me by Dr. Hooker.⟩

CHAPTER V

THE STRUGGLE FOR EXISTENCE AS BEARING ON NATURAL SELECTION

INTRODUCTION

A fortunate change of ink after the first 18 folios of the manuscript of chapter five reveals significant details in its history. Darwin started writing this chapter under the title 'On Natural Selection' and only later decided to add 'The struggle for existence' as the main theme. The original ink, now brown, is clearly distinguishable from the black of the later additions, notably in the title of the chapter, the added last sentence on folio 8: 'This present chapter will be devoted to the Struggle for existence,' and the slip of paper with the revised beginning of the direct discussion of this theme (fol. 9 A).

Although in the original brown ink version Darwin placed 'War of nature' as an alternative to 'Struggle of nature' as a rubric for this section, and began it in the Hobbesian vein, 'all nature is at war,' and although, through Erasmus Darwin he knew the even harsher Linnaean image of 'One great slaughter-house the warring world!'[1] he later changed his rubric to 'struggle for existence'. This he could interpret more broadly than war between organisms to include the physical environment as well: 'A plant on the edge of a desert is often said to struggle for existence' (fol. 30 A').

The latter statement is in an interpolated addition equalling about three to four full folios of text giving an extended definition of the term struggle for existence. This interpolation and another one of half a dozen pages mainly on intraspecific competition are the only extensive revisions of the manuscript for this chapter. I have found no sure evidence of their dates of writing, but it is tempting to relate them to the entries for this chapter in Darwin's Pocket Diary. Here he originally wrote under 1857: 'Feb. 27 Fin^d Ch. 5 Struggle for Existence' then later cancelled the date to change it to March third. The earlier date might well be for the completion of the original draft and the later for the completion of Darwin's revising.

The further accumulation of notes and observations in the portfolio for this chapter continued, even long after the completion of this draft. In the spring of 1857 soon after he completed the chapter draft, he began a series of observations on the effects of enclosure on portions of the heath at Farnham, Surrey, near Moor Park, once the residence of Sir William Temple and occasionally of Jonathan Swift, and later of Dr. Lane, whose water-cure Darwin took several times. These observations relate closely to those given in this chapter on folios 48 and 49 on the heath at Maer Hall, near Stoke-on-Trent, Staffordshire, the estate belonging to Josiah Wedgwood, Darwin's uncle and father-in-law. Some of these Farnham Heath observations are

---

[1] The Temple of Nature (London, 1803) Canto IV, line 66. Cf. Linné: 'lanienam & Bellum omnium in omnes', in Politia naturae, Cap. II, 15, but omitted in Brand's English translation, and Linné: 'bellum omnium perpetuum in omnes, et horrenda laniena', in Föreläsningar öfver Djurriket..., ed. E. Lönnberg (Uppsala, 1913), p. 1.

included in the appendix to this chapter. The corresponding published passage is on pages 71 and 72 of the first edition of the *Origin*. Even in May 1862, a year after the publication of the third edition of the *Origin*, Darwin recorded observations on the struggle for existence on heath land, which, by labelling 'Ch. V', he still associated with the Natural Selection manuscript rather than with the *Origin*, for which the chapter number would be three.[1]

THE STRUGGLE FOR EXISTENCE
AS BEARING ON NATURAL SELECTION

[completed March 3, 1857]

1/In treating of the variation of our domestic productions it was shown that the changed conditions of their existence had some direct effect on them, as food on size, heat on their hair &c., but that indirectly the effect was more potent in tending to render their whole organisation plastic, or less true to the parental type. This view of the organisation being thus rendered plastic in various ways, as if, though of course not really, by mere chance, is, I think strongly supported by the many facts given in our third chapter, showing how sensitive the reproductive functions generally are to changed conditions. It was further shown in the first chapter, that Selection by man, whether intentional or un-intentional, combined with the strong principle of inheritance, played a most important part in adding/2/up very slight variations in a given direction.

In the last chapter we have seen that in all organisms in a state of nature there are at least individual differences, & in some a considerable amount of variation. It would be strange, inasmuch as variability in main part is due to changed conditions, if this were not so, as Geology consists of the history of the many changes which the earth and its inhabitants have undergone. ⟨& from these changes its inhabitants must suffer or profit. No one who has studied Lyell's Principles of Geology will dispute this. Look to our last epoch, within which the far greater proportion of the now living beings have existed, & reflect over how a vast an expanse of land in Europe & both Americas the sea flowed & left its shells & boulders: reflect on the prodigious changes of climate evidenced by the long intercalated glacial period:/3/all those organisms which were so situated that they could not emigrate must have suffered almost every possible change which their organization could withstand; indeed far more, & there must have

---

[1] C.D. MSS. vol. 46.1, item 53.

been much local extinction. Occasionally a living being must get into an island or other isolated site, where it would be exposed to new conditions & yet might survive, like the very many productions naturalised by man's intervention. Some reasons were given in the first chapter for supposing that abundant food might be one main cause of variation under domestication; & I think I shall be able to show hereafter that the species, which are now most vigorous, ranging furthest & abounding most in individuals are those which vary most; & thus we may believe are the best nurtured.

Let the cause be what it may, organisms/4/in a state of nature are in some degree variable;⟩[1] But mere fluctuating variability, or any *direct* effect of external conditions (to which subject we shall return) are wholly inadequate to explain the infinitude of exquisitely correlated structures, which we see on all sides of us. Look at the Anteater with its great claws & wonderful tongue; or at the Woodpecker, or the Hawk which may swoop down on it, or at the wood-boring beetle on which it preys, or at what we consider the humblest creature, the parasite so admirably formed to cling to its feathers./

5/The most credulous believer in the "fortuitous concourse of atoms" will surely be baffled when he thinks of those innumerable & complicated yet manifest correlations. In quite simple cases, as in seeds furnished with hooks so as to be transported by animals, the believers in such a doctrine might, perhaps, adduce the case of the cultivated Teazle, believed by many botanists to be a mere variety, & yet so well adapted, that it cannot be imitated by man's art, for a special purpose; & he might say as chance in this instance has favoured man, so in other cases it might favour the plant. But no one I should think could extend this doctrine of chance to the whole structure of an animal, in which there is the clearest relation of part to part, & at the same time to other wholly distinct beings. It is superfluous to give examples: every animal if we know it well, could suffice; but the/6/instances are more obvious in some cases than in others, as perhaps in those given, or as in those insects, which have their structures specially adapted to lay their eggs in the larvae of other particular species of insects; others again being adapted to lay their eggs in special plants together with a marvellous poison ⟨which no chemist can understand or imitate⟩, which will cause the tissues of the plant in question to develop a gall of fixed form, serving as food to the insect, & appearing like a prison, but out of which the prisoner in due time knows full well how to escape.

---

[1] [For the rest of this cancelled passage see appendix.]

No theory of the derivation of groups of species from a common parent can be thought satisfactory until it can be shown how these wondrous correlations[1] of structure can arise. I believe such/ 7/means do exist in nature, analogous, but incomparably superior, to those by which man selects & adds up trifling changes, & thus brings his pigeon or canary-bird or flower up to a preconceived standard;—or gets one breed of dog to point to his game & another to retrieve it, in a manner which no wild animal would follow;— or gets the wool of one breed of sheep to be good for blankets, & another for broad cloth. If those slight variations of structure, which we see occurring in beings in a state of nature & which from our ignorance we attribute to chance, or changed conditions, if these could be selected & added up, not for man's good, but for that of the being in question, in such case the structure of one part might be adapted to another part, or to some distinct organism/ 8/& the whole being might be harmoniously modified. And for myself I am fully convinced that there does exist, in Nature, means of Selection, always in action & of which the perfection cannot be exaggerated. I refer to that severe, though not continuous struggle for existence, to which as we shall immediately see all organic beings are subjected, & which would give to any individual with the slightest variation of service to it ⟨at any period of its life⟩ a better chance of surviving, & which would almost ensure the destruction of an individual varying in the slightest degree in an opposite direction. I can see no limit to the perfection of this means of Selection; & I will now discuss this subject,—the most important of all to our work. This present Chapter will be devoted to the Struggle for existence./

9 A/*The Struggle for existence*.[2] All Nature, as the elder Decandolle has declared with respect to plants, is at war.[3] When one views the contented face of a bright landscape or a/9/tropical forest glowing with life, one may well doubt this; & at such periods most of the inhabitants are probably living with no great danger hanging over them & often with a superabundance of food. Nevertheless the doctrine that all nature is at war is most true. The struggle very often falls on the egg & seed, or on the seedling,

[1] [Above the word correlations in the MS., Darwin pencilled in 'co-adaptation' as an alternative term.]

[2] [Cf. appendix for Darwin's earlier versions regarding this key phrase and for discussion.]

[3] [Quoted by Lyell sans cite: *Principles*, 9th ed. p. 670, in the chapter in Theories respecting the original introduction of species, where C.D. marked his copy. See Augustin P. de Candolle, art. 'Géographie botanique', in *Dictionnaire des sciences naturelles*, vol. 18, p. 384.]

larva & young; but fall it must sometime in the life of each individual, or more commonly at intervals on successive generations & then with extreme severity. This struggle & destruction follows inevitably in accordance with the law of increase so philosophically enunciated by Malthus.[1] In a country undergoing no great change, on a long average the numbers of all the species cannot increase; & unlike man, other organisms/10/cannot artificially increase their means of support, which must determine the extreme limit of their numbers. Yet all living beings, if not destroyed, even the slowest breeders, tend to increase in geometrical proportion, & often at an enormous ratio.

Everyone must have seen statements of the number of eggs & seeds produced by many of the lower animals & plants.[2] To illustrate geometrical progression one meets in works on arithmetic calculations such as, that a Herring in eight generations, each

---

[1] Essay on the Principle of Population 1826. Franklin & many others have clearly seen & exemplified the great tendency to increase in all the lower animals & plants. [See Franklin, B., observations concerning the Increase of Mankind...Boston, 1755.]

[2] I will copy out a few instances of numbers of eggs & seed. Mr. Harmer in Phil. Transact. 1767, p. 280, weighed the whole & portions of roe & counted in this portion the number of eggs. The number differed considerably in different individuals.

Carp	203,109	and	101,200	lowest number
Cod	3,681,760			
Flounder	1,357,400	and	133,407	do
Herring	36,960	and	21,285	do
Smelt	38,278	and	14,411	do
Lobster	21,699	—	7,227	do
Prawn	3,806		3,479	do
Shrimp	6,807		4,090	do

(N.B. These observations on the F. Water fish are confirmed by independent calculations by C. F. Lund in Acts of Swedish Academy Vol. 4.)

Astacus gammarus   12,440 [Linné] Brand Amoen Acad. p. 343

Holothuria 5000 ova in one night Sir J. Dalyell

Doris, 600,000—counted by myself. Journ. of Researches p. 201. [*Powers of the Creator.* vol. I (London, 1851), p. 52]

Bombyx mori 500. [Anon.] Silliman's Journal. Vol. 18, p. 282.

Wasp   the Rev. Prof. Henslow counted 300 females in one nest in Autumn

Ascaris lumbricoides, sixty-four million. Carpenter Comp. Phys. p. 590. This is the greatest number, I recollect to have seen; & it is almost inconceivable.—

Plants

Helenium 3000 seeds ⎤
Zea mays 2000   ⎥   Linnaeus [p. 93] in
Papaver 3200   ⎬   Brands Amoen.
Nicotiana 4300   ⎦   Acad. vol. 2.
                               p. 409

Wild carrot, (a very fine one) according to my calculations had 40,000 seeds Wild parsnip according to Rev. Prof. Henslow had 2250 seeds one which I gathered, had I fully believe 12,000 seeds.

fish laying 2000 eggs, would cover like a sheet the whole globe, land & water: Linnaeus in the Amoenitates Acad. says that an annual plant producing a single flower with only two seeds (& no plant nearly so barren exists) in twenty years would yield one million plants.[1] The great engineer Vauban calculates that from one sow...[Sentence left incomplete]. Buffon ranks fifteen animals as less fertile than man (a statement which I rather doubt); &/11/ yet man in the United States, has doubled in 25 years. The Elephant is supposed to be the slowest breeder of all living creatures; & I have seen it stated that were this not so, elephants would overrun the world! The elephant is supposed not to breed till ⟨20⟩ perhaps 30 years old; its length of life is not known, but as one of unknown age when taken lived according to Dr. Falconer 120 years, I think it will not be an exaggerated statement to take ⟨80⟩ 90 years as the possible duration of life & that each pair produces ⟨four⟩ three pair of young: in this case from one pair there will be at the end of 500 years 5,111,514 elephants alive: or if we assume that the pair produced eight young there would be above fifteen millions alive. Hence we can plainly see that it is not from want of fertility that this animal, the least fertile of any, does not overrun the world.[2]

But we have far better evidence than calculations of the possible increase, namely the actual increase of many animals & plants under favourable circumstances. The marvellous increase of several of our domestic animals where run wild in different parts/12/of America have repeatedly been quoted:/12 v/for instance the great herds, some even of 8000, seen in Cuba only 27 years after the discovery of that island:[3]/12/Nothing has astonished me so much in this respect as to find in Sarmiento's Voyage that in only 43 years after the horse was first imported[4] into Buenos Ayres, where it immediately ran wild, it was in possession of the Indians at the Straits of Magellan, 1200 miles to the south. We have similar facts in New S. Wales: thus in 1788[5] 29 sheep & 5 cattle were introduced;

---

[1] ['On the Increase of the Habitable Earth' pp. 94–5 in Select Dissertations from the Amoenitates Academicae...Trans. F. J. Brand, London, 1781.]
[2] [See ULC vol. 46.1 f. 35 for calculations and comment re increase of elephants.]
[3] Lyell's Principles of Geology 9th Edit. 1853, p. 685. Robertson [II, book 8, p. 394] quotes for increase of cattle in S. America Oviedo, ap. Ramusio III, 101 Hackluyt III, 466 & 511 Churchill Collect III, 47 & v, 680, 692 Feuille, I, 249 Acosta, Lib III c. 33. [Robertson's citation for Acosta should read 'Lib. iiii cap. 33'. To identify all these citations, the editions used by Robertson would have to be identified. For example, his reference to 'Churchill Collect. III, 47' seems to fit pp. 43–4 of the 1st ed., i.e. Ovalle, *Chile*, book 1, ch. 21.]
[4] In 1537, see Rengger, Natur. der Sä[u]gethiere von Paraguay S. 334.
[5] [Report given in Sydney Smith's Works, vol. I, p. 324.]

29 years afterwards the numbers were for sheep, 170,920, & for cattle 44,753; & no doubt many must have been slaughtered in the interval. In 1418 a single female rabbit was turned out in the island of Porto Santo[1] in a few years 3000 were killed at one time; & 36 years afterwards Cada Mosto in his voyage/13/speaks of them as innumerable;[2] nor is this wonderful as it has been calculated[3] that one pair might produce 1,274,840 individuals in four years. Equally striking & well known are the many facts, showing the astonishing increase of many native animals, when two or three favourable seasons have followed each other consecutively: thus during the famous drought of 1826–28 (inclusive) in La Plata the whole country literally swarmed with mice, which disappeared with the returning wet. In Germany a similar increase of field mice was accompanied by an astonishing increase in stoats &c. which preyed on them. It would be superfluous to give the cases amongst my notes of the enormous increase of Birds, fish, frogs, snails & insects, when turned out in new countries: the one island of Mauritius[4] would afford striking instances in all these classes except fishes; & for fish we may turn to N. America. Bees & wasps taken from Mauritius have come to swarm, as I am informed by Capt. Moresby on the miserable /14/coral Chagos islets.

Of the rapid ⟨& often overwhelming⟩ increase of plants run wild, innumerable instances could be given. America over large districts has been peopled by plants from the old World & in La Plata to a quite overwhelming extent: on the other hand there is scarce a region of the world which has not got now widely extended colonists from America since the time of Columbus: in India, as I am informed by Dr. Falconer, three of the commonest plants from Cape Comorin to the Himalaya are of American origin. In the island of P. Santo.[5] . . ./

15/In the foregoing cases, & innumerable others could have been added, we cannot account, at least in any great degree, for this wonderful ⟨observed⟩ rate of increase, by the law of fertility having been suddenly altered in each species. In the higher animals, the period of gestation & suckling, the number of young produced at a birth, the length of natural life, would almost certainly remain constant; probably the animals would breed at a little younger age & oftener when better fed than in their native country; but

[1] By J. G. Zarco in Kerr's Collect. of Voyages. Vol. 2, p. 177.
[2] [For Cada Mosto see Kerr, *Voyages*, 2: 205.]
[3] Fothergill Philos. of Nat. Hist. p. 137.
[4] Sonnerat's Voyage aux Indes vol. 2. p. 83. I could add other instances.
[5] [Darwin left this sentence incomplete and the rest of the folio blank except for the pencilled memorandum: 'St. Helena'.]

this could hardly apply in all cases as in short-lived animals & annuals. No one will maintain that the American Parkinsonia has spread over all India, or that the European cardoon & thistle have overwhelmed the plains of La Plata, owing to their producing more seed than in their aboriginal land. Undoubtedly the great increase must almost exclusively be due to all, or nearly all the young surviving & breeding, with the old likewise still surviving & breeding. The result of geometrical progressions invariably strikes one with surprise. The observed rate/16/of increase in the foregoing instances could not possibly be continued for centuries, for neither earth nor ocean could hold the product:/16A/Nor is it probable that the cessation of increase or actual decrease as with the mice of La Plata, would be in any high degree influenced by lessened fertility; for I think the young would perish, before the old were starved to the degree as not to breed; & in the case of the domestic animals run wild they would hardly spread into districts, already stocked with native animals, so unfavourable to render them in any marked degree sterile. Indeed according to Mr. Doubleday's theory, in which for reasons given in our third chapter, [See ch. 3, fol. 99] I do not believe, but/16/which has found several advocates, organic beings when pressed for food, breed the more freely, causing the struggle for life to be more fearful.

In a state of nature, all plants annually produce seed, excepting a few which propagate at a great rate by suckers &c., & still fewer which are just able to live in the extreme arctic regions & on high mountains, where they have to struggle not against other living beings but against cold. All or nearly all animals pair in a state of nature excepting apparently a few males in excess, & a few barren individuals. Had this not been so, it could/17/hardly fail to have been observed in our game-birds & other carefully observed wild animals.[1] The time of pairing, I believe, always falls at a

[1] With respect to barren birds, which are not at least in the case of Solan Geese, young individuals, it seems that they are not very rare in sea-fowl. See Wilson's Voyage round Scotland Vol. 2, p. 77. For the excess of males see the fact given in regard to Partridge by White of Selbourne in Letter xxix. But there are other facts mentioned in the same letter in regard to both males & females of sparrows & owls, quickly getting a new mate, when one has been shot, which are of difficult explanation. This fact has been particularly observed in the case of the Magpie: Jenner (in Philosoph. Transact. 1824. p. 21) relates the case of a pair of Magpies with a nest, of which seven were successively shot, but the widow or widdower was again immediately paired: in another case given by Macgillivray (British Birds. vol. i, p. 571) six females were successively shot on the same nest of eggs. As many nests, especially conspicuous nests like that of the magpie, are annually taken, one may conjecture, that a bird having a nest, offers an irresistible attraction to either sex of a nestless pair, to break their marriage vow.

period when the animal is at full vigour; though no doubt it is of still more consequence that the young should be produced at a time when food is superabundant & the other conditions of life favourable: hence it is in itself highly probable that nearly all animals pair annually or biennially according to the period of gestation. We have seen how great has been the actual increase of horses & cattle, in short periods, though many must have been slaughtered or killed by accidents; & these animals, when compared to the great mass of living beings must be considered as extremely slow breeders: we know the actual rate of doubling of man, a still slower breeder; & we have seen the possible increase of the supposed slowest breeder, the elephant, if allowed to live & breed at its natural rate, even for a few centuries, whereas we have to consider hundreds of thousands of years. Therefore I consider nothing can be/18/more certain, than that every single species on the face of this earth would rapidly swarm to an incalculable degree, if many individuals were not continually destroyed at some period of their lives from the egg or seed upwards, either during each generation or at short intervals in the successive generations.

Checks to increase in animals. What are the checks to this ⟨possible, & as we sometimes see the actual⟩ tendency to a high rate of increase in every living thing? This is a most difficult & curious question, which cannot be completely answered in any single instance. This subject of the Police or economy of nature has been ably discussed by many authors from the time of Wilcke[1] nearly a century ago to the present day when it has been ably handled by Sir Charles Lyell. A volume would be required to treat the subject properly, & I can give here only a few of the leading facts, which have most struck me. The checks are often of a very unexpected nature. Let us look first at our domestic animals/19/ become feral in America, about which we might expect to know most. Though both cattle & horses multiplied greatly in La Plata when left on the desertion of the colony in 1537 to themselves, & ⟨though⟩ subjected to the attacks of Indians; yet at no time have they run wild in Paraguay; & both Azara & Rengger[2] clearly show that this is owing to the greater number of a certain fly, there, which lays its eggs in the navel of the newly born young. In parts of Brazil, cattle can hardly be kept even in a domestic state, whole

---

[1] [Carl von Linné, praeses, H. C. D. Wilcke, respondent, 'On the Police of Nature', pp. 129–64 in F. J. Brand's *Select Dissertations from the* Amoenitates Academicae.]

[2] [Rengger] Naturgesch. der Säugethiere p. 335, & 360. [Felix d'Azara, *Voyages dans l'Amérique méridionale.* I, p. 215 and Azara *Essais sur quadrupèdes*, II, p. 368.]

herds perishing from exhaustion in the dry season from the multitude of ticks (Ixodes) with which they are infested:[1] in another part they failed from the attacks of blood-sucking bats on the calves.[2] In La Plata, where these causes do not come into play, great droughts are almost periodical, & ⟨horses &⟩ cattle of all kinds perish actually by the million, more especially by rushing by thousands into the great rivers, & from drinking saline water.[3] These droughts destroy myriads of wild/20/animals, & even birds, whereas we have seen that during these very same periods mice swarm to an incalculable degree.—I may add that everyone has heard of the terrible destruction of sheep in Australia from the droughts: so it is in India, & Dr. Falconer tells me in places where formerly one man could kill 30 or 40 Deer in a day, for some years after a great famine & drought, hardly a single deer could be got. But to return to the cattle, further south in the Falkland Islands, there are no droughts, or injurious flies, or ticks or bats, & the cattle are magnificent animals & have multiplied greatly; but, as I am informed by Capt. Sulivan,[4] who has kept cattle in these islands, every few years a hard winter like the 1849 destroys numbers, & even those that survive in the following spring are so much weakened that many die of diseases & get lost in the bogs.—The Horses there do not suffer so much from the snow, as their instinct teaches them to scrape the ground with their hoofs; but oddly enough they have multiplied/21/far less than the cattle, & here were left to eastern end of the island; though the western is the more fertile:[5] the Gauchos can account for this only from the stallions constantly roaming from place to place & compelling by kicks & bites the mares to desert their young: Capt. Sulivan can so far corroborate this statement that he has several times found young foals dead, whereas he has never found a dead calf.[6] Horses here deteriorate in size, & they are apt to grow lame from the boggy soil, so climate here, no doubt, aids in checking their increase[7] but the fact of their not spreading seems to show that

[1] Gardner's Travels in Brazil p. 295, 388.
[2] In parts of Demerara Fowls cannot be kept from the same cause, Waterton's Wanderings p. 163, 4th Edit.
[3] Darwin Journal of Researches p. 134.
[4] [Bartholomew James Sulivan, admiral and hydrographer was one of Darwin's shipmates on the 'Beagle'. See ULC vol. 46.1 fols. 17–18 of second numbering, for Darwin's notes, dated March, 1856, on Sulivan's information.]
[5] Journal of Researches p. 191.
[6] [Sulivan MS. letters to Darwin. C.D. MSS. vol. 46.1 fols. 73 v–74, (undated portion of letter) and fol. 81 v from letter dated Jan. 13, 1844.]
[7] It is possible that in this case the Horses' fertility may be somewhat lessened: for in the Shetland Isl'd (Fleming's Philosophy of Zoology. Vol. 2. p. 10) the Pony does not reach maturity till its four year, is not vigorous beyond its twelfth; &

the check falls chiefly on the young. I may add that Rabbits, though very numerous in certain parts of the Falklands likewise have not spread: what the check is here, I have no idea; or what the check is in Jamaica[1] where the Rabbit is feral but has not multiplied.

There can be no doubt that carnivorous animals keep down the numbers of the animals on which they prey. It is worth noticing the number of domestic animals destroyed in single Kingdom./ 21 A/In the year 1823 in Livonia there were destroyed by the wolves 1800 horses, 1800 cattle, 15,000 sheep, 2500 goats, 4000 pigs, 1200 fowls, 673 geese &c &c.[2] The number destroyed, however, must often depend on complex relations: to give a single instance, according to Nillsson[3] wolves have of late increased in Halland & foxes decreased; & this it is believed/22/is chiefly owing to the wolves running down & devouring the foxes, as has often been witnessed; but they can do this only on open plains, so that the proportional increase & decrease of wolves & foxes here depends indirectly on the presence of trees.[4] We are perhaps apt to lay too much stress on the amount of food as determining the numbers of any species; for it seems well ascertained that game in any district, even in this our highly cultivated country, where so few hawks or carnivorous animals are seen, can more certainly be increased by the trapping of vermin than any other means.—But there are some few animals which are probably never, either whilst young or old, destroyed by beasts of prey as the elephant; & yet they do not increase to the extent, which their degree of fertility would soon permit: in this case the check is no doubt periodical famines & droughts which we have seen occur in India; & when weakened they would be very apt to perish in morasses, as seems to have happened with the fossil mastodons of N. America. On the coast of Africa, Capt. Owen[5] gives a curious account of the

breeds only biennially. The dampness of the climate probably is the deteriorating agency, for Wrangell (Expedition to the Polar sea p. 28) states that in the extreme climate of N. E. Siberia, the Horse is serviceable even at 30 years old. With respect to the wild stallions killing their foals, the same thing has been observed in Australia, see Haygarth's Bush Life, p. 76.

[1] Gosse's Sojourn p. 441.
[2] [Anon.] Silliman's Jour. v. 20, p. 177. Rev. encyclop. Sept. 1830.
[3] Lloyd Field Sports of N. Europe Vol. I, p. 395.
[4] A beast of prey must often prevent other animals from haunting districts in which they could live and might prefer.
[5] Surveying Voyage Vol. 2, p. 274 [Contrast with the note on this same passage which Darwin gives in his *Journal of Researches* (1845) p. 133, where he *correctly* quotes Owen as writing: 'A number of these animals had some time since entered the town in a body, to possess themselves of the wells, not being able to procure any water in the country. The inhabitants mustered, when a desperate conflict

sufferings of the/23/elephants, which in a body fairly took possession of a town for the sake of the water & drove out the inhabitants who numbered about three thousand!

I will give a few other instances of checks to increase from
apparently trivial causes. The ferret cannot be kept in W. Indies[1]
owing to the chigo or sort of flea, which burrows in their feet. In
[   ] the half-wild dogs invade each other's districts when
pressed for food, fight & ⟨wound each other⟩ flies lay their eggs
in the slight wounds & cause their death. Everyone has heard how
Rein-deer[2] are forced to migrate in vast bodies & annually perish
in multitudes owing to the mosquitoes. Dean Herbert seems often
to have been perplexed[3] why certain animals do not increase: he
instances the toad, of which such myriads are often seen, showing
that they do not perish in the egg-state, & as no animal preys on
the toad, he asks why they do not increase infinitely: I can adduce
one check, namely a maggot of some fly, which breeds in their
nostrils, & which destroys thousands in Surrey, as I have seen, &
in parts of Kent, as I have been informed by Mr. Brent. But the
Dean might have asked with still more force/24/why the natter-
jack, (Bufo rubeta), which lays eggs enough to people the earth
in a few generations, is confined to a few spots in England, where,
however, it is common as on Gamling-gay Heath. What animals
can seem less concerned with each other than a cat & Humble-Bee;
yet Mr. [H. W. Newman][4] shows that field mice are the most
powerful enemies to the Bee, & the cats determine the number
of mice, as everyone knows in his house, & hence he believes that
Humble-bees are apt to abound near villages, owing to the
destruction of the mice. From the facts given in our third chapter,
I cannot doubt that the number of seed produced by certain
flowers will be determined by the part which Bees play in their
fertilisation; & on the number of seed to a certain extent depends
the number of the plants; & on them the number of certain other
insects & on them certain birds ad infinitum. To attempt to follow
the mutual action & reaction in any one case, would be as hopeless

ensued, which terminated in the ultimate discomfiture of the invaders, but not
until they had killed one man and wounded several others.']
[1] Gosse's Sojourn p. 447.
[2] Wrangell's Travels p. 48. [i.e. Wrangel, *Expedition to Polar Sea*].
[3] [Darwin left a space between brackets here in his manuscript for a reference to
be supplied later.]
[4] [Darwin left blank spaces here in his manuscript for the name of his authority
and for the reference. In the published version he supplied the name H. Newman.
See *The Origin of Species*, 1st ed, p. 74. H. W. Newman, 'On the Habits of the
Bombinatrices', *Entomol. Soc. London, Trans.* N.S., Proceedings Section, p. 88,
1850–51.]

as to throw up a handful of ⟨sawdust⟩ feathers on a gusty day & attempt to predict, where each particle ⟨of sawdust⟩ would fall./

25/This subject is so important for us, that I must be excused for making a few more remarks. Our British Birds are probably the best known wild animals. Take the case of the familiar Hedge sparrow (Accentor modularis), which that acute observer, Mr. Waterton,[1] says will not increase in numbers, however carefully protected. If not killed it could probably live at least seven years:[2] it generally has two broods of about five eggs, but let us suppose that only every other pair rears any young, we will say only two pair. We thus seem to allow a fair amount of destruction at an early age; yet if we suppose that in Mr. Waterton's grounds there were at one time eight pair, the above rate of increase would yield at the end of the seven years, when the eight old pair would die, 2048 birds; but we have just seen, that though carefully protected by man they do not increase at all. It cannot in this case be any difficulty in finding a place for a nest; & I sh'd think hardly more than three out of four nests would be taken by cats; & only one out of four nests are supposed to be preserved/26/in the above calculation. That in many other cases the loss of the nest is a most important check we may infer from the wonderful increase of Magpies & some other comparatively rare birds in Mr. Waterton's park,[3] where in one year 34 pair of Magpies bred & reared 238 young ones. The Hedge sparrow in a garden near a house can hardly suffer much from Hawks & the smaller wild carnivores, which are so influential in checking the increase of game-birds. I doubt whether the young birds, during the first few months suffer greatly; at least with the Robin everyone must have noticed their numbers in their mottled plumage. ⟨& in our migratory birds, as White long ago observed in his letters, the check must fall on the young birds which leave us, for what we imagine to be a more favourable climate, for comparatively few of those which migrate return to us.⟩[4] The domestic cat is I believe a potent enemy, which with other occasional causes of death must prevent any great increase in numbers; but I believe nearly all our Birds

[1] Essays on Nat. Hist. 2 Series p. 95.

[2] In the N. American Journal of Science vol. 30, p. 81. It is said [by J. Bachman] that the same pair of Saxicola sialis built its nest in one place for 10 successive years;—a Muscicapa fusca for 9 years; a Turdus for a longer period; Falco borealis for 12 winters. Eckmarck Amoen. Acad. noted the same home in starling for 8 years [see: Linné, 'On the Migration of Birds']; a Motacilla & Kestrel for 6 years. In Montagu's Ornith. Dict. [p. 217] it is said that a Goldfinch lived in confinement for 23 years.

[3] Essays p. 269. [1st series, essay on Magpie.]

[4] [Selborne, part 2, letter 16.]

do go on increasing/27/in numbers, till there comes a severe winter, which greatly reduces their numbers & sometimes exterminates them in certain districts.[1] After the winter of 1854,[2] judging from the number of nests in my shrubberies & from the number of birds on my lawn, I estimated the decrease at four-fifths compared with previous years. In the summer of 1855, butterflies & moths abounded in an extraordinary manner, which some naturalists at the Entomological Society attributed, I believe rightly, to the lessened destruction of the caterpillars by birds: the little Tomtit (Parus coeruleus) has been observed to feed its young with caterpillars 475 times in the day.[3] With man we consider an epidemic which destroys ten percent as frightful; but in ⟨this⟩ above case with the birds it seemed to me that the destruction had been at least 80 per cent.

With the higher animals, as soon as the young can provide for themselves they are generally driven away by the old: in their forced wanderings many probably perish; but some no doubt find a home, in spots where the destruction has been above the average, more especially after any unusually fatal period. The Rev. L. Jenyns informs me that in Swaffham, during twenty years, sparrows/28/& Rooks were unmercifully killed ⟨for a reward offered per head⟩, but the most careful observer could observe no diminution in their numbers during this period: no doubt the spare birds from the surrounding parishes flocked in; but what would have become of these birds had not there been room made for them in Swaffham? undoubtedly they would have wandered away, some few have found a home & the others have perished during the first severe winter. In all cases, probably, the destruction is unequal in different parts of the whole area inhabited by the species; but this does not alter the final result; Nor is it applicable to the endemic species of small insulated regions: we may go in imagination from spot to spot, ⟨& everywhere the rate of increase is far higher than what can possibly be supported⟩ & we may fancy that here & there the conditions ⟨of life⟩ are so favourable that all survive to their full term of life; but if this be so the destruction must be very heavy in other spots, for, as repeatedly remarked the rate of increase in every living being is so high that

---

[1] [John L. Knapp] Journal of a Naturalist, p. 182.

[2] Severe winters destroy not only the inhabitants of the land, but of the sea; both certain species on the coast, as described by Hugh Miller (Royal Physical Soc. of Edinburgh Feb. 28th 1855), but likewise on banks under the water: thus in 1829–30 Kröyer (Eding New Phil. Journal 1840. p. 25) says eight million oysters were computed to have been destroyed by the frost.

[3] Macgillvray's British Birds Vol. 2, p. 438.

the earth could not hold the product. In animals capable of much locomotion, & inhabiting a continent or the ocean, it is likely that many wander to the/29/extreme confines of their natural range & there perish in larger numbers than elsewhere. But how rarely could this be ascertained!/29v/A pair of sparrows bred for the first time in 1833 in [the] island of Colinsay, one of the Hebrides, but in 1841, no descendants could be seen.[1]/29/Richardson[2] speaking of the extreme northern range of the American Antelope, says that almost every year a small herd lingers on a piece of rising ground not far from Carlton-house; but few or none "survive until the spring, as they are persecuted by the wolves, during the whole winter." So again with Arctic Fox, he says "Most of those which travel far southward are destroyed by rapacious animals; & the few which survive to the spring, breed in their new quarters, instead of returning to the north. The colonies they found, are, however, soon extirpated by their numerous enemies."

In those animals which produce an astonishing number of eggs, the destruction probably chiefly falls on the eggs, as is known to be the case with Fish, from other fish, water-beetles &c. But when the old can protect their young few are generally produced as with the larger carnivorous birds: the Lion, however, produces several young at a birth, but when the/30/Lioness is hunting for food, it is asserted the hyaenas prey on her young. In very many other cases the check falls not on the egg, but on the young: thus Smeathman[3] thinks that "not a pair in many millions" of the Termes or white ant "lays the foundation of a new community," common ants being the chief destroyers. In other cases, of which instances have been given, the very young do not seem especially to suffer: thus White of Selbourne long since remarked in his sixteenth Letter [to Barrington] that in our migratory birds those returning yearly, from what we imagine to be a more favourable climate "bear no sort of proportion to the birds that retire."/

30/A/As in this chapter I repeatedly use the expression of struggle for existence; I may here remark that I employ it in a very large sense./30 A'/Carnivorous animals prowling for their prey in a time of dearth may be truly said to be struggling for existence; so when seeds are sown so thickly that all cannot grow, they may

[1] Wilson's Voyage round Scotland Vol. 1 p. 368. [Wilson here wrote of Stornoway not Colinsay.] I may add that Partridge hens have been turned out here but they became extinct. In 1841 Rooks bred for the first time in this island. Will they hold their own?

[2] Fauna Boreali-Americana [part 1] p. 88, 263.

[3] Philosoph. Transact. 1781, p. 167.

be said to struggle, though not voluntarily against each other. A multitude of animals are directly dependent on other animals & on plants; & plants on the nature of the station inhabited by them; & here the idea of dependency seems quite distinct from a struggle. But a plant on the edge of a desert is often said to struggle for existence; this struggle consisting in the chance of a seed alighting in a somewhat damper spot, & then being just able to live; so it may metaphorically be said that carrion-beetles struggle for existence, when fewer animals die than usual in any district. In many cases when an animal depends on another or on a plant/30a/it destroys or injures it to a certain degree; & here more strictly there may be said to be a struggle. Again another idea comes into play, for it may be said to be chance, which seeds in the capsule of any one plant shall be devoured by a bird or insect, but it may metaphorically be called a struggle which individual plant of the species shall produce most seed, & so have the best chance of leaving descendants;/30a'/& again it may be called a struggle whether the plant or the bird ⟨or insect⟩ which feeds on its seeds gets the upper hand. A minute parasite which is absolutely dependant on an animal, cannot be said to struggle with it; yet its numbers will generally be dependant on the vigour of the animal which it will sometimes injure, & with the increasing vigour of the animal the weaker parasites will perish; so that here there may be said to be a struggle between parasite, & parasite & the animal; as there likewise will be which parasite or which carrion feeding beetle shall lay most eggs & so have the best chance of getting into another animal's body or feeding on its carcass.

I hardly know any living being which is more *dependent* on others, & which seems less subject to a struggle in the strict sense of the word then the Misseltoe; for it depends on certain trees for support, on certain insects for fertilisation, & on certain birds for diffusion; yet even here, when several seeds are dropped close together there must be a struggle which shall grow; there may be said to be a struggle which plant/30 B/shall produce most seeds with most tempting pulp for the thrushes; & lastly there may be said to be a struggle between parasite & tree, for the latter will suffer severely from too many misseltoes. In many of these cases, the term used by Sir C. Lyell of "equilibrium in the number of species"[1] is the more correct but to my mind it expresses far too much quiescence. Hence I shall employ the word struggle, which

---

[1] [E.g. section heading, *Principles*, 9th ed., p. 670 and page heading 1st. ed. Vol. II, p. 134.]

has been used by Herbert & Hooker &c.,[1] including in this term several ideas primarily distinct, but graduating into each other, as the dependency of one organic being on another,—the agency whether organic or inorganic of what may be called chance, as in the dispersal of seeds & eggs, & lastly what may be more strictly called a struggle, whether voluntary as in animals or involuntary as in plants./

30 B'/To return to our subject, it is difficult to realise that every animal is kept down by a severe "struggle"; yet it accords with, & aids us in understanding, much that is passing around us. Lighten the pressure on any one organism in the slightest degree, quite inappreciable by us, & its numbers will instantly increase. Why are some species rare or quite absent in one district, & abundant in another, under, as far as we can judge, similar/31/conditions. Innumerable instances could be given; & several even within the limits of England; as the absence of the Nightingale in Devonshire, water-wagtails (Motacillae) & carrion-crows in certain districts: during 15 years I have only twice seen a swift (Cypselus) in the parish in which I live; yet how common a bird over nearly all England. We can perceive why the sparrow & partridge have increased in numbers in some districts with extended cultivation; but who can explain why during the last 20–30 years the Missel-thrush (Turdus [   ]) has increased in Ireland, Scotland in England, as I have likewise myself noticed. Why did the Robin (Sylvia rubecula) decrease & finally disappear in the year in parts of Belgium. A small wading bird (Pelidna [   ]) has increased of late considerably on the shores of the United States. In New S. Wales as Mr. Sutton stated before the Geographical Society some parrots have greatly decreased, & some disappeared; others equally conspicuous as the white cockatoo have remained in about the same numbers, & others as the Blue Mountain parrot have increased. No doubt if we had accurate accounts in past centuries, we sh[d] have endless cases of great changes/32/in proportional numbers: I will give only a single instance from Prof. Nilsson.[2] A large Bat (Vespertilio noctula) is now common in Sweden, having appeared about the year 1825, & was quite unknown to Linnaeus; but it seems from the bones found in parts of the walls of the

---

[1] [Herbert, 'Local Habitations and Wants of Plants', *J. Hort. Soc.*, 1 (1846), 47. Hooker, *Flora Indica*, i, 41, 42. Aug. de Candolle, article 'géographie botanique', *Dict. sci. nat.*, vol. 18 (1820), pp. 384, 386. Alphonse de Candolle, *Géographie botanique* (1855), p. 453.]

[2] Report Brit Assoc. 1847. p. 79. Prof. Nilsson gives other curious instances: the water wagtail, (Motacilla alba) was very numerous 30 years ago, then it vanished & now again has reappeared.

Cathedral, which it now again haunts that about 700 years ago it was also very common. Lastly it is the common rule, that a species is abundant within what has been called its metropolis, & towards the confines of its range both in longitude & latitude becomes, often rather abruptly, rarer & rarer, till it disappears; & there seems to be no difference in this rule, whether or not the beings be locomotive: yet as it can exist towards the confines of its range, & as its fertility certainly usually then lessened, how is this? In all these cases, namely of a species abundant in one district & rare or quite absent in an adjoining one,—in their increase or decrease in numbers,—we shall feel little surprise, if we steadily look at the average number of every single species in its most favoured site, as determined by a severe struggle, of which in no one case can we perceive/33/all the elements: the merest grain in the balance will then determine whether the range should be lessened or increased.

The manner in which the diverse checks act & react must be exceedingly complicated. When there is no compensation there will be a steady but slow decrease in numbers: thus "the fur-trade even when best managed has always been a decaying trade," & post has to be pushed beyond post into the interior: so it has been with whaling; but how different our game. Neither partridges, or grouse or hares are fed, & yet how many hundred thousands are annually killed with no decrease in the stock: no doubt they could be exterminated as the capercailye has been: with our game man compensates by the destruction of vermin, & he kills many which would otherwise have perished during the winter. Let not a gun be fired or a trap set in England for the next 20 years, & I think it may safely be predicted that there would be less game, almost certainly not more. For instance/34/Bruce remarks[1] that in Abyssinia Boars, ⟨foxes⟩ & Hares are held unclean & are not hunted, but yet they do not increase in numbers; & he accounts for this by the number of Hyaenas; but whether Hyaenas would destroy many hares may be doubted.

Whatever the number of a species in any country may be, the average being determined by a complex struggle, that number will steadily decrease, if we add without any compensation the least additional cause of destruction, until the species becomes extinct. But the rate of decrease will be very slow: if we have 1000 individuals & we destroy on an average ten per cent more every year at the period when the number is least than were heretofore destroyed, it will take 298 years to reduce the numbers

[1] Travels. vol 5, p. 84.

to fifty. But often with the decreasing numbers of the organism destroyed, the numbers of the destroyer will be diminished, & the check thus lessened & its action almost infinitely prolonged. It may well happen that a large additional number of a species might be destroyed without in the least lessening/35/their average numbers; for the destruction may fall before an habitually recurrent period of dearth, which would have in any case thinned their numbers: it is even quite conceivable that such destruction might increase the minimum average, for more food might thus be preserved against the period of dearth, as for instance in dry countries, in which the herbage withers up & serves as natural hay. Many other considerations might have been added showing how complex the action & reaction of the checks to increase must be.

Besides the many & complex checks tending to cause a decrease in the numbers of a species; an inordinate increase, under the most favourable conditions, is prevented in some cases at least, as in our game, by mysterious epidemics, which seem connected we know not how, with the closer aggregation of many individuals of the same kind.—

The great difficulty, which at least I have experienced in fully realising the struggle for life covertly going on around us: I think is partly due to our familiarity with our domestic/36/animals. We see how easily they are reared, how long they live & how seldom they perish from accident; & we overlook our care of them whilst very young & that we artificially preserve food for them & so prevent recurrent famines; but the millions annually slaughtered over Europe, with the stock still kept up, ought clearly to show us what destruction there must be with the allied animals in a state of nature. Nor ought we to feel the least surprise at our not being able to point, how, when & where the check falls on any animal in a state of nature: for the case of man, incomparably the best known, (& in some respects more simple, though in others as in the moral ⟨restraint⟩ check of Malthus or as Laing[1] more correctly calls it the prudential restraint, very much more complicated) shows how ignorant we are. Without careful statistical tables: how little could we have judged of the different rates of increase, & expectancy of life amongst different ranks, at different times, in different countries & even within the limits of the same town.[2]

[1] [See Samuel Laing, *Journal in Norway...*, 2nd ed., 481. (Conclusions at very end of book), and his *Notes of a Traveller*. First Series, 158, (ch. x, section on 'Checks on Over-population'.)]

[2] Mr. Neison has shown (Statistical Soc. March 17th 1845) that in the same town the expectancy of life with mature men of different trades differs by 50 per cent.—

37/*Mutual Checks of Animals & Plants.* We have considered as yet almost exclusively the manner in which animals check the increase of other animals. But plants & animals are even more importantly related; as are plants with plants. This subject is so important for us, in several ways, that I must be excused for entering into some details, but they shall be few. All animals live on plants either directly or indirectly; & their breath is the plants' chief food; so that the relation of the two kingdoms on a grand scale is very obvious. But it is probably much more precise than it at first appears. One at first supposes that grass-eating animals devour all plants nearly alike; but of Swedish plants it has been ascertained[1] that oxen eat 276 kinds & refuse 218; goats eat 449 & refuse 126; swine eat 72 & refuse 271,[2] &c. Southward of La Plata, I was astonished, as others have been,[3] at the change effected in the appearance of the plains by the depasturing of the cattle; & could not for some time believe but that there must have been a change in the geological nature of the country. What plants the many small/38/rodents live upon is seldom known, but every one must have heard of the destruction of whole plantations by mice, & rabbits &c. I have heard it remarked that all, or nearly all our spinose & prickly plants are liked by the larger quadrupeds; the spines being an evident protection to them; & I have sometimes fancied that the very common prickliness of the bushes on desert plains was chiefly due to the greater protection from animals requisite for any bush to live, where the vegetation was scanty. It has, also, been shown in detail by Forskahl[4] that those plants which are not eaten by cattle are attacked in an extraordinary degree by insects; from 30 to 50 species sometimes preying on a single plant: I presume a plant preyed on by both insects & quadrupeds would be exterminated.

I will not do more than allude to the enormous amount of injury, even to extermination, effected by insects on plants; on which subject copious details are given by Kirby & Spence.[5] Land mollusca are, likewise, potent enemies to many plants, especially when young, as every gardener knows: and early on a dewey/39/ morning in what extraordinary numbers they sometimes swarm! In all these cases the relation is obviously mutual: the increase

[1] Stillingfleet Tracts. 1762. p. 361, on authority of Hasselgren in Amoen. Acad. [See Linné, 'Swedish Pan', in Stillingfleet Tracts (where name is mis-spelled.]
[2] [Sic. Stillingfleet *Tracts* p. 361 gives the figure 171.]
[3] Journal of Researches p. 118.
[4] [Linné, 'The Flora of Insects', in Brand's *Select Dissertations*, pp. 361, 366, 367.]
[5] [*An Introduction to Entomology*, I. Letter 6 is on injuries caused to living plants by insects.]

or decrease from any cause of plant & animal mutually affecting each other.

But animals serve plants, as well as destroying them; & in destroying some plants they invariably favour others. In how many ways do they transport their seeds! Even when they devour the seeds if one out of a thousand escapes, it may be of the utmost importance to the plant; of which I shall presently give a curious instance. Though Bees devour much pollen, they are indispensable to the fertilisation of some plants, & generally most useful: different plants are visited by different kinds of Bees; & some by none, but which absolutely require other insects in order to produce seed. Worms I believe[1] play an important part for plants in turning up the ground, & in burying seeds. I have often thought when seeing the quantities of manure collected under the most shady tree in a field during hot weather that even this in the great war of nature/40/might make a sensible difference in the vigour & spreading of a tree: on the other hand, Lieut. Breton[2] says he has known in Tasmania that trees which were flourishing have actually perished as soon as the land was depastured; & he suspects that this is caused by the ground being bared & thus dried.

At St. Helena the upper plains, to an extent of 2000 acres were originally wooded, & it seems pretty well made out that the goats & swine which were introduced[3] in 1502 & soon multiplied, destroyed all the young trees; & that by degrees the old ones perished of age; so that 220 [years] afterwards it is said "the old trees have mostly fallen"; & now the upper plains are covered with grass without a single tree. Some of the trees are known to be now absolutely extinct. In the surface soil, I collected eight kinds of land-shells, now extinct; & their extermination & that of many insects has likewise been in all probability, indirectly due to the goats. To give one more example: near Inverorum [Inveroran?] in Scotland, I saw a whole hill-side covered with young birch-trees so nearly of the same age, that I enquired why so useless a/41/tree had been planted; but was told that about ten years before the district had ⟨been⟩ converted from sheep-pasture into a deer forest; & that sheep devour young birch-trees, but that deer do not. The growth of the birch, would certainly greatly alter the vegetation on the whole bank; & with the plants, the insects would change; & with them, the birds, of which I shall presently give an instance. It is not too strong an expression to say that the

---

[1] Geological Transactions Vol. [   ]. [Charles Darwin, 'On the Formation of Mould', *Geol. Soc. London, Trans.*, 2nd ser. 5 (1840) 505–9.]
[2] Tasmanian Journal Vol. 2. 1843 p. 136.   [3] Darwin, Journal of Researches p. 489.

introduction of a single mammal might change the whole aspect of a district, even to the minutest living details.

On the struggle between plant & plant: the struggle here is not so obvious, but not less certain. Plant does not actually prey on plant, excepting in a few root & branch parasites. Nearly all plants, however, are favoured by the decay of others; and this is indispensable to those which live in peaty earth. In very many cases, also, shade is indispensable or highly favourable: but in plants growing in the shade of others there is some, though perhaps slight reciprocal action, for such plants must rob their protectors of some nutriment;/42/as we see in the greater vigour of our orchard fruit-trees,when the ground is kept bare beneath them. Plants, also, often offer protection to the seedlings of others; & as Stillingfleet[1] has remarked how often do we see a young tree springing out of a furze or thorn bush on a common which has protected it from the attacks of cattle, ultimately to be overshadowed & destroyed by it.

Generally the struggle between plants is like that of those quadrupeds in the same country, which devour nearly the same kind of food. We have evidence of the struggle on a grand scale in the many thousand hardy plants which can be perfectly preserved by simple weeding in our Botanic & common gardens & shrubberies, but which never spread beyond our gardens or spread to perish./ 42 v/Long ago Gouan was in the habit of sowing near Montpellier many foreign seeds likely to grow, several of which succeeded for some years;[2] but Mr. Bentham informs me that he searched in vain, & all are now extinct: the ground here is sterile & bare, & we must suppose the native plants in the long run beat the foreigners in the spots where both could grow./42/It is instructive to observe how frequently foreign plants spring up for a year or two in the rubbish thrown from a garden; but how certainly in a/43/few years, more or less, they are overwhelmed by our native weeds. The foreigners languish, perfect few seeds; & of these seeds, few germinate; & the seedlings are generally smothered./43 v/ Rothof[3] sowed 39 kinds of hardy garden & agricultural seeds on earth thrown out of a ditch in a bog in process of being reclaimed, & only seven came to maturity; eleven seemed capable of ripening their seed; twelve germinated but did not thrive & nine did not

---

[1] Tracts p. 74. [Carl von Linné, praeses, Isaac J. Biberg, resp., 'The Oeconomy of Nature'.]
[2] [Gouan, Antoine, *Herborisations des environs de Montpellier*...(Montpellier, 1796). pp. ix–x, 227–42. cf. Candolle, *Géogr. Bot.* pp. 799–800.]
[3] Acts of Swedish Academy Vol. 4.

germinate./43/In our uncultivated banks & woods, far more seedlings of our native plants spring from the ground, than can possibly come to perfection; this may be conspicuously observed with some of our trees. We see the same fact in our crops; for thin-seeding requires good farming,—that is land with many weeds must be thickly sown, to give the right number a chance of succeeding. In our gardens we can raise common culinary plants with certainty; but sow the same seeds in any number on an adjoining grass field, where there would be nearly the same animal enemies, & you will not raise a plant. Preoccupation of the ground, no doubt, is most influential against chance seeds; but its power has been, I think, sometimes over-rated: all plants in a state of nature undergo a kind of rotation of crops, exhausting one spot & springing up in another, being supplanted & supplanting others: in a coarse meadow the patches of Dactylis &c. which are not browsed, if marked, will be found to change their place; so that if/44/the seed of a plant fitted to overmaster the others, be annually sown it will at last find a proper site. And the many naturalised plants in every land from the even chance seedlings will not rarely intrude on a preoccupied surface. Seeing on what a nice balance of power a plant can become naturalised, it is no wonder that the most skilful Botanist cannot in the least predict, as was remarked to me by Dr. Hooker, what plant will become naturalised in a given country, though he may safely assert that some will not.

No one will question that there is a limit of heat & cold, dampness & dryness, beyond which a plant cannot survive; but it seems that few plants reach this extreme limit. This may, I think, be inferred from what they can endure in our gardens; but more especially as once or twice in a century we have a winter of extreme cold or a very chilly or dry or wet summer; & yet I have not seen any record of a zone of dead plants having been observed towards the confines of their natural range. But what havoc an extraordinary winter will make in our gardens & more especially in our shrubberies! It may be inferred from this, that owing to the struggle between plant & plant, hardly any species reaches/45/very near its extreme climatal limit. In arctic regions & on lofty mountains, where each plant has to struggle against few other living beings, but against severe conditions; zones of dead trees have been observed, as by Ledebour on the Altai, & by Hearne in N. America, who describes a band of dead and blasted stumps upwards of 20 miles in width beyond the living wood.[1]

[1] Ledebour, in Hooker Bot. Miscell. vol. 2, p. 251. Hearne's Journey to the Northern Ocean p. 101.

In the arctic regions & on high mountains very many plants become much stunted; & though I have not met with any precise observations on this head, I think it would certainly have been noticed had this often happened with plants at their lower limits on mountains & at their southern limits in the lowlands: of this latter case I have noticed only one instance,/46/namely the Sugar-maple which in the southern United States is said[1] not to attain above the third of the height which it does in Canada: on mountains, also, I have met with only one instance, namely in the Beech, which is stated [   ] to be stunted below the level of [   ] on the [   ]. Again when the northern range of a plant does not fall near the Arctic regions, it seems seldom to become stunted at its northern limit: as several British plants do not range beyond Northumberland & Durham, I asked Mr. Story to attend to this point for me, & he has sent me a list of 32 plants in this predicament observed by himself & friends; & it appears that only three or four of these are at all dwarfish. Trees,[2] however, seem more commonly to suffer I presume, from being more exposed to the winter temperatures: & several of our British trees become dwarf in Scotland; & so it is according to Kalm[3] with the Sassafras & Tulip-tree in the United States./

47/These several facts are explicable if we look at plants as not actually limited by climate, but by struggling with other plants under conditions beginning to be unfavourable; for the struggle would be severer in proportion to the number of enemies or opposed species, & these would be more numerous on the lower than on the higher slopes of a mountain, & in the southern than in the northern half of our colder temperate regions.

No one has written more forcibly on the struggle between plant & plant than the experienced horticulturist, the Dean of Manchester. Mr. Herbert shows[4] most clearly that those plants which live in sterile & peculiar soils often do not live there or under this or that degree of moisture, because they prefer it, but because they can thus "get a poor livelihood in peace & quiet" and their "enemies cannot grow to choke them." Speaking of some Crocuses confined to sterile hills in the Ionian islands he says that when secured from their native soil & transported into a garden they acquired ten-fold vigour. There are many cases on record[5] ⟨besides

[1] Kalm Travels in N. America Vol. 1, p. 142.
[2] Alph. De Candolle Geograph. Bot. p. 72.
[3] Travels in N. America Vol. 1. p. 142–Sir G. Grey in his Expedition Vol. 2, p. 262 says that the Xanthorroea, though not a tree, declines in health & growth in proceeding northward.
[4] Journal of Hort. Soc. on the local habitation of Plants vol I. p. 46.
[5] Alph. De Candolle Géograph. Bot. p. 428, 453, 455.

the striking/48/ones given by Herbert⟩ of the same species growing
in very different situations in different countries,—as Herbert
instances the Orchis monorchis & militaris in England on chalk
Banks, & in reed-beds on the edge of Lake of Brienz,/48 v/the
common milkwort (Polygala vulgaris) in England on dry upland
pastures, in Zante on alluvial & very moist meadows./48/—Such
cases are probably in main part though not exclusively due to
other plants more vigorously occupying the sites in one or both
countries which the species in question would most enjoy; for
with so flexible a constitution there would be few sites on which
such plants could not exist.

To show how one plant can influence others, & like-wise many
animals, I am tempted to give one very common case. In Staffordshire on the estate of a relation, where I had ample means of
acquiring all particulars, there was an extensive barren heath,
never touched by the hand of man; but on one side several hundred
acres had been planted about 25 years before with larch & Scotch-
fir, nothing whatever having been done, except small holes having
been dug, & the whole enclosed. The effect on the native vegetation
was quite remarkable in the very great change in the proportional
numbers of the plants found on the Heath; & in the presence of
12 species (not counting grasses & carices to which I did not
attend) not growing/49/on the Heath; of these twelve, three had
never been observed elsewhere in the neighbourhood by a relative
who had attended pretty carefully to the botany of the district.
The change in the insects must have been even greater; for six
insectivorous birds were extremely common in the wood & were
not to be seen on the Heath; where two or three other insectivorous
species lived, but did not frequent the plantations. I was interested
by one particular: young oaks were springing up of all ages by
hundreds, in parts at the distance of a mile from any oak-tree,
here & there actually appearing as if they had been sown broad-
cast; but I was assured that this never had been the case; & the
woodmen told me that there was not the least doubt how they
came there; that they had *repeatedly* seen rooks dropping acorns
in their flight across the woods: there was no rookery near, & the
line of flight would take the birds across the heath where there
were no oaks, so that this ⟨curious &⟩ most efficient means of
dispersal must have been wasted for centuries, until the decay of
the leaves of the fir-trees & the growth of other plants had made
a bed on which the acorns soon after being dropped could germinate.
I have given instances/50/to show what an effect the introduction
of a single quadruped can indirectly produce on the vegetation

of a country; & here we see that the introduction of a tree, with no other change whatever, can produce as great an influence on other plants, birds & insects.

Make the ground quite bare, as on a railway cutting, & it may be almost said to be chance by what plants it will be at first covered, being dependent on the nature of the soil, the kinds of plants growing near, the means of diffusion & number of their seeds & the direction of the wind; but in a few years, notwith-standing that the number of the seeds of the first occupants will probably have been increased a million-fold, the proportions will greatly change, & ultimately become the same as on adjoining old Banks. Many curious accounts have been published of the change of vegetation when a N. American forest has been burnt or cut-down & then left to nature. This has been called rotation; & it seems pretty clear[1]/51/that in our meadows & woods, when not suddenly destroyed that there is a real rotation, like that followed by farmers & probably dependent on the same causes, viz chiefly exhaustion of the various chemical elements in the soil required in different proportions by the different families of plants. The same principle probably comes into play in causing the beautiful diversity of plants in our meadows & woods: the good farmer every fifth or seventh year plants the same crop on the same field; but nature raises her crops altogether in exact proportion to what the soil can support, each kind slowly changing its place, with this great difference that she is not the determined enemy of any bird, insect or slug, & cares not what or how many plants overmaster the others. But when a forest is burnt down, whilst still in full vigour, & a very different vegetation, as is invariably the case, springs up, it seems doubtful whether this should be called rotation in the above sense; the change would rather appear to be due to what seeds are ready in the ground, or quickest brought there; on the rate of growth of the seedlings & their immunity from animal attacks. In these cases, the trees/52/ reassume in the course of ages the same beautiful variety in the same exact proportions as in the surrounding virgin forest: this has been noted in many parts of the world, as over the ancient American ruins in Central & North America.[2] On how many & complex contingencies must this wondrous battle prolonged over centuries have been determined by which each species has recovered its rights!

[1] Alph. De Candolle. Géograph. Bot. p. 448, 472.
[2] An Enquiry into the Origin of the Antiquities of America by J. Delafield. [p. 55 seems most apt.]

It is indeed a wonderful conflict, on which I cannot cease marvelling. Causes appearing to us most trifling are potent. In the Staffordshire Heath formerly alluded to, a small portion had been broken up & attempted to be cultivated, for two or three years; but had utterly failed & was planted with fir trees at the same time with other parts of the heath; & 25 years afterwards, the under-growth was so different that the lines of separation could be most easily traced. In walking over the most barren heath where four or five plants held absolute sway, I have often been surprised to see a line of turf along small pathways: is this owing to the heath being mechanically destroyed? or do/53/animals follow the paths & occasionally, though rarely drop a little manure? Manure may be directly injurious to the Heaths; but I have noticed in a neglected field of my own, that manuring caused a marked decrease in the hard-heads (Centaurea nigra); yet this plant certainly likes manure, but the more vigorous growth of other plants must have checked its increase. In this same field I have observed in different summers, an obvious difference in the proportions of the several plants; showing how rapidly a slight change in season allowed one species to increase over another. So again in old meadow land, which has been ploughed years ago, the same species may be observed in the slightly damper furrows & slightly dryer ridges, but in different proportions: in this (& other such cases) there can be no doubt that the plants growing both in the furrows & on the ridges, could for a time cover the field, if all the other plants were exterminated, but that having to struggle with other plants, the slightest difference in dampness, determined the proportional numbers in either case./

54/The old divine Jeremy Taylor says, "Tell me why this turf this year brings forth a daisy, & the next year a plantane."[1] No one can answer. But let it not be called chance. The chemist may throw a dozen salts into solution & may hope to predict the result; the naturalist cannot do this with the living beings dispersed by ten thousand ingenious contrivances all round him; but when we see the virgin forest reassuming its beautiful variety apparently in the same exact proportions, over the ancient Indian ruins, we must see how little of what we call chance has to do with the final result. This struggle, this war of nature, becomes only in the least degree intelligible to us, by keeping steadily in mind that each plant would cover the ground for a period if left to its natural powers of increase; for no one will doubt if four-fifths of our British plants were suddenly exterminated, the remaining fifth would soon decently clothe the land./54 v/One may wonder why

---

[1] [Taylor, Jeremy: 'Of Modesty', in *Holy Living*.]

any one or half-dozen of the most vigorous plants in England, annually producing thousands of seeds, growing in all sorts of ordinary stations, existing here in the middle of their range & therefore well capable of bearing somewhat more heat & cold, damp & dryness, why such plants do not monopolise the whole surface. But assuredly/54/every single plant, even the most vigorous & predominating in its nature, is habitually destroyed in multitudes at some period of its life from the seed upwards, either annually or at recurrent periods, by means, which we very/ 55/seldom can perceive; the only difference between the weak & strong being that proportionally to the number existing at any time the weak one has been destroyed during former generations, / or has been prevented increasing, more than the one called strong.

In considering the facts now given, & many similar ones known to any naturalist, one caution is perhaps necessary. Although certainly the most different organisms very often act & react on each other in the most complicated way; yet from such cases exciting our surprise we may perhaps be led to attribute too much to this mutual action from remote parts in the polity of nature. That part of the complex term struggle for existence, which is more correctly expressed by dependency, generally relates to organic beings remote in the scale of nature; & individuals of the same species are hardly ever dependent on each other, excepting in their sexual, parental & social relationship. But we have seen how dependency graduates into a struggle for existence. On the other hand that part of the idea, more correctly expressed by the word struggle, applies in its fullest force between individuals of the same species. When we remember that individuals/55 a/of the same species, whether animals or plant, live on nearly the same food & are exposed to the same dangers & difficulties, it is in itself probable that the struggle will be here most severe at some period of life. Probably it will be nearly equally severe between the individuals of two varieties, when they meet, & secondly between closely allied species or between organisms, however different in structure, if they have nearly related habits & encounter each other. /55 a v/What can be more remote than a locust & a ruminant quadruped, yet they must often powerfully affect each other. In the cases of rare species, having few individuals thinly scattered, we may infer that the struggle, as far as organic beings are concerned, is chiefly with ⟨other⟩ distinct species ⟨or conditions of existence⟩./55 a/And lastly the struggle will often be very severe with the external conditions of existence independently of the co-inhabitants of the district.

We have some evidence how powerfully allied species affect each

199

other: every one has heard how the Norway Rat has exterminated the Black Rat under the most different climates & circumstances of all kinds from the Polar circle[1] to within the Tropics, in the New & Old world: in New Zealand[2] the Black Rat had previously almost expelled a previously introduced species: in Färoe[3] "the decrease of the mouse has been in proportion to the increase of the Rat," so that the common mouse, which was the earlier inhabitant, has been almost exterminated./55 b/Even with varieties of our domestic animals it has been found by experience[4] that other breeds of sheep cannot exist on the mountains of Cumberland with the Herdwick breed, "for they stand starving best." If one species of Swallow were to increase we might expect that other Swallows would suffer more than other Birds; & so it seems to be, for with the late curious increase in parts of the United States of the Hirundo fulva, the Barn swallow has decreased.[5] When the red-legged Partridge increases, the common Partridge decreases; so it has been observed with the Pheasant & black-grouse. Again Fish with allied habits must chiefly affect fish; & thus the shad (Clupea sapidissima) has increased in the Hudson, in parts full twenty-fold, owing to the erection of a dam, & the consequent decrease chiefly of another species of Clupea.[6] In Russia the small Asiatic Cock-roach (Blatta asiatica) has everywhere driven before it the great cock-roach.[7]/55 c/ The [   ] Leech exterminates the [   ] when placed in the same pond. And to go to the other extreme of the scale how fatally does civilized man cause the extermination of savage men.

I have said that the struggle is often severe between organic beings & their conditions of existence, independently of the co-inhabitants: this chiefly holds good on the confines of life, as in the extreme arctic regions or on the borders of a desert like the Sahara. When animals & plants actually perish from cold or drought, there cannot be said to be any struggle between the individuals of the same species; but between the constitution of each & the destroying element. But more generally, the cold or drought for instance, kills by lessening the food, & then there may be most truly said to be a struggle between the individuals of the same species or of species with allied habits. To give one instance to show how during such periods one variety may indirectly

---

[1] Lloyd Field Sports of N. Europe vol. 2 p. 321.
[2] [Darwin left a blank space here for a reference he never supplied.
[3] Landt Description of Faroe p. 213.
[4] The Northmen in Cumberland by R. Ferguson p. 22. 1856.
[5] Dr. Brewer in N. American Journal of Science. vol. 38 p. 392.
[6] Mr. Adams in N. American Journal of Science Vol 20. p. 150.
[7] Pallas Travels in Russia vol. i. p. 16.

master another: in La Plata, during/55d/the great drought, the cattle perish chiefly from famine & the Niata breed would be utterly exterminated, if not protected, for from the peculiar shape of the jaws they cannot feed on twigs of trees so well as the common cattle when all the dried up herbage has been consumed;[1] but if there were no bushes whatever in the country probably the Niata cattle would pass through the ordeal as well as the common breed; both with greatly reduced numbers.

Hence, I think, we may conclude, that as a general rule, the struggle for existence in its strictest yet never simple sense is most severe between the individuals of the same species, & next between the individuals of two distinct varieties, or species, or even classes if their habits are somewhat allied. In all cases, the struggle being ruled & modified by multiform relations./

55e/*Facts apparently opposed to there being a severe struggle in all nature*:—I will now give the few cases which alone have seemed to me to throw doubt on the struggle for existence. Perhaps the most striking is the existence of species, even locomotive species as mammals, confined, without any physical barrier & with no difference in conditions appreciable by us, to a very small locality, but there very abundant; for it might be argued that if there be such a power of increase, & as the species is abundant in the locality in question, showing that the conditions of its existence are there favourable, why does it not spread.—Many instances in all classes could be given of facts of this nature: Mr. Bentham has often insisted to me, how remarkable it is that certain plants should be found in a single spot, as the Pyrenees & no where else in the world; & should there be abundant; & therefore apparently not like a species on the point of extinction.[2] Some local species have been known to exist in the same place/56/for one or two centuries.[3] But by far the most remarkable case of this nature on record, is that of certain species & even varieties of land-shells in Madeira & P. Santo, are positively stated by Mr. Wollaston[4] to swarm on

---

[1] [In the MS. the caret for the insertion of this final clause, added between the lines, is placed thus: 'feed ∧ on twigs.']

[2] Bartram in his Travels (p. 466) speaks of "a singular and unaccountable circumstance" namely that he found a Franklinia (Gardenia) alatamaha growing plentifully over two or three acres in E. Florida, but that he never met with elsewhere. Mr. Wollaston (Variation of Species p. 153) gives plenty of cases of common insects, though extremely local insects, in Madeira.

[3] Al. De Candolle. Geograpn. Bot. p. 471.

[4] On the Variation of Species p. 132. Helix Wollastoni is one of the most striking cases, & the *varieties*, as so considered by Mr. Wollaston, of H. polymorpha obey the same law.

certain hillocks on these islands, where they are also found fossil, & that they occur no where else either fossil or recent in the whole group, which has been thoroughly well investigated. The superficial calcareous beds in which these very local land-shells occur, include a few extinct species, & I am informed by Sir C. Lyell that the island has undergone considerable change since their deposition: hence we must conclude that these land shells, each on its own site, has swarmed probably for several thousand years, & yet have just held their own place & have never spread!/

57/In cases like these latter in which each district has a representative species, filling as far as we can perceive the same place in the economy of nature, the difficulty is, perhaps, not quite so great as it at first appears; for let us take one of those common land shells,/57 v/which we positively know, from the extraordinary numbers occasionally appearing in favourable seasons, can rapidly increase, & is therefore habitually kept under by checks of some kind; & let us suppose it to inhabit two points [,] hillocks a few miles/57/apart, I should think that probably the inhabitants of those two hillocks were the lineal descendants of the first colonists, without having in many cases been at all intermingled; for although no doubt the checks would fall much heavier at some times on the inhabitants of the one hillock than on the other; yet if they were not wholly exterminated on the one, the rapid power of increase common to these & almost all the lower animals, together with their slow power of travelling, would allow the survivors of the hillock which had suffered most to breed up their numbers before they could be invaded by the inhabitants of [the] other hillock, though they would be to a certain extent by the inhabitants of the intermediate low land; but during another season the lowlands might be invaded by highlanders. The result would be different with slow breeding animals having rapid powers of travelling as with birds, or plants having seeds easily blown by the wind. Thus far I can admit, the weight of /57 bis/slow diffusive progress, to which Mr. Wollaston[1] attributes so much importance. The result would, also, be very different if the land-shell inhabiting one hillock was a variety having the smallest advantage over the individuals in the intermediate tract & on the other hill, for then it would surely spread; but in the Madeira case we may suppose that each species or variety long inhabiting its own hill is at the very least as well adapted to the conditions (I do not mean mere climatal conditions) there occurrent as to the conditions of the other hill.

Those cases in which a plant is absolutely confined to one

---

[1] Variation of Species p. 125, 130, 153.

small area, & is there very abundant, without close representative species in other adjoining districts, seems to offer more difficulty. On a less striking scale, the same difficulty is often encountered, namely in plants being very abundant on one spot, but not found anywhere/58/else in the district or even Kingdom, & yet without any perceptible difference in the conditions: These, however, are exceptional though not very rare cases, the common rule apparently being that very local plants or animals[1] are not numerous in individuals. But the fact which has struck me the most, is that given by Alph. De Candolle, that some few "social plants" are social[2] to the extreme limits of their range, or are not thinly scattered as might be expected, & when consequently we must suppose that the conditions have begun to be unfavourable. If social plants could help each other like some social animals, from which the term social has been borrowed, there would be no difficulty, for then as far as they could range, they would range in company. But there seems to be no essential difference, only one in degree, between a social plant, & one numerous on any one site. Al. De Candolle has shown in his admirable discussion on this subject,[3] that most social plants are thus inhabiting peculiar or unfavourable sites as salt-marshes, heaths, arctic regions, beneath water &c., & where consequently as only few plants can grow there peculiarly adapted plants grow together in great numbers. Hence, also, in islands, inhabited/59/by only few species, they are very apt to be social; as they are wherever the conditions are very uniform. But the fact which has seemed to me to show that there is no essential difference between very common plants & social plants, is that some naturalised plants are social in their adopted country,—as is eminently the case with the cardoons & thistles on the plains of La Plata, & not, as far as I can make out, in their native home. Nevertheless it seems to me that many plants, both those commonly called social, & those abounding in numbers in some one spot & not elsewhere found in the neighbourhood or even in the whole world, may be said, in a somewhat strained sense, to help each other, so that if they did not live in numbers, they could not live all.—

It follows from the doctrine of the struggle for existence that every plant is checked in its increase in the seed, seedling, or mature state. For simplicity let us suppose in any plant that the *main* check falls on the seed, owing to its being devoured by some

[1] Al. De Candolle. Geograph. Bot. p. 470.
[2] Geograph. Bot. p. 462. M. De Candolle instances the Cistus & Lavenders &c on the plains in the south of France: some alpine plants: & forests of trees in the Arctic regions.   [3] Geograph. Bot. p. 460.

bird or insect: the argument will be just the same if applied to the seedling & we/59a/suppose a great loss by slugs or other animals. We must bear in mind that in all probability that this will not be the *sole* check; a certain percentage of seed, for instance, perishing by not getting buried &c. Now from a thousand/60/ plants of the same kind growing together, there will be a far better chance of many seeds being preserved than from a dozen plants,— that is as long as the increase of the bird or insect which preys on the seed is checked by some other agency & is not determined by the seed of the plant in question: if with the increase in seed the numbers of its devourers increased in the same ratio, then it would make no difference in the proportion saved whether there were a thousand or a dozen plants; but if the devouring birds or insects could not thus increase, owing to the want of food in winter, or owing to being preyed on by other animals &c., & this would very often be the case, then there would obviously be more seed saved from the thousand plants than from the dozen.—We see this often practically illustrated; a farmer notices a peculiar ear of wheat, & plants the seed in his garden, but it is notorious that without he carefully protect his dozen wheat plants, he will hardly save a seed owing to sparrows: I have seen this occur & in the same year: I raised some hybrid Radishes & ⟨with all sorts of protection⟩ had the greatest difficulty/61/in saving a few seed out of thousands of pods from the attacks of another bird, the green-finch.—Yet in a large plot of seed Radishes or in a field of wheat, plenty of seed can be secured. Beyond a doubt, there would be great difficulty in a small colony of radishes or wheat establishing itself in my garden, supposing that they could sow themselves. In animals we have seen the same thing occur in small colonies of foxes & antelopes naturally establishing themselves as described by Sir John Richardson, in N. America, though these instances occurred near the limits of their range.

Another & quite distinct cause may come into play in deter-mining that a social plant could not exist beyond the limit in which the conditions were so highly favourable, that large numbers could grow together: in dioicous plants there must be at least two individuals near each other, & if the fertilisation of the plant be due to the wind, & not to insects, bearing in mind that they will be planted by chance, it seems almost necessary that there should be a good many together in order to be thoroughly fertilised & produce their full complement of seed.[1] Now we have seen in

---

[1] I have previously shown in our third chapter that many trees are dioicous & monoicous, & they are apt to be social.

the third chapter that there is good reason to believe that many/ 62/plants are what Sprengel called dichogamous; & when the fertilisation is not aided by the voluntary flight of insects, these could seed well only when growing in masses: I believe many Grasses are in this predicament, namely depending to a great extent on other individuals for their fertilisation; & are not visited by insects; & grasses are commonly social.

From these two considerations, more especially the first one, (& it is likely there are other considerations overlooked by me) I think we can to a certain extent see why a plant/62 v/may, or rather must, exist socially in numbers together, even near the confines of its range, if it can exist at all: we can, also, see why a plant or animal may exist in/62/large numbers in one spot & not spread; for when once established in numbers it might escape destruction by its enemies, but when thinly scattered in colonies, ⟨owing to the severe struggle going on⟩ all might easily perish. Hence this fact which seems at first paradoxical, & is so if we look chiefly to climatal or soil conditions as of predominating influence, ceases to be paradoxical when we look at all organic beings as periodically struggling for existence with their utmost energy against their enemies. Authors have often spoken of the occupation of the soil, as a powerful/63/element in distribution: in the strict sense of the word, if we remember that plants undergo a natural rotation & that seeds are disseminated in a multitude of ways, I think it can have very little influence: in the sense above given, namely that plants or animals when once established in numbers, by their very numbers escape destruction, I have no doubt this occupation is potent.

Another class of facts seemed at one time to me opposed to there being a severe struggle in nature; namely animals having recovered in a state of nature from severe injuries, as evidenced by the fossil Hyaena[1] which had part of its upper jaw entirely worn away; or by the famous Mylodon described by Owen with a fractured skull. Mr. Couch caught a cod-fish with no eyes, yet in good condition.[2]/63 v/Mr. Blyth mentions two nearly blind Indian crows; but these very singularly were fed by other members of the flock.—Rengger describes rickety Jaguars with short legs as not very uncommon in Paraguay./63/Lame birds have been noticed for several years building in the same nest. Birds, more especially rooks, have not very rarely been observed with their upper & lower mandibles crossing & distorted; & this has been observed even in the case of a /64/Woodpecker (Picus erythro-

[1] [Buckland] Phil. Transact. 1823. p. 85.   [2] Transact. Linn. Soc. Vol. XIV p. 72.

cephalus) which one would have thought would have most severely suffered from such a malconformation.[1] All these cases show only that the struggle for existence is periodical & not incessant, of which fact we have plenty of other evidence: in the first very severe winter the rooks with the crossed bills would no doubt be cleared off.—

⟨In some cases the term struggle is not very appropriate; for instance in the Misseltoe (Viscum   ); as it can hardly be said to struggle with any other beings, though evidently *dependent* on them: if it increased in an inordinate degree it would greatly injure the few trees on which it can grow: it would probably be actually exterminated if the Thrush genus which it helps to feed became extinct; & Kölreuter has shown that its fertilisation is dependent on certain insects:[2] probably deficient means of dispersal is a principal check in this case.⟩/

65/Finally I must allude to an opinion, which I have repeatedly seen advanced, but probably without deliberation;—namely that the numbers of any species depend on the number of its eggs or seed, & consequently not on a struggle for existence at some period of its life or its parents' lives./65 v/This belief has probably arisen from the larger animals, which can seldom be supported in very great numbers in any country, producing few young; but most of them can protect their young; nor is this relation invariable, as we see in the Crocodile, & amongst Birds in the ostrich./65/The number of the eggs is no doubt one element in the result but by no means one of the most important. How many rare fish there are existing in very scanty numbers, yet annually producing thousands of ova! Years ago I was struck with this in finding a large sea-slug (Doris) at the Falkland Isl[d], very rare & yet on calculating the number of the eggs of one individual, I found six hundred thousand. The Condor lays only two eggs & yet in parts it is quite as common, (for I have seen between twenty & thirty take flight from one cliff) as the American Rhea, which lays between twenty & forty eggs & even more: but we need not go so far, the Kitty-wren, (Sylvia troglodytes) lays on an average just twice as many eggs as the other British wrens or Sylviadae, yet we see no corresponding relation in numbers.[3]/65 a/The Picked

---

[1] Mr. Blackwall Researches in Zoology has collected several cases, p. 173–6: see also [Sheppard & Whitear] Transact. Linn. Soc. Vol. 15, Part I, p. 9.

[2] [Koelreuter (Erste) Fortsetzung (1763), p. 72.]

[3] [On the verso of the manuscript sheet, fol. 65, ending here Darwin wrote: 'Put a remark that fertility is most important in *rapidly* increasing but not in final results. This is crucial difference. In the ultimate number no doubt other elements are far from unimportant.']

Dog-fish (Squalus acanthias) actually swarms on many coasts & yet is said to lay only six eggs; whereas the Cod-fish sometimes lays above three million & a half.[1] Again many Diptera increase at such a rate, that Linnaeus has stated that three flies of Musca vomitoria would devour a horse as quickly as a Lion:[2] yet there are other flies, which produce only a single egg, or rather pupa, ⟨at a birth & probably⟩ in their whole life, and yet such flies/66/ are by no means rare, as all who have had their horses tormented by the horse-fly, (Hippobosca) must well know. Amongst plants, I have looked through lists, in which a few of the most abundant plants of a country are marked, & have often noticed amongst them the bearers of the fewest seeds. But the most conclusive evidence of all may be derived from fossil tertiary shells; we have numerous cases of a shell formerly rare & now common in the same region, or the reverse case; & I presume no one will imagine that these shells laid a different number of ova at the two periods. There is an old Eastern fable that the locust lays ninety-nine eggs, that if it laid the hundreth it would overrun the world; this fable is probably as false as it is old.

Upon the whole none of the facts, which seem at first to deny that all organic beings have at some period or during some generation to struggle for/67/life are of much weight; on the other hand the several remarks & illustrations given in the foregoing pages, imperfect as they are, appear to me conclusively to show that such struggle, often a very complex nature, does truly exist. I have found myself that much reflexion is necessary fully to realise this struggle & dependence of one being on another: our great ignorance of the complete biography of any one single plant or animal makes us slow to believe in the multiform & often extremely obscure checks to their increase. Look at any piece of wild ground, & notice that hundreds, often thousands of seeds annually produced by each plant & disseminated by a hundred ingenious contrivances; —think of the number of eggs produced by each insect, worm & snail,—each animal strives to live, each plant will live if it can,— & yet the average number cannot possibly long increase: go from spot to spot, till you reach the confines of life, & the same story is

[1] Yarrell British Fishes vol. 2. p. 401; Fleming's Philosophy of Zoology vol. 2 p. 356.
[2] [Darwin probably remembered this from Lyell's *Principles of Geology*, for in his copy of the 9th ed. (1853) an X is marked in the margin of p. 673, where this statement is attributed to Linnaeus (on the authority of Kirby and Spence— *Introduction to Entomology*, see 1815 ed. I, 250—who give no source reference) whereas the same assertion in Erasmus Darwin's *Temple of Nature* (London, 1803), Additional Notes IV, p. 17, is not marked in Darwin's copy. Linnaeus added this statement in the '12th edition' of the *Systema Naturae* (Stockholm, 1767) to his entry for *Musca vomitoria* (see Tom. I, pars II, p. 990).]

predetermined. Everywhere, the rate of increase,/68/if unchecked, will be geometrical; whilst the means of subsistence on the long average will be constant; & we know in our slow-breeding larger domestic animals, how large & rapid the result of this ratio has been in an unstocked country. We must regret that sentient beings should be exposed to so severe a struggle, but we should bear in mind that the survivors are the most vigorous & healthy, & can most enjoy life: the struggle seldom recurs with full severity during each generation: in many cases it is the eggs, or very young which perish: with the old there is no fear of the coming famine & no anticipation of death. Philosophical writers, such as Lyell, Hooker,[1] Herbert &c. have most ably endeavoured to make others appreciate the struggle & equilibrium of life, as clearly as they do themselves; & I should not have discussed this subject at length, had it not been in many ways of great importance for us; & had I not occasionally met with good observers of nature, who by such remarks, as that the number of the individuals of a species was determined by the number of its eggs:—or that when an island partly subsides into the/69/ocean, it will become (as if not already) crowded in an extraordinary degree with living beings,—show as it seems to me, an entire ignorance of the real state of nature. Nature may be compared to a surface covered with ten-thousand sharp wedges, many of the same shape & many of different shapes representing different species, all packed closely together & all driven in by incessant blows: the blows being far severer at one time than at another; sometimes a wedge of one form & sometimes another being struck; the one driven deeply in forcing out others; with the jar & shock often transmitted very far to other wedges in many lines of direction: beneath the surface we may suppose that there lies a hard layer, fluctuating in its level, & which may represent the minimum amount of food required by each living being, & which layer will be impenetrable by the sharpest wedge./

70/*Corollary on the relation in Structure of organic beings.* It follows almost necessarily from what we have seen of the struggle for existence, dependent on the habits of animals & plants, that the structure of each organic being stands in most intimate relation to that of other organisms. For habit generally goes with structure, not withstanding that in most great families, a few species having the same general structure can be picked out with habits in some degree aberrant. It is very important in order, as I believe, to

---

[1] Hooker & Thomson Flora Indica: see the remarks in Introduction p. 41. [cf. notes 2, 3 p. 175, and 1 p. 188.]

understand many facts in geographical distribution, the steps towards extinction, & the principle of natural selection, fully to appreciate how intimately visible structure, by which we discriminate species from species & genus from genus, is related to the structure of other organic beings. Obviously every living being has its constitution adapted to the climate of its home; but this seems to produce scarcely any visible difference in structure:/70 v/ thus in every kingdom we have a few species keeping identically the same structure under the most opposite climates—look at Poa from Equator to T. del Fuego, up to limit of snow in Cordillera./ 70/Thus species of such tropical genera as the Elephant & Rhinoceros, inhabited during the glacial epoch very cold countries, with no essential difference in organization; for their woolly covering however important for their habits cannot be /71/looked at as an important difference in structure. It has often been noticed that many tropical families of plants send out one or two species, having of course the structures of their family, into the cool temperate regions; on the other hand, such northern genera as the Rose & willow have each a species inhabiting the hottest plains of India.[1] I presume that many highly succulent & vascular plants are so far related to a hot climate that they could not exist where severe frost would burst their textures; but it would seem that much caution is required in drawing all such conclusions. For instance seeing the vast number of Heaths at the Cape of Good Hope, & hearing[2] that every family of modest size, even leguminous & compositous plants, there have some & often many species with heath-like foliage, it would appear a safe induction that heath-like leaves were related to a dry & moderately hot climate; yet our heaths inhabit damp & cold mountains. We find animals & plants/ 72/inhabiting the most abnormal stations, as hot & sulphureous springs & deep caverns into which a ray of light never penetrates, & yet not displaying any great difference in structure from species of the same genera inhabiting ordinary stations.

Whether an animal or plant lives, breathes or moves on land, air or water certainly influences the structure in a most important manner; but even in these cases there is a secondary & perhaps equally important relation to the coinhabitants of the same element. Whether an animal feeds on vegetable or animal food, plainly influences structure, though here the relation is between organic beings, either alive or dead, & often of a special nature. Moreover if we run over in our mind the various structures of the commoner

[1] Hooker, Himalayan Journal vol. 2. p. 255.
[2] Drege & Meyer. Zwei Pflanzengeograph. Doc. Flora 1843 B. 2 p. 26.

animals, we shall see that the manner of obtaining their prey or food & of escaping danger from other living beings is almost equally influential on their structure.

As the relation of plants in structure to other organic beings is not so obvious as in animals, I will briefly run through the life of a plant in/73/the abstract, & which will serve as a summary for parts of this chapter. Beginning with the flower, which has its dangers from flower-feeding beetles &c, I cannot doubt from the facts given in our third chapter, that the beauty of the corolla, the scent in night-bloomers, the positions of the nectary & of the stamens & pistils to each other stand in many cases in direct relation to insects of special genera & classes. When the seed is matured, animals in multitudes prey on it; & it will escape destruction by its size, hardness, defences, chemical nature or mere number. Its dispersal in some cases depends partly on hooks or on agreeable pulp: even the down of a thistle is perhaps important to it, in as much as the ground is thickly covered by other plants & thickly sown every year: under this same relation to other plants, the period & rapidity of germination will be all important. So again the amount of nourishment surrounding the embryo within the seed, we may believe is given to certain plants that in their earliest days they may succeed in struggling with other plants. The seedling has its special enemies as has the mature plant, which/74/sometimes defends itself from animals by prickles, more often by its chemical composition, & which often gains the day over other plants by rapid growth or mere height, at the same time protecting & shadowing other plants, & feeding them with its decayed leaves.

One set of plants will allow another set to live only on some bare chalk banks, though not perfectly suited to them; but the relation of different plants to each other growing on the same plot of ground must be equally important. Cut a piece of turf & look at the inextricable mass of roots, each growing rapidly in the line where it can find food: it is like a battle between voracious animals devouring the same prey. The power of each plant in an entangled mass to get its food apparently will depend on their different periods of activity & on the depth & manner of growth of their roots. Each plant requires certain inorganic bases & a certain amount of moisture; but this in many cases will depend as much on other co-existing plants as on the nature of the soil; for even with regard to moisture one sees in hot summer how the grass though shaded is often dried up under a tree. To give one example,—/75/the turnip can beat many weeds ⟨from overshadowing them⟩ by its rapid growth, & so as farmers say cleans the ground; this rapid growth, I may add, apparently stands in relation to the enormous destruction which this plant suffers during its early state alone from the Haltica & saw-fly. The turnip is said to contain but a small percentage of the salts of phosphorus, yet farmers find it adviseable to give it phosphate of lime, owing to its rapid growth, rather than to wheat which has ultimately to assimilate a larger percentage of phosphorus, but is a slower grower. So that amongst plants struggling together in a soil very poor in phosphorus, it is quite possible that one requiring much phosphorus might beat another requiring but little of this substance.

From these several considerations I think we may safely conclude that a plant or animal if naturalised in a new country, under exactly the same conditions of climate & soil as in its native country; but associated with a different set of organic beings, would in fact be generally placed under quite as new conditions as if the climate had been somewhat modified. Under an extremely different climate it would not/76/become naturalised. It would probably be quite unimportant to the naturalised organism, whether the greater number of its compatriots were to it new or old forms; those which stood in some relation to it would alone be important, & then in the highest degree; & these influential forms might be as different as possible in the scale of nature, but more commonly those having somewhat similar habits & therefore often systematically related would be the more important. We may put the case in another point of view; let us in imagination wish to alter the structure or constitution of any being so that its numbers might increase: on the confines of its range we should have to change its climatal constitution & in doing this we should not have, judging from analogy, much to alter its structure; even in the midst of its range, as we see the proportional numbers of the inhabitants of a country are changed according as the season is wet or dry &c., we might in some cases increase its numbers by a similar change: always having to do this without deteriorating in the slightest degree its multiform relations to the other inhabitants of the same place. But these relations are so numerous, so complex & so important that we may believe that it would/77/ probably be easier to make some slight change in structure in respect to the other co-inhabitants in order to allow its numbers to increase. How totally ignorant we are how this could be effected, we shall immediately perceive, if we ask ourselves what we should alter. In the case of single species of a Family or Order inhabiting a country, or in such cases as the Misseltoe, we can perceive that the altered structure would have to stand in relation to beings systematically far removed: we may imagine a greater power of penetrating the bark of the apple, or the berries being rendered more attractive to birds might aid the misseltoe to increase in numbers but it is all the wildest conjecture. Very commonly the altered structure would have to be in relation to nearly allied forms & here the difficulty of imagining a favourable change of structure is even greater. Would mere increase in size & strength prevent the black rat from yielding in so many quarters of the world to the Norway rat: this is quite doubtful, at least the great size of the occidental Blatta has not saved it from its puny Oriental congeners. What change could we make in the Barn-swallow of the United States to allow it to withstand the inroads of the allied Hirundo fulva? And so we may continue to puzzle ourselves in infinitely numerous cases./

78/I have discussed this subject at some length, for it seems to me most important under many points of view, that we should fully realise our ignorance, & never forget, that though the constitution of each being is necessarily related to the climate of its country, yet that not only in animals but in plants, much, probably far the greater part of the structural differences between species & species stands in the most direct yet generally unperceived relation to the other organic beings of the same country.

CHAPTER VI

ON NATURAL SELECTION

INTRODUCTION

Darwin's Pocket Diary records two periods of work on his chapter on natural selection, namely March 1857 and the spring of 1858. The first draft, completed on March 31, 1857, was written on sheets of gray wove foolscap which are distinguishable from the bluish gray paper used for the later interpolations and revisions. The outline of this original form of the chapter appears in the original table of contents made before the later revisions:

Aside from one comment on a separate slip of paper dated June, 1858 (fol. 53 A, printed here as note 1, p. 263), the dating of the additions and revisions of the manuscript is uncertain. The valuable clarification on folio 21, 'By Nature, I mean the laws ordained by God to govern the Universe', Darwin might have added above the line at almost any time, after he had completed the lines where it is inserted. Still probably most if not all the additions and revisions were made sometime in the period between April 14 and June 12, 1858 which according to his Pocket Diary, Darwin devoted to 'Divergence and correcting ch. 6.' Darwin had mentioned divergence in the original draft of the chapter but only briefly, whereas in the revised version he devoted over forty new pages to this topic. The following table gives the numbers of all the pages or scraps whose pale blue gray colour indicates they were written later than the original draft:

12 A	37 v	49
13 v	38	51 (bottom half starting
16v	38 v	'slowness of selection')
17 v a & B	39 A	to 62
26*	40	64 to 76.
26a to 26 nn	40 A	

The revision of the later part of the chapter added to its length so that the original folio 58 later became folio 77, and (after cancellation of a final sentence running on to the original folio 59 and the addition of a new concluding sentence) now ends the chapter.

ON NATURAL SELECTION

[completed March 31, 1857]

1/How will the struggle for existence, which we have discussed in the last chapter, act? Annually during thousands on thousands of generations, multitudes have been born more than can survive to maturity. The least possible weight will turn the balance which shall live & which die. Look at the young in the same litter or nest, something must determine which shall live & procreate its kind. If two beings were absolutely alike in all respects, during the whole course of their lives, it might be truly said to be chance, which of the two should come to maturity & procreate their kind. But such absolute identity can hardly be predicated of any living beings; & certainly, as has been seen in the fourth chapter, there is a considerable amount of variability in nature. A large proportion of the variation, which does occur, may be quite unimportant for the welfare of any particular organism, & such variation would not in the least be affected by the struggle for existence. On the other hand, any variation, however infinitely slight, if it did promote during any part/2/of life even in the slightest degree, the welfare of the being, such variation would tend to be preserved or selected. I do not say that it would be invariably selected, but that an individual so characterised would have a better chance of surviving.

If we reflect on the infinitely numerous & odd variations in all parts of the structure of those few animals & plants, on which man may be said to have experimentised by domestication, & again on the many, though slight variations which have been noticed in a state of nature, it would be most strange if in the course of thousands of generations, not one variation added to the welfare of some varying organic being; in thinking of this we should bear in mind how multifarious, singular, & complex the relations for each living being are in habits & structure to other organic beings & to climate, both for securing food & escaping many dangers, during the various stages of life. Again we should bear in mind that whole treatises have been written, showing what numerous, what trifling, what strange peculiarit[i]es are inherited, or tend/3/ to be inherited, that is appear in some of the offspring or reappear in their descendents. An individual, therefore, which from having some slight profitable variation, was preserved or naturally selected, would in many cases, tend to transmit the new, though slight modification to its offspring. Moreover the causes, which

214

from their extremely complex nature we are forced generally to call mere chance, which produced the first variation in question would under the same conditions often continue to act; & assuredly these causes would be eminently likely to act on individuals having some inherited tendency ⟨however slight⟩, in this same direction: so that the cause of the variations & inheritances would act & react on each other, thus giving fresh & fresh opportunity for natural selection to seize on & preserve whatever modification of structure habit, or constitution, was in any degree useful./

3 v/On the other hand, any modification if in the slightest degree injurious would be rigidly destroyed. In the struggle for existence, during the long course of generations, individuals thus characterised, would have a very poor chance of surviving. Even if the injurious modification from the nature of the conditions, or from a strong principle of inheritance, appeared again & again, it would be rigidly rejected again & again./

3/I can hardly imagine any change in structure habits &c so slight that it might not be useful to an individual/4/of a species, & hence be selected. It seems at first to be simple chance which individual insect shall fall a prey to a bird; yet birds are guided by their eye-sight; & we so often see leaf-eating insects are green or those living on bark, mottled-brown, we may believe that a slight change in the shade of colour, might in the long run cause such individuals better to escape destruction & leave offspring with the same inherited tint. Colour is thought an unimportant character by naturalists; but when we see as it has been fancifully[1] said that "the ptarmigan is lichen in summer & snow in winter, that the red-grouse is heather, & black-grouse peaty earth"; & when we remember the main check to the increase of our game birds, is owing to birds & beasts of prey, I can see no reason to doubt that in birds varying in colour as does the red-grouse, that the finest tints of colour might be selected owing to such individuals suffering less. Such selection would perhaps/5/the more readily be effected with birds & insects when they invaded a new district, or slightly changed their habits, which certainly occurs, as we see with insects attacking our exotic plants. I observe in many German & French pigeon-books, that people are cautioned not to keep

---

[1] See, also, some good remarks on the colour of those birds giving them a better chance of escape from birds of prey by Mr. C. St. John in his Tour in Sutherland-shire, Vol. 2, p. 179. [As Prof. John L. Brooks has pointed out to me, Darwin's next quotation about the ptarmigan derives from Blyth's article, 'An attempt to classify the "Varieties" of Animals...' *Mag. Nat. Hist.* 8 (1835), 51. Note that Darwin's statement that 'the ptarmigan is lichen in summer' derives from Blyth's phrase, 'lichen rock', rather than from the 'mossy rock' of the original statement by Robert Mudie.]

white pigeons, as they suffer much the most from hawks. Nor let it be said that the occasional destruction of individuals of a particular colour could have no influence on the colour of the whole body; for it is well known how effective is the destruction of any lamb with a tinge of black in keeping the flock pure.

Again take a beast of prey, pressed for food owing to the destruction by a dearth of the animals on which it feeds; what a trifle will determine which shall survive; the least superiority in power of scent, a shade of colour so as to be less conspicuous, (I have noticed that a prowling white-piebald cat is far easier seen by birds than a tabby), the power of springing an inch further may well determine its success,/6/when life depends on success: in such cases one meal lost may be the turning point; here it may be truly said that the last straw breaks the camel's back./6 v/And success will depend not only on the vigour of the moment, but often on the condition in which the animal has been able to keep itself during several previous months./6/Or again look at the surprisingly large annual destruction of shrews by cats either by mistake or for sport, as shown by the number found killed but not devoured on our gravel-walks: supposing for the moment that this destruction is a main check to the increase of shrews, may we not believe that an individual born by chance with an inheritable stronger odour, & so a little more repugnant to the prowling beast of prey would have a better chance of escaping; & from this *individual* others still more offensive might be selected, till a shrew was formed with an odour as insufferable to man & beast, as that of some foreign allied animals.

A sudden or great variation most rarely, some will/7/say never, occurs in nature; but if it did, & were profitable it of course would be selected; but small modifications, let them appear ever so trifling, if in the least influential on the welfare of the being, I can see no reason after the most careful consideration to doubt would tend to be preserved or selected. They would, also, tend to be inherited; & slight modification might thus be added to slight modification in any given direction useful to the animal;—just as in our domestic animals & plants modifications useful to man have been added together & rendered permanent by artificial selection./

7a/Natural selection may act at any time of life; for variations appearing at one period tend, as we have seen, to reappear at the corresponding period: thus peculiarities in the caterpillar or coccoon of the silk-moth are inherited; & any modification in a caterpillar or coccoon useful to it, might be naturally selected & made permanent/7a v/and so it might be through however many stages of existence, or alternations of generations to use Steenstrup's expression,[1] any animal may pass. Thus also,/7a/the embryo might be modified by selection in relation to the mother's womb: in Yorkshire according to that excellent writer Marshall[2] big-buttocked calves were selected, until they were found to destroy many cows during calving, & thus a deviation of this kind, if left to nature, would be soon eliminated: on the other hand if this deviation were useful in any way to the embryo, or to the calf after birth, no doubt in the course of time the parental structure might become modified by selection to allow of such births; for facility in parturition is undoubtedly hereditary./

7a v/In the Tumbler pigeons, the beak has been rendered so short by long-continued selection, that Mr. Eaton[3] says he is "convinced that better head & beak birds have perished in the shell than ever were hatched, the reason is that this amazingly short-faced bird cannot reach the shell with its beak, & perishes in the shell if the Fancier does not extricate it." But by long-continued selection a shell thinner at the right end might be naturally obtained, for we know that the eggs [of the] common Hen often vary in thickness./

7a/So again any modification in either sex separately, whether useful to that sex alone, or in functional relation to the other sex, or to the flock or to the young, might be selected & become attached to that sex alone; for/[Folios 7b to 7k are missing]

8/creation of each living thing endowed with a small limit of variability, or with the theory of a great amount of slow modification; & it will be the object of this work in the latter chapters to make this comparison. But for the present, in order to explain my principles, I must assume that there is no limit or no close limit to variation during the long course of ages.

From what we have seen in the first chapter the main cause of variability seems to lie in a change of the conditions of existence, perhaps aided by abundant food. That many countries have suffered great changes during the same geological period, that is within the period of existence of the majority of the same species, no geologist will dispute. Reflect for a moment on the vast changes of climate & of the level of the sea, during the glacial epoch,— a mere sub-division of one geological period. Now let us/9/take the case of a country subjected to some climatal or other change;

[1] See the wonderful facts given in Steenstrup's most interesting work, translated and published by the Ray Society.
[2] [*Rural Economy of Yorkshire*, II, 183.]
[3] A Treatise on the Almond Tumbler, p. 33.

the proportional numbers of its inhabitants will be altered & organic beings better adapted to the new climate will flow in from the surrounding countries, as they certainly did into Europe during the glacial epoch. But if the country were cut off by some impassable barrier, from the adjoining warmer or colder or dryer countries, as the case might be, or if one supposed country was an island, then new beings could not immigrate, & fewer of the old inhabitants would be exterminated for there would not be new beings to take their place; the majority would suffer & then would decrease in numbers; but some few, which were previously just able to reach so far south (supposing for the moment that the change was from warmer to colder) under the new conditions would be favoured & would increase in numbers. Bearing in mind how intimately each organism is related to other organisms, & even to the proportional individual numbers of each, for one organic being in large numbers may well be far more influential for good or evil to another, than if in small numbers, there can, I think, be no doubt/10/that in our imaginary country the selections of nearly every inhabitant would be seriously disturbed both by the change of climate, & more especially by the changed proportions of the other inhabitants & by destruction of some few./10 v/Each being would be placed under conditions, such as the world had never exactly seen before./ 10/Moreover the changed conditions ⟨of existence⟩ would tend to make some of the organic beings more variable than heretofore. Under such circumstances, it seems to me that it would be quite extraordinary, if in some few at least of the slightly varying organisms, no profitable variations better fitted for the new & complex combination of conditions occurred./10 v/A very slight modification would often suffice to give some advantage between the struggling inhabitants; for we have before seen, that the severest struggle, leading even to the extermination of one, often lies between closely allied & therefore very similar species of the same genus.—If any such profitable modification did/10/occur, I cannot doubt but that it would be slowly though steadily selected; & the variety thus selected would gain strength & increase in numbers.

Under the above circumstances, which though imaginary must repeatedly have occurred in the world's history, the conditions would probably be most favourable for some rapid selection & consequent modification of forms; nevertheless I think we may conclude that there does not exist a land in which the process may not be going on slowly. Everywhere organic beings present individual differences,/11/& some few more marked variations. No country can be named in which all the inhabitants are perfectly adapted to its conditions of existence: this may seem a rash assertion, but I think it can be fully justified. Each being in its native country no doubt is adapted to its conditions of existence as perfectly as the other coinhabitants, in proportion to the average number of the individuals of its kind; but not one country, still less not one island can be named which does not possess many organic beings naturalised thoroughly well as far as we can judge.[1]/11 v/M. Alph. De Candolle has insisted strongly on this fact of the universality of naturalised plants & has drawn the foregoing inference from it. The number of naturalised plants in Europe & N. America is probably in great part due to great changes effected by agriculture; but I think Sir C. Lyell has shown that the action of man on other organic beings though more potent, does not differ essentially from that of any other animal when introduced naturally into a new country.[2] In the case of many plants naturalised in the uncultivated parts of many islands, man has probably in no ways influenced the conditions. No one will assert that the existence of the cat, rat &c in New Zealand, of introduced monkeys in Cape de Verde Isl[d]—of horses & cattle in La Plata &c &c is due to changes effected in the natural state of their countries through man's intervention./11/Now does this not show, that in the natural polity of each land there were places open, which could be filled by other beings more perfect, not by any ideal standard, but by actual proof, in relation to the previous inhabitants & to the climatal conditions of that land? Nor let it be said that individual differences are so slight that the most careful selection could make no sensible change by adding them up during a long course of ages; for man, even during mere scores of years, has certainly thus acted on differences so slight/12/as to be inappreciable except by an eye long educated. Therefore I conclude that there is no land, so well stocked with organic beings, or with conditions so unvarying, but that in the course of time, natural selection might modify some few of the inhabitants & adapt them better to their place in the great scheme of nature. I may here add that hereafter we shall show good reason for believing that it is not the oppressed & decreasing forms which will tend to be modified, but the triumphant, which are ⟨increasing in numbers, extending their range, & coming into new relations,⟩ already very numerous in individuals, widely diffused in their own country & inhabiting many countries, which are most variable & so will be most apt to be modified & so become under new forms still more triumphant./

[1] [Pencilled addition:] so has Bunbury in Linn. Transact. [21 (1854) 188–9.]
[2] [*Principles*, 9th ed. (1853) p. 664.]

12 A/*Illustrations of the Action of Natural Selection.* In order to make it clear how I believe natural selection acts, I must beg permission to give one or two imaginary illustrations./12/Let us take the case of a wolf, which preys on various animals, securing some by craft, some by strength & some by fleetness; & let us suppose that the fleetest prey, a deer for instance, had from any change whatever increased in numbers, or other prey had decreased in numbers during that season of the year, when the wolf is hardest pressed for food; I can under such circumstances see no/⟨12⟩13/reason to doubt that the swiftest & slimmest wolves would in the long run be preserved & selected; always provided that they retained strength to master their prey at this period or some other period of the year when compelled to prey on other animals./13 v/I can see no more reason to doubt this, than that the Breeder can greatly improve the fleetness of his greyhounds by long-continued & careful selection./13/The same process would tend to modify the deer in order to escape the wolf slowly rendered fleeter; though it might happen that some other & incompatable modification might be more important to this animal, as getting food during some other season. Even without any change in the proportional numbers of the animals on which the wolf preyed, a single cub might be born with an innate tendency either of instinct or structure leading it to pursue certain prey; nor can this be thought very improbable seeing that of our cats, one naturally takes to catch rats & another mice, & according to the excellent observer Mr. St. John[1] one to bring home winged game, another hares & rabbits, & another to hunt on marshy ground & almost nightly to catch woodcocks & snipes, how if any innate slight change of habit or structure benefitted our wolf, it would be more likely to survive & procreate many young, than the other wolves; & some of its young would/⟨13⟩14/probably inherit the same tendency, & thus a new variety might be formed, which would either supplant or coexist with the parent form. Or again with our wolves, those inhabiting a mountainous district might readily be led chiefly to hunt different prey from those on the lowlands; & from the continued selection of the best fitted individuals in the two sites two varieties might slowly be formed, which would, cross & blend where they met, but to this subject of intercrossing we shall soon have to return; I may add that according to Mr. Pierce there are two varieties of the wolf in the Catskill Mountains in the ⟨United States⟩,[2] one with a light grey-

---

[1] Wild Sports & Nat. History of the Highlands. 1846, p. 40.

[2] ['A Memoir on the Catskill Mountains...', *Amer. J. Sci.*, 6 (1823), see p. 93.]

hound like form which pursues deer, & the other more bulky with shorter legs & which more frequently attacks the shepherd's flocks.

If the individual numbers of a plant depended chiefly on the wide dispersion of its seed, so that some might fall on a proper site, any plant which had its seed furnished with pappus a little better adapted to be wafted; or with pulp more agreeable to Birds, would have a better chance of being dropped where it could germinate & reproduce its kind; & I can see no reason why nature should not thus select the most dispersable seed than that gardeners should be able to go on selecting varieties having more & more differences in seed, pod, or fruit./

15/Let us now take a more complex case: some plants excrete a sweet juice apparently for the elimination alone of something injurious from their sap, as in the case of the glands at the base of the stipules of some Leguminosae; & this juice is greedily sought by insects. Let us suppose the juice to be excreted at the inner bases of the petals, & insects in seeking the juice would be apt to get dusted with pollen, & carry it on to the stigmas of other flowers of the same kind, & so cross them: this, as we have every reason to believe, would make more vigorous seedlings which would have the best chance of surviving; & some of these seedlings would probably inherit the nectar-excreting power; & those individual flowers which excreted most nectar would be most visited by insects, & oftener crossed, & so in the long run would gain the upper hand. In order to increase the amount of nectar, the nectaries & with them the petals might become modified, as well as the position of stamens & pistils in relation to the particular insect which visited the flower; some insects like ants being of not the slightest service to the plant; others as Bees being very useful in fascilitating intercrosses. We might have taken for our example, insects devouring pollen instead of nectar, & as pollen is formed for a definite object its destruction appears at first a simple loss to the plant; yet if a little was occasionally or habitually carried to another plant, owing to the visits of the pollen-devouring/16/insects, & a cross thus effected, although nine-tenths of the pollen were destroyed, it might still be a great gain to the plants; & those individuals which produced more & more pollen & had larger & larger anthers would be selected. Indeed this process of selection of larger & larger anthers might be carried on, merely that some of the pollen might escape destruction, without any indirect advantage being gained by the pollen being robbed, in the same manner as many plants probably produce thousands of seeds, in order that a few may escape destruction.

When our plant had by natural selection been rendered so attractive to insects, that unintentionally on their part they regularly carried pollen from flower to flower; & how effectually they do this, ⟨the result of Kölreuter's artificial fertilisation of flowers, the same number being left to insects, clearly shows⟩ I could easily show by many striking facts; then another process might commence. No naturalist doubts the advantage of what has been called the "physiological division of labour"; hence we may believe that it would be an advantage to a plant to produce only male organs in one flower or one whole plant, & only female organs in another. If then an individual plant tended to fail, in either sex in the different flowers of the same individual, or on all the flowers on different individuals; nor does this seem/17/very improbable,/17 v/as it can be shown that the two sexes in the same flower are sometimes rendered sterile in different degrees, when the plant is exposed to changed conditions of life, & as we see in/17/nature how many gradations there are between dioicous, monoicous & polygamous plants; then if this incipient division of labour profited the plant in the least degree, it might be increased by natural selection, until one plant had separated sexes.

Lastly let us turn to nectar-feeding insects in our imaginary case: let us suppose that the plants of which we have been slowly increasing the nectar by continued selection was a common plant, & that certain insects depended in main part on its nectar for food. Now/17 v/I could give many facts, showing how eager Bees are to save time, & to visit flowers as rapidly as possible—for instance their habit of cutting holes at the bases of flowers, which they can enter with a little trouble—bearing this in mind,/17/I can see no reason to doubt that an accidental deviation in the size or form of the body, far too slight to be appreciated, or in the curvature or length of the proboscis &c might profit a moth, fly or Bee, so that an individual so characterised would more rapidly obtain food & so/18/have a better chance of living & leaving descendents with a tendency to a similar slight deviation of structure./18 v/For instance the tube of the corolla of the common red & tall incarnate clovers do not on a hasty glance appear very different in length; the Hive-bees can easily suck the nectar out of the latter, but not out of the common red clover; so that whole fields of the plant offer precious nectar on which the welfare of the community depends, in vain to our Hive-bees. On the other hand I have elsewhere experimentally shown that the fertility of clover depends in the closest manner on the visits of Bees, which by moving parts of the corolla push the pollen on to the stigmatic surfaces./18/Thus

I can understand how a flower & Bee might slowly become either contemporaneously or one after the other modified & adapted in the most perfect manner to each other.

I am well aware that the doctrine of natural selection exemplified in the above imaginary examples, is open to the same objections, which were at first launched out against Sir Charles Lyell's noble views on "the modern changes of the Earth, as illustrations of geology", but we now very seldom hear the action of the coast-waves, for instance, called a trifling & insignificant cause, as applied to the excavation of gigantic valleys or to the formation of the longest lines of inland cliffs. In our imaginary examples, it may be observed that natural selection can act only by the preservation & addition of infinitesimally small inherited modifi-cations each profitable to the preserved being; but as modern geology has almost banished such views as the excavation of a great vally by a single diluvial wave, or cataclysms desolating the world, so will Natural Selection, if it be a true principle, banish the beliefs of/19/the continued creation of new organic forms, & of any subsequent, great & sudden modifications in their structures.

We must add to the effects of natural selection, the *direct* action, probably very small & almost certainly slow, of climatal conditions; —we must not forget, & I believe this to be of very wide application, that in the modification of one part, either during the same or during an earlier period of life, other parts will be altered according to the complex & unknown laws of the correlation of structures, for instance a selected modification of the larva would almost certainly influence mature forms;—we must allow something in the higher animals for the effect of habit & disuse, of which again the action must be always slow;—but over all their causes of change, I ⟨fully believe⟩ am convinced that Natural Selection is paramount.

Comparison of nature's selection with man's selection. From the facts given in the two first chapters, it cannot be doubted that man can do, & has done, much in the modification of animals & plants by the artificial selection of variations. But he labours under great disadvantages: he selects only by the eye & acts therefore on external characters alone: he cannot perceive slight constitutional differences; nor the course of every/20/nerve & vessel: he can by no means tell whether all parts & organs are correlated perfectly, but only so far that life & tolerable health are preserved. Far from allowing each being to struggle for life; he protects each to the utmost of his power, both during youth & times of dearth & from all enemies. Instead of selecting steadily from generation to generation, he only occasionally selects; & his judgement is often bad or capricious: he & his successors never go on selecting for the same precise object for thousands of generations. Even when most carefully selecting he sometimes grudges to destroy an animal, imperfect in some respect, as it comes up to his standard in some other respect. Each being is not allowed to live its full term of life & procreate its kind, according to its own capacity to exist. He does not always allow the most vigorous males to be the fathers of their breed. He often begins his selection with some striking abnormal form, differing widely from anything observed in nature, & of no use to the being selected. From migrations, changes of agriculture &c, he often unintentionally changes the conditions to which his products are/ 21/exposed; or intentionally crosses them with individuals brought from another district or country, as was done in the darkest ages. He selects any peculiarity or quality which pleases or is useful to him, regardless whether it profits the being & whether it is the best possible adaptation to the conditions to which the being is exposed: nor does he regularly exercise the selected peculiarity: he selects a long-backed dog, or long-beaked birds & trains it to no particular course of life;—he selects a small dog or bird & feeds it highly;—a long limbed animal & exercises its fleetness only occasionally or not at all like the Italian greyhound. And lastly, to repeat, he can judge by external characters alone, & not from the perfect action & correlation of the whole organisation during the whole course of life.—

See how differently Nature acts! By nature, I mean the laws ordained by God to govern the Universe.[1] She cares not for mere external appearance; she may be said to scrutinise with a severe eye, every nerve, vessel & muscle; every habit, instinct, shade of constitution,—the whole machinery of the organisation. There will be here no/22/caprice, no favouring: the good will be preserved & the bad rigidly destroyed, for good & bad are all exposed during some period of growth or during some generation, to a severe struggle for life. Each being will live its full term & procreate its kind, according to its capacity to obtain food & escape danger. Nature will *never* select any modification without it gives some advantage to the selected being over its progenitors under the conditions to which it is exposed. Every selected change will be fully & regularly exercised. Nature will not commence with some half-monstrous & useless form; but she will act by adding up deviations so slight as to be hardly or not at all appreciable by

---

[1] [This sentence was added above the original line of text.]

the human eye. Natural conditions remain constant for enormous periods, or generally change very slowly, so will the consequent variability be slight, & the selection very slow. Nature is prodigal of time & can act on thousands of thousands generations: she is prodigal of the forms of life,/23/if the right variation does not occur under changing conditions so as to be selected & profit any one being, that form will be utterly exterminated as myriads have been. No complications are too great for nature: a contingency happening once in a thousand generations may lead to the extermination of a variety: she can gradually select, either simultaneously or successively, slight changes adapting the selected variety to a score of other beings, most widely apart in the great scale of nature.

Can we wonder then, that nature's productions bear the stamp of a far higher perfection than man's product by artificial selection. With nature the most gradual, steady, unerring, deep-sighted selection,—perfect adaption to the conditions of existence,—the direct action of such conditions—the long-continued effects of habit & perfect training, all concur during thousands of generations. Here we meet with no hereditary useless monsters. All who have reared animals & plants believe that trueness is dependent on long-continued & careful selection, & on exposure to the same conditions. How incomparably truer, then, must nature's varieties, called by us species/24/when strongly marked, be, when compared with the varieties reared by man. Now trueness or the absence of variability, is the most important characteristic mark of a species in contrast with a variety, second only to the sterility of hybrids, & not second to this in the eyes of some, as Gaertner & Herbert whose studies would naturally have led them to attribute the greatest importance to the laws of breeding. If we admit, as we must admit, that some few organic beings were originally created, which were endowed with a high power of generation, & with the capacity for some slight inheritable variability, then I can see no limit to the wondrous & harmonious results which in the course of time can be perfected through natural selection.

It may, perhaps, be here worth notice, that amongst barbarous nations, there will be little intentional/25/selection, & the animals in great degree will be left to struggle for life without aid under conditions nearly constant; & it has been remarked that in such cases the breeds approach much more closely in character to true species, than amongst civilised nations.

Seeing what man has done in a few thousand years, I have sometimes wondered that nature considering the perfection of her means has not worked quicker, than geology teaches us to believe she has in the modification of organic beings. But from what has gone before, & from what will presently follow, we may see that there are most powerful retarding agencies always at work.

The forms produced by natural selection, if quite modified, will be called species, if only slightly different, will be called varieties; if no further variation occurs in the right direction by which the variety may be further profited, I can see no reason why a variety may not remain in that state during an enormous lapse of years/ 26/; & we have seen in the fourth chapter, that some varieties such as the land-shells in the calcareous superficial beds of P. Santo certainly are of high antiquity.

But that a variety should remain constant during whole geological periods is excessively improbable; for we have seen in our 5th Chapter in how important a manner the structural differences of each organism is most intimately related to those of the other coinhabitants of the same district; & as all these are struggling for supremacy, & will hence constantly *tend* to be modified & become improved, if one variety be so fixed as not to vary at all in a fitting direction, & so become through natural selection adapted to those other changing organic forms to which it is related in the polity of nature, it will be exterminated./[1]

26*/*Extinction.*—The general subject of extinction will be discussed in a future chapter on palaeontology. But extinction must be here noticed, as bearing in a very important manner on the theory of Selection. As man in any country improves his breeds, he neglects the less improved & these gradually disappear. Hear Youatt[2] on the cattle of northern Yorkshire: at the commencement of the 18th century the ancient black cattle were the only breed. To them succeeded the long-horns, which by degrees spread over the whole northern & midland counties; but much valued as they were, they were after a time "swept away, as if some by some strange convulsion of nature". For they had to give way to the short-horns, & these for the last century have maintained their ground; & no doubt will do so, until some better breed be formed, if better can be. So it has been with innumerable varieties of our cultivated plants; "old sorts being fairly beaten out by new & better ones."[3] Thus it has been, & thus it will be, with man's

---

[1] [See appendix for short cancelled passage. Fol. 27 is gone, replaced by fols. 26 and 26 nn.]

[2] Cattle. Library of Useful Knowledge, 1834, p. 248, 199.

[3] [Anon.] Gardener's Chronicle, 1857., p. 235.

productions. In nature, the same species existing in two now separated areas, might become modified in one or both, & the resultant forms might continue, whilst/26 a/separated, to exist for any length of time. Such forms are often called by naturalists representative or geographical species, races or varieties: they are maiden knights who have not fought with each other the great battle for life or death. But, whenever from the union of the two areas, they meet, & come into competition, if one has the slightest advantage over the other, that other will decrease in numbers or be quite swept away. But as we see in a vast number, perhaps in a large majority of cases, that the varieties of the same species, & the species of the same genus, inhabit the same country, or divisions of it not separated by impenetrable barriers, generally the varieties as well as the species will have come into competition with each other & with their parents from an early period or even from the very commencement of their formation; and as a form can be selected by nature solely from having some advantage, at least in the spot where the selection is going on, over its parent form; the parent will be almost infallibly there exterminated by its own offspring.

Hence, we may, I think, safely conclude, that/26 b/natural selection (like man's selection) almost necessarily entails a nearly proportional amount of extinction;—one species whilst forming beating out another, & one even the finest variety, if having any kind of advantage over another, taking the place of & exterminating the less favoured & less modified variety. It is in each country, a race for life & death; & to win implies that others lose.

Principle of Divergence.—This principle, which for want of a better name, I have called that of Divergence, has, I believe played a most important part in Natural Selection. To seek light, as in all other cases, by looking to our domestic productions, we may see in those which have varied most from long domestication or cultivation, something closely analogous to our principle. Each new peculiarity either strikes man's eye as curious or may be useful to him; & he goes on slowly & often unconsciously selecting the most extreme forms. He has made the race-horse as fleet & slim as possible & goes on trying to make it fleeter; the cart-horse he makes as powerful as he can: he selects his Dorking-fowls for/ 26 c/weight & disregards plumage; the Bantam he tries to get as small as possible, with elegant plumage & erect carriage: a pigeon has been born with slightly smaller beak, another with slightly longer beak & wattle, another with a crop a little more inflated than usual, another with a somewhat larger & expanded tail &c; his eye is struck & he goes on selecting each of these peculiarities, & he makes his several breeds of improved tumblers, carriers, pouters, fantails &c, all as different or divergent as possible from their original parent-stock the rock-pigeon; the intermediate, & in his eyes inferior birds, having been neglected in each generation & now become extinct. It is the same with his dress, each new fashion ever fluctuating is carried to an extreme & displaces the last; but living productions will not so readily bend to his inordinate caprice./26c v/⟨Moreover, far more *fancy*-pigeons will be kept, (I do not mean those kept as food) after they have become broken up into very distinct breeds, than when fewer & more similar birds existed; for each fancier likes to keep several kinds, or one fancier keeps one kind & another becomes famous for another breed.⟩

26 c/Now in nature, I cannot doubt, that an analogous principle, not liable to caprice, is steadily at work, through a widely different agency; & that varieties of the same species, & species of the same genus, family or order are all, more or less, subjected to this influence. For in any country, a far greater number of individuals descended from the same parents can be supported, when greatly modified/ 26 d/in different ways, in habits constitution & structure, so as to fill as many places, as possible, in the polity of nature, than when not at all or only slightly modified.

We may go further than this, &, independently of the case of forms supposed to have descended from common parents, assert that a greater absolute amount of life can be supported in any country or on the globe; when life is developed under many & widely different forms, than when under a few & allied forms;— the fairest measure of the amount of life, being probably the amount of chemical composition & decomposition within a given period. Imagine the case of an island, peopled with only three or four plants of the same order all well adapted to their conditions of life, & by three or four insects of the same order; the surface of the island would no doubt be pretty well clothed with plants & there would be many individuals of these species & of the few well adapted insects; but assuredly there would be seasons of the year, peculiar & intermediate stations & depths of the soil, decaying organic matter &c, which would not be well searched for food, & the amount of life would be consequently less, than if our island/ 26 e/had been stocked with hundreds of forms, belonging to the most diversified orders.

Practice shows the same result; farmers all over the world find that they can raise within the period of their leases most vegetable matter by a rotation of crops; & they choose the most different plants for their rotation: the nurseryman often practices a sort of simultaneous rotation in his alternate rows of different vegetables. I presume that it will not be disputed that on a large farm, a greater weight of flesh, bones, and blood could be raised within a given time by keeping cattle, sheep, goats, horses, asses, pigs, rabbits & poultry, than if only cattle had been kept. In regard to plants this has been experimentally proved by Sinclair[1] who found that land sown with only two species of grass, or one kind of grass with clover, bore on an average 470 plants to the square foot; but that when sown, with from 8 to 20 different species, it bore at the rate of about 1000 plants, "& the weight of produce in herbage & in hay was increased in proportion." It is important to observe that the same rule holds for different & not very distinct varieties of the same species when sown together; for M. L. Rousseau, a distinguished practical farmer, on sowing fifteen varieties of wheat/ 26f/separately, & the same kinds mixed together found on actual measurement that the latter "yielded a much heavier crop than that obtained on far better land on which the unmixed wheats were grown for the purpose of the comparative trial."[2]

We see on a great scale, the same general law in the natural distribution of organic beings; if we look to an extremely small area supposing the conditions to be absolutely uniform & not very peculiar./26f v/Where the conditions are peculiar & the station small as compared with the whole area of the country, as Alpine summits; Heaths salt-marshes, or even common marshes, lakes & rivers, &c.—a great number of individual plants are often supported, belonging to very few species: so it is with Fresh-water shells; so it is with the marine inhabitants of the arctic seas. But even in these cases, though the individuals appear to be very numerous compared with the species, yet even in these cases, the coinhabitants belong to very different types; for instance Dr. Hooker has marked for us all the plants in Britain, which he thinks may be called truly aquatic: they are, [   ] in number, & they belong to [   ] genera and to [   ] orders.— With respect to the number of individuals to the species, we shall have to return to this subject in our chapter on geographical distribution, & I will here only say that I believe it mainly, but not wholly, depends, on the manufacturing, if I may so express myself, being

---

[1] The author of Hortus Gramineus Woburnensis, in Loudon's Gardener's Mag. Vol. I., 1826, p. 113.

[2] Gardener's Chronicle & Agricult. Gazette, 1856, p. 859. See, also, p. 858. and 1857, p. 179 [Samuel Taylor 'The Thick and Thin Sowing Discussion'. pp. 178–9.]

small in size (& sometimes in duration); that is that the number of individuals is small in comparison with the numbers of individuals of the commoner species which inhabit ordinary stations: for we have seen in our 4th Ch. that it is species which most abound in individuals which oftenest present varieties, or incipient species./ 26 f/Supposing the conditions to be absolutely uniform & not very peculiar or unfavourable for life, we seldom find it occupied by any two or three closely allied & best adapted forms, but by a considerable number of extremely diversified forms. To give an example, I allowed the plants on a plot of my lawn three feet by four square which was quite uniform & had been treated for years uniformly, to run up to flower; I found the species 20 in number, & as these belonged [to] 18 genera & these to 8 orders & they were clearly much diversified./26 f v b/The most remarkable exception to this rule, under conditions not apparently very peculiar, is one given by Mr. C. A. Johns[1] who says that he covered with his hat, (I presume broad-brimmed) near to Lands End six species of Trifolium, a Lotus & Anthyllis; & had the brim been a little wider it would have covered another Lotus & Genista; which would have made ten species of Leguminosae, belonging to only four genera! The wretched soil of Heaths, though covered thickly with one or two species of Erica, supports very little life, as judged by their extremely slow growth, & yet, selecting the very worst spots, I have very rarely been able to find a space two yards square, without one or two other plants, belonging to quite different orders, not to mention a good crop of Cryptogams.

To show the degree of diversity in our British plants on a small plot, I may mention, that I selected a field, in Kent, of $13\frac{1}{2}$ acres, which had been thrown out of cultivation for 15 years, & had been thinly planted with small trees most of which had failed: the field all consisted of heavy very bad clay, but one side sloped & was drier: there was no water or marsh: 142 phanerogamic plants were here collected by a friend during the course of a year; these belonged to 108 genera, & to 32 orders out of the 86 orders into which the plants of Britain have been classed. Another friend collected for me all the plants on about 40 uncultivated, very poor, acres of Ashdown Common in Sussex; these were 106 in number, & belonged to 82 genera & 34 orders; the greater proportional number of orders in this case being chiefly owing to the presence of water & marsh plants on the Common: the vegetation was, however, considerably different in other respects, no less than nine of the 34 orders, not being found on the field of thirteen

---

[1] Phytologist, Vol. 2, p. 908.

acres in Kent.—/26 f/To give another example of a small area having singularly uniform conditions of life; namely one of the low & quite flat, coral-islets having a wretched soil, composed exclusively of coral-debris, but with a fine climate; for instance Keeling Atoll, on which I collected nearly every phanerogamic plant, & these consisted of 20 species[1] belonging to 19 genera & to no less than 16 different orders!/

26 g/The extreme poverty of the floras of all such islets may be partly due to their isolation & the seeds arriving from lands having different Floras, but chiefly to the poverty & peculiarity of the soil; for coral-islets, when lying close to large volcanic groups, have an almost equally poor & closely similar flora: the extreme diversity of the plants, the twenty in the case of Keeling islands, belonging to sixteen orders, can, I think, only be accounted for by the fact that of all the plants of which the seeds have been borne across the sea in the later periods of the natural colonisation of the island, those alone, which differed greatly from the earlier occupants, were able to come into competition with them & so lay hold of the ground & survive.

As with plants so with insects. I may premise that entomologists divide the Coleoptera into 13 grand sections, & then into families, sub-families &c. Mr. Wollaston[2] carefully collected during several visits all the Beetles on the Dezerta Grande, a desert volcanic islet about four miles long, & in widest part only three-quarters broad, lying close to Madeira; & he found 57 species, belonging to 47 genera; & these to all 13 grand sections, except two, which being aquatic forms, could not exist on this waterless islet. Again on the Salvages, an extremely small volcanic isl[d]. between Madeira, & the Canaries, six beetles were collected, & these/26 h/belonged to six genera, to six Families, & to three of the grand Sections![3] As a general rule, I think we may conclude, that the smaller the area, even though the conditions be remarkably uniform, the more widely diversified will its inhabitants be: for to this very diversity, the power of supporting the greatest possible number of living beings, all of which are struggling to live, will be due.

There is another way of looking at this subject; namely to

[1] Described by the Rev. Prof. Henslow in Annals of Nat. Hist., 2. Ser., Vol. I, 1838, p. 337.
[2] Insecta Maderensia. 1854.
[3] In the volcanic Galapagos Islands in the Pacific, I carefully collected all the Coleoptera during several weeks; but omitting two probably naturalised species, I got only 24 species, which have been described by Mr. Waterhouse in Annals & Mag. of Nat. History Vol. 16. 1845, p. 19.—The 24 species belong to 18 genera, to 17 families & to 10 out of 13 grand sections. So here again we see the same rule as in other cases in the text. ['Lundy Island' added in pencil.]

consider the productions naturalised through man's agency in several countries; & see what relation they bear to each other & to the aboriginal productions of the country, i.e. Are they closely allied to, that is do they generally belong to the same genera with, the aboriginal inhabitants of the country? Do many species of the same genus become naturalised? If we looked only to the inorganic conditions of a country, we might have expected that species, belonging to genera already inhabiting it, & supposed on the common view to have [been] adapted by creation for such country, would have formed the main body of the colonists: or/ 26i/the many species of certain favoured genera would have been the successful intruders. On the other hand, the principle of diversity being favourable to the support of the greatest number of living beings would lead to the expectation, that land already well stocked by the hand of nature would support such new forms alone, as differed much from each other & from the aborigines. Alph. De Candolle[1] has fully discussed the subject of naturalisation: He shows that 64 plants have become naturalised in Europe (excluding species from neighbouring regions) during the last three centuries and a half; & these 64 species belong to 46 genera & 24 orders; of the genera, 21/46 are new to Europe.[2] Again in N. America, 184 species have become naturalised & these belong to 120 genera & to 38 orders; of the genera, 56/120 are new to N. America.[3] A list of the naturalised plants in Australia & on many islands would give similar, but much more striking results. The number of new genera naturalised in Europe & N. America, reciprocally from each other, is the more remarkable when we consider how much allied the two floras are; & that a very large proportion of the/26k/naturalised plants inhabit land, cultivated nearly in the same manner, which would favour the introduction of allied forms & many forms of the same groups. Hence, I think, we may conclude that naturalised productions are generally of a diversified nature; & as Alph. De Candolle has remarked native

---

[1] Geographie Botanique, p. 745, 759, 803.

[2] In some respects small areas, not including in the sub-regions many indigenous representative species, are best for comparing the native ⟨indigenous⟩ & naturalised productions. De Candolle gives a list (p. 645 et seq.) (in large type) of 83 plants, which he considers as certainly naturalised in Great Britain: these belong to no less than 71 genera: & of these 31/71 are new to Britain. The indigenous genera include on an average about 2.8 indigenous species: the naturalised only 1.1.

[3] Dr Asa Gray seems to consider many more plants as naturalised, than does De Candolle for in his Manual of the Botany of the Northern United States (2nd Edit.) he gives a list of 260 naturalised plants, belonging to 162 genera, of which no less than 100 are new to America. The naturalised genera include on an average 1.6 species: the indigenous include 2.6.—

floras gain by naturalisation, proportionally to their own numbers, far more in genera than in species.

If we turn to animals, we find, though our data are very scanty, the same general fact: no where in the world have more mammals become well naturalised than in S. America (cattle, horses, pigs, dogs, cats, rats & mice); & yet how extremely unlike is the native mammalian Fauna of S. America to that of the Old World!

The whole subject of naturalisation seems to me extremely interesting under this point of view, & would deserve to be treated at much greater length. It confirms the view that in natural colonisation, for instance in that of a coral-islet, diverse forms very different from the few previous occupants, would have the best chance of succeeding. It shows us, & by no other means can we form a conjecture on this head,/261/what are the gaps or still open places in the polity of nature in any country: we see that these gaps are wide apart, & that they can be best filled up by organic beings, of which a large proportion are very unlike the aboriginal inhabitants of the country. Consequently we might perhaps from this alone infer, that natural selection by the preservation of the most diversified varieties & species, would in the long run tend, if immigration were prevented, to make the inhabitants, more & more diversified; though such modified forms would for immense periods plainly retain from heritage the stamp of their common parentage.

The view that the greatest number of organic beings (or more strictly the greatest amount of life) can be supported on any area, by the greatest amount of their diversification is, perhaps, most plainly seen by taking an imaginary case. This doctrine is in fact that of "the division of labour", so admirably propounded by Milne Edwards,[1] who argues that a stomach will digest better, if it does not, as in many of the lowest animals, serve at the same time as a respiratory organ; that a stomach will get more nutriment out of vegetable or animal matter, if adapted to digest either separately instead of both. It is obvious that more descendants from a carnivorous animal could be supported in any/26m/country: if some were adapted, by long continued modification through natural selection, to hunt small prey, & others large prey living either on plains or in forests, in burrows, or on trees or in the water. So with the descendants of a vegetable feeder more could be supported, if some were adapted to feed on tender grass &

---

[1] [Milne-Edwards, *Introduction à la Zoologie générale*...Paris, 1851 see p. 35, pp. 55–7, and art. 'Organisation' in *Dict. class. hist. nat.*, vol. 12, Paris 1827, pp. 332–44.]

others on leaves of trees or on aquatic plants & others on bark, roots, hard seeds or fruit.—

Perhaps I have already argued this point superfluously; but I consider it as of the utmost importance fully to recognise that the amount of life in any country, & still more that the number of modified descendants from a common parent, will in chief part depend on the amount of diversification which they have undergone, so as best to fill as many & as widely different places as possible in the great scheme of nature. Now let it be borne in mind that all the individuals of the same variety, and all the individuals of all the species of the same genus, family &c, are perpetually struggling to become more numerous by their high geometrical powers of increase. Under ordinary circumstances each species will in the briefest period have arrived at its fluctuating numerical maximum. Nor can it pass this point, without/26 n/some other inhabitants of the same country suffer diminutions; or without all the descendants of one species becoming similarly modified in some respect so that they better fill the place of their parent-species; or without (& this would be the most effectual) several varieties & then several species are thus formed by modification, so as to occupy various new places, the more different the better, in the natural economy of one country. Although all the inhabitants of the country will be tending to increase in numbers by the preservation through natural selection of diverse modifications; but few will succeed; for variation must arise in the right direction & there must be an unfilled or less well-filled place in the polity of nature: the process, moreover, in all cases, as we shall presently see, must be slow in an extreme degree.

Let us take an imaginary case of the Ornithorhynchus; & suppose this strange animal to have an advantage over some of the other inhabitants of Australia, so as to increase in numbers & to vary: it could, we may feel pretty sure, increase to any *very great* extent, only by its descendants becoming modified, so that some could live on dry land, some could feed exclusively on vegetable matter in various stations, & some could prey on various animals, insects fish or quadrupeds. In fact its descendants would have to become/ 26o/diversified, somewhat like the other Australian marsupials, which, as Mr. Waterhouse has remarked, typify in their several sub-families, our true carnivores, insectivores, ruminants & rodents. Moreover it can, I think, hardly be doubted, that these very marsupials would, profit by a still further division of physiological labour; that is by their structure becoming as perfectly carnivorous, ruminant & rodent as are our old-world forms; for it may well be doubted (not here considering the probable intellectual infirmity of the marsupialia in comparison with the other or placentate mammals) whether many marsupial vegetable feeders could long exist in free competition with true ruminants, & perhaps still less the carnivorous marsupials with true feline animals. And who can pretend to say that the mammals of the old world are diversified & have their organs adapted to different physiological labours to the extreme, which would be best for them under the conditions to which they are exposed? Had we known the existing mammals of S. America alone, we should no doubt have thought them perfect & diversified in structure & habit to the exact right degree; but the vast herds of feral cattle, horses, pigs & dogs,/ 26 p/at least show that other animals, & some of them as the horse & solid-horned ruminants, very different from the endemic S. American mammals, could beat & take the place of the native occupants.

In Chapter IV we have seen on evidence, which seems to me in a fair degree satisfactory, that on an average the species in the larger genera in any country oftenest present varieties in some degree permanent, and likewise a greater average number of such varieties, than do the species of the smaller genera. It is not that all the species of the larger genera vary, but only some, & chiefly those which are wide-rangers, much diffused & numerous in individuals. In the same chapter we also saw that/26 p v/the species in the larger genera are thought by highly competent judges to be more closely related together, being clustered in little sub-groups round other species, than are the species in the smaller genera; & this closer affinity & grouping of the species in the larger genera, & the fact that there is no unfailing test by which to distinguish species & varieties,/26 p/all to a certain extent confirm the view that varieties, when in some degree permanent, do not essentially differ from species, more especially from such species, as are closely allied together. Hence I look at varieties as incipient species./

26 q/I have lately remarked that the formation of new varieties & species through natural selection almost necessarily implies (as with our domestic productions) much extinction of the less altered, & therefore less favoured, descendants from the same original parent-stock, whose places they occupy in the struggle for life. Hence, though the larger genera may be now varying most, & must, according to our theory, have varied largely, so as to have become modified into many specific forms, yet such large genera must have suffered a large amount of extinction, & very many intermediate & less modified forms have been wholly swept away.

Diagram I .

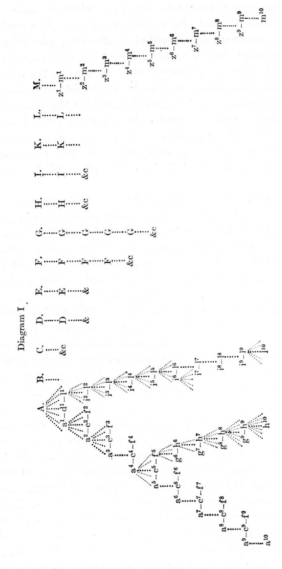

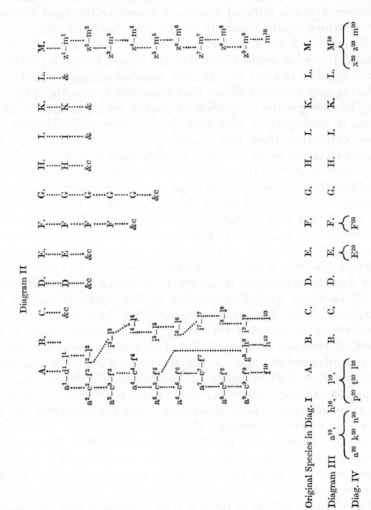

Nevertheless, I think we may infer that in any given country, on the whole, there will have been rather less extinction, proportionally to the whole amount of extinction within any given period, amongst the larger than amongst the smaller genera. For the species which vary most & thus give rise to new species, are chiefly the very common & much diffused species, & therefore the most favoured forms, which would naturally be the least liable to extinction; & such common & much diffused species tend to belong to the larger genera. Indeed it seems to me that the simple fact of a number of allied species, beyond the average number of allied species, inhabiting any country; shows that there/26r/is something in common in such groups of species, or genera, which is favourable to them, & consequently that they would suffer proportionally less from extinction than the smaller genera. Therefore, from the species of larger genera tending to vary most & so to give rise to more species, & from their being somewhat less liable to extinction, I believe that the genera now large in any area, are now generally tending to become still larger. But what will be the end of this? for we do not find in nature genera of indefinite size, with innumerable species. Here in one way comes in the importance of our so-called principle of divergence: as in the long run, more descendants from a common parent will survive, the more widely they become diversified in habits, constitution & structure so as to fill as many places as possible in the polity of nature, the extreme varieties & the extreme species will have a better chance of surviving or escaping extinction, than the intermediate & less modified varieties or species. But if in a large genus we destroy all the intermediate species, the remaining forms will constitute sub-genera or distinct genera, according to the almost arbitrary value put on these terms,—according to the number of intermediate forms which have been destroyed,—and/26s/according to the degree of difference between the extreme species of the original genus. Nevertheless the modified descendants from the common parent-stock, though no longer forming what is called the same genus, may still go on becoming more & more numerous, & more & more diversified.

The complex action of these several principles, namely, natural selection, divergence & extinction, may be best, yet very imperfectly, illustrated by the following Diagram, printed on a folded sheet for convenience of reference.*/26s v/This diagram will show the manner, in which I believe species descend from each other & therefore shall be explained in detail: it will, also, clearly show

* Fold-out version can be found at the end of the book.

several points of doubt & difficulty;/26s/Let A to M represent the species of a genus, numerically large compared with the other genera of the same class in the same country, & arranged as naturally as can be done, so that A & M are the two most distinct forms in all respects. The unequal distances of the letters may represent the ordinary way in which the species, even when as in this imaginary case all are closely related together, yet stand unequally related in little sub-groups. This genus may have one, two or even more varying species. Any of the species may vary; but it will generally be those species which are most numerous in individuals & most diffused; & this shows that such species have already some advantages over the other inhabitants of the country. From our principle of divergence, the extreme varieties of any of the species, & more especially of those species which are now extreme in some characters, will have the best chance,/26t/after a vast lapse of time, of surviving; for they will tend to occupy new places in the economy of our imaginary country. I do not mean that any of these points are of invariable occurrence, but that in the long run such cases will prevail. The extreme species A and M will differ in very many respects; but for convenience sake we may look to any one character, & suppose A the most moisture-loving & M the least moisture-loving species.

We will first take the simplest case. Let M inhabit a continuous area, not separated by barriers, & let it be a very common & widely diffused & varying plant. From the fact of M. being very common & widely diffused, it clearly has some advantages in comparison with most of the other inhabitants of the same country; but, we will suppose, that it might become still more common, if retaining the advantages which it already has, it could endure still more drought. It is a varying species; & let z^1–m^1 represent numerous, extremely slight variations of many kinds, produced at intervals, of which m^1 alone is a more drought-enduring variety. As m^1 tends to inherit all the advantages of its parent M, with the additional advantage of enduring somewhat more drought, it will have an advantage over it, & will probably first be a thriving local variety, which will spread & become extremely common & ultimately, supplant its own parent. We may now repeat the process, & let the variety m^1 vary in a similar manner; perhaps/26u/many thousands of generations may pass before m^1 will produce another variety m^2, still more drought-enduring & yet inheriting the common advantages of m^1 & M; but if this should ever occur, the same results, as before, will follow; & ultimately, by repeating the process, there may be produced m^{10}, which may either be, according to the amount of difference thus acquired, a very strongly marked variety, or a sub-species, or good species, enduring far more drought than M & probably with correlated differences of structure. In each stage of descent, there will be a tendency in the new forms to supplant its parent, though probably, as we shall see, very slowly, & so ultimately cause its extinction. But if M had originally inhabited a country separated by barriers into distinct districts, in one or more of which the varieties M^{1-10} had never originated or had never been able to enter, M and m^1 & ultimately m^{10} might be living contemporaneously, but separated:/26u v/or, again, if m^{1-10} had been produced, capable of enduring more drought, but not at the same time enduring an equal amount of moisture with the parent M, both parent & modified offspring might coexist: the parent (with perhaps a more restricted range) in the dryer stations, & m^{1-10} in the very dryest stations./

26u/It should always, be borne in mind that there is a wide distinction between mere variations & the formation of permanent varieties. Variation is due to the action of external or internal causes on the generative systems, causing the child to be in some respects unlike its parent; & the differences thus produced may be advantageous or disadvantageous/26 v/to the child. The formation of a permanent variety, implies not only that the modifications are inherited, but that they are not disadvantageous, generally that they are in some degree advantageous to the variety, otherwise it could not compete with its parent when inhabiting the same area. The formation of a permanent variety must ⟨can⟩ be effected by natural selection; or it may be the result, generally in unimportant respects, of the direct action of peculiar external conditions on all the individuals & their off-spring exposed to such conditions. We shall best perceive the importance of the difference by glancing at our domestic breeds: in our truest breeds, innumerable slight differences are continually occurring & can be detected by measurement, but only those differences which improve the breed in the often fanciful eyes of the Fancier are rendered permanent by the animals so characterised being carefully preserved, matched & largely bred from; all other slight differences being lost, by the animals not being largely bred from, & from indiscriminate crossing. If, however, the process of selection were continued for a long time by two Fanciers, under very different conditions of climate or food, some subordinate differences would probably arise between the two lots, owing to the direct action of such conditions. Now in our diagram, the letters z^1–m^1, z^2–m^2 &

c represent all sorts of successive slight variations, of which m$^{1-10}$, the most drought-enduring varieties alone have been naturally selected & been rendered permanent.

This natural selection has been possible, owing to there having been/26 w/a place in the economy of our imaginary country, which the descendants of M, from inheriting all or some of the advantages over the other inhabitants which made M a very common species, could seize on, when rendered more drought-enduring.

With respect to the process by which each new & improved variety supplants its parent, this must often have gone on in two slightly different manners, differing, however, only in degree. In those animals which are highly locomotive & of which two individuals unite for each birth, there can only seldom have arisen as we shall hereafter see, within the same continuous area, especially if of not very large size, distinct varieties, for they would become blended by such free crossing. In such cases, modifications must be effected quite insensibly by the natural selection of mere individual differences; nearly in the same way as many of our domestic breeds throughout whole districts have been insensibly changed from their ancient state. So that in our diagram the letters m$^{1-10}$ may represent in the case of the higher animals, not recognizable varieties, but mere ideal steps in a real, yet insensibly gradual, change of structure. In organic beings which do not cross freely/26 x/& which are more stationary, & which are capable of propagating at a great rate, a variety might easily be formed in one spot (more especially if in some slight degree isolated) & might not spread & supplant its parent-stock, until it had become developed by the continued natural selection of similar extremely slight or individual differences into a distinct & plainly recognizable variety./26 x v/I am inclined to think from the frequency of local varieties, though the subject must remain very doubtful, that this latter process has been a very common one, for a variety would often be unable to supplant its parent, until it had become considerably modified so as to have a decided advantage over it. For instance in the imaginary case of the varieties m$^{1-10}$ which are supposed to inherit all the characters of M, with the addition of enduring more drought; these varieties would inhabit stations, where M could not exist, but in the less dry stations m$^{1-10}$ would have very little power of supplanting their parent M; nevertheless during unusually dry seasons m$^{1-10}$ would have a great advantage over M & would spread; but in damper seasons M, would not have a corresponding advantage over m$^{1-10}$ for these latter varieties are supposed to inherit all the characters of their parent. So there would be a tendency in m^{1-10} to supplant M, but at an excessively slow rate. It would be easy to show that the same thing might occur in the case of many other new characters thus acquired; but the subject is far too doubtful & speculative to be worth pursuing. I will only add that/26x/with the most freely crossing & locomotive animals, when inhabiting an area, separated by barriers only to be passed after geographical changes or through some most rare accident, a similar process must often have occurred; for in such cases, distinct & plainly marked varieties might have been insensibly formed in the different districts by the selection of mere individual differences; & when these districts became united, so that the varieties could mingle & come into competition, the best variety would supplant the other varieties or the parent-stock.

To return to our diagram. I do not suppose the process generally to have been so simple as represented under M, where a simple variety m^{1-10} in each stage of descent has been naturally selected. We have seen that not only more species, especially the very common species, in the larger genera in any country present varieties in some degree permanent, but that each such species on an average tends to present a greater number of varieties, than do the species, especially the rarer species, in the smaller & less flourishing genera. As varieties from a species tend to inherit the advantages which/26y/made the parent common, these varieties will ultimately tend to be common & to vary; moreover they descend from a variable stock, & are still exposed to the conditions which made their parents vary, hence for this cause they will be liable to vary. Consequently there will be a tendency in the original varying species, after a vast number of generations to produce an almost infinite number of varieties; but our principle of divergence explains how the most diversified varieties will generally have decided advantages over the less diversified & intermediate varieties, causing their extinction & thus reducing the number of varieties living at any one time. These remarks are illustrated in our diagram under A., which species, after many generations represented by dots, is supposed to have varied largely, & to have produced these varieties a^1, d^1 l^1 in some degree permanent; of these, again after many generations & much variation; the two extreme varieties a^1 and l^1, are supposed to have produced other varieties in some degree permanent; of which the extreme varieties have again reproduced others, represented finally by a^{10} & l^{10}. In the diagram I have been able to represent only one other branch proceeding from f^5, & giving rise to a third variety h^{10}

which being the extreme form in its own branch has the best chance of surviving/26 z/& seizing on some place in the natural economy of the country inhabited by the genus.

By continuing the process represented in the diagram, the forms marked a^{10}, h^{10}, l^{10}, may be made different in any degree, till they would be universally be [sic] ranked as good species; & the number of such new forms would continually tend to increase. These new species will generally have supplanted, perhaps by a very slow process their several parents in each stage of descent & their original common parent A,—that is if formed in one continuous area, or as soon as they came into competition with each other if formed in different areas. The original species A. was supposed to be the most moisture loving plant; & if for simplicity sake we imagine a^1 more moisture loving & l^1 less moisture loving, but inheriting some of the advantages which made A in the great & complex battle for life a very common species; & the offspring of these varieties to be continually selected on the same principle, a^{10} will have been rendered so moisture loving as to have become semi-aquatic, & l^{10} far less moisture loving than A; & in the third branch, h^{10}, about as moisture loving as A, for it has descended from f^5 which was more moisture-loving than A, and subsequently has become less so. Not that I at all suppose the diversity is ever thus confined to one point; for as a^{1-10} becomes moisture-loving & as l^{10} becomes less moisture-loving both would under the extremely complex conditions to which all organic beings are exposed, come to be exposed to new dangers & /26 aa/would have to gain some other advantages over other organic beings with which they would have to compete. So that in love of moisture & in many other respects, a^{1-10}, h^{1-10}, l^{1-10} would come to differ or diverge more & more from each other & their original parent-stock.

A little reflexion will show the extreme importance of this principle of divergence for our theory. I believe all the species of the same genus have descended from a common parent; & we may call the average amount of difference between the species, x; but if we look at the contemporaneous varieties of any one species, the amount of difference between them is comparatively extremely slight & may be called a. How thus can the slight difference a be augmented into the greater difference x; which must on our theory be continually occurring in nature, if varieties are converted into good species? The process feebly illustrated in our diagram, I believe, explains this; namely the continued natural selection or preservation of those varieties, which diverge most in all sorts of respects from their parent-type, (but still largely inheriting those advantages which made their parents generally dominant & common species) so as to fill as/26 bb/many, as new, & as widely different places in the economy of nature, as possible.

A glance at Diagram 2. will perhaps render this plainer. The varieties a^{1-10}, l^{1-10} may be here again for simplicity be looked at as more & less moisture loving plants; & everything is the same as in diagram I (the third branch h^{6-10} cannot be introduced) except that it is left to mere chance in each stage of descent, whether the more or less moisture loving varieties are perserved; & the result is, as graphically shown, that a^{10} & l^{10} differ in this respect; & so in other respects, hardly more than did the first varieties (a^1 l^1) which were produced.

In regard to the difference between varieties & species, I may add that varieties differ from each other & their parents, chiefly in what naturalists call unimportant respects, as size, colour proportions &c; but species differ from each other in these same respects, only generally in a greater degree, & in addition in what naturalists consider more important respects. But we have seen in Ch. IV, that varieties do occasionally, though rarely, very slightly in such important respects; and in so far as differences in important physiological characters generally stand in direct relation to different/26 cc/habits of life, modifications however slight in such characters would be very apt to be picked out by natural selection & so augmented, thus to fit the modified descendants from the same parent to fill as many & as widely different places in nature as possible. We shall, also, see in a future chapter that a large part of the differences in structure between species may be accounted for by the mysterious laws of correlation; by which, I mean, that when one part is modified, (or the whole animal at one age, as with the larvae of insects) other parts necessarily become altered through the correlated laws of growth. That there is no obvious & unmistakeable difference between the differential characters of species & varieties, is plainly shown by the number of debateable forms in the best known countries, which are ranked by one good naturalist as true species, & by another as mere varieties.

Our principle of divergence has another very important bearing. In the diagram, A. has given rise to three new species, & M to one. The other species of the genus, B to L, are supposed to have/26 dd/transmitted unaltered descendents. Hence, even supposing that A & M have been supplanted as I believe will usually have been the case, by their modified & improved descendants, the genus will have become not only more divergent in character

(a^{10} more aquatic than A; & m^{10} more drought-enduring than M.) but numerically larger. The original species A to M were supposed to be closely allied, but yet to exhibit traces, as is so general, of being divided into sub-groups. The sub-groups, after the formation of the new species, will be slightly altered & increased in number; for a^{10} & h^{10} will be closely related together from common descent from f^5, & closely yet less closely with l^{10} from descent from their common ancester A.; and they will all differ as much, generally more from B, than did A. So again m^{10} having constantly diverged from the characters of M will now stand more distant from L, than M originally stood. This is represented in the Diagram III. And from the reasons already given, I believe there will be a constant tendency in the modified descendants of A & of M to go on thus producing more & more new specific forms & thus more & more modified or divergent.

What will be the limit to this process in nature? Though many genera are large, they do not include an indefinite number of species. I believe that there is no limit/26 ee/to the number of species tending to be formed from the most favoured forms in any country (or those which have any [sic] the greatest advantages over the coinhabitants), except the number of species which the country is capable of supporting; but such modified descendants, or new species, after a long period will have to be ranked not in the same genera, but in distinct genera, families or orders. For if we suppose the process illustrated in diagram I. to have long continued & the modified descendants of A to have become extremely much multiplied and diversified in many ways, they will tend to take the places of & thus exterminate the species B.C.D. &c, which originally were nearest related to A. & were not then such common & flourishing species. So if M had left several modified & divergent descendants, it would have been with L, K. &c./26 ee v/It may be here worth observing, that although the new species in taking the place of the old (their great uncle) may have acquired through natural selection, some of their characters; this kind of resemblence would be called by naturalists that of analogy, & the real affinity of the new species would be with their real parents: thus l^{10} might come to simulate some of the character of B, from occupying its place in nature yet the real affinity of l^{10} would be with A.—/26 ee/Continue this process, & all, or nearly all the original species (A to M) will become extinct. In Diagram IV. this is represented, E & F alone now having descendents, whether or not modified. And the final result will be, that we shall have two large groups of modified descendants,/26 ff/coming from the two species, generally the extreme species, (A & M) of the original genus, and differing as much as natural selection could make them from each other & from their two parents, which at the first start differed much: assuredly these two new groups of new species would be ranked in different genera, which would be very distinct, if all the original intermediate species from B to L. had been exterminated, but somewhat less distinct if some of these species (as represented in Diagram IV.) had left descendants, whether or not modified.

Now for a moment let us go back many stages in descent: on our theory the original twelve species A to M are supposed to have descended & diverged from some one species, which may be called Z, of a former genus. But now, according to the result given in the last paragraph, Z will have become the ancestor of two or three very distinct groups of new species; & such groups, naturalists call genera. By continuing the same process, namely the natural selection of generally the most divergent forms, with the extinction of those which have been less modified & are intermediate, Z may become the ancestor of two very distinct groups of/26 gg/genera; & such groups of genera, naturalists call Families or even Orders. But to this subject, we shall have to return in our Chapter on Classification.

I have previously remarked that there seems to be no limit to the number of modified descendants, likely to proceed from the most favoured form in any country,—the most favoured always tending to diverge in structure & take the place of & exterminate the less favoured & intermediate forms,—except the total number of species, which the country is capable of supporting. But it may be objected that as natural selection, extinction & divergence must have been going on since the dawn of Life, why have we not an infinite number of species, almost as many species, as individuals? We shall presently see that natural selection can act only with extreme slowness. Nor do we by any means know that the maximum number of species, which any country would be best fitted to support, has anywhere been as yet produced: the fact that there is no country which does not support several, often many, organic beings naturalised by man, without, as far as we know,/26 hh/having caused the extinction of an equal number of the indigenous productions, renders it probable that such countries were capable of supporting a greater number of specific forms than nature had supplied them with. Even the Cape of Good Hope, which is apparently the richest district in the world in different kinds of plants has received, as I am informed by Prof. Haney from [   ] to [   ] naturalised species. Many geologists, indeed believe that the number of species in the world has gone on increasing from the earliest geological days; but I am sorry to own that the evidence on this head seems to me quite insufficient./

26 hh v/It might indeed be argued from the enormous list of shells, found in the eocene Paris basin, & even in the ancient Silurian system of Bohemia, as so admirably worked out by Barrande, that at these periods & in these places, a greater number of species existed than anywhere at the present day. But it may be doubted how far such comparisons are in any instance trustworthy; for we have reason to suppose that the duration of each sub-division of each geological formation is so enormous, that it is not fair to compare all the species found in one such sub-division with all existing within an area at the present day. Barrande's "colonies" show, according to Sir C. Lyells explanation of them,[1] what changes of climate or currents must have taken place within certain definite periods: the Glacial epoch within what may be called the present period, should teach us caution, for far lesser changes than the glacial epoch, not easily to be detected in ancient geological formations, might alternately bring in & displace, & apparently mingle many organic beings, which never really co-inhabited the same area./

26 hh/But if the time has not yet arrived, may it not at some epoch come, when there will be almost as many specific forms as individuals? I think we can clearly see that this would never be the case. Firstly, there would be no apparent benefit in a greater amount of modification than would adapt organic beings to different places in the polity of nature; for although the structure of each organism stands in the most direct & important relation to many other organic beings, and as these latter/26 ii/increase in number & diversity of organisation, the conditions of the one will tend to become more & more complex, & its descendants might well profit by a further division of labour; yet all organisms are fundamentally related to the inorganic conditions of the world, which do not tend to become infinitely more varied. Secondly as the amount of life & number of individual beings, whether or not much diversified, also primarily depends on such inorganic conditions; if there exist in any country, a vast number of species (although a greater amount of life could be supported) the average number of individuals of each species must be somewhat less than if there were not so many species; & any species, represented by but few individuals, during the fluctuation in number to which

---

[1] Supplement to Manual of Geology, 2nd Edit. 1857. p. 34.

all species must be subject from fluctuations in seasons, number of enemies &c, would be extremely liable to total extinction. Moreover, whenever the number of individuals of any species becomes very small, the ill-effects, as I believe, of close inter breeding would come into play. Lastly we have seen in our Chap. iv & shall presently again see, that the amount of variations, & consequently of variation in a right or beneficial direction for natural selection to seize on & preserve, will bear some relation within any given period, to the number of individuals living & liable to variation during such period: consequently when the descendants from any one species have become modified/26 kk/into very many species, without all become numerous in individuals, which [we] see hardly ever to be the case with all the species of the same genus or family, there will be a check amongst the less common species to their further modification: the lesser number of the individuals serving as a regulator or fly-wheel to the increasing rate of further modification, or the production of new specific forms.

Subject to these restraining influences, I can see no limit to the number of modified descendants, which might proceed from the most favoured forms, whatever they may be, now living in the world. If we return to look to the future, as far into the remotest futurity as the Silurian system lies in the remote past, our theory would lead to the conclusion that all organic beings which will live at that far distant period, will be descendants from a very few of our contemporaries; perhaps from not so few, owing to the increasing complexity of the organic world, as our existing organisms have descended from; for our animals have descended, according to our theory, from four or five ancestral types & our plants from apparently still/26 ll/fewer; & if we rashly dare trust to mere analogy, all our plants & animals from some one form, into which life was first breathed.

Taking a more modest glance into futurity, we may predict that the dominant genera, now abounding with common & widely difused species, will tend to be still more dominant for at least some considerable lapse of time, & will give rise to new groups of species, always diverging in character, & seizing on the places occupied by the less favoured forms, whether or not their near blood relations, supplanting them & causing their extermination. The great & flourishing genera both of plants & animals, which now play so important a part in nature, thus viewed become doubly interesting, for they include the ancestors of future conquering races. In the great scheme of nature, to that which has much, much will be given.

Finally, then, in regard to our principle of Divergence, which regulates the natural Selection of variations, & causes the Extinction of intermediate & less favoured forms, I believe it to be all /26 mm/ important as explaining why the average difference between two species of the same genus, the parents of which by our theory once existed as mere varieties, is greater than the average difference between two such varieties. It bears on, & I think explains, the classification or natural affinities during all times of all organic beings, which seeming to diverge from common stems are yet grouped like families within the same tribes, tribes within the same nations, & nations within the same sections of the human race. We shall, also, hereafter see that these views bear on palaeontology & explain why extinct forms either fall within existing groups, or, as is so frequently the case, are in some slight degree intermediate between them.

The relation of all past & present beings may be loosely compared with the growth of a few gigantic trees; that is if we suppose that from each of the innumerable twigs, innumerable buds are trying to sprout forth, & that the other buds, twigs & branches have the best chance of growing from getting more light. The buds & twigs may represent existing species, & all beneath their living extremities may represent extinct forms. We know that the twigs proceed from lesser branches, these from larger & these from main limbs, from the trunk, & that the several branches & limbs are of very unequal/26 nn/sizes; & this grouping of the branches may represent the natural classification of organic beings. In our living trees we can trace in the gnarled & leafless branches the connecting links; but so imperfect are our palaeontological records, that we can only here & there find a form which may be called a forked branch, with its two arms directed towards two now distinct groups of organisms. As we know that the gnarled branches were at successive periods tender twigs crowded with buds, so we may believe that every organic class, whether or not now having lineal descendants on the earth, swarmed at each stage of descent under diversified forms of life. Many a smaller & larger branch, & even some main limbs have utterly perished, from being over topped by the ever diverging budding twigs; so it has been with whole groups of organic beings. Here & there a branch is still alive, carrying only a few twigs & buds; & these will represent the organic groups having few species & fewer genera, which are now on the road to extinction. As buds give rise by growth to fresh buds, & these, if vigorous, branch out & give rise to many a diverging branch still branching out, & causing the death of many a feebler twig &

branch on all sides & beneath, so by generations I believe it has been with the great Tree of Life, which fills the crust of the earth with fragments of its dead & broken branches, & covers with its ever living, ever diverging & marvellous ramifications, the face of the earth./

27,28/Long ere this, a crowd of difficulties will have arisen in the reader's mind, overwhelming my theory of natural selection, more especially when applied to organs or beings widely different in the same great classes. Some of these difficulties are indeed great enough almost to crush my belief; but many, I think, are only apparent. Is it possible to believe that the eye with its admirable correction for spherical & chromatic aberration, & with its power of adapting the focus to the distance, could have been formed from the simplest conceivable eye, by natural selection? Is it possible for the instinct of a bee, which produces a cell constructed on the highest geometrical principles, to be thus perfected? I confess that my mind recoils from such an admission; yet, reflecting on the known gradations in so wonderful an organ as the eye amongst existing animals,—a mere/29/small fraction of those which have lived,—I can see no logical impossibility; & as far as probability is concerned, a safe conclusion can be drawn, as it seems to me, only from the general phenomena of organic beings, as indicative whether each being has been simply created or has been produced by the common laws of generation with superadded modification. But these questions, & likewise the general subject of instinct shall be discussed in separate chapters.

What shall we say of small & apparently trifling organs, yet most useful to the animal possessing them, as the eye-lash, or a tail serving as a fly-brush; could these have been produced by natural selection, which is in fact selection for life & death? But I have already shown how cautious we should be in deciding what trifle may turn the nicely-suspended balance of life in the great struggle for existence. Again how could a swimming animal be turned into a crawler, or a walking animal into a flyer: how/30/ could they live in an intermediate state? Undoubtedly nothing can be effected through natural selection except by the addition of infinitesimally small changes; & if it could be shown that in cases like the foregoing, transitional states were impossible, the theory would be overthrown. This being so, it may be further asked, do we not meet in certain members of a class organs, which, as far as we can see, are absolutely new creations, & which cannot be some other part or organ modified by natural selection in accordance with the laws of morphology? We shall see that such cases are surprisingly few & hard to find.

Again it has often been urged that if species were subject to change all nature would be in confusion & the limits of no species distinct; but this argument depends on the assumption that the change is rapid & that many species are simultaneously undergoing change. If species were as distinctly defined, as some authors pretend, systematic/31/natural history would be a far less difficult subject, that those authors will find if they will take up for description almost any group, especially a varying group of species; but to this subject I shall immediately recur. So again it has been said, if species were subject to change, we should find plain evidence of such change in our collections of fossil remains; but the force of this objection, in main part, lies in the supposition that the records of geology are as ancient as the first commencement of life, & that they are far more perfect than some of our most experienced geologists have shown good reason for believing that they are in truth. I will here only ask those who make this objection, can they believe that at some future geological epoch, fossil remains will tell that which we do not now know, namely what were the exact steps by which the various British breeds of sheep & oxen have descended/32/from some one or two parent stocks. It should be remembered we do not mean forms intermediate between horse and tapir, but between both of them & some unknown common parent.

Lastly why do two species when crossed, either yield few or no offspring, & these more or less sterile, & why do those varieties which we may safely conclude are descended from a single species yield abundantly fertile offspring? To this important subject I will devote a chapter. And all the foregoing great difficulties, & some curious special cases shall be stated in detail, as fairly as I can, & be discussed. That some difficulties remain to be mastered will not be thought surprising by those who will make allowance for our ignorance on what is daily passing round us in the living world; & our incomparably greater ignorance of the many past worlds which have teemed with life.—/

33/*Causes favourable & unfavourable to Natural Selection.*—Having given a pretty full outline of my theory, it will be necessary to discuss as well as we can, though very imperfectly, the circumstances, favourable or the contrary to natural selection. We have seen that variability is the foundation. The variation, whatever its cause be, must be inherited or tend to be inherited to be of any use. Certainly this tendency is very strong & applies to the most trifling changes; but it often fails; & the offspring instead of taking after their parents resemble their grandfathers or more remote ancestors. We see this repeatedly perhaps oftenest, at least most plainly, where strongly marked varieties are crossed; but in all cases it must tend to retard natural selection.

Again the variation must be in the right direction to profit the individual, otherwise it will not be selected. I do not here refer to the direct effects of climatal conditions, for these must be quite unimportant, in relation to the numberless exquisite co-adaptations of each organic being to other inhabitants of the area./

34/I am inclined to believe that in the polymorphous or protean groups of species, as they have been called, mentioned in our Ch. IV which we meet with in every great class, we see more fluctuating variability,—perhaps the very tendency to vary being inherited,—the variation being of no use in any one direction to the being in question, & therefore with no one character steadily selected, augmented & rendered nearly constant.

The expression of variation in a right direction implies that there is a place in the polity of nature, which could be better filled by one of the inhabitants, after it has undergone some modification: the existence, therefore, of an unoccupied or not perfectly occupied place is an all important element in the action of natural selection. I do not doubt, as previously remarked from the number of naturalised productions, that everywhere such open places ready to be filled exist; but it is obvious that such places or gaps will be more frequent, & it may be said wider, in districts favourable for life, but yet not thickly stocked with various forms. Districts subjected to some physical change & cut off from free immigration will be thus circumstanced;—for instance part of a continent separated by a desert or mountain-barrier, into which after climatal changes,/35/the other inhabitants of the continent cannot freely enter; or better still a volcanic island, rising from the ocean at first with few or no inhabitants, but receiving an occasional stray colonist. Now both Mr. Wollaston & Alph. de Candolle[1] have strongly insisted that isolated areas are the chief scenes of what they consider, like most naturalists, as the actual creation of new species & likewise of varieties. It is not, I may add isolation in the abstract which seems to affect organic beings; for the very same spot may easily be isolated for one set of beings & not to another: thus Madeira is not isolated for birds for annually birds are blown there from the mainland, & there is only one endemic or peculiar bird & that not a very distinct species: from what we know of the habits of land-molluscs this

---

[1] Variation of Species, p. 70. Bot. Geograph. [See II, 1092–4, 1125.]

island must be closely isolated for them, & a large majority of the species are endemic; whereas there is not a single endemic sea-mollusc, & these, little as we know of their means of dispersal, can hardly be so completely isolated as the land-molluscs: again coleoptera are seldom strong flyers, & therefore would be here more isolated/36/than the other orders of insects, & Mr. Wollaston tells me that he believes that there are far more endemic species of Coleoptera than in the other orders. We have seen in the last chapter that birds, for instance, in the struggle for existence would be apt to come more into competition with other birds, than with other animals; & so land-molluscs with land-molluscs, & beetles with beetles: consequently a few beetles or land-molluscs (whether we suppose them the remnants of an ancient population before the island was severed from the mainland, or as I think far more probable, occasional stray colonists) placed by themselves in this island would find themselves in a far more disturbed condition & with more places opened to them in their own scheme of nature, than would those other animals, which found themselves associated with all or nearly all their old compatriots with whom they had long struggled in their native land.

Isolation by itself will apparently do nothing; we can find on mountain summits, & in the lowlands innumerable instances of plants & insects with not another individual of the same species within a distance of many miles, & which we have no reason to doubt have long remained there, & yet are absolutely identical with the same species/37/from elsewhere. Isolation under a some-what different climate introduces another element of change; but the fact which must strike every naturalist is that isolation under the same climate seems to have been eminently favourable to the production of new forms. The climatal conditions of Madeira could probably be paralled on the shores of Europe, as closely as the habits of most species require, judging from their ranges on the mainland; yet, as Mr. Wollaston has shown, those islets swarm with peculiar endemic Coleoptera & Land Molluscs. We see the effects, of isolation under the same climate in the numerous endemic species, both with whole groups & in the separate islets of the Galapagos & Sandwich & Canary archipelagoes, & in the West Indies, as far as some of their productions are concerned./ 37 v/In our chapter on Geographical Distribution, I shall enter on some details showing how extremely rich isolated islands are in endemic species in relation to their areas, as compared with an equal area on the most favoured mainlands. In the case of some of the above archipelagos it is known, & in/37/other cases it is highly probable, that the inhabitants, excluding those peculiar to the archipelago, are differently grouped to what they are in the mainland, & differently on the separate islets, so that a colonist would be exposed in each to a somewhat different set of competitors. But to this subject, also, we shall have to return in our chapter on Geographical Distribution./38/From the foregoing considerations I conclude that the association of an organic being in any country with a different set of those beings, with which it comes into the most direct competition or dependence, as eminently favourable for natural selection for acting on whatever variations may occur, & so seizing on & filling up new places in the economy of that country. I look at this as so important as to be second only to variability, the basis on which the power of selection rests. Now an organic being could be particularly liable to become associated with new competitors, either when first by chance entering an isolated region into which few of its compatriots had entered; or when living there, after climatal or other changes had destroyed many of the inhabitants, & the isolation of the spot had checked free immigration of new & better adapted inhabitants. In this way, I think, isolation must be eminently favourable for the production of new specific forms. It must not, however, be supposed that isolation is at all necessary for the production of new forms;[1] when a species spreads widely it will almost universally become associated with new competitors & there will often be some advantage gained by the selection of some modifications/38a/in its structure. I do not doubt that over the world far more species have been produced in continuous than in isolated areas. But I believe that in relation to the area far more species have been manufactured in, for instance, isolated islands than in continuous mainland.

The rate at which selection can act, depending on the chance production, as we must call it, of slight favourable variations; it might well happen, that of two forms undergoing modification, the one might beat out the other, if brought soon into competition; whereas if time had been allowed the other might have gained through selection some advantage, by which it could have held its own, when thrown into competition with the first. In this way, I can believe isolation may have played an important part in allowing two varieties from the same species to be considerably modified, before they are enabled to struggle with each other for existence./

[1] See on this subject some excellent remarks by Dr Hooker in his Review of A. De Candolles Geographie Botanique in a note in Hooker's Journal of Botany, vol. VIII p. 83. [p. 153]

38 a A/Isolation, moreover, comes into play in lessening the amount of inter crossing, but here we are launched on a sea of doubt. That the majority of animals have their sexes separated or when united require the concourse of two individuals for the production of young is certain; & I think it has been shown in the third Chapter that occasional crosses will take place both/39/ with plants & animals far oftener than would at first be anticipated: but facts do not allow us to say that such occasional crossing is of universal occurrence. In those few cases, moreover, in which intermediate forms have been observed between two strongly marked varieties or reputed species, unfortunately we hardly ever know whether they are due to crossing, or to the intermedial action of external conditions & of the powers of natural selection. But as two individuals of most animals & some plants habitually unite for reproduction; this crossing will obviously retard, perhaps obliterate, the process of selection by dragging back the offspring of a selected variety towards its parental type. Let us suppose a stray gravid female or a pair of any animals to reach a small isolated island; if their offspring instantly varied & the old died, there would be no crossing, but such an improbable supposition may be quite disregarded; but if after several generations when the island was pretty well stocked some of their offspring slightly varied in any favourable direction; these would be selected or preserved, & though they would in all such cases be apt to cross with the parent-form;/39 A/yet the offspring from such crosses would have a stronger inherited tendency to vary in nearly the same favourable manner, as did the first variety; & natural selection would by preserving such individuals continually augment the tendency; until all the individuals might become insensibly modified in the same favourable manner. Just in the same way as a large herd of cattle may be modified by crossing even with a single bull of an improved shape & by the continued selection of the crossed offspring most like the Bull; & this would be much facilitated if the conditions of the country had ⟨any⟩ the weakest tendency occasionally to produce animals of the desired character./

40/I am inclined to believe, that wherever very many individuals of a freely crossing & highly locomotive animal existed, the retardation of any selected modification from crossing would be so strong, that it could hardly be overcome, without indeed the tendency to vary in some particular direction was extremely strong. Hence I infer that some degree of isolation would generally be almost indispensable. This isolation may result from the nature of the area; or from the varieties as soon as produced, keeping to a certain extent separate; & we shall immediately see that some partial separation of varieties, can & does take place in nature. That isolation from locality is important with highly locomotive, freely crossing animals, I infer from the fact, that with birds & mammals, the varieties & close & very doubtful species, (not here considering mere monstrosities, such as albinoes &c) generally inhabit distinct areas.

On the other hand, with organic beings, such as most plants, which do not cross for each birth or which are not highly locomotive so as to cross with individuals over a wide area, or which when favoured can increase at a great rate, I can well believe that a small body of any selected variety might be/40a/more quickly formed & hold their own against the ill effects of crossing, without being completely isolated. Though in such cases, isolation, at least partial isolation at first, would be favourable to their natural selection.

I have just taken the case of the selection of a variety of a freely-crossing animal, on an isolated island; if we suppose the same process to be going on, in some favourable spot, but open all round to the inroads of the parent or unaltered form, there would be crossing with the parent form, not only at first in the actual birthplace of the variety, but all round its confines, where there might be neither the same tendency to vary nor the same place in the polity of nature open & ready to be filled up by the selected forms; in such cases, the retardation from crossing would be extreme./

41/In all these cases of crossing, we should remember the facts given in the third chapter, which convinced me that the offspring from two varieties have a greater amount of vigour & fertility, which would give them an inherent advantage, however slight, over the parent forms; tending thus to obliterate the variety, but on the other hand leaving descendants with some inherited tendency on which the same original cause of variation, we may believe, would be very likely to react. I am tempted to give an illustration of the effects which I should attribute to isolation in regard to crossing: in Madeira there are 20 land-birds[1] which breed in the island, & of which only one does not inhabit Europe or Africa; but besides these twenty, 26 stray species from the continent have been observed. Of these stragglers about 17, as I am informed by Mr. Harcourt, appear every two or three years, & some of

---

[1] Excluding Grallatores; see Mr. E. Vernon Harcourt's excellent paper on the ornithology of Madeira in Annals & Mag. of Nat. History June 1855. I am infinitely obliged to Mr. Harcourt for having given me much valuable information on this subject.

them almost annually, & occasionally in little flocks, which has been noticed in the case of the starling, rook &c.—This being/ 42/the case, it seems most improbable that individuals of those species which breed on the island, should not likewise be occasionally blown there from the continent, although of course it is almost impossible to prove this. Therefore I should infer that the Birds of Madeira have not undergone modification, in the first place because the small island is well stocked with the same species, which have long struggled together in other & not very dissimilar lands, & secondly because, any slight tendency to change, which I believe would occur as the conditions cannot be identically the same, is checked by an occasional cross with quite unaltered forms having no tendency to vary;—the crossed offspring having greater vigour & hence a better chance of surviving.

What a contrast is presented by the Galapagos Islands, situated in a most tranquil climate, without any storms to blow birds from the mainland, which is nearly twice as far off; in this considerably larger group we have 26 land-birds, of which 25 are endemic or peculiar to the archipelago! Of these 26 species, 8 belong to one endemic genus Geospiza, & five others belong to three sub-genera closely allied to Geospiza; there are three closely allied mocking-thrushes, & two tyrant-flycatchers; so that I imagine that there were only/43/14, perhaps only 11 original stray colonists, which arrived at different periods, & which had to fill the places in the economy of nature, occupied by 20 birds in the very much smaller island of Madeira; hence I suppose that nearly all the birds had to be modified, I may say improved by selection in order to fill as perfectly as possible their new places; some as Geospiza, probably the earliest colonists, having undergone far more change than the other species; Geospiza now presenting a marvellous range of difference in their beaks, from that of a gross-beak to a wren;[1] one sub-genus of Geospiza mocking a starling, another a parrot in the form of their beaks. In this archipelago, moreover, there could be little retardation, or none, from crossing with unaltered forms from the continent.

I have remarked that in animals of which two individuals unite at each act of reproduction some degree of separation must be if not actually necessary, yet most advantageous. This may arise from a selected individual with its descendants, as soon as formed even into an extremely slightly different variety, tending to haunt a somewhat different station, breeding/44/at a somewhat different season, & from like varieties preferring to pair with each other.

---

[1] Zoology of the Voyage of the Beagle, Pl. 36 to 44.

The following facts show that this is possible. After matching for experiment the most distinct breeds of Pigeons, the birds, though paired for life, seemed to me to show plainly a liking each for its own kind, so that I was led to ask Mr. Wicking, who has kept a larger stock of various breeds together than any man probably in Britain, whether he thought the different breeds, supposing that there were plenty of males & females of the same kind together, would prefer to match together; & he without having any theory unhesitatingly answered that he was convinced that they would:/ 44 v/it has, moreover, often been remarked that the Dovecot pigeon, the ancestor of all the breeds, seems to have an actual aversion to the several fancy breeds.[1]/44/It has been asserted[2] that sheep of different breeds turned out together tend to separate, one sort taking to the more upland another to the lowland pastures: in the Shetland Islands[3] two breeds of sheep have long kept distinct, the one haunting the mountain summits, the other the lower lands. In the Falkland Islands, Capt. Sulivan assures me/ 45/that the herds of white & brown cattle tend to keep separate, though neither are quite pure: the white haunt the mountains, & contrary to what might have been expected, they breed about a month earlier than the brown. In the New Forest[4] the herds of brown & pale-coloured deer have long kept separate, without intermingling. We have seen in the Catskill Mountains[5] two varieties of the wolf hunting different prey. In N. America, Sir John Richardson[6] says that "there are two well-marked & permanent varieties of the Caribou deer that inhabit the fur-countries; one of them *confined* to the woody & more southern districts, & the other retiring to the woods only in the winter & passing the summer on the Barren Grounds": so that these annual migrations are different; the woodland variety retiring more inland in September, the other more southward. So in Tasmania, Mr. Gould informs me that there are two very slightly different varieties of [   ] one of which migrates & the other does not. Many instances could be given of

[1] The Dovecote by the Rev. E. S. Dixon, p. 155.
[2] [A group of four notes are missing at this place in the manuscript. Darwin published a revised version of the text on fols. 44–5 in *The Variation of Animals and Plants under Domestication*, 1st ed. (London, 1868), II, pp. 102–3. On the MS., he pencilled a 'U' over the portions used. It seems likely that he transferred this sheet of notes to his later MS. The citations are supplied from the published text, ch. 16, notes 6 & 7: 'For the Norfolk sheep, *see* Marshall's "Rural Economy of Norfolk", vol. ii. p. 136.']
[3] ['See Rev. L. Landt's "Description of Faroe", p. 66.']
[4] [Darwin attributed statement to Bennett: 'White's "Nat. Hist. of Selbourne", edited by Bennett, p. 39.']
[5] [Note missing.]   [6] Fauna Boreali-Americana, p. 239, 250.

Birds of the same species inhabiting the same country, some of which migrate and some do not & which can be distinguished by very slight differences. In all such cases there would be some tendency for varieties having such different habits to keep distinct./

46/We have seen in the fourth chapter how the Common Ravens in Faroe drive away the pied Ravens, though sometimes pairing with them: the hooded & common crow haunt different districts which must check their crossing, for when they meet they often cross; but here we have to do with forms considered as species by most ornithologists. So again in India reputed species of Coracias, as I am informed by Mr. Blyth, intermix & blend on the confines of their range. So do, to give one instance in insects, the Carabus purpurascens of Western Germany & the eastern C. violaceus; at least where they meet there is a reputed third species C. exasperatus, which presents varieties undistinguishable from the two foregoing species.[1]

In the case of plants, as there is reason to suppose that in the majority of cases or at least in many cases only an occasional cross occurs, there will be less retardation in natural selection from this cause; more especially as any favoured variety might rapidly increase, & hold its own, on exactly the same principle, that seed-raisers cultivate large plots of the same variety in order to get pure seed & lessen the ill-effect of an accidental cross. A variety might, also, easily affect a slightly different station & seed/47/at a different period on a hill-top for instance as is known often to be the case. Indeed there are innumerable instances of varieties of plants occupying particular sites or whole districts in the midst of the range of the species: thus the Centaurea nigricans, which Prof. Henslow, as we have seen has proved by culture to be only a variety of C. Nigra, occupies Hampshire to the exclusion of the common forms. The primrose & cowslip are sometimes found mingled though generally affecting slightly different stations. Although there can be little doubt that crossed varieties of plants will have an advantage from their inherent vigour; yet we shall see in our Chapter on Hybridism that there are some few curious but well ascertained facts showing that between certain varieties the pollen of one far from having a prepotent fertilising power on the other variety, is less influential. This leads me to remark, that although facts are greatly wanted to support the hypothesis, that sterility may supervene between varieties slowly formed by natural selection, I think I shall be able to show in the same chapter that

[1] Erichson's Report in Ray Soc. Reports 1841–1842. p. 161.

this is not in itself very improbable. At least I shall be able clearly to show that the difficulty in crossing species & the sterility of their off-spring, by no means follows laws, as if simply/48/ordained to keep species distinct. On the hypothesis that sterility at last supervenes between varieties formed by nature, & called by us species, there will obviously be not the least difficulty, where this has happened in keeping such varieties for ever distinct: But on this hypothesis it may be very important that two varieties during the early formation until converted into species should be isolated or kept apart.

If in opposition to the general facts, given in the third chapter, there do exist organisms, of which two individuals never, or only at intervals of thousands of generations, unite or cross, then these cannot be kept uniform by intercrossing & selection cannot be thus retarded. In such cases the formation of new varieties & species will be stopped only from the absence of a new place in the polity of nature, from the want of variability, the variations not being inherited, the offspring taking after its grandfather or more remote ancestor, instead of its parent./

49/The number of the individuals of any species must form one important element in the formation of new species through natural selection. Several considerations incline me to lay considerable stress on this. We have seen in Ch. IV, on evidence which seems to me satisfactory, that it is actually the common species abounding with individuals which oftener present varieties; & I there gave the obvious reason, that when many individuals existed there would be a better chance within a given period of variations arising, which might in some way prove beneficial to a selected variety. Just in the same way, as an/49 A/agriculturist with a large stock of animals to work on, will have a better chance of gaining a prize for the standard of perfection than will one having only a few animals to select from: so again it is nurserymen, who raise large crops of our different flowers, who generally succeed in getting new & prettier varieties. As in each country all the variable forms are striving through selection to get the upper hand, there is not unlimited time for any one; & if/50⟨55⟩/any particular form be not modified it will run a good chance of being left behind in the race & being thus exterminated.

On the other hand a large number of individuals will apparently be injurious by favouring intercrossing with the selected forms. But we have not facts enough to guide our conjectures on these complex points: it may be that varieties, even amongst organisms which do not freely cross, generally arise on a small spot, partially isolated, in the midst of the range of the parent-species; & that they remain there till so much modified, as to spread largely by overcoming the parent form; sometimes crossing with it on its confines with selection continually acting on the crossed forms./

50 v/The supplanting of a parent-species by a variety, which inherits all the characters & advantages of the parent, with some superadded advantages, will generally be an extremely slow process, as already explained in a former part of this chapter; for instance a variety more capable of enduring drought or resisting some insect will have an advantage over its parent only in the dryer spot, or chiefly where the enemy abound, yet during fluctuations of seasons, when very dry or when the hostile insect is unusually abundant, it will everywhere have an advantage, & tend to spread & supplant its parent./

50⟨55⟩/As perfectly isolated spots, such as islands, are often small, selection will be here retarded by the fewness of the individuals; but at the same time the competition will be less severe & there will be less danger of the extermination of a new variety from their being fewer forms to give rise to other new & victorious varieties or species. The greater number of open places in the polity of nature in islands, especially if stocked by chance colonists only at long intervals, could probably more than counteract the evil from the fewness of individual numbers;/51⟨56⟩/Certainly, oceanic islands abound out of all proportion to their area, with endemic forms, in comparison with continents; but for reasons hereafter to be given, I suspect that the formation of species through nat[ural] sel[ection] has been slower. Considering the whole world, from the fewness of the completely isolated spots, & from the difficulty of the subsequent diffusion of new forms therein produced, such isolated spots, will probably not have played a very important part in the manufacturing of species.

Slowness of Selection.—From the various considerations now advanced, we can see that the formation of new species must be an extremely slow process. New places in the polity of nature for the occupation of the modified descendants of any species can be formed in most cases only at an extremely slow rate. Such new places will be due to physical changes, which will act either directly on the habits or requirements of the inhabitants, or, in a more important manner, indirectly, by causing the extermination or change in proportional numbers of some of the species; by immigration, also, not only will new places be opened to the immigrating species, but the economy of many of the old inhabitants may be thus most seriously affected. All such changes will generally occur either very slowly or at long intervals. Secondly we require for the formation of new species, variability, & repeated variation of/52/ the most diversified nature, in order that changes of structure may occur in the right direction. Variability will depend on the conditions, more especially on changing conditions, to which the organic being is exposed; & the amount of variation will in part depend on the number of varying individuals. Selection acts only by the addition of infinitely small & numerous variations in some given & advantageous direction; & the process will be stopped by want of inheritance in any such characters & retarded by inter-crossing.

I can well believe that many will exclaim, that these causes are amply sufficient wholly to stop all modification through natural selection: I do not believe so; but the result must be judged of by the general phenomena of nature. That changes will usually be extremely slow, I fully admit; & I am convinced that a fair view of the geological history of the world accords perfectly with an extreme degree of slowness in any modification of its inhabitants./

53/*On the absence of intermediate forms or links between species of the same genus.*—One of the most obvious difficulty on our theory, is if two or more species have descended from a common parent, & have been so slowly modified by numerous small changes, why do we not see all around us, or find embedded as fossils in the earth, innumerable varieties or the finest links closely connecting in an unbroken chain such species? This subject must be discussed here at some length, & likewise in our chapter on palaeontology. That such links must, on our theory, have existed, or do now exist, I fully admit. With respect to the nature of the links it is difficult always to keep clear of one source of deception, namely the expectation of finding *direct* links between any two species which we are considering: an example from our domestic breeds of pigeons will make what I mean clear; if we take a carrier & Fan-tail pigeon & consider their origin, we have not the least reason to expect graduated links between them, namely birds with longer beaks slightly covered with wattle & at same time with tail slightly expanded; but what we should find, if we had records of every bird kept by fanciers, during the last few thousand years, would be varieties intermediate in character between carriers & the rock-pigeons, & between fan-tails & rock-pigeons:/53 v/The rock-pigeon, being in its general characters intermedial between these two breeds, though not having a long-beak covered with any wattle, or having its tail at all expanded.[1]/53/So again, still more strong, if we look to two species/54/remote in character; for instance the Horse & Tapir; from not having any idea, what on our theory, was their common ancestor, it is hardly possible to avoid the conclusion that numerous forms directly intermedial between these two must have existed: whereas it might well happen that the common ancestor was fully as unlike in many of its characters a horse or a tapir, as these two animals are from each other, yet being in its general organisation intermediate between them, though, perhaps much more nearly resembling one of these two genera, than the other.

From what we have already seen in this chapter, it seems probable that each variety, whether arising insensibly from the slow modification of the whole parent-stock, or when formed in a separate area, or on some one spot within the same area with its parent, & subsequently spreading, will tend in the long run to supplant & exterminate its parent-stock; for its formation is due to some new advantage gained under the conditions to which it is exposed, & it will generally largely inherit the advantages of its parent. This process will be continually repeated. In all these cases we could obtain a chain of intermediate gradations, only by discovering fossil remains of extinct forms; for of those living at one time & within one area we should see only the parent-stock and one or two varieties, which if destined to become triumphant will increase in numbers & range & so ultimately supplant the parent; the parent, I may add,/55/being ranked as the variety, as soon as its range became less than that of the conquering variety. In the cases of insensible modification we should not at any one time see within the same area, a variety recognizably different from the parent, only mere individual differences.

Why in those classes, of which fossil remains are capable of being preserved & have been abundantly discovered, we do not find innumerable links connecting recent with extinct species, will be most conveniently discussed in our chapter on palaeontology. I think several fairly good reasons can be assigned. I will here only add that the whole force of the difficulty rests on the assumption that our geological records are not only nearly continuous in time, but during each period nearly continuous in space; for otherwise varieties, which seem at first to be so frequently local

---

[1] [Fol. 53 A] June 1858. I doubt whether I have got *intermediate* links yet clear. An animal rarely ranges over whole continents from climate—if it ranges to some extent then it will get into new conditions, but they will change rather abruptly, & only few cases—so we ought not to expect infinite gradation at same time only over moderate area, over which climate will let it range.

could only rarely be preserved. We should, also, remember that the definition of the term species is arbitrary; if an extinct form be found to a certain extent intermediate in character between two existing species, as is of such frequent occurrence; this may be fairly viewed on my theory as one of the intermedial links; the extinct form may have been the actual ancestor of our two species, or/56/more probably it may be an early & less modified descendant of the common ancestor, either in the direct line of descent of one of the two species or in a collateral & extinct line;[1] but all naturalists would rank our in some degree intermediate fossil as a distinct species, without they likewise discovered every intermediate grade between it & one of the living species; but that this should be asserted obviously requires the collection of very many specimens, which generally must have been embedded at slightly different periods & over a considerable area: supposing moreover this to have been effected, as occasionally has been the case, nothing more is thought about it; it is only the case of two forms at first ranked by our palaeontologist as two species & subsequently proved by a second palaeontologist to be merely varieties. Conchologists now doubt whether certain sea-shells, living on the shores of N. America & Europe should be ranked as species or varieties; when the present day has become a miocene or eocene epoch is it probable that the palaeontologists of that far future epoch will find fossilised intermediate links between these now living & doubtful forms. He who does not expect this, has no right, as far as I can see, to expect now to find all the fossil links between a recent & closely allied fossil shell./

57/Looking now to the present time alone, if we travel for instance southward over a continent; we find at the point whence we start many species very common, but as we travel southward some of them become, more or less abruptly, rarer & rarer, till they disappear; but as they disappear, other closely allied or representative species, apparently filling nearly the same place in the economy of nature, take their place, at first being rare, & then becoming more or less abruptly, common. The two species, both comparatively rare, often commingle in neutral territory which

---

[1] What I mean may, perhaps, be best understood by turning to the diagram printed at p. [236–7] Let a^{10} & l^{10} be two now living forms with all their ancestors extinct. If A should chance to be discovered it will be strictly intermediate, though it might in many of its characters far more resemble a^{10} than l^{10}; if a^2 were found, it would be a lineal ancestor of a^{10}, & it would be largely intermedial between a^{10} & l^{10}, for it had diverged but little from the parent-type A. So it would be with f^3 or i^3, for they are early & collateral descendants from A, which have become extinct & transmitted no descendants.

is *narrow*. Every naturalist must have been struck with very many such cases amongst the birds & mammals of large continents: it may be observed with plants in ascending mountains, & with shells, as discovered by the dredge, in the descending depths of the sea.[1] Why in such neutral or border territories without any barriers dividing them into sub-regions and under apparently quite intermedial conditions do we not commonly find intermediate & graduated forms, connecting the two species, which are supposed by our theory to have originally descended from a common parent. That we most rarely find such forms is most certain; the two species, even selecting the most locomotive & freely crossing animals, on comparison, will be found in every single respect as distinct, as if specimens had been taken from the metropolis of each species. This/58/for a long time, formerly appeared to me a most serious difficulty; but the difficulty is largely due, as I believe, to common yet erroneous views on several points in nature.

In the first place we should be very cautious in concluding that because a continent is now continuous, it has remained in this state during the whole period of existing species. How many extensive areas have been greatly elevated within the period of existing shells; & what wonderful changes of level are shown by erratic boulders now scattered over the low-lands & mountain-summits, & which have been borne on ice-rafts over the sea. What an enormous amount of recent depression of level may be inferred from the structure of living coral-reefs. Even when we have no direct evidence, the form of the land sometimes leads to the conclusion, as in the case of the southern extremity of Africa, which is so extraordinarily rich in species, that it formed at no very remote epoch a large archipelago of islands. It is probable that very many single volcanic islands have within the recent period existed as a group of islets; like those forming the little Madeira group which are inhabited by many distinct species & distinct varieties. Even when there has been no change of level, desert tracts may formerly have intervened, where the land is now continuously fertile. If we look at some of the larger volcanic/59/islands, or read Mr Webb & Berthelot's account of Teneriffe, we shall see that some of the valleys are almost as perfectly separated for some organic beings from each other by lofty spurs as if divided by arms of the sea.

In such isolated fragments of land, groups of the same species

---

[1] See Prof. E. Forbes numerical observations on this head in his Report. Brit. Association on the Aegean Sea 18 [43] p. [174 & passim.]

might become differently modified, for they would be associated, especially after any changes in climate &c, with different sets & different proportional numbers of competing associates; & in such cases there could be formed no graduated intermedial links by crossing; nor would the more important conditions of life, in relation to other organic beings, graduate insensibly away between one of the isolated fragments of land and another. After reelevation, if the new forms had been sufficiently modified not to cross & blend with each other, each would spread as far as it could, & would mingle in the intermediate territories with other forms proceeding from different birth-places. On each of the once isolated spots & there alone, we ought to find if our geological records were perfect, intermedial links between the new forms and the states under which they formerly existed.

Nor should we forget the facts, already given, of varieties of the most freely crossing animals, sometimes keeping apart, or breeding at different seasons &c, which would greatly lessen or prevent the formation/60/of intermediate links by crossing, though it would not often lessen the function of such links in relation to the intermediate state of the conditions of life. Unfortunately in those cases, in which intermediate varieties have been found, connecting two races or two closely allied species, we hardly ever know, whether they have originated from crossing or from the direct & graduated action of climate, or from natural selection having fitted those intermediate forms for intermediate conditions of life. And our ignorance on this head greatly adds to the perplexity of this whole subject.

Although I believe the former broken & isolated state of parts of now continuous areas, & in a lesser degree the voluntary separation of the varieties of the higher animals, have played a very important part in the formation of species since become commingled, or just meeting in a border territory; yet I do not doubt that many species have been formed at different points of an absolutely continuous area, of which the physical conditions graduate from one point to another in the most insensible manner. But here lies a source of deception; we are so much struck with the evident manner in which the heat or moisture graduates away, in going from one latitude to another, that we can hardly avoid/ 61/overlooking the more important relations of organic beings to each other. We have every reason to believe, from what we see in gardens & manageries, that almost all organisms can withstand more heat, cold, moisture or dryness, than they are exposed to within their natural range; the definite limit to the range of most species, under gradually increasing unfavourable conditions, being the presence of other competing forms better adapted to such conditions. So that in going for instance southward, the decreasing numbers & final disappearance of any species, is not by any means wholly due to the extremely gradual change of climate, but to the sudden presence of other competing forms, or the sudden absence of others, on which our species may chiefly depend for food; & the relation of the prey or food will again depend on other organic beings; all nature being bound together in an inextricable net-work of relations./61 v/A change in climate is very obvious, but the struggle for existence, depending on many contingencies & chiefly on other organic beings often far removed in the scale of nature is extremely obscure; & it is most difficult to keep this steadily in mind. Hence we have no reason to expect that in going southward that any one species ought to be insensibly modified in relation to the slowly changing climate, but chiefly in relation to each new set of those organic beings, with which it comes into the most direct competition or stands in some relation; & the zone with really intermediate conditions, will depend in chief part on the range of other organic beings. As we see that the range of most organisms is in some degree defined, the species becoming, generally within a rather narrow space, rare & then quite disappearing, the zone with really intermediate conditions for any two species will generally be narrow, & therefore cannot support any vast number of varieties intermediate between such two species.

It comes to this that if the majority of the living forms in any country, as every one can see with care, are defined in their character and do not insensibly blend together, then the relations in range & in all other respects of any one form undergoing modification will tend to be defined: if organic beings had been in a wholly preponderant degree related to climate alone, then the range & specific modification of any form undergoing modi-fication would have been related in an indubitably [?] ⟨clear⟩ manner to the insensibly changing climate./

62/Whether we ought, on our theory, to find many cases of two species closely connected by intermediate links in the narrow zone, which is really intermediate in all its relations to the two bordering species, must depend on whether at the same period many species are undergoing modification & on whether inter-mediate varieties, when once formed are likely to endure for long periods. Every fact in geology seems to show that species change very slowly & therefore I conclude that but few species are undergoing modification at any one period; but as the process by our theory is excessively slow, some such cases ought to occur in every large area. I believe that they do, & in our Ch. IV several cases have been given of varieties connecting two forms, which have been considered by several naturalists as good species. The cases on record are probably few compared with those which exist in nature; for varieties or sub-species or species (for there is no rule to follow in knowing what to call such forms) seem to be scanty in individual numbers, & hence would be observed, generally, only in countries which have been well worked.

The truth of intermediate varieties being individually rare is of importance to us. Mr./63/Wollaston[1] has stated his opinion that this is the case, & he informs me that it is founded upon his observations on insects & land-molluscs; and from his immense experience in collecting, few naturalists have a better right to express an opinion. I applied to Mr. H. C. Watson & to Dr. Asa Gray for their opinions on this head; as from their critical knowledge of the floras of Great Britain & the United States, everyone would place great confidence in their judgment. Both these botanists concur in this opinion, & Mr. Watson has given me a list of twelve nearly intermediate varieties found in Britain which are rarer than the forms, which they connect. But both these naturalists have insisted strongly on various sources of doubt in forming any decided judgement on this head./

64/Therefore, as it seems to me, we ought to expect to find only some few cases of intermediate varieties, inhabiting a narrow zone between the areas inhabited by any two species which they closely link together. But it may be asked, if varieties intermediate in character between two bordering species are ever once formed in such narrow intermediate zones, why do they not endure for as long a time as the species which they connect? & if they did so endure, cases of linking varieties could hardly fail to have become in the course of time with species after species undergoing modification far commoner in nature than they seem to be. I think some sufficient reasons can be assigned why they should not last for very long periods. As they inhabit a narrow zone (for we have seen zones with really intermediate conditions must generally be narrow) they can hardly be, & do not seem to be, numerous in individuals, so that they would be in some degree liable to extinction from great fluctuations in seasons, or any extraordinary increase of enemies. They are, also, bordered on each hand by forms adapted to the somewhat different physical conditions, to greater heat or

---

[1] Variation of Species p. 5.

cold, moisture or dryness &c, to the coinhabitants of the bordering regions, so that if during [a] few successive seasons the temperature became higher or lower &c, they would/65/be liable to invasion on either hand; & if they had not great powers of endurance or of migration, or if any slight obstacles intervened to migration, they would be liable to be wholly extirpated. Moreover in the case of any two species having moderately wide ranges & commingling, as is so often the case, in a narrow border territory, if we suppose this border territory to have been once peopled by a chain of intermediate links connecting the two bordering species, we can see that these latter from having wider ranges would be more abundant in individuals, than the intermediate forms in the narrow intermediate zone; and on the principle already explained of a large number of individuals greatly favouring the production of favourable variations, one or the other of the two bordering species would have a better chance of being modified or improved so as to seize on the place of the intermediate links, & perhaps even to invade the territory of the other bordering species.

Finally, then, I suppose, that a large number of closely allied or representative species, now inhabiting open & continuous areas, were originally formed in parts formerly isolated; or that the varieties became in fact isolated from haunting different stations, disliking each other, breeding at different times &c, so as not to cross./66/That amongst those organisms, of which two individuals rarely (or never) unite for reproduction, that varieties have arisen on some one spot & from having some advantage over their parents either during occasional times or at all times has spread (perhaps sometimes crossing on their confines) & have supplanted their parent-forms; & this would be most readily effected in small & isolated districts. That amongst organisms of all kinds, I suppose, that many species have been formed on different points of open & continuous areas, of which the physical conditions change insensibly, & that in such cases linking varieties have been formed, but that these would not tend to be infinitely numerous & spread over a wider space, for they would by no means be related solely to the insensibly changing climate, but in an equally or more important manner to the somewhat definite ranges of certain other organic beings. Such linking varieties (whether produced by the action of natural selection or of external influence in an intermediate degree, or by crossing) seem, as might have been inferred from their theoretically restricted range, not to be abundant in individuals; & hence, I believe, would be apt to be exterminated by fluctuations of seasons, extraordinary increase of enemies &c,

and by the inroads of the bordering species, which/67/they link together. And lastly, I believe that these bordering species would have a better chance, owing to their greater individual numbers, of being modified & improved, so as to seize on the places of the intermediate & linking varieties. I am well aware, that if I wished to treat my subject as a mere advocate, it would have been better to have slurred over all these complex actions & contingencies, which apparently must affect the formation of new species, & of the relative importance of which I cannot judge; but my object is to point out all difficulties, as plainly, as lies in my power./

68/*Summary of Chapter.*—During the severe struggle for existence, to which all organic beings, owing to their high rate of increase, are exposed, during some period of their lives or during some shortly succeeding generations, Natural Selection acts by the simple preservation of those individuals which are best adapted to the complex contingencies to which all are related. Natural Selection can seize on plainly marked variations or on the slightest modifications, on mere individual differences even though inappreciable by the human eye, if in any way whatever advantageous to the individual, from its egg state to as late a period as the powers of generation last & can transmit any new character. As pecularities are often, probably generally, inherited at corresponding ages, it can modify the egg or seed, the larva, or pupa, without causing any change in the adult form except such as necessarily follows from correlation of growth. As peculiarities are often inherited by the corresponding sex, it can modify each sex in relation to the other; and the individuals of the male sex may be modified by sexual selection, enabling them to struggle for supremacy with other males, like natural selection modifies both sexes that they may struggle for supremacy/69/with other & distinct organisms. Sexual selection will also aid natural selection in giving most offspring to the most vigorous males, under whatever conditions they live. Natural Selection will scrutinize every habit, instinct, constitutional difference, every organ external & internal, will preserve the good, & rigidly reject the bad. It may pause in its work for thousands of generations, but whenever a right & fitting variation occurs, without error & without caprice natural selection will seize on it. From the several reasons already assigned, the process in all or nearly all cases will be excessively slow.

The greater the variability the better the chance of favourable variations. Individual differences seem to be of almost universal occurrence; a larger amount of variability apparently depends mainly on changed conditions of life. The chance of favourable variations occurring will, also, stand in some close relation to the number of the individuals of the varying species. External conditions will, also, act directly on the individuals differently exposed & so modify them to a certain/70/limited extent: as will, also, use & disuse; but to these subjects we shall have to recur in a future chapter.

Intercrossing will prevent or retard the process of natural selection; but here we are involved in much doubt. Those animals, which move much about & unite for each/70 v/birth will thus be kept truest to their parental type; or if undergoing change will be modified in an insensible manner, without any recognizable variety being formed at any one period. It may, however, be otherwise in those cases, in which varieties of the most freely crossing animals,/70/from their very first commencement, haunt some distinct station or breed at different periods &c. Those organisms which rarely cross, & which are capable of increasing at a quick rate, may be formed on some one spot, & thence spread with little retardation from intercrossing.

The direction in which natural selection will act & its very power to effect any thing will mainly depend on there being places in the natural economy of any country not filled up, or not filled up as perfectly as possible. And this will depend on the number, nature, & relations of the other inhabitants of the region, in a far more important manner than on its physical conditions. Look/71/ at the woodpecker or the Bee or almost any other animal or on plants (though here the relations to other organisms, as we have seen in our last Chapter, are less plain, though not less certain) & see how clearly their structure is related to other organic beings: a woodpecker or bee may inhabit the hottest or coldest, the dampest or driest regions, yet how essentially similar is its whole organization. Hence I infer that the association of an organism with a new set of beings, or with different proportional numbers of the old inhabitants, as perhaps the most important of all elements of structural change. If a carnivorous or herbivorous animal is to be modified, it will almost certainly be modified in relation to its prey or food, or in relation to the enemies it has to escape from. Change of climate will act indirectly in a far more important manner than directly, namely in exterminating some of the old inhabitants or in favouring the increase of others. The immigration of a few new forms, or even of a single one, may well cause an entire revolution in the relations/72/of a multitude of the old occupants. If a certain number of forms are modified through natural selection, this alone will almost certainly lead to the modification of some of the other inhabitants. Every where we see organic action & reaction. All nature is bound together by an inextricable web of relations; if some forms become changed & make progress, those which are not modified or may be said to lag behind, will sooner or later perish.—

When a district is isolated, so that after any change in its physical conditions, new beings cannot freely immigrate, or enter only by a rare accident, the relations between its inhabitants will assuredly in time become greatly disturbed. Hence I infer that isolation would be eminently favourable to the production through natural selection of new specific forms. Isolation will also to a certain extent lessen the retarding influence of intercrossing. It will facilitate the supplanting of the parent type by its modified offspring, & lastly it will give time for a variety to be sufficiently changed so as/73/not to blend with, and to hold its own against, other varieties formed elsewhere, with which it may hereafter be thrown into competition.

As each new variety is formed through natural selection, solely from having some advantage over its parent, each new variety will tend to supplant & exterminate its predecessor. In regard to the intermediate links by which each new species must once have been closely connected with its parent, we could expect generally to find such only amongst fossil remains. In those cases however in which a species, ranging over a continuous area, is at the present day in the act of breaking up into two or more distinct species, we ought to find intermediate links in that narrow border territory which is really intermediate in all its organic & inorganic conditions; but we have no reason to expect to find many such cases, & we do find some. The intermediate links in such border territories, from reasons already assigned, would be liable to early extermination.

As a general rule we have seen that widely diffused species, abounding with in-/74/dividuals, & belonging to large flourishing genera, are those which vary most. Of the varieties descended from any one species, the most divergent, or those which differ most from each other & their parents in all respects, will in the long run prevail, for they will be enabled to fill more & more widely different places in the polity of nature. It follows from this that the amount of difference which at first may have been very small between any two varieties from the same species, in each successive set of new varieties descended from the first two, will steadily tend to augment as the most divergent or different will generally be preserved. From reasons already given, namely from the number of different places in the polity of any country not being indefinitely large, and from the individual numbers of each species necessarily being small where very many species exist, which will render such poor species liable to accidental extinction, and will check further modification,—the number of species inhabiting any country will not increase indefinitely; and as the most divergent are those which are the most likely to succeed, the intermediate forms, whether called varieties/75/or species of the same genus or of distinct genera, will tend to disappear.

The groups already large being those which vary most, & the principle of divergence always favouring the most extreme forms, & consequently leading to the extinction of the intermediate and less extreme, will taken together give rise to that broken yet connected series of living & extinct organisms, whose affinities we attempt to represent in our natural classifications. For all organic beings during all time seem to have been related to each other like twigs diverging from the same branch, branches from the same limb, and limbs from the same main trunk representing the common ancestor of a whole class of organisms, with many an intermediate branch and limb now lost.

Finally, then, in regard to the several contingencies favorable to natural selection, I am inclined to rank changed relations or associations between the inhabitants of a country from opening up new places in its polity, as the most important element of success. The amount of variability, which is largely contingent on the/76/number of individuals, as of secondary importance; though perhaps time being given for each new variety to be perfected before being thrown into competition with other varieties, may be almost equally important. A diminished amount of inter-crossing is probably the least important element. But the subject is far too much involved in doubt for us to be enabled to weigh ⟨to strike any balance between⟩ these several contingencies./

76 A/Thus far I think we may with safety conclude that a large tract of land, stocked with nearly similar species, if by subsidence converted into a group of islands, like those of the great Malay archipelago, would in the course of time be eminently favourable for the production of new forms; & such archipelagoes are known to be extraordinarily rich in species. In the course of time, after our supposed subsidence, we might expect the destruction of some species through climatal changes and the occasional introduction of stray colonists; oscillations of level when downwards would cause more destruction, when upwards would extend the area, &

make new stations;—/77⟨58⟩/and these combined causes would act powerfully on the relations of the inhabitants to each other & would thus open new places in the polity of nature for natural selection to fill. Such changing conditions would also add to the variability of many of the organisms. In such large islands, there would be plenty of individuals to act on; intercrossing, at least on the confines would be prevented; & time would be allowed for the varieties in all the islands to be strongly marked & perfected so as to have a better chance of escaping annihilation, when thrown into competition with other & more favoured varieties, formed elsewhere. Those organisms which were originally common to the whole region, before the first great subsidence, might become converted into new forms, whether called varieties or species, in each separate island, or in some of them remain unaltered, according to the nature of the organic forms with which they had to struggle in each island after it had undergone physical changes. If we now suppose our archipelago, through renewed elevation, to be reconverted into continuous land; then of the several forms produced from the same parent-species in each former island, some would probably remain on the spot to which they had been adapted, some would spread, & if only slightly different might become blended by crossing with other varieties, or they would exterminate them, or if sufficiently distinct might live commingled with them. But in the case both of varieties & species, the most divergent, or those which had become most modified so as to fill the most diverse new places, would have the best chance of surviving.

CHAPTER VII

LAWS OF VARIATION:
VARIETIES & SPECIES COMPARED.

INTRODUCTION

The writing dates for chapter VII are not clear from Darwin's Pocket Diary, which lumps it together with the following chapter. Having finished chapter VI at the end of March, 1857, Darwin presumably commenced writing chapter VII at the beginning of April; for, to judge from the identical four illustrative facts appearing both in a paragraph in Darwin's letter of April 8, 1857, to Hooker[1] and on folio 8 of chapter VII, it seems likely that both passages were written at nearly the same date. Probably by June 5 Darwin's writing had almost reached folio 105, for on that date he mentioned in a letter to Hooker[2] that: 'I have been so much interested this morning in comparing all my notes on the variation of the several species of the genus Equus and the results of their crossing.' This topic, the next to the last discussed in the chapter, forms the subject of folios 105 to 113, and Darwin presumably started to write it up soon after he had reviewed his relevant notes.

Darwin had probably completed chapter VII by July 5, 1857, when he mailed to T. H. Huxley a fair copy of folios 41 to 44 together with the following letter:

Down, Bromley, Kent
July 5 [1857]

My dear Huxley

Will you be so kind as to read the two enclosed pages as you said you would, and consider the little point therein referred to. I have not thought it worth troubling you with how far and in which way the case concerns my work, the point being how far there is any truth in MM Brullé and Barneoud. My plan of work is just to compare partial generalisations of various authors and see how far they corroborate each other. Especially I want your opinion how far you think I am right in bringing in Milne Edwards view of classification. I was long ago much struck with the principle referred to: but I could then see no rational explanation why affinities should go with the *more or less early* branching off from a common embryonic form. But if MM Brullé and Barneoud are right, it seems to me we get some light on Milne Edwards views of classification; and this particularly interests me. I wish I could anyhow test M. Brullé's doctrine; as in Vertebrates the head

---

[1] ML, no. 56.   [2] L & L, II, 101–2; NY, I, 459.

consists of greatly altered Vertebrae, according to this rule, in an early part of the embryonic development of a Vertebrate animal, the head ought to have arrived more nearly to its perfect state, than a dorsal or cervical vertebra to its perfect state. How is this? I have been reading Goodsir, but have found no light on my particular point. The paper impresses me with a high idea of his judgment and knowledge, though, of course, I can form no independent judgment of the truth of his doctrines. But by Jove it would require a wonderful amount of evidence to make one believe that the head of an elephant or tapir had more vertebrae in it, than the head of a Horse or Ox. Many thanks for your last Lecture. How curious the development of *Mysis*!

yours very sincerely

CH. DARWIN

Do you know whether the embryology of a Bat has ever been worked out?[1]

14 Waverley Place, July 7, 1857

My dear Darwin—

I have been looking into Brulle's paper, and all the evidence I can find for his generalization (adduced by himself) is contained in the extract which I inclose—Let us dispose of this first— Paragraph No. 1. is true but does not necessarily either support or weaken his view, which rests on paragraph No. 2.—Now this paragraph is a mass of errors—You will find in my account of the development of *Mysis* that the antennae appear before the gnathites are any of them discoverable—& Rathke states the same thing with regard to *Astacus*—and I believe it to be true of *Crustacea* in general.

The second statement, that the legs do not appear until the buccal appendages have taken on their adult form is equally opposed to my own observations & to⟨all⟩ those of all who have worked in this field.

It would have been very wonderful to me to find Brullé resting such a generalization on such a basis, even had his two affirmations as to matter of fact, been correct. But as they are both *wrong*— one can only stand on one's head in the spirit—

Next as to the converse proposition marked 3). It is equally untrue—Mouths antennules backwards The appendages in *Mysis*

---

[1] C.D. MSS. box 145, pp. 152. W. R. Dawson, ed., *The Huxley Papers a descriptive catalogue*...(London, 1946), p. 26, lists this as Darwin letter no. 22 and gives the date July 5, 1857.

& in *Astacus* appear in regular order from before backwards wholly without respect to their future simplicity or complexity—and, what is still worse for M. Brullé, the ophthalmic peduncles, which as you know well are the most rudimentary & simple of all the appendages in the adult make their appearance at the most very little later than the mandibles & increase in size at first out of all proportion to the other appendages

M. Brullé bases his whole generalization upon what he supposes to occur in the *Crustacea*—whereas the development of both *Astacus* & *Mysis*—affords the most striking refutation of his views Tant pis pour Brullé!

And now having *brûler*'d Brullé (couldn't help the pun) I must say that I can find no support for his generalization elsewhere—There are two organs in the *Vertebrata* where developmental history is especially well qualified to test it—the Heart & the Nervous system—both presenting the greatest possible amount of variation in their degree of perfection in different members of the vertebrate series—The heart of a Fish is very simple as compared with that of a Mammal & a like relation obtains between the brains of the two—[Darwin's comment: 'Good'.] If Brullé's doctrine were correct therefore the Heart & Brain of the Fish should appear at a later period relatively to the other organs than those of the Mammal—I do not know that there is the least evidence of anything of the kind—On the contrary the history of development in the Fish & in the Mammal shews that in both the relative time of appearance of these organs is the same or at any rate the difference if such exist is so insignificant as to have escaped notice—

With regard to Milne Edwards views—I do not think they at all involve or bear out Brullé's. Milne Edwards says nothing, [C.D.: ' ?See to this'.] as far as I am aware about the relative time of appearance of more or less complex organs—I should not understand Milne Edwards doctrine as you put it, in the paragraph I have marked: he seems to me to say that, not the *most highly complex*, but the most *characteristic* organs are the first developed —Thus the chorda dorsalis of vertebrates—a structure characteristic of the group but which is & remains excessively simple, is one of the earliest developed—The animal body is built up like a House —where the Judicious builder begins with putting together the simple rafters—According to Brullé's notion of Nature's operation he would begin with the cornices, cupboards, & grand piano.

It is quite true that "the more widely two animals differ from one another the earlier does their embryonic resemblance cease" but you must remember that the differentiation which takes place is the result not so much of the development of new parts as of the *modification of parts already existing and common to both of the divergent types.*—

I should be quite inclined to believe that a more complex part requires a longer time for its development than a simple one; but it does not at all follow that it should appear *relatively* earlier than the simple part. The Brain, I doubt not, requires a longer time for its development than the spinal cord. Nevertheless they both appear together as a continuous whole, the Brain continuing to change after the spinal cord has attained its perfect form. The period at which an organ appears therefore, seems to me not to furnish the least indication as to the time which is required for that organ to become perfect

You see my verdict would be that Brullé's doctrine is quite unsupported—nay is contradicted by development—so far as animals are concerned—& I suspect a Botanist would give you the same opinion with regard to plants—

Ever yours faithfully

T. H. HUXLEY

[Passage copied by Huxley from Brullé, *Ann. sci. nat.* ser. 3 *zool.* 2 (1844), 282–3:]

1) En suivant, comme on l'a fait dans ces derniers temps les phases du developpement des Crustacés, on voit que les pièces de la bouche et des antennes se manifestent avant les pattes; celles ci ne se montrent que par suite des developpements ultérieurs —2) De leur côté, les antennes sont encore fort peu développées que les pièces de la bouche le sont déjà plus; enfin c'est lorsque les appendices buccaux ont revêtu la forme qu'ils doivent conserver que les pattes commencent à paraître. Il en résulte donc cette conséquence remarquable [sic], *que les appendices se montrent d'autant plus tôt que leur structure doit être plus complexe.* On trouve, en outre, dans ces développements divers une nouvelle preuve de l'analogie des appendices. Ainsi les pattes n'ont pas de transformation à subir elles ne se montrent que quand les autres appendices ont déjà revêtu la forme de mâchoires ou d'antennes. 3) Donc dans un animal articulé *les appendices se montrent d'autant plus tard qu'ils ont moins de transformations à subir*: c'est le complement de la loi précédente. On peut par conséquent juger du degré d'importance et de complication d'un appendice par l'époque même a laquelle il commence à se manifester

[As immediate reactions to Huxley's letter, Darwin
jotted down in pencil the following:]

VII, 41 A/There is only one point in which I cannot follow you.
—*Supposing* Barneouds I do not say Brulles remark were true
& universal, i e that the petal which has to undergo the greatest
am't of development or modification begins to change the soonest
from the simple & common embryonic form of the petals; then
I cannot but think it w^d throw light on Milne Edwards proposition
that the wider apart ⟨more different⟩ the classes of animals, the
sooner do they diverge from the common embryonic plan.—which
common embryonic plan, may be compared to the *similar* petals
in the early bud.—the several petals in *one* flower being compared
to the distinct, but similar embryos of the different classes.—I see
in my abstract that M. Edwards speaks of the most perfect &
important organs being first developed & I sh^d have thought that
the char[acteristic] organs w^d be developed.

These comments Darwin developed in his reply of July 9, 1857, in which
he thanked Huxley and mentioned his decision that he would 'not allude to
this subject, which I rather grieve about, as I wished it to be true; but, alas!
a scientific man ought to have no wishes, no affections—a mere heart of stone.'[1]

Also as if to reject this discussion on Brullé, Darwin took his pencil, altered
the numbering of folio 45 to read '40 to 45', and he half cancelled the corresponding entry in his table of contents.

LAWS OF VARIATION:
VARIETIES & SPECIES COMPARED

1/We have seen in our first & fourth chapters that changed conditions
of existence, especially if accompanied with excess of food, seems
to be a main cause of variation. But it must be owned that we
are profoundly ignorant in regard to the first cause of variation.
We do not know, whether the change in the conditions must be
in some degree abrupt to cause much variation as ⟨I think⟩ may,
perhaps, be inferred from such changes alone affecting the fertility
of organisms; or whether a much slighter change over a longer
period would not be equally effective. We can assign no sort of
reason why one organism varies greatly under domestication, &
why another varies hardly at all: why in a state of Nature, most,
but not all the species of certain whole groups are excessively
variable; & we do not even know whether this latter sort of protean
variation is the same as ordinary variation. Ignorant as we thus

[1] ML, no. 52.

are in regard to the primary cause of variation, yet when varieties do appear, we/2/can sometimes, in a very dim & doubtful manner point out some of the laws governing the changes in structure, as was attempted in the first chapter. Here I shall further treat on this subject; & compare domestic varieties with those naturally produced, & both together with the forms called by Naturalists species.

If it can be shown, even partially, that species differ from each other in a similar manner & apparently according to similar laws, as do varieties, it strengthens our view, that species are only strongly marked varieties with the intermediate gradations lost. The old cosmogonists believed that fossil shells, resembling but not identical with living shells, had been created within the solid rock; & they asked why God should not have thus formed them? The paleontologist would probably now reply, that we see in the fossil & living shell plain evidence of similar structure, & therefore he would affirm that their origin & formation must have/3/been alike. So I believe that the similarity of the laws in the formation of varieties, & in the so-called creation of species, indicates that varieties & species have had a like origin; & not that the one has been due to the nature of surrounding causes, & the other to the direct interposition of the Hand of God.—

The laws which obscurely seem to govern variation, & which were briefly alluded to in our first Chapter, together with some others not then mentioned, may be grouped under the following heads. (1) The immediate action of the ⟨external⟩ conditions of life. (2) The effects of habit & disuse (3) The correlation of growth, namely the manner in which the modification of one part affects another part, either through quite unknown relations, or by such relations as that called by Geoffroy St. Hilaire balancement or compensation, by which the large development of one part is supposed to cause the reduction of another; or by such as the early arrest of development in a part,/4/—the period, at which any modification supervenes, any early change of structure affecting parts subsequently developed;—multiple parts strongly tending to vary in number;—homologous parts varying in a like manner & tending to cohere &c.—(4) Parts developed in any species in an extraordinary manner ⟨& rudimentary parts⟩ tending to vary. (5) Distinct species presenting analogous variations; & a variety of one species, resembling in character another species: reversions to ancestral forms. (6) The distinctive characters of varieties more variable than specific characters; specific characters more variable than generic: secondary sexual characters variable./4a v/Lastly, varieties occuring most frequently amongst those species, which are most closely allied that is those which fall into the larger genera—⟨also amongst the more common species, (or those which are the most vigorous in any region & are consequently most abundant in individual members?) also amongst those which have widest ranges.⟩ It, also, seems that the species in the larger genera, are apt not only to be the most variable but to have the widest ranges & to be the most abundant in individuals. From the facts to be given under the last head we gain, if the view that varieties & species do not essentially differ be true, a slight but deeply interesting prophetic glance into the far future of the organic world; we can dimly see whither the forms of life are tending; where about in the great scale of Nature new species will arrive, & where old forms will tend to disappear./

4 bis/*The immediate or direct action of external conditions.* When we find that certain individuals of a species placed under peculiar conditions, are all or nearly all affected in some particular manner, especially if all are soon affected, & more especially if the modification does not seem of any use to such individuals, so that probably it is not the result of selection, then I should be inclined to attribute the effects to the direct action of the conditions of existence. But it is most difficult to eliminate ⟨the power of⟩ selection:/4 bis'/thus we have reason to believe that climate produces some immediate & direct effect on the woolly covering of animals; but when advantage of this is taken by man & a long-wooled animal is produced by artificial selection, it would be wrong to attribute such wool to the immediate action of climate; & so it would be in the case of natural/4 bis/selection.

From the facts given in the first chapter, I think we may in some case attribute greater size, early maturity, & the nature of the hairy covering &c to the immediate action of food & climate./

5/The time of flowering in plants, & of breeding in animals no doubt is affected by climate; & a more curious difference has been observed in a Lizard, namely that it is oviparous in dry Northern Chile & viviparous in humid Southern Chile.[1] But in such cases we can seldom, perhaps never, separate the various elements of change; we cannot tell whether it be cold or damp or lessened or different food which has produced any given result. The wretchedly dwarfed & often distorted state of the shells in the Baltic may be safely attributed to the brackish waters; for the shells grow more perfect as they approach the open sea. Few Naturalists, however,

---

[1] M. Gay, Annal. des Science Nat. Zoolog 2ᵈ series. Tom v. p. 224.

would rank such shells, or the stunted plants on a lofty mountain, as varieties: But I can hardly see where to draw a line of separation: I presume that it is assumed that these dwarfed states are not hereditary; & this would be a valid distinction; but we have previously seen how difficult it is even to conjecture what is inherited in a state of nature./

5 bis/In some cases of shells having an immense range, as that of the common Buccinum undatum from the North Cape to Senegal,[1] which presents a perfect series of intermediate grades between the extreme northern & southern forms; I presume that the modification may be attributed to temperature: but in cases, where we have a strongly marked variety, at the northern & southern ends of the range, with a narrow zone inhabited by an intermediate form, of which I have observed marked examples with cirripedes, it would be rash to attribute the difference to climate, for natural selection probably has come into play & according to my views is in the act of making two species. In regard to colour, Forbes[2] says "it is easy for the practised conchologist to distinguish specimens of the most painted shells, gathered on the southern coasts of England, from those taken on other parts of our shores:" So it is in a marked degree with the tints of certain shells, specified by him, which range from the shallow laminarian zone into great depths./

6/In the case of insects, if we read the accounts given by Oswald Heer[3] & Wollaston on the changes which the same species undergo in ascending mountains, & in approaching the pole, generally but by no means always becoming darker-coloured we can hardly avoid attributing the change to climate. So again, Mr. Wollaston[4] clearly shows that residence near the sea-coast tends to make insects lurid, & affects them in various ways. In regard to Birds, it will suffice to quote Mr. Gould,[5] whom no one will accuse of running varieties together, & he says that birds of the same species are brighter coloured in the interior of continents than near the coast, which he attributes to the greater clearness of the atmosphere far from the sea.[6]/6A/It is well known that in animals with fur, the skins are much more valuable, the further North they are

---

[1] [Deshayes,] Annales des Sciences. Nat. 2ᵈ series Zoolog. Tom v. p. 291.

[2] Report Brit. Assoc. on British Marine Zoology 1850 p. 254.

[3] Quoted by Bronn. Ges[ch]ichte der Natur B. 2. s. 96.—and Mr. Wollaston on the Variation of Species p 39–41.

[4] Ib. p 57–64.

[5] Zoolog. Soc. Meeting May 8. 1855.

[6] From the character of the *species*, not *varieties*, inhabiting very dry districts, as the Galapagos Archipelego,—the deserts of Peru & Northern Patagonia, it would appear as if dampness was an element in the bright colouring of birds & insects.—

collected.[1] In plants several cases are on record of the same individual or all its seedlings changing in a few generations, without the aid of selection, the tint of its flowers when brought from its native home into our gardens.[2]/6 A v/Cold seems to lessen the intensity of the colours of flowers, as is asserted to be the case with some on high mountains, & as has been observed by the Dutch cultivators with their Hyacinths.[3]/6 A/Moquin Tandon gives some instances of plants acquiring by variation more fleshy leaves, when growing near the sea.[4] It has often been asserted that the same plant is more woolly when growing on mountains than on lowlands, & Moquin Tandon[5] asserts that this change occurred with several species from Pyrenees when placed in the Botanic Garden at Toulouse: but Dr Hooker informs me that the Anthyllis vulneraria is glabrous in the Alps & woolly on hot dry banks:/7/ moreover Dr Hooker after tabulating some Alpine floras does not find that in truly alpine species the proportion of woolly plants to be large. He is inclined to believe that dryness has a stronger tendency to produce hairs on plants.[6]

Most of these variations are apparently of no service to the organisms thus characterised, & therefore not having been affected by selection, may be wholly attributed to the immediate action of the conditions of existence. Small & unimportant as are the modifications, it deserves notice, that they almost invariably tend in the same direction with the characteristic differences of the species peculiar to the districts under comparison. Thus, how incomparably more beautifully coloured are the sea-shells of the Tropics compared with those of the cooler temperate regions. It is, also, well known that shells confined to great depths are almost colourless. Alpine species of Coleoptera are generally dark-coloured; & Mr. Wollaston expressly states as every collector must have noticed that beetles confined to the sea-coast are generally "lurid-testaceous or pale brassy"./7 bis/Species of plants living near the sea frequently have fleshy leaves; those of dry & hot countries woolly leaves; those in tropical regions brilliantly coloured flowers. Arctic quadrupeds are thickly clothed with fur. The species of birds, which are confined to the interior of continents, according

---

[1] Bell's British Quadrupeds on the Ermine Stoat: see Bronn's Ges[ch]ichte der Natur. B. 2. s. 87.

[2] Dr. Hooker on the Climate & Vegetation of the Sikkim Himalaya p 49. in regard to a Rhododendron; & see, also, Mr. H. C. Watson's account of the Azorean Myosotis.

[3] Moquin-Tandon Eléments de Teratologie p. 42.

[4] Eléments de Teratologie Végétale p. 73.   [5] Id. p. 62.

[6] Moquin Tandon gives, also, several facts in corroboration of the same view. Ch. 2.—p. 65.

to Mr. Gould, are more beautifully coloured than those which inhabit the coasts & adjoining islands. In all these cases, the species, which according to our views are only strongly marked varieties, are naturally affected in the same manner, but in a stronger degree, as the forms admitted by naturalists to be mere varieties.[1]

In some cases the action of external causes, which I have called immediate, from its influencing apparently without selection, all the individuals exposed to it,/8/seems indirect in its influence; by which, I mean, that very different conditions will produce the same result. Thus Dr. Harvey, the highest possible authority on sea-weeds, says[2] that the Fucus vesiculosus at the Canary Islands, where the heat is too great for it, appears under a nearly similar form, as in the Baltic where it is injured by the brackish water & mud; & he adds that no one "would be prepared for the fact that the heat of the tropical sea would exercise the same trans-forming power on a particular plant as the mud & fresh-water of a colder climate." In other cases, also, it would appear that an organism presents a nearly similar range of variation under what-ever condition it is exposed: thus to give a very trifling instance, the common Polygala has blue, white & purple flowers in the cold humid island of Faröe in 62° n.[3] in England & southern Europe. The Juncus bufonius which ranges from the arctic regions to the equator "in every region seems to present the same variations in its size & branching."[4] These cases, which I believe to be not common, though Dr. Hooker thinks a good many could be collected,/ 9/lead us back to the perplexing facts of polymorphous species & genera, discussed in the fourth Chapter; they show us how ignorant we are on the subject of variation, & how prepotent an influence, the organisation of the species has on the causes, whatever they may be, of variation.

Upon the whole, I think, we must attribute some effect to the immediate action of external conditions; but I am inclined to think it is very little. Innumerable instances could be given of organisms of all kinds exposed to an immense range of climatal & other conditions,[5] & yet not varying in the least, & although, as

---

[1] [Here Darwin added in pencil:] 'If Buckman did not use selection, here allude to his facts as strongest evidence of direct action of food & cultivation.'

[2] Sea-side Book 1849 p. 66. [Cf. ML no. 56: Darwin letter of April, 8, [1857] and L & L II, 90–1 Darwin letter of April 12 on theme of this paragraph.]

[3] Landt, Description of Feroe p. 180 see Herbert in Hort. Journal vol I p 48 (?) or ask Bentham Lat. of Zante.

[4] D. Don on Indian Junci in Linn. Transactions vol. 18. p. 324.

[5] Perhaps one of the most striking cases, is that given by Göppert (Wiegmann's Archiv fur Naturgesch. 1837 p. 210) of unaltered species growing on hot soil

Mr. Wollaston has remarked, we ought by no means to infer because these causes have no influence on one species, they will have none on another; yet I think we may to a certain extent be guided by ⟨the frequency of such cases of non-variation.⟩ As I consider those forms which are ranked by most Naturalists as independently created species, as only strongly/10/marked varieties, the high degree of generality of the fact, that the tropical & temperate, & temperate & arctic zones, are inhabited by species, often closely allied, ⟨of the same genera,⟩ as strongly confirmatory of the view, that climatal conditions have no great influence on organisation; but to those, who look at species as independently created, these latter facts will have no weight.

Acclimatisation.—Though climatal conditions may have no great influence on organisation or visible structure, yet it is notorious that the great majority of organic beings are adapted, within moderately narrow limits, to the climate ⟨of the regions⟩ which they inhabit. When, therefore, a Naturalist meets an animal with a very wide range, for instance the Puma in the reeking hot forests of Central America, on the dry deserts of Patagonià, in the damp cold woods of Tierra del Fuego & up to the limits of eternal snow on the Cordillera, he is much surprised; for he is accustomed to meet for instance, one species confined to the Tropics, another to the temperate & another to the cold regions; his surprise is, also, increased, from falsely attributing (as I believe) far too much weight/11/to the relations between climate & visible structure; climatal conditions are manifest; but the more important conditions determining each creature's power of getting food & escaping dangers are obscure in the highest degree. Nor must we overrate the degree of adaptation in the constitution of each living being to the climate of its own restricted home: when a new plant is introduced from a foreign land, until actual trial we cannot closely tell what range of climate it will endure. Even plants confined to certain islands, & which have never ranged, as far as we know beyond the narrow confines of their home, are found to endure very different climates: look at the Snowberry tree (Chiococca

above burning coal; & other similar cases given by Humboldt in regard to certain grasses on the edges of hot-springs. Many plants have enormous ranges (see Hooker Introduct. New Zealand Flora p. x) & remain unaltered; some range from the base of the Himalaya & other mountains up to an immense height. A land-shell, the Nanina vesicula ranges from the hot plains of India up to 10,000 feet (Huttons Chronology of Creation p. 202) on the Himalaya, where a Toad has an immense range (Hooker Himalayan Journals vol. 2. p. 96)—For wide range of insects see Mr. Wollaston's excellent discussion, p. 29–31 in his Variation of Species.—It would be easy to accumulate innumerable examples.—

racemosa) how difficult to eradicate from our shrubberies, who would have ever supposed that it had been naturally confined to the West Indian islands? ⟨Those who think each species created, as we now see it, will. Must we say that such island plants were created for the prospective chance of the island becoming joined to the mainland & then the plants in question spreading?—⟩

Nevertheless there can be no question that very many, probably most organic beings are pretty closely adapted to their own & no other climate; & if the species/12/of the same genus are descendants from one common parent, many of them must in the course of ages have become accustomed to very different climates. Is this possible? I think the following facts, though few from the nature of the case, show that plants at least do become in some degree acclimatised. Dr Hooker states[1] that he has found a great difference in the hardiness of individuals of several Himalayan plants, depending upon the height at which the seeds were gathered: he instances seedling Pines, which taken at the height of 12,000 feet, were hardy in England, whilst those from 10,000 feet were tender; & so there is a great difference with the Rhododendron arboreum according to the height at which the seeds have been collected. Mr. Thwaites, the curator of the Botanic Garden at Ceylon, whose accuracy is well known, writes to me, that he finds "that individuals of the same species are acclimatised to different elevations,—being more & more impatient of cultivation at any station, according as they have been transported to it, from stations of greater & greater altitude." Again Mr. H. C. Watson has cultivated a variety of a British Lysimachia brought home from the Azores, & found it was decidedly tender.[2]/

13/I think, also, that there can be little doubt that the varieties & sub-varieties of our domestic animals & plants become in a slight, though very slight degree, acclimatised each to its home: I infer this from the caution incessantly given in works ancient & modern of agriculture, ⟨in all countries,⟩ even in the old Chinese Encyclopedias, not rashly to change the breed of any animal or race of plant from one to another district, more especially in wild mountainous districts./13 v/The horses from Algiers stood the climate of the Crimea better than those of Europe: Merino-sheep from the Cape of Good Hope "are far better adapted" to India than the same breed from England:[3] the cactus introduced into India from Canton, Manilla, Mauritius & the Kew Botanic gardens were undistinguishable to the eye, but the Cochineal insect perceives a

[1] Flora of New Zealand Introduct. p. xi.   [2] Phytologist. vol 2. p. 976.
[3] Royle Productive Resources of India. p. 153.

great difference, for it will flourish only on the Indian plants, supposed to have been formerly imported by the Portuguese.[1]/ 13/Different dogs have extremely different capacities for standing heat, but then their probable origin from distinct species renders this case of no value. No one, I presume doubts that the Negro & Laplander have very different constitutions in regard to climate.

Again we have some instances, but here also from the nature of the case but few, of animals naturally extending their range, though we do not know how far the individuals actually become acclimatised to their new homes: thus Audubon gives several instances of Birds, which undoubtedly/14/have extended their range much further northward during late years in the United States.[2] Thus, also, there can be little doubt that owing to the introduction of cattle, a vulture (Cathartes atratus) in S. America now ranges many hundred miles further south than it originally did three centuries ago.[3] The innumerable instances of plants, not cultivated by man, & of some few animals ([) insects for instance (]) not domesticated, which have been naturalised through his agency in many countries under different climates show clearly that organic beings can adapt themselves, whether or not becoming acclimatised, to new conditions. Look at the common mouse & rat which have run wild on the hottest & dryest volcanic & coral islets under the equator, & in Faroe in the north & at the Falkland Islands in the south; it is opposed to all probability that these species had aboriginally nearly so wide a climatal range. The Fallow-deer is feral in Barbuda in the West Indies, & can live on the shores of the Baltic; but it is superfluous to give other instances.—

These facts lead me to believe, that many organic beings by slowly extending their range, can become acclimatised. Whether the acclimatisation is/15/effected by mere habit, or by the natural selection of individuals born with a constitution, fitted either to greater heat or cold, it is impossible to say: probably both actions concur. The spreading of any organism, in those cases in which there is no physical barrier, will depend, mainly, on the nature of the other inhabitants, that is whether there be any place which it can seize in the polity of nature. If there be such place animals & plants will, sometimes extend their range, even although the climatal conditions are in some considerable degree unfavourable

---

[1] Id. p. 59.

[2] [See Appendix for a group of Darwin's reading notes etc. attached here.]

[3] Zoology of the Voyage of the Beagle [Part III, Birds] p. 7. The Rio Negro is about 500 miles south of Monte Video, where according to tradition they did not formerly exist, having come there from still further north.—

to them, as we see with the Elephant reduced in size in India north of Lat [   ]; & with the Capercailzie,[1] in Northern Scandinavia; & with the dwarfed trees in the northern parts of Scotland & the United States. But the spreading will, also, depend upon how closely the organism has become rigidly acclimatised to the conditions of its native home. Nearly all our domestic animals & some plants have great climatal flexibility of organisation, as we see in their cultivation & in their becoming feral under such different climates; & in their generally retaining perfect fertility under sudden & great changes of climate. Although in many cases we do not know/16/what were the parent forms & what their natural ranges, or how many aboriginally distinct species are now blended together in our domestic races; yet if we look at the whole body of our domestic productions or even if for instance we run through the shorter catalogue of our domesticated Birds—there can be no doubt that they live under a much greater diversity of climate than do an equal number of organisms taken at haphazard in a state of nature. The arguments given towards the close of our second Chapter have convinced me that our domestic productions were not aboriginally selected from having this constitutional flexibility, though doubtless they are far more useful from possessing it; half-civilised man could neither know, nor would he care, whether the animal which he was taming or the plant which he was cultivating was thus constituted; he would not care for this more than did the Laplander when he domesticated the Rein-deer, or the inhabitants of the hot deserts of the East when he domesticated the Camel. Hence then, I conclude, from the very general, though as we have just seen, not/17/universal constitutional flexibility of our domestic productions, either that organisms in a state of nature possess this same quality far more generally than we should expect from their natural ranges, or that the simple act of domestication gives this constitutional capacity for bearing climatal changes in a high-degree. It may be doubted, whether if the wild parent-form or multiple parent-forms of the Horse, the goat the Fowl &c the maize, tobacco, rice, wheat &c were suddenly carried from their wild native state into the various climates under which the domestic races now flourish, they would be prolific & healthy. If this doubt be correct & an organic being subjected to domestication or change of some kind, has its constitutional adaptation to special climate so far broken down, that

---

[1] L. Lloyd Field Sports of the N. of Europe Vol I p. 284. in Lapland this bird seldom weighs more than 9 or 10 pounds, whereas in the southern parts of Sweden it not seldom exceeds 17 pounds in weight.—

it acquires a *general* degree of flexibility, then we can perhaps understand a statement insisted on by M. Alph. De Candolle, which long appeared to me very strange;—namely that with the progress of knowledge, plants in a state of nature are found to divide themselves into two opposed categories, "les unes locales et ordinairement très locales, les autres très répandues."[1] ⟨For according to this notion, as soon as a plant begins to spread, it would be in predicament of a domesticated production & would gain flexibility of organization & might spread very far.—⟩/

18/Finally then I conclude that most animals & plants are capable of spreading beyond their present confines, when no physical barrier is opposed to their progress; the main & general check being the presence of other & better adapted organic beings; a second check being their native acclimatisation but that this may be overcome by habit & natural selection; & that when overcome, the being tends to gain a general degree of flexibility of organisation, allowing it to spread very widely, as far as climate is concerned; its means of obtaining food & escaping danger being then the sole but powerful checks to extension. On this view, such facts as the former existence of a rhinoceros & elephant adapted to a glacial climate—the wide extension of man himself,—of his domestic productions & of those accidentally transported by him —are not exceptions to a general law: it is only that these animals have lost their special acclimatisation & have regained their normal constitutional flexibility./

19/*Effects of use & disuse on structure.*—That constant action will increase the size of a part & that this increase becomes hereditary, I think can hardly be doubted from the facts given in the first chapter for instance the size of the mammae in our cows & goats when habitually milked, the more muscular stomach of owls & gulls fed on vegetable matter; & the great weight of the bones of the legs of the domestic duck &c. On the other hand from disuse parts decrease in size, as we see in the wings of the duck & of the Cochin China fowl. (?) Nor is this at all surprising because as we have seen parts become visibly more developed, or atrophied from accidents & operations, during the life of an individual.

---

[1] Geographie Botanique p. 484

    (a) [On verso of this folio, Darwin pencilled the following remarks: 'Col. Sykes. Fowl from India, *native home*, bred readily in *this* country—*screw loose*—we must say that act of domestication by itself in a being never transported to other country gives flexibility to endure climate!

    (a) A screw loose—this fact of when adapted & enabled to beat two sets of organisms is enabled to beat many more sets, *must* be far more important element —yet above must come into play.']

In a state of nature, the same variety cannot be observed during very many generations; the conditions of existence when they change change most slowly; ⟨& if a sensible modification did occur in any form, that form would naturally be considered as a distinct species,⟩ hence we cannot recognise the effects of use & disuse in varieties in a state of nature. But if we look at species, as only strongly/20/marked varieties, we frequently meet with structure analogous to that resulting from disuse under domestication. Thus the great logger-headed Duck[1] of Tierra del Fuego, which so much surprised the old voyagers, & which I have often watched, cannot use its wings more than a fat Aylesbury duck, & is under any extremity incapable of flight. Feeding, as it chiefly does in the great beds of floating kelp, it does not require wings to escape from danger, to which it would hardly be more exposed, than the ocean-haunting Penguins. The islands of Mauritius, Bourbon, Rodriguez, of North, South & Middle New Zealand, & of Philip all have had birds, incapable of flight; & when we remember that no beast of prey inhabited these islands, & that ground-feeding birds usually take flight only to escape dangers, I should attribute, their almost wingless state to disuse./

21/In New Zealand, the birds incapable of flight, belong, as we know from Prof. Owens wonderful discoveries, to 3 or 4 very different orders; & therefore I should infer that at least so many birds had colonised these islands ages ago, & had since given birth to the score of birds in this state now inhabiting these islands.[2] But as several of these belong to the ostrich family it may be supposed that one at least of the original colonists, arrived, we know not how, at these islands in an already almost wingless state. But in regard to the other almost wingless birds of New Zealand & of the other specified islands, it seems to me probable that they arrived by flight & that their wings since became almost atrophied from disuse in their new & protected homes. In ostriches which inhabit continents & great islands, as we see that they/21A/ can escape danger by their fleetness, & in close quarters by their dangerous kicks, quite as well as any small quadruped, disuse together with the increasing weight of their bodies may well have rendered them incapable of flight. The fact of so many birds with imperfect wings inhabiting oceanic islands, naturally leads us to/

[1] Micropterus brachypterus *Eyton.* Zoology of Voyage of Beagle. [Part III] p. 136.
[2] Nov 21/57 conversation with Owen I think 3 types Rallidae—Aptornis either distinct or a Parrot—& Dinordinae [sic], which includes Apteryx [?]—If there could be winged Dinordinae—these might have come by flight—If Dinordinae close to Rallidae or other winged Birds then perhaps always wingless—Though I should think even Struthioidae were once winged.—

22/Mr. Wollaston's[1] remarkable discovery of the frequently apterous condition of the Beetles at Madeira; for no less than 200 species out of the 550 coleopterous inhabitants of this island, have their wings in various stages of reduction & are incapable of flight; & this undoubtedly is a wonderfully large proportion./22 v/The more wonderful, as winged Beetles would during the whole existence of Madeira as an island have had a better chance of getting there than aboriginally wingless species; just on the same principle that many European birds have by their wings reached Madeira; & that the only mammals existing there are the winged Bats. We see clearly the tendency in the beetles of Madeira to be wingless in the fact mentioned by Mr. Wollaston, that 17 genera here have wingless species, which genera usually have winged species in other parts of the world. Moreover of the /22/29 endemic genera, that is genera strictly wholly confined to the island, no less than 23 have *all* their species incapable of flight! Still more remarkable is Mr. Wollaston's conviction, & no one can be a more capable judge, that some few of the very same species, common to Europe & Madeira, are wingless on this island & winged on the continent; & he gives full details in regard to three of them. Here, then, I may add we have another case of varieties in a particular locality marking the species, which are exposed to the same conditions; or as I should look at the case we here have permanent & strongly marked varieties, called species, very naturally possessing the same character with the less-strongly marked forms, called by naturalists/ 23/varieties.

In regard to the origin of the apterous condition of the Madeiran coleoptera; as Mr. Wollaston repeatedly remarks, that the Beetles on the more exposed rocks lie concealed during the almost incessant winds, & immediately appear in numbers, when the winds lull & the sun shines, something may, perhaps, be attributed to the mere disuse of their wings just as with the males of the silk-moth. But I am inclined here to lay far more stress on the principle of selection with its antagonist action of destruction. Beetles from not being powerful flyers are very liable to be blown out to sea, as I have ⟨repeatedly⟩ witnessed, & this would naturally happen far oftener on a small island than on a continent; therefore on an island active individuals with a strong tendency to use their wings would be oftener destroyed, & sluggish individuals with their wings reduced in size, however little the difference might be, would in the course of ages be oftener preserved, & would leave offspring with the same inherited tendency; & this process ultimately,

---

[1] Insecta Maderensia p. xii. and Variation of Species. p. 82 to 87.

through continued selection, might render the beetles quite safe from being blown to sea, by rendering their wings rudimentary. As the danger would be obviously greater, in the smaller & more exposed islets, I have ascertained through Mr. Wollaston's kindness,/24/that on the Dezertas, a mountainous rock near Madeira, four miles long & about three-quarters in breadth, there are 54 Beetles; & that of these, 26 are winged & 28 wingless, which is a proportion one-fourth larger, than the Dezertas ought to have had in accordance with the proportions of the winged & wingless coleoptera in the whole archipelago./24 v/In working out the proportions, the insects believed by Mr. Wollaston to have been introduced by the agency of man have been left out on both sides.—On the Dezertas, however, the number was only three. If I had contrasted the Beetles on the larger island of Madeira itself, with those on the Dezertas alone, the proportions would probably have been greater than that given in the text./24/From the Salvages, a little rock, between Madeira & the Canary islands, six Beetles are known to Mr. Wollaston, & four of these are apterous: at Kerguelen island, Dr. Hooker found only one beetle & one moth, & both were apterous.

Any beetle which from not being a ground-feeder or which absolutely required wings for any purpose, would on the principle above explained run great risk of utter extinction; without indeed its conditions of life were so highly favourable that it could bear great occasional loss from being blown to sea. Now one of the most remarkable features in the entomology of Madeira, strongly insisted on by Mr. Wollaston[1] is the entire absence or extreme rarity of certain whole Families &/25/great genera of Coleoptera, which abound in species on the mainland of Europe under a similar climate. Thus to take the Families alone of Cicindelidae there is not one species; of the following great groups only one in each up to the present day, has been discovered, namely Buprestidae, Elateridae, thalerophagous Lamellicorns, Telephoridae, Oedemeridae, Silphidae & Pselaphidae. No one but an entomological collector will fully appreciate this most remarkable fact. In considering this list it occurred to me that these very Families (the remark does not apply to all the genera) were exactly those which from their habits of life do actually use their wings far more than other Coleoptera: accordingly I enquired from Mr. Wollaston whether this was not the case, & he has gone through the whole list &, with the exception of the Pselaphidae, says that undoubtedly it is so. Therefore I think we may with some safety conclude that

---

[1] Insecta Maderensia p. x.

a vast majority of those Beetles, the habits of which did not allow them to subsist without wings & therefore did not allow them to become apterous through selection & disuse have been exterminated: & this conclusion supports the former one on the origin of the apterous species./

25A/On the other hand, in those classes of insects which are not ground-feeders & are rapid & powerful flyers, this very power might save them from utter destruction, by allowing them to battle against the wind. Such insects might even have their wings enlarged by natural selection; and Mr./25/Wollaston[1] says he is by no means certain that this is not actually the case with the Lepidoptera & some flower-feeding beetles, which if they are to live at all, must have wings & use them. Hence I can see no difficulty in two directly opposite processes going at on the same time with different members of the same great class; some having their wings reduced by selection & disuse, others having them increased, —just as Pigeon Fanciers during the few last centuries have decreased & increased the length of beak of the tumbler & carrier pigeons, both derived from the same stock. The turning point will have been when an insect first arrived on the island, whether, according to the nature of its food, its individual numbers were increased by its flying less & so running less chance of being blown to sea; or flying better so as to conquer the winds./

26/Such, I believe, to be the explanation of the conditions of the wings of the insects on Madeira; but it must be plainly confessed, that doubt is thrown on it, from the fact, discussed by Mr. Westwood[2] that in many parts of the world, there are insects belonging to various orders, of which individuals may be taken either winged or perfectly apterous; of this fact the common Bed-bug is a well known instance. It has been thought that the wings are developed during hot seasons, but the evidence seems to me hardly sufficient. The facts anyhow plainly show that there is something in regard to the wings of insects, which we do not in the least understand.

Loss of tarsi.—

We will now turn to another somewhat analogous case: Kirby has remarked[3] that in certain Scarabaeidae, (dung-feeding beetles) the anterior tarsi of the males are generally broken off: he examined seventeen specimens in his own collection " & not a single one had

---

[1] Variation of Species p. 87.

[2] Modern Classification of Insects. Vol [II, pp. 473, 158, 431.] Also Mr. Wollaston Variation of Species p. 43–45.—

[3] Introduction to Entomology vol. 3. p. 337. [Actually p. 338.]

a relic of the anterior tarsi;"/26 v/& in Onitis apelles they are so rarely present, that the tarsi in this beetle have been supposed by some authors not to exist./26/I remember formerly, when largely collecting in this Family, having made the same observation; & Mr. F. Smith of the British museum tells me that he, also, has observed it. This frequent, & almost habitual loss of a portion of the/27/front limbs of the males is not common to all the genera, having the same general habits, for it is not observed in Copris or Onthophagus./27 v/I do not suppose that the tarsi are lost by the males fighting: at least in Lethrus, in which the males are known to fight furiously, the tarsi were quite perfect.—/27/If mutilation were inheritable, as many authors believe,—if cutting off a dog's or cat's tail tended to make them produce tailless offspring,—then we might have expected some result from this almost habitual loss of the tarsi; but I cannot believe in mutilation being inherited. Nevertheless so constant a loss clearly shows that the anterior tarsi are of not much service to the insect & therefore probably are not much used; & disuse, I do not doubt causes atrophy & is inherited. Now in the genus Onitis above referred to & likewise in Phanaeus,[1] members of the Scarabaeidae, the tarsi are "very slender & minute", & may be said to be quite rudimentary; indeed in the Brit. Mus. I could not find any specimen of Phanaeus with tarsi, & in another genus, Ateuchus, (which includes the great sacred beetle of the Old Aegyptians) it is well known that the tarsi of the front legs are absolutely deficient/27 v/It would be easy to bring forward cases of the atrophy or entire disappearance of parts *apparently* from disuse; but as these occur in all the individuals of the species, & as I cannot illustrate them by analogous losses merely in individuals or varieties, I have not given them in the text. Many parasitic Crustaceans have their limbs atrophied when attached for life to fishes. In another totally distinct Kingdom, there is a striking case in as much as it occurs in nearly full-grown individuals in the Pholas lamellata; this shell has been described as a distinct species, but has been shown by Mr. W. Clark[2] to be the half grown animal of Pholadidea papyracea, which after it has domed its shell, does not any longer require its foot for boring, & consequently the whole large muscular foot is "depauperated &

---

[1] M. Brullé (in Annal. des Science. Nat. 2 series Zoolog. Tom 8. p. 284) asserts that in Phanaeus the males are deprived of tarsi, whereas the females almost always have them. He adds that in Onitis, the females of some of the species have tarsi, whilst in other species neither the males or females have them. I do not know whether M. Brullé was aware of the frequent *accidental* loss of the tarsi in several other coprophagous genera.—

[2] Annals & Mag. of Nat. History vol 5. 1850. p. 12: see also Dr. Fleming's British Animals [p. 451.]

finally obliterated."/27/Hence I am inclined to attribute the very small size or loss of the tarsi in these beetles, wholly to disuse./

28/*Blindness*.—I have one more class of facts of the same nature to bring forward. It is well known that moles & some allied genera owing to their subterranean habits have either very small yet perfect eyes,/28 v/as in European mole, in which the eyelids are hidden under thick fur, & are one-third of the size of the head of a middle-sized pin;/28/or, eyes excessively minute, & fairly covered over by the hairy skin, so that if they have any vision at all, it must be confined to the dimmest perception of mere light.—The burrowing Aspalax, (a Rodent & therefore belonging to another order of animals) is in the same predicament; its eye being excessively minute & covered not only by skin, but by a tendinous expansion. Now in S. America there is a very common rodent, the Tucu-tucu (Ctenomys Braziliensis), more subterranean in its habits even than the mole: I heard of a Spaniard who had often caught them, & without my making any remark, he stated that "invariably very many are found blind": he procured me some specimens, which I kept alive, & one of them was evidently stone-blind; I preserved it in spirits & Mr. Reid dissected the eye, & found that the blindness had apparently been caused by inflammation of the nictitating membrane. As blindness tends to/ 29/cause atrophy & as diseases of the eye are believed to be strongly hereditary (especially with horses), I can see no difficulty in believing that the eyes of the Tucutuco might be reduced by disuse & disease to the state of those of the Aspalax: yet as inflammation of the eyes must be injurious to any animal, & as the Aspalax can live in its blind state, it may well have been that the absolute closing of its eyes was effected by the continued selection of smaller & smaller eyes & more closely shut eyelids.—

It is well known that in the deep caves of Styria there are many blind insects, & crustacea arachnidae & a reptile the Proteus: in the caves of Kentucky there are, also, blind insects crustaceans, fish, & a Rat./29 v/The various stages of abortion of the eyes in these Kentucky animals is very curious: some have no trace of an eye, some have a rudiment, & the Crustacean has the footstalk for the eye without the organ,—it has the stand for the telescope without the instrument./29/Now as the existence of useless eyes could hardly be injurious to these animals, I should attribute their blindness to simple disuse. Although it is trenching on a distinct subject, I may remark, that many of the cave animals of/30/Europe & No. America, though exposed to closely similar conditions of existence, are except in their blindness very little allied.[1] According to my views, these animals were not created in their respective caves, but American animals must have got into the Kentucky caves, & European animals into those of Styria, slowly penetrating, century after century into the profounder abysses, & gradually have become blind by disuse: they would, also, become modified in any other way, through selection gradually fitting them for their new & dark homes. Now in regard to the Kentucky caves, Prof. Dana informs me that the Crustacean is[2]/

31/In the discussion on the Madeiran insects, I remarked that it was quite possible that natural selection might at the same time be enlarging or reducing the wings of different insects of the same class. In the caves of Kentucky I think we have evidence of something analogous in regard to the eyes of the animals; the contest, however, being here between selection enlarging & disuse alone reducing these organs. The blind cave Rat, instead of having rudimentary or no eyes, has eyes of an immense size; & Prof. Silliman Jun. who kept this animal alive, thought that after a period & when accustomed to the light, it acquired some slight degree of vision. Now if we may suppose that this animal did not habitually live in the utterly dark parts of the caverns, we may suppose according to our principles, that the individuals with infinitesimally larger eyes & a more sensitive optic nerve had been continually selected, until some American rat from the outside world, had been converted into this strange inhabitant of darkness, with its large [?] eyes, blue fur & long moustaches.[3]/

32/In the depths of the ocean, & in deep & dark wells some Crustaceans as Calocaris & Niphargus are blind.[4] Now though I am not aware that any Fish inhabiting very deep water is normally blind, yet it seems to bear on the above facts, that the Gadus lota[5] at the depth of 100 fathoms has its air-bladder frequently atrophied, often accompanied by total blindness. On the other hand, it has been "remarked that fishes which habitually descend

---

[1] Trans. Entomolog. Soc [   ].
[2] [Here Darwin broke off in mid-sentence. On the lower half of the sheet he pencilled the following memoranda: 'Fish & Rat.—
     In the caves of Styria I have failed in finding out the affinities of the insects, but one or two are even thought to be only varieties of European insects.—Proteus has American & European species Look in Dict Class. for range of each genus & write to Dana to ask']
[3] [Darwin left a pencilled question mark at the end of this sentence of which the last half dozen words were scrawled in pencil.]
[4] E. Forbes. Report Brit. Assoc. 1850 p. 254.
[5] Prof. Jurine in Mem. de la Soc. d'Hist. Nat de Geneve Tom 3. p. 149.

to great depths in the ocean have large eyes ".[1] And one most remarkable fact is on record, ⟨which is worth giving, though of a most perplexing nature.⟩ M. Eudes-Deslongchamps gives with great detail two cases[2] of eels taken from wells about 100 feet in depth, which had their eyes of immense size, so that their upper jaw in consequence projected over the lower. But here comes the remarkable fact the first specimen was shown to Agassiz, & he/ 33/ thought it was specifically identical with the common Eel. One of the wells was within the precincts of a prison; & it seems impossible to conjecture how the eel got in; & it seems, moreover, quite incredible that such an alteration could have supervened during one generation: it is, also, most improbable that there should be a race of subterranean eels, for, I believe it is well established that the eel invariably breeds in the sea. Surrounded with difficulty as this case is, we apparently have in the large eyes of these eels, & in the blind Gadus from the deep parts of the lake Leman, a parallel case to the opposite condition of the eyes of the Kentucky cave-fish, crustaceans &c contrasted with the large eyes of the cave-Rat./

34/*Correlation of growth*.—In the first chapter I briefly alluded to several laws, appearing to govern variability. These laws are most imperfectly known; & I will here recapitulate them, adding a few remarks, more especially in regard to a comparison of the structure of those forms, recognised as varieties, & those which are generally supposed to have been formed by distinct acts of creation. Physiologists admit a principle, called "nisus formaticus", which repairs, often in a wonderful manner, accidental injuries; & I think we may infer, that if any part were greatly increased or altered in form by continued selection, this "nisus" would give corresponding size to the vessels & nerves &c, without the direct aid of selection though this might always come into play.

I alluded in the first chapter to the mechanical action, attributed by Vrolik, to the shape of the bones of the pelvis of the mother on the head of the human embryo in different races./34 bis/In various groups of Birds, the form of the kidneys differs remarkably, & M. St. Ange[3] attributes these differences to the varied shape of the pelvis, which would seem to have acted mechanically on them; & the form of the pelvis probably stands in direct relation

---

[1] Sir John Richardson Encyclop. Brit. [8th ed.] art. Fish. ['Ichthyology'] p 219.
[2] Mém de la Soc. Linn. de Normandie vol. 5 1835. p. 47, and vol 7. 1842. [p. xxix.]
[3] Annales des Science. Nat. 1 Ser. Tom 19 p. 327.

to the different powers of locomotion. So again in Snakes, Schlegel[1] has remarked that the varied positions of the heart & of the lungs, the riband-like liver with the gall-bladder removed from it, the anomalous position of the kidneys & organs of generation, all stand in direct relation to the shape of the body, formed for crawling, & to the manner of swallowing: how much of these remarkable modifications ought to be attributed to direct selection acting on slight variations in these important organs, & how much to the indirect, & almost mechanical action of changes in the form of the body & of the mouth, it would be very difficult to say.

In our first chapter I showed that Isidore Geoffroy St. Hilaires law of the multiple parts whether physiologically important or unimportant varying much, in number, holds good both in regard to varieties/35/& to species; I presume that this stands in relation to a greater or less amount of plastic matter, out of which the multiple organs have to be developed, having been accumulated at an early embryonic age.

Homotype ⟨Homologous⟩ parts tend to vary in a similar manner, owing, it may be supposed, to their similarity at an early embryonic period; or one part tends in its variation to imitate another part of ⟨the same homotype⟩ nature. Thus the great anatomist Meckel, has insisted, as stated by Isidore Geoffroy Saint-Hilaire[2] "que les muscles du bras, de l'avant-bras et de la main ne s'ecartent presque jamais de leur type normal par le nombre, et la disposition de leurs parties, sans tomber dans les conditions qu'offrent dans l'état régulier, les muscles de la cuisse, de la jambe & du pied; et reciproquement."

Homologous parts both in animals & plants seem to have a strong abnormal tendency to cohere or unite;[3] & the variations thus caused, can often be so closely paralleled by normal structures, that it is difficult to believe that the parallel is accidental./

35a/Moreover it would appear that multiple parts are especially apt to be variable in form as well as in number. M. Isidore Geoffroy[4] insists on this; & M. Moquin-Tandon[5] observes that "les organes répétés le plus de fois sont aussi ceux dont le developpement est le plus variable." As this "vegetative repetition", to use Prof. Owen's expression, is a sign of a low or little specialised organisation, the foregoing remarks on the variability of multiple parts seems

---

[1] Essay on Serpents, Engl. Translat. p. 26
[2] Hist. Gen. des Anomalies Tom. I p. 635.
[3] Isi. Geoffroy. Hist. Gen. des Anomalies. Tom I. p. 541, 545.—For plants see M. Moquin-Tandon Elements de Tératologie Vegetale. 1841. p. 248, 267.
[4] Hist. des Anomalies Tom I. p. 60, 638 650. Tom. 3. p. 456.
[5] Teratologie Végétale p. 124.

to fall under an observation often made by naturalists that the lower animals are more variable than the higher. And with plants, Dr. Hooker remarks[1] that "variations in the floral organs are apparently more likely to occur the less the individual parts deviate from the normal type, the leaf; as if the more complete adaptation to a special function rendered them less liable to casual variation." Or as the/35B/case may be put, as long as an organ had to act in many ways, its exact form would probably not signify; just as a knife for cutting all sorts of things, may be almost of any shape, but a cutting tool for some particular object had best be of some particular shape; so with an organ as it began to be specialised through natural selection for some particular end, its particular structure would become more & more important; & this same natural selection would tend to keep the form constant by the rejection of accidental deviations, excepting indeed such few as tended to improve the organ; ⟨& these it would seize on;⟩ whereas until the exact shape or structure of the organ became important for its function natural selection would hardly come into play in checking any slight fluctuations in its form./

36/There can be no doubt, that many parts of the organisation of every living thing are correlated together, so that if one part changes, another part will tend to change, by a bond which we can sometimes see dimly but often not at all. Some instances were given in the first chapter of variations thus related; for instance hair & teeth believed by most physiologists to be of an homologous nature in the so-called Turkish or naked dogs: now if we take a general survey of the mammiferous Kingdom, the two orders which are most anomalous in their teeth, namely the Edentata & Cetacea, are certainly most anomalous in their dermal covering; as we see in whales, contrasted with seals, & in the armour of the armadillo Mylodon &c & Ant-eater. I presume the remarkable fact of the seedling Cuscuta not having cotyledons, though germinating in the soil, stands in direct correlation with the mature plant being parasitic on the elaborated sap of other plants & so not requiring leaves.

As I have said the bond of correlation is often quite hidden from us; remember the blueness of the eyes & deafness in cats,—the nakedness of young pigeons & their colour—/37/constitutional differences & complexion &c. So in the gravest & in unimportant monstrosities Is. Geoffroy[1] remarks "que certaines anomalies co-existent rarement entre elles, d'autres fréquemment, d'autres

---

[1] Flora Indica Introduct. p. 29. & Ranunculaceae p. 2.
[2] Hist. des Anomalies &c Tom 3. p. 402

enfin presque constamment, malgré la difference très grande de leur nature, et quoiqu'elles puissent paraitre *completement independantes* les unes des autres ". In looking at organic beings in their normal state one incessantly sees throughout whole groups of animals & plants, having quite different habits, two parts of their organization having no apparent connection, yet almost identical throughout all the species: but it is most difficult in such cases to know whether there is any correlation in the parts. The mere fact of the community of structure in the two parts throughout many allied forms is no proof whatever, according to our theory, of any correlation of growth, for it may be wholly due to community of descent. And in the ancient parent of the allied forms, the two parts may have acquired their present structure & apparent connexion, from having been independently modified for separate purposes through natural selection. ⟨Just as the Fancier is now making by artificial selection the beak of his tumbler-pigeons very short, & the feet very small, without, perhaps, there being any correlation whatever in the growth of these parts.⟩/37 v/But it would be rash even in this case positively to assert that there was no correlation; for it is well known that acephalous monsters are especially liable to have imperfect feet./

38/On the other hand, when in a group of species, the same part or organ differs in each, such differences are very generally, perhaps universally, accompanied by at least slight differences in the surrounding parts. Thus Prof. Owen/38 v/remarks[1] that "he knows of no analogy in the whole mammalian series that would justify a belief" that the lower jaws should not be different in two genera, characterized by a difference in the number of their teeth./38/Such differences in the connected parts, when slight & apparently unimportant in function, may in all probability be attributed to correlation of growth.

As we can hardly suppose that internal & structural differences in the fruit on the same individual plant can be of use to the species, we must attribute the differences in the pericarps,—in their shape, their appendages, & even in the ovary itself with its accessory parts—[2] of the central & marginal florets of many compositae, to some correlation of growth. Possibly it may be a case of compensation, yet there does not seem to be any direct/38 b/ relation between the state of the fruit & the presence or absence

[1] Proc. Geolog. Soc. 1842. p. 692.
[2] H. Cassini in Annal. des Sciences Nat. 1 series. Tom. 17. p. 387.—C. C. Sprengel in his Das Entdeckte gives figures Tab xx of the achenium in central & ray florets of Picris (Helminthia) & Tussilago; Thrincia offers another instance.

of the ray-like corolla in the outer florets. Possibly the differences may be related to the mutual pressure of the flowers: at least the Decandolles[1] are inclined, in the case of certain states of Dianthus polymorphus, to account for the abortion of the anthers & the greater length of the style "to the lateral compression of the flowers in the cymes." But it seems extremely doubtful whether this explanation can be applicable to the differences in the internal structure of the seed, which has been observed in the inner & outer flowers in some Umbelliferae: thus in Hasselquistia[2] the seeds of the ray-flowers are orthospermous & those of the disc coelospermous; & analogous differences have been observed in the Coriander; it is, I may add, to show how important these differences of this kind are that Decandolle has founded on them the classification of the order./38 b v/It is by no means the Umbelliferae with the densest heads, which have the corolla most frequently developed in the external flowers; & in the carrot it is the central flower which is developed in an unusual manner. Perhaps, this whole class of facts are in some way related to nutrient flowing more freely to the central or exterior florets, & may be connected with causes which tend to produce peloria in the line of the axis. But in some instances I suspect, that C. C. Sprengels view that the exterior florets are developed & one bit of calyx in Mussaenda to make flower conspicuous to insects.[3]/

38 b/Certain Leguminosae bear on the same plant flowers of two different kind, & with the/38 c/flowers, as I am informed by Mr. Bentham, the pod sometimes differs. Ad. de Jussieu has described[4] two different kinds of flowers borne by certain species of Malpighiaceae; one flower of the ordinary sort, the other without a corolla or with a mere rudiment of it, two ovaries without a style &c; so that Jussieu remarks in these degraded flowers "the greater number of the characters proper to the species, to the genus, to the family, to the class disappears, which thus laugh at our classifications". Nothing is known of the use or meaning of the two kinds of flowers borne by these & other plants, but I presume that the internal & structural differences in the imperfect flowers, which, however, seed as well & often better than the perfect flowers, can be of no service to the plant, & must be due to some correlation of growth.[5]

To give an instance of a correlation, which I should attribute

---

[1] Mém. Soc Phys. de Genève. Tom. 9 p 78.
[2] Tausch in Annal. des Sciences Nat. 2 series Bot. Tom IV. p. 41.
[3] [See appendix for later memoranda on a folio (watermarked 1858) also numbered 38.] [4] Archives du Mus. d'Hist. Nat. Tom. 3. p 82.
[5] [Here Darwin later pencilled: 'Nectary & petiole [?] of column in Pelargonium'.]

wholly to natural selection, & not to the laws of growth;—winged seeds are never found[1]/39/in an indehiscent fruit; or, as I should put the case, seeds could become winged through natural selection only in fruit which opened, so that the seeds which were blown furthest got an advantage over those less fitted to be acted on by the wind, & thus gradually became winged; & this could never happen through natural selection in a fruit which did not open.

Those who have studied monstrosities believe,[2] that any affection of a part developed during the early life of the embryo tends to modify other parts of the organization subsequently developed. This seems so natural that it can hardly be doubted; & hence the later formed structures as they are necessarily subjected to the influence of all previous abnormal changes, are the most liable to monstrosities & variations. On the same principle monstrosities of axis of the plants almost always affect the appended structures.[3] We may infer from these considerations that the same cause tending to produce a monstrosity or variation would produce different results according to the period at which it acted on the embryo. Perhaps we may to a certain extent understand those sudden & great variations,/40/called by horticulturists 'sports', whether in the bud or seed, by supposing that a modification takes place at a very early age of development & greatly disturbs the whole organisation. I think there can be no doubt that in those animals, which live an independent & active life in their larval condition, any great modification at this period would sensibly alter the structure of the mature animal; & as many insects, when mature, live for a very short time, & never even feeding, have nothing to do but procreate their kind, much of the difference between species & species, may well in many cases be almost wholly due to correlations with their larval condition: on the other hand modifications in the mature state will almost necessarily have been preceded by modification at an earlier age. It must not, however, be supposed that a great amount of change, caused by the continued addition through natural selection of small changes, of any one organ, or at any one period, necessarily causes a correspondingly great change in all other parts of the organisation; or at all other periods of life; for I think the facts given in the first chapter on the changes due to selection under domestication, show that such is not the case./

[1] Alph. De Candolle in Annales des Scienc. Nat. 2 series. Bot. Tom XI. p. 281.
[2] Isidore Geoffroy St. Hilaire Histoire des Anomalies. Tom 3. p. 392 Andral was strongly of the same opinion.—
[3] Moquin-Tandon Elements de Teratologie Vegetale. p. 113.

41/M. Brullé[1] in a memoir on the embryonic transformations of the Articulata[2] insists "qu'un appendice se montre d'autant plus tôt, qu'il doit acquerir un development plus complet". In another part he strongly reurges the truth of this proposition, & asserts that the converse holds good. It would almost appear according to this view as if more time were required for the growth of a part which has to undergo greater embryonic modifications, & that consequently its development had to commence earlier. M. Barneoud[3] has shown something analogous in plants having irregular flowers; for he finds in an Aconite, in Orchidaceae, Labiatae & Scrophulariaceae, that at a very early age the petals are equal & similar; "mais bientôt on remarque entre elles, une difference de grandeur d'autant plus forte et plus précoce/42/que la fleur est plus irreguliere a l'etat adulte". ⟨So that in these cases, the parts which have to undergo most modification from their archetype, grow quicker than the less modified parts.⟩

Prof. Milne Edwards[4] makes a different but somewhat analogous comparison: he does not compare parts in the same individual developed from similar & homologous elements, but the same functional system in quite different groups of animals; & he seems to think that according as the organs in question are most developed in any class, the earlier they appear in the embryo in that class: thus he contrasts the circulatory system in the Vertebrata, in which it is so highly perfected, with the same in Annelids. Indeed the main basis of all affinities, so strongly insisted on by Milne Edwards in this paper & elsewhere,[5] seems to hang on the same principle,—namely that the more widely/43/two animals differ from each other, the earlier does their embryonic resemblance cease; thus a fish on the one hand, & mammals together with birds on the other hand branch off from the common embryonic form at a very early period, whereas mammals & birds being more closely related to each other than to fish, diverge from each other at a later period. This seems to accord with M. Brullé's principle

[1] [Darwin had a fair copy made of the text running from the top of fol. 41 to the first paragraph ending on fol. 44. A few slight additions or changes he made on the fair copy are incorporated in the text given here. The fair copy was sent to Huxley, whose answering letter and Darwin's additional comments are in the introduction to this chapter. Presumably because of Huxley's criticism of Brullé, Darwin later wrote at the top of fol. 41: 'Do not copy this Heading or pages' and, changed the number of fol. 45 to read: '40 to 45' evidently intending to omit fols. 41 to 44.]

[2] Annales des Sci. Nat. 3 Series Zoolog. Tom. 2. p 273, 282–283.

[3] Annal. des Scien. Nat. 3 series Bot. Tom. 6. p. 270, p. 287.

[4] Annales des Sci. nat. 3 Series Zoolog. Tom. 3. p 176.

[5] Annal des Science. Nat. 3 Series Zoolog. Tom. i p. 65.

that the more each part is changed from the common archetype the earlier it is developed; for as a fish differs in nearly all its organization from a mammal, more than a bird differs from the mammal, the fish as a whole would have to be differentiated at an earlier period than a bird. So with Mr. Barneoud's case, if we look at an irregular flower at a period between its earliest condition & maturity, the more irregular & modified petals from having grown at a quicker rate may be said to have been earlier developed. I presume that actual time is not referred to in any of these cases; only relative time one organ being compared with another; for, as is well known, the heart of the chick arrives at the same stage/ 44/of development with that of a mammal in a far shorter actual period of time.[1]

If the foregoing principle be really true & of wide application, it is of importance for us; for then we might conclude that when any part or organ is greatly altered through natural selection it will tend either actually first to appear at an earlier embryonic age or to grow at a quicker rate relatively to the other organs than it did before it had undergone modification.: consequently, as we have seen in the case of monstrosities this early formation will tend to act on the other & subsequently developed parts of the system. This same principle would, also, probably play an important part in the following so-called law of balancement or compensation of growth./

40 to 45/*Compensation or Balancement*: Geoffroy St. Hilaire & Goethe brought forward about the same period this law, which has been admitted by some naturalists & utterly rejected by others: it seems to me that there are the gravest difficulties in proving its truth, & yet I must think that it holds good to a large extent. Goethe puts the case under a clear point of view, when he says[2] "the budget of Nature is fixed; but she is free to dispose of particular sums by any appropriation that may please her. In order to spend on one side, she is forced to economise on the other side." That this sort of compensation holds good with the modifications which our domestic productions have suffered, I can hardly doubt after the facts given in our first chapter; for instance in plants rendered sterile & seedless by their artificial treatment the nutriment goes to the enlargement of the fruit./

46/In monstrosities this law seems, also, to hold: Isidore Geoffroy

---

[1] [See appendix for long note removed from this MS. and placed in port folio C 40 f.]

[2] Pictet on the writings of Goethe, translation in Annals & Mag. of Nat Hist. vol. 2 p 318 1839.

St. Hilaire gives the following example as the best out of hundreds, "dans lequel l'antagonisme de développement m'a semblé aussi evident que possible. Il existait en effet du côté gauche un rein & une capsule surrénale de grandeur ordinaire, et du côté droit, un rein extremement petit et une capsule tres-volumineuse."[1] M. Moquin-Tandon gives several cases of this same law in monstrosities in the vegetable kingdom.[2]

But the question which here more immediately concerns us, is whether we can discern this law in the structure of species in their normal condition. The case of the ribs being so numerous & the limbs absent in serpents has been advanced as one of compensation of growth; & it may be so,/47/but as, according to the principles of this work, a part may be diminished by disuse, & another neighbouring part augmented by use or still more effectually by continued natural selection (for instance the greatly lengthened palpi & antennae in the eyeless cave animals), I do not see how such results are to be distinguished from compensation of growth. Nevertheless so many cases of apparent compensation of growth can be advanced, that I conclude there must be some truth in the law. For, as Mr Waterhouse has remarked to me, it would appear that when any part is greatly increased, adjoining parts or organs do not retain their usual or typical size, but are actually diminished. The large size of the canine teeth & the smallness of the premolars in the Carnivora may be given as an instance. The great size of the thorax & the small size of the abdomen in the Brachyurous Crustaceans & the exactly converse case in the Macroura, have been advanced as cases of compensation: Adouin, for whose opinion/48/one must entertain the highest respect, insists most strongly[3] on the mutual relation in development of the three divisions of the thorax, in the several great orders of insects.

The following great Botanists seem to believe in the law of compensation, not merely in monstrosities, but in plants in their normal state; De Candolle, the elder, Richard Moquin-Tandon & Auguste de Saint-Hilaire. This latter Botanist, (no relation of the founder of the law) gives as instances of compensation, the expansion of the petiole, & the abortion of the limb in many leaves;—the great development of the bracteae when flowers are not developed as in the crown of the pineapple;—& a crowd of instances in which the doubling of the organs in one whorl seems to cause the abortion

[1] Hist. des Anomalies Tom I. p. 276. This case is quoted from M. Martin St. Ange.
[2] Elements de Teratologie Vegetale. p. 155–160.
[3] Annales des Sciences Nat. Tom I (1 series) p. 111. & 416.

of the organs in the succeeding whorls.[1]/49/Moquin-Tandon, besides cases quoted from Decandolle & some monstrosities[2] brings
forward as a case of balancement, the elongated peduncles & bright-
colours of the rudimentary flowers in the Feather Hyacinth; & again
the development of the corolla & the abortion of the stamens & pistils
in the outer flowers of the Snow-ball-tree (Viburnum opulus); &
something of the same kind would appear to hold good in the outer
florets of many Compositae & some Umbelliferae. Ach. Richard[3]
believes that the great size of the bulbillas in certain Crinums
causes the pericarp in these species to be almost rudimentary./

50/If one could feel thoroughly convinced of the truth of this
law of compensation, it would be important. On our view of
species having arisen like varieties we could understand its action.
⟨& we need not call in fresh creations to play the part of the laws
of simple growth.⟩ The order of the development of parts would
probably be seen to be a very important element in change; if
compensation be as powerful a law as many have thought, for the
first developed parts would be apt to rob & so cause the deterioration
of subsequently formed parts. It would, also, I believe, throw some
light on rudimentary organs & parts. I have sometimes been
inclined to think that the supposed law of compensation might be
put under a simpler form; namely that nature, like a careful
manufacturer, always tries to be economical in her materials; &
if any part or organ can be spared, whether or not any adjoining
part be in consequence largely developed, it is spared, & matter
so/51/saved. Animals belonging to very different classes, when
parasitic within other animals & thus protected, offer instances
of this truth: I am thinking of two Cirripeds namely Proteolepas
& the male of Ibla which live within the sacks of other cirripedes,
& in both of these & in no other member of the class the entire
capitulum or carapace is absent & thus saved. In many such
cases, I doubt whether it can be truly said that any other part or
organ has been, either as cause or effect, developed in excess; but
the less nutriment required, owing to some parts of the body
under changed circumstances being through natural selection less
& less developed, might be of service to any creature in the severe
struggle for life to which all are exposed: just as on the same
pasture a greater number of animals in a moderately thin state,
could be kept alive, than of animals with a thick layer of fat./

[1] Lecons de Botanique 1841. p. 145, 199, 619. Again in Annal. des Sciences Nat.
Bot. 2 series Tom I. p. 333. he advances this law in relation to stamens & petals
in certain genera.
[2] Elements de Teratologie p 157.
[3] Annal. des Sci. Nat. Tom 2. 1 series, p. 15.

52/A part normally developed in any species in an extraordinary degree or manner, in comparison with the same part in allied species, tends to be highly variable.—

Several years ago, Mr. Waterhouse[1] published a/52A/remark to nearly this effect; Professor Owen, also, seems to have come independently to a similar conclusion. I was formerly much struck with Mr. Waterhouse's remark, for I could see no reason why in a species, if looked at as an independent creation, a part developed in any highly peculiar manner or to an extraordinary size should tend to be eminently variable: on the other hand if a species be only a strongly marked variety, the cause of this variability, we shall see, is not of very difficult explanation./

53/I must here premise that our apparent law, which we are here going to discuss relates only to parts differing greatly from the same parts in species if not actually congenerous at least pretty closely allied: nor do I suppose that the rule is of universal application. To give an imaginary example, the wing of a Bat is a part developed in a highly remarkable manner in comparison with the front-legs of other mammals, but our law would not here apply: it would apply only to some one Bat having wings developed in an extraordinary degree, or manner, compared with other closely allied Bats,. When several species within the same genus differ remarkably one from the other in some part or organ, which is uniform throughout the rest of the same Family, then according to our law, the part or organ in question should tend to be variable in the species of the genus. Our supposed law is applicable to any character, although attached exclusively to either male or female sex, if the character be very remarkable in comparison with the same part in the corresponding sex of the/ 54/allied species. Moreover as all secondary sexual characters, whether or not developed in any especial manner, may be considered as in some degree a departure from the typical structure of the group to which the species in question belongs,[2] for instance the male Turkey, Fowl & Pheasant all depart a little more from the typical structure of the Gallinaceae than do the females; so does the female common glow-worm depart far more from the typical structure of the Lampyridae than does the male,—hence

[1] A. Nat. Hist. of the Mammalia. 1848. vol. 2. p. 452, note 1, "As a general rule where any species is characterized by a maximum of development of certain parts, those parts are more subject to variation in the different individuals of the species than are parts which approach more nearly to the normal conditions."

[2] W. [actually John] Hunter's Animal Oeconomy Edited by R. Owen p. 47.— Westwood in Entomolog Discourse has made primarily [?] same remark. [Possibly in Addresses to the Entomological Society, 1851–53, London 1851–53.]

it seems to be conformable to our law, that all secondary sexual characters should be more variable, as I believe they are, than the characters common to the two sexes.

Before giving a list of the more striking facts, which I have accidently met with, I must remark that the cases implying *extraordinary* development cannot be very frequent; & secondly that it is very difficult to collect facts of this kind: I have experienced this myself, & have seen it in others, namely that it is scarcely/ 55/possible, on being asked, to call to mind relations of a complicated kind without going deliberately through every species in a group with which one must be thoroughly familiar. Having been struck with Mr. Waterhouses remark before I undertook the classification of the Cirripedia I attended to it & was astonished at its wide application; so that I generally found some most striking & remarkable character in a species of far less use for classification than I had anticipated owing to its surprising variability.[1] Moreover from Cirripedes being hermaphrodite, the cases are the more valuable, as clearly showing that the law holds good without any relation to sexual distinctions. As Birds are generally remarkably constant in their structure, I have also particularly attended to those few cases in which, in comparison to closely allied birds, some part presents a very unusual character, & we shall immediately see how apt these characters are to be universally variable. These cases of Birds, together with my own experience with cirripedes, have/56/mainly convinced me that there is much truth in our supposed law.—/56 v/On the other hand I have been led to doubt its truth from not having noticed any analogous remarks in Botanical works, & I believe in the present state of Natural History Botanical generalisations are more to be trusted than those deduced from Zoology. I applied to Dr. Hooker on this subject, who after careful consideration, informs me that though some facts seem to countenance the rule, yet quite as many or more are opposed to it. In plants one large class of cases, namely secondary sexual characters are not present. Moreover, as Dr. Hooker has remarked to me, in all plants there is so much variability, that it becomes very difficult to form a judgment on the degrees of variability: in a Bird having a beak of unusual structure we are at once struck at any variation, as the beak in other birds very seldom varies; but with a plant, how difficult to judge whether an abnormal leaf or petal varies more than leaves or petals of ordinary forms!

In parts developed to a great size, a source of deception should be

[1] Monograph of the Balanidae. Ray Soc. 1854. p. 155.

here noticed; namely that the variation, even if not really greater than in other species in which the same organ is of the usual size, would be far more conspicuous. But this source of doubt does not apply to parts developed not to a great size, but in an unusual manner./

56/Naturalists have repeatedly remarked that every part of the living frame can be shown to be variable in some or another species: hence, as a mere coincidence, I should have expected that some few instances would have occurred of parts developed in any remarkable manner, being likewise variable in the same species. But it must be remembered that instances of parts developed in great excess or very differently from the same part in allied species are not numerous; & secondly that the cases of variability in organs which are usually constant in form (of which fact we have several instances in the following list) are decidedly rare; therefore the improbability is very great of variability, itself rare, being a mere chance concomitant, of unusual development, also rare in the same part or organ in the same species. Hence, I think, we may infer that there is some direct relation between the variability/57/& the unusual, though normal, development of the same part. I may here add that many Naturalists believe that variability is related to the slight functional importance of the part: ⟨I do not myself believe in this doctrine;⟩ it is therefore worth notice that when a part or organ is developed in a remarkable manner in a particular species, the most obvious inference is that it is of at least as much, probably of more, importance to the species in question, than the same part or organ where less developed in the allied species; & yet, as we shall immediately see, it is nevertheless generally highly variable./

58/The Hystrix cristata has a skull readily distinguished by "the enormous size of the nasal bones," but these bones, & "the highly arched upper surface of the cranium "are subject to considerable variation ".[1]

The *male* Narwhal has, perhaps, the most anomalous teeth of any mammal, & here we have variability in the length, of the tusk & sometimes the second incisor is developed into a short tusk.[2]/

[1] Waterhouse. Nat. Hist. of Mammalia vol. 2. p. 452. [Darwin here jotted in pencil the following:] Zoolog. Soc. beginning of 1857 Owen Length of arm of ourang-outang longest arm & very variable in length—Owen has some other cases about teeth I feel sure ⟨Falconer tusks of Elephants but the sexual Elements certainly variable but not confined to a species or genus⟩ To give a few instances of male *sexual* characters variable—Tusks of Elephants ⟨Narwhal⟩ Mane of Lion if Persian be same species—Deers Horns [1 or 2 words illegible.]

[2] Scoresby. Arctic Regions Vol. I. p. 490.—Kane's Arctic Exploration vol 1 p. 455. I have looked in Penny Encyclop. & Dict. Class & can find out no other particular of variability. [Note continued at foot of p. 310.]

59/The Wax-wing, Bombycilla garrula, is very remarkable from the wing-feathers being tipped by scarlet horny points which differ a little in the male & female: Macgillivray adds "the principal variations have reference to the wax-like appendages to the secondary quills".[1]

The Chimney Swallow, Hirundo rustica, differs from many of its congeners, by its forked tail, which is much shorter in the female: Macgillivray says that it exhibits little variation, except in the tinge of the red on its breast "& in the lateral tail-feathers being more or less elongated."[2]

The Oyster-Catcher, Haemantopus ostralegus, certainly has a remarkable beak, & Macgillivray says "considerable differences occur in the size of the bird, & especially in the length & shape of the bill."[3]

The Cross-Bill, Loxia Europaea, has a most singular bill, as its name implies: several ornithologists have been struck by its great variability: Macgillivray says "the variations which I have observed in adult birds are not remarkable, excepting in regard to size, & especially in that of the Bill, which varies considerably/60/in length, curvature & the degree of elongation of lower mandible." He then gives various measurements showing how remarkably great the variations are in this important & generally constant part of the Bird's structure:[4]/60 v/The upper mandible, moreover, sometimes crosses from the right & sometimes from the left; & this variation is the more remarkable, as certain muscles are unequally developed on the two sides, in accordance to the side to which the upper mandible crosses over:/[5]

60/The long-legged Plover or Himantopus forms a small genus with closely allied species, quite remarkable from their extraordinary length of the legs compared with their nearest allies. Mr. Gosse[6] has carefully attended to the measurements of the legs in H. nigricollis, & he finds no two birds with exactly the same length of leg, there being as much as half an inch in length difference between the extreme specimens. This bird is likewise remarkable by its

[Reading note sheet, marked '12' in ochre:]
Gleanings from the Menagerie of Knowsley Hall 1850.
J. E. Gray p. 55. Though Horns in Deer (a sexual character) are of service for separating the species into groups. & though they have largely been used for specific distinction, yet Dr. G. finds, ['] yet it has been found that animals of same herd or even family & sometimes even the same specimen under different circumstances, in succeeding years have produced Horns so unlike in size & form, that they might have been considered as belonging to different species ['] See *good* case in Sir J. Richardson Fauna. p 241 [part i. Re variation in horns of Barren Ground Caribou, *Cervus tarandus* var. *arctica*.]

[1] British Birds vol. 3. p. 535.   [2] British Birds vol. 3. p. 560.
[3] British Birds vol. 4. p. 155.   [4] British Birds. Vol. i. p. 423.—
[5] Yarrell in Zoological Journal. vol. 4. p. 459.   [6] Birds of Jamaica p 388.

bill being slightly upturned; but Mr. Gosse finds this character well pronounced in only one out of 16 or 18 specimens.—/

60 bis/In Trochilus polytmus the curvature of the beak seems in some degree a sexual character, being according to Mr. Gosse[1] plainest in the female; but the curvature "varies in the individuals, ⟨& I possess several females whose beaks are more curved than in T. Mango."⟩

One of the species of Chamelion [sic] (C. bifurcus) is most extra-ordinary from its nose being divided & produced into two horn-like protuberances; but H. Schlegel[2] says that "the nasal prominences are subject to variation."

In the genus Cygnus, the trachea in some species follows the usual course, in others it makes the most remarkable convolutions, entering the breast-bone, & these convolutions differ greatly in some of the species; in the Whooping Swan "the diameter of the trachea & the extent to which it enters the crest of the sternum varies"; in Bewicks Swan, also, the trachea is not constant, the horizontal loop being sometimes absent, & in some specimens it does not differ from that of the Whooper.[3]

61/*Cirripedes.*—In Conchoderma the valves are very abnormal in shape & astonishingly variable but then they are in some degree rudimentary. One species of Concoderma [sic] differs from all other Cirripedes in having curious ear-like appendages to the capitulum & these, also, are very variable. Alepas cornuta differs from the other species of the genus in having horn-like projections on the capitulum, & these are variable in shape & position. Balanus laevis differs from all other cirripedes in having the basis filled up with a cancellated structure; the extent to which this is effected is very variable & very often there is no trace whatever of this remarkable structure. In Chthamalus antennatus the third pair of cirri (legs) is *very* remarkable in having one of the rami wonderfully elongated & apparently developed to act as an antenna; but this elongation of the one ramus & the number of its segments, are marvellously variable; & the arrangement of the spines, which are of functional importance & generally constant, was equally or even more variable, being arranged/62/on two distinct plans. Acasta sulcata is unique in having the pedicel of the fourth cirrus developed into most beautiful, curved, prehensile teeth; but of this remarkable structure there was not a trace in some specimens

[1] Birds of Jamaica p. 98. [2] Essay on Serpents Engl. Translat. 1843. p. 216.
[3] E. Blyth in Calcutta Review 1857. p. 155. and Macgillivray British Birds vol. 4. p. 651, 665.

from the same district which after the most careful examination I am fully convinced belong certainly to the same species: moreover similarly anomalous teeth on the lower segments of the cirri were also *highly* variable. These teeth are not mere spines but actual modifications of the margin of the limb: their presence caused also the abortion of the usual moveable spines. As I look at a strongly marked variety as not essentially differing from a species, I may advance as an illustration of our law, a variety of Balanus balanoides which I at first described as a distinct species: in this the segments of the posterior cirri had ten pairs of main spines—a number quite unparalleled in any other cirripede whatever; but on examining many specimens I found the number varying from seven to ten pairs!/

63/Lastly I may advance the case of the opercular valves in Pyrgoma & in the too closely allied genus Creusia: the opercula valves, I may premise, are of the highest functional importance, & stand in direct relation to the most important muscles in the animal's body. These valves present very slight differences in most of the genera of sessile cirripedes; but in Pyrgoma they differ in the most striking manner in the different species: I had not sufficient specimens in most of the species to ascertain whether they varied much as they ought to do according to our law; but in/63 v/Pyrgoma cancellatum, the ridge giving attachment to the great & important adductor scutorum muscle is developed in the most wonderful & abnormal manner, & it is variable: in P. dentatum our/63/law is fulfilled in a more marked manner: the scutum in this species has a special ledge greatly developed;—it has the articular ridge developed into a unique tooth-like projection;—the whole outline of the tergum is most unusual & on the inner side there is a unique tooth; now all these extraordinary conformations varied/64/in so wonderful a manner, that it is no exaggeration to say that the *varieties* differed far more from each other in these important parts of the structure than do the other *genera* of sessile cirripedes in the same parts. Creusia spinulosa might be added to this list; but the variation in the opercular valves was so great & so hopelessly perplexing that after weeks of labour I had to give up in despair the determination of what to rank as species & what as varieties.[1]/

65/*Insects.*—We will now turn to insects, & give some illustrations

---

[1] See my monographs on the Lepadidae & Balanidae (p 155) published by the Ray Soc. Under the heads of the Genera & Species, above specified, full details are given.—

from several of the great orders. One of the most striking cases
has been given to me by Mr. Wollaston, namely that of a beetle,
the Eurygnathus Laterillei [actually Latreillei][1] the *female* of which
presents "the extraordinary anomaly" of its head being immensely
more developed than that of the male; & Mr. Wollaston believes
that the case is unparalleled in the whole vast order of Coleoptera:
now this, though serving as a well-marked specific character, is
so excessively inconstant that "scarcely two females have their
heads of exactly the same size"; in some there being only a tendency
in this direction, in about two-thirds of the specimens, the head
being "literally immense"./65 v/The females of some species of
Dyti[s]cus, which normally have their elytra deeply furrowed in
a very remarkable manner, are sometimes quite without these
furrows; yet such females have been caught in connection with
the males.[2]/65/Mr. Wollaston believes that the Harpalus vividus
is the only species in this great genus, which has its elytra connate,
"but this character, anomalous as it is, is far from uniform". In
the whole genus Scarites the mandibles are in both sexes remarkably
developed, compared with other Carabideous genera, & Mr. Wollaston informs that "in size they are imminently variable."—In
the/66/Stag-Beetle, & indeed generally in the Lucanidae, the
mandibles in the males are enormously developed & are eminently
variable not only in size but in the form of the terminal teeth;
yet the mandibles, as Mr. Stephens[3] has well remarked in ordinary
cases "are dwelt upon almost with mathematical nicety." The
astonishing variability of so important an organ as the mandibles
& of some other organs in this & in many of the following cases, is
rendered very striking if the same part or organ be compared in
a set of females of the very same species, where they will be found
to be almost absolutely identical in form.

Mr. White showed me a series of specimens of a magnificent
Chalcosoma from the Philippines in the British Museum, in which
the females were absolutely similar, but the males exhibited the/
67/most surprising series of varieties in the curious horns on the
thorax & head; these horns being five or six times as large in some
specimens than in others & with great diversity in the teeth: so it
is in Megasoma & many Dynastidae. So again in the males of many
Scarabaeidae & of some Cetoniidae. To turn to another quite
distinct group of Coleoptera; the males of some Staphylinidae are
horned, & the horns are very variable, as in Bledius.

[1] Insecta Maderensia. p. 20.—
[2] Westwood, Modern Classification of Insects. vol. 1 p 104.
[3] Illustrations of British Entomology Mandibulalata. vol 3. 1830 p 367.

The whole snout is much elongated in the male Attelabus & in some Curculionidae, & is in them very variable as I am informed by Mr. Waterhouse. In the male of the Truffle beetle (Leiodes) the thighs are much incrassated &, here again, as I am informed by Mr. Waterhouse they are very variable.—In the males of Choleva, the trochanters of the hind legs are liable to great variation.[1] In the carrion feeding Necrodes littoralis, the males have incrassated & dentated femora, increased [sic. Stephens says "incurved"] tibiae/68/dilated tarsi, sculptured thorax, costated elytra, & everyone of these points is highly variable![2]

To turn to another Order, the Homoptera: the Umbonia spinosa was pointed out to me by Mr. White as having most singular spinose projections of the thorax, in both sexes, & these are highly variable. Again in Fulgora or the Lantern-Fly & in the Fulgoridae the forehead is most singularly dilated into a muzzle, sometimes even equalling the whole rest of the body in length! this strange projection differs greatly in the different species,[3] is not confined to either sex, & is very variable in several species, as I saw in specimens shown me by Mr. White.

Lastly to take one other great order, the Hymenoptera, in which I am indebted to the highest authority Mr. F. Smith, for the following striking illustrations of highly abnormal characters in several species, being, as heretofore, highly variable. Both sexes of the Chrysis ignita[4] are highly peculiar from the apex of the abdomen being armed with four teeth;/69/but these are so variable in length as well as in position, as to assume nine distinct types of form, & are occasionally nearly or quite absent! The male of the Andrena longipes, in some examples, but not in all, has an enormously large head in comparison with that of the female: in another species, Andrena fulva, large males have a long acute tooth at the base of the mandibles, but in smaller specimens this is reduced to a mere tubercle; & this form was consequently described by Kirby as a distinct species. In a male Saw-fly, the Tenthredo femorata, a series of specimens "exhibits a wonderful difference in the development of the posterior femora." On the other hand the females of the two following Bees have a peculiarity very remarkable & confined to two or three species in their respective genera: namely in the Osmia fulva, two stout horns on the front of the head, & these vary greatly in length & shape,

---

[1] Andrew Murray, Monograph of the Genus Catops 1856 p 14–.
[2] Stephens Illustration of British Entomology vol 3. p. 367.
[3] Westwood Modern Classification of Insects vol 2. p 428.
[4] Entomologists Annual 1857 [p 31.].

being somewhat bifurcated when large & wedge-shaped when small; & secondly in the Nomada lineola, two teeth on the labrum, & these vary so much in length/70/that the varieties have been described by Kirby as distinct species.[1]

In considering the foregoing facts, & others might have been added, we see that they fall under three heads, namely of some striking peculiarity being eminently variable, when attached to both sexes, or when attached exclusively to the male sex, which is the commonest case, or exclusively to the female sex. The cases seem to me too numerous & striking to be accounted for by the mere chance coincidence of variability & unusually great development; more especially when we bear in mind how remarkably constant in character many of the very same organs or parts are when not developed in any extraordinary manner in the other species of the same groups. Our laws seem to hold equally/70A/ good, & when all the species of a group differ somewhat from each other in some part, as with the opercular valves in Pyrgoma; as when a single species differs somewhat in some part from its congeners.—indeed the cases do not essentially differ from each other. As genera are mere conventional groups, I should have expected that when a set of genera were closely allied but yet differed from each other in some one organ to a marvellous degree that this organ or part would have been variable in the species of such genera. This, I think does happen sometimes, but certainly very far from always. Thus amongst the Homopterous Insects, we have numerous closely allied genera differing from each other in certain parts in the most extravagant & grotesque manner conceivable,—with ball-spines, bladders, lanterns such as a child might draw out from his fancy—& yet, as I saw with Mr. White in the British Museum, these astonishing peculiarities did not vary much in the species of the several genera./

71/Mr. Waterhouse believes that the extreme diversity in the development of the mandibles & horns in the Lucanidae & Dynastidae is related to the manner in which the insect in its larval state has been nourished; & it deserves notice that in some species, for instance in the Angosoma centaurus there is, as I have been informed by Mr. White, much less variability than in the allied species. In the Lucanidae & allied families, the existence of males presenting a wide range of varieties in their secondary male characters, from extreme development to a close approach to the

---

[1] [Later note on separate slip of paper:] F. Brauer advances in favour of the law as given in the Origin (Verhand. Zoolog. Bot. Gesell. in Wien. 1867 Dec. 4) some highly peculiar characters in the wings of the dimorphic females of the Neuropterous genus Neurothemis.

female condition, is so very general, that a collector is not satisfied, as I am informed by Mr. Waterhouse, until he possesses a complete series of this kind for each species; & this fact perhaps does indicate that there is here something quite unknown & different from the other cases of variation & abnormal development, One is at first strongly tempted to explain all these cases of variability in the secondary male characters by the hypothesis of a great diversity in the virile force of the males; on the same principle that the horns of deer are affected by emasculation, by the amount of food,/72/or by unnatural conditions as confinement on ship-board;/72 v/and I think this explanation may be true to a large extent. We must, however, be cautious in inferring loss of virile powers from loss of the secondary male characters; to give one instance; Sebright Bantam has not sickle-feathers in the tail, yet a writer in Poultry Chronicle, shows that one thus deficient, was the father of an unusual [?] number of chickens.[1]/72/But an analogous hypothesis, would be rather bold when applied to the several cases of variation in remarkable developments, characteristic of the female sex: in the Eurygnathus we should have to suppose that about one-third of the females, namely those with small heads nearly like those of the males, were in some degree sterile. Moreover this view is clearly inapplicable to abnormal characters in no way connected with sexual function, & common to the two sexes as in many birds & as in the hermaphrodite cirripedes, which have afforded us so many instances of parts unusually developed being highly variable.

But now let us turn to what we know in regard to domestic varieties: we have seen in our two first chapters that fancy breeds, —those which the fanciers are now improving ⟨by selection⟩ to their utmost,—are much more difficult to breed true or vary more in the admired & selected points, than breeds which have long inhabited any district without/73/particular care having been paid to them; I refer of course to pure breeds alone & not to fluctuating mongrel breeds which would necessarily be variable from crossing. For example compare the head & beak of the common & improved Tumbler, of the common & improved Carrier, with these same parts in any old [?] breed, as the Fan-tail, in which these points have not been much attended to, & observe what an astonishing range of variation the head & beak present in the two former cases. The cause seems obvious, namely that in each of the later generations, individuals with certain admired points most strongly

---

[1] [Additional pencilled comments:] Hewitt says they are generally deficient in virile force Hen-tailed Game Cocks show no loss of virile powers.

developed have been selected, so that the particular characters in question, though the difference in each generation may have been so slight as to have been scarcely appreciable, have not been fixed by strict inheritance during a long course of centuries. Moreover we have seen in our first chapters that new characters, or those in course of improvement through Selection often become, from quite unknown causes, attached in a greater or lesser degree to one sex, far/74/most generally the male sex,—take for an instance the wattle in the improved carrier Pigeon; furthermore it would appear that those characters which become under domestication attached to one sex, are eminently variable.

Now if we look at species as only strongly marked & very permanent varieties, & consequently at all the species of a small group, as the descendants from some one form,—like the fancy pigeons from the Dovecot,—then those parts in which all the species agree will have been inherited by them for an enormous period, & ought to be thoroughily fixed in the breed: in such cases it will make no difference whether or not the part is developed like a Bat's wing, in an extraordinary manner: on the other hand any part developed in an extraordinary degree or manner compared with the same part in the closely allied species, according to our theory, will have undergone an immensely long course of modification through natural selection within a comparatively recent/ 75/period; for as natural selection acts only by the addition of successive extremely small changes, & as the part in question is developed in an *extraordinary* degree or manner, the process of addition must have required a very long time to have produced the given result; & all this must have taken place since the several species branched off from the common parent stock & therefore long subsequently to any considerable change in the other parts of their organization. Consequently, in accordance with the analogy of our improved domestic breeds, we might have expected that such parts or organs would be the least strictly inherited, with a strong tendency to reversion to the aboriginal parent form. Moreover we might have expected from the same analogy, that some of the comparatively late & extraordinary developments would have become attached to either sex, generally to the male sex, without as far as we can see profiting either sex; & furthermore we might have expected that such secondary sexual characters would have been highly/76/variable,—all facts which seem to hold good in nature. Nor we must forget that Sexual Selection, by which the variations in the secondary characters confined to the males alone, & useful to them in their struggle for the females are added up & accumulated is less rigid than ordinary selection; the less successful males generally leaving some offspring; so that those secondary sexual characters which are of use to the male, would be less rigidly scrutinised & sifted than the characters on which the life or death of the individual male & female depended. On the other hand, if we look to the generally accepted doctrine of each species having been produced by an act of creation, I can see no explanation of the several facts given in the present section, showing that secondary sexual characters, especially if developed in an extreme degree, & generally that all parts developed in any very extraordinary manner, are apt to be highly variable./

77/*A part so little developed, as to be called rudimentary, tends to be highly variable.*—

The subject of Rudimentary organs will be treated of in a separate chapter; I refer here to this one point of variability, as standing in relation to our last proposition of parts developed in an extraordinaryly great degree being variable. The cause, however, I believe to be different: organs become rudimentary through disuse (aided, perhaps, by the principle of compensation & often by natural selection) & through the effect of disuse becoming hereditary at a period of life corresponding with that of the disuse. Disuse shows of course that the part in question is not useful to the Species, & therefore natural selection cannot come into play to keep fixed a part when become useless & rudimentary, namely by destroying all injurious departures from one fixed type. The continued existence of a rudimentary organ depends wholly on the strong principle of inheritance, as we shall, hereafter,/78/attempt more fully to explain. On the other hand, a part developed in an extraordinarily high degree is as I suppose variable, from not having become strictly inheritable,—from natural selection not having had time sufficient to overcome the tendency to reversion & to regulate its own work of adding up very many small successive modifications.—/[1]

79/*Monstrosities: arrests of development.*—As monstrosities can not be clearly distinguished from variations, I must say a few words on some of the conclusions arrived at by those who have studied the subject. Geoffroy St. Hilaire & his son Isidore[2] repeatedly

---

[1] [Pencilled comment later cancelled:] ⟨The diverse branching of horns in Lucanidae & the nine types of abdominal points in Chrysis, shows not *all* reversion—there is fluctuation as well as reversion.⟩

[2] Principes de Phil. Zoolog. 1830 p. 215. And Histoire des Anomalies 1836 Tom. 3. p. 437 et passim.

insist on the law that monstrosities in one animal resemble normal structures in another. So in the vegetable Kingdom M. Moquin-Tandon says, "Entre une fleur monstrueuse et une fleur normale il n'y a souvent d'autre difference que l'état accidentel de la première et l'état habituel de la seconde. La monstrosité est donc en general, l'application insolite a un individu ou a un appareil, de la structure normale d'un autre individu."[1] As the resemblance between a monstrosity & a normal structure is generally not very close, & as the comparison is often made with forms remote in the scale of nature, & as when all within the same great class/80/are included, a vast field for comparison is opened, I cannot avoid the suspicion that some of the resemblances given are simply accidental. But I imagine no one would account for all the resemblances on the doctrine of chance. To give two or three of the best instances from Mr. Isidore Geoffroy;—in the pig,—which has the snout much developed & which is allied, but, as Owen has shown, not so closely as we formerly thought to the Tapir & Elephant, a monstrous trunk is developed oftener than in any other animal: the frequent monstrosity of three, four or even a greater number of breasts in woman seems to stand in relation to the fact of most mammals having more than two mammae: Carps are very subject to a curious monstrosity causing their heads to appear as if truncated, & an almost exactly similar but normal structure is met with in the species of Mormyrus, a genus of fish belonging to the same Order with the carp.[2] Notwithstanding such facts, &/81/many others could be given from the animal & vegetable Kingdoms, I cannot believe that in a state of nature new species arise from changes of structure in old species so great & sudden as to deserve to be called monstrosities. Had this been so, we should have had monstrosities closely resembling other species of the same genus or family; as it is comparisons are instituted with distant members of the same great order or even class, appearing as if picked out almost by chance. Nor can I believe that structures could arise from any sudden & great change of structure (excepting possibly in rarest instances) so beautifully adapted as we know them to be, to the extraordinarily complex conditions of existence against which every species has to struggle. Every part of the machinery of life seems to have been slowly & cautiously modelled to guard against the innumerable contingencies to which it has to be exposed.—

[1] Elements de Teratologie vegetale p. 116 p. 342. The same view is taken by M. Auguste St. Hilaire in his Morphologie Vegetale p. 818.
[2] Histoire des Anomalies Tom I. p 285. Tom 3. p. 353. p. 436.

As all vertebrate animals, for instance, pass/82/through nearly similar embryonic changes, we can see that arrests in the development of any part,—a doctrine on which M. Isidore Geoffroy lays much stress—will account for a certain degree of resemblance of many monstrosities to the normal structure of other animals, even when very remote in the same great class. A very frequent monstrosity in plants having irregular flowers, such as Snap-dragons, is their becoming regular; & as such flowers are known to be regular in their early bud state, I presume that this monstrosity would be admitted to be an arrest of development; as an instance of how all monstrosities are governed by laws, it may be added that the flowers nearest the axis are much the most apt to become regular;[1] thus I have seen a Laburnum tree with the flowers at the end of each raceme open & not having the proper papilionaceous structure.

Other monstrosities appear caused not exactly by arrest, but by abnormal development; thus in the case of a monstrous number of mammae or digits, it may be surmised that in the embryo of all vertebrate/83/animals there is a tendency at some very early age to produce several mammae or digits, & that this tendency from quite unknown causes occasionally becomes fully developed in animals normally having only one or some small number. There are other monstrosities connected with the doubling of parts, the union of distinct embryos &c, to which we need not here allude. And there are other monstrosities, apparently not to be explained by arrests of or increments of development, which are common to various animals & plants in the same great classes, & which I presume can be understood only on the supposition of similar abnormal conditions acting on organic structures having much in common,—so created according to the common belief, but according to our views due to inheritance from a common/84/though sometimes immensely remote stock. I will only further remark that according to these same views, a part or organ may in one creature become normally reduced in size or quite atrophied from disuse during successive generations, in another it may suddenly become so in a monstrosity by an arrest of development;—again in one creature a part may by long-continued natural selection become greatly increased in size or number, in a monstrosity it may suddenly be thus increased by abnormal development; but the possibility of this diverse origin of similar parts, through normal & through monstrous formation, evidently rests on the common embryonic structure of the two forms; & how organisms remote in the same

---

[1] Moquin-Tandon Teratologie Vegetal p. 189.

great classes come to have a similar embryonic structure will be treated of in a future Chapter.—

M. Isidore Geoffroy Saint-Hilaire makes one generalisation which concerns us & well deserves notice;—[1]/85/namely, that the more an organ normally differs in different species of the same group, the more subject it is to individual anomalies: thus taking the case of monstrous deplacements of organs, he affirms, that "Les organes qui se déplacent le plus fréquemment sont aussi ceux qui présentent des déviations plus considérables de la position *normale*."

We will now proceed to some remotely analogous considerations in regard to varieties & species./

86/*Distinct species present analogous variations; & a variation of one species often resembles the normal structures of an allied species:* or more commonly resumes the general character of the group to which it belongs.

In the first Chapter I gave a few instances of variations produced under domestication, resembling in character distinct species; & as some of these might be called inherited monstrosities such case can hardly be distinguished from those alluded to in our last section. Our present section relates more especially to varieties produced under nature or in organisms not much affected by domestication. But I think the bearings of our present discussion will be best shown by first giving an illustration from trifling variations in that group of domestic varieties, which I know best, namely pigeons. In all the main breeds there are analogous sub-varieties, similar in colours,—in having feathered legs, & turn-crowned heads; in several of the breeds & in sub-varieties of others, the lesser wing-coverts are chequered with white & the primaries white. None of these points have any direct relation to the aboriginal parent breed, the Rock-Pigeon; yet, I think it/87/ cannot be doubted that these analogous varieties are due to the several breeds having inherited a like organization from a common source; this organization having been acted on by similar organic & inorganic causes of change: just as we know that children of the same family often show a remarkable parallelism in symptoms when suffering from disease.[2]

Some of these variations as feathered feet, chequered wing-coverts &c are fixed in & characteristic of certain breeds & sub-breeds; therefore when such character appears for the first time

[1] Histoire des Anomalies Tom. i p. 281, 418, 650 &c.
[2] Sir H. Holland [   ].

in a breed, the sub-breed thus characterised presents an analogy to other breeds properly so characterised. On the other hand when a character of the above kind is lost; or to give another instance when a blue Pouter which ought to have all its primaries white is "sword-flighted" that is has some of the first primaries coloured,—or when a Turbit which should have a white tail throws a dark tail[1] (of which Mr. Tegetmeier has had an instance) these/88/are not new variations, but the partial reversions to the parent *breed* but not to the parent *species*. Of reversions to the aboriginal species I have given an excellent instance in my discussion on Pigeons, in the fact that all breeds occasionally throw blue birds, & that these always have the two black bars on the wing, generally a white rump & a white external web to the exterior caudal feathers, —all characteristics of the aboriginal Rock Pigeons. It deserves, as we shall presently see, especial notice that these just specified characters are frequently brought out by crossing two Pigeons neither of which are blue, or probably have had a blue bird in their race for several ⟨many⟩ generations: why the disturbance caused by a cross should have this effect we are perfectly ignorant. In respect to all cases of reversions to ancestral characters, I may revert to the only hypothesis which appears to me tenable; namely that in such cases the child does not in truth resemble its ancestor a hundred or thousand generations back more than its immediate father, but that in *each* generation there has/89/been a tendency to produce the character in question, & that this tendency at last for causes of which we are profoundly ignorant overmasters the causes which have for so long rendered it latent. This does not seem to me more surprising than that the merest rudiment or vestige of an organ should be inherited for numberless generations. Those who explain an abnormal & monstrous number of mammae in a woman from the fact of the number of mammae in vertebrate animals being generally greater than two, will admit that a tendency, as well as an actual rudiment, may be inherited for any length of time. Under this point of view reversion to an ancestral form is only an arrest of development,—or the appearance in the mature state of a character which ought to have been passed through in an earlier stage.

Supposing that we had reason to believe that all the breeds of pigeons had descended from one stock, but did not in the least

[1] I may remark that in crossing various breeds I have clearly noticed that colour sticks to the caudal feathers than to any other part, & secondly to the few first primaries: I have repeatedly noticed in crossing black & white birds of very different breeds, the few first primaries are black, succeeded by white feathers.

know what its characters were, or the ancient character of any of the breeds we should be quite perplexed to conjecture, when an individual was born with a turn-crown whether this was a case of reversion of a character formerly attached to the breed, or a new variation analogous to what had/90/at some former period appeared & become fixed in some other breed. In the case, however, of the blue birds, as so many characters appear together without, as far as we can see, any necessary correlation,/90 v/& as these characters arise from a crossing of distinct breeds,—a cause wholly unlike what must aboriginally give the blue colour—/90/we might have pretty safely inferred that the black wing bars, white rump &c were due to reversion. But whether, or not, we could tell which characters were due to reversions (either to the aboriginal species or to some subsequent but ancient breed) & which to new variations analogous to those already existing in other breeds or sub-breeds, we should without hesitation put all down to a community of organisation from common descent./90 v/Those who believe, as I do, that our Fowls are all descended from the *Gallus Bankiva* have an analogous case in so many breeds, as was remarked to me by Mr. Tegetmeier, having sub-breeds with their feathers edged or laced & other sub-breeds with their feathers transversely barred or pencilled. ⟨This latter character may be derived from the hen of the G. Bankiva (though transferred to the Cocks of some of our breeds) & may be ranked as a case of reversion;⟩ It is doubtful whether either class of colour-marking can be attributed to reversion but both the lacing, & pencilling are variations analogous in one sub breed to another, & likewise to some other quite distinct species of Gallinaceae./

90/Now let us turn to nature; we have frequent instances of distinct species & strongly marked natural races, presenting analogous variations. Thus many Foxes, as C. lagopus, fulvus & vulpes present crucigerous varieties;[1] the American &/91/European Bears both sometime have young with a white collar.[2] So in the British brambles, our best authority[3] says "nearly allied species are apt to sport in parallel varieties........so that the species being ascertained, the same designation & very nearly the same description will characterise the variety in each case": these nearly allied species are themselves looked upon by many Botanists as strongly inherited races.[4] An excellent observer in Sweden, Anders-

[1] Sir J. Richardson Fauna Boreali-Americana p. 84, 93.
[2] Id. p. 15.
[3] Dr. Bell Salter in Henfreys Bot. Gazette vol 2. p. 114.
[4] See the account by Mr. Ed. Lees on Brambles coming true from seed, in Phytologist vol. 3. p. 54.

son[1] describes a set of varieties (which have been described as species) of Carex ampullacea & vesicaria, which "present a perfect analogy of every form in one species to those of the other." As I look at species as only strongly marked varieties, I may adduce one other, but distinct case, namely the remarkable correspondence, as insisted on by Prof. Fries[2] between/92/particular series of the American *species*, of Hieracium, with those of Europe. Now all such cases, of parallelism of variation would be ordinarily accounted for by the species having been created with a nearly similar organization: following the analogy of our domestic productions we should attribute it to community of descent.

If there had been a permanently crucigerous species of Fox (as some believe there is), then the crucigerous variety of another species would have been a case of variation analogous to a distinct species; so would it have been if our supposed crucigerous species had produced a variety without the cross. As we do not know the ancestors of organisms in a state of nature, whether ranked as varieties or species, we can very seldom tell whether their varieties, when resembling in character other species of the genus, are variations for the first time in the breed, or reversions to a state through which the species in question had formerly passed. But in some very few cases we can form from indirect evidence a conjecture on this head./92 v/Thus the British Stoat (Putorius ermineus) may be called a variety in as much as it does not regularly turn white in winter: it has inhabited this country since the Glacial period,[3] & during that period, analogy from other countries can leave no doubt that it was always white in winter.[4]/

93/I will now give in small type such cases as I have collected illustrative of one or a few species varying in a manner closely analogous to other species of the same group.—I may recall to mind my former remark on the difficulty of collecting such cases excepting by an author himself carefully going through the group with which he is most familiar. I think that the following cases are too numerous & precise to be accounted for by mere chance, more especially as the comparison is always made with allied

[1] Henfrey Bot. Gazette vol 2. p. 251.
[2] Bot. Gazette vol 2. p. 185.
[3] Owen, British Fossil Mammals p. 116.
[4] Blyth remark in Loudon [*Mag. Nat. Hist.*] vol. 8. p 50—makes me doubt the case —see about Irish Hare & then give up case [;] if Hare turns allude to it—Bell seems to think common in North—Will not do to quote, without Variable Hare offers a good case [Separate note slip:] Stoat Case Loudon Magazine Vol. 7. 504–591 [vol.] 5. 294—718 ['Zoophilus'] [vol.] 8. 51 [Blyth]. The Stoat turning white in Cornwall strongest case. [The first three references are to short communications of paragraph length.]

species, generally closely allied species. ⟨It has several times occurred to me in reading an account of a set of species of the same
genus which have differed in some remarkable character, that
I have truly predicted that I should find this same character
described as variable in the individuals of some of the species.⟩
In the list of cases of parts greatly or abnormally developed being
highly variable, in so far as the variations bring back the species
to the common type of the genus they might have been here
introduced. All the cases, alluded to in Chapter 4 of varieties
intermediate between two species, whether the one or both vary,
in fact come under this head. And of variations of what systematists
consider/94/trifling characters, as colour, size, proportion being
analogous to other species of the same genus, innumerable instances
could be given,—indeed a large part of the difficulty in identifying
species seems due to varieties approaching in character other
species—but in such cases it seems hardly possible to distinguish
mere general variability, from variations having some direct
relation to the structure of other species of the group. But as far
as they go they confirm what I must consider a law, namely that
variations whether in some degree permanent, or occurring only
in single specimens of one species often assume the character of
another species of the same group./

95/Prof. Vaucher shows (in Mém. Soc. Phy de Genève Tom. i. p. 300)
that the modes of gemmation are constant in each genus with some few
exceptions; for instance, the species of Syringa bred in two ways; & in the
common Lilac the two modes of gemmation are sometimes seen even in the
same bush.—

Decandolle states (in Mem. Soc. Phys. de Geneve Tom. 3. Part ii. p. 67)
that in the Lythraceae some of the very natural genera have some species
with petals & some without; & that in Peplis portula the *individuals* indifferently have petals or none.—

In the Primulaceae, according to M. Duby (Mem Soc. Phys. de Geneve.
Tom. x. p 406) Samolus is the only genus in which the ovary is adherent to
the calyx; and in another genus in this family, namely Cyclamen, some
individuals of one of the species have the ovary partially adherent./

96/Oxalis buplevrifolia has simple & lineal leaves unlike all the other
species of the genus, but Aug. St. Hilaire found some individuals with this
enlarged petiole surmounted by the three usual, though here small leaves.
(Morphologie Vegetale p 143)

The alternation & opposition of the leaves is respectively constant throughout many great Families, but in the Salicaria & Polygleae the species, have
either alternate or opposite leaves; & both often are found even on the
same individual. (Aug. St Hilaire Morphologie Vegetale p. 183)

The torsion in the aestivation of the corolla was thought to be uniform in
the Gentianaceae, but in Gentiana Moorcroftiana & Caucasica it is different
from the rest of the family & in individuals of the latter species, it is found

to vary in the individuals (Decandolle in Annal. des Science Nat. 3 Series Botany Tom I p. 259)/

96 bis/In the Malpighiaceae, A. de Jussieu (Archives du Museum d'hist. Nat. Tom 3. p 86) says the leaves are always opposed, with the single exception of Acridocarpus; & even in this genus one may sometimes remark a tendency, especially in the lower leaves, & even a complete return to opposition

Decandolle has seen a variety of Geranium pratense (Mém Soc. Phys. de Genève Tom 1. p 443) with the two upper petals white & the three lower blue; "on retrouve ici dans les Geraniées à fleurs régulières, cette tendence a la disparité des petales si remarkables [sic] dans plusieurs Pelargoniens [sic]." Geranium & Pelargonium, as every one knows, are very closely related genera.— /

97/Aug. St. Hilaire (Mem. du Mus. d'Hist. Nat. Tom XI. p 49) says that in the different species of Helianthemum one may observe "toutes les nuances possibles entre le plancenta [sic] purement parietal et des loges parfaitement distinctes", & that "dans l'*H. mutabile* une lame, plus ou moins large, s'etend entre le pericarpe et le placenta"; so that in this species the degree of division of the ovary into lodges seems to be variable.

In the Compositae the presence of a ray to the outer florets is generally a constant character; but in Senecio, for instance, some of the species have a ray & some not; & of those species which have not, as the *S. vulgaris*, varieties are found having a small ray; on the other hand in the species ordinarily having a ray, as S. Jacobaea, varieties are sometimes found without a ray./

98/W. Herbert (Amaryllidaceae p 363) says that by crossing Calceolaria arachnoeides [*sic* in Herbert] which is purple with C. corymbosa, which has small purple specks on the yellow corolla, flowers were produced to the surprise of cultivators broadly blotched with dark & even blackish purple "but the subsequent discovery of a Chilian biennial species, which I shall call C. discolor, blotched with a reddish purple in a manner somewhat similar, shewed that such an arrangement of the colour was a natural variation of the genus, which the cultivator might therefore have expected, if all the natural species thereof had been previously known."

Nerine curvifolia fertilised by a hybrid curvifoliâ-pulchella, produced seedlings, of which one produced a young crimson leaf, "such a remarkable seminal variation brings curvifolia in closer affinity with N. marginata, which is distinguished by a red margin to the leaf". (Herbert Amaryllidaceae p. 412)/

99/Gaertner says (Bastarderzeugung p. 50) that Lychnis diurna when growing in dry places sometimes has sharp teeth on the sides of the petal, as is the case with Lychnis flos cuculi. I may add/99 v/that I have seen a seedling Spanish Pink D. Hispanicus with its petals so deeply cut & the point so much elongated, as to call to mind the petals of Dianthus superbus.—/

99/Azalea with five stamens & Rhododendron with ten stamens, of which half are of inferior power are closely allied genera as everyone may see; & W. Herbert says (Journal of Horticultural Soc. vol 2, p. 86) he has raised seedlings of A. Pontica & Indica frequently producing 7, 8, or 9 stamens. Azalea nudiflora in N. America presents "numberless varieties some of them exhibiting ten or more stamens". Asa Gray Manual. p. 257 2nd Edit.

The northern species of Gladiolus present, according to Herbert, (Journal of Hort. Soc. vol 2, p 90) "a strange diversity of seed"; there being a winged or foliaceous margin in some of the *species*, which totally disappears in other *species*; & in G. *communis* some *varieties* have it curtailed, & some almost obsolete.

Of the Oak genus some species are evergreen & some diciduous [sic]; & the varieties of Quercus cerris are so variable in this respect, that Loudon (Arboretum et Fruticetum vol 3. p. 1846) says its varieties "may be arranged as deciduous, sub-evergreen & evergreen."—So it is with the genus Berberis. (Hooker Flora Indica p. 218)/

100/Moquin-Tandon (Teratologie Vegetale p 138) says he has found a plant of Solanum dulcamara in which all the upper flowers had two or three stamens "beaucoup plus longues et plus grosses que les autres", & in S. tridynamum & Amazonium three stamens are habitually much more developed than the others.

Dr. Hooker. (Journal of the Linnean Soc. vol. 2. p 5 Bot.) believes that the Lobeliaceae ought to be included as merely a tribe of the Campanulaceae. For in the Lobeliaceae, "even the irregular corolla affords no good mark, for some states of the Wahlenbergia saxicola (one of Campanulaceae) have an oblique corolla, & unequal inclined anthers, of which two have the connective produced into an appendix, thus imitating a prevalent feature of the Lobeliaceae." The coincidence of these several imitative characters deserves attention.—[1]/

101/The American wolf is generally esteemed a distinct species from the European: Sir J. Richardson says (Fauna Boreali-Americana. Quadrupeds p. 76) "the black mark above the wrist which characterises the European wolf is visible in some American wolves but not in all".

The Didelphys Azarae has a broad black stripe on the forehead; & the D. crancrivora [actually cancrivora] has an indistinct dusky line; the D. Virginiana has *occasionally* "a small dusky stripe on the forehead (G. R. Waterhouse,—Marsupialia in Naturalist Library 1841. p. 84)

The genus Timalia (allied to the Thrushes) according to Swainson (Fauna Boreali-Americana. Birds p 31) stands in a group in which the bill is either notched or entire,—a character generally of high importance; & in Timalia pileata some individuals "have the bill perfectly entire, some slightly, & others distinctly notched; all apparently being old birds, full plumaged, & not differing in the slightest degree in other respects"./

102/Yarrell has stated that the Little Ringed Plover (Charadrius minor) can always be distinguished from Ch. hiaticula by a dusky spot on the inner web of the outer tail-feather; this feather being in C. hiaticula wholly white, but Mr. Garrett & Thompson have shown that this spot does occur *in some specimens*. (Nat. Hist. of Ireland: Birds: vol. 2. p. 103

The position of the Spleen differs much in various serpents, "so as sometimes to occur at a distance from the pancreas & isolated at the posterior surface of the stomach"; and H. Schlegel (Essay on Serpents. Engl. Translat. 1843 p. 55) says he has observed *individual* variations in this respect.—/

103/Mr. Wollaston remarks (Variation of Species p. 62) that "it is almost diagnostic of the genus Gymnaetron that its representatives should be thus (ie with blood-red dashes on the elytra) ornamented typically, or else that those species which are normally black should, *when they vary*, keep in view, as it were, this principle for their wanderers to subscribe to".

---

[1] [Note slip:] Hooker's Misc. 3/109.

Mr W Wilson says the Andromeda polifolia like Vaccinium vitis idea *sometimes* has the stamens attached to the corolla. Lindley puts these genera in *remote* nat. Fam.!!

Mr. Waterhouse informs me that the Pachyrhynchus orbifer one of the splendid Curculionidae of the Philippine Archipelago, which is the most variable of the genus, in its variations typifies the regular markings of the other species. So again in varieties of Cicindela campestris the golden marks became united as in C. sylvicola; & on the other hand in varieties of this latter species the marks become disunited as in C. campestris.

The classification of the Fossorial Hymenoptera was mainly founded by Jurine on the neuration of the wings, & this has been adopted by all subsequent writers; but in Typhia [i.e. Tiphia] & more especially in Pompilus, there is a considerable difference in this respect between the species, & in some of these species there are even *individual* variations (Shuckard on Fossorial Hymenoptera 1837, p. 48, 40, 43.)/

104/I will now give some examples from cirripedes. Acasta fenestrata, & in a lesser degree A. purpurata present a very remarkable character (Darwin Balanidae p. 305) in the shell being perforated by six clefts or holes in the lines of suture; & we have a similar character in some *varieties* of A. sulcata. In the different species of Pyrgoma, the opercular valves on each side are sometimes quite separate & sometimes so perfectly calcified together that even the line of juncture cannot be distinguished; & in P. milleporae the degree of union (p. 368 idem) is very variable in different individuals. In Bal. improvisus (p. 251 id.) certain varieties have their terga closely imitating the form of the same valve in the allied B. eburneus: so again a very remarkable variety of the common Balanus balanoides is the form of the tergum & in the parietes being tubular makes a close approach (p. 275 ib) to the very distinct B. cariosus. Certain varieties (p. 453 ib) of Chthamalus stellatus & cirratus have the anterior ramus of the third pair of cirri elongated & antenniform, prefiguring, as it were, the remarkable structure of this same cirrus in Ch. antennatus./

104 v/⟨Chthamalus Hembeli presents a unique character in the walls of the shell, when old, growing inwards & replacing the basis (Darwin Balanidae p. 450)⟩/

105/Lastly I will give in rather more detail the case which has interested me most, & which combines several considerations. The common Donkey ⟨sometimes is destitute, even when not an albino, of the characteristic transverse stripe on the shoulders⟩//[1]...[a double shoulder-stripe]/105A/is said to have been seen in the Koulan of Pallas, now generally admitted not to be the parent of the domestic ass.[2] The Hemionus is well known to be characterized by not having the cross shoulder stripe, but a trace of this stripe is asserted to appear occasionally.[3]

106/The Quagga, though strongly banded in the front part of

---

[1] [The information in the cancelled passage is repeated in *Variation*, I, 63, where Darwin adds the note: 'One case is given by Martin, "The Horse", p. 205.' The manuscript folio is sheared off at this point. The broken continuity of the text is restored from the corresponding passage in the first edition of the *Origin*, ch. v, p. 163. See also *Variation*, II, p. 43, on Hemionus.]

[2] Dict. Class. d'Hist. Nat. Tom 3. p. 563. [Desmoulins, art. 'Cheval'.]

[3] Horses, Naturalist Library by Col. H. Smith p. 318. Also E. Blyth in the India Sporting Review 1856. p. 320.

the body is without stripes on the legs; but one individual which Lord Derby[1] kept alive had a few distinct zebra-like transverse bars on the hocks.

Again in the Horse, dun or mouse-coloured or eel-back ponys & horses invariably have (I believe) a dark stripe down the spine, as in the Hemionous, sometimes a transverse shoulder stripe, as in the Donkey, & sometimes dark zebra-like bars on the legs as I have myself seen. I have heard of cream coloured horses with the dorsal stripe in India & others with the transverse shoulder stripe in S. Wales & in other parts of the world. A friend has likewise informed me that he had a brownish horse with the spinal & shoulder stripe. I have been informed by two other friends that they have seen Roans with the spinal stripe. Chesnut horses, also, of very different breeds not rarely have a dark & well defined stripe down the back. Col. Hamilton Smith, who has given numerous most curious facts on this subject,[2] believes that the Dun Ponies have originated in a distinct, wild race or species: they are found in Iceland, commonly (as informed by a friend) in Norway/107/ Spain, near the Indus, & in the great islands of the Malay Archipelago, everywhere characterized by the longitudinal stripe, occasional shoulder stripe & bars on the legs. It is a very ancient race, & existed (together with cream-coloured horses ⟨which we have seen also have the dorsal & shoulder stripe⟩ in the times of Alexander & are either truly wild or feral in the East, & were so at no very remote period in parks in Prussia. It is admitted by Col. Smith that duns appear occasionally in herds of variously coloured horses, but he would account for all such cases by a cross at some time from his dun-stock; I suspect, considering the wide range & antiquity of this colour & its occurrence in wild breeds, that it might be argued with much probability that this was the aboriginal colour of the ⟨aboriginal parents⟩ of all our domestic horses. However this may be, the shoulder stripe & bars on the leg are now only an *occasional* appearance./107 v/It seems to me a bold hypothesis to attribute the spinal stripe in roan, cream-colour & chesnut horses to a cross at some time with a Dun./107/ In regard to the chesnut colour said to be strongly inheritable Col. Smith, who admits so freely various wild stocks, doubts about there having been one of this colour as it is characteristic of every breed; & Hofacker[3] shows that chesnuts/108/are bred from both

---

[1] Gleanings from the Menageries of Knowsley Hall 1850. p. 71 a splendid work by Dr. J. E. Gray.

[2] Horses, ⟨p. xi Preface⟩ p. 109, 156 to 163, ⟨275⟩ 280, ⟨288⟩.

[3] Ueber die Eigenschaften &c 1828. s. 12. [Darwin later noted in pencil:] I must see to this, per[haps?] I translated [?] wrong??

parents of different colour. The stripe is only occasionally present; it has been seen in common chesnut horses, in the heaviest dray horses. I have seen it in a remarkably small pony from India. Hence I believe that the chesnut colour & probably the Roan itself are variations, & the dorsal stripe an occasional concomitant of these colours.

Here then in the horse, Donkey, & other equine animals we have several cases under domestication & in a state of nature, of variations analogous in one variety to another variety & to allied species. Remembering in how remarkable manner in pigeons the blue colour & allied tints with black wing bars &c were brought out by crossing the most distinct breeds, let us see what is the result of crossing the various species of the Horse genus. But first, let me remark that it would appear that the Dun Ponys & chesnut Horses with these asinine marks often appear from the crossing of two breeds of the Horse: this certainly is the case with the so-called Kutch or Kahteawar breed[1] "which are generally greys or light duns & almost invariably have the zebra marks on the legs with list down the back"; & these are bred from a Kutch mare & an Arab sire; & it is asserted[2]/109/that Arabs are never duns. Now for crosses between species: Rollin [sic][3] asserts that the common mule between ass & Horse are particularly liable to the zebra marks on the legs. ⟨Burchell's zebra (E. Burchellii) is not striped on the legs, but hybrids between it & common ass in two instances were plainly barred on the legs[4]⟩. In Lord Morton's famous case[5] of the hybrid from a male Quagga & a *chesnut* mare (not thorough bred), & in the two subsequent colts from the same mare & a black Arabian, the bars across the legs were "more strongly defined & darker than those on the legs of the quagga, which are very slightly marked": indeed it can hardly be/110/said that the Quagga has ordinarily any bars on the leg. Lastly & this seems even a more curious case than the last in regard to our present subject, the Hemionus differs from the Ass in having the spinal stripe but not the cross shoulder stripe & with the legs without any trace of bars but a hybrid figured in that splendid

---

[1] Col. H. Smith. Horses p. xi on the authority of Major Gwatkin, Stud-Master.

[2] Col. H. Smith p. 211.

[3] [Blank space for reference. See Roulin, *Acad. sci., Paris Mém. divers savans* 6 (1835), 338.] Mr. Martin in his History of the Horse p. 212 gives a figure of a Spanish mule with the strongest zebra marks on whole length of legs: especially front legs; I have seen a fine cream-coloured mule with all four legs strongly barred.

[4] Martin Horses p. 223. See also the splendid drawings in Dr. J. E. Gray Gleanings from the Menagerie of Knowsley Hall 1850.

[5] Philosophical Transactions. 1821. p. 20.

work, the Knowsley Menagerie, has all four legs with transverse bars; *there are even some zebra-like stripes near the eyes*, & on the shoulder there are three short transverse stripes. This last character reminds one of the variety of the common Ass & Koulan with a double shoulder stripe. Dr. J. E. Gray further informs me that he has seen a second hybrid quite like the one figured. Here we see most plainly what an extraordinarily strong tendency there is for the bars to come out in crosses between those species of Horse, which have naturally plain legs.

I will only further remark that in Hybrids from the zebra & ass or Hemionus in which as the one parent has striped legs, stripes on the legs might be expected, it is clear that the stripes are more plainly developed on the legs than elsewhere as may be seen in two figures in the Knowsley/111/Menagerie, & still more plainly in a Hybrid figured by Mr. Geoffroy & F. Cuvier[1] in which there are hardly any stripes on the legs. In one of these hybrids between Ass & Zebra there is a double cross shoulder stripe./111 v/In two hybrids from the common ass with plain legs & Burchells Zebra, the legs are barred quite as plainly, perhaps rather more plainly, than in Burchells Zebra.[2]/111/Again in the offspring from a Bay mare & a hybrid Ass-Zebra, the bars on the legs are to be seen, & I was assured at the Zoological gardens were extremely conspicuous when the animal was young. It may be noticed in connection with dun Ponys that in several of these hybrids, dun or slate-like tints prevail.

What shall we say on these facts? Those who believe in the independent creation of species—& if there does exist such a thing as a species distinct from a permanent variety, undoubtedly these equine animals offer perfect examples—will say that they have been created with an organization so much in common, that under certain unnatural conditions & crosses, characters appear which mock those in animals created in other & remote countries: they/ 112/will have to admit that the bars on the legs of the zebra were so created & more strongly inheritable than the bars on the body; but that the similar bars occasionally appearing on the ass or on the several above hybrids are due to variation. It seems to me far more satisfactory to follow the striking analogy of domestic pigeons & attribute all the cases to one common cause, viz community of descent. Let it be remembered that the races of domestic Pigeons differ more from each other in external appearance than do the several equine species; & that in all the races when from

[1] Hist. Nat. des Mammiferes 1820. Tom i.
[2] Gleanings from the Knowsley Menagerie; the skin there figured I have seen in the British Museum: see also Martin on the Horse p. 223.

simple variation or crossing a blue tint appears (comparable to the dun in Horses) almost invariably the black wing-bars appear (comparable to those on the legs of the horse, ass &c) often accompanied by other characters, as white rump &c (comparable to shoulder stripe &c). But although these colours & markings appear in the several breeds of Pigeons the form of head & body &c do not alter; & so it is with the equine animals when they become occasionally striped & barred.[1] From the facts previously given, it is possible that the bars & stripes on the several equine animals might be analogous variations from the/113/several species having inherited a common organization, but the concurrence of several characters & more especially the characters being brought out by crossing—a cause wholly unlike that which produces the bars on the aboriginal parts seems to me clearly to indicate reversion to ancestral character;[2]—this ancestral character being latent in the young of each generation & occasionally brought out when the organization is disturbed by a cross or other cause: hence, probably, it is that the stripes on the legs of the common Donkey are said to be plainest in early youth, as they were in the complex cross of Ass, Zebra & Horse. It is to my mind very interesting thus to get a glimpse into the far past, millions of generations ago, & see a dun-coloured animal, with dorsal & transverse shoulder stripe, barred legs, & striped body, the common parent of the Quagga, Burchell's Zebra, the Hemionus, Ass, & Horse.—

Finally I think the fact of varieties of one species often assuming some character of another species as shown in the several foregoing instances,—though it is in most cases impossible for us to conjecture whether the variation be an old character reappearing from reversion, or a new one appearing in any creature for the first time but like what has previously appeared in a collateral relation owing to like causes acting on a like organization—accords well with the view that the several species of the same group, like the varieties of the same species, have descended from a common parent./

114/*Characters distinguishing varieties are more variable than those distinguishing species & specific characters are more variable individually than characters distinguishing genera or higher groups.*—

This proposition will sound, I apprehend almost like a truism to the systematist. In regard to the variability of the character

---

[1] Mr. Martin in his History of the Horse (p. 97) has well remarked that the dun or eel-back Ponys are asin[in]e only in colour & not in form.

[2] This seems to have been the opinion of Rollin in [Acad. Sci., Paris. *Mém. divers savans* 6 (1835), 338.] & of the Rev & Hon. W. Herbert, who in his work on the Amaryllidaceae (p. 340) alludes to the Dun Pony with dorsal stripe.

of varieties, nothing need be said, for it is self-evident. In regard to specific characters being more variable than generic many will at once assert that differences in the less important parts distinguish species, & in the more important parts, genera; & that the less important from affecting the welfare of the individual are more variable, than the more important parts. That this includes part of the truth I do not doubt; but in our future Chapter on classification, we shall see that some most competent judges consider that the importance of a character under a systematic point of view is not related (as we see in embryonic *rudimentary* parts) to its physiological importance but simply to its presence throughout many different forms, or in the case of species to its non-variability throughout many individuals. In animals, I think there can be no doubt that the parts more immediately connected with the habits of life, & those more immediately exposed to external agencies, as the dermal appendages are individually the most variable parts. But characters even of this latter/115/kind often present the highest degree of generality. Look at the presence of feathers common to the whole great class of Birds: if the Ornithorhynchus had been clothed with feathers instead of hair, its place in the system of nature would not have been altered, but naturalists would have been far more surprised at the fact, than at certain important parts of the skeleton making some approach to that of a bird: and why, except from the generality of mere dermal appendages such as feathers being characteristic of the whole class of Birds & of that class alone?

We see the truth of our proposition in colour size & proportion of parts being the most general diagnostic characters of species, & notoriously the most variable individually. But when any the most trifling character is common to many species of a group we are surprised to find it variable in that group. If all the many species of a genus of plants had yellow flowers, we should be more surprised at one varying into red & yellow flowers, than if about half the/ 116/species had red & half yellow. But why should this be so if we look at each species as an independent creation? But if we look at yellow & red-flowered species of the same genus as having descended from a common parent, it implies that there has been variation in this very respect since the period when the species, first as mere varieties, branched off from a common stock; & as most genera have not a very high geological antiquity the period cannot have been in a geological sense very remote. I believe that it takes an enormous period of inheritance to render any character perfectly true or free from reversion; and as the descendants of a common stock will generally retain much in common, the same causes which at an early period caused the parent to assume red & yellow flowers will be apt still to react on their offspring.

In the fourth chapter I attempted to show that every part of the organization in some group or other was occasionally variable. —But we require something more precise for our theory: in as much/117/as all the species of the same genus are supposed to have descended from a common parent, it is implied that all the diagnostic characters between such species have varied within the very group in question, & within the period since they branched off from their common parent. But the very fact of the existence of a set of species, that is according to our theory strongly marked varieties, implies that the variation must have commenced long ago to allow of the accumulation of slight differences through natural selection, & therefore we have no right to expect invariably to find evidence of variation in the diagnostic characters at the present day. Yet we ought sometimes, perhaps often to discover such evidence, owing, as just stated, to new character apparently requiring an enormous period to become thoroughily fixed, & likewise to similar causes still acting on a similar organization tending still to produce variability in the same parts. Consequently all the facts above given of varieties of one species imitating in character another/118/species,—whether trifling characters not enumerated or those somewhat more striking cases which have been tabulated, & all cases of varieties & close so-called species intermediate in their whole organization, are of especial value in establishing the probability of our theory. Under this same point of view the facts before tabulated of parts or organs extraordinarily developed in single species in a group, tending to be highly variable, may likewise be looked at as valuable, as showing within the group itself, the possibility at least of specific changes.

M. Isidore Geoffroy Saint Hilaire's proposition, before stated, that parts or organs which differ most in the same group are most subject to monstrosities, may be here alluded to. According to our theory, such parts & organs have varied much since the group of species originated, & as variations may be called slight monstrosities, we can to a certain extent understand how such parts should be particularly liable to great & sudden variations or monstrosities.

It would be tedious to enter into more details; but/119/I believe another & related proposition could be established, namely that in animals presenting secondary sexual characters, the allied species generally differ in the same points in which the sexes differ./

334

119 n/I will give a few facts, which have led me to this conclusion: I could easily have added others. In most Coleoptera the joints of the tarsi offer characters of highest value: in the Engidae, however, they exhibit numberless differences, "*even in the sexes of the same species*" (Westwood Modern Classification. vol. I. p. 144). In the Hymenoptera Terebrantia, "the antennae are very variable (ie differ) in the number & form of their joints both in the various species & in the sexes of the same species." (Ib. vol 2. p. 89.) we have analogous facts in the curious growing of the elytra of the females & in the different species of Dyticus (Ib. vol I. p. 104). Shuckard in his essay on the Fossorial Hymenoptera shows that in certain genera, as Tiphia, the neuration of wings, a character of highest importance, differs in some of the species & in the sexes of certain species. The mandibles in the Lucanidae, & the horns in the Dynastidae differ in the males of the different species. In Deer the Horns, so eminently sexual, differ greatly in different species: in sheep in which they are more of a sexual character than in cattle, as the wild females have them either small or not at all, they vary far more in the several domestic races, or quasi-species than do the horns of cattle. The tusks of Elephants, a sexual character, differ greatly in the several allied genera & sub-genera & even in the races of the Indian Elephant. In Gallinaceous birds, the length & curvature of the tail is eminently a sexual character, & if the female of the allied genera & sub-genera be/119 n v/compared the length of tail differs remarkably in the several species.—The naked & carunculated head is a specific character in the Turkey & only sexual in the allied Ceriornis./

119/According to our theory, secondary sexual characters are due to variations becoming primarily attached (as we see in our domestic races) to one sex & if found useful to that sex alone, being augmented & perpetuated by sexual selection; & as the part in question is thus supposed to be variable (and in a former section of this chapter I think it has been shown that secondary sexual characters are eminently variable in the individuals) we might naturally have supposed that variations of the same kind would have affected the several species of the group,—which species we look at as descendants from a common parent, just as much as we do at the male & female of the same species. Hence I believe that individual variability of any part or organ, differences in the same part in the two sexes, & likewise in the several species of the same group are all facts closely connected together & explicable on our theory./

120/*Summary.* In former chapters we have seen that Naturalists have no means, no golden rule, by which to distinguish varieties, whether produced under domestication or in freedom, from species. Looking to the productions of the best-known countries, & taking the hig[h]est authorities, we often find the widest differences in opinion which form to denominate as species, & which as varieties.

Isolated districts are equally favourable for the birth of varieties & species. In this chapter we have seen that although the conditions of life, as food, climate &c, seldom appear directly to cause any great modifications in structure, & must be quite important in regard to all those beautiful adaptations of one organism to another; yet what slight changes the external conditions of life do produce, are analogous to those characteristic of the species exposed to the same conditions. The fact of an organism varying in a like manner under widely different conditions, shows how inferior in importance the direct & immediate effects of such conditions are in comparison to the beings own organisation./

121/There seems no great difficulty in believing that those organic beings, which are so well endowed as to be enabled to beat their competitors in the struggle for life, & thus spread, should soon become acclimatised through natural selection & habits to a new climate; & if we admit this, some facts in geographical distribution & in the history of our domestic productions are explained. Though we cannot actually trace in organisms in a state of nature, the effects of disuse on structure; yet if we admit that species are mutable we can explain by disuse certain peculiarities of structure in relation to the habits of the species, as wingless birds & insects on islands, & eyeless animals in dark caves & subterranean burrows: but in some such cases, it is highly probable, that natural selection may have played a part, either in reducing the structure, or in a directly opposite way by enlarging & perfecting the organ, whichever tendency was at first most profitable to the individual.

We have seen in this chapter that the growth of the whole organic structure is correlated by many obscure laws,—as compensation, the tendency/122/in homologous parts to vary in a like manner & to cohere subsequently so that if one part should be modified by accumulated variations other parts would in consequence be modified: when flowers on the same individual plant habitually & normally differ & we see internal structural differences in their seeds, we are moved to attribute such differences to some unknown laws of growth. Similar laws of correlation, are common, as far as we can judge, to the production of varieties, & to the so-called creation of species.

Parts developed in an extraordinary manner in a species, as compared to its nearest allies, seem to be highly variable: but why should this be so, if species have been independently created? But if, in accordance with our theory, we attribute such extraordinarily developed parts to a long course of natural selection within recent times,—and this will generally have been the case, as natural selection can act only with extreme slowness, & we are comparing organisms closely allied in blood by descent & yet differing greatly/123/in some one respect,—then we can understand the great variability of such parts, on the same principles that the parts recently & greatly modified by artificial selection are the most variable in our domestic productions. Rudimentary parts are likewise highly variable; & why should this be so, if these rudiments were created, as we see them, in their present useless condition? Why should one species in varying so often assume some of the characters of a distinct, though allied, species? Why should the ass or dun-coloured horse be often born with stripes like those on a zebra: why should the hybrid from the ass & hemionus, both with plain legs, be conspicuously striped on the legs & even slightly on the head? Why should a variety of Geranium resemble in the colouring of its petals a Pelargonium? And a score of similar questions could be asked. If the ass, horse & zebra have descended from a common ancestor, like our domestic breeds of the Pigeon, we can to a certain extent understand the reason; but on the view of their independent creation, these facts/124/seem to me a mere mockery; & I could nearly as well believe that fossil shells had been created within the solid rock, mocking the live shells on the beach.

We admit as a truism that the distinctive characters of Varieties are apt to be highly variable; but why should the characters distinguishing species, be more variable than those, even when functionally unimportant, distinguishing genera; ⟨or what is the same thing, why should the characters differing in two closely allied species be more variable, than the characters, sometimes the very same characters, distinguishing two more different sets of species:⟩ why, for instance, if one plant has a blue flower & another closely allied species a red flower, should their colour be more likely to vary, than in two species of the same Family one taken out of a genus with all the species blue flowered, & the other out of a genus with all the species red flowered? According to our doctrines, the existence of sub-varieties presupposes/125/a previously existing parent variety, from which they have inherited very much in common; the existence of two or three closely related species presupposes a previously existing parent species, as does the existence of all the several species in any genus, from which parent they have inherited much in common, but less than in the case of sub-varieties. Hence it follows that the characters, by which the sub-varieties of one variety, the two or three species of the same sub-genus, & all the species of the same genus, resemble in each case their parents, must have been inherited during a longer period than those characters in which the sub-varieties & the species differ from each other. And we have reason to suppose that mere length of inheritance tends to render characters more fixed; so that the characters inherited from the more ancient parent will tend to [be] more fixed or less variable, than the characters by which the member[s] of the same group differ from each other; that is the distinctive characters of varieties will tend to be more variable than those of species, & the distinctive characters of species more than generic. Moreover the forms which have varied recently will often remain exposed to the same causes, which first produced the changes in their structure; & hence the same parts will often be again affected & so kept variable./

126/Why, again, in animals are the secondary sexual characters when strongly displayed so variable? ⟨if each species be an independent creation?⟩ Such sexual characters, according to our view do not differ essentially from strongly marked differences between species in all other respects most closely allied; & we have just seen that such differences tend to be highly variable from reasons already assigned. Sexual characters, moreover, have generally been accumulated by sexual selection, which is less rigid than the struggle for life & death. Sexual characters have become attached to one sex alone, whereas ordinary specific characters have become attached to both sexes; but our theory looks at all the species of the same genus as the descendants of a common parent, with as much certainty as it does at the males & females of the same species. Hence it is not surprising that naturalists have so often described the sexes of the same species, as distinct species & even as distinct genera.—

Ignorant as we are on the primary causes of variation, yet as far as we can obscurely see, the laws governing variation/127/are the same as those concerned in the production of species. Therefore, I conclude that the facts given in this chapter, as far as they can be trusted, support our theory that Varieties & Species have had a like origin;—& not that Varieties are due to the laws impressed on nature & Species to the direct interposition of the Creator.—

DIFFICULTIES ON THE THEORY OF NATURAL SELECTION IN RELATION TO PASSAGES FROM FORM TO FORM

INTRODUCTION

On September 30, 1857, having finished chapter VIII the previous day, Darwin wrote Hooker that he was already looking over note 'scraps' for his next chapter and added in a postscript: 'Though I work every day, my last two chapters of rough M. S. have taken me exactly six months! Pleasant prospect!'[1] The Pocket Diary entries corroborate this, for the one following that of March 31, for the completion of chapter VI is: 'Sept. 29th finished Ch. 7 & 8; but one month lost at Moor Park.' Yet Darwin's work during these six months was by no means concentrated solely on writing. Numerous letters to Hooker concern his extensive collection of statistical materials from voluminous regional floras as evidence for his discussion of the thesis that 'wide ranging and common and much diffused species tend most to vary', later written out in the supplement to chapter four. Besides this major project, his letters mention more incidentally activities such as his experimental observations on his 'weed garden' and on the mechanics of transport of young fresh water molluscs by birds along with formulating considered advice to Hooker on the selection of recipients for the Royal Society medals. And his September 30 letter also mentions to Hooker that: 'We have lately been taking a very extravagant step & are building a new dining room & bredroom over; so are in the midst of bricks & rubbish.—'

DIFFICULTIES ON THE THEORY OF NATURAL SELECTION IN RELATION TO PASSAGES FROM FORM TO FORM[2]

[Completed September 29, 1857]

In the sixth chapter I briefly alluded to many grave difficulties, enough at first sight to overwhelm our theory of natural selection. In this chapter we will consider those connected with the absolute necessity of all passages having been extremely gradual from one

---

[1] C.D. MSS., 114, letter no. 210.

[2] [Darwin noted in pencil at top of this sheet: 'Be careful in use of word transitional as = intermediate']

living being into another, or of one part or organ into another. That this is absolutely necessary follows from all causes of variation apparently acting only slowly, & more especially from our paramount principle of natural selection inevitably thus acting by the addition of numerous slight modifications,—each modification profiting the selected or preserved individual. I fully admit that the difficulties in relation to these passages are many, & in some instances extremely great; & I wish here only to consider whether the difficulties are actually insuperable & enough by themselves to overthrow our theory, whether or not other facts support it./

2/We will commence with cases of intermediate & possibly transitional habits. It has been asked how for instance a land carnivorous quadruped could be converted into an otter; for how could it have subsisted during its transitional state? A far more difficult question would have been, how could a Bat have lived, before its wings had become perfected? ⟨by long continued natural selection?⟩ I may premise that it is immaterial to us, whether slight changes of structure supervene first leading to changed habits, or whether habits & instinct change slightly first, & the animal is benefitted in subsequent generations by slight selected modifications of structure in relation to the already slightly changed habits. In the chapter on instincts we shall see that mental attributes vary slightly, can be selected & are inherited, like corporeal structures.

Let us take the case of the Bat, which is one of the most difficult, that has/3/occurred to me. What were the stages by which probably an insectivorous & terrestrial animal could have acquired the capacious wings of the Bat;—every single, slight, intermediate grade being so useful to the animal in that state, that it was enabled to conquer in the struggle for life, to which it must inevitably have been exposed? We cannot answer this question even by conjectures. The earliest known, Eocene Bat apparently was as perfect, as one of the present day: if our geological records really make any approach to perfection, this would be a fatal objection. But the whole subject of the value of the evidence from fossil remains will have to be discussed in a separate chapter; & I will here pass by this apparently fatal objection, applicable in many other instances, & consider our more immediate subject, namely the possibility of transitional stages between a Bat & an animal not capable of flight. On this difficulty/4/I think we ought to be extremely cautious in laying much stress.—

Look how amongst Birds, the most perfectly winged animals,

we have the Penguin, which uses its wings exclusively as paddles for diving, & as front legs on the land; (as I have witnessed),—the logger-headed Duck (Micropterus brachypterus Eyton) as flappers on the surface of the water, & never as wings,—the ostrich as sails—& the Apterix is destitute of wings capable of any use. Yet all these Birds are enabled to hold their own place in the great struggle for life; & no one will doubt that their wings are most useful to the Penguin, & Logger-headed Duck, & I presume of some use to the ostrich perhaps in its first start to escape a beast of prey. Many birds use their wings as paddles for diving & for flight, others flap along the surface of the water or run with expanded wings before taking flight: is it not thus conceivable that by continued selection organs used exclusively for diving or flapping or sailing on land might come to serve exclusively for flight? But had some half-dozen genera of Birds become utterly extinct, he would have been a bold man, who would have said/ 5/that Birds might have flourished on our earth which did not fly, but used their wings solely as sails, flappers, paddles, or front legs. I am, of course, far from pretending to indicate what were the transitional grades, by which Birds came to fly; it might have been through some wholly distinct line of change: the penguin may be, as the logger-headed Duck probably is, the degraded descendant of a perfectly winged Bird. All that I want to show is, that as far as habits are concerned & judging only from now living animals, great transitions are possible.

Seeing that we have flying Birds Mammals, & formerly flying Reptiles, & seeing that so eminently an aquatic animal as a crab can by a contrivance to keep its branchiae moist exist on dry land, it is conceivable, that the so-called flying Fish,—which can glide to such great distances through the air, turning & rising slightly in its rapid course by the aid of its fluttering fin[1]—could have been converted into a perfectly flying animal: had this been the case, & our present flying-fish/6/unknown, who would have ventured to have even conjectured that the sole use of an early transitional stage of the pectoral fin was to escape danger in the open ocean. Certain fish use their pectoral fins for splattering over the mud, for jumping & even for climbing trees; if fish had become, like land-crabs & onisci, terrestrial animals, how easily ancient transitional uses of the pectoral fins, might have baffled all conjecture.

Amongst mammals, we have squirrels with the tail forming a flattened brush, & we have others/6 v/with "a peculiar wideness

---

[1] Owen's Lecture Fishes. p 170.

in the posterior part of the body & a fulness of skin of the flanks being an approach to the forms of a true flying squirrel[1] & these latter have a wide membrane connecting the front & hind legs together & in one species a slight fold of skin uniting the base of the tail to the hinder thighs. All these contrivances aiding the animal to glide great distances through the air from the top of one tree to the base of another./6/There are, also, gliding insectivorous opossums with the flank-membrane developed in different degrees. In the Galeopithecus, or flying Lemur, which was formerly ranked amongst Bats, the membrane extends from the corner of the jaw & includes all four legs & the tail: the membrane on the flank has a muscle for extending it: the rather long fingers of all four hands are also connected by skin: its habits are imperfectly known, but it is said to descend trees "par une sorte de vol/7/ retardé."[2] ⟨& to be even partly aquatic in its habits.⟩

The fact that each animal lives by a struggle,—that each would increase inordinately if not checked at some period of its life,—is constantly eluding us; so that we find it difficult to realise that in course of thousands of generations the power of gliding a few inches further through the air may make an important difference to an animal in escaping dangers or getting food. For myself I can see no difficulty in the means of gliding through the air in squirrels having been perfected through natural selection from a mere flattened brush-like tail to a wide flank membrane; & amongst Lemurs (though all such supposed intermediate forms are extinct) to the enormously developed membrane of the living Galeopithecus; & ⟨in some other unknown & extinct tribes of animals,⟩ even amongst other extinct animals to the wonderfully perfect wings of the Bat. The graduated structure amongst squirrels & the almost intermediate condition of the Galeopithecus between an aerial & terrestrial animal ought at least to/8/make us very cautious in supposing that numerous animals constructed on every intermediate type between a Bat & land quadruped, could not formerly have flourished in the great battle of life. Who would have ever supposed/8 v/that at the present day there should be a Bat feeding chiefly on frogs & occasionally on fish;[3] or that the

---

[1] Sir J. Richardson Fauna Boreali–Americana. Quadrupeds p. 191. It is the Pteromys petaurista which has the base of the tail united to the thigh, see Dict. Class. Hist. Nat. Art. Pteromys. [In vol. 14, art. by Isidore Geoffroy-Saint-Hilaire.]

[2] Dict. Class. d'Hist. Nat. Tom. 7. p. 122. [Desmoulins, art. 'Galéopithèque Galeopithecus'] Buckland Bridgewater Treatise.

[3] Mr. Blyth gives an account of these habits in the Megaderma lyra in India, in Annals & Mag. of Nat. History. vol. 15.—1845. p. 463.

frugivorous Pteropus, when put on a floating raft, should take to the water & "swim pertinaceously after a boat."[1]/

8/To return to the objection which has actually been made that a land-carnivorous animal could not be changed into an Otter, for it could not live during the transitional state. The genus Mustela is closely allied to Lutra or the otter, & indeed was made into one by Linnaeus. Some species of Mustela occasionally haunt the water, & the common Polecat has been known to lay up stores of half-killed frogs; the N. American Vison-Weasel (Mustela vison allied to the M. lutreola of N. Europe) has webbed feet, a flattened head, short ears, close fur & a tail all like an otter: it can dive well, & preys on fish: but during the winter when the water is frozen, it hunts mice on the land:[2] here then we have an animal allied to the otter, wholly aquatic during part of the year & partly terrestrial during another part. Can it, then, be said that there would be any great difficulty, as far/9/as transitional habits are concerned, in converting a polecat into an otter. The possibility will rest on there being a place open in the polity of nature, which would allow of a polecat living & increasing in numbers, if rendered more & more aquatic in habits & structure. On the same principles an otter could be converted into a seal-like animal; not, perhaps, now when seals actually exist & well fill their place in nature, but before a seal had been formed. It might well happen through natural selection, that an aquatic animal should be converted into a terrestrial animal, retaining perhaps a trace of its former webbed feet; & subsequently have some of its descendants refitted to inhabit the waters.

Numerous instances could be given, like that of the Vison-weasel, of diversified habits in the same individuals or in the same species both under the ordinary conditions of its life & under peculiar circumstances; & similar differences could be given amongst different, but closely allied species of the same genus. I will give only a very few cases chiefly amongst birds. In S. America,/10/a tyrant-flycatcher[3] may often be seen hovering over one spot & then proceeding to another like a Hawk, but its stoop is very inferior in force: at other times it remains motionless on the edge of a piece of water, & like a Kingfisher dashes at any small fish near the margin. In Tierra del Fuego, there is an Owl (Ulula rufipes) which I have reason to believe preys chiefly on the lower

[1] Mr. Lay on a Pteropus from the Bonin Islands in Zoological Journal vol iv. 1829. p 459.
[2] Sir J. Richardson Fauna Boreali–Americana Quadrupeds. p. 49.
[3] Saurophagus sulphuratus: Zoology of the Voyage of the Beagle p. 43.

marine animals, & I have seen its stomach gorged with the remains of large crabs. The Woodpecker with its peculiar feet, stiff tail, strong wedge-like beak & long tongue, has often & justly been adduced, as a perfect instance of the adaptation of a bird to prey on insects concealed in the bark & wood of trees: but on the wide grassy plains of La Plata, where not a tree exists, the Picus (Chrysoptilus) campestris Licht. feeds *exclusively* on the ground: even in its colouring in the peculiar undulating flight, & loud cry it resembles pretty closely our common green species.[1] A North American woodpecker[2] has the extraordinary propensity of catch-ing/11/flies on the wing! & some other N. American species feed largely on fruit. In our own country the Titmouse genus (Parus) are properly insectivorous; yet everyone may continually hear in autumn, the loud hammering, of the Parus major, like that made by the especially adapted Nuthatch, as it breaks with its beak the kernels of the yew-berries held on a branch of a neighbor tree. One more instance,—Hearn[e] states that the black bear fishes for small crustaceans in the sea, by swimming about with its mouth widely open, so that here a terrestrial quadruped almost mocks a whale in its occasional manner of getting food![3]

⟨In such cases, if under changing conditions one of the diversified lines of life were especially favoured, it does not seem very difficult to believe, that the structure of the descendants from a parent form having very different habits might become greatly modified through natural selection.⟩/

12/When an animal or plant is introduced into a new country, and vast numbers have been thoroughily naturalised, some slight changes in its habits of life can hardly fail often to occur; & likewise indirectly in some of the aborigenes of the country. Look at our many introduced quite new plants, how they are preyed on by our native animals, sometimes almost to the exclusion of their original food.[4] Within two years after planting Berberis dulcis a bush very unlike the common barberry I found its twigs covered with Aphides on which Coccinellae were preying; its flowers were visited by Bees, visits indispensable, I believe, to its fertilisation, & its fruit

---

[1] Zoology of the Voyage of the Beagle p. 113.

[2] Picus varius. see Mr. T. Macculloch in the Boston Journal of Nat. History vol 4. p. 406.

[3] [C.D. MSS., vol. 48, note slip no. 8: 'Hearne's Travels p. 370 The black bear catches fresh water insects *by swimming with mouth open* "*like whales.*" These insects are in wonderful numbers. So that they are driven together into the bays of the lakes to the thickness of two or three feet| & make dreadful smell. These insects of two kinds. All the bears stomachs distended.']

[4] Loudon's Magazine of Nat. History vol. v 1832 p. 154. Thus the caterpillar of the Death's Head sphinx is very rarely found except on the Potato or jasmine.

was devoured by the Robin, which would disseminate its seed. In Tasmania, I found the dung of the introduced quadrupeds, so different from that of any native animal, supporting numerous beetles; & this was likewise the case even in the island of St. Helena,[1] where there was no native quadruped; yet Entomologists know that stercovorous beetles are usually restricted in their habits. Innumerable/13/parallel instances could be given.

The changes produced by civilised men in many countries must have sensibly affected the habits of life of the native animals & plants. How many insects there are in Britain, which, as far as known, subsist exclusively on artificial or foreign substances. There are many egregious plants, which are scarcely ever seen except on cultivated land, though probably most of them are foreigners.

Thoroughily wild animals of the same species, when inhabiting districts of a very different nature can hardly fail to differ somewhat in habits. Some land birds are common to Tierra del Fuego, absolutely covered by inpenetrable forests, & the Falkland Islands, where not a tree exists. In our own country, the Scotch Highland Fox has considerably different habits from the lowland Fox of England. Even with our domestic animals, individuals show a tendency to different habits of life,—of which I/14/have already given instances in the Cat. Very many/14 v/cases are on record of the inhabitants of the sea gradually becoming accustomed to brackish & even fresh water; & of the reverse change; which circumstance, considering how fatal an abrupt change of this nature is to most animals is surprising.[2]

14/In all cases of naturally diversified or of changed habits, I can see no great difficulty in structure being modified through the natural selection of variations better fitted to some one of the diverse habits, or to a new & changed habit of life. If our reason tells us that the structure of the woodpecker is admirably adapted for its insectivorous life on trees, & that another structure is generally best for the capture of insects on the ground or in the air, then I can see no reason why the ground or aerial woodpecker might not have their structure still more modified, than it actually is, from the typical structure of the genus Picus, & so be improved for their aberrant habits of life; or on the other hand why birds having in some degree the habits of these aberrant woodpeckers might not be modified in structure till they acquired the strictly arboreal habits of the typical woodpeckers. To take one more

[1] Darwin Journal of Researches p. 490.
[2] See Bronn. Ges[ch]ichte der Natur B 2. p 55–58.—Macculloch Highlands & W. Islands. [IV] p. 377. Many other references could be easily given.

extreme case, that of the Black Bear seen by Hearn: if its habit of catching small crustaceans by swimming with widely open mouth/15/became, from the crustaceans being always present with the loss perhaps of other prey, highly important for its sustenance, then as slight variations in the size & shape of the mouth would almost certainly occur during millions of generations, some of these would almost certainly aid ever so little individuals in this strange way of fishing; & these individuals, as all cannot possibly live, would have a better chance of living, & thus such slight variations would be continually added up through natural selection, till an animal, which we should think monstrous, was produced, thoroughily aquatic & with a mouth perhaps proportionally as large as that of a whale. Who would not think it monstrously improbable, if he had never heard of a whale, that so gigantic an animal could subsist by sifting with its huge mouth the minutest animals from the waters of the sea!

Facts do not tell us, as far as I can see, whether habits generally change first, corresponding structures being/16/subsequently selected; or whether structures modified through variation, generally first leads to perhaps ⟨nearly simultaneously⟩ changed habits. In the case of no organic being can we pretend to conjecture through what exact lines of life its progenitors have passed. We may use our knowledge of the habits of existing animals as a guide to conjecture; and somewhat further, in as much as it is probable that amongst the many living & greatly diversified descendants of some ancient & extinct form, some would retain the habits not greatly modified of their several progenitors at different stages of descent. Really to know the transitional states, by which the habits & structure of any one animal have been acquired, we ought to study the long line of its direct ancestors alone, & neglect all collateral branches. How little chance there can be, of one ever knowing, even very imperfectly, all the lineal ancestors of any one form will be seen, when we come to consider the real poverty, under the guise of richness, of our geological records.

It may, however, be here noticed, that when the habits of life of/17/any species or group of animals, is undergoing any great change, as from swimming or gliding to flying, although assuredly every single transitional form has to subsist under a most severe competition for life, yet we ought not to expect the new habit with its corresponding structures, to be developed under many subordinate forms, each with numerous individuals, until it had arrived at a high degree of perfection. To arrive at this, according to our principles of natural selection an enormous period of time would be required. On our former imaginary case of Fish being rendered true flyers, we could hardly expect that they would give rise to a whole class of subordinate groups, fitted for various subordinate stations, until they had obtained through the slow action of natural selection the power of flying perfectly; the intermediate & transitional states, we might expect, would be comparatively few in number at any one period, as we see at the present day with our so-called flying-fish. ⟨So it would probably have been with Lemurs,—the/18/Galeopithecus being looked at as a transitional form—if Bats had not existed, & Lemurs had been developed by natural selection into true flyers.⟩ Hence it seems probable that mere transitional states between very different lines of life, would seldom be largely developed at any one period; for this would not happen till the changing form—changing from having some advantage over its compatriots—could fill its new place in nature with a high degree of perfection. The perfected descendants would generally cause, by the very principle of natural selection or the struggle for life, the extinction of their less perfect progenitors. Hence, also, the chance of finding fossil remains of the progenitors of any organism, during its transitional & less perfect state, would be so much less in proportion as they had been developed under fewer subordinate forms & under fewer individual numbers./

19/We have as yet considered only the possibility of transitional habits, & the difficulties which they seem to oppose to our theory; but some of these same facts may be fairly viewed as supporting our theory. He who believes that each species has been independently created, must feel surprise, at least I remember formerly having felt great surprise, at an animal manifestly adapted for one line of life, following another & very different line. I will take again my illustration from Birds. It cannot be doubted that the general configuration of a Goose is for an aquatic life; & the meaning of webbed feet is unmistakeable; but there are long-legged geese,/ 19 v/which run like gallinaceous birds, & seldom or never enter the water: thus Mr. Gould informs me he believes that the Cereopsis goose of Australia is perfectly terrestrial, & I am told at the Zoological Gardens that this bird & the Sandwich Island goose seem quite awkward on the water:/19/in S. America the Upland Goose (Chloephaga Magellanica Eyton)[1] never frequents the water, except for a short time after hatching for the protection of its young; the feet of this goose are well webbed. The long-legged

---

[1] Zoology of the Voyage of the Beagle Birds p. 134.—Capt. Sulivan has given me further particulars on the habits of these birds.—

Flamingo (Phoenicopterus) has webbed feet, but lives on marshes & is said seldom even to wade except in very shallow water. The Frigate-bird (Fregata aquila) with its extremely short legs, never[1]/20/alights on the water, but picks up its prey from the surface with wondrous skill; yet its four toes are all united by a web.[2] The web, however, is considerably hollowed out between the toes; & tends to be rudimentary: as is the case likewise with the Cereopsis goose: I notice this, because it connects our present class of facts with rudimentary organs, hereafter to be discussed: in the foregoing cases the function of the webbed feet may be said to have become rudimentary without a corresponding change in structure.

On the other hand there does not exist a more thoroughily aquatic Bird, than the Grebe (Podiceps), but its toes are only widely bordered by membrane. The Water-Hen (Gallinula chloropus) may be constantly seen swimming about & diving with perfect ease; yet its long toes are bordered by the narrowest fringe of membrane; other closely allied birds belonging to genera Crex, Parra &c can swim well & yet have scarcely any trace of web; & their extremely long toes seem admirably adapted to walk over the softest swamp & floating plants; yet the common corn-crake (Crex pratensis)/21/belonging to one of these very genera, having the same structure of feet, haunts meadows & is scarcely more aquatic than a quail or partridge.

Several of the cases already given, such as that of the Ground-woodpecker may be looked at under our present point of view: I may add that on the plains, inhabited by these woodpeckers, I saw parrots (Conurus Patachonicus) which build in banks & which can never alight on a tree as one does not exist, & yet have their feet, strictly adapted for perching with two toes in front & two behind: on these plains there were also tree-frogs,[3] with their toes enlarged into little sucking discs for climbing. Numerous other instances could be given; in the S. United States, the Swallow-tailed Hawk, (Nauclerus furcatus), a true hawk, has very long wings & a forked tail, & it lives, by catching whilst on the wing, insects. In the quiet creeks of Tierra del Fuego, I was particularly struck/22/with the habits of the Puffinuria (Pelecanoides) Berardi:

---

[1] Zoology of the Voyage of the Beagle p 146. Vigors Linn. Trans vol 14 [pp. 418–20.]

[2] Mr. Westwood, (Modern Class. of Insects. vol 2. p. 272) has remarked with surprise that certain parasitic Bees, which have no use for their jaws, have these tools as strong, as in working species. Again Mr. F. Smith showed me specimens of another hymenopterus insect, Scolia, in which the legs are eminently & typically fossorial, but which from being parasitic certainly does not use its legs for burrowing. [Query added in pencil: 'How in the *Drone* Hive Bee?']

[3] Hyla agrestis: Bell, in Zoology of Beagle Reptiles. p. 46.

this bird in its manner of swimming, of flying, though rarely in a straight line by the rapid beating of its short wings, then dropping suddenly as if struck dead & diving to a suprising distance, would by anyone be mistaken for an auk or grebe: but the structure of its nostrils & beak & other characters show that it undoubtedly is one of the Petrels,—those most aerial of Birds which hunt the surface of the wide ocean for their prey: here then we see a bird taken from a family having most widely different habits, adapted to fill the place of the Auk of the Northern hemisphere, which are not found in the south./22 v/In this case there has been a considerable change in the form of the body & power of respiration; the wings have been greatly shortened, the tail altered in shape & hind toe lost. I will now give one more instance of an entire change in habits, with no sensible correlated change in structure;—I refer to the Water-Owzel (Cinclus aquaticus), a member of the common Thrush family; it is sub-aquatic in its habits, using its wings for diving, & its feet for grasping stones under the water; & yet the acutest observer would never have foretold this singular manner of life from the most careful examination of its structure./

22/All such facts must seem strange, as long as we look at each species as independently created: it will be said that a bird belonging to one type of structure has been adapted by the Creator to another line of life; but this seems to me only /23/restating the case in dignified language. The theory of natural selection implies that every single animal in each region tends to increase in number with a geometrical power, & so may be said to strive to gain subsistence anyhow it can, & to fill any place in the economy of nature which it can seize: bearing in mind the many & complicated contingencies to which each animal must be exposed in the long course of its existence, & remembering that the world is not open from end to end for immigration, (as we see proved by the many productions naturalised by man's aid) it seems to me perfectly natural on our theory, that occasionally an animal of a wholly different class or occupation should intrude into that of another species or group of species which laboured under some disadvantage however slight. How far its structure would become modified in relation to its new habits, would depend on how far any change would be advantageous to it, & whether variations in the right direction had occurred, & on how long a time selection had been at work accumulating such variations. On these principles, it/24/ is not surprising that there should be webbed geese living & running on the dry land, & webbed Frigate-birds never alighting on the water,—that there should be woodpeckers & tree-frogs where there is not a tree,—that corn-crakes should live in meadows instead of swamps—that there should be diving thrushes, & petrels with the habits of auks.—

Differing Organs of extreme perfection & complication. Although in our consideration of the possibility of great changes in habits corresponding changes in the whole bodily structure have generally been implied; yet it will be adviseable to look at some special cases of particular organs. What shall we say of the eye? Is it conceivable that this transcendant organ with its power of adjusting its focus to different distances & of letting in more or less light,— with its nearly perfect correction for chromatic & spherical aberration, could have been formed by the accumulation, through natural selection, of infinitesimally/25/slight variations, each useful to its possessor. I confess that no language at first seems too strong to condemn the absurdity of such a notion.—

To judge our theory according to its own principles, we ought not to compare the variously perfected eyes in any one group, one with another; but the eye of each species only with the eyes of all its lineal progenitors; so that if we look to the eyes of many species in the group, we should have to look to many lines of ascent converging up to one common parent. This is impossible; & all that we can do, is to look at the eyes of all existing animals within each great class, as a guide for judging how far a transition from one stage of perfection to another stage is possible; at the same time never forgetting how small a fraction the living are compared with the extinct,—almost infinitely small, as I believe. Let us briefly consider the eyes of the Articulata: we have as the/ 26/lowest grade, an optic nerve, coated with pigment sometimes having a kind of pupil, but without a lens or any other optical mechanism. I need hardly say that we have in this work nothing to do with the origin of nervous sensibility to light, any more than with the origin of life. From this rudimentary eye, ⟨as it must be called,⟩ which cannot possibly distinguish figures, & can only perceive light from darkness, there is an advance towards perfection by two fundamentally different contrivances. Firstly, stemmata (or the so-called "simple eyes") which have a crystalline lens, with a cornea & more or less perfect vitreous body—that is the essential parts of the eyes of the higher animals—, & which act by the rays from each point of the object viewed converging on different points of the retina. Secondly "compound eyes", formed of numerous, diverging, transparent, narrow cones, separated by pigment, & which act simply by excluding all the rays from each point of the object viewed, except the pencil which comes in a line perpendicular to the convex retina; so that/27/a separate & distinct image of each separate point of the object is made at the base of each separate transparent cone. Hence Müller[1] the discover[er] of this principle of vision, calls these compound eyes, a mosaic dioptric instrument. In the Articulata we have numerous grades of perfection in the eyes: besides endless difference in form proportion & position of the transparent cones, & in number up to 20,000 in a single eye, there are cases, as in Meloe, in which the facets of the cornea are "slightly convex both externally & internally, that is lens-shaped": in many crustaceae there are two corneae, the external smooth, & the internal divided into facets, within the substance of which, "renflemens lenticulaires paraissent s'etre développés"; but sometimes these lenses can be detached in a layer distinct from the cornea. The transparent cones are usually attached to the cornea, but not rarely they are detached & have their free ends rounded, &/28/in this case they must act, I presume, as converging lenses, & not simply as tubes excluding all oblique rays. Prof. Milne Edwards thinks that the transparent cones of the compound eyes are homologous with the crystalline lenses of the stemmata or simple eyes; & that behind the transparent cones there is apparently a vitreous substance: on this view the lenses in & beneath the cornea of the compound eye is a structure superadded to that observed in the stemmata. Altogether Müller divides the compound eyes into the three main classes with seven sub-divisions of structure: he makes a fourth main class of "aggregates" of stemmata or simple eyes each of which contains the essential parts of the simple stemmata, namely a lens & globular vitreous humour; & he adds "this is the transition form between mosaic-like compound eyes, unprovided with concentrating apparatus, & the organ of vision with such apparatus."

Seeing the numerous gradations & diversity in the eyes of the Articulata, numbering probably at least a hundred thousand in kind, & that/29/the eye of each is good for its habits, then if the eye varies ever so little, & I know of no reason to doubt this, I can see no great difficulty in believing that amongst the Articulata, the eye might be perfected through natural selection from a simple optic nerve to the most complex of compound eyes having numerous transparent cones, a double cornea &, the inner one having both

---

[1] Elements of Physiology, translated by Baly vol. 2 p. 1110 et seq. All the facts here given are taken from this great work, & from Milne Edwards, Hist. Nat. des Crustacés. Tom i. 114 et seq, who has particularly attended to the vision of the Crustacea.

facets & lenses. If we here encounter no greater difficulty than in the case of other structures, then if we look even at the transcendantly perfect eye of the eagle, though we have hardly any guide for judging on the probable transitional stages, yet I think the difficulty is not actually fatal to our theory of natural selection. Nor ought we feel much surprise at our entire ignorance of the transitional stages by which the eye of the higher animals has been perfected; in as much as the links by which the Vertebrates have reached their present position at the head of the animal Kingdom seem to be wholly lost; and amongst existing animals, we/30/are far from understanding the contrivances by which some of the highest perfections of the eye are gained, as for instance the adaptation of the focus to different distances.

A large part of the great difficulty which I have felt in persuading myself that so inimitable an organ as the eye could be perfected by natural selection, has arisen from our constant & almost involuntary habit of comparing the eye with the microscope or telescope. We know that these beautiful instruments have been produced by the long-continued efforts of the highest human intellects; & we naturally infer that the eye has been formed by a somewhat analogous process. But may not this inference be presumptuous? Have we any right to suppose that the Creator works by the same means as man? If we must compare the eye to an optical instrument, we ought ⟨according to our theory,⟩ to take a thick layer of transparent tissue, with nerve sensitive to light beneath, & then suppose/31/⟨that from external causes,⟩ every part of this layer to be continually changing slowly in density, so as to separate into layers of different densities & thicknesses, placed at different distances from each other & with the surfaces of each changing from flat to various degree of convexity; & further we must suppose that there is a power always intently watching each slight accidental alteration in the transparent layer, & carefully selecting each, which may in any way or degree tend to produce a distincter image ⟨at one end⟩ under the circumstances in which the instrument is used; each of the many new states of the instrument being multiplied by the million; & each preserved till a better is produced, the old being then destroyed. In living bodies, variation will cause the slight alterations, generation will multiply them almost infinitely, and natural selection or the struggle for life will pick out with unerring skill each improvement. Let this process go on for millions on millions of years, & during each year on millions of individuals of many kinds, & may we not believe that a living optical instrument might be formed, as much superior to one of glass, as the works of the Creator ⟨Nature⟩ are to those of Man ⟨Art⟩./

32/In regard to other organs of extreme perfection, for instance the ear, analogous remarks may be made as on the eye. No doubt there will be in all such cases, many & wider gaps in the known transitional stages, which we cannot bridge over even conjecturally; but the question here is, are the gaps so wide & impassable that they are fatal to our theory, whatever other evidence can be advanced in its favour. In the case of hearing to give one instance of a great difficulty: in a genus of little pelagic Crustaceans, called Mysis, the auditory organs are seated in the caudal plates or swimmers at the posterior extremity of the body.[1] According to our theory it would at first appear necessary that these organs should have been moved by infinitely small & numerous variation from the front of the head, where the auditory organs occur in other crustaceans, to the end of the tail; & the possibility of this might be thought to be in some/33/favoured, as I found the case in cirripedes (a sub-class of Crustaceans) placed on the sides of the body about half way between the posterior & anterior extremities of the body; but this would be a false view, as Prof. Huxley has shown that in Mysis the acoustic nerves run to the posterior abdominal ganglion, whereas in other crustaceans they run, as far as known, to the first cephalic ganglion; & it seems impossible to effect such a change by slight transitions. This difficult case can apparently be got over only by hypothetically supposing that hearing is nothing but sensibility to a common vibration carried to an extreme pitch; & hence that a nerve of common sensation might in any part of the body be perfected so as to perceive the finest & most rapid vibrations in air or water. An analogous case occurs in vision: that excellent observer, Quatrefages has shown that some Annelids, which can swim & crawl tail first have eyes at both extremities of the body; & there is another annelid with a pair on each segment of its body. Now it has long been known that some lowly organised animals, have no eyes, yet seem to distinguish/34/light from darkness; & it has been supposed that their bodies are generally sensitive to light; but Müller[2] has well remarked that this is quite hypothetical that these animals may perceive only the heat or other influence concomitant with the light.—But our theory almost requires that in low animals, like the annelids & planariae with eyes in diverse parts of their bodies,

---

[1] This curious discovery was made by Frey & Leuckart, & has since been confirmed by Prof. Huxley. see Medical Times & Gazette. 1857. p. 354.
[2] Elements of Physiology Vol 2. p. 1123.

that an ordinary nerve of sensation may be rendered specially sensitive to light.—/

35/*Transitions in organs: several kinds of transitions possible.* ⟨*Organs without known transition states & changes of function in organs.*⟩ As natural selection can act only by the accumulation of slight variations, it may naturally be asked, do not absolutely new organs appear in species, or in small groups of species within a class? That this is of rare occurrence is shown by that old saying ⟨of Linnaeus?⟩ become a canon in natural history, "Natura non facit saltum." In recent days one of our highest authorities has insisted how prodigal nature is in variety, how niggard in innovation.[1] And in the same spirit, a great Botanist says "Nature, as we have seen a thousand times always proceeds by transitions."[2]

Before giving a few cases of real & apparent difficulty from the absence of transitional states in understanding how an organ could have reached its present condition, I must make a few remarks on the kinds of possible transition. But first I must admit that those naturalists who speak so strongly/36/of nature not moving by leaps, seldom, probably never, mean to go as far as our theory requires, namely the existence at some period of transitional states ⟨between the same organ in any two members of the same class⟩ as fine as those between an admitted variety & its parent species. Thus to give a single instance, if we look at the family of Humming birds, we shall find a pretty close gradation in the length & form of beak, & although there are considerable gaps in the series, most naturalists would say that nature had here proceeded by transition; but, as Mr. Gould showed me, there are very many forms, for instance in one strange form with the beak bent almost rectangularly downwards, & another with it upturned; & in these two cases there are hardly any transitional forms. Such cases, however, do not seem to me to offer any real difficulty, that is if we admit that the living members of a group bear but a very small proportion to the extinct, as follows inevitably from the working of natural selection. Those naturalists who would lay much stress on so simple a case as this, will long ago have rejected our theory, so no more need be said on it./

37/In considering the possibility of transitions of an organ from one state to another, we should bear in mind that a part having a nearly similar structure may perform in the same individual or in two individuals functions wholly different. Secondly that two

---

[1] Milne Edwards. Introduct a la Zoolog. Generale 1851 p. 9 & 10.
[2] Aug. de S. Hilaire Lecons de Botanique. 1841 p. 508.

widely different organs may perform simultaneously in the same individual the same function; so that whilst one of these organs was continued or perfected through natural selection in its function, the other might come to be used for some quite distinct purpose. Thirdly that organs & the use of parts change normally in the same species with age, or when placed under different conditions; & in these cases it does not seem difficult for the organ to retain throughout life either one or the other of its states.

I will now give a few facts illustrating these remarks; & they will show how cautious we ought to be in assuming in almost any case that a passage from one state of an organ to another is impossible; or/38/that an organ apparently quite new in its class is not some other part changed in function. In such cases the extinction of a few forms would often utterly baffle us in conjecturing through what stages an organ has apparently passed.

Prof. Milne Edwards has often insisted[1] how frequently in the lowest animals, the same fluid & apparently the same tissues serve for digestion, nutrition & respiration;[2] thus the Hydra has been turned inside out; the outer surface then serving for digestion, & the inner ceasing to digest & no doubt respiring. This same naturalist, as well as others often insist on the advantages of a division of physiological labour; for instance that a surface will digest better if it has not at the same time to act as lungs, or that a stomach will digest vegetable matter more effectually, if it has not, also, to digest flesh; thus/38 v/it presupposes says von Baer mere prejudice not to rank the stomach of a Ruminant above that of a man./ 38/Owing to this advantage from division of labour natural selection will always tend, where habits permit, to specialise organs. In such cases as the Hydra & many lower animals, the/39/same tissue perform multiple functions;—thus also in many crustaceans, the limbs act as swimmers & branchiae:[3]/39 v/in the Loach (Cobitis) the whole alimentary canal acts of course for its proper end, but likewise in aid of the lungs, "as this fish swallows air & voids carbonic acid"[4]: in the larva of the Dragon-fly, water is taken into the intestine by the anus & its oxygen absorbed for respiration; & I may add by the violent expulsion of this water, the animal progresses./

---

[1] Annales des Scien. Nat. 3 series Zoolog. Tom 3. p. 264 and Introduct. Zoolog. Generale.

[2] Dr. Carpenter in his admirable Principles of Comparative Physiology 4th Edit. p. 131. shows that 'in cases where the different functions are highly specialised, the general structure retains, more or less, the primitive community of function which originally characterized it.'

[3] Milne Edwards Introduct. Zoolog. Gen. p. 64.

[4] Owen, Hunterian Lectures: Fish p 281.

39/But we have, also, many instances of distinct organs in the same individual simultaneously performing the same function. Many articulate animals have stemmata & compound eyes which are organs constructed on a fundamentally different principle; the compound eyes not necessarily having any lens or concentrating apparatus. In respiration double organs are common: many animals breathe wholly by their skin, or as in Nereids by a highly vascular portion near their legs,[1] aided more or less by branchiae. Even in frogs it has been experimentally proved that the skin largely aids the lungs. The Proteus & other perennibranchiate reptiles, have at the same time both lungs & branchiae. Certain spiders have both a pulmonary sack & tracheae:[2] ⟨one species of Nemoura (an insect allied to May flies) has branchiae & tracheae;⟩ tracheae act by carrying air to the diffused blood, & branchiae or lungs by bringing the blood to the water or air, & so are fundamentally different. A few genera of Gasteropod Mollusca have a pulmonary sac combined with branchial organs.[3]/

40/The proper function of the swim-bladder in fish is explained by its name, but in some fish it becomes divided by vascular partitions & has an air passage or ductus pneumaticus into the oesophagus, & certainly aids respiration[4] but these fish have, also, branchiae. There can be no doubt[5] that the lungs of the higher vertebrata are homologous or "ideally similar" with the swim-bladder of fish; & according to our theory the progenitor of all the vertebrate animals having lungs, had a swim-bladder, ⟨& that the transition was effected by⟩ the swim bladder having been perfected for respiration through natural selection;—the ductus pneumaticus having become the wind-pipe or trachea,—whilst the branchiae have been atrophied. As the Branchiae became useless for respiration, they might have been slowly converted for some other purpose. Thus the wings of all insects are believed by many entomologists to be homologous with the dorsal branchiae & scales of aquatic annelids; & therefore according to our views there has

---

[1] Milne Edwards Introduct. Zoolog. Gen. p. 63.

[2] Dugès in Annal. des Sc. Nat. 3 [actually 2] series Zoolog. Tom 6. p. 182.—

[3] Owen, Hunterian Lectures, Invertebrata 2nd Edit. p. 560. The Ampullaria of which I was shown drawings by Mr. Woodward offers an excellent instance of these double organs.

[4] I may just allude to Webers curious discovery of the swim bladder in certain fish being brought into connection with the organ of hearing by a chain of little bones & cavities, & so aiding this function. Indeed in some fish (Owen, Hunterian Lectures. Fish p. 210) as the Cobitis barbatula, the swim-bladder apparently subserves no other function.—
See the most interesting account of the use & homologies of the swim-bladder in Prof. Owen's Hunterian Lectures on Fish p. 273–281.

been, during a long course of descent & a great change of habits, an actual conversion of branchiae into organs of flight though we cannot even conjecture what were the transitional stages. On the other hand the ramified aquiferous/41/vessels of Annelids, by which the circumambient water circulates through their bodies, are believed by Prof. Huxley to be homologous with the air-tracheae or respiratory organs of insects.[1] We see, perhaps, in the double row of lateral sacs in the leech & earth-worm a transitional stage, for these sacks are considered by Prof. Owen[2] as the first morphological step towards tracheae; but their chief office is to excrete mucus.

I will give only one more instance of a ⟨very perfect⟩ morphological or homological transition between organs quite distinct in function; indicating, as I believe in all such cases, a real conversion through natural selection during a long line of descent. In most pedunculated cirripedes there are on the inside of the sack, two small simple folds of skin (called by me ovigerous fraena) with a row of minute glands on their edges. These glands secrete a substance which becomes attached to the ova & thus prevents their being washed out of the sack; but in some few cirripedes of this family which either live embedded or have a more perfect shell, this safe-guard does not seem to be required, & the fraena have no glands, but are/42/larger. In another closely allied family, with perfectly enclosed shells, the fraena have no glands, but are very much larger, & are plicated & sub-plicated, so as to expose a vast surface to the constantly renewed water of the sack; & here these folds of skin have been considered by everyone, as branchiae, ⟨which undoubtedly they are,⟩ although we have but to look a very short distance in the same sub-class to see them serving exclusively as a bridle to retain the eggs.

In the several examples now given, we have seen in respiration alone, the whole skin, or a part, the legs, alimentary canal, mucus-sacks, ovigerous fraena & the swim-bladder either aiding or actually converted into true breathing organs; and in the case of insects, branchiae probably converted into wings. In several of the instances we have also seen two distinct organs simultaneously serving the same office in the same individual.

In all the vast number of animals undergoing metamorphoses,

[1] [Darwin memorandum on separate sheet:] Ch. 8 Huxley says he is inclined to think that aquiferous tubes are homologous of tracheae, *but from no other reason except they carry circumambient fluid through body.*—Thinks mucus-sacks may be same—I doubt whether I had better quote at all—⟨He disbelieves that Branchiae of Squillae can be considered as a new organ.⟩

[2] Hunterian Lectures. Invertebrata 2nd Edit. p. 239.

in which the organs at two periods of life are extremely different, it seems quite possible through natural selection to carry on the state during either term of life into the other term. Field mice (Arvicolae)/43/differ from true mice by their molar teeth having fangs; but Mr. Waterhouse tells me that the teeth of old field-mice have been so often observed with fangs that this structure seems almost normally to supervene with age in some species. The two broods in certain annually double-brooded butterflies & moths, differ sensibly in size & colour of which fact Mr. H. Doubleday has given me striking instances. Most parasitic plants are parasites, & most climbing plants are climbers from their earliest days; but the Cuscuta or dodder germinates in the ground becomes parasitic, & its roots then perish: certain shrubs[1] become climbing lianas only after having grown to a height./43 v/Some species of Atriplex[2] bear on different flowers on the same plant seeds of very different size, colour & smoothness. The same thing occurs, though in definite positions in the flower & seeds of the ray & center in some Umbelliferae & Compositae. The position of the ovule is an important character generally uniform in large groups of plants, but in Buttneria & a few other cases, the same ovarium has one ascendant & the other suspended.[3] Moreover Al. Brongniart gives a case[4] of an erect ovule becoming, during maturation, suspended./43/Certain grasses[5] have fibrous roots when growing in moist soils & bulbous when in dry: the immersed & surface leaves of Ranunculus aquatilis differ in a surprising manner: the common Holly when old generally has its lower leaves prickly & the upper smooth. Although these facts do not throw the least light on how a particular state at a certain age, or time of year or under certain conditions is acquired; yet they are worth notice as showing the possibility of a kind of transition by the loss of one of states, different from/44/ordinary transition.

That an organ should acquire a particular state at one time of life if useful to the species, presents no particular difficulty, as we have seen that there is a tendency for a variation, or accumulated amount of variation at any period of life to be hereditary at a corresponding period; & we may perhaps hypothetically extend an analogous view to a variation in connexion with some

[1] A. de Jussieu. Archives du Mus. d'Hist Nat. Tom III. 1843. p. 102. Monographie de la Fam. Malpighiacées.

[2] Mr. J. Woods in Henfrey's Bot. Gazette vol. I. p. 328.

[3] Aug. St. Hilaire in Annal. des Scien. Nat. 1 series Tom 6. p. 134 and in Mem. du Mus. d'Hist. Nat. Tom x. p. 156.

[4] In Rhamnus, Annal. des Scienc. Nat. 1 series Tom x. p. 324.

[5] Alopecurus geniculatus & Phleum pratense, Hopkirks Flora anomala p. 22.

peculiar conditions; /44 v/ thus the presence or absence of wings in certain insects is believed by several entomologists to stand in relation to the temperature of the season./44/Individual plants raised by florists, as certain ⟨Hollyocks⟩ Dahlias?, have been noticed always to produce flowers of two colours; a variety called the heterophyllous oak produces leaves of several shapes; these tendencies might become strictly hereditary, especially if aided by selection from the two forms being in any way useful to the plant. Seedling Hollies differ greatly in prickliness & the tendency is known to be hereditary; suppose the natural conditions tended to make all the leaves smoother (& luxuriance & starving seem to have a direct action on thorns & prickles); then any natural seedling with all its leaves smooth would be unprotected from grazing animals & would be destroyed ⟨& would not reproduce its kind⟩, but if smooth only in its upper leaves it might perfectly reproduce its kind, & thus a variety or species be produced with leaves of two kinds, owing to natural selection caring only about the lower leaves./

45/*Real & apparent cases of difficulty in the transition of organs.*— By the foregoing cases it will have been seen that an organ may pass through the most extraordinary changes in function & form; this having been apparently facilitated, sometimes, by one organ performing two or more functions, & being then specialised & modified for one function; sometimes by two distinct organs performing the same function, the one being continued for the same & the other being either atrophied or transferred to another office; & sometimes by an organ normally having two states at different ages or under different circumstances, one of the states being preserved & the other lost. Probably many examples might be collected of a part or organ, which, from our not knowing of any intermediate grade, we should be very naturally led to look at as created for some new & special end. But, considering how small a proportion the living bear to the extinct, I have been much surprised at the difficulty, which I have found in collecting many good examples of such/46/cases. It should, however, be here noticed that if we look to an organ in a very isolated being, as the duck-like bill of the ornithorhynchus; or to an organ common to the greater part of a great class, as to the swim-bladder in fishes, the web-secreting organs in Spiders & a thousand such cases, we are very seldom able to indicate intermediate states, & therefore are not able even to conjecture how such structures could have been produced through natural selection. But this on our theory could hardly be expected for isolated beings are supposed to be isolated by extinction; and in the case of an organ common to the whole or greater part of a larger class, in order to find its intermediate stages we should generally have to ascend far in time (the natural selection of many diverse forms always implying a vast lapse of time, & the extermination of numerous less perfect forms) to about the period when the whole or greater part of the class branched off & inherited from a common parent the organ in question. And to ascend very far back in time & to find the intermediate stages, by which an organ common to a whole large class was produced, would require infinitely more perfect geological records, than we can hope to acquire. We can only hope to do this, when intermedial states happen to have been handed down by inheritance to the present day.—/47/But cases of extreme difficulty, judged even by the principles of our theory, undoubtedly do occur. Prof. Milne Edwards[1] who admits that new organs are occasionally though rarely created, adduces the branchiae of the higher Crustaceans as an instance of an organ, not formed by the modification of any preexistent part. But I must think the case lately given of the branchiae of cirripedes, a sub-class of Crustacea, ought to teach us extreme caution: there has been much extinction amongst the Lepadidae or pedunculated cirripedes & if a few more forms had become extinct, no one could have ever told, that the branchiae of the Balanidae were not a new & special creation.

Most naturalists look at the poisonous glands in venomous snakes as specially created organs, & not as modified salivary glands, which their position would indicate; for their intimate structure is wholly different. Here then, apparently, is a case in point. But as we know that many innocuous snakes have channelled or grooved fangs, which convey into the wounds made by them a copious supply of saliva from the large glands at their bases;[2]/48/and as I have been informed by Dr. Andrew Smith, that a bite of *such* snake ⟨(Coluber rhombeatus)⟩ caused him immediate pain more than could be accounted for the mere prick, I must believe that saliva[3] now in some degree injurious & no doubt useful to the even so-called innocuous snakes could by natural selection be slowly converted into a poison, as deadly as that of the most venomous snakes, entailing with it a change in the intimate structure of the gland.

[1] Introduct. a la Zoolog. Generale. p. 61, 65 &c.
[2] H. Schlegel, Essay on Serpents, translated by Dr. Trail 1843, p. 42, 47.
[3] Dr. Smith, also, informs me that all the Dutch Colonists assert that the Boomslange (Bucephalus Capensis) a snake without any proper poison gland, causes the death of small animals, which it bites.

In another snake (Tropidonotus rudis),[1] we have an extraordinary structure, namely the points of certain processes of the vertebrae are tipped with enamel, & penetrating the oesophagus, apparently serve as teeth./48 v/No intermediate structure is known; but, Prof Owen tells me that by passing the finger down the gullet of other snakes, homologous process can be felt, pointing downwards; & he thinks is quite possible that they may aid in forcing prey down the oesophagous. ⟨The point of the ribs, I may add in serpents, certainly aid in their progression⟩/

48/I had thought that the case of the Surinam Toad (Rana pipa Linn.) was quite isolated, here the male glues the eggs of the female on her back;/49/the skin of which swells & rises so as to form cells. In these cells the eggs are hatched & the young pass their tadpole state. But I find that in a common French Toad, well called *Bufo obstetricans*, the male helps to deliver the female, & then attaches them to his own thighs. Moreover lately in the same quarter of the globe inhabited by the Pipa, a [Here Darwin left a blank space in the manuscript to allow several lines for an example he never supplied.] Amongst insects I think it likely that instances of apparently quite isolated structures might be found:/ 49 v/although the highest authority, Kirby & Spence,[2] say "there is a regular & measured transition from one form to another, not only with respect to beings themselves, but, also, to their organs —no new organ being produced, without a gradual approach to it." Can a regular transition be shown in the case of the/49/ wonderful musical instrument of the male Cicadae with its double membrane, powerful muscles & two apertures like those of a violin?[3] The Bombardier Beetle (Brachinus) in England & as I have seen on the banks of the Plata, curiously defends itself by crepitations of an acrid fluid & smoke-like gas; but many other allied beetles squirt from their tails/50/an intensely acrid fluid (as I know for I have received a discharge from Cychrus in my eyes), but not so volatile as to turn into gas, & therefore not accompanied by a crepitation. The sting of a Bee or wasp is an admirable weapon, but homologous[4] (or rather identical as in the Hive Bee the eggs pass through it) with the ovipositor of other Hymenoptera; & the ovipositor in the Ichneumonidae is known to be occasionally used as an organ of defence, causing "a painful irritation", & for driving prey out of concealed places./50 v/Wasps & Bees & Ants

[1] Schlegel on Serpents p. 45.
[2] Introduct to Entomology. vol. 3. p 474.
[3] Westwood Modern Classification of Insects. vol. 2. p. 422.
[4] Westwood Modern Classification of Insects vol. 2. p 77, 117, 141. See also [Lacaze-] Duthiers in Annales des Sciences Nat. 3 series. Zoologie Tom. [   ].

use their stings solely as an organ of defence & battle; whereas the Fossorial Hymenoptera almost exclusively use theirs for half-killing insects, & storing their nests with semi-animate prey. And here we see in existing Ichneumons both uses of the sting shadowed forth in an ovipositor: but I shall return to this subject.—[1]/50/ ⟨Kirby confirms the fact of Ichneumons stinging.⟩

The separation of the two sexes at first seems a difficult case. On our theory it requires that the early progenitors of every class should have been hermaphrodites,—a view countenanced even in regard to mammals, by the rudimentary mammae & womb[2] of the males. In plants we can trace numerous/51/intermediate steps between hermaphrodite & unisexual flowers. In animals very few such stages are known; but in all the many cases in which hermaphrodites couple or are mutually necessary to each other, as in the oyster in which the male & female elements are matured at different times, it is not very difficult to believe that in some individuals the male & in some the female power might become less & less potent, so as ultimately to abort. And in the Hydra & certain corals one individual is sometimes exclusively male, sometimes exclusively female, but generally hermaphrodite.[3]

I have given this case of the sexes, that I might allude to the Complementary males of cirripedes, which show in how unexpected a manner nature can effect a transition. Nearly all cirripedes are hermaphrodites, though belonging to the great class of Crustacea, in which the sexes almost universally are distinct; but in two genera[4] I found the sexes quite/52/distinct;—several minute males, fourteen in one instance, being attached parasitically on to one female. It may be asked how was the separation of the sexes in cirripedes effected? I venture to assert, if two other small genera had become extinct, this question could never have been answered. In these two genera, some of the very closely allied species have the sexes distinct (in one instance the female carrying in two pouches a pair of minute, mouth-less, short-lived males, which when dead are succeeded by another pair), whereas other species are hermaphrodite, but with the male organs rather feebly developed; and these hermaphrodites are aided by a succession of minute short-lived males. From these males being paired not

---

[1] Westwood in Loudon. Mag. & Nat. History. vol. 6. p 414. & Modern Class. of Insects. vol 2. p 88. & 150.

[2] [Baly] Supplement to Müller's Physiology. 1848 p. 111.

[3] Owen, Lectures on Invertebrata. 2 Edit. p 125 p. 137.

[4] Cryptophialus & Alcippe: the two other genera alluded to, are Ibla & Scalpellum. A detailed account of these facts is given in my two volumes on Cirripedes published by the Ray Society.

with females, as in every other known case in the animal kingdom, but with hermaphrodites, I have called them Complementary Males. How easily in these two genera a separation of the sexes could be effected: we have but to make the male organs in the hermaphrodite already feebly developed, still/53/more feebly developed so as to abort; & the males are already parasitic on the females, & will then, ceasing to be complementary, assume the full dignity of the male sex.

As chemical compounds are definite, it seems at first almost impossible that a substance in one plant should change by gradual transition into a chemically different compound in another plant. That the proportions of different compounds in the same species change most readily under culture is well known; as in the case of wheat & the opium-poppy. But Prof. Christison has shown[1] that Oenanthe crocata produces a virulent poison in England, but is innocuous in Scotland: that Hemp yields a peculiar gum-resin only when grown in hot countries; and Dr. Stenhouse[2] has shown that the same species of lichen from different regions of the world, contains somewhat different chemical & crystalline substances, which are used for dying.—/

54/The electric organs of Fish,—those wonderful organs which, as Owen says, "wield at will the artillery of the skies"—offer a special difficulty. Their intimate structure is closely similar to that of muscle;[3] but it is most difficult to imagine by what grades they could have arrived at their present state.[4] Nevertheless the fact, ⟨recently discovered⟩ that Rays[5] which have never been observed to discharge the feeblest shock, yet have organs closely similar to those of true electric fishes, shows that we are at present too ignorant to speculate on the stages by which these organs, now affording such a powerful means of defence to the Torpedo & Gymnotus, may have been acquired. But the special difficulty in this case lies in the fact that the Electric fishes, only about a dozen in number, belong to two or three of the most distinct orders or better sub-classes of Fish.[6] This curious/55/ subject of closely similar organs occurring in organic beings, which are remote, in the scale of nature, that is, according to our theory,

---

[1] Gardeners Chronicle 1857. p. 518.   [2] Philosoph. Trans 1848. p. 63.

[3] Owen, Hunterian Lectures: Fish. p 217.

[4] Dr. Carpenter in his Principles of Comparative Physiology (IV. Edit.) has an interesting discussion on the Electric organs of fishes: compare p. 465–470, & 471.

[5] Dr. Stark Proceed. Royal Soc. Edinburgh. Dec. 2. 1844. On Jan. 6. 1845. Mr. Goodsir read paper on same subject, & shows that the organ in the Ray is the middle & posterior of the caudal muscle, greatly modified.

[6] Valenciennes in Archives du Mus. d'Hist. Nat. Tom 2. 1841. p. 44.

which have branched off from a common progenitor at an immensely remote period & therefore can hardly owe this similar anomalous organ to community of descent will be hereafter considered.

Neuter Insects.—We now come to our last & by far most difficult case of transition, namely the existence amongst wasps, bees, ants & termites of neuters, or sterile females, which often differ in structure & instincts from their parents, & which cannot themselves propagate their kind. We here encounter an accumulation of difficulties. I shall be compelled incidentally to allude to the subject of Instincts, which will hereafter be treated of in a separate chapter, but I shall here as far as possible confine myself to corporeal structure; the remarks, which follow are, however, all applicable to instincts. Of the difficulties, firstly, we have the fact of Neuters occurring in Bees &c, belonging to the Hymenoptera, & in White-ants/56/or termites, belonging to the Neuroptera, that is to a distinct order of insects: this case is parallel with that of electric organs in fishes of distinct orders & will presently be discussed. Secondly, how could the females have been rendered sterile by the agency of natural selection? Thirdly, when formed, how could they possibly come to have a different structure & instincts from their parents? This latter most curious difficulty will best be understood by an example. In certain Ants the neuters consist of two kinds, as in Eciton,[1] the soldier-neuters have enormous, peculiarly curved jaws & instincts, greatly different from the jaws & instincts of the other working neuters, and of the fertile females & males: in another species of Eciton, the soldier-neuters have large heads & instincts likewise different from the three other occupants of the same community. Now supposing that these soldier-neuters had been ordinary male or female insects, I should have boldly said, that first a slight enlargement of the jaws or head had been favourable to an individual, that this had consequently flourished & propagated its kind; that of its offspring, those with the largest jaws or heads had ⟨been selected⟩ survived, & that this/57/process had been continued, until great protruding jaws or heads had been attained. But in neuter ants, which are absolutely sterile, how is this possible? Granting that in an individual neuter a very slight enlargement of the jaws or more bellicose instincts had been of use to it or its community, & that it or they had in consequence benefitted by the better chance of

---

[1] I am greatly indebted to Mr. F. Smith of the British Museum, one of the highest authorities on Hymenoptera, for much valuable information on all points in the following discussion.

surviving; yet the neuter could leave no offspring to inherit the peculiarity,—to vary again, & again have the favourable variation selected & propagated. How then could the great jaws & peculiar instinct have been produced by the accumulating power of natural selection? I confess that when this case first occurred to me, I thought that it was actually fatal to my theory; but we shall presently see, that though a very grave difficulty, it cannot in my opinion be considered as absolutely fatal. The case, moreover, is of great interest, for it clearly shows that the Lamarckian doctrine of all modifications of structure being acquired through habits, & being then propagated, is false; for whatever may have been the habits of life/58/of our neuters, they never leave, (at least in ants) offspring to inherit the effects of habit or practice. For my own part, though I do not doubt that use & disuse may affect structures & be inherited, yet long before thinking of this case of neuter-insects I had concluded that the effects of habit were of quite subordinate importance.

First we will consider the simple fact how it is possible that communities of insects should come to possess sterile females or neuters: this is not a special difficulty, but only one of the same class, as that of any organ in a highly peculiar condition. There is some gradation not only in the numbers of the neuters in different species, but likewise, apparently in the degree of sterility: some ants have but few neuters, whereas in the Hive Bee the neuters have been estimated at up even to 40 or 50 thousand to one Queen. In ants there is no reason to suppose that the neuters ever lay eggs, though Huber[1] has seen them coupled with males, an act which always causes their death./

59/In the Hive-Bee, the neuters occasionally, though rarely lay eggs, which invariably produce only males. In Wasps & Humble-Bees the early-born females are small, ⟨do not differ in structure from the Neuters⟩ do not survive to winter like the large females, & the eggs they lay yield only males. These small females are said by Huber[2] to be attended by a small number of males, & therefore it might be thought that they had been fertilised; but this, probably is not the case anymore than with those neuters of the Hive-Bee, which have been known to lay eggs.[3]/59 v/Lastly in one of our common Wasps[4] the workers produced late in the

[1] Kirby & Spence Introduct to Entomology vol 2. p. 51.
[2] Kirby & Spence. vol. 2. p. 117.
[3] Dzierzon & Von Siebold True Parth[en]ogenesis Engl. Translat. 18 57. Those authors, have, also, made the wonderful discovery that in the Queen Hive Bee it is exclusively the unimpregnated eggs which produce males.
[4] Vespa Germanica. F. Smith in Zoologist [—See I (1843) 161–6.]

autumn are larger than those produced earlier in the summer & almost seem to be graduating into the state of fertile females./

59/In our third Chapter numerous facts were given showing how readily organic beings under changed conditions, not unfavourable to life & health were rendered sterile. When a cow produces twin calves, & one is female, she is a free-martin & always sterile. In the male Lucanidæ, or stag-beetles we have seen that collectors are not satisfied till they possess a series/60/ from mandibles developed to an enormous size to mandibles differing little from those of the female: this I believe to be caused by the amount & kind of food which the larva have obtained, & one must suspect that it stands in some relation to the virile power of the males. Male & female Brachyourous Crustacea differ in the width of the abdominal segments; but in some species intermediate individuals are not rarely found[1] & these females are believed to have been rendered sterile from some unknown cause/60 v/In Lepidoptera Mr. Newman[2] has given good reason for believing that the females in the autumnal broods, when two broods are not normally produced, are utterly sterile: this has been observed in France & England; & the high authority & experience of Mr. Doubleday is adduced in support of this remarkable fact./

60/Now let us suppose that these Crustacean, Lepidoptera or Lucanidae were truly social; many males & females (as with wasps & humble-bees) living & working together for the common good: in this case it seems not improbable, owing to the vast fecundity of the lower animals, that a certain number of females, working like the others, but without any waste of time or vital force from breeding, might be of immense service to the community. If this were so, & we see it is so with social insects, then natural selection would favour those communities, in which some of the individuals/ 61/had been exposed to conditions, or eaten food which had rendered them in some slight degree less fertile than the other individuals. In the social Hymenoptera we have to suppose that in long past ages some of the larvae were fed in the early part of the summer on some peculiar food or otherwise treated so as to have been rendered slightly less fertile than the other larvae. Then natural selection or the struggle for life, would ensure the continuance or the increase of the same treatment, so that the degree of sterility or the number of sterile individuals might be increased.

Now for our great difficulty of how neuters, not having progeny can be modified through natural selection, so as to fit their various

---

[1] [Unfilled space left for citation.]   [2] Zoologist 1857. p. 5764.

offices in nature. With wasps & humble-bees the large females alone survive the winter; & in the spring in their solitary state they perform all the duties of their neuters which are subsequently produced, & do not differ from them essentially in structure; Hence the neuters of the different species of wasps & Humble bees might be modified by inheriting any selected modifications in the females/61 v/just as bullocks of the different breeds of cattle, ⟨& capons of different breeds of poultry,⟩ differ from each other & slightly from the perfect males of their own breeds./

62/In the Hive-bee, the queen differs greatly from the neuters in instincts & in many important points of structure, as in the mouth, shape of sting, absence of wax-secreting pockets, & of the several curious contrivances for collecting & carrying pollen.[1] Now in most of these respects the Queen differs not only from her own neuters, but from the typical character of most social Bees; & it might be argued on our theory that the Hive neuters have retained by inheritance from an early progenitor certain normal characters which the Queen had lost. It deserves notice that the cuckoo-like species of bees, which lay their eggs in other bee's nests, have lost similar points of structure, either through disuse, or through natural selection, or both combined. That in the Queen Hive-Bee the loss of certain parts or any modification of structure should become attached to the female sex alone, is not at all surprising as we have in previous chapters seen; but here we have the truly astonishing fact that these hereditary losses are correlated with a particular treatment of the larva: this we know from/63/the fact that larvae which would certainly have become neuters & therefore would have had wax-pockets & the corbicula on their hind-legs &c can by a particular line of treatment be turned into Queens without wax-pockets & pollen-collecting instruments. According then to our theory, the neuter or sterile females of the Hive-bees owe those characters which they have in common with other social Apidae to ancient inheritance; the fertile female having lost them either by disuse or through natural selection, but always in correlation with a certain line of treatment during the larval state.[2] This seems a very bold hypothesis; but then there are two kinds of neuter Hive-bees; one larger, with a more capacious stomach, much greater power of secreting wax & which does not build; the other

---

[1] Kirby & Spence Entomology vol 2. p. 131.
[2] It is remarkable that the males when fed on royal jelly are not affected; but when they are hatched in workers-cells, they are believed to be rendered smaller: Kirby & Spence Entomology. vol. 2. p 126, [actually 127] 161.—

smaller, a nurse & builder.[1] How can we make this fact accord with our theory of natural selection? But first let us take the case of neuter-ants, in which analogous facts are more strongly displayed.—/

64/In Ants the case is reversed as compared with Hive-Bees, for their neuters, in differing from the fertile females, differ in an extraordinary manner from the typical structure of their sub-order, —namely in being always wingless, in the very peculiar shape of their thorax, in the frequent absence or very rudimentary condition of the ocelli & indeed in some genera (as Ponera) in being destitute both of ocelli & compound eyes.[2] But it deserves notice that in one allied family of non-social Hymenoptera, the Mutillidae, the females are wingless, destitute of ocelli, & the thorax is often singularly like that of a neuter ant:[3] hence, perhaps, it may be inferred that there is some correlation[4] between these points, so that if one were modified, the other points would tend to follow./ 64 v/The neuters, also, differ from the fertile females in size, in the shape of head, & of the mandibles, sometimes in the number of joints (Pseudomyrma) of the antennae, & in the form (Crypto-cerus) of the abdomen.[5] /64/Considering the terrestrial & sub-terrestrial habits of ants, the Lamarckian doctrine that they have lost their wings & ocelli by inherited disuse seems very tempting, but how utterly false; as it is just the wingless individuals which can never leave offspring! As queen-ants, like the large female wasps & Humble-bees, at the first foundation of a /65/community do all the work, any selected modification in them would be transmitted to their neuter offspring; but how could these neuters have acquired, through natural selection, a structure so widely different from that of their mothers? Moreover the neuters in closely allied species of the same genus, which we by our theory believe to have all descended from a common ancestor, also, of course, differ from each other. But the difficulty comes to a climax when we remember that amongst the neuter-ants of the same identical species we have in several genera two kinds extraordinarily different in structure & instincts; as in the case of Eciton already alluded to, in which the soldier neuters have enormous jaws & the working neuters, whom they guard ordinary jaws; & in another

[1] Kirby & Spence vol I. p. 492.
[2] Westwood Modern Class. vol 2. p 218, 235 and F. Smith on British Formicidae Transact. Ent. Soc. vol 3. Part 3. p 110, 113, 115.
[3] Westwood Modern. Class. vol 2. p. 213.
[4] See remarks on this subject by Mr. Westwood in Annals & Mag. of Nat. History. vol 6. (1841) p. 81.
[5] F. Smith in Entomolog. Transacts vol 2. p. 215 & vol. 3 p 156. [See Appendix for letter from Smith about forms of Cryptocerus workers.]

species of Eciton & likewise in Atta the soldier neuters have heads twice as big as those of the working neuters; In a Mexican genus[1] there are ordinary working-neuters & others never quitting the nest, with the abdomen swollen into a great, almost diaphanous sphere nearly five times as large in diameter as that of the common workers: these inactive neuters serve as mere distillers of a sac-charine fluid, which is stored up in a sort of comb. Lastly in the Driver ants of W. Africa (Anomma arcens, Westwood) there are, according to the Rev[d] T. Savage, three classes/65 v/of workers, differing in size, in their manner of biting, & in the work which they perform: and all are absolutely blind.[2]/

66/Grave as these several difficulties are, do they overwhelm our theory ⟨of natural selection⟩? Let us turn to our best guide the process of selection by man in our domestic productions. Man almost invariably selects from external appearances & breeds from the individual which he approves of: but let us suppose that he cooks & tastes a cabbage or radish & finds it very fine flavoured: that individual plant is utterly destroyed; but let him sow seed from several plants of the same stock in separate beds; of these seedlings let him cook & taste some out of each bed; & let him again save seed from the bed which produced the best-flavoured plants; & so repeat the process: in time, I cannot doubt he would get his desired variety true without ever having bred from a selected individual, only from a selected family. Breeders of cattle, like the famous Bakewell[3] who have attended to the grain of the muscle & to the fat & lean being well marbled together must have followed this plan, of breeding from the family to which the slaughtered animal belonged. To give another hypothetical illustration: the oxen or castrated animals of the Craven/67/cattle have horns not only much longer than those of the Bull, but even than of the cow: now I have such confidence in the principles of inheritance & in man's power of selection, that I fully believe by carefully noticing which families produced oxen with the longest horns, a stock might be reared, which not having themselves very long horns, yet when castrated would invariably produce oxen with extraordinarily long horns: this seems to have been effected, I presume accidentally, as far as size of body is concerned, in the oxen of the Devonshire & Herefordshire breeds,[4] which oxen are

---

[1] Myrmecocystus Mexicanus M. Wesmael in Bull. Acad: Royales: Bruxelles. Tom. 5. p. 766.

[2] Transactions of Entomological Soc. vol 5. p. 9 & 16.

[3] See, Marshall's account of Bakewell's proceedings in Youatt on Cattle. p 191.

[4] Youatt on Cattle, p. 17. The oxen of the Devonshire cattle are much larger in body than the Bulls; & the Bull is very much larger than the cow. In the Hereford-

of an extraordinary size. This principle of selection, namely not of the individual which cannot breed, but of the family which produced such individual, has I believe been followed by nature in regard to the neuters amongst social insects; the selected characters being attached exclusively not only to one sex, which is a circumstance of the commonest occurrences, but to a peculiar & sterile state of one sex.

Now to take the case of neuter ants, which neuters differ more from their parents than in other social insects; if the absence of wings/68/was any advantage to them, & we may fairly suppose that it would be so, seeing that the queen ants & termites tear off their own wings as soon as they found a colony, then I believe that those males & females which happened to produce neuters with their wings aborted, or with them ever so little less, would have a slight advantage; & of the myriads every year born these males & females would have a better chance of surviving & procreating neuters with less & less wings, till wingless neuters were produced. Judging from the Mutillidae, with the loss of wings, the thorax, owing to the laws of correlation of growth, would be modified, & possibly even the head & ocelli.

No doubt the process of selection would be retarded in an extreme degree by its action being indirect,—that is on the family alone; the individuals themselves born with any useful variation never leaving offspring. Had it not been for neuter insects, I am bound to confess that I never should have supposed that this process could have been as efficient, as our theory requires it to have been in the case of neuter insects. Extremely slow as such selection must/69/be; we have, at least for Bees & Termites, a superabundance of time, for fossil Bees have been found in Jurassic strata & Termites actually in beds of the Carboniferous age !¹

According to all analogy, neuters from the same parents would not all present the same variation, or the same in the same degree. In the case of Ants, for instance, analogy would lead us to infer that some few might be born with wings slightly smaller or the jaws slightly larger but that the other neuters in the same nest would retain their uniform character. Therefore, it may be urged we ought to have, or have had, communities presenting intermediate grades; more especially as there can here have been no

---

shire cattle, the ox is, also, a very large animal; & this cannot be simply accounted for by the effects of castration; as in the Durham or short-horn oxen, I am assured, there is no such inequality of size compared with the bulls & cows. In regard to the Horns of the Craven oxen see p. 197 of Youatt's Work.—

¹ Lyell's Manual of Geology. 1855. p. 389: Pictet's Paleontologie 1846 Tom. IV. p. 109.

crossing between the several neuters to keep them all uniform in structure. But I believe, first taking the case in which all the neuters have been altered from the maternal type, that variation is generally so insensibly slow, that without a comparison could be instituted between/70/the neuters at the present day with those belonging to the same species, a thousand or more generations ago, no difference could be perceived: I suppose that social communities are profited, first by a few neuters having wings, jaws or other organs different from the same parts in the fertile females in so slight a degree as to be imperceptible by us; & that the tendency in the parent to produce such neuters is increased by natural selection, until all the neuters are thus characterised;/ 70 v/and that subsequently the amount of difference is augmented first in a few neuters, & then again extended to all the neuters; & so onwards./

70/But if it were an advantage to the community that only certain proportion of the neuters should have, for instance, larger bodies or jaws than the other neuters, then I can see no insuperable difficulty in believing that by selection parents could be formed, which would produce a certain number of sterile females with big bodies or jaws & a certain number retaining their former small bodies or jaws. To give an instance from the vegetable kingdom, in which an analogous difference has appeared suddenly, & with long continued selection might perhaps be rendered hereditary by seed: there is a grape.[1]/71/which produces almost regularly on the same bunch small round and large oval berries,—a character I may add, considered by Odart as usually amongst the most constant in the vine. In the case of the two kinds of neuters in the same nest,—the acme of difficulty on our principle of natural selection— there must have been communities presenting during a long period grades between the large bodied or jawed individuals & the small: it must have taken an extraordinary length of time for selection acting only on the parents to produce a defined line of demarcation between the two sets of neuters in the same nest,—between the warriors & workers, for instance, in Eciton. Considering how very few social insects are well known I am surprised that I am able to adduce on the highest authority some instances of intermediate grades in the same nests. Mr. F. Smith informs me that in the nests of Formica flava, though there are large & small neuters, they so graduate into each other that it is impossible to separate them into two distinct bodies. In F. sanguinea, the neuters if viewed in mass may be divided into two bodies, differing

---

[1] Count Odart. Ampélographie 1849. p. 71.

considerably from each other in size & colour;[1]/72/& their instincts are slightly different, for if on a hot midday when all are in their nest, the bank be struck, the large neuters alone come to the surface as defenders. But Mr. Smith tells me that if all the neuters in a nest be carefully examined, a considerable number will be found graduating from one extreme to the other. Again in the Driver-ant of Africa/72 v/which presents the unique case, as far as I know, of *three* classes of neuters in the same nest; the largest warriors are thrice as large as the least workers, & differ in some other trifling respects; but Mr Westood[2] expressly states that "there seems indeed to be a regular gradation in the size from the largest to the smallest," & "I must confess that I can discover no distinct character to separate the largest individuals from the others."/

72/In these cases, we have intermediate forms between the neuters in the same nest; but if we compare the amount of difference between the two classes of neuters in the nest of different social insects, even within the limits of the same genus as in Eciton, we, also, find a gradation, as might have been expected from selection having in some communities produced a far greater difference between the neuters, than in others. In Humble-bees the neuters seem generally to vary more in size but not in colour than the males or females: Mr. Smith[3] states that in *B. muscorum*, which includes seven reputed species of Kirby, the several varieties sometime coexist in the same nest, but generally some one is preponderant: Mr. Newport/73/moreover, seems to have found[4] in some species two classes of neuters like those in the Hive Bee; but the difference must be very slight, as this does not seem to have been noticed by other observers. In the Hive-Bee[5] the two

---

[1] Transact Entomol. Soc. vol 3. P. 3. p. 102. In F. fusca, I may add, (p. 105) the difference in colour between the two-sized neuters is not invariable; the smaller neuter having "*usually* much paler legs & antennae" than the larger neuter.

[2] On Anomma arcens: Entomological Transactions. vol. 5. p. 16.

[3] Catalogue of British Hymenoptera: Apidae—1855. p. 213. Compare size of the neuters & others in the several species of Bombus.

[4] Westwood Modern Class. vol 2. p. 279.

[5] The Italian Bee (Apis Ligustica) is now considered by capable judges as only a variety of the common Hive-Bee, & this view is rendered extremely probable by their perfect fertility together. They differ considerably in colour, & the Italian Bee is more industrious, flies quieter so that von Berlepsch (Bienen-Zeitung 1856 p. 4 and Dzierzon's article p 61) says he could distinguish hives of the two kinds with his eyes shut, & it stings *much* seldomer: but as it is expressly stated that the Queens differ in colour & in seldomer stinging, this case, though very interesting as showing that strongly-marked varieties can arise in Bees, does not throw light on the difference between the workers & the Queens, or between the two classes of workers. It seems, however, that in the ancient Roman times (True Parth[en]-ogenesis by von Siebold Engl. translat. p 71. and Bienenzeitung 1856 p. 4) and

kinds of neuters differ very slightly in size, considerably in instincts & in the development of the wax-pockets; but in the latter respect the difference is not strongly defined, as the nursing-neuters "do secrete wax, but in very small quantities:[1] ⟨occasionally, what apiarians call Captain or black-bees appear in a hive; & here we have the groundwork for the production through natural selection of a third class of neuters, should such Captains prove in any way useful to the community[2]⟩ Even in British Ants alone we have some with all the neuters in the same nest quite uniform in size & structure; others with neuters of two classes differing slightly in size & not apparently in habits[3] as in Formica flava; others in size & colour & somewhat in habits as in F. sanguinea, and accompanied by slight differences in shape of thorax as in F. nigra. In many ants the neuters are quite destitute of the ocelli, which are present in the perfect sexes; & in the smaller neuters of F. flava the ocelli are "only distinguishable under a high microscopic power", whereas in the larger neuters of this species they are "distinctly visible",[4] but yet far smaller than in the males or females./74/In some of the species, also, of Eciton the two classes of neuters differ only slightly in size, whereas in other species of this genus, & of Atta & of Myrmecocystus & of Cryptocerus we find the most astonishing differences, in heads, jaws & abdomen.[5]

According to our theory it might easily happen that parent-ants after having produced two forms of neuters, should through natural selection come to produce more & more of one form till none of the other were left. I infer that this has actually come to pass with the Polyergus rufescens[6] which from making slaves of the neuters of other species (how this truly wonderful instinct could have been acquired, will be discussed in a future chapter) has no working neuters, but only warriors, or slave-takers, which have jaws incapable of building a nest./

75/I have discussed this case of neuter social insects at great length, for it/75 v/is by far the gravest difficulty, which I have encountered; so grave, that to anyone less fully convinced than I am of the strength of the principle of inheritance, & of the slowly accumulating action of natural selection, I do not doubt that the

at the present day common dark-coloured neuter-bees appear amongst the golden Bees even in Italy: this, however, may possibly be due to crosses owing to the common Bee having been anciently introduced into Italy.

[1] Kirby & Spence Entomology vol I. p. 493.
[2] [Pencil note:]? Bevan, Westwood [*Classification of Insects*] 2/279. perhaps old.
[3] Kirby & Spence, Introduct. Entomology vol. 2. p. 51.
[4] F. Smith in Transact. Entomolog. Soc. vol. 3. Part 3. p. 108.
[5] Entomological Transactions Mr. Smith on Brazilian Ants. vol. 3. p. 161.
[6] Westwood Modern Class. vol. 2. p. 219, 232.

difficulty will appear insuperable. But I have now done my best to show how I reconcile with our theory, the facts/75/(as far as corporeal structure is concerned) of the differences of the neuters from the fertile females, & of the two classes of neuters from each other;—namely by the continued selection, not of individuals which have varied in some way profitable to themselves, but of the stock which has produced any profitable variation; the variation having become correlated to a certain state of one sex ⟨favourable to the community⟩. A division of labour is possible in communities of man, through his intellect, his traditions & artificial instruments; but in communities of insects, which have almost unvarying instincts & for instruments, only their jaws & limbs, a division of labour seems to be possible only by the production of sterile individuals; for had the different workmen been capable of breeding together, the several castes would have been blended together & so lost./

76/I have now finished, as far as seems necessary, the subject of the transitions of organs. We have seen some cases, as that of the eye, most difficult from its transcendant perfection; some from no transitional stages being known, and some from our not seeing as with Electric fishes, how any transition is possible; but I think facts enough have been given to show how extremely cautious we ought to be in ever admitting that a transition is not possible. Considering the number of forms which undoubtedly have been ⟨exterminated &⟩ utterly lost, I am much surprised that we have not encountered very many more cases of extreme difficulty in attempting to show how one organ or part may be slowly converted into an apparently quite distinct organ.

Finally it seems to me highly important to bear in mind that he who believes that each species has been independently created, can only say that it has so pleased the Creator never or most rarely to introduce a new organ. Or he may mask his ignorance, & say with Milne Edwards[1] that the "law of economy"/77/is almost as paramount in nature, as the law of "the diversity of products". But on our theory of gradual modifications through natural selection, the law of economy is only the law of descent, the canon "*Natura non facit saltum*" becomes scientifically explicable.—

Similar & peculiar organs in beings far remote in the scale of nature.
—I have already alluded to the remarkable case of Electric organs occurring in genera of fish, as in the Torpedo & Gymnotus almost

---

[1] Introduction Zoolog. Generale. Chapter i.

as remote as possible from each other: but the organs differ not only in position, & in the plates being horizontal in one & vertical in the other, but in the far more important circumstance of their nerves proceeding from widely different sources.[1] I have also alluded to another very remarkable case, namely both ants, belonging to the Hymenoptera & termites belonging to the Neuroptera having communities, served by sterile females; the fertile females, I may add in both cases, losing their wings, as soon/78/a new community is founded: but, according to the prevalent belief, there is a wide difference in the two communities in the larvae of the termites being the workmen. The luminous power of certain insects is a rare & curious property; but in the Lampyridae it is the under surface of the abdominal segments, in Elater two spots on the hind part of the thorax, which shine.

The eye of the cuttle fish contains all the essential parts of the same organ/78 v/in the Vertebrata, belonging to a different Kingdom: a cornea, crystalline lens, & vitreous humour, corpus ciliare & retina are said to exist,[2] but it seems that neither the cornea, or the iris, are homologous, that is different parts are worked in for the same end; & the structure of the retina is extremely different.[3] To give a case of parts of little importance; in the Echidna, one of the most aberrant of the Marsupialia, & in the placental Hedgehog, we see the body protected by very similar spines.

In the Vegetable Kingdom, Orchis & Asclepias belong/78/to the two main divisions of phanerogamic plants, yet they present a curious resemblance in their means of fertilisation; in both, & in no other plants, the pollen-mass is attached by a footstalk to a sticky gland[4] which when touched by an insect/79/is drawn out, & is thus carried on to the stigmatic surface: moreover, according Aug. St. Hilaire[5] the sticky gland with its footstalk, which becomes during growth united to the anther, are developed in both cases in a similar manner. The leaves converted into pitchers in Sarracenia & in Nepenthes is another instance of a nearly similar structure in plants far from closely allied.

According to our theory when we see similar organs in allied beings we attribute the similarity to common descent. But it is impossible to extend this doctrine to such cases, as those just given of the Orchis & Asclepias, the Torpedo & Gymnotus, the Echnida & Hedgehog &c,—excepting in so far that community of descent, however remote the common ancestor may have been, would give

[1] Owen: Hunterian Lectures, Fish. p. 214.   [2] Müllers Physiology p. 1117.
[3] Carpenter Principles of Comp. Physiology 4th Edit. p. 730.
[4] R. Brown. Transact. Linn. Soc. vol 16. p. 685.
[5] Lecons de Botanique p. 448.

something in common to the general organisation. Just in the same way as in our last Chapter we have seen that the occurrence of similar monsters in the most diverse members of the same great class may be attributed to a like organisation from common descent, being acted on by like abnormal/80/causes of change. In the case of the eye of the molluscan Cephalopod & of the vertebrate animal, I do not pretend that we have one single fact (without it be the resemblance of the germinal vesicle) to induce us to believe that the members of these two great Kingdoms have had a common descent. It is not, I think, at all surprising that natural selection should have gradually given a fish & a whale something of the same forms, from fitting them to move through the same element; just as man in a small degree has given by his selection something in common to the form of the grey-hound & race-horse. A similar doctrine, I infer, must be extended to the above given remarkable cases of similar, though very peculiar & complex structures, in beings remote in the scale of nature. Such cases are not common; & in some of them the parallelism, as we have seen in the electric organs of fishes & in the eye of Cephalopod & mammal is not absolutely strict. Men, without communication have sometimes simultaneously hit upon the same curious invention: here man's intellect, which is nearly the same for all, may be compared with the power of selection which is the same throughout nature; & the general state of knowledge, the groundwork of all man's inventions, may perhaps be compared to that degree of general resemblance in organisation, which the members of the same great class have derived from common, but immensely remote, ancestors./

81/*Organs of little importance modified by Natural Selection.*—As natural selection acts solely through life & death by the preservation of slight favourable variations & the destruction of less favourable ones, the formation or modification of organs of apparently extremely little importance to the life of the individual has often seemed to me fully as great a difficulty, as the formation through such means of the most perfect & complex organs./81 v/In the case, however, of those animals, which possess will & choice, we must not forget "sexual selection", which may modify parts of little general importance, namely such as favour the struggle between male & male, or such as serve to charm the females; & characters gained by sexual selection amongst the males seem not rarely to be transferred to some extent, as shown in our sixth chapter, & possibly sometimes to a large extent, to the females.

In as much as assuredly we do not/81/really know the entire
economy of any one being, we may sometimes attribute importance
to characters which are of little or no service to the individual;
sometimes we may place to the account of natural selection that
which is wholly due to the laws of growth; & probably still oftener
we think that of little importance, which in truth is of the greatest
in the struggle for life.

Thus if we had known only the green woodpecker, we might
have said that its colour was of service to it in escaping dangers
in the woods, but the many black, white & crimson woodpeckers
show that probably this would have been a false view: seeing how
over the whole world Kingfishers, both male & female, are brilliantly coloured, we might naturally attach some importance/82/
to their colours in relation to their fish-taking habits; but a closely
allied & similarly coloured bird, the Dacelo Jagoensis inhabits
deserts, far from water & preys on lizards & grasshoppers.

Seeing how absolutely necessary whiteness is in the snow-covered
Arctic regions to the prey-seizers & the preyed, we might attribute
the absence of colour to a long course of selection; but it may be
that whiteness is the direct effect of intense cold; & that the
struggle for life has only so far come into play that coloured animals
would in the arctic regions live under a great disadvantage. So
again, the curious recurved hooks on the tips of the branches of
the Java Palms (   ), which are so strong & effective that the
natives use a branch as a thief-taker, are quite necessary to this
trailing plant that it may climb the lofty forest-trees; & hence
we might attribute (& perhaps truly) the formation of these hooks
to a long course of selection; but the many curiously formed
thorns & hooks on trees, which can apparently be of no use to
them from their height, may lead to the conclusion, that such
hooks are simply due to unknown laws of growth; & that in the
Java/83/palm the plant has become a trailer so as to take advantage
of the already formed hooks, & not the hooks slowly formed to suit the
changing habits of the plant.—The open sutures in skull of the just-
born mammal/83 v/which allow the bones to close together so as to
facilitate birth, have often been advanced as a case of special
adaptation; but as the sutures are equally open in the skull of the
young bird or reptile, which has to come only out of an egg, we
see that this structure must be due to some quite independent
cause; & being present has only been taken advantage of in the
birth of mammals./

83/Probably we oftenest err in attributing too little importance
to slight points of structure in the struggle for life. Looking at

the tail of the Giraffe, which seems quite like an artificially constructed fly-flapper I thought at first that surely this instrument
could never have been modified & adapted for its humble end,
through natural selection; but when I remembered Bruce's account
of the torments suffered from flies by the largest & thickest-skinned
pachydermata in Abyssinia; & when I remembered that the
extension of the introduced quadrupeds in S. America, is in many
cases ⟨absolutely⟩ governed by insects, I felt that it would be
rash in this case to put limits to the powers of long-continued
selection. Again I doubted whether the form or size of the external
ear could be modified by natural selection; but how all-important
is hearing to the Hare, & we know in domestic rabbits how
prodigiously the ears have been increased by the fancier's selection,
so that rabbits have been exhibited, with the two ears from tip to
tip [   ] inches in length: /84/ sportsmen, also, know how injurious
it is to crop the ears of terriers, which have to enter burrows; &
cruel gamekeepers crop the ears of cats, for when this is done they
will hardly enter a wood. Again I thought that such an apparently
small point of structure, as the eye-lashes, could never have been
formed or modified by selection: yet at times when the struggle
for food is most severe, what a momentary difference in vision
must often determine which shall survive & which perish; what
a trifling difference may often determine which individual shall
escape some beast of prey or other danger. But why nocturnal
marsupials should not have eye-lashes would, I suppose puzzle
anyone to account for.[1] Vultures which wallow in putridity have
the skin of their head naked; whether this adaptation is due to
selection, I will not pretend to conjecture; & we should remember
that the head of the clean-feeding Turkey-cock is naked like that
of the Turkey-buzzard.—

In all cases of organs of apparently trifling importance, we
should bear in mind that selection may act on them from their
concurrence in a more or less perfect state with other advantages
or disadvantages; for when the chance of life is/85/trembling in
the balance from some quite distinct cause, an extremely slight
difference, as more or less protection from insects or temporarily
better vision, might well determine which way the well-poised
beam should strike; for of those annually born a few alone can
leave offspring. Moreover a part or organ, though of secondary
importance to most animals, may be of the highest to some having
particular habits, as the external ear or eye-lashes to a burrowing
animal; & under such conditions the organ might readily be

---

[1] [Pencil note:] ?? Owen cannot remember having made any such statement.

perfected by natural selection, & subsequently inherited by numerous descendants modified in other respects, to whom the organ was of less importance but yet useful in its perfected state. Even in this latter case natural selection might be enabled to check any decidedly injurious deviations from the perfected state; as for instance the eye-lashes growing inwards, which causes to man much suffering & weakness of vision; quite enough almost to ensure the destruction of an animal which had to provide for itself in a time of dearth.—/

86/Several distinguished writers[1] have of late protested against the utilitarian doctrine that every part of every organic being is of use to it: they seem to think that nature plays with her work for mere variety sake or for beauty. Are we to believe that infusoria are exquisitely sculptured for man to admire them through the microscope? This protest against utilitarianism seems to me rather rash, as assuredly we do not know the whole life, its dangers & advantages, of any one single being; if we did, we could say why one is rarer & one commoner in any country. In the structure of each being, very much must be attributed to the correlation of growth,—that is when one part is modified for the good of the organism, other parts will in consequence be likewise in some degree modified: very much, also, must be attributed to inheritance from ancient progenitors, as we see in an exaggerated degree in rudimentary/87/organs. But in every case, according to our theory, the structure of the ancient progenitor could ⟨must⟩ have been modified or acquired, solely through its own good. So that all structures in all beings, making allowance for the correlation of growth to a larger but unknown extent, & making some allowance for the direct action of food & climate, must either have been useful to a progenitor or be now useful to the present descendant. The doctrine that structure is developed for variety or beauty sake would, if proved, be fatal to our theory.—

Looking again, not to the separate parts or organs, but to the whole individual, one is sometimes tempted to conclude, falsely as I believe, that nature has worked for mere variety: thus when we hear[2] that Mr. Bates collected within a day's journey, in a quite uniform part of the valley of the Amazons, 600 different species of Butterflies (Gre[a]t Britain has about 70 species), one may at first doubt whether each is adapted to its own peculiar & different line of life; but from what we know of our own British Lepidoptera/

[1] Prof. Huxley. Royal Institution Feb. 15 1856. p. 6.—The Rev. C. Kingsley, Glaucus., [: cf. pp. 100–1.]
[2] A. Wallace, Narrative of Travels on the Amazons. 1853. p. 469.

88/we may confidently believe that most of the 600 caterpillars would have different habits, or be exposed to different dangers from birds & hymenopterous insects. Mr. Wallace in his interesting Travels[1] seems to doubt the strict adaptation even of very differently constructed birds; for he lays much stress on the fact of having repeatedly seen the ibis, spoon-bill & heron feeding together on precisely the same food; & so with pigeons, parrots, toucans &c. But until it can be shown that these birds feed throughout the year on exactly the same food, & are throughout their lives from the nest upwards exposed to the same dangers,—for to want or danger each must be sometimes exposed, otherwise each would increase inordinately—the fact of their feeding together for a time or even for the whole year, seems to me to tell as nothing against the strictest adaptation of their whole structure to their conditions of existence.

On the other hand natural selection will produce nothing on the whole injurious to the species; no part or organ, though subject to the acutest suffering, will be actually formed, as Paley has remarked, to give pain. But /89/natural selection will not necessarily produce absolute perfection, as judged of by our poor reason. Each organism must be sufficiently perfect in all its parts to struggle with all its competitors in the same country; but by no means with all existing beings, as we see in the lessened numbers & even extinction of indigenous animals when others are introduced. We may err greatly, but can we call the sting of the Bee or Wasp perfect, when its use causes the insect's death by the tearing out of its viscera; the Bee, as I am informed by an apiarian, seeming conscious of its fate & never returning to its hive.[2] But if this fatal power of stinging, though it causes the loss of one member (but a member which does not breed) be of use to the community, it satisfies the requirements of the principle of natural selection. If we look at a Bee as an independent creation, this fact of death ensuing from the instinctive use of its own weapon must appear, as was long ago remarked/90/by John Hunter,[3] very singular: but on the principle of inheritance we can perhaps, understand how the two barbs came to be retro-serrated, so that their withdrawal is so difficult; for the two very same organs are serrated in the same manner in very many members of the same order, for the sake of sawing or boring holes for their eggs, in a manner which

[1] Id. p. 84.
[2] Bevan, Honey Bee. 1827. p 278. gives the best account of the act of stinging of the Bee which I have met with.
[3] Philosophical Transactions 1792. p. 191.

has mostly justly excited the admiration of every observer[1] Hence I infer that the ancient progenitors of Bees & Wasps used their ovipositors as boring instruments & that their eggs were laid with an acrid secretion; the boring instrument having since been converted in the neuter bees & wasps exclusively into an organ of defence; the acrid fluid having been intensified into a virulent poison. If in any member of the order, the sting gradually came to be habitually used for any purpose, then I do not doubt that natural selection, by always/91/favouring those individuals, which could easiest withdraw their less-strongly barbed stings, could make the spicula as smooth as in the Sphegidae,[2] which require their frequent use in order to half-kill their prey as a store for their larvae.

If we admire the female tiger savagely defending her young, or the hen-bird facing a hawk even to her own destruction, can we equally admire the Queen-bee always trying with the utmost fury to sting to death her own just born rival daughters:[3] We are accustomed to maternal love, but here we have instinctive, inveterate maternal hate; but both are the same, if useful to the community, to the unconscious & unpitying power of natural selection. We may err greatly, but can we call the drone or male Hive bee a perfect creation, whose sole function is to unite with the female; this union inevitably causing its death? If in most insects, we admire the means by which the male finds the female, —as by that almost incredible power of scent in moths which so often leads the male even down a chimney into a chamber in which the female is confined—or which leads some other moths to find/92/& know their females, which never leave their cocoons & remain in a rudimentary & almost monstrous condition; if we justly admire this, can we equally admire the production, in order to fertilise two or three queens, of some 2000 drones, utterly useless in the hive, not even collecting their own food; not even serving as scavengers like the male wasps, & slaughtered before their natural term of life[4] by their own nearest relations.

If in very many plants we admire the manner in which insects are tempted to visit the flowers, so as to carry the pollen exactly on to the stigmatic surface,—as for instance in Orchis or Asclepias,

1 Westwood Modern Classification of Insects. 1840. vol. 2. p. 77, 117, 141. Also M. [Lacaze-] Duthiers in Annales des Scienc. Nat. 3 series. Zoolog.—Tom. [   ].
2 M. Fabre in Annal. des Scienc. Nat. 3 series Zoolog. Tom vɪ. p. 161 [series is 4, not 3] I have examined the sting of Pompilus & I could see no trace of Barbs.
3 Kirby & Spence Introduct to Entomology vol 2. p 142.
4 Desborough. on the Duration of Life in the Bee: Transactions of Entomolog. Soc. vol. 2. Part v. p. 156.

or the Kidney Bean, in which latter the Bee always alights on the left side, where the stigma lies exposed,—can we look at this end as attained with equal perfection by the pollen being blown by chance, as in our coniferous trees, on to the ovules; for this is effected by the elaboration of dense clouds of the precious granules which are wasted to such an incalculable degree that buckets-full have been swept off the decks of ships at sea. In the Dionaea we may admire the beautiful contrivance, by which the leaf-appendage/ 93/closes like a steel rat-trap & catches insects,—beautiful at least for the plant, if those be right who believe that it is manured by the dead insects. But what shall we say of the terrific waste of insect-life by the varnished & sticky-buds of the Horse-chesnut & other plants; the scales of which are soon blown far away by the wind with the almost innumerable insects sticking on them; on one large tree with thousands & thousands of buds, there seemed to be on an average at least four insects sacrificed on each bud. But in all these cases, if the animal or plant can successfully struggle with its competitors, the principle of natural selection is satisfied.

As in nature selection can act only through the good of the individual, including both sexes, the young, & in social animals the community, no modification can be effected in it for the advantage of other species; & if in any organism structure formed exclusively to profit other species could be shown to exist, it would be fatal to our theory. Yet how often one meets with such statements, as that the fish in the Himalayan rivers are bright-coloured, according/94/[to] an excellent naturalist, that birds may catch them! How the fish came to be bright-coloured I can no more pretend to explain than how the Gold-fish, which Mr. Blyth ⟨informs me he⟩ believes to be a domestic variety of a dull-coloured Chinese fish, has gained its golden tints, or than how the Kingfisher, which preys on these fish, comes to be so brilliantly coloured, without, as far as we can see, any direct relation to its habits. A great physiologist supposes that glow-worms shine that birds may find & devour them! The aphis excretes a sweet fluid, highly useful to ants, & necessary, I presume, to those species which keep the root-feeding aphides in their subterranean nests; but must we infer from this, that aphides were created for the sake of the Ants? An acute observer supposes that the nectar of flowers was created specially for insects; but here there is reason to believe that it serves as an excretion for the plant, & besides in many cases is indispensable by tempting insects for their fertilisation.

How often one sees it stated that insects produce innumerable larvae, & plants innumerable seeds, that animals/95/may feed on them; or that a surprising number of plants, as Wrangell has remarked, bear edible berries in the tundras of arctic Siberia that birds may be there supported: but is it not more reasonable that the innumerable seeds & larvae are produced that some may escape destruction, & in the wretchedly barren Siberian tundras, may not the dung of the birds be almost indispensable to many plants, or at least as good for them, as the pellicle of guano with which some agriculturists coat their seeds, or as the so-called albumen with which nature coats not the outside of the seed, but the embryo within. One author supposes that plants with pitcher-like leaves were created that animals might drink out of their contained water; but the Sarracenia grows in bogs where water abounds. One more of the many instances which could be given, will suffice: it is commonly believed that the Rattle-snake has been created with fangs to destroy & the rattle to warn its prey! In this instance, I may just remark, that in a venomous allied S. American/96/Trigonocephalus, I observed that it constantly vibrated, especially when irritated, the last inch of its tail, with sufficient force, to make a slight noise when gliding amongst dry stalks of grass: I presume that no one would think that this habit was of any more use,[1] either to other animals as a warning or to itself for any object, than the vibration of its tongue, or the curling of a cat's tail when angry: now let us suppose that the little bead with which the tail of this snake, like that of many others, is terminated, were not annually moulted with the rest of the skin, but adhered only slightly to the new & larger bead formed with the new skin, we should then have the actual structure, manner of formation & vibratory movement of the rattle in the true rattle-snake; & our new rattle would be of no more use to the Trigonocephalus, & no more created to warn other animals, than its vibrating tongue or the curling of the tail in the cat or enraged lion.

Finally, although within the same class species having a nearly similar structure may be adapted to the most diverse habits, I believe that each single species has had its whole structure formed through natural selection, either in ancient time for the good of its progenitors, or/97/more recently for its own individual good; every modification, however, having been subjected to the laws of the correlation of growth & to the direct action of ⟨the conditions of existence, as⟩ food & climate. This conclusion seems to me to accord sufficiently well with the famous principle enunciated by

---

[1] If of any use it is more likely to serve to paralyse by fear or fascinate its prey.

Cuvier "celui des conditions d'existence, de la convenance des parties, de leur coordination pour le rôle que l'animal doit jouer dans la nature."[1]

Before summing up this chapter, I may remark that if our theory be extended to the utmost limits, which facts of any kind permit, nothing is easier than to make the whole appear to oneself quite ridiculous;—namely by asking whether a rhinoceros & gazelle, an elephant & mouse, a frog & fish, a bird, lizard & mammal could possibly have descended from a common progenitor. Involuntarily one immediately looks out for a chain of animals *directly* connecting these extreme forms. One forgets for the moment, that these/98/great groups have been perfectly distinct for enormous geological periods; some of them, almost if not quite as distinct at the earliest period of which we possess any fossil records, as at the present day; & therefore if intermediate forms ever did exist, they would all, or nearly all be, assuredly now utterly lost. To lessen in some degree the ridiculous impression of the foregoing question, one ought to think of such animals as the Ornithorhynchus, which though an indisputed mammal, presents in its skeleton & other parts some few plain resemblances to reptiles & birds. When mentally comparing a rhinoceros & gazelle, one ought to bear in mind that Cuvier & all our elder naturalists considered the Pachyderms & ruminants as the two most distinct orders of Mammalia; but now Owen has so connected them by Eocene forms, that he has made them into one great group. Look at the mud-fish (Lepidosiren annectens), which is so intermediate in structure, that although the greatest living authority considers it to be certainly a fish, many highly competent judges class it as a reptile: if then there be any truth in our theory, it would not be ridiculous to suppose that the Lepidosiren could/ 99/be modified by natural selection into an ordinary fish, or into a reptile. The case is almost parallel with that often encountered by philologists: to one who knew no other language, dead or living, besides French & English, how absurd would the assertion seem, that *evêque* & *bishop* had both certainly descended from a common source, & could still be connected by intermediate links, with the extinct word "episcopus". Let it not be supposed that I wish to underrate the extreme difficulty of extending my theory to its utmost limits. I feel it in every sense. The utmost which I wish, is to deprecate mere ridicule,—a tempting but faulty weapon for the discovery of our universal aim, Truth.—

[1] Quoted from Geoffroy Saint-Hilaire Principes de Philosoph. Zoolog. 1830. p. 65.

Summary. I think facts enough have been given,—on the unexpected transitions in the ways of life in animals of the same class,—on the diversified habits in the same species or in closely allied species, and on the changes of habit in the same species when placed under new conditions—to show how extremely cautious we should be in admitting that any animal, a bat for instance, could not have been formed by the modification/100/of another animal with totally different habits. On our theory of changes in habit or structure, due to the struggle for life common to every species, we can understand such cases, as birds with webbed feet never haunting the water, which must seem strange if every different species is viewed as an independent creation. So, also, with separate organs, I think facts enough have been given to show, what extraordinary changes in function may be effected; these changes being often facilitated by the same organ performing two wholly different functions, or changing its function during growth, or by two organs simultaneously performing the same function. Seeing the gradation in nature even in so perfect an organ as the eye, each stage being useful to its possessor, it does not seem actually impossible that such organs should have been modified & perfected by natural selection. From our ignorance of the entire economy of any one being, we ought to be very cautious in concluding that any part is too insignificant to have been formed by this same principle; seeing that the part might have been perfected during the life of an ancestral species to which it was of the highest importance, & seeing that natural selection might slightly act on the variations/101/of a most insignificant organ, when accidentally concurrent with other advantages & disadvantages. Even the extraordinary difficulty of neuter insects differing in structure from the fertile females, & being divided into castes in the same nest, can hardly be considered actually fatal to our theory, if we consider what man could probably effect & indeed has effected under somewhat analogous circumstances by his feeble powers of selection. Nor can the rare cases of closely similar, but not strictly homologous organs, in organic beings far remote in the scale of nature, be considered as fatal; for the same means of natural selection acting on nearly the same materials, might sometimes hit on the same result. Considering the vast number of extinct forms, it is surprising that far more numerous cases cannot be readily found, of organs without any known transitions serving to indicate the probable steps by which they were formed. The extreme rarity of the appearance of any quite new organs in a class, is an astonishing fact, as long as we look at each of the

13   **385**   SCD

innumerable/102/living & extinct species, as independent creations, but gives great support to our theory of gradual modification. Organic beings seem to be perfect only in that degree required by our theory, namely to be enabled to struggle with all competitors in their native country. If we trust our reason, which fills us with the most lively admiration for very many adaptive contrivances, others, like the sting of the Bee or the wasted pollen of coniferous trees, can hardly be considered as equally perfect. We have no good reason to believe that any organic being has been created for the good of another species, though so many mutually profit by each other. The doctrine that each part in each species has been formed (subject to the laws of growth) either for the good of its progenitors or for its own good, accords sufficiently well with Cuvier's principle of the "conditions of existence", & seems to fulfil all that we really see in nature.

CHAPTER IX

HYBRIDISM

INTRODUCTION

In a Christmas day letter of 1857 Darwin wrote to Hooker: 'I have just finished a tremendous job, my chapter on Hybridism: it has taken me 3 months to write, after all facts collected together!'[1] This confirms the Pocket Diary entry: 'Sept. 30th to December 29th on Hybridism.—'

The manuscript for this chapter so clearly reveals two distinct stages in Darwin's writing, and so many of the rejected sheets of the earlier version have been preserved that it would offer abundant material for a special study of Darwin's procedure in rethinking and revising his text here. The earlier and later drafts are easily distinguishable by their colours, for the former is written on sheets of gray foolscap and the latter on pale lilac sheets. At the University Library Cambridge, see for example, folios such as 19 which has a passage on gray paper which was sheared off the foot of the folio numbered 4h in the earlier draft, and pasted on the newer lilac folio 19, and folio 38 on which the top five lines having been sheared off from the top of the gray sheet originally numbered 20 in the earlier draft were pasted on another lilac sheet. In all, the following lavender folios of the newer version have gray paper pieces cut from the earlier version and pasted on for the revision: 19, 23, 24, 25, 34, 38, and 45. Here in the text as printed the beginnings and the ends of these older gray cuttings have been marked by a vertical bar. In addition seven gray sheets from the earlier draft were taken over bodily to be incorporated in the newer draft, having their original folio numbers cancelled and replaced by the numbers appropriate to their places in the sequence of the later draft, namely: 20 ⟨5⟩, 21 ⟨6⟩, 30 ⟨13⟩, 31 ⟨14⟩, 32 ⟨15⟩, and 33 ⟨16⟩. Then in section C. 40. g of the Darwin MSS., there are about three dozen rejected manuscript sheets of gray foolscap representing more of the earlier version of this chapter, namely original folios numbered: 3, 3A, 3E, 3D.1–3D.3, supplements a to h to folio 4, 13, 17–18, 25–36, 48, 53–59, and 61–64.

Even after the publication of the *Origin of Species*, Darwin continued to make notes which he labelled for this chapter ix of his Natural Selection, rather than for the corresponding chapter viii of the Origin. For example, also in the section of the Darwin MSS. marked C. 40. g. there are note slips which Darwin marked 'Ch ix' and dated: 'Dec 5–59', 'Aug 1860', and 'Ap 19, 61'. If intended for the *Origin*, such notes would have been marked chapter viii for the corresponding chapter there.

Later Darwin began to remove material from the manuscript of this chapter in order to use it in other works he was getting ready to publish. In the margin of folio 136 he scribbled the pencil note: 'Used in Dom. Animals', he wrote 'used' on folio 21 v, and elsewhere to signal similar use he marked passages with an encircled 'U' on the following folios: 21–23, 38, 99, 104, 116, 129–132, and 134–135 v; while he marked a passage on folio 119 with an encircled '2'. More drastically, he sheared off from other folios parts

[1] C.D. MSS., vol. 114, letter no. 218.

which are now missing, presumably because he attached them to the now missing manuscript for *Variation under Domestication*. The wording immediately preceding or following these passages cut away and now missing from the Natural Selection manuscript usually so closely parallels passages published in *Variation* that the continuity for this chapter can easily be restored from the published text of Variation. The sources for these restorations are given in the relevant footnotes. The following sheets have been thus cut up: 21, 38, 40, 40 v, 40A, 80, 119 v, and 132.

HYBRIDISM

1/This important subject concerns us under the following five heads. Firstly: are species invariably sterile when crossed, & are the resultant hybrids likewise sterile? Undoubtedly they are very generally infertile in some degree. But besides the extreme difficulty of deciding in some cases what forms to rank as species & what as varieties, we shall see that there is so insensible a gradation from utter sterility to perfect fertility that it is most difficult to draw any distinct line of demarcation between the two;—more especially as other quite independent causes often simultaneously tend to give some degree of infertility. In some very few cases it is, I think impossible to withstand the evidence that forms which are universally admitted to be good species are quite fertile together & produce quite fertile offspring.

Secondly: are those forms which from their known descent or other reasons must, in accordance with common usage, be called varieties, invariably quite fertile together & produce quite fertile offspring?/

2/This question may be answered by an almost universal affirmative; even in the case of varieties differing in an extreme degree from each other. But we shall see from a few experiments, carefully conducted by hostile witnesses, that the fertility of varieties when crossed can hardly be considered as absolutely universal. Nevertheless the extreme rarity of any, even the slightest degree of infertility between the most distinct varieties more especially in the animal kingdom is one of the greatest difficulties opposed to the theory of species being only strongly marked & constant varieties; a difficulty far more grave in my opinion than the sterility of crossed species.

Thirdly: do the several laws governing the degree & kind of infertility in the first cross & in the hybrid offspring, when these latter are paired inter se, or with one of their pure parents or with distinct species, indicate that species were created with this tendency to sterility in order to keep them distinct; or does the sterility seem to be an incidental consequence of other differences in their organisation? I think the numerous facts, which we shall give, clearly point to this latter alternative.—/

3/Fourthly: can the sterility of one species when fertilised by another & of their hybrid offspring be in any degree explained; so that the view of their sterility being only an incidental consequence on other differences be, at least partially, supported? I think that some little light can be thrown on this subject by the analogy of what often takes place, when organic beings are placed out of their natural conditions of existence.

Lastly: independently of the question of fertility, do the off-spring of two species & of two admitted varieties, when crossed, follow the same laws in their variability, in their resemblances to their parents, & in other such points? I believe it can be shown that they do.—/

3 v/I may premise that the whole subject is extremely compli-cated & that it is scarcely possible to make any universal proposition on any one head. On many points it seems to make great difference whether the forms experimentised on, have been long cultivated or domesticated./

3/*Sterility of species when crossed & of their hybrid offspring.*—The sterility of two pure species, when first crossed & that of their hybrid offspring has not always been kept sufficiently distinct. It does not seem a priori improbable that there should be difficulties in the union of two distinct species; we might imagine, for instance, that in plants the pollen tube/4/of one did not grow sufficiently long or in the right direction to reach the ovule of another species &c; though in truth the obstacle is probably always of a more recondite nature. But when the germ has been fertilised, & a healthy, long-lived hybrid is produced, it seems a far more wonder-ful fact that it should remain throughout its life utterly sterile. It is generally supposed that species have been created with this quality of being sterile one with another in order to prevent the many varied forms in nature becoming blended in extricable [sic] confusion. And this at first seems extremely probable; for no doubt if species did blend together, much of the perfect adaptation, —that division of labour—by which each species is excellently fitted for its own particular line of life would be lost; & consequently a lesser amount of life be supported in any given area. It is, also, generally supposed that the hybrids themselves have been rendered sterile in order that when formed (& undoubtedly they are occasionally formed in a state of nature) they should not perpetuate

themselves; but on the view/5/of sterility having been impressed
on species by direct creative action, it seems rather strange that
it should not have been impressed with sufficient strength to
prevent the production of a hybrid in any case.

If it could be proved or rendered highly probable that sterility
in the first cross or in the hybrid offspring was a specially created
endowment, it would be to us a fatal difficulty. By our theory
this sterility, whether or not we can throw light on its origin, must
be looked at as an incidental concomitant; like, for instance, the
greater or lesser facility with which one kind of tree can be budded
on another. This must be so, for sterility cannot have been produced,
at least in the case of the hybrids themselves, by natural selection,
as sterility obviously could not be favourable to them. In the
case of sterility between species & species, in as much as this is
favourable to them by keeping their characters pure & unmixed,
it is just/6/possible that the tendency might have been acquired
through natural selection; but I know of no fact leading to this
conclusion; whereas I do know of facts leading to the view that
it is an incidental concomitant of other differences.

The important service rendered by sterility in keeping the forms
in nature distinct, perhaps, leads us to overrate its importance as
a criterion of species. To explain what I mean: different species
of trees graft on each other with different degrees of facility, &
though trees in forests occasionally become naturally grafted
together, no one would look at this difference in facility, as an
endowed quality to prevent the more distinct kinds from becoming
inarched in a forest. Yet if it could be shown that invariably
different species of trees could not be grafted together or grafted
with difficulty, whereas all varieties could invariably be grafted
with perfect facility, this quality, though quite unimportant to
the plant, would be nearly or quite as good a criterion & as
valid an objection to our/7/theory, as the sterility of species when
crossed.

⟨What we have to show in order to render the facts here treated
of, not utterly subversive of our theory, is nearly the same as in
the case of any peculiar organ, namely to show how sterility
could first arise, to show that it is variable in degree & that there
is a gradation in different species from a lesser to greater degree
of sterility. And all this, I think, can be done.—⟩

I will first treat of Plants & will subsequently make only a few
comparative remarks on animals; for Hybridism has been attended
to with infinitely more care amongst plants than with animals.
Kölreuter & C.F. v Gärtner almost devoted their lives to this

subject; & the care, the conscientious accuracy & the astonishing amount of labour exhibited by them is admirable. Next comes the Hon. & Rev. W. Herbert (Dean of Manchester), who experimentised during even a longer period, but who never kept or published such systematic records; but had one advantage in/8/having large means at his disposal & in being one of the most skilful of horticulturists. Besides these three great authorities, we have Andrew Knight, Sageret, Lecoq & Wiegmann & many others.

I may premise that I have used the term mongrel for the offspring of two *reputed* varieties, & that of hybrid for the offspring of two *reputed* species./8 v/Hybrids are designated by the names of the parent species combined by a hyphen; & the first name in the mother; thus Dianthus armeria-deltoides, means a hybrid form D. armeria fertilised by the pollen of D. deltoides./

8/By "reciprocal crosses" I mean the *union* of species A the father & *species* B the mother, & on the other hand of B as father with A as mother/8 v/: for instance Dianthus armeria-deltoides & D. deltoides-armeria are hybrids from reciprocal crosses.—/8/ By "reduction", I mean the process by which the off spring of A & B, whether species or varieties, is brought, by repeated crosses in successive generations with either A or B, nearer & nearer to that form.

Kölreuter, whose admirable labours have been confirmed by every subsequent observer, concludes that all species whatever, when crossed, are in some degree sterile; but then he cuts the knot, for when in 10 cases he finds two reputed species quite fertile together he assumes that they are varieties. Unfortunately it is/9/not now possible always to know what plants, he really experimentised on;[1] but it is probable that several of the ten would be considered by the best authorities as truly only varieties.

Gaertner, after his truly vast experience, comes emphatically to the same conclusion, namely that two distinct species are *never* perfectly fertile together: he even disputes the entire fertility[2] of Kölreuter's ten cases & will not admit that they are varieties; but as from his table it appears that he has tried only three of them, I do not see what right he has to come to this conclusion.

[1] Thus the plants now corresponding with the Hibiscus manihot & vitifolius of Linnaeus, (Syst. 2 Edit) which Kolreuter experimentised on & found quite fertile together, Dr. Hooker tells me appear to be very distinct forms & have even been ranked in distinct genera. It seems, also, difficult to make out what is meant by Sida crista minor & major: Dritte Fortsetzung p. 114, 118.—1766.—

[2] Versuche & Beobachtungen Ueber die Bastarderzeugung 1849. p. 414 & 579 et passim. The three which he has tried are Datura stramonium & tatula: D. laevis & stramonium: & Malva sylvestris & mauritiana.

The laborious plan followed in every instance by Gaertner to measure the fertility of species when crossed with other species (& likewise of their hybrid offspring) was to take the *average* number of seeds in both pure parents growing naturally (& this is not quite so difficult[1] as might have been anticipated as I have found by trying in a few cases), & then to take the *maximum* number of seeds ever produced by the crossed species. Gaertner took the maximum in order to eliminate the acknowledged ill effects[2] of the pollen not being always applied at exactly the right time or not often enough at successive periods,/10/and of the plant being cultivated in a pot & placed not in a greenhouse but in a chamber, & lastly of the early castration of the anthers. He admits[3] that in order to get the proper maximum, many flowers in successive years should be experimentised on. Hence it is much to be regretted that he did not take for his standards of comparison the same species artificially fertilised with their own pollen & treated in every way like the crossed species. But I suppose the labour would have daunted the almost dauntless Gaertner. To test the ill effects of the processes, I have gone through the Table, & have picked out all the cases,[4] in which Gaertner actually did artificially fertilise plants, 20 in number, with their own pollen, or with that of another plant, universally admitted, & even by Gaertner himself, to be a mere variety; & these latter are 13 in number. Thus we have altogether 33 cases; & out of them 16 are marked as having had less than full fertility & 17 as producing the full number of seed. Hence the necessary treatment lessens the fertility of every other plant, when artificially self-fertilised./

11/Now admitting that the number of species, which are quite fertile when crossed is extremely small, the effects of the treatment alone would reduce the number by half. Moreover Gaertner himself admits that to get the proper maximum in crossed plants, many flowers should be experimentised on during successive years; & this has been done in comparatively few cases.

---

[1] Bastardz. &c. p. 207, 211.

[2] Bastardz. &c. p. 212. See also Gartner's Beiträge zur Kenntniss der Befruchtung 1844. p. 332 p. 365. at p 600 there is case in Tropaeolum showing the good of successive applications of pollen.—In some cases, as was also found by Kölreuter, culture in pots tends to increase the fertility of crossed plants & Hybrids.—To give one single instance of the ill effects of artificial fertilisation, taken by hazard from Gaertner Bastard. p. 385: Lychnis vespertina naturally yields 210–230 seeds, but fertilised with own pollen artificially it yielded as maximum only 192 seeds.

[3] Bastardz. p. 210, 214.

[4] I have made the case as favourable as possible to Gaertner by not counting those cases of artificial self fertilisation in which as in the Leguminosae there is great difficulty in the operation, as I have myself found. Nor have I counted some cases in which he utterly failed, as this would indicate that there was some

Although these considerations seem to me to throw some doubt on the universality of Gaertner's statement that species when crossed are *never* equally fertile with the pure species; they do not in the least make me doubt the high generality of his conclusion; for he experimentised on many hundred plants, & he asserts that he never once[1] got the full & normal number of seeds. But the case already given in our fourth Chapter of the very numerous experiments made during four years by Gaertner on no less than 170 flowers of Primula veris & acaulis & on Anagallis arvensis & coerulea, in which genera there is no apparent difficulty in effecting a cross, nor have other experimentisers found any difficulty, must give rise to serious misgivings; for Gaertner only twice succeeded in getting any good but scanty/12/seed from Primula, & none from Anagallis. As I cannot doubt, at least in the case of the primrose & cowslip, that they are only varieties; & as Gaertner failed either wholly or nearly so in crossing them, one may well ask in how many cases he may have failed in a lesser degree?

I will give two or three examples of the results obtained by Gaertner by counting the seeds. In the several species of Dianthus the normal number in a capsule varies from 80 to 120; whereas in the many species which he cross-fertilised he obtained only from 2 to 54.[2] For this genus & for Verbascum & Lychnis he gives the following decimal table;[3]

12 bis/As with the first cross between pure species, so with their hybrid offspring, Kölreuter & Gaertner maintain that they are invariably sterile./12 bis v/I may here remark that even in the most sterile hybrids, the pistil, ovary & even ovules appear to the eye perfect, but the ovules will not form an embryo; so it is with the stamens, but the pollen is manifestly imperfect as may be seen by everyone who has ever examined a hybrid. With hybrid animals in like manner the spermatozoa are imperfect,

---

fundamental error in the operation; thus I have not included amongst the varieties, Primula veris acaulis & elatior or Anagallis coerulea & arvensis, as these, moreover, are not considered as varieties by Gaertner. I may add in respect to the Leguminosae, that Gaertner crossed 32 flowers of the common Pea with the pollen of undoubted varieties & did not in one single instance obtain full fertility, nor did he with the Kidney Bean.—I have repeatedly tried to cross the varieties of the Sweet Pea, & have always failed, except [   ]. Andrew Knight succeeded with the common Pea, as I have also succeeded.

[1] There are some contradictions between the text & Table, which I cannot reconcile: thus in table it would seem that he once got the full number of seed from Lychnis diurna ♀ & L. vespertina ♂. So Matthiola annua ♀ & glabra ♂, & reciprocally are marked in table as fully fertile; but the contrary is stated in text. p. 102, 197. There are, moreover, some similar contradiction in regard to the fertility of some hybrids in the genera, Malva, Lychnis, Lobelia, & Verbascum.

[2] Bastardz. s. 195.   [3] Bastardz. s. 216, 219.—

	Proportional numbers of seeds.
Dianthus barbatus naturally fertilised with own pollen	1.0000
Crossed with pollen of	
D. superbus	0.8111
japonicus	0.6666
armeria	0.5333
chinensis	0.2600
collinus	0.2333
deltoides	0.2222
carthusianorum	0.1111
virgineus	0.0111
&c &c	
diutinus	0.0033
Lychnis diurna nat. fert. with own pollen	Proport. no. of seeds 1.0000
crossed with pollen of	
L. vespertina	0.7777
Cucubalus viscosus	0.2222
L. flos cuculi	0.0021
Silene noctiflora	0.0011
Verbascum Lychnitis nat. fert. own pollen	Proport. no. of seeds 1.0000
crossed with pollen of	
V. phoeniceum	0.8061
— nigrum	0.6336
— blattaria	0.6224
— thapsiforme	0.4081
— austriacum	0.3877
— macranthum	0.2653
— thapsus	0.2142
— pyramidatum	0.0306

though the microscopical structures of the testis, even in so sterile an animal as the common mule, present no imperfection: the ovules also are to all appearance perfect in the female common mule.[1]/12 bis/In those very few cases in which Kölreuter obtained fertile hybrid plants, he holds that their parents should be considered as varieties; & as with his first crosses Gaertner disputes

[1] M. Coste is the authority for ovules, see Colin Traite de Phys. Comp. 1856. Tom 2. p. 530. For microscopical structure of testis & state of spermatozoa see, Lallemand in Annal. des Sc Nat 2 series. Tom 15 p. 52. p. 298 &c. For state of ovules &c. in plants, see Gaertner Bastardzeugung s. 262.

their entire ⟨perfect⟩ fertility. That hybrid plants are very generally in some degree sterile, & that they stubbornly retain their sterility, I cannot in the least doubt; & I will give a few of the cases, which have most vividly impressed this conclusion on my mind. That hybrid plants are universally sterile, I cannot admit, from facts presently to be given. Indeed if the first cross between two species be ever quite fertile as I believe it to be; nothing is known to make one suppose that its offspring would be sterile.

Even in hybrids when crossed during successive generations by the pollen of either pure parent, although the progeny in each generation gradually assumes the characters of the pure parent, & acquires fertility, yet perfect fertility is the element last acquired.[1] A hybrid plant may thus come perfectly to resemble in external appearance one of its pure ancestors & yet be/13/utterly sterile ![2] For example, the hybrid called by Gaertner, Nicotiana paniculato-rustica, which means that the gr-gr-grandmother was pure N. paniculata, all nearer relations having been pure N. rustica, (or in the language of breeders having only 1/16 of blood of N. panicu-[la]lata) differed in no respect from N. rustica, except in producing less seed.[3] Kölreuter[4] found the very same thing in the successive crosses between these same two species; but made reciprocally, so that in the fourth generation, the plant could not be distinguished from N. paniculata, but its pollen was not so good, especially in the autumn. Again Kölreuter[5] found that the hybrid Mirabilis jalapa-longiflora fertilised by the pure M. longiflora, produced plants more sterile than their hybrid mother. These plants (which were 3/4 M. longiflora & 1/4 M. jalapa) produced with their own pollen seven seedlings, of which some were quite sterile. But three of the seven produced altogether 15 plants which were very sterile. But one of the fifteen produced nine seedlings; the seeds of these nine seedlings seemed nearly worthless. Here then we have a high degree of/14/sterility continued down to the gr-gr-grandchildren (self-fertilised in each generation) of a hybrid which was fertilised by one of its own pure parents.

In the cases just given, the first hybrid had been fertilised, either in one or in all the succeeding generations by the pollen of one of the two parent species. In hybrids fertilised from the first by their own pollen, Gaertner repeatedly states that he has never known the fertility to increase in the successive generations, even in the case of the most fertile hybrids; but he has often known it

---

[1] Gaertner Bastardz. s. 450, 459.   [2] Gaertner Bastardz. 449, 460.
[3] Bastardz. s. 447.   [4] Dritte Fortsetzung s. 47.
[5] Compare Nova Acta Petropol. for 1795, p. 324. and 1797 p. 373, 375.

to decrease; so that in a late generation the hybrid could not be fertilised even by the pollen of either pure ancestral species.[1] In the successive generations of self-fertilised hybrids, occasionally a seedling is produced extremely like one of its pure ancestral species; but such seedlings are not more, generally less, fertile than the first hybrid.[2] Gaertner gives a full account of the successive generations of Dianthus armeria-deltoides: this hybrid yielded seed for ten generations; having sown/15/itself in his garden for the first six or eight; at each generation it yielded less & less seed, & at the tenth its fertility was quite lost.[3]

I will abstract two analogous cases from Kölreuter: Two hybrid plants of Mirabilis jalapa-longiflora, self-fertilised produced 16 seedlings (grandchildren of the two pure species), most of which were very sterile; but one produced nine seedlings. Of these nine, four were slightly fertile & altogether yielded ten plants, which were excessively sterile, only one having produced anything, namely three seedlings. These three were the gr-gr-gr-grandchildren of the two pure species.[4]

Kölreuter found the cross between Mirabilis jalapa & M. dichotoma, nearly as fertile as the pure species so that he says he should have doubted whether the parents ought to have been considered as distinct species, had it not been for the portentous stature of the hybrids. One of the hybrids thus raised, & self-fertilised produced 28 seedlings, of which 14 were more fertile than their hybrid parent & some of them even more fertile than their pure grandmother, M. jalapa; but the remaining 14 were considerably less fertile, Kölreuter then took eight of the most fertile of these 28 hybrids,/16/& raised from them 34 seedlings: of the 34, (which were gr-grandchildren of the two pure species) only one produced an abundance of seed, & nine were excessively sterile. So that we have seen in some of the hybrids of the second generation a marked increase of fertility (in opposition to Gaertner's statement), but found in all except one of the third generation a high degree of sterility.[5]

I have given only a few examples, but it is impossible to study the work of Kölreuter & Gaertner, without coming to the conviction that the fertility of hybrids, when self-fertilised during successive generations, rarely, perhaps never, increases; on the contrary it generally decreases. But this latter fact, I think, is

[1] Bastardz. s. 418–421. [2] Bastardz s. 439. [3] Bastardz. s. 553.
[4] No record is given of the fertility of these three last plants. Compare Nova Acta Petropol. 1795. p. 332, & 1797. p. 373, 381, 392, & 403.
[5] Nova Acta, 1793. p. 394; 1795 p. 316; 1797, p 383–389.

perhaps partly due to an independent cause. Gaertner repeatedly states[1] that hybrids, even the less fertile kinds, if *artificially* fertilised with pollen of their own hybrid sort for some generations, sometimes decidedly improve in fertility. This is a very surprising fact, considering that, as we have lately seen, the artificial process of fertilisation lessens the fertility of about half the pure species experimentised on,—those flowers/17/which presented any peculiar difficulty for the operation having been excluded from the enumeration. But I think the increased fertility from artificial fertilisation may be explained in the case of hybrids, by the undoubted good which always follows from a cross with another individual of the same kind, as shown in our third chapter. When a plant is artificially fertilised it is castrated at an early period, & the pollen from another individual, or at least another flower must necessarily be used during each successive generation. On the other hand when a hybrid is allowed to seed spontaneously, it will have to be isolated in a green house or chamber, in order to prevent accidental crosses from either pure parent or allied hybrids, which the experimentiser will generally possess; & hence the visits of insects will be checked or quite prevented & the pollen will not then be carried from flower to flower or plant to plant of the hybrid; or if the hybrid is grown in a garden,[2] there will seldom be, owing to the trouble of making hybrids & their sterility, a large bed of the same kind. Consequently the hybrid will generally be fertilised by its own individual pollen, & far from receiving the benefit of a cross in each generation, it will suffer from the undoubted/18/ ill-effects of breeding in & in. As in hybrids we already start with ⟨a strong tendency to⟩ sterility, I think the close interbreeding of carefully guarded hybrids will account in part for their increasing sterility; but not wholly, for the increase in some cases is too rapid, being observed even in the second generation[3]. As, however, Gaertner found the fertility of even the less fertile hybrids was actually improved by the process of artificial fertilisation, which we know is so often injurious & can hardly fail to be injurious in some degree, I can hardly think that the fertility of any hybrid has been fairly tested for successive generations until a large bed of it has been left growing in the open air, freely exposed to the visits of insects & the other means by which nature habitually crosses the individuals of the same species.

[1] Bastardz. s. 418, 421, 554.
[2] It must be owned that Dianthus armeria-deltoides, before alluded to, was grown in a garden for six or eight generations, but it is not said whether there were many plants of this hybrid.—
[3] Gaertner Bastardz. s. 421.

Let us now hear the results arrived at by W. Herbert, the third greatest Hybridiser who ever lived. He agrees generally in the closest details with Kölreuter & Gaertner, with one important exception, namely that he attributes much more fertility/19/in many cases both to the first cross between species & to their hybrid offspring. He says[1] "it is certainly not correct as a general law, though some have stated it, that the number of seeds in one pericarp is smaller in hybrid than in the case of natural impregnation; it is true in some cases, & the reverse occurs in others." This difference may be partly due to Herbert having accidentally experimentised on more favourable groups of plants; to his having in many cases raised at some time whole beds of hybrids; partly, perhaps, to his not having so closely observed the slight shades in sterility; but chiefly I am inclined to think to his great skill as a horticulturist & to his having ample means in numerous green & hot-houses: for it is certain[2] that hybrids are more sensitive in their fertility to their conditions of life than are pure species.

I will now give some of Herberts principal facts: ⟨47⟩ | he asserts,[3] that the hybrids from "the yellow Linaria genistifolia & the purple L. purpurea, & from Pentstemon angustifolium & pulchellum are both perfectly fertile, sowing themselves about the garden". Again the hybrid from Lobelia siphylitica & fulgens "reproduces itself abundantly".|/20/|⟨5⟩ Gaertner tried the cross between these same two Linaria & absolutely failed: he seems, also, to have found the hybrid between the two Lobelias much less fertile.—Herbert[4] states that the hybrids between Petunia nyctanigenaeflora & phoenicea are "not only fertile but seed much more freely than either parent":—⟨Here we see that Herbert has apparently tested the fertility of the hybrid by Gaertner's plan, namely by actual comparison of the number of seeds produced &⟩ in this instance there could have been no error by fertilisation through one of pure parent species, for the hybrid was forced & set its seed before any other |/20A/Petunia came into flower. Gaertner tried this cross, but with him the fertility did not come up to that of the parent species.[5]

Of the species of Gladiolus,/20/|⟨5⟩ Herbert[6] remarks that there can scarcely be two more dissimilar than *G. cardinalis* & *tristis*; "yet the produce of these intermixed is fertile, & where the third species *G. blandus* has been, also, admitted into the (compound)

---

[1] Amaryllidaceae p. 354.
[2] Gaertner Bastardz. s. 10, 32, 384.
[3] Amaryllidaceae p. 345.
[4] Amaryll. p 379.
[5] Bastardz. s. 388, 719. The reciprocal crosses (s. 177) between these species are not of equal fertility.
[6] Journal of the Horticult. Soc. vol 2. 1847 p. 88.

union, it is fertile in the extreme, incomparably more so than the pure G. cardinalis [''']: this, I may add makes the case very singular, as complicated crosses of three or more species are usually very sterile.

In Hippeastrum, Herbert says[1] that the species when crossed produce "offspring invariably fertile". In Crinum[2] Herbert had/ 21⟨6⟩/"a pod from C. capense fertilized by C. revolutum in which *every ovule* produced a seedling plant, which I never saw to occur in a case of its natural fecundation"!/21 v/Here it is impossible that the species self-impregnated in its wild state could have been more fertile, even if we assume that it is normally as fertile. This last case leads on to perhaps the most extraordinary fact recorded on hybridism, namely those cases in which a plant is *less* fertile with its own pollen, than with the pollen of a distinct species; though its own pollen is proved to be good by fertilising other species. Herbert was led to make | the following experiment from having observed during several years (in a letter to me in 1839 he says he had *then* made the observations during five seasons) that every hybrid Hippeastrum[3]/21⟨6⟩/when ferti[li]sed by the pollen of some other hybrid Hippeastrum yielded much more seed than with its own pollen. He was thus led to try an analogous experiment on a pure species, namely on a bulb of Hippeastrum aulicum, lately imported from the Organ Mountains of Brazil; this bulb[4] // [produced four flowers, three of which were fertilised by their own pollen, and the fourth by the pollen of a triple cross between *H. bulbulosum, reginae*, and *vittatum*; the result was, that "the ovaries of the three first flowers soon ceased to grow, and after a few days perished entirely: whereas the pod impregnated by the hybrid made vigorous and rapid progress to maturity, and bore good seed, which]/21 A/vegetated freely." Herbert adds "this is a strange truth, & the more remarkable from the difficulty of obtaining cross-bred seed at all in the genera which are most nearly related to Hippeastrum, namely Habranthus & Zephyranthes".—/

22/Gaertner has observed[5] analogous facts occasionally occurring in Lobelia: thus in two instances, the ovaries of L. fulgens could not be fertilised by their own pollen, though they set seed with the pollen of L. syphilitica & of L. cardinalis; & yet the pollen of those flowers of L. fulgens was good, for it fertilised L. syphilitica. So it likewise was with Verbascum nigrum./23/|Kölreuter more-

---

[1] Amaryll. p. 345. [2] p 351.
[3] Amaryllidaceae p 371. Journal of Hort. Soc. vol 2. 1847 p. 19.
[4] [The manuscript is sheared off at this point. The corresponding passage in *Variation*, II, 139, is here substituted for the missing portion.]
[5] Bastardz. s. 357.

over[1] described long ago a similar case in Verbascum phoeniceum, which was fertilised by 4 other species, but yielded no seed to its own apparently good pollen.|In Passiflora[2] also, it has been found that the plants could be much more easily fertilised by the pollen of a distinct species, than with its own.—

In these several curious cases, more especially in those which are only occasionally occurrent, we must suppose that the plants are in an abnormal state; though they are not to the eye in any way imperfect; & both pollen & ovaries are quite capable of performing their proper function when exposed to the action of a distinct species. We may attribute this result in part to the good always resulting from a cross. But it is a most singular fact that the self-fertilising power of a plant should ever be actually inferior to its perfect capacity simultaneously for hybridising & being hybridised./

24/Lastly we must allude to the numerous & complicated crosses, though their history is very imperfectly known, now carried on for many years by horticulturists, amongst the species of Azalea, Rhododendron, Calceolaria, Fuchsia, Rosa, Petunia & Pelargonium. In this latter genus, according to | Herbert,[3] the first great step "was the production of the plant called, *ignescens* by the inter-mixture of the group to which *betulinum, citriodorum* &c belong, with a tuberous rooted scarlet one. The fertility of that plant set wide the doors to innovation but the stream is confined within certain limits." | Very many of the beautiful varieties of Pelar-gonium, are extremely sterile; but this seems often quite indepen-dent of their hybrid origin; some varieties having become barren & some having come fertile after a few years culture.[4] The species & varieties of Calceolaria have been crossed, as Herbert remarks *ad infinitum*: he states[5] that even the/25/hybrid from *C. integrifolia*, a woody shrub, & *C. plantaginea*, as humble & herbaceous as a plantain, though at first sterile, during the second year "reproduced itself as perfectly as if it were a natural species from the mountains of Chile". The other great American genus of Fuchsia has likewise been crossed in the most complicated manner. | Yet there is no difficulty in getting abundance of seed from several of the varieties & so it is with Petunia.—One of the best seeders

---

[1] 2 Fortzet. p. 10. & 3 Fort. p. 40.
[2] H. Lecoq De la Fecond: et L' Hybrid: 1845. p. 70. Likewise M. Mowbray in Transactions of Horticultural Soc. vol 7. p. 95.—Bosse has made same observation, Gaertners Bastardz. s. 64.
[3] Journal. Hort. Soc. vol 2. p. 95.
[4] D. Beaton in Cottage Gardener 1856 p. 44, 55, 61, 94, 109.
[5] Journal Hort. Soc. vol 2. p. 86.

amongst the highly cultivated & generally sterile Races is of hybrid origin.[1]

Everyone has seen the splendid results of the most complicated crosses between the several species of Azalea & Rhododendron[2]: Mr. Gowen who raised some of the early crosses at Highclere assures me that some of them yielded numerous self-sown seedlings./ 25 v/I applied to Mr. C. Noble of Bagshot, so well known for the numerous splendid Rhododendrons raised by him, on the degree of fertility of his hybrids, & he has given me the names of several, the offspring of R. arboreum & maximum, & of altaclerense (itself a hybrid from Pontico-catawbiense fertilised by arboreum) & catawbiense, which he says he is sure produces as many seeds as any pure species. He adds that the kind raised in great numbers as stocks for grafting, is a hybrid from R. Ponticum & Catawbiense, & that this "seeds as freely as it is possible to imagine."/25/These facts, though many of them are not known with scientific precision, are important because it might have been inferred from Kölreuter's & Gaertners experiments that the successive generations of/26/ both simple hybrids & hybrids reduced one or two steps towards either parent form, *invariably* became more & more barren, but with what is known of the history of these several genera of highly cultivated plants, it is scarcely possible to believe in this conclusion. A steady & quickly increasing degree of sterility would have struck nurserymen & horticulturists. These facts moreover, strengthen my previous remark, that the only fair way of testing, as nature would test, the fertility of hybrids in successive generations, is to have numbers of the same kind growing in the open air, & allowed freely to cross.

Reviewing all these facts on the fertility of the first cross between two species & of their hybrid offspring, the precise observations of Kölreuter & still more of Gaertner demonstrate that in the great majority of cases no union whatever can be effected between two species; & that when effected their fertility & that of their offspring is very generally impaired to a serious degree. On the other hand, the fact that when Kölreuter found what he considered/27/perfect fertility, he at once ranked the two forms as varieties;—that in Gaertner's experiments the fertility not rarely approached pretty closely to that of the pure species, & that the necessary artificial fertilisation is shown to be in about

[1] Cottage Gardener 1856. p 206—Mr. Appleby on the Petunia—P. Phoenicea crossed with P. violacea have produced all the pink & purple vars—crossed with P. nyctaginiflora have produced the white vars.

[2] Herbert, Amaryll. p. 359. do in Hort. Journal vol 2. p. 86.

half the cases decidedly injurious;—that Gaertner failed in some
cases in which Herbert succeeded; that he failed almost entirely
in crossing primrose & cowslips & entirely in crossing the blue
& red Anagallis,—that Herbert & others in some cases found
undoubtedly distinct species when crossed not only fertile, but
actually more fertile than the pure species, every ovule in the
pericarp in one instance setting;—that in some few cases plants
have hybridised other species, & been hybridised by them, far
more readily than they could be self-fertilised;—and lastly attaching some little importance to the unscientific experiments
tried on so large a scale by Florists; it seems to me impossible to
admit, that species when crossed & their offspring are invariably
sterile even in a slight degree, or that the sterility invariably
increases in successive generations./

28/*On the difficulty in distinguishing species from varieties by the
test of fertility.*—Forms known to have descended from a common
parent are universally admitted to be varieties; but this can
seldom be told except with cultivated plants, & in other cases, in
order to decide whether to rank a plant as a species or a variety,
we must rely on the opinions of the best & most cautious Botanists,
who, however may of course be easily mistaken. If we followed
Kölreuter's simple rule & called all plants, which were quite
fertile together, varieties, it might be thought that we should at
least arrive at a decided result; but this is not so, for we have
seen that the two most laborious & careful experimentisers who
ever lived, often come to a diametrically opposite conclusion on
this head: and this alone almost suffices to show that, *practically,*
fertility will not serve to distinguish varieties from species.

We will now briefly consider those cases, in which Gaertner
found considerable fertility in the first cross & in the hybrids, but
in which on counting the seeds, he ascertained their *maximum*
was less than the *average* yielded by the pure species growing
freely under the most favourable conditions. By comparing, in the
case of closely allied & more or less doubtful species,/29/the
evidence from fertility with that which can be derived from any
other source, & with the opinions of the best Botanists, we shall
see in how curious & instructive a manner, the evidence is almost
equally doubtful, & graduates away on both sides./

29A/*Matthiola annua, glabra & incana*: Gaertner experimentised on the two
first species of Stocks. In the table at end of column (Bastardz. p. 706) their
union is marked as yielding the full & normal number of seeds, but in the
text (p. 102 &c) they are expressly said not to be perfectly fertile. Mat. glabra

fertilised by annua yields more seed than the reciprocal cross. The hybrids from this cross are said not to vary (p. 168) like the off-spring of crossed varieties; but yet (p. 247) some slight variation is /30⟨13⟩/admitted: Gaertner attributes the greatest importance to variation in the first offspring as determining whether forms are to be ranked as species or varieties; but I cannot think, and we shall hereafter have to discuss this point, that this is of so much importance. With respect to the hybrid Mat. annuo-glabra (p. 388) its fertility is said to approach very nearly but not to equal, that of its pure parents. Kölreuter (Dritte Fortsetsung p. 116) crossed Mat. annua & incana & reciprocally; he obtained perfectly fertile capsules, & the hybrids raised from them, are expressly stated to be as fertile as the pure species ever are: Gaertner doubts this, but he does not appear to have tried these two forms. Now Robert Brown, ⟨a host in himself,⟩ & a few other Botanists as Spach consider Mat. annua, glabra as only varieties of *incana*, whereas most Botanists have treated them as species. As far as fertility serves for a test, we see the two greatest authorities divided. I should trust most to Gaertner on the point of fertility; & everyone I suppose on Botanical grounds would prefer leaning on the opinion of Robert Brown.

Datura stramonium & tatula: These species when united reciprocally are not according to Gaertner (p. 197) equally fertile: the hybrid off-spring do not vary (p. 168); Gaertner (p. 385) asserts that the hybrid D. stramonio-tatula gave at most 220–280 good seeds, whereas the two pure species give respectively 800 & 600 seeds. Gaertner (p. 273) lays great stress on the fact that these two species when crossed with *D. quercifolia*, yield very different hybrids, which Gaertner does not believe is the case when two varieties of one species are crossed with another species: but this conclusion/31⟨14⟩/seems grounded exclusively on some experiments with the varieties of Nicotiana, hereafter to be discussed, & which seem to me to be contradicted by some experiments made by Kölreuter on the same genus.—/31 v/Gaertner, also, crossed D. laevis with stramonium, & the union (p. 687) yielded less than the normal number seeds. On the other hand,/31/Kölreuter (Zweite Fortsetsung p. 125) tried reciprocally Datura stramonium & tatula & found the hybrids thus produced as fertile as the pure species: he also got hybrids from D. stramonium & inermis or laevis (Acta Acad. St Petersburg 1781. Part II. p. 304) & these he calls "foecundissimae," yielding 400–500 or even more seeds in each capsule; so that he concludes all these species are varieties. Now Asa Gray in his Manual (2 Edit [p. 341]) (See in his large Flora whether he enters in details) considers D. stramonium & tatula as only varieties; & the late Dr. Bromfield an excellent observer states (Phytologist Vol. 3. p 597) that he traced in the U. States every grade between *D. stramonium* & the purple *D. tatula*. Are we then to throw over such excellent observers as Asa Gray & Bromfield & believing Kölreuter was mistaken, follow Gaertner & Linnaeus & deduce that these two forms are true species. Or shall we believe that the lessened fertility of Gaertners hybrids, careful & conscientious as he seems to have been, was due to the treatment of the plant; in an analogous manner to the fact recorded by Kölreuter (Acta Acad 1781 p 303) that the hybrid Datura ⟨inermi⟩ laeviferox when planted in the open air yielded from 120–130 seeds in each capsule, but in pots only from 60–70?/32⟨15⟩/I confess my opinion of Gaertner's circumspection is so high that, though I do not believe that Kölreuters hybrid Daturae were as little fertile as Gaertner's yet I do believe, considering that with Gaertner, the hybrid D. stramino-tatula yielded only about 1/3 of the seed of the pure species, that there really is some lessened fertility; nevertheless it may be questioned, as we shall see from facts presently to be given, whether these forms should be considered as anything but varieties.

Lychnis diurna & vespertina: Gaertner made very numerous experiments on these plants during several years. In the table at the end it seems that he *occasionally* got the normal number of seeds from L. diurna fertilised by L. vespertina; & from its resultant hybrid; as well as from the reciprocal cross & its resultant hybrid. But in the text (p. 218) he arranges the species of this genus according to their sexual affinity, as deduced from counting their seeds, whence it would appear that L. diurna fertilised with the pollen of L. vespertina yields as a maximum only $\frac{777}{1000}$ of the average number of seed of L. diurna; & secondly that L. vespertina fertilised by the pollen of L diurna, yielded $\frac{810}{1000}$ of the average seed of L. vespertina; Hence, also, we here see that the reciprocal crosses are not of equal fertility. With respect to the fertility of the hybrids, L. diurno-vespertina (p. 385) yielded the maximum of 125; the pure L. diurna yielding 150–180, & L. vespertina through artificial impregnation with its own pollen giving 192 seeds, but when naturally fertilised, from 210–230 seed. On the other hand,/33⟨16⟩/the reciprocal hybrid L. vespertino-diurna, artificially impregnated with its own pollen, gave (Kenntniss der Befruchtung p. 598) a maximum of 234 seed, which is a greater number than occurs in either pure parent./33 v/Gaertner further states (p. 68) that in L. diurna after self-fertilisation, the pistil & petals turn colour in half an hour, but when dusted with the pollen of L. vespertina, they do not change their colour till from one to one & a half hour—facts which show in L. diurna a quicker fertilisation by its own pollen than by that L. vespertina./33⟨16⟩/Gaertner gives without any details, one marvellous statement (Bastarderzeugung p. 515) namely that the hybrid resulting from the union of the two reciprocal hybrids Lychnis diurno-vespertina ♀ & L. & L. vespertino-diurna ♂ is absolutely sterile! I have only to add that Gaertner remarks (Bastarderzeugung p. 577) that the hybrids from Lychnis diurna & vespertina in their variability are analogous, to the products of two varieties crossed. Hence we see that the general evidence, adduced by Gaertner is in favour of these two forms being species, though it seems from the Table at the end of the book that the fertility of the first cross & of the hybrids occasionally mounts up to the full normal number; & one case is given in detail of hybrids being fertile in excess.—Now if we look to the opinion of Botanists we find Linnaeus, Sir James Smith, ⟨Hooker⟩ & Prof. Henslow consider them, as mere varieties, whereas/34/most botanists think them distinct. I have cultivated L. diurna for three generations & could not observe in its variations any approach to L. vespertina; & as far as I can judge from the various published statements the two ought to be considered distinct. (C. C. Sprengel has some remarks on this subject in his Geheimniss der Natur s. 260: Gaertner discusses the subject at length: Tausch shows in the Flora 1833 B 1. s. 225, that L. diurna sometimes produces a white coloured variety.)

|Gaertner crossed several times in several ways, Cucubalus alpinus, C. Behen latifolius, C. Behen angustifolius, C. Italicus, C. pilosus & C. littoralt & never obtained the full & normal complement of seed, though the fertility seems to have been generally only one degree under it. This was the case even with the two admitted varieties of C. Behen.|/34 v/Yet, as I am informed on the high authority of Mr. Bentham all these forms are in his opinion & in that of many Botanists, only varieties of Silene inflata. All such cases of

course remain subject to the doubt whether Gaertner experimentised on properly named plants./

34/In all the several Floras which I have consulted, Dianthus glaucus is considered a variety of D. deltoides: Kölreuter (Dritte Fortsetsung s. 94) sowed seeds of the former & ocasionally raised plants closely resembling D. deltoides: further he (Nova Acta Petro: 1785 s. 284) crossed these two/ 35/forms, & raised many plants "in summo gradu foecundae". Nevertheless Gaertner (Bastardz. s. 539, 414) thinks Kölreuter is in error, & concludes that these are true species.

I need not here do more than to recall to mind that Gaertner after the most persevering effort concluded that, Primula veris, acaulis, elatior, (the primrose, cowslip & oxlip) are good & distinct species, from being highly infertile one with another; & that Anagallis arvensis & coerulae, tried on 19 flowers were absolutely sterile! I must believe that these experiments failed from causes analogous with those which prevented his entire success in his crosses of the garden varieties of the common Pea & Kidney Bean.—

These facts suffice, I think, to show that when forms in nature approach each other so closely that Botanists are divided whether or not to rank them as species, their fertility when crossed, & that of their hybrid offspring, approaches so closely to the normal value, that it is most difficult to decide the point by the test of fertility. Anyhow we see two observers, the most experienced of any, having the same theoretical opinions on the independent creation of species & who grudged no labour to arrive at the truth, often coming to directly opposite conclusions./

36/*On the infertility of varieties when crossed.*—I remarked at the beginning of this chapter, that the perfect fertility, even in most cases the decidedly increased fertility, of the most distinct varieties when crossed, was the gravest difficulty in our present subject. It is notorious, for instance, that the several varieties of cabbage, though so widely distinct in general appearance, are perfectly fertile together. So it is with our different breeds of cattle, dogs & poultry; but we shall, as yet, confine our attention to plants. I shall now show, taking the best evidence which can possibly be obtained,—though I admit that the evidence here is not un-impeachable, any more than in regard to the sterility of undoubted species when crossed,—that some forms which are generally or universally recognised as varieties, are not perfectly fertile together. These cases, though necessarily few in number, as the crossing of varieties has very seldom been carefully attended to, are, I think, of great importance in our present subject.

Gaertner fertilised 13 flowers, on different plants, //[1] [on a dwarf

---

[1] [Fol. 37 was cut up, and only the bottom portion, now numbered 36 v, was preserved with the manuscript. The missing text is replaced by the corresponding

maize bearing yellow seed[1] with pollen of a tall maize having red seed; and one head alone produced good seed, only five in number. Though these plants are monoecious, and therefore do not require castration, yet I should have suspected some accident in the manipulation had not Gärtner expressly stated that he had during many years grown these two varieties together, and they did not spontaneously cross; and this, considering that the plants are monoecious and abound with pollen, and are well known generally to cross freely, seems explicable only on the belief that these two varieties are in some degree mutually infertile. The hybrid plants raised from the above five seed were intermediated in structure, extremely variable, and perfectly fertile.[2] No one, I believe, has hitherto suspected that these varieties of maize are distinct species; but had the hybrids been in the least sterile, no doubt Gärtner would at once have so classed them.]

36 v/|Gaertner made most numerous experiments on many species of Verbascum, & with nearly all the species, he tried both the white & yellow varieties of *V. lychnitis* & *blattaria*; & he asserts most distinctly in two of his works[3]/38⟨20⟩/that the white-flowering *varieties* of Verbascum crossed with the white-flowering *species* bear more seed than when yellow-flowering varieties are crossed with white flowering species. of this genus. So again similarly-coloured varieties of the same species are more fertile together | than when differently coloured varieties of the same species are crossed. That these really are varieties, no one has doubted; & Gaertner actually raised[4] one variety from seed of the other. The serial arrangement of the species & varieties according to their sexual affinity or number of seeds yielded, was ascertained by experiments on no less than nine species repeatedly crossed by both the yellow & white varieties of the above two species.[5] In one instance alone, Gaertner enters[6] into minute details on this head: //[7] [but I must premise that Gaertner, to avoid exaggerating the degree of sterility in his crosses, always

passage in ch. 16 of *The Variation of Animals and Plants under Domestication*, 1st ed (London, 1868) II, 105.]

[1] [Surviving MS. note, slightly more detailed than printed note:] Zea minor semine luteo: Bastardz. s. 87, 169. From the table at the end, it, moreover, appears that altogether this experiment was tried on 22 flowers.—

[2] ['"Bastarderzeugung", s. 87, 577.']

[3] Kenntniss der Befruchtung p. 137; and Bastarderzeugung p. 92 & p. 181.

[4] Bastardz. s. 307.

[5] Some errors have crept into the Table; for the degree of fertility assigned to each species & variety does not always perfectly accord with their serial arrangement.

[6] Bastardz. s. 216.

[7] [Fol. 38 is sheared off at this point. The continuity of text is supplied from *Variation under Domestication*, II, 106.]

compares the *maximum* number obtained from a cross with the *average* number naturally given by the pure mother-plant. The white variety of *V. lychnitis*, naturally fertilised by its own pollen, gave from an *average* of twelve capsules ninety-six good seeds in each; whilst twenty flowers fertilised with pollen from the yellow variety of this same species, gave as the *maximum* only eighty-nine good seed; so that we have the proportion of 1000 to 908, according to Gäertner's usual scale. I should have thought it possible that so small a difference in fertility might have been accounted for by the evil effects of the necessary castration; but Gäertner shows that the white variety of *V. lychnitis*, when fertilised first by the white variety of *V. blattaria*, and then by the yellow variety of this species, yielded seed in the proportion of 622 to 438; and in both these cases castration was performed. Now the sterility which results from the crossing of the differently coloured varieties of the same species, is fully as great as that which occurs in many cases when distinct species are crossed. Unfortunately Gäertner compared the results of the first unions alone, and not the sterility of the two sets of hybrids produced from the white variety of *V. lychnitis* when fertilised by the white and yellow varieties of *V. blattaria*, for it is probable that they would have differed in this respect.] // 40 v/⟨It may be noticed that Gaertners strong wish to draw a strong line of demarcation between species & varieties, must have made him unwilling to admit his own curious discovery of this sexual affinity in Verbascum of varieties to each other & of varieties to distinct species, with flowers of the same colour.⟩ //

40/I am enabled to give one other & slightly different case & founded on much better evidence than the last. Kölreuter describes minutely five varieties of //[1] [the common tobacco,[2] which were reciprocally crossed, and the offspring were intermediate in character and as fertile as their parents: from this fact Kölreuter inferred that they are really varieties; and no one, as far as I can discover, seems to have doubted that such is the case. He also crossed reciprocally these five varieties with *N. glutinosa*, and they yielded very sterile hybrids; but those raised from the *var. perennis*, whether used as the father or mother plant, were not so sterile as the hybrids from the four other varieties.[3] So that the sexual

---

[1] [The text is again sheared off, but it can be partially replaced from *Variation*, ii, pp. 108–9.]

[2] ["'Zweite Forts.,'" s. 53, namely Nicotiniana major vulgaris; (2) perennis; (3) Transylvanica; (4) a sub-var. of the last; (5) major latifol. fl. alb.']

[3] ['Kölreuter was so much struck with this fact that he suspected that a little pollen of *N. glutinosa* in one of his experiments might have accidentally got

capacity of this one variety has certainly been in some degree modified, so as to approach in nature that of *N. glutinosa*.] . . . . . .

40A//infertility between the varieties of the same species, as tested by crosses with an extremely distinct species, seems to me a particularly interesting case.

In the varieties of the Maize, Verbascum & Gourd, it will have been observed that the fertility seems to have been slightly lessened only in the first cross, & not in the mongrel offspring: had, indeed, the yellow & red-seeded Maize, for instance,/41/in addition produced barren offspring, they would have been unanimously ranked as distinct species. In Nicotiana, however, certain varieties of one species, when crossed with a very distinct species, did produce hybrids more sterile, than when another variety of the very same species was used in the same cross as either father or mother. Although the infertility of hybrids is in itself, I think, a much more remarkable fact than the infertility of the first cross, yet this latter fact has been universally acknowledged as a test of equal value for discovering the essence of a species; indeed in some respect it seems of higher value, as more directly tending to keep the forms in nature distinct. We shall, also, presently see that in crosses of undoubted species there is by no means a uniform relation between the difficulty of the first cross & the sterility of the hybrid offspring.

To our short list of varieties in some degree infertile together, or having different degrees of fertility when, as in Verbascum & Nicotiana, crossed with other species,/42/those several Botanists who believe that the several forms before specified in the genera, Matthiola, Datura, Lychnis, Cucubalus (Silene) & Dianthus, are not species, but varieties, will have to add them to the list.

On the other hand some may say that not only the forms just alluded to, but that the reputed varieties of the Maize, Verbascum, Gourd & tobacco are true species. But even Gaertner with his strong predisposition to call the finest forms species, did not venture to do this: in the case of Verbascum no botanist considers the yellow & white flowered forms of V. blattaria & lychnitis as species: Gaertner's statement that he raised one variety from the other would, also, have to be dis-believed. Moreover on this view,

mingled with that of *var. perennis*, and thus aided its fertilising power. But we now know conclusively from Gäertner ('Bastarderz.,'' s. 34, 43) that two kinds of pollen never act *conjointly* on a third species; still less will the pollen of a distinct species, mingled with a plant's own pollen, if the latter be present in sufficient quantity, have any effect. The sole effect of mingling two kinds of pollen is to produce in the same capsule seeds which yield plants, some taking after the one and some after the other parent.']

it must be admitted in the case of the maize & tobacco that the hybrids raised from crossing these supposed species are perfectly fertile.

With much more apparent probability, others may say that the difficulty/43/in crossing the varieties in the four genera must have been entirely caused by some want of skill or by the injurious effects of the necessary manipulation. But in the Zea & Cucurbita no manipulation was requisite, as the sexes stand in separate flowers. And if in these several cases, the lessened fertility has been thus caused; & if we add to these cases, those of the primrose, cowslip, & oxlip, of the anagallis & all leguminous plants, it cannot possibly be any longer pretended that we have evidence worth anything on the infertility of a vast number of related forms, which are universally acknowledged to be distinct species, but which when crossed are in some degree fertile together.

As for myself I believe in the very general infertility of even closely related species when crossed; & further I am forced to believe that the forms generally called varieties, even those which have originated under culture, are occasionally in some slight degree infertile together. Believing this, & as it has, I think, been shown in the first section of this Chapter that some few undoubtedly distinct species are perfectly fertile together, I conclude, that not only the test of fertility practically fails, as we have seen by comparing the results arrived at by Kölreuter & Gaertner; but that theoretically the forms called species cannot in all cases be distinguished from those called varieties either by their fertility when first crossed; or by the fertility of their offspring./

44/*Laws & circumstances governing the infertility of crossed* ⟨species⟩ *plants & their hybrid offspring.—*

My object in treating this subject at some little length, is to see how far the facts indicate that the infertility is a quality, especially created, in accordance with the common view, to prevent species mingling in nature.

When the results obtained by the almost innumerable experiments tried on various plants are compared, we find a perfect gradation from absolute sterility to perfect fertility,—even to fertility, according to Herbert, beyond the natural degree. In plants belonging to different Families, the pollen of one when placed on the stigma of another, has no more effect than so much inorganic dust: this, also, is not rarely the case[1] with species even of the same genus. The first evidence of some sexual affinity between

[1] Gaertner. Bastardz. s. 96.

two plants, is a little shorter persistence of the corolla or calyx in a flower when dusted by the pollen of another species, than when simply castrated.[1] Even the pollen itself behaves differently when laid on the stigma of an allied plant, to what it does on one not at all sexually allied to it; & so does the stigma itself.[2]/45/Gaertner, also, describes a curious gradation[3] in the quicker & quicker withering & change of colour in the corolla, & in the more & more perfect development of the pericarp & external parts of the seeds, in accordance with the closer sexual affinity of the crossed plants. We then come to plants, in which after trials prolonged over many years, one or two out of thousands apparently perfect seeds, will germinate[4]. From this low degree of fertility, in which a single seed is occasionally found to germinate, a perfect series can be most easily shown, in the increasing proportional number of good seeds, up to nearly perfect fertility: but facts enough have already been incidentally given on this head.

Hybrid seedlings from plants with very little sexual affinity are sometimes weakly & tender, & cannot be raised with the greatest care:[5] but generally hybrids, as we have seen in our third chapter, are more hardy, vigorous, precocious & of larger stature than their pure parent species. Herbert, I may add | has clearly shown[6] that some kinds of Narcissus now cultivated in our gardens must have been formed by hybridism between two & three centuries ago, & have been propagated ever since by | offsets./

46/As in the first cross, so in the hybrid offspring a perfect gradation from sterility to fertility can be shown. But even in the most sterile hybrids, the pollen of either pure parent[7] will generally cause the flower to endure longer than it otherwise would have done: so that even in these hybrids, the sterility can hardly be considered absolute, though such hybrids never have yielded, & probably never could yield a single seed, which could germinate.

The degree of facility in effecting a cross, & the fertility of hybrids are both much influenced by the more or less favourable conditions to which they are exposed.[8] Besides this extreme susceptibility to external circumstances, it is most clearly proved by Kölreuter & Gaertner that the degree of innate fertility in the

---

[1] Gaertner Bastardz. s. 189.

[2] R. Brown Linn. Transact. vol. 16 p. 708 Gaertner Bastardz. s. 9, 19, 110.

[3] Bastardz. s. 68, 102.

[4] Gaertner Bastardz. s. 8, 101, 138.

[5] Herbert has given striking cases in Horticultural Journal vol 2. p. 11. also Amaryllidaceae p. 360; See also Gaertner's Bastarderz. p. 520 p. 548; & Kölreuter in Nova Acta Petrop. 1794 p 391, and 1795. p 325.

[6] Hort. Journal vol 2. p 20.

[7] Gaertner. Bastardz. s. 412, 533.   [8] Gaertner Bastardz. s. 10, 32, 384.

same hybrids is excessively variable. Hybrids raised one year from the same parents will be far more fertile than others raised another year.[1]/47/Both simple hybrids in the first & succeeding generations, & hybrids in course of reduction to either parent form[2], when raised in these several cases from the very same capsule, will differ extremely in fertility[3]. So strongly marked is this variability in the fertility of hybrids from the same parents & under the same conditions that Gaertner has remarked[4] that this quality is to a large extent contingent on the *individual*, as well as on its parentage. Again, in regard to the first cross between two pure species, sometimes individuals are found[5] which obstinately refuse to cross; & he concludes that many *individuals* must always be tried, before the sexual affinity of two forms can be determined.

The fact that no one has crossed plants belonging to distinct Families, & that in bigeneric crosses, the genera have generally been closely allied[6], & the cases before given of the high or very nearly perfect fertility of the crosses between species which are so closely/48/related that botanists have doubted whether to call them species or varieties, all show that commonly there is a pretty close parallelism between systematic affinity, & the fertility both of the first cross & of the hybrid offspring. But this parallelism is very far from being invariable or uniform.[7] Every single experimentiser has been struck with surprise at the numerous cases of most closely allied species[8] which cannot be made to unite or which produce utterly sterile offspring, & on the other hand at the very great dissimilarity of some forms which unite most easily. In the same Family, the species of one genus, for instance Dianthus, will cross very easily & yield unusually fertile hybrids, whereas in another genus, as in Silene, the most persevering efforts of Kölreuter & Gaertner[9] failed in producing hybrids between the

---

[1] Gaertner Bastardz. s. 385, 391. Thus Aquilegia vulgaris-canadensis sowed itself freely one year; but similar hybrids in a subsequent year would produce scarcely any good seed.

[2] Kölreuter 2 Forts. s. 98: Gaertner Bastardz. s. 461.

[3] Kölreuter 1 Fortsetz. s. 14. Gaertner Bastardz. s. 366, 554.

[4] Bastardz. s. 143, 406.   [5] Gaertner Bastardz. s. 165.

[6] The following are some cases of bigeneric crosses: Rhododendron & Azalea.—Rhododendron & Rhodora, a remarkable cross.—Lychnis & Silene (Gaertner)—Hymenocallis & Ismene (Herbert), Gloxinia & Sinningia,.—Brunsvigia & Vallota (D. Beaton) not considered good genera.—Cereus, & Epiphyllum not considered good genera by many Botanists.—Cereus (Gaertner Bastard. 179) with Echinocactus & Melocactus, genera usually admitted.—

[7] Gaertner Bastardz. 121, 168, 408. Gaertner says, (s. 194) the parallelism is less strict with the fertility of the hybrid than with the facility of making the first cross.

[8] Gaertner Bastardz. s. 174, 164 gives a whole catalogue of such forms.

[9] Bastardz. s. 140, 195, 197.

very closely allied species. Even within the limits of the same genus, for instance Nicotiana, in which the many species have been more largely/49/crossed than in almost any other genus, Gaertner found one species, N. acuminata, which is nearly related to the other species, yet absolutely failed to fertilise or be ferti[li]sed by no less than eight other species![1] Analogous facts were observed by Kölreuter in Digitalis. Herbert[2] remarks that Crinum Capense, Zeylanicum & scabrum are very similar in their general appearance & yet produce excessively sterile hybrids; whereas one of them namely C. Capense yields when crossed with C. pedunculatum fertile offspring; yet the latter is as unlike C. Capense "as perhaps any two species of any known genus" & indeed has been put by some authors into a distinct genus. So again Herbert[3] asks how it comes that all the forms of Hippeastrum are excessively fertile together, whilst, in a closely allied genus Habranthus (or rather in the opinion of most botanists a mere section of the same genus) every attempt to cross the species has entirely failed. Numerous similar cases[4] could be added. Although we may predicate that forms very remote in our systematic classification certainly will not unite; yet/50/assuredly systematic affinity does not unlock the law regulating the fertility of the first cross & of the hybrid offspring. We shall, also, find this conclusion strongly corroborated when we come to compare reciprocal crosses.

Gaertner has shown[5] that external differences even when strongly marked, in the form & size of the flower, fruit, seed, pistil, pollen & cotyledons, do not always prevent a union between two species. So it is with remarkable differences in general habit, for a shrubby & herbaceous calceolaria have crossed, as have two species of Erythrina, one of which bears its flowers directly from the root, & the other is almost arborescent & blossoms from the exits of its leafy branches[6]. In the Cactae we have still more surprising unions with fertile offspring, as far as external shape is concerned; namely between the prickly & angular Cereus speciosissimus, to whip-like C. flagelliformi & the unarmed C. (Epiphyllum) phyllan-

---

[1] Bastardz. s. 147.
[2] Amaryllidaceae. p 343.
[3] Hort. Journal. vol. 2. p. 10.
[4] Lindley in his Theory of Horticulture p. 336 [330], gives the currant & gooseberry, apple & pear, as cases of close species which will not unite. Herbert in Hort. Journal vol 2. p 82 insists on the impossibility of crossing the very closely allied species of crocus & iris [crinum], & contrasts this with the facility of crossing extremely unlike species of Narcissus. So again (ib. p. 95) in the genus Pelargonium, between certain species, there is a "secret insuperable bar"—
[5] Bastardz. s. 180, 183, 275.
[6] Herbert. Amaryllidaceae p. 364, & Hort. Journal vol 2. p. 102.

thoides: several species, also, of Cereus have been crossed with Melocactus & Echinocactus.

Herbert believed that the difficulty in effecting a cross/51/ depends on some constitutional difference between the species; if by this, differences not externally visible are only meant, it certainly is true. But if we take a more common acceptation of the word & imply such differences, as the duration of life, period of flowering, adaptation to climate, this view cannot be considered as generally holding good, though it apparently does so in some cases. Thus the tender Indian Rhododendron arboreum has been crossed with the hardy R. Ponticus, & even with R. Dauricum[1], which flourishes under the intensely cold climate of Eastern Siberia. In the crosses between Rhododendron & Azalea we see evergreen & deciduous bushes united. And Gaertner[2] has shown that annual, biennial & perennial species can be united. A marsh & woodland species of Crinum, have produced a hybrid, as I was told by Herbert.

A remarkable discovery made by Kölreuter, shows I think even better than the above special cases, that neither systematic affinity, or resemblance in general habit or in constitution will account for the capacity of some species to unite & for the incapacity of others. I allude to reciprocal crosses. There are very many/52/cases, in which species A can be easily fertilised by the pollen of B, whilst B absolutely resists or receives with great difficulty the pollen of A. Thus Kölreuter found that Mirabilis jalappa fertilised by the pollen of M. longiflora produced a good many hybrids, which self-fertilised yielded seeds "numero non adeo exigeo" [sic]; whereas the reciprocal cross (i.e. M. longiflora fertilised by M. jalappa), was tried during fourteeen years more than 200 times, & yet utterly failed.[3] I will give only two striking cases from Gaertner:[4] Nicotiana Langsdorfii will fertilise four other species, but cannot be fertilised by them: the common & Canadian Columbines naturally yield nearly the same number of seed; but A. vulgaris fertilised by A. Canadensis gives us a maximum 151 seed, whereas the reciprocal cross yielded as a maximum only 29 seeds. In sea-weeds, Mr. Thuret[5] has shown that Fucus serratus could quite easily be fertilised by F. vesiculosus, whereas he never once could effect, after repeated trials, the reciprocal cross./

[1] Herbert Amaryll. p. 359.
[2] Bastardz. s. 143.
[3] Nova Acta Petrop. 1793, p. 391. Kölreuter gives other nearly as striking cases in Lycium & Linum, in Acta Acad. 1778. p 219 & Nova Acta 1783. p. 339.
[4] Bastardz. s. 147, 195, 199: see his general remarks on this head s. 176.
[5] Annal. des Scienc. Nat. 4 series Tom 2 & 3.

53/Gaertner found that this unequal fertility in reciprocal crosses was extremely common in all intermediate & lesser degrees: it could even be detected between species, very closely related to each other, as in Matthiola annua & glabra, Datura stramonium & tatula, which some botanists consider to be only varieties.[1] It is also an important fact that the hybrids raised from reciprocal crosses themselves have not equal, occasionally even very unequal fertility[2]. These facts are the more remarkable as the hybrid offspring of reciprocal crosses are generally so like as to [be] undistinguishable in appearance;[3] /53 v/yet in their inner most nature they cannot be identically the same, for Gaertner[4] found that reciprocal hybrids were reduced at different rates, when repeatedly crossed with one of the parent-species./

53/Now it seems impossible in these numerous cases to say that species A absolutely refuses to be fertilised, or is fertilised with great difficulty by the pollen of B, on account of any systematic or, in the common sense, constitutional difference, whilst the very same individual B can be easily fertilised by the pollen of this species or individual A. We are driven to look to/54/some difference in the sexual relations of the two species, which Gaertner calls their (Wahlverwandtschaft) elective, & which I have called their sexual affinity: by this expression we must include the relation of the pollen to the stigma, as well as to the ovule, & of the stigma to the pollen; & possibly even the relation of the hybrid embryo to the seed, formed by the pure mother-plant, as long as it is nourished by it.

⟨If in these reciprocal crosses, there had been only occasional instances of entire or almost entire refusal to be fertilised on one side, we might have been tempted to explain the fact by some physical obstacle, either in the length of the pollen-tube & structure of the pistil, to the act of impregnation; but such a view does not seem to accord well with the graduated differences in the fertility of reciprocal crosses. It may possibly be that in some cases average equal number of ovules are really fertilised in the reciprocal crosses, but that a greater number perish, when nourished in the seeds of one parent species than in those of the other parent-species: this point does not seem to have been attended to by microscopical dissection. But on this view, how are we to account for the unequal fertility of the hybrids themselves, when raised from reciprocal crosses.⟩/

---

[1] Gaertner Bastardz. s. 177, 197   [2] Gaertner Bastardz. s. 407.
[3] Gaertner Bastardz. s. 201, 223–5. It is a singular fact that both Kölreuter & Gaertner found marked exceptions to this rule of the similarity of reciprocal hybrids in the genus Digitalis.   [4] Bastardzeugung s. 459, 465.

54a/I may give one other statement by Gaertner in regard to the hybrids from reciprocal crosses, which it requires almost more than my faith in his accuracy to credit:—namely that a hybrid, even a very fertile hybrid, when crossed with a reciprocal hybrid, is quite sterile. Thus the hybrid Lychnis diurno-vespertina & the hybrid L. vespertino-diurna which differ from each other only in their reversed parentage are both very fairly though unequally fertile, but when these two hybrids are crossed, they produce no offspring. He gives five other instances, & appends an et cetera to his list; & concludes by saying "we have found these hybrids absolutely sterile".[1]/

55/If sterility had been ordained simply to prevent the confusion of specific forms, it might have been expected, I think, that there would have been an uniform relation between the difficulty in effecting the first cross between any two species, and the sterility of their hybrid offspring. Such relation does hold good to a large extent: if two plants can be crossed very easily, more especially if they can be reciprocally crossed very easily, their hybrid offspring are generally pretty fertile; & conversely if they cannot be easily crossed reciprocally the hybrids are generally very sterile. But there are strong & curious exceptions to all such rules.[2] The hybrids, moreover, as we have just seen from the very same two species, when crossed reciprocally, often differ, even considerably in fertility. As a general rule hybrids when self-fertilised, & even when fertilised by the pollen of either pure parent species, yield far less seed than did their parent in the first cross by which they were formed:[3] but some few hybrids, as from between the species of Datura & Dianthus yield more seed than does the first cross. On the other hand, there/56/are many most striking cases of species which can be united with facility, whose offspring are excessively or even absolutely sterile: thus Nicotiana suaveolens fertilised by N. glutinosa yielded no less than 256 good seed in one fruit; but the hybrids raised from these seeds were absolutely sterile[4]. So again the closely related species of Verbascum unite so easily that this not rarely happens without any artificial aid, yet the hybrids raised from these species are excessively sterile.[5] Within the same genus Dianthus some species unite very easily, but produce hybrids most sterile; whereas other species can be united with the utmost difficulty, but produce hybrids very fairly fertile[6].

[1] Gaertner Bastardz. s. 515.   [2] Gaertner Bastardz. s. 200, 406, 407.
[3] Gaertner Bastardz. s. 13, 425.   [4] Gaertner Bastardz. s. 194, 405.
[5] Kölreuter Dritte Fortsetz. s. 37, 42, 4: Gaertner Bastardz. s. 580, 591.
[6] Kölreuter 2 Forts. s 108. 3 Forts. s. 108.

Hence we must conclude that the fertility of the first cross & of the resultant hybrids, certainly in several cases follows widely different laws./

57/I will now give a few other of the best ascertained facts, to show on what special, curious & complex laws, the fertility of first crosses & of hybrids depends. Hybrids always yield more seed to the pollen of either parent-species, than to their own; & the pollen even of a third & quite distinct species is sometimes more effective than their own.[1] During the reduction of a hybrid to either parent-form, which I may add requires more or fewer generations according to the species & even according to the individuals experimentised on, the fertility is extremely variable, but gradually increases (with a few exceptions) as the hybrids assume the character of either pure parent; yet it is sometimes seriously impaired, after the hybrid perfectly resembles the pure parent form. A hybrid reduced by the use of the pollen of the mother-species acquires fertility in the successive generations quicker, than when the same hybrid has been reduced by the pollen of the father-species.[2] In all cases the male sexual organs suffers first; that is the pollen sooner suffers & during reduction is more slowly reperfected, than the capacity of the ovule for fertilisation.[3]

58/Hybrids in the first generation generally all resemble each other, but occasionally single seedlings are produced differing considerably from the rest, & these are called by Gaertner "exceptional types"; they closely semble either the father or mother species, & are almost always quite sterile; even when the other hybrids from the same capsule have considerable fertility.[4] These "exceptional types", also, sometimes appear in the succeeding generations of hybrids, whether fertilised by their own pollen or once by that of either pure parent; & as in the first generation, these exceptional types have diminished or even quite destroyed fertility.[5]

Again hybrids are usually nearly intermediate in appearance between their parents; but some species regularly produce what

[1] Gaertner Bastardz. s 425–427.
[2] Gaertner Bastardz. s 419–455.
[3] Gaertner Bastardz. s. 350, 355, 435. Kölreuter has made the same observation on the greater liability of the male organs than the female organs of hybrids to suffer.
[4] Gaertner Bastardz. s. 244. In some hybrids which I raised between the common Carnation & Spanish Pink, one plant was extraordinarily like the pure Spanish Pink, but it was not more fertile than the other quite sterile hybrids.
[5] Gaertner Bastardz. s. 439, 442. Kölreuter gives, also, strong instances of this law.

Gaertner calls "decided types", which take much more after one parent-species than the other. This is caused by the prepotency of one species over the other./59/These "decided types" are with some few exceptions sterile. This stands in close relation with the rule that when species cannot be easily crossed reciprocally their offspring are sterile; for it is found that those species which yielded "decided types" cannot be reciprocally united: thus if the hybrid offspring of A & B be nearly intermediate in appearance, then A can be fertilised by B, & B by A; but if the offspring takes decidedly after either parent, then a reciprocal cross between them can seldom be effected[1].

In the foregoing cases, we see that close resemblance in hybrids to either pure parent, if it appears exceptionally in only a few of the hybrids, or if it appears in a very "decided" manner in all the hybrids, is connected with lessened fertility;—a fact which assuredly would never have been anticipated. It stands, moreover, in direct opposition to what usually takes place in the gradual & regular reduction of hybrids by the application of the pollen of either pure parent-species in each successive generation; for in this case as the hybrids gradually approach the pure parent-form, they acquire/60/fertility. ⟨We see, also, in the foregoing cases how little necessary relation there is between fertility & external resemblances⟩.

Several cases are known of species which will not unite with each other, but will both unite with a third & distinct species.[2]

A species, when crossed with several other species of the genus, may have a very strong power of transmitting its likeness to all its hybrid offspring: a species may, also, have a remarkable power of fertilising the other species of the same genus; but these two powers are quite distinct & by no means necessarily go together.[3]

In nearly all cases in which three or four species have been united, Gaertner found the hybrids to be excessively variable & extremely sterile. But/61, 62/this sterility in complex crosses is not invariable, as Herbert & others have shown[4] in the genera Gladiolus, Crinum & Rhododendron even when four or more species have been united./62 v/In some of the hybrid Rhododendrons, raised by Messrs. Standish & Noble[5] no less than six species have been blended together by successive crosses into a single hybrid, namely Rhododendron campanulatum, maximum, Ponti-

---

1 Gaertner. Bastardz. s. 221, 286.
2 Gaertner Bastardz. s. 202, gives examples in Nicotiana & Dianthus.
3 Gaertner Bastardz. s. 289.   4 Herbert Hort. Journal. vol. 2. p 19, 88.
5 Journal of Hort. Soc. vol. 5. p. 274.

cum, purpureum, Catawbiense & arboreum,—species coming from the most distant quarters of the world, & having the most different climates; most of these complex Rhododendron hybrids seem to be very fairly fertile; & some of them extremely fertile, as has been previously mentioned./

62/Taking a general review of the facts now given on the infertility both of first crosses & of hybrids,—we see a most insensible gradation from absolute sterility to high or perfect fertility, —we see the fertility not only eminently susceptible to external conditions, but independently of conditions innately variable in an extreme degree, so as sometimes to depend to a large extent merely on the individual selected;—we see that the infertility does not closely follow mere external, or systematic, or constitutional (in common sense) differences; we see this very plainly in reciprocal crosses in which there is a very general, & sometimes an enormous, difference in the result, solely owing to one of the two species having been used as father or mother; nor is this difference confined to the first cross, but affects the fertility of the hybrid offspring;— again we/63/see that the fertility of the first cross & of the hybrid offspring by no means always runs parallel;—we see several other curious facts, the pollen of the mother-species giving fertility during the reduction of a hybrid, sooner than that of the father-species;—the male sex failing easier than the female;—the extreme sterility of exceptional types i.e. of hybrids which suddenly assume the appearance of either pure parent, & of decided types or those which have not intermediate structure but regularly take after either parent;—& other such odd cases.—Now do these several laws & facts,—which it should be observed include all the known principal facts in hybridism, look as if they had been specially ordained for the simple purpose of keeping specific forms in nature distinct? I think that their complexity & singularity give a decided negative to this question. The several laws & facts seem to me to be incidental on other & unknown differences in the sexual organs & products of the two species which are crossed. And differences in the sexual organs & products/64/will stand in some relation, but by no means necessarily in a close & uniform relation, to systematic differences, which imply the sum of all the differences of all kinds.

I may illustrate what I mean by these laws of infertility being *incidental* on the sexual differences of species, by the action of an artificial poison, which, from what we know would certainly be in some degree different in widely different plants, but would probably be nearly alike on members of the same genus; & this action might be called incidental, for as the poison did not exist naturally the species would not have been created, or modified by natural selection, so as to have different powers of resisting it. The action moreover, would be incidental in differences in the absorbent & nutritive systems, wholly inappreciable by us. But another illustration already alluded is so apposite, that it is worth giving in rather fuller detail: I refer to grafting & budding./

65/The capacity for grafting, ⟨like that for hybridism,⟩ is limited but less closely than of hybridism, by systematic affinity. The ash, Lilac olive, Phyllirea, Chionanthus & Fontanesia, belong to the same order, & though so very different in appearance, can be grafted together; but the Jasmine, as Von Martius has remarked, cannot be grafted on them, "which confirms the propriety of separating these two orders"[1]. Many species of Pyrus can be grafted together; & the Pear can be grafted on three other genera, namely the white-thorn (Crataegus) Quince & Cotoneaster; but it would be an error to suppose that all the species in the same Family can be grafted together, or grafted with equal facility; the apple though so closely allied to the pear takes on it with much difficulty.

Although a small bush can sometimes be grafted on a tree, as Cytisus purpureus on C. alpinus; although plants of the most widely different external shape can be grafted together as various Cacteae; though a plant from a warm climate will take on one from a cold, as with species of Rhododendron,—a shrubby plant on an herbaceous one, as the tree Paeony on the common, a deciduous tree on an evergreen, as the common cherry which/ 66/will even fruit on the Portugal laurel; yet it has been observed[2] that trees of very different size, or which grow at very different rates or have wood of very different hardness or saps of different nature or flowing at different periods, cannot be grafted together or take with much difficulty. But who can explain why the pear takes so infinitely easier on the quince, classed in a distinct genus; than on the apple a species of the same genus; or why the pear takes easier on the apple, than the apple on the pear;[3]/66 v/or again why the common Gooseberry cannot after repeated trials

---

[1] Lindley The Vegetable Kingdom 3 Edit p. 616. I have myself had the Lilac grafted on the common Ash, & reciprocally the ash on Lilac.

[2] N. C. Seringe Flore des Jardins p. 250: Loudon in Encyclop of Gardening p. 650 asserts that the Acer platanoides on account of its milky sap will not receive other maples.

[3] Loudon's Gardener's Mag. vol. 1, p. 200. Diel has made the same observation as quoted by Gaertner, Bastardz. s. 632. For the Currant & Gooseberry case, see [Godsall,] Gardener's Chronicle 1857. p. 757.

be grafted on the currant, though so closely allied systematically & apparently in constitution; whereas, on the other hand, the currant will take on the gooseberry./66/By these latter facts we are reminded of the unequal reciprocity of Hybrids.

Thouin names[1] three species of Robinia which when grafted on other species are generally quite barren or yield but very few seeds whilst ungrafted they often seed pretty copiously; on the other hand some species of Sorbus when grafted on/67/distinct species yield twice as much fruit, as do seedlings on their own root. We can, I apprehend no more account for this latter fact, than for some of Herbert's crosses yielding more seed than the same species naturally self-fertilised; nor can we account for the barrenness of the Robinias any more than for the barrenness of many hybrids.

Although we may probably account for Peaches succeeding best on plum stock by the hardiness of the plum's roots in our climate; & for the pear often succeeding best on the quince by its over luxuriance being checked; & for certain apples flourishing best when grafted on the Paradise variety, owing to the fibrous root of the latter not penetrating so deeply into poor soil; and for late & early Varieties of the Walnut[2] not taking kindly on each other owing to their sap flowing at different periods. Yet who can explain why one variety of the Pear succeeds far better than another variety on the Quince; & why, as it is positively asserted[3] some varieties will not succeed at all on the quince./

68/Why do certain varieties of the Apricot, & of the Peach, prefer certain varieties of the plum?[4] Why will not the Golden variety of the common Lime take on its own species, but freely on the distinct American species?[5] Sageret, moreover, gives reasons for believing that *individual* stocks have a repugnance to receive the grafts of certain varieties,[6] in the same way as we have seen the *individual* plants resist being hybridised: and the rare cases of varieties crossing with different degrees of facility with each other & with distinct species, have been paralleled by the foregoing not rare case of varieties grafting with different degrees of facility.

In drawing this parallel I am very far from wishing to make it

---

[1] Annales du Museum. Tom XVI. p. 214.
[2] Loudon Encyclop. of Gardening p. 650.
[3] Loudon's Gardener's Mag vol. 3. p. 380. All horticulturists have remarked on some degree of diversity in the different varieties of the Pear. see Sagerets Pom. Phys. p. 65 p. 222. A. Knight in Hort. Transact. vol 2. p. 203.
[4] Sageret Pom Phy. p. 321, 346.
[5] Loudon's Gardener's Mag. vol. 6 p. 317.   [6] Pom. Phys. p. 222.

appear that grafting & crossing are allied processes: many species will graft together with the utmost facility which cannot be crossed: the mere cohesion of the cells in grafting, & the intimate fusion of the two cells in sexual union must be fundamentally different.[1] But I think facts enough have been given to show that the capacity both for grafting & for crossing is limited, but by no means wholly,/69/governed by systematic affinity; & that in many cases we can assign no cause why certain forms will not graft or will not cross. We have seen, also, that varieties are not exempted from differences in their capacity for grafting & crossing. I presume that no one would say that the capacity for grafting with its somewhat complex & obscure laws, is a specially created endowment; I presume that all will admit the capacity is incidental on differences in the constitution, more especially of the vegetative tissues, in different species. So do I conclude that the capacity for crossing is incidental on differences, more especially in the sexual relation (taken in its largest sense) of the species subjected to experiment.

I may add that in the Robinias & in some other cases which might have been added, the barreness is incidental on the grafting, as the power of being grafted is incidental on other contingencies. The lessened fertility, also, from close interbreeding must be looked at as incidental on wholly unknown laws, for it does not seem probable in the case of plants & the lower animals that this should be a special endowment. So again the very frequent sterility from changed/70/conditions of existence, which was so fully treated in the third chapter I should look at as only incidental; for this tendency to sterility could not have been acquired by natural selection; it may well be doubted whether it would have been created, as its only use would be to keep organic beings within certain limits & this would apparently be superfluous as climate & the struggle for life would be amply sufficient. ⟨Nevertheless there are a few cases known of plants extending sometimes by the accidental dispersal of the seeds, into conditions where they are rendered sterile; but more cases could be given⟩ Moreover several cases could be given of plants living in profusion, where they do not increase ⟨propagate⟩ by seminal reproduction.

Causes of the sterility which is *incidental on Hybridism.*—Very little light can be thrown on this subject. The following remarks will apply to animals as well as plants. There is clearly a fundamental difference between the sterility of Hybrids & that of the first cross

[1] Gaertner in his Bastardz. s. 606–633 has an excellent discussion on this subject.

between two pure species. In hybrids the sexual functions ⟨elements⟩ are deteriorated, as can be plainly seen at least in the pollen & spermatozoa. In first crosses between two pure species both sexual elements/71/are of course perfect; but either the pollen never reaches or does not penetrate the ovule, or reaching it does not cause an embryo to be developed; or an embryo is developed but perishes at an early age. When the pollen of a plant is placed on the stigma of a distantly allied genus, both the pollen & stigma are often, as before mentioned in some degree affected by their affinity, but the pollen-tubes do not properly penetrate the stigmatic tissues. Here we have the first of the three apparent causes of sterility. Thuret[1] in attempting to cross distinct genera of Fucus, saw the antherozoids cling to the *naked* spores, but no germination ensued: here we have the second cause. But in some very rare cases Thuret observed the commencement of germination, the spore subsequently perishing;[2] & this is our third cause./71 v/ That the early death of the embryo is in some cases is one very potent cause of the little fertility between two species when crossed, I cannot doubt from some facts communicated to me by Mr. Hewitt, who has had the largest experience during many years in making hybrid Gallinaceae. Mr. Hewitt has had in one year above 300 eggs from crosses between various pheasants & the common cock pheasant & Fowl; & he assures me that he has "opened hundreds of eggs, containing partially formed embryos of hybrid pheasant-fowls"; yet these two birds are so sterile together that only 3 or 4 per cent of the eggs produce chickens. Again out of 55 eggs from the hen Silver Pheasant, fertilised by the Gold Cock pheasant, he got only three hybrids, but on opening the bad eggs, he found "that many had germinated."—/

71/The extreme difference of fertility sometimes observed, & in a lesser degree often observed, in reciprocal crosses amongst plants especially makes us feel how ignorant we are on the whole subject: are we to suppose that, though A can be readily fertilised by B, the pollen of A cannot reach the ovule of B; or reaching it that it does not cause an embryo to be developed or being developed that the embryo perishes at an early age? These points have/ 72/not been investigated as far as they might have been by microscopical dissection. Considering the double & unnatural condition of hybrid embryos, & that they have to be nourished within

---

[1] Annal des Sciences Nat. 4 Series. Tom. 2. & 3.
[2] See Gaertner's account (Bastardzeugung s. 101) of his "Fructificatio subcompleta," in which only a small withered embryo is developed: when seeds of this character very rarely germinate, the seedlings are weak & soon perish.—

a seed or egg formed by one of the pure parents & therefore of a somewhat different nature to their own, & considering how easily the young both of plants & animals are affected by unfavourable conditions, & lastly considering that hybrid seedlings & young animals raised from between very distinct species are often tender & delicate, it does not seem improbable that in many cases the cause of infertility in first crosses lies in the early death of the embryo. But opposed to this notion, is the general health & vigour of hybrids when once produced. When two plants very remotely allied are crossed the probable cause of their infertility seems to be that the pollen does not reach the ovule; between plants more closely related, that the pollen does not cause an embryo to be developed, & perhaps in not a few cases that the embryo is developed & then perishes. But why the pollen-tube should not penetrate the stigma of a remotely allied plant, or why if penetrating it should not develop an embryo, is no more explicable, & apparently is no greater a difficulty, than why some trees can & some cannot be grafted on others./72 bis/⟨Supposing that the fertilisation is equally effective on both sides, it might well happen that the hybrid embryos, might perish at an early age in very different numbers from being nourished in the two cases by different mother-species. Moreover, although the offspring from reciprocally crossed plants are with few exceptions identical in external appearance, which makes the unequal number in which they are produced the more surprising, yet their inner nature, & consequently perhaps their liability to perish, must in some degree different, for Gaertner has shown[1] that they are capable of being reduced to either pure parent forms in a different number of generations.⟩

With regard to the sterility of hybrids themselves which are imperfect in their sexual functions,—a fact which in itself has always seemed to me much more remarkable than the difficulty in effecting a first cross—I think we can perhaps see our way a little more clearly. In our third chapter, numerous facts were given showing that slight changes of condition & crosses between closely allied forms or varieties were good for all organic beings & their offspring; but that changes of conditions beyond certain limits affected the reproductive system in an especial & injurious/ 73/manner, independently of general health: therefore it completes the parallel that crosses beyond certain limits of affinity should injuriously affect the reproductive system. In this same chapter I showed in detail how similar in many respects the sterility superinduced by unnatural conditions was to that caused by

[1] Bastardz. s 228, 459, 465.

hybridism; I will here recapitulate only the more important points. In both cases the sterility is often quite independent of general health; how healthy & how sterile is the common mule! In both cases the sterility occurs in various degrees: in both the male element is the most liable to be affected, but sometimes the female more than the male.[1] In both, the tendency goes to a certain extent with systematic affinity; for whole groups of animals & plants are either greatly or very little affected by unnatural conditions of the same kind, without our being able to assign any adequate reason; & whole groups of species tend to produce sterile hybrids; but there are often marked exception in both cases in the same groups. No one can tell till he tries, whether any particular animal will breed under confinement, or any plant seed freely under cultivation; nor can be tell till he tries whether any two species of the same genus will produce more or less sterile hybrids. Lastly when organisms are placed during several generations under conditions not natural to them, they are extremely liable to vary, which/74/is due, as I believe, to their reproductive system having been thus specially affected though in a lesser degree than when entire sterility is caused: so with hybrids, when they can breed, their successive generations are eminently liable to vary, as every experimentiser has observed.

Seeing how similar the results are in these two apparently very different cases, let us compare the secondary causes: in the one case the structure & constitution of the organism remain the same, but the conditions of life to which it is exposed have been changed, & hence results sterility: in the case of a hybrid, the conditions of its existence may remain unchanged, but its constitution & all the laws of its growth from its earliest days, from being compounded of two distinct forms, can hardly fail to have suffered disturbance, whatever may be the conditions of life to which it may be exposed; & hence sterility is the result. There can be but very few species in nature with their whole constitution & laws of growth so similar that the blending of the two would not cause a disturbance, different from, but we may suppose as great, for instance, as the giving a plant a little too much water during one season of the year, which we know will in some cases not in the least effect its general/ 75/health or prevent its flowering, but will render its pollen utterly impotent.

As the double & compounded nature of a hybrid is inherited by its offspring, it is not surprising that the infertility (subject, however, like the whole rest of the organisation to variation) should

---

[1] Gaertner Bastardz. s. 357. 360.

be likewise inherited: the gradual increase of infertility which has
not rarely been observed in the successive generations of hybrids,
I am strongly tempted to explain in large part for reasons already
assigned, to the evil of close interbreeding; & we know that their
infertility is highly susceptible to unfavourable conditions.

During the gradual reduction of a hybrid by successive crossings
with either pure parent, the stain of the mixed constitution
⟨foreign blood⟩ is gradually washed out, & fertility is acquired,.[1]
But no light as far as I can see, can be thrown on the very singular
fact that amongst hybrids, "exceptional types" (or those which
suddenly & abnormally closely resemble either parent-form) &
"decided types" (or those which normally closely resemble either
parent-form) are almost always extremely sterile; without, indeed,
we might suppose that whilst the large part of the organisation
took closely after the other, then I think we could understand
how there would be a greater disturbance in the machinery of
life, than if every part of the machine had a more strictly intermediate structure./

76/It will have been seen that I would explain somewhat
differently the two cases of the sterility of hybrids themselves
& of the difficulty in effecting first crosses;—namely in the case
of hybrids by their double & heterogeneous nature producing
closely analogous results to what changed conditions do when
acting on pure species; & in the case of first crosses, either by
obstacles perhaps of various kinds to the act of fertilisation—
(somewhat analogous to those in making grafts) or sometimes by
the early death of the embryo. Although the cause of sterility in
first crosses & in hybrids seem to be, almost necessarily, somewhat
different, I do not think that it is surprising that there should be
a considerable degree of parallelism in the results; for in both
cases the sterility is related to the amount of difference between
the parent-forms. Moreover we should bear in mind that even
within the limits of the same genus, the parallelism is by no means
always close between the number of hybrid offspring produced
by a first cross, & the fertility of the hybrids when obtained. How
many cases there are, as with the common mule, in which there
is no great difficulty in producing the hybrid, the hybrid itself
being excessively sterile; & in plants there are a good many cases,
exactly the converse. Even between the very same two species,
when crossed reciprocally, we have seen that there is sometimes
the utmost difference in the number of offspring obtainable,

[1] [Darwin later pencilled 3 reversed question marks in the margin beside this next
sentence.]

according as one or other is used as father or mother; whereas in the hybrids themselves obtained from such reciprocal crosses there is only occasionally/77/a considerable ⟨slight⟩ difference in fertility. ⟨There is another important point of difference between the hybrid offspring of first crosses & the offspring from a first cross (taking of course species which have not been affected by culture) are generally, as we shall presently see, very uniform in character, for their parents have not been exposed to conditions tending to make them sterile & variable, whereas the offspring from hybrids which are newly [?] domesticated [?] in their successive generations, are eminently variable, like the offspring of species which have suffered during successive generations from conditions not natural to them.⟩

Finally we have seen that sterility occasionally ensues when two species are grafted; in a lesser degree from the close inter-breeding of individuals of the same species; in a marked manner from exposing organic beings to conditions of life different from those to which nature had adapted them; & lastly from crossing species, I believe, as soon as we can explain why unnatural conditions make a pure species sterile without necessarily affecting its health, then & not till then, we shall understand the sterility of hybrids;—but that something different (& different in different cases) is required to explain why few or no progeny is produced from a first cross between two pure species. When our physiological knowledge is so perfect that we can explain why trees of different orders can never be grafted on each other, we may perhaps hope to explain the more difficult problem why the pollen of a given species will not penetrate or penetrating will not fertilise the ovule of another species./

Animals

78/In regard to the sterility of animals when first crossed & of their hybrid offspring, I shall discuss only a few points, & chiefly in comparison with plants. Carefully conducted experiments have seldom been made; & we have not here excellent Treatises[1] like those on hybrid plants.

---

[1] In the several compiled lists, which have been published, little care seems to have been taken in sifting evidence. Several cases are recorded, as due to hybridism, which I can hardly doubt have been simply monstrosities. The statement by Hellenius that he crossed a common ram & female deer (Cervus capreolus) quite recently has been admitted; not withstanding that he found the hybrids perfectly fertile inter se! But I believe there is no doubt that the mistake arose from the Mouflon the reputed aboriginal of the sheep) being called a *Roe* in Sardinia; see Observations on Zoology by C. L. Bonaparte. [Wagner] in Ray Soc. 1841–42 p. 85. Again M. E. de Selys-Longchamps says (Bull. Acad. Royale de Bruxelles

With animals the will & instinct come into play preventing or checking first crosses; but their importance has, I think, often been greatly exaggerated. No doubt some cases, as in the experiments recorded by Buffon & Hunter[1] two species have shown a strong aversion to cross; but then, quite independently of any sexual relations, distinct kinds of animals often dislike each other./ 78 v/In one of the cases of aversion given namely, wolf & dogs other experimentisers have not observed any aversion,—& it is perfectly well known that a tame dog from Parry's ship coupled with a wild she-wolf.[2]/78/If we lay on one side some few tribes of animals, as the Ruminants, which breed very readily under confinement, it is hardly an exaggeration to say that in menageries hybrids are produced almost as easily as pure-bred species: let anyone look over the Reports of the Zoological Soc. for a number of years, in which both pure breed & cross-bred births are recorded, & he will see that this is true. How rarely do the Fringillidae breed in confinement,—yet at least nine species, belonging to three distinct genera have produced hybrids with the Canary: no species has been crossed oftener with the canary than the/79/Siskin (Fringilla spinus); yet instances of siskins breeding in confinement are extremely rare.[3] I could give many cases of birds kept with others of the same species almost in a state of nature,[4] yet pairing with distinct species. Under strict confinement this is still com-

Tom XII [see pp. 341–3]) that the Baron de La Fresnaye procured *seven* hybrids from Anser cygnoides ♂ & Anser Canadensis ♀: one coupled with canadensis & two others with other species all were sterile. Is not this the foundation of M. Chevreul's statement (Annal des Sc. Nat. 3 series Bot. 1846. Tom 6. p. 188.) that the Baron produced hybrids from these two species & that "il est remarquable que leurs hybrides se soient reproduits deja *jusqu' a sept fois*."? Morton, changes the A. canadensis into A. cinereus & says that Chevreul himself has seen "the progeny extend through seven generations".

[1] Animal Oeconomy p. 310. The she-wolf had to be held. But I have known the same thing to be necessary in choice fancy spaniels, which had long been closely interbred; the female having thus lost sexual passion.

[2] It is scarcely possible to read M. Mauduyt little Treatise Du Loup et de ses Races 1851 without believing that in the Pyrenees crosses between free wolves & Dogs not very rarely take place.—See Supplement to Parry's Voyage 1819–p. CLXXXV: for additional case see Franklins narrative vol v. p. 140. [? See Appendix 5, p. 664] See also Pallas. on this subject.

[3] Mr. Milne says he has never known or heard of more than one instance: Loudon's Mag. of Nat. Hist. vol. 3. p. 440. See my Chapter 3 on the difficulty in making the Fringillidae breed in confinement.—

[4] Waterton says. Essays on Nat. History 2 series p. 42, 117. that a Canada Goose, living with 23 other birds of the same species, paired with a solitary Bernacle Gander, though of so different a size, & produced young. A wigeon (Mareca penelope) associated with others of same species, paired with Pintail (Dafila acuta). Loudons Mag. of Nat. History. vol. 9. p. p 616.—I have heard of many similar cases. Indeed the Rev. E. S. Dixon (Ornamental & Domestic Poultry p. 137)//

moner: thus a female Bonnet monkey in the Zoological Gardens, I was often assured by the Keepers preferred the male of any other species to her own, & she produced a hybrid with the Rhesius monkey.—

The well authenticated cases, also, of hybrids bred between species, both in a state of nature, show that there cannot have been any strong aversion between them; though I do not doubt that generally these crosses have been caused by an inequality in the sexes of the pure species. Thus five species of wild Grouse have produced hybrids together; about 18 cases in Great Britain alone are now on record of hybrids produced between the Black hen grouse (Tetrao tetrix) & the cock pheasant & reciprocally.[1] Several cases are known amongst Ducks; & distinct species of insects have often been caught in union.[2] With rare exceptions, all that can be said with truth is that the sexes of distinct species, when very remote do not in the least excite (though even here/80/some strange anomalies have been recorded) each other passions; & that when more closely related, they excite each other, but in a less degree than natural;[3] but not apparently that they cause more natural aversion than do the same sexes of the two species when confined closely together.—

With respect to the degree of fertility of first crosses in comparison with the fertility of the hybrid offspring from such crosses, few facts seem accurately known; & indeed from several causes can be made out only with much difficulty, more especially with those animals which produce only one or two young at a birth. The best evidence known to me refers to the common mule; in rearing which it has been found that only [   ] conceptions follow from 100 unions; whereas with the mare & horse [   ]

---

[1] [Spicer] Zoologist vol 11–12—1853–1854. p. 3946 [4294].

[2] A Gold Pheasant turned out & free in the woods at Henley Park produced hybrids with the common Pheasant; [Lowcock,] Annals & Mag. of Nat. History vol 6. 1841. p 73.—Numerous cases are on record of the Carrion & Royston crow pairing, see ch. 4.—Several hybrid Ducks have been shot at different times. see [Fennell] Loudon. Mag. of Nat. History. vol. 9. p. 616. See Macgillivray British Birds vol 1. p 398 on the breeding of Goldfinch & Green linnet in nature.—I can see no reason to doubt that the Black bird & Thrush have produced hybrids together: see [Berry] Loudons Mag. of Nat. History. vol 7. p. 599: this fact is corroborated by another case in Macgillivrays British Birds vol 2. p. 92. It would appear from Bechstein (Naturgesch. Deutschlands B I s. 950) that hybrids have been naturally produced from the Black & Brown Rat. For insects see Bronn's Ges[ch]ichte der Natur B. 2. p. 164. [Shuckard,] Annals of Nat. Hist. vol 7. 1841. p 526.; Westwood Transact. Entomolog. Soc. vol. 3 p. 195.

[3] Mr. Hewitt (Poultry Book by Tegetmeier. 1857 p. 123) says that after a domesticated Cock Pheasant, has become attached to a Hen of the common Fowl, the introduction of a female pheasant "will estrange all feelings of affection, which had before-times been indulged."

per cent of the union are fruitful: Azara, moreover, states that the mare ceases to produce at an earlier age to the male ass than to the horse[1] //

81/In crossing Gold or silver pheasants with the common Pheasant "most of the eggs prove barren", & the chickens are difficult to rear, & when reared are almost invariably sterile[2]/81 v/The hybrids from the Cock Pheasant & Common Hen are I believe universally quite sterile & do not even show any sexual passion; & there is considerable difficulty in producing them, for Temminck[3] asserts that out [of] 100 eggs, only two or three young can be raised: Mr. Hewitt informs me that out of above 300 eggs from these two birds, he raised not above a dozen hybrid chickens: in the same laying, however, if a single egg proves to have been fertilised, several can generally be hatched.[4]/81/On the other hand, hybrids from the common Duck & the Musk-Duck. (Anas boschas and Cairina moschata) are utterly sterile & even without any passion; but yet can be raised with great facility, & are raised in large numbers in the U. States for the table, as I was informed by Dr. Bachman; so that, as I infer, most of the eggs must be fertile/ 81 v/but even in this case Mr. Garnett of Clitheroe who has raised many of these hybrids & from reciprocal crosses, informs me that the proportion of good eggs is not so great as with the Common Duck.—/81/Mr. Brent, tells me that in his crosses between the canary-bird, gold-finch, Linnet & green-linnet (Loxia chloris) he has often had the full number of eggs & every egg fertile: the hybrids from these birds will often breed with either pure parent, & some of them very rarely inter se. Mr. Brent, also, in crossing the Stock & common Pigeons (C. oenas & livia) found both eggs fertile, but the young were very difficult to rear, & ⟨with him were⟩ sterile. A Pomeranian bitch with a dog-wolf produced ten puppies.[5]/

82/From these few facts I presume that there is, as with plants, some pretty close relation between the facility of getting offspring from first crosses & the fertility of the hybrids when raised; but I much doubt, more especially from the case of common mule & musk duck, whether the relation is uniform. The fertility of the first cross, when utterly barren hybrids are produced, does not

---

[1] [The text is sheared off here; the information from Azara is to be found in his *Quadrupèdes*, ii, 349.]
[2] Mr. Hewitt: Poultry Chronicle 1855. vol. 3 p. 15 & Temminck Hist. Nat. Gen. des Gallinacees vol. 2. p. 323.
[3] Hist. Nat. Gen. des Gallinacées Tom 2. p. 314.
[4] Mr. Hewitt in the Poultry Book p. 125.
[5] Pennants Quadrupeds 3 Edit. vol. 1, 1793 p 238.

seem so much impaired as with plants. ⟨The frequency [of] several cases of very young hybrids being difficult to rear & of the first-laid eggs being addled perhaps indicates that the fewness of the progeny is in ⟨large⟩ part due to the deaths of the embryos at an early age.—⟩

With animals, it is difficult to decide whether in first crosses as with plants there is much or any unequal reciprocity, for here instinct comes into play. The reason why male Fringillidae alone are generally paired with female Canary-birds, is that hens of wild species, if not taken quite young from the nest, will seldom receive a male of any kind; & they will not build a nest, or use one when made for them.[1] The greater facility of getting mules than hinnies I have heard attributed to a difference in a sexual instinct between the male ass & horses; & this perhaps accounts for the much greater frequency of crosses between the domestic dog & she wolf, than reciprocally./82 v/It is said further that though the he-goat crosses readily with the sheep; but ⟨yet⟩ that the Ram will not produce with the she-goat.[2] Mr. Fink has found the reverse to hold good; namely that a Ram crosses more readily with she-goat than reciprocally.[3] Mr. Garnett of Clitheroe informs me that the common Drake will seldom have any intercourse with the Musk-Duck, whereas, as is well known, the Musk-drake takes with perfect readiness to the common Duck; but when they are paired, Mr. Garnett tells me that he has not observed any difference in the number of young produced./82/But Bechstein who was well aware of the foregoing causes of difficulty, states that the male/83/House sparrow succeeds better with the female tree-sparrow (Fringilla domestica & montana) than reciprocally.[4] I could add other facts pointing to the same direction, but many more facts are wanted to draw any definite conclusion on reciprocal crosses amongst animals. ⟨Hybrids themselves are only so far imperfect, as far as the most careful examination shows, that the spermatozoa in Mammals & Birds are in the same state as in pure species in the intervals of rut.[5]⟩ Gartener[6] advances some evidence, ⟨but hardly sufficient in my opinion,⟩ showing that in hybrid animals; as with hybrid plants, the male sex fails easier than the female: I shall presently give a very striking case of this

[1] Bechstein, Stubenvogel 1840 4 Edit. s. 247.—Mr. Brent has made the same remark to me.
[2] Lucas on authority of Bomare. Héréd: Nat. Tom 2. p. 185.
[3] Communication to Board of Agriculture vol ɪ. p 280.
[4] Stubenvögel s. 210, 224.
[5] ⟨Annales des Sci. Nat. [   ] Gaertner Bastardz s. 340, 382.⟩
[6] Bastardz. s. 340, 382.

in hybrid Yaks. I do not know whether it is anyways connected
with this tendency that male hybrids, ⟨as remarked by Buffon &
others & I believe truly,⟩ are oftener produced than females.[1]/
83 v/A perfect gradation in the degree of sterility of first crosses
& of the hybrids themselves could be given; from unions, which
as between the Guanaco & Goat[2] never produce offspring, to cases
where a single instance is known of offspring produced after
repeated couplings as in case of Peacock & Guinea-fowl, to such
cases as the common mule, where they are habitually produced;
& so with hybrids themselves, as we shall see in the cases presently
to be given in more detail./83/The fertility of first crosses & of
hybrids seem to be much affected by favourable conditions of all
kinds; the common mule is said[3] to conceive more frequently
in hot countries[4]: I strongly suspect that the hybrids from Anser
cygnoides & cinereus, are more fertile, as we shall presently see, in
India than in Europe; the only known case/84/of these hybrids
breeding inter se in Europe was effected by Mr. Eyton[5] taking
a male & female hybrid from different hatches, & thus in a slight
degree lessening the ill-effect of the closest interbreeding. Age,
either very slightly too great or too little, interferes most seriously
with the fruitfulness of first crosses: we have seen that according
to Azara mares fail to produce to the male-ass, earlier than to the
horses. Mr. Brent informs me that it is an axiom with Canary
fanciers never to put a hen bird over four years old to a male
Gold-finch, as there would be no hope of produce. On the other
hand more hybrids can be raised from between pheasants & common
Hens, in their second year than in their first:[6]/84 v/Mr. Hewitt
tells me that eggs from these two birds, laid later than April &
May, *invariably* failed to produce chickens./84/So again a Canada
goose crossed by a Bernicle gander[7] for the two first years laid
barren eggs, but in the third two young were hatched out of seven
eggs. In the case of the hybrids themselves from the Canary &
other finches, in the few instances in which they have bred the

---

[1] Flourens de la Longevité Humaine 1855. p. 156. [Darwin added in pencil (one
wonders when): 'See my paper on Species of Primula for additional facts.' See
Darwin, 'On the Character and Hybrid-like Nature of the Offspring from the
Illegitimate Unions of Dimorphic and Trimorphic Plants.' *Linn. Soc. London.
J. Bot.*, 10 (1869), 433–4.]

[2] Dict. Class. de Hist. Nat. Tom 3. p. 448. [Desmoulins, art. 'Chameau'.]

[3] Gaertner Bastardz. s. 381 & Morton on Hybrid Animals in Edin. New Phil.
Journal vol 43. p. 264.

[4] But an apparently well authenticated case of the mule breeding in Scotland is
given in Smellies Edit. of Buffon 3 Edit. vol. 8 p 18.—[in note by Wm. Smellie.]

[5] Charlesworth's Mag. of Nat. Nistory 1840. vol 4. p. 90.

[6] Bechstein Naturgesch Deutschlands B. 3. s. 434.

[7] Waterton's Essays on Nat. History 2 series p. 42, 117.

eggs produced the first year have been observed to be sometimes either very small or the young birds to be very weakly; but in the following years stronger hybrids have been produced.[1]

As with plants, a hybrid can be reduced or absorbed by successive crossing to the form of either pure parent form, of which I shall give a good instance under Phasianus versicolor.[2] I have seen a triple cross amongst quadrupeds, namely a hybrid from a mare by a hybrid ass-zebra. Mr. Gould tells me that Phasianus, versicolor, torquatus & colchinus have blended together in the woods of Norfolk./

85/Systematic affinity, though limiting the possibility of hybrids being produced, certainly does not, any more than with plants, absolutely regulate this capacity. But it is very difficult to draw any just conclusion with animals on this head; for under confinement so many pure species either will not unite, or uniting are quite sterile. How, for instance, can we compare the capacity for hybridising between the several Families of Carnivora, whilst the plantigrades, though freely coupling, so rarely breed in confinement[3]; or how can we compare hawks & gallinaceous birds, whilst the former have never been known to breed when tamed. We are almost driven to look to animals in a state of nature in order to judge of their tendency or capacity to produce hybrids; but then in a state of nature we can seldom form any opinion, on the degree of sterility of their hybrid offspring. By the foregoing remark I do not wish to doubt the common opinion that the Gallinaceae are eminently capable of hybridisation;[4] I believe this to be case; but how much of this capacity to attribute to several species having been domesticated, & to most of the species breeding readily under confinement, & to the social habits of many, I know not. I strongly suspect that the great Pigeon Family, though several

---

[1] Bechstein Stubenvögel 4 Edit. s. 248. Mr. Brent informs me that he has had these small eggs from a hybrid canary–goldfinch, fertilised by pure Goldfinch.

[2] The hybrid Ph. versicolor & colchicus was reduced to the pure form of P. versicolor by two successive crosses with P. versicolor: i e P. colchicus was absorbed in three generations. Mr. Flourens, shows (Longévité Humaine 1855. p. 145.) that in four generations the Jackall was reduced into the Dog & the hybrid Yak seems to be reduced in three generations. From these facts animals would appear to be reduced by crossing with each other at a rather quicker rate than plants.

[3] See chapter 3. p. 76 et seq.

[4] The Anatidae have almost or quite equal capacity; Swans, Geese, Ducks of various sub-genera have crossed very freely: see Selys-Longchamps, in Acad. Roy. de Bruxelles. Bull. Tom. xii. no. 10.—Nineteen cross were enumerated by M. Bartlett before the Zoolog. Soc. 1847. April 13. Amongst the Fringillidae Bechstein (Naturgesch. Deutsch. B. 4. s 468) enumerates nine species belonging to genera Fringilla (with its sub-genera) Loxia & Emberiza which have yielded hybrids together. These several Families include, with the exception of Pigeons, nearly all the Birds which have been domesticated.—

have crossed, are /86/much less capable of hybridisation than the
Gallinaceae. In a state of nature the grouse-genus (Tetrao) seems
particularly inclined to cross; more especially Black-Game (T.
tetrix) which has crossed with the pheasant, Fowl Capercailzie,
Red Grouse and willow grouse, as may be seen in the following
table; but we have no reason to suppose that the hybrids are in
any degree fertile.

The pintail Duck (Dafila acuta) seems to have strong tendency
to cross with the Wigeon & the common Duck,[1] though these
three Ducks are placed by modern systematics in distinct genera;
& the hybrids of the Pintail & common duck are fertile, in an
unusual degree; the hybrids again from Anser cygnoides & cinereus
are far more fertile than hybrids between other & apparently
much more closely related geese. Many cases could be given of
very sterile hybrids produced by the crossing of species apparently
very closely related; as between the common & collared Turtle-
doves; between Gallus bankiva & Stanleyi; between the common
& changeless ⟨Polish⟩ swans[2]/

87/As far as our present imperfect state of knowledge serves,
I conclude with Gaertner that the laws regulating the fertility of
first crosses & of hybrids amongst animals are nearly the same as
with plants; but my impression is (notwithstanding the obstacle
sometimes opposed by the will & instinct of animals) that first
crosses are more easily effected between animals considerably
remote in the scale of nature, than between equally remote plants:
of course nothing can be more vague than the comparison of the
differences in plants & animals; but if we assume (though it would
be a very bold assumption) that the genera of birds, for instance
amongst the Rasores, are of at all equal value with the genera in
any order of plants, then I think my remark holds good. On the
other hand, I believe that amongst animals, the hybrids themselves
are generally more sterile than with plants./87 v/We shall see some
proof of this conclusion by comparing in the following table the
very wide limits within which amongst the Rasores, first crosses
have been effected, with the very general sterility of the hybrids

---

[1] On the fertility of Hybrids of Pintail & Common Ducks see Proceedings. Zoolog
Soc. 1831. p. 158: and [Fennell] Loudon's Mag. of Nat. History vol 9. p. 616.
I have heard of other instances & statements that the hybrids have bred inter se.
At the Zoological Gardens I saw hybrids, which I was informed by the Keeper
were descended from half-bred pintail by a Duck & then by a Pintail; so that
these hybrids had 10/16 of pintail in them & 6/16 of common Duck.

[2] Mr. Yarrell informed [me] that a hybrid from these two closely related swans was
quite sterile at Lord Derbys: I shall refer presently to the sterile hybrids from the
two very closely related species of Gallus: for the evidence about the Turtle-doves,
see Ch. 2. of this work.

produced. This latter fact is, also, strongly brought out in the few cases, immediately to be given of the highest degree of fertility observed in any hybrids. This conclusion, moreover, seems quite conformable with the analogy, which I have drawn between the sterility of hybrids themselves (not of first crosses) & that in pure species from changed conditions of life: for the complex organization of an animal might well be more disturbed by being blended with that of another species, than in the case of a plant; & we have seen that very many animals if confined in their own country are rendered sterile, where with plants this is rare without they suffer from some very marked change of conditions./

87/I have given the following Table of all the well authenticated crosses which I have heard of in one order of Birds, the Rasores; in order that those who have not attended to the subject, may see how numerous the crosses have been, & between what different forms./

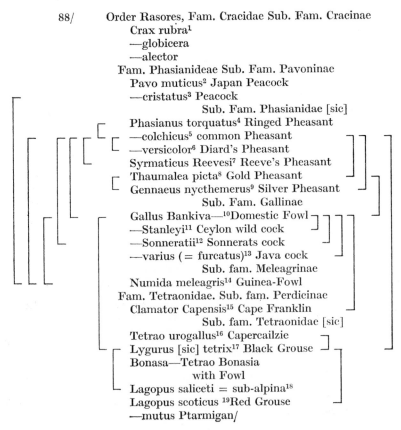

88/   Order Rasores, Fam. Cracidae Sub. Fam. Cracinae
     Crax rubra[1]
     —globicera
     —alector
     Fam. Phasianideae Sub. Fam. Pavoninae
     Pavo muticus[2] Japan Peacock
     —cristatus[3] Peacock
     Sub. Fam. Phasianidae [sic]
     Phasianus torquatus[4] Ringed Pheasant
     —colchicus[5] common Pheasant
     —versicolor[6] Diard's Pheasant
     Syrmaticus Reevesi[7] Reeve's Pheasant
     Thaumalea picta[8] Gold Pheasant
     Gennaeus nycthemerus[9] Silver Pheasant
     Sub. Fam. Gallinae
     Gallus Bankiva—[10]Domestic Fowl
     —Stanleyi[11] Ceylon wild cock
     —Sonneratii[12] Sonnerats cock
     —varius (= furcatus)[13] Java cock
     Sub. fam. Meleagrinae
     Numida meleagris[14] Guinea-Fowl
     Fam. Tetraonidae. Sub. fam. Perdicinae
     Clamator Capensis[15] Cape Franklin
     Sub. fam. Tetraonidae [sic]
     Tetrao urogallus[16] Capercailzie
     Lygurus [sic] tetrix[17] Black Grouse
     Bonasa—Tetrao Bonasia
     with Fowl
     Lagopus saliceti = sub-alpina[18]
     Lagopus scoticus [19]Red Grouse
     —mutus Ptarmigan/

1 Temminck Hist Nat. Gen. des Gallinacees vol 3. p. 13–21. The hybrids raised from these species are sometimes sterile & sometimes fertile with the pure parent species.

2 On the authority of Mr. Blyth in Rev. E. S. Dixon's Dovecot & Aviary 1853 [sic.] p. 88. They perished when a few days old. See [Blyth] Indian Sporting Review New Series vol. 2. p 253.

3 Mr. Mitchell. Sec. Zoological Soc. saw at Amsterdam a hybrid between the Peacock & Guinea Fowl & brought home a drawing of it. A clergyman in England informs me that he has seen his Peacock though having a Pea-Hen frequently unite with a Guinea-hen; but no offspring was produced—

4 The hybrids from P. torquatus & colchicus there is good reason for believing are quite fertile see text infra.

5 The account of the hybrid Pheasant-Fowl, being fertile with pure Pheasant in Proc. Zoolog. Soc. 1836. p. 84—does not at all satisfy me for I saw the Birds But Dr. P. Lucas quotes Bomare (L'Heredité Naturelle Tom. 2. p. 307) for the same fact of the occasional fertility of these Hybrids with the pure pheasant; but I doubt the account, for Mr. Hewitt has had much extra-ordinary experience in making these hybrids; never once saw them show any sexual feeling.—

6 With respect to the high fertility of the hybrids between P. versicolor & Colchicus, see text infra.

7 Hybrids from the male of this Bird & the common pheasant were raised during two seasons in Zoological Garden. A male hybrid was paired with hen common & hen Rees pheasant ⟨(Penny Encyclop. vol. 18, p 61)⟩ but was quite sterile. M. S. Report sent me from Zoological Society.

8 For crosses between Gold & Silver, each with the common see Proc. Zoolog. Soc. 1836. p. 84—& See also Is. Geoffroy Saint Hilaire Essais de Zoolog. Generale 1841. p. 493. The hybrids are said to be always sterile, but Temminck (Hist. Gen. des Gallinacées vol 2. p. 323) says that one hybrid from Gold & common Pheasants produced with pure P. colchicus a sterile hybrid grandchild of the two pure species. I give the cross between the Gold Pheasant & common Fowl on the authority of the M. S. return from Zoological Society: this hybrid was quite sterile. Mr. Hewitt has repeatedly tried to make this cross, & likewise between the Silver Pheasant & common Hen without success, though the birds coupled freely.—

9 Mr. Hewitt informs me that he raised three hen hybrids from the cock Gold Pheasant & Hen Silver Pheasant.

10 [There is no note with Darwin's reference number here, but on verso of fol. 88 he added:] Temminck Gallinacées 2 p. 75 says without particulars that turkey & Hoccos & Turkeys & cocks unite. Append this at note.

11 Layard in Annal of Nat. Hist vol XIV. 2 series 1854. p. 63. The one hybrid raised was quite sterile. This is a very remarkable fact seeing how close this species is to G. Bankiva.

12 Mr. Blyth raised many hybrids from this bird & a domestic Hen from Aracan, but these were quite sterile inter se & with the domestic cock or Hen. But several years ago I myself saw at the Zoological Gardens young birds, which were the *offspring of hybrids* inter se from a Sonnerat cock & Bantam Hen; some of these had returned most closely to the pure Sonnerat & others, as I was told, to the pure Bantam: there could hardly be any mistake here; for they had then only one Sonnerat cock; & the fact of the young (or grand-children) returning towards both pure species strongly corroborates their asserted parentage.—I was told by the curator Mr. Miller at the Zoological Gardens that hybrids have been raised between this Fowl & the Pheasant.

13 Wagner's Report on Zoology for 1843–44 in Ray Soc. for 1847. The hybrid from this species with common Fowl has been called G. aeneus. They are commonly raised in Java as I am informed by Mr. Crawford, but are believed to be always sterile.

87A/The Families & genera are arranged in accordance with Mr. G. R. Gray's classification generally acknowledged as one of the best. The Brackets, imply that hybrid offspring has been produced by the two forms so connected. The degree of fertility of the hybrids, is given in the notes, where nothing is said, nothing is known & in this case generally, their sterility may be safely inferred./

89/I will now discuss in some detail the degree of fertility of certain hybrids in successive generations.

Dogs—Wolves Jackalls Foxes: that hybrids between dogs & wolves & Jackalls possess a certain degree of fertility is notorious ⟨I think evident from the frequent practice amongst savages, as has been alluded to in a previous chapter of universally crossing their dogs with wolves⟩: M. Flourens (Longévité Humaine 1855. p. 143, 156) has made laborious experiments on a large scale with these hybrids having raised no less than 294 individuals. He finds that when bred inter se the dog-wolves invariably became sterile at the third generation, & the hybrids of the Dog & Jackall at the fourth generation./ 89 v/⟨He raised no less than 294 hybrids between these three species.⟩ This increasing sterility in the successive generations is curiously parallel to the same fact, as observed by Gaertner, in plants./89/But nothing is said to show that care was taken to prevent the ill effects of close interbreeding: if the hybrids were all raised from one dog & wolf & one dog & Jackall, & all were placed under the same conditions of life, then I think, starting with some degree of sterility, it is not at all surprising that their sterility should have been increased by the interbreeding, to such a degree that Mr. Flourens could not rear any offspring beyond the fourth generation. I saw a female hybrid from a Dog & Jackall in the Zoological gardens which even in this its first generation was so sterile that it only comes periodically into an/

---

14 Morton describes their hybrid in Proc. Philadelphia Acad. of Nat Sciences. Quoted in Annals of Nat. Hist. vol 10 [19], 1847. p. 210 The Zoological Society once possesed specimen of this hybrid, which was quite sterile. The cross from between Guinea-Fowl & common Pheasant, I give on authority of M. S. return from the Zoological Society: this hybrid was quite sterile.

15 Temminck Hist Nat. Gen. de Gallinacees vol 3 p. 301. The hybrids are said to be always sterile. This remarkable cross is confirmed by Mr. Blyth, on the authority of Major W. Sherwill in the Indian Sporting Review new S. vol 2. p 241.—

16 The reappearance of Tetrao medius of some authors in Scotland after the re-introduction of the Capercailzie is the best & most curious proof of its hybrid origin from this Bird & the Black cock. See [J. Wilson] Proceedings Royal Soc. of Edinburgh December 19. 1842 [vol. 1, p. 395.]

17 Lloyd states (Field Sports of N. of Europe vol I. p. 314) on the authority of Nilsson that the Black-cock has crossed with the Fowl, but the chicks survived only a few days; but Nilsson is so excellent an authority that his statement, may, I should think, be trusted.

18 Bronn Ges[ch]ichte der Natur B. 2. s. 166 and Wagner Report on Zoology for 1844 in Ray. Soc. for 1847 [p. 293.] ⟨I presume that the translation is correct & that the term Ptarmigan is meant.⟩ The Hybrid is the Tetrao hybridus lagopides of Nilsson.

19 There seems no reason to doubt from Mr. Macgillivrays account in his British Birds vol I. p 162. that hybrids from the Black & Red Grouse of Scotland have been produced in a state of nature.

90/imperfect state of heat. Hybrids from the Fox & Dog will breed with the Dog, as shown in Pennant's Quadrupeds 3 Edit. vol I. p 239. & in Herbert's Amaryllidaceae. p. 338.

Goat & sheep. The fertility of the hybrids from these two animals, classed in distinct genera, has been exaggerated by some authors, see for instance ⟨(Nott & Gliddon, Types of Mankind 1854. p 379)⟩/90 v/It is stated that Goats & sheep are habitually crossed in Chile (Molina Historia Geograph. del Reyno de Chile. 1788. P[art.]. I. p 376); & similar crosses have been effected in Europe.—/90/M. Chevreul (Annal des Sci. Nat. 3 ser. Bot. Tom. VI. p. 188) affirms on the authority of Mr. Gay who must have had excellent opportunities of ascertaining [?] the truth that the male hybrid from the male Goat & female sheep are paired with female sheep, & that the offspring, with 3/4 of the blood of the sheep, are propagated inter se, but that the character of the fleece deteriorates, so that after some generations they have to take a fresh cross with the half-blood or first hybrid. This seems a very remarkable degree of fertility, for it implies that after several generations the hybrids with one-fourth of the goats blood in them are fertile with the hybrid having half goats blood.

Bactrian & common Camel. There seems no doubt that hybrids are commonly raised (Burnes Travels in Bokhara vol. 3. p. 154) between these species though so remarkably distinct in structure & constitution. Eversmann enters into details & clearly distinguishes the Bactrian from common camel, &/ 90 Bis/states that the hybrids have either one or two humps, & he expressly adds that these hybrids are fertile (Quoted in C. Ritter's Erdkunde. B. 8. Th. I. s 655, 659: I was guided to this reference by Nott & Gliddon Types of Mankind [p. 380]); but it is not expressly stated whether these hybrids are fertile inter se or only with either pure parent-species. The old traveller Olearius, as quoted by Desmarest in Encyclop. Method, says these hybrids are sterile, like the common mule. The great facility with which crosses can be made between the several forms of Llamas (Auchenia) & the *reputed* fertility of the hybrids, renders Eversmann's statements somewhat more credible./

91/*Muntjac Deer.* At Lord Derbys the Cervulus vaginalis from the Malayan archipelago & C. Reevesii from China bred together (Gleaning from the Menagerie Dr. J. E. Gray 1850. p. 65); & Mr. Thompson the intelligent Curator (& now curator of the Zoological Gardens) assures me that he is certain that the hybrids were perfectly fertile inter se, & that a herd was thus reared. No doubt these are close species; & Kölreuter in such cases would have at once called them varieties.

Bos grunniens (Yak) *& taurus var. Indicus.* Various hybrids have been produced between several species of oxen & Buffaloes, but all, as far as I can find out are quite sterile, with the exception of those between the two above-named apparently very distinct species, which indeed by many naturalists are ranked in distinct genera. These hybrids are raised in large numbers in the Himalaya. The case is interesting, in as much as the two parent species have remarkably different constitutions, the yak enduring extreme cold &

the Indian cattle extreme heat: the yak, moreover, browses on very different plants.—Mr. Schlagintweit has lately stated (British Association: Dublin 1857)[1]/92/that these hybrids are perfectly fertile together & that he examined some bred inter se to the seventh generation! But I must think that this is incorrect; for Dr. Falconer has given me the following precise details obtained from the best possible authorities in Little Thibet: the male Yak is habitually crossed with the female cow, but the reciprocal cross can also be made: the female hybrid (called Bsohn) can be fertilised by the male yak; & the female offspring (with 3/4 of Yak blood, called Gurmoh) can be crossed again with the male Yak; & the progeny (namely great-grand-children of the two pure species first crossed) are then quite fertile; both on the male & female side, & can be crossed inter se or with either pure parent. But in the earlier stages of reduction, though the female hybrids are fertile with the pure Yak, the males are impotent./

93/*Fringillidae.*: Bechstein states (Stübenvogel 1840. p. 248) that the hybrids from the canary birds with the Goldfinch, & with the siskin (F. carduelis & spinus) have certainly interbred inter se. I have heard of another case. Capt. Hutton, a good observer says (Calcutta Journ. of Nat. Hist. vol 2. 1843. p. 530) he has himself reared the young from a pair *inter se.* of mule birds from the canary & linnet. (F. canaria & linota), which young (grand-children of the pure species) "again in turn produced & reared a brood". The cases, however, of hybrids from the Canary & other Finches breeding inter se are undoubtedly rare..But the experiment, cannot be considered as fairly tried until, some of the species with which the canary-bird is crossed breed quite readily in confinement, which occurs with none of them. I have seen it stated but have failed to find the original account that Vielliot found the hybrids from the F. citrinella & canaria, which are very closely allied species, perfectly fertile./

94/*Hybrid Pheasants*: I am assured by Mr. Thompson (see also the Sale Catalogue [by E. S. Stanley]) formerly curator at Lord Derbys that he is certain that the hybrids from P. versicolor & Colchicus (manifestly distinct species) bred quite freely inter se, & that their progeny (grandchildren of the two pure species) again had offspring. The hybrids were also reduced by crosses with pure P. versicolor, & the great grand-children of the two pure species are undistingushable from P. versicolor (Report Zoolog. Soc. April 1855. p. 17): In the summer of 1856 I saw the latter birds & two of the hens had laid together 24 eggs, & 15 young had been reared from them.

Phasianus torquatus has of late years been turned out in several places into our woods, & has certainly crossed with the P. Colchicus. These species are closely allied, but I believe are universally admitted to be distinct./94 v/ Temminck (Hist. Nat. Generale des Gallinaceas vol. 2. p. 326) distinctly states that the hybrids from them are fertile inter se. We have, also, indirect evidence on this head: the/94/kind of pheasant, called the ring-necked is not uncommon in some parts of England, & is stated by Selby to have spread within a comparatively short time over nearly the whole of Northumberland. (Montagu's/95/Ornitholog. Dict. Rennies Edit. p 370). From different

---

[1] [See Schlagintweit, 'Notes on Some of the Animals of Tibet and India', *Brit. Ass. Advanc. Sci., Rep. for 1857* (1858) p. 107.]

statements which I had read, I had always considered the ring-necked pheasant as a simple variety not due to crossing until Mr. Gould was so good as to show me the several points of resemblance between the common ring-necked pheasant & an undoubted hybrid from P. colchicus & torquatus. Besides the white collar, a trace of a white line over the eyes, a slightly more fulvous tinge on the flanks, & sometimes a trace of green on rump, & the bars on the tail all seem to indicate a cross from P. torquatus, as is likewise the conclusion arrived at Mr. Blyth after careful examination of the subject; it seems highly improbable that these several characters should have all concurred from simple variation. If this be so, the case is interesting, for as we have seen that the ring-necked pheasant has rapidly increased in some districts, its fertility must be very great; & although it is not at all known what is the exact proportion in which the two species are blended in the English ring-necked pheasant; yet there can hardly have been repeated crosses for many generations with the pure P. colchicus, otherwise the characters of P. torquatus would have been much more completely obliterated./

96/⟨Pintail & common Duck (*Dafila acuta & Anas boschas*). These two ducks seem to cross very readily, & several cases are on record of these hybrids, breeding readily with the pure parent species (Proceedings Zoolog. Soc. 1831. p. 158: Loudons Mag. of Nat. History. vol. 9 p 616): but Mr. James Hunt, the intelligent Keeper at the Zoological Gardens, showed me several years ago a lot of young birds, which he knew to be the offspring of a pair of these hybrids inter se:—I have heard, also, but more vaguely of another parallel instance.⟩

96/*Chinese & common Goose* (*Anser Cygnoides & Cinereus domesticus*). These birds are so distinct that most ornithologists place them into distinct genera; many cases are on record of hybrids from these birds breeding readily with the pure parent species;/96 v/but they are generally quite sterile inter se; & one case is on record (["Zenas,"] Poultry Chronicle vol. 3. 1855 p. 487) of a hybrid with 3/4 of pure China-goose blood in them, being sterile inter se. Nevertheless/96/Mr. Eyton once succeeded by pairing birds reared from different hatches, in rearing no less than eight young birds from half-bred hybrids inter se, though both the parents were young birds, which would naturally have lessened their prolificness. (Charlesworth Mag. of Nat. Hist. vol I. 1837. p. 358. & vol 4. 1840. p. 90)/

97/In India over large tracts of country, in the N. W. Provinces, in Assam, & near Calcutta, the geese as I am informed in letters from Mr. Blyth[1] & Capt. Hutton[2] are of a mixed breed, clearly intermediate in all their characters, even in their voice, between A. cygnoides & cinereus: Such acute & experienced ornithologists as these two observers could not possibly have been mistaken on this head. In many of these districts neither pure species is kept. Therefore there can be no doubt that, as I am assured by these gentlemen that these geese breed inter se; & Mr. Blyth says he believes they are fully as prolific as the common English Goose. It is indeed I think, obvious that their prolifick-ness, must be great; otherwise the breed would not be commonly kept for

---

[1] [See Blyth's 'Notes for Mr. Darwin', C.U.L. Darwin MSS. vol. 98, fol. 106.]

[2] Mr. Sundevall (Annals of Nat. Hist. vol. 19. 1847. p 171) noticed these crossed geese near Calcutta. Pallas in the Act. Acad. St. Petersburgh. 1780 p. 83. speaks of the hybrids from these two genera as being very prolific.

profit. We may perhaps attribute the much greater fecundity of their hybrid when bred in India than in Europe, to the difference of climate, & probably in large part of the numbers raised,—the ill-effects of close inter-breeding being thus wholly removed./

98/I have now given in some detail all the cases which I have collected on the degree of fertility of the most fertile animal hybrids. But I think that hardly one of these cases has been sufficiently investigated. The scantiness of the facts plainly shows how rarely there is with hybrid animals any approach to perfect fertility. In considering, however, the subject we should always bear in mind that the experiment is never fairly tried, without both parent-species breed perfectly well under domestication or confinement, & without both are placed under favourable conditions & without several hybrids, not related to each other, are raised at the same time, so that the ill effects of close interbreeding in the successive generations may be excluded. Very few cases are on record in which all these conditions have been fulfilled. Nevertheless, the hybrids from Phasianus Colchinus with P. torquatus & with P. versicolor, & in India from Anser cygnoides & cinereus probably make a very close approach to perfect fertility, or perhaps are perfectly fertile together.

Most naturalists now believe that many of our domestic/99/ animals, as dogs, cattle, sheep &c., have descended, each from several aboriginally distinct species. In some cases, this seems to me the most probable view. Those who admit this view, must suppose either that there once existed several distinct species, which were capable of uniting & of producing perfectly fertile hybrids, which we have now domesticated around us; or they must suppose, in accordance with the view first broached by Pallas,[1] that species originally infertile together, become quite fertile through a long course of descent under domestication. That the making of the first cross should be facilitated by both species having been thoroughly domesticated seems extremely probable; but I know of no actual facts to support this view, except the statement by Dureau de la Malle who has so closely studied classical literature, that the common Mule was produced with more difficulty in the time of the Romans than at present:[2] on the other hand, Gaertner[3]/100/could perceive no difference in the facility of hybridising cultivated & wild plants & on asking

---

[1] Act. Acad. St. Petersburgh. 1780 Part II. p. 84, p. 100.

[2] Annales des Sc. Nat. Tom 21. (1 series) p. 61.

[3] Bastard[er]zeugung s. 11, 12. [At top of next folio, Darwin added in pencil: 'Aegilops being more fertile N. Hist. Review.']

W. Herbert he expressed to me the same opinion: but neither of these botanists have experimentised on the very same species in its wild & cultivated state. Even if the first cross between two animals could be effected more easily, when both were domesticated, I know of no fact whatever countenancing the view that the fertility of the hybrids thus to be produced, would be greater, after the parent species had long been domesticated, than at first. Nevertheless, I must confess that there seems to me much probability in this hypothesis. Believing as I do, that our dogs, for instance, have descended from several distinct wild stocks: analogy prevents my believing that if their wild stocks had been caught & paired that their offspring would have been as fertile as are our mongrel dogs:[1] but how much of the infertility in this hypothetical case would have to be put down to the wild parent stocks not breeding readily under confinement, & the many hybrids/101/not having been raised so as to prevent the ill-effects of continued inter-breeding, it would be hard to conjecture. If this hypothesis could be proved true, it would throw considerable light on the history of our domesticated animals; & would be interesting for our theory as it would show that the sterility of hybrids was a varying quality, which would in some degree lessen its importance as a diagnostic character between Species & Varieties.

Fertility of crossed Races ⟨*Varieties*⟩ *in Animals, & their Mongrel Offspring.*—As I have already more than once remarked, after seeing the almost universal lessened fertility of even very closely related species when crossed, the perfect, nay very generally increased, fertility both of the first cross between the most widely different varieties & of their mongrel offspring is a highly remarkable fact. In plants I have been able to give a few cases pretty well authenticated of some degree [of] sterility in crossed varieties: with animals I can advance no satisfactory evidence on this head. From the facts/102/ given in our sixth (?) chapter, I do not doubt that animals of the same variety sometimes prefer pairing with each other; but this, though an interesting fact, & bearing on the reputed repugnance of distinct species to unite, does not concern their fertility when crossed./102 v/Nor can physical obstacles from

---

[1] Mr. Blyth ['Zoophilus'] has remarked (Calcutta [actually India] Sporting Review vol 2. new series p. 133) that the N. American wolf & the Canis latrans keep distinct in their native state, & yet it is believed that the Indian dogs, which I think there can be hardly a doubt have descended from these two wild species, mix readily together; & Richardson has described (Fauna Boreali-americana [I] p. 80) an intermediate race.—For an interesting discussion on this subject, see Nott & Gliddon, Types of Mankind p. 384.—

great difference in size between two varieties be strictly considered as causing lessened fertility. Thus it is well known that bitches paired with dogs of large size, often die during parturition. I presume very unequal size would sometimes interfere with the union of varieties; though A. Knight got offspring from a Dray stallion & Norwegian Pony[1] & chickens from a Cochin cock & Sebright Bantam Hen were exhibited at Manchester[2]/102/When we hear that certain domestic breeds of native American dogs[3] do not pair or even readily associate with other breeds; when we hear it said, but on what grounds I know not,[4] that certain breeds of dogs, are more fertile when crossed together than other breeds, the explanation probably is in the case of the American dogs, & perhaps in the latter case, that these dogs have descended from primordially distinct species, & not that any degree of relative sterility has been acquired during domestication. The same explanation probably applies to Bechstein's statement[5] (if to be trusted) that dogs of the Spitz breed can be easiest crossed with Foxes: in plants, however, I may remind the reader that we apparently have good cases of varieties of the same species uniting with different degrees of /103/facility with distinct species. Believing as I do that some of our dogs are descended from the European wolf, & seeing that the hybrid from the wolf & dog shows some sterility, I should have been tempted to surmise had the experiment been made with a breed like the Hungarian sheep-dog, which is extremely like the wolf, that the dog had become so much modified by domestication that the fertility of its offspring when crossed with the wolf had become impaired.

Many naturalists believe that the Llama & Alpaca are only varieties of the wild Guanaco: Mr. W. Walton, who has particularly attended to these animals[6]/104/says the two first named often breed together, but that their offspring are quite sterile & are hence called "Machorras". From other statements which I have heard, I doubt the fact, but supposing it to be true, the inference would be that those naturalists are right who view the Llama & Alpaca as distinct species, now utterly exterminated in their wild state. A good authority says that the first cross from the Long- & Short-horned cattle is excellent, but that in the third or fourth generation "there is much uncertainty whether the cows will hold

---

[1] A. Walker on Intermarriage p. 205.   [2] Poultry Chronicle vol. 2. 1854. p 446.
[3] Rengger Saügethiere von Paraguay 1830 s. 153, on the Hairless dog.—Gosse's Sojourn in Jamaica p. 339. on the Alco or Mopsy Dog.
[4] Gaertner, Bastardzeugung s. 577.
[5] Naturgeschichte Deutschlands. 1801. B. 1. s. 638.
[6] The Alpaca. 1844. p. 29.

to the bull; & full one-third of the cows among some of these half-breds fail of being [in] calf."[1] /104 v/In the case of all our domestic animals, though crosses are very frequently made, yet from their extreme uncertainty, they are very rarely propagated for many generations, so that it is quite possible that a slight degree of infertility might long remain undiscovered. But in the case here given it so happens that the cross has been systematically made, & according to Mr. J. Wilkinson[2] a half-breed has been fully established; & this makes me doubt Youatt's statement; for any marked degree of infertility would surely have prevented any one establishing the half-breed./104/Supposing that this remarkable statement could be trusted, I do not doubt that some naturalists would immediately argue that our Long- & Short-Horns have descended from two distinct species.—

I have given the foregoing details to show how much inherent & almost insuperable difficulty there must always be, from our ignorance of the/105/history of our domestic breeds, in this subject.—If any two breeds had ever become so different as to be in the slightest degree sterile together; scarcely any amount of evidence would convince naturalists that both had descended from the same parent stock.

The case of perfect fertility between varieties, which has struck me most, is that of Pigeons: I have myself largely experimentised on the fertility both of simple & the most complicated crosses between the most distinct breeds; & I have given my reasons for fully believing that all are descendents of one species. Compare a Pouter, Tumbler, Carrier, Fantail & Barb, which produce together quite fertile mongrels, & see in how marked a manner they differ from each other, in comparison with Gold, Silver & common Pheasant, or with the Java & common Fowl. (Gallus domest. & varius), from which it is often difficult to rear any young/106/& these when reared are utterly sterile! Such cases as this of the Pigeons are very surprising & seem to stand in direct opposition to our view that species do not essentially differ from varieties. But there are some considerations which make the case not quite so contradictory, as it at first appears. In the first place, it has, I think been clearly shown, in accordance with Gaertner's conclusion, that the power in any two species of easily producing hybrids & these more or less fertile does not strictly run (as well seen in the different results from reciprocal crosses) with their systematic affinity, that is with the amount of resemblance which can anyhow

[1] W. Youatt. Cattle 1834. p 202.
[2] in his well known "Remarks addressed to Sir J. Sebright" 1820. p. 38.

be detected by the eye: hence we should err if we were, under the view of varieties not differing essentially from species, to infer that because a Powter-pigeon apparently differs more from a Tumbler than the common pheasant does from the Silver, that there ought necessarily to be fully as great difficulty in getting fertile hybrids from the two pigeons, as from the two pheasants./

107/Secondly, those many naturalists who believe that most of our domestic animals are descended each from several aboriginal species, & who, therefore, must believe either that perfect fertility between species when crossed is very far from uncommon in nature, or is a quality readily acquired under domestication, will feel little surprise at the fertility of varieties when crossed. On the view, indeed, of sterility being commonly lost between species when under domestication, it would be most strange if sterility were, also, to supervene between varieties under raised domestication.

Lastly: sterility from hybridisation, like that from other & quite distinct causes, must be looked at, not as a specially endowed quality, but as incidental on other & unknown differences in the sexual organisation of the species which are crossed,—as we see very plainly in the frequent & great differences in the results of reciprocal crosses. Now man both in his unconscious & in his methodical selection of varieties useful or pleasing to him, selects almost exclusively by the eye; he has neither the power or the wish to affect by continued selection those/108/obscure & inner constitutional differences, on which the sexual affinity of distinct species seems to depend. Moreover man does not select each variety in exact relation to the conditions to which he exposes it; nor does he keep the conditions as constant as possible; nor is his selection uniform in direction & extremely slow. How differently nature acts! She keeps her conditions uniform or nearly uniform for thousands & tens of thousands of generations: if she modifies her productions she modifies them most slowly & uniformly only for the good of the selected variety. And who can say what may be the difference in the results of these two kinds of selection? In a somewhat analogous manner, as species which are so generally adapted by nature to a certain limited climate, would appear, as before shown, when domesticated by man to lose to a large extent this close adaptation, & as the varieties raised by him acquire only in a very limited degree such kind of adaptation; so it seems not very improbable that species domesticated by man may lose, as some have thought, this tendency to sterility when crossed, & the varieties/109/raised by him not acquire it, or only, as with certain plants, in a very slight degree. Hence I conclude that it is not so surprising, as on our view of the nature species it must at first appear, that varieties formed under the care of man, should not have become modified in their sexual organisation, in that mysterious manner on which their greater or lesser power of crossing with other forms & of producing more or less fertile offspring depends, in the same remarkable manner as is so generally & eminently characteristic of Species./

110/*Comparison of Hybrids & Mongrels, independently of their fertility*: *Plants* We must now compare the hybrid offspring of species & the mongrel offspring of varieties, & see how far they resemble or differ from each other, independently of their fertility. This subject is of some importance for us; for it would be strange if the union,—that most mysterious problem in physiology—of two varieties, produced by natural causes & the offspring of two species, supposed to have been separately created, yielded offspring, which followed the same laws in their likeness to their parents, & in other respects; & if this can be shown to hold good, in all essential particulars, the fact may be used as an argument that the origin of species & varieties is essentially alike. For several reasons I will discuss this subject separately & first for Plants & then for Animals.

As some authors have considered hybrids as monsters, & beyond the pale of law, it will be well to premise that Gaertner & Kölreuter have clearly shown that this is quite a false view. ⟨Undoubtedly the hybrid progeny of two/111/species follows, at least approximately, the same laws of resemblance as do the individual offspring of the two sexes of the same species.⟩ When two not-cultivated species have been repeatedly crossed, their hybrid offspring of the first generation are as a general rule found closely to resemble each other;[1] & when rarely "exceptional types" are produced, which abnormally closely resemble either parent-species, such types when reproduced at long intervals from the same two species, are alike. Moreover, when these exceptional types are sufficiently fertile to propagate themselves, their offspring generally revert to the normal hybrid type of the two species in question.[2] Nor, I may add, do malconformations[3] occur oftener in hybrid plants than with pure species: even their reproductive organs, as we have seen, are only functionally imperfect./

112/Gaertner with his immense experience has compared in

[1] Gaertner Bastardzeugung s. 234.
[2] Bastardzeugung s. 238, 424.   [3] Bastardzeugung s. 518 557.

detail[1] hybrids & mongrels: it is evident that he would gladly
have seized on any difference in the progeny of species & varieties,
for such is the clear tenour of his whole admirable work. He
reduces, here leaving out the question of fertility, the main differences to two,—namely that mongrels, especially in the first
generation, but likewise in the succeeding generations are much
more variable than hybrids in the first & even in the succeeding
generations: secondly, that mongrels evince a much stronger
tendency to revert to either pure parent-form than do hybrids; but
this latter difference in fact is only a part of that first specified
of greater variability & less antiquity. As a general rule, I think
there can be no doubt that mongrels of the first generation are
much more variable than hybrids of the first generation. But
"exceptional types"[2], which are nothing but strongly marked
variations of a definite kind, do occasionally occur amongst hybrids
of the first generation[3]: and Gaertner further admits that lesser
variations do likewise sometimes occur, but he adds, & the observation is an /113/important one, that he has noticed these lesser
variations only in hybrids from species which have long been
cultivated in gardens./113 v/In fact there cannot be a doubt that
hybrids of the first generation from between two species both long
cultivated often differ greatly from each other as in hybrids from
distinct species of Rhododendron, Passiflora, Fuchsia &c. I saw
at Spofforth[4] two hybrids between Rhododendron & Azalea raised
by Herbert from the same seed-capsule, & they differed greatly
in appearance[5]/113/It is, also, an important observation that these
lesser variations in the first generation have been principally
observed by Gaertner[6] in hybrids between species so closely related
that they have been thought by some authors to be varieties,
though ranked by Gaertner by the test of fertility, as true species.
In this greater variability in hybrids from between very closely
related species we plainly see a gradation towards the strongly
marked variability of mongrels in their first generation. Moreover,

[1] Bastardzeugung s. 582.
[2] [Darwin later pencilled: 'Are not these only reversions?']
[3] Bastardzeugung s. 238.
[4] [At Spofforth, Dean Herbert gardened and experimented with hybrids; see his
Amaryllidaceae, p. 359.]
[5] Herbert Amaryllidaceae. p. 359: Mr. Sabines account of hybrid Passiflora in
Hort. Transact. vol IV. p. 261, 266. In Hybrids from Fuchsia coccinea fertilised
by fulgens, Mr. Thwaites [probably in letter to Darwin] says "scarcely could
any two be found so much alike as to be undistinguishable": he adds a truly
remarkable case of one single seedling, half of which more resembled the one
species, & the other half the other species.
[6] Bastardzeugung. s. 247, 249, 577.

it is well to observe that there are causes [?] of variability in hybrids, which seem to escape all law; thus in the following combinations, Dianthus barbatus fertilised by the hybrid D. barbato-carthusianorum gives many more varieties than the reciprocal cross of D. barbato-carthusianorum by D. barbatus; so again with these distinct species, Lobelia fulgens fertilised by L. cardinali-syphilitica yields a more variable hybrid progeny, than L. cardinali-syphilitica fertilised by L. syphilitica.[1]/

114/Now considering these facts, can it be considered as a surprising or important difference that mongrels in the first generation should be more variable than hybrids? In the first place the far greater number of varieties have been produced by long-continued cultivation, & this we have seen makes true hybrids variable in their first generation. Secondly, according to Gaertner, hybrids are more variable from between closely related species than from between those which are very distinct; & of course varieties are closely related to each other./114 v/Thirdly, variability in itself is certainly inherited, & as varieties are in many cases only recognised as varieties from this very quality, it would be strange if their mongrel offspring were not commonly thus characterised./

114/But we do not know that mongrels, especially from varieties in a state of nature are universally highly variable in the first generation: we must remember that *extremely* few experiments have been systematically made & recorded on varieties: I find in Kölreuter's works[2] a few cases of crosses some made reciprocally, between several varieties of Mirabilis, Matthiola & Nicotiana; & no mention is made of any extreme degree of variability, as from the tenour of his works/115/might most safely have been expected, had such variability occurred.

Turning now to the generations succeeding the first in hybrids, every observer has been struck with their extraordinary variability, & some authors have even thought this a more important characteristic of hybrids than their lessened fertility. Gaertner freely admits this variability in the successive generations of hybrids whether fertilised by their own pollen, or by that of either pure parent-species: Kölreuter used the strongest expressions on this head, as does W. Herbert.[3]/115 v/This highly remarkable & unexplained difference in the degree of variability of hybrids in the first generation, compared with that in the successive generations

---

[1] Bastardzeugung s. 445, 507, 513.
[2] Zweite Fortsetzung s. 56, 126, 128: Journal de Physique 1782. p. 285: Nova Acta Petrop. 1795, p. 333 & 1797. p. 393.
[3] Gaertner's Bastardzeugung s. 518: Kölreuter in Nova Acta Petrop: 1794. p. 391: Herbert Amaryllidaceae. p. 348.

is quite conformable with the view of the cause of ordinary variability (independently of crossing), which as stated in the first chapter seems to me by far the most probable; namely that the reproductive system is affected by the conditions of life to which either one or both parents have been exposed, in the same manner but in a lesser degree, as in those many cases in which actual sterility supervenes from cultivation or confinement.—For in hybrids of the first generation from between two species which have not been modified by cultivation, the reproductive system of neither parent has been in any way affected; whereas in the successive generations from hybrids, we well know from the lessened fertility of the latter how seriously their reproductive system has been generally affected./

115/If mongrels in their successive generations are more variable than hybrids in the corresponding generation, which I think probably is the case, at least with cultivated varieties, the difference is only one of degree; the comparison moreover can only have been vaguely made for I know of only one case on record, in which the offspring of two varieties have been carefully observed/116/for several, in this one case for four generations[1]. Nor is ⟨the⟩ rule universal that the successive generations of hybrids are highly variable; for Gaertner has given[2] five cases in which the progeny kept constant, & one, namely the offspring of the hybrid Dianthus armeria-deltoides was observed even to the tenth generation. In the same manner it would appear that occasionally, though very rarely, the mongrels keep true; I am assured by an intelligent nurseryman that "Dale's hybrid turnip" has every appearance of being a hybrid, & that it does not vary; & Mr. Beaton,[3] has remarked that "Melvilles most extraordinary cross between the Scotch Kale & an early cabbage is as true & genuine as any on record". In these cases there may probably have been some selection in the early generations; but had the variability been as extreme, as it generally is, no one would/117/have had patience to have raised a true mongrel race.

With respect to Gaertner's statement that mongrels show a greater tendency to revert to either pure parent state, nearly all the foregoing remarks are applicable, as indeed this reversion is only a form of variability. Even if proved strictly true & I must

---

[1] Kolreuter on varieties of Mirabilis in Nova Acta Petrop. 1797. p 393.
[2] Bastardzeugung s. 553. Dean Herbert showed me some hybrids from two species of Loasa, which had kept constant for several generations.
[3] Cottage Gardener. 1856 p. 110; Wiegmann in his Bastardzeugung s. 33 says that seedlings from his mongrel cabbages, as a general rule, retained their blended nature.

repeat my remark on how few mongrels have been carefully observed during several generations; it would be only a difference in degree, for Gaertner gives many cases of true hybrids reverting to ancestral forms/117 v/; & it deserves notice that this tendency to reversion was observed much oftener with cultivated than with wild species. Nevertheless it is/117/an anomalous circumstance that according to Gaertner the most fertile hybrids which according to our view, differ least from varieties do not show this tendency to reversion.[1] I may give one remarkable instance from Kölreuter[2] in the offspring of the hybrid Mirabilis Jalapa-longiflora, fertilised by several varieties of M. Jalappa, in which *some* of them, though having in their composition only one quarter of M. longiflora, yet more closely resembled that species, than their hybrid mother./

118/With the exception of the new specified differences, though these seem to me to be of extremely little importance, hybrids & mongrels in nearly every other respect apparently have the closest resemblance,—the degree of fertility, as before, not being here considered. Both are remarkably luxuriant, hardy & precocious. Both generally come alike from reciprocal crosses.[3] Both follow nearly the same rules in their resemblances to their two parents: Gaertner[4] has taken much pains in classifying the resemblance of hybrids to their parent species, both in the first & successive generations; he makes three classes but which he fully admits, blend together in an inextricable manner; (1st) hybrids very nearly intermediate in their whole structure; (2nd) hybrids, (& these are extremely common) resembling, but not identical with, one parent in one part, & the other parent in another part; (3rd) hybrids decidedly resembling one of their parents. In the first generation any particular hybrid may generally be classed under one of these heads, but in the succeeding generations the same hybrids often break into all three classes: so it is with mongrels. Gaertner asserts that hybrids following the first or intermediate class of resemblance are generally/119/raised from between very closely allied species; & he further remarks that mongrels com-

---

[1] Bastardzeugung s. 236, and 420–446, 474.

[2] Dritte Fortsetzung s. 53, 59.

[3] The only cases, carefully recorded of reciprocal crosses amongst varieties, which I know of, are given by Kölreuter in Mirabilis, Matthiola & Digitalis thapsi & purpurea. (Dritte Fortsetzung s. 126, 128: Nova Acta Petrop. 1795 p. 333. 1797. p. 393: Journal de Physique 1782 p. 285). I should not expect the rule of the similarity of reciprocal crosses to hold strictly with long cultivated plants (see Wiegmann, Bastardz. s. 10, 11 on mongrel cabbages) for Gaertner. (Bastardz. s. 223) says it is more particularly true in regard to hybrids from species growing in their wild & natural state.

[4] Bastardz. s. 277–94, 580.

monly belong to this same type.[1] But I must think, from what I have myself seen in mongrel cabbages & raddishes, & from some of Mr. Knights descriptions of mongrel apples & grapes[2] that with long cultivated plants the variation of the mongrels is so excessive that seedlings even from the same pod might be generally ranked in all three classes.

Gaertner has clearly shown that certain species possess a prepotent power over other species with which they are crossed (distinct from their fertilising power) of impressing their likeness on their hybrid offspring.[3]/119 v/To give one single instance from Gaertner of Nicotiana //[4] [*paniculata* and *vincaeflora* are crossed, the character of N. *paniculata* is almost completely lost in the hybrid; but if N. *quadrivalvis* be crossed with N. *vincaeflora*, this latter species, which was before so prepotent, now in its turn almost disappears under the power of N. quadrivalvis. It is remarkable that the prepotency of one species over another in transmission is quite independent, as shown by Gärtner, of the greater or less facility with which the one fertilises the other.]/ 119/This difference of prepotency has not been proved to exist in varieties, owing as I believe, to the fewness of the experiments tried; but it holds good, as we shall see, with the varieties & apparently even with the indi[vi]duals of animals, & I cannot doubt that it does also hold with plants; for this prepotency is closely connected with the power of one species reducing another/ 120/by successive crossings; & the power of one species reducing another in fewer or more generations depends not only on its specific difference, but on that of the variety or even on the indi[vi]duals used, & likewise on whether the species has been long cultivated.[5]/

120A/Some of the special cases of resemblance of hybrids to their parents are curious. If we put aside the species having a prepotent power of transmitting their likeness, then in complex crosses of two species, the appearance of the hybrid depends on the proportions in which the parent-species have been blended together. Thus in Dianthus barbato-barbatosuperbus, in D. barbato-superbobarbatus, & in D. barbatosuperbo-barbatus we have two species differently mixed, but in the same proportion, namely with three-fourth of D. barbatus in each & they closely resemble each other. (Gaertner Bastardz. s. 504) But the result is wholly different when a hybrid is fertilised by the

---

[1] Bastardzeugung s. 282, 578. This was, also, Mr. Knights opinion p. 39 Treatise on the Culture of the Apple & Pear.
[2] Philosophical Transactions 1799 p. 201.
[3] Bastardz. s. 290 256.
[4] [The addendum sheet is sheared off at this point, presumably for use in *Variation under Domestication* (II, 67), from which the continuity of text is supplied.]
[5] Gaertner Bastardzeug. s 458, 461, 465.

pollen of a third pure species (for instance Lobelia fulgenti-cardinalis fertilised by L. syphilitica), for in this case the triple hybrid always closely resembles (though having only half blood) its pure father, so closely that it might often pass as a mere variety of it: Kölreuter says he was almost as much astonished at one such case in Nicotiana, as if he had seen a cat born with the form of a lion. (I Fortsetz. s. 42. Gaertner Bastardz. s. 511. gives several cases. M. Regel quoted in Journal de la Soc. Imp. d'Horticulture vol 1855. p. 251. makes a similar remark on some hybrid Achimenes.) On the other hand, in the exactly reversed case, namely of a pure species fertilised by the pollen of a hybrid from two other species (for instance, Lobelia fulgens fertilised by L. cardinali-syphilitica), the triple hybrid does not take after its pure mother or after its hybrid father (Gaertner Bastardz s. 507)./

120/Gaertner[1] adduces, the fact that one species can be made by repeated crossings to reduce or absorb another, as an "unequivocal proof" that species have fixed limits. This seems to me singular reasoning, for/120 v/Gaertner assuredly would not have disputed that one variety might be reduced by another; and/120/ supposing that the case had been exactly the reverse, namely that it had been found impossible to reduce by crossing one species into another, might not this with much greater force have been advanced as a proof of the aboriginal & immutable difference of the two species? This argument was indeed used to me/120 v/by an acute observer, on my telling him of a case, where the effects of a single cross from the Malaya breed of Fowls was occasionally perceptible in a stock of poultry after an interval of 40 ⟨thirty⟩ years: he argued from the stain of the Malay blood being so permanent that it could be only due to its being an aboriginally distinct species. On the other hand M. M. Boitard & Corbié[2] have argued that because in crossing certain breeds of Pigeons, their characteristic features are lost even in the first generation & cannot be recovered without extreme difficulty, by crossing the mongrels repeatedly with pure birds, that such breeds are true species! So we see how this argument may be turned round & round to do any duty./

120/At one time I had thought it probable that if a variety produced by culture were crossed with an unaltered & distinct species, that the artificial variety would have less power/121/than the unaltered species of impressing its likeness on their mutual hybrid offspring. But for some few crosses made by Kölreuter[3] between the varieties of one tobacco with another species, & between several long-cultivated forms of Dianthus with wild species, this does not seem to be the case; & a hybrid from between

---

[1] Bastardz. s. 475.　　　　[2] Les Pigeons. 1824 p 198.
[3] Fortsetzung s. 29. Dritte Fort. 72, 79, 83, 87, 103: in the Zweite Fortset. s. 116, there is, however, one somewhat opposed case.

a variety of one species & a second unaltered species seems as often to come intermediate as from between two unaltered species. Gaertner lays great stress[1] on the fact that when two distinct, but closely allied species are crossed with a third species, the two sets of hybrids are very distinct from each other, even more distinct than the two closely allied pure species are from each other. On the other hand if two varieties of one species are crossed with a distinct species, he asserts that the two sets of hybrids differ very little from each other. In regard to this latter statement, no other facts are given, but two sets of crosses from between several varieties of two species of tobacco with/122/two other distinct species. Now Kölreuter[2] also crossed several varieties of tobacco with a distinct species & he expressly states that the hybrids differed as much from each other as might have been expected from the difference of the varieties, & more than this could hardly have been expected. This same result seems to have followed crosses of differently coloured varieties of Verbascum with distinct species & of two varieties of Digitalis purpurea with D. lutea[3]/

123/*Comparison of hybrid & Mongrel Animals, independently of their fertility.*—In comparing hybrids & mongrels together, & both with their two parents, we meet with great difficulties, besides those necessarily inherent in all such comparisons. In animals we have quite commonly secondary sexual differences, so that a hybrid or mongrel has to be compared more or less with both sexes of both parents;/123 v/& in hybrids, owing I presume to their sterility, the secondary male characters are developed late in life & apparently not fully at any period; for instance Mr. Hewitt informs me that he has never seen even in old hybrid Pheasants & fowls, full-sized spurs./123/In the next place, differently from in plants, the progeny from reciprocal crosses between two species, or two races, is generally unlike, & this greatly complicates the case. As with plants, one species or one race is prepotent over another in transmitting its likeness to its crossed offspring; but with animals the prepotency seems often to run in one sex, which probably accounts for/124/the very frequent dissimilarity of the offspring from reciprocal crosses. But besides this general pre-

---

[1] Bastardz. s. 273, 581.   [2] Fortsetzung s. 31. Zweite Forts. s. 56.
[3] Journal de Physique 1782. p. 291. Dritte Fortsetzung s 6, 35. Acta Acad. Petrop. 1781. p. 249, 257 [Darwin cancelled the next sentence, before completing it: '⟨Even Gaertner would not have disputed that a common or wild cabbage, if crossed for instance with a Kohl-Rabi & Brussel Sprouts (as I have seen) would yield remarkably different⟩' He later pencilled: 'Probably here Laburnum case.']

potency certain parts or characters, in certain species or even genera, appear to be more readily transmitted by one than the other sex, perhaps most frequently by the male sex; & they may be transmitted to both sexes, or only to one sex; & that sex may either be the same or different from the one which transmits the character.[1] Moreover some authors believe that the age & vigour of the male influence the character of the crossed offspring, but I have not met with any satisfactory evidence on this head. Altogether it is not easy to exaggerate the complexity of the subject, & I will in consequence make only a few remarks on some of the points of comparison between hybrids & mongrels.

Isidore Geoffroy Saint-Hilaire has stated[2] that hybrids from between two species generally present fixed & constant characters/ 125/ partly those of the father & partly those of the mother: on the contrary that mongrels are either intermediate like hybrids, or resemble entirely one of their parents. Two somewhat different considerations are here excluded, namely the resemblance of crossed offspring to their parents, & their homogeniety [sic] with respect to resemblance. I think there can be no doubt that hybrid animals, exactly as with plants, are either intermediate in structure, or take more after one parent in one part & the other parent in another part, or are altogether more like one parent than the other. But hybrid animals, in the first generation perhaps hardly ever so closely resemble either parent as do mongrels: Bechstein, however,[3] says that hybrids from the Canary & Fringilla spinus, always have both the colour & form of the latter. With respect to homogeneity, hybrid animals from between the same species as with plants not long cultivated, seem generally to be alike; but there are marked exceptions as in the offspring from a Dog & Wolf for instance those described by Wiegmann[4] two of which resembled the ordinary wolf hybrid, but a third took closely after the pointer: in a flock of hybrids from the common & Chinese goose, I saw some with/126/black & some with yellow beaks like one or the other parent; & the Rev W. D. Fox informs me that in some other hybrids which he had seen there was considerable diversity in the degree of resemblance to either parent goose. Hybrids from between the Canary & linnet are said to differ; & from between the common Pheasant & P. torquatus Temminck says some are

---

[1] Numerous facts confirming these propositions may be found in Dr. Prosper Lucas' work on L'Hérédité Naturelle. I shall have to give illustrations on several of them.

[2] Dict. Class. d'Hist. Nat. Tom x p 121.—1826; & subsequently in other publications, as Essais de Zoologie Generale 1841 p. 516.

[3] Stubenvögel 1840. s. 239.   [4] Bastardzeugung s. 21.

like one parent & some like the other, & some intermediate./126 v/ Of two hybrids from the Guinea & Common Fowl, examined by Morton, "one looked more like the fowl, the other had much stronger resemblance to the Guinea Fowl."[1]/126/But in all these cases one or both of the parents have been more or less domesticated./126 v/Hybrids, however, from the Carrion & Hooded crow resemble in their colour either parent or an intermediate.[2] Hybrids from the Capercailzie & Black-Game differ in size & colour from reciprocal crosses, but this is not sufficient to account for the amount of difference; for /126/out of twenty of these hybrids, and all of the male sex, not two are said to be quite alike.[3]/

127/So extremely few cases are on record of hybrid animals breeding inter se, that it is not known what rules the successive generations of hybrids would follow in their resemblance to their parents; but there is a high degree of probability that they would vary like the successive generations of hybrid plants, & take after either one or other parent-species. In the offspring from the hybrids from a hen Bantam & Gallus Sonneratii, which I saw, some resembled in an extraordinary degree the pure grandfather G. Sonneratii; & I was told that others had taken closely after their grandmother, the Bantam Hen.

Now let us turn to mongrels: the general rule seems to be that they are in some degree intermediate, between their two parents, & homogenious [sic][4] It is notorious that breeders, who have once crossed two breeds of cattle or sheep can gradually foretell what the character of the/128/offspring will be *in the first generation*, which shows that there is no such great variability. I have crossed in Pigeons Barbs & Fan-tails, Pouters of two sub-breeds & Fantails, & in this latter case reciprocally, & I have been surprised at the similarity of the mongrels, even in colour; as I have found likewise to be the case from several other crosses. It is chiefly in dogs, pigs, Fowls & other animals producing many at a birth, that such great diversity has been observed, even in the young produced at the same time; but I can hardly persuade myself that there is not

---

[1] For Guinea Fowl, see [Morton] American Journal of Science 2 Ser. vol 3. p. 204. For the Pheasants see [Temminck] Hist. Nat. Generale des Pigeons et des Gallinacees Tom 2. p 330. I give the linnet case on the authority of Dr P. Lucas Tom. i p. 211; but this diversity in hybrid canaries I do not believe [,] from enquiries which I have made [,] to be common.

[2] Naumann, as quoted in Bronn's Ges[ch]ichte der Natur. B 2. s. 172.

[3] L. Lloyd. Field Sports of the North of Europe vol. i. p. 285; on the authority of Mr. Falk: Latham, also, has in his Synopsis (Supplement) noticed the great variability of these hybrids.

[4] See Colins Traite de Phys. Comp. des Animaux Domestique. Tom 2. p. 356, who has well treated this subject.

some exaggeration, when the young have been said to *perfectly* resemble one of the parents. Mongrels, bred inter se, after the first generation, no doubt present the most extraordinary diversity & reversions to their pure grand-parents, as I have myself seen from the very uniform pigeon-mongrels when bred inter se. Occasionally, however, characters immediately become fixed in a mongrel breed;/128 v/Boitard & Corbie[1] who have had immense experience in crossing Pigeons, assert that from a Pouter & Runt //[2] ["a Cavalier will appear, which we have classed amongst pigeons of pure race, because it transmits all its qualities to its posterity."]/128/The Editor of the Poultry Chronicle[3] bred some blueish fowls/129/from a Black Spanish & Malay hen, & these remained true to colour "generation after generation"[4] By the aid of some selection several intermediate mongrel breeds of sheep, as the Oxford & Shropshire Downs have been firmly established, & amongst cattle a breed, before mentioned from Wilkinson between Long & Short-Horns.—

With respect to the rules of resemblance of hybrids & mongrels to their parents, it deserves notice that very many attempts have been made to give laws such as that the Father gives external characters, & the mother internal or vital organs &c &c None of these rules, if widely extended to all animals seem to hold good, as has been ably shown by Dr. P. Lucas & Gaertner[5] by merely contrasting the diversity of the rules given by different authors, who have had ample means to form an opinion. Similar rules have been enounced for plants, & have I think been conclusively shown by Gaertner not to be true. Prepotency of one species or race over another, has been generally confounded with the influence of sex. If we confine our view to the races of one species, or perhaps even to the species of the same group, some such rules may/130/hold good; for instance it seems in crossing reciprocally different breeds of poultry, that the cock very generally gives colour;[6] & in sheep that the Ram gives the character of the fleece & horns, & the Bull its horns or absence of horns. But what

---

[1] Les Pigeons. p. 37.

[2] [The addendum sheet is sheared off at this point. The continuation is supplied from *Variation under Domestication*, ii, p. 97.]   [3] Vol. i. 1854 p. 101.

[4] [Here Darwin noted a pencilled addendum on the verso: 'I crossed Penguin [ducks] & Black B[uenos] Ayres & the offspring kept not perfectly but very nearly true of a brown colour, a few darker again to blue, with center white mark on breast & even the bill.' Cf. *Variation* ii, pp. 97–8.]

Bastardzeugung s. 264–266. L'Heredité Nat. Tom. 2. B[ook] 2. Ch. i. I could add other rules to those given by these authors. In not a few cases examples have been given of crosses, without the manifest necessity of a reciprocal cross having been made.   [6] Cottage Gardener 1856 p. 101, 137 for Poultry.

concerns us, is that I have never observed that a different rule has been given for hybrids & mongrels/130 v/and I think we may safely follow Lucas[1] that the same wide & diverse rules of resemblance are common to the crossed offspring between species, varieties & individuals of the same race./

130/One case, however, seems to occur frequently with mongrels, almost in accordance with Is. Geoffroy's remark, and which as far as I am aware has not been noticed in hybrid animals from between species in a state of nature; or only in a very slight degree as in case of carrion & hooded crow; namely either the perfect transmission or entire absence of some marked character of one of the parents in the mongrel/131/but intermediate states also appear: so it is sometimes with the condition of the hair. The dwarfed & turnspit like structure of the Ancon sheep when crossed with others seems to have either not at all or almost perfectly transmitted their characters. Piebald animals, & such cases as the mongrel offspring from the Dorking & other fowls, having five toes on one foot & four on the other—the cross from the ⟨solid⟩ whole-hoofed & common pig, which with Sir R. Heron had two feet whole & two normally divided—are probably due to this same difficulty of fusion in certain characters.[2]/131 v/Black, white & other coloured varieties of several kinds of animals have been observed in a state of nature, far oftener than piebald individuals, which shows the same tendency for certain colours to appear, independently of crossing, fully developed or not at all./131/I strongly suspect that characters which refuse to blend have first appeared suddenly & perfectly developed: I do not believe that any structure slowly acquired through selection, whether artificial or natural, can be transmitted in this entire or quite negative manner/

132/*Prepotency.* As with plants, one species of animal seems to be prepotent over an other in impressing its likeness on its hybrid offspring. This according to Flourens[3] is the case with the Jackall over the Dog, & seemed to me to be so with one of these hybrids which I examined. I cannot doubt that this is strongly the case with the ass over the horse; the prepotency here running more

---

[1] L'Hérédité Naturelle Tom. 2. p 179–184.

[2] [Re solid-hoofed pigs, Darwin in *Variation*, ii, p. 92 cites letter from Heron to Yarrell. For extent, see Darwin MSS. c. 40 g. The original text, later cancelled, here continues: 'In certain animals in a state of nature, when there has been no crossing, as with squirrels & hamsters, wholly white or black young appear far oftener than piebald or intermediate tints.' Unknown except from *Variation*, ii, p. 92, note.]

[3] [Note missing. In *Variation*, ii, p. 67, Darwin cites: 'Flourens, "Longévité Humaine", p. 144, on crossed jackals.']

strongly through the male ass. The Pheasant preponderates over the fowl, in those hybrids which I have seen.[1] ⟨but it is most difficult to form any accurate judgment on this head.⟩ In races of our domestic animals, numerous instances can be given of prepotency of one over the other. Godine[2] has given a very curious case of a strongly characterised goat-like race of Cape Sheep; the Ram of which was crossed with 12 other breeds, & the offspring were so like the ram, that they could hardly be distinguished from it; the lambs, however of two ewes, of this breed, when crossed for six successive years with a Merino-ram, resembled most closely merinos.[3] Sturm states positively from repeated experiments that a Holland Bull //[4] [was much less prepotent than a Holland (i.e. Friesian) cow in crosses with Swiss cattle.]

133 & 134/Prepotency seems, also, to be characteristic of individuals, of either sex, of the same race; for we can understand in no other way the manner in which marked features are transmitted in certain families, after marriages with different females. So amongst our domestic animals, certain individuals have been notorious for transmitting their characteristic qualities.—/

134 Bis/I have met with several observers who have expressed a strong opinion that when an ancient or naturally formed breed or species is crossed with a modern domesticated one, that the characters of the former preponderate in a cross. That a character which has long been inherited should continue truer than one recently acquired is almost self evident; but when we cross one breed with another, the separate question of prepotency (as we have just seen in the case of Trumpeters) comes into play. Ancient breeds may be prepotent over modern breeds when crossed with them; & I have met with some facts which countenance this view; ⟨but many others seem strongly opposed to it.⟩/134 Bis v/This opinion may have partly arisen from a difficulty or impossibility of improving old established breeds in wild mountainous countries

---

[1] [Note missing. In corresponding passage in *Variation under Domestication* (II, 68), Darwin mentions Mr. Hewitt's descriptions of hybrids between pheasant & domestic fowl. Cf. Wm. Wingfield and G. W. Johnson, *The Poultry Book*....ed. W. B. Tegetmeier, London, 1856, pp. 165–7.]

[2] [Darwin's note slip for this reference is lost. Essentially the same statement is given in *Variation*, also without a source reference. There again the name is given as Godine in the text, but in the index the passage is listed under Godron and this seems more probable. Considerable search in the works of Dominique Alexandre Godron has not led me to the source of the information Darwin gives.]

[3] [Here Darwin marked the point of insertion for an addition to the text; this addendum is now missing, but see corresponding passage in *Variation*, II, 66.]

[4] [The text is sheared off at this point, and folio 133 is missing. From the clues in *Variation*, ch. 14, II, 66., it seems Darwin had in mind the passage in Sturm, *Ueber Racen*, p. 107.]

by crosses with improved artificial breeds; but the unfavourable conditions for tender animals here come into play. On the other hand, I have met with several facts which seem strongly opposed to the foregoing supposed rule. Thus,/134 Bis/the almost monstrous Ñata cattle, of S. America before alluded to, have arisen within the last three centuries, but yet are prepotent over other cattle: no breed is more modern or artificial than the Improved Short-horn, yet I observe in all accounts from the continent, that no breed is more potent in impressing its character on other/135/native breeds, & hence partly its great value for exportation. Drooping ears are no doubt due to domestication, yet in a hybrid from a Jackall & Terrier, which I saw in the Zoological Gardens, though the Jackall preponderated, yet the ears drooped; & this hybrid wagged its tail./135 v/Mr. Hewitt[1] describes hybrids from the Cock Pheasant & five differently coloured breeds of the common hen; & these hybrids differed greatly from each other, in colour, showing that the Cock pheasant had no marked degree of prepotency at least in colour over these several domesticated varieties./135/The Penguin Duck is an almost monstrous race, but in some mongrels from this bird & the common Duck, & in some remarkable hybrids with the Aegyptian goose in the Zoological Gardens (Tadorna Aegyptiaca), the upright & singular gait of this breed seemed to me equally to prevail.

It is notorious that both a species & race can be reduced or absorbed by repeated crossings with a distinct species or race. The number of generations required probably differs in different species, as we have seen to be the case with plants; & this probably accounts for the great diversity of opinion of breeders on this head; some saying that 12 or 13 or even 20 generations are required, others more commonly saying (as with plants) that 8 or 9 amply suffice. It is certain that the rate of reduction differs according as the male or female of the reducing race/136/be used;[2] & this naturally follows from prepotency running in animals more in one sex than the other. I know of no facts showing that one strongly marked race can reduce another either more quickly or more slowly than one species can another. The great grandchildren of the common pheasant were reduced to the perfect appearance of P. versicolor by three crosses; & so it has been with a mongrel Fantail-pigeon, reduced by a pure Fan-tail, but as Boitard &

---

[1] Poultry Book by Mr. Tegetmeier 1856 p. 124.

[2] Note used in Domestic Animals Chapter 15. Crossing [See *Variation* ii, 88, note 9: 'Sturm, "Ueber Racen, &c.," 1825, s. 107. Bronn, 'Geschichte der Natur,' b. ii. s. 170, gives a table of the proportions of blood after successive crosses. Dr. P. Lucas, "l'Hérédité Nat.," tom. ii. p. 308.']

Corbie have remarked three or four more generations would be required to make sure of the purity of the offspring.

Since Lord Morton's famous case of the Quagga & Arabian mare, it has been universally admitted that the subsequent offspring of a female mammal is affected in an incomprehensible manner by a first cross from a distinct species. And there is copious evidence that this is likewise the case between different races of animals & even different individuals of the same race.[1]

Finally it seems that the same rules hold generally good for crossed plants & animals whether distinct species, or varieties or merely individuals of the same race are crossed in regard to their resemblance to their parents, their variability/137/prepotency & reduction. ⟨The same rules seem to hold good for hybrids & mongrels & for the offspring of individuals of the same race when they present recognizable differences.⟩ The chief difference between plants & animals seems to be that prepotency, or an extraordinary power of impressing resemblance & consequently of reduction, commonly differs in animals to a large extent in the two sexes of the same race or species.

The chief difference between mongrels & hybrids, whether in plants or animals, seems to be that in the first generations hybrids are generally uniform in character, but this is not universally the case; nor are all mongrels very variable./137 v/Domestic races of animals often have characters which have originally appeared in a sudden & monstrous manner, & these I suspect are frequently transmitted either perfectly or not at all to their mongrel offspring; & this seems rarely to be the case with hybrids from between two species, neither of which have been modified by domestication./ 137/In the succeeding generations, mongrels probably are more variable than hybrids in the corresponding generation but this does not seem to me, considering their origin, to be at all surprising; occasionally, though very rarely, both hybrids & mongrels keep true in their successive generations. ⟨Lastly when races are crossed, those characters, which, as I suspect have originally appeared suddenly, are much oftener transmitted either perfectly or not at all to their crossed offspring than in the case of crossed species, of which the characters have not been formed in this sudden and monstrous manner.⟩/

138/*Summary on Chapter*.—Weighing all the evidence given in this chapter, I think we must conclude that the first cross between the forms, called by naturalists species, & their hybrid offspring

---

[1] Used in Dom. Animals [See *Variation*, I, 403, ch. XI, n. 137 (note 151 in 2nd ed.): Lord Morton, '"Philos. Transact.," 1821, p. 20.']

are with rare exceptions sterile in some degree. But when closely related forms are tried, the sterility so graduates away, that the two best observers, who ever lived differ diametrically whether or not they are perfectly fertile together. The attempt to measure fertility by so nice a process as counting the number of seeds is seriously interfered with [by] the ill-effects of manipulation & the seclusion of the specimen & culture. We probably see the importance of the latter, in the difference of the results obtained by Gaertner & Herbert when experimentising on the same two species. Gaertner's failure to obtain full fertility between many forms, ranked by all the best botanists as varieties must shake our confidence in his conclusion that species are universally in some degree sterile together. The increasing sterility of hybrids when naturally self fertilised for successive generations may I think be safely attributed in large part to the ill-effects of close inter-breeding; for it seems otherwise impossible to understand, how artificial fertilisation, in itself injurious, should aid, as Gaertner asserts it does, their fertility./

139/But it is, I think, impossible to admit that species when first crossed & their offspring are invariably sterile together even in the slightest degree, after Herberts repeated observations on, for instance Crinum, in which he found that every crossed ovule produced a seed, which never happened with natural fertilisation. Nor should we pass over the apparently perfect fertility of several florists flowers, as in the genus Rhododendron, which have of late years been crossed in so complicated a manner: in these cases alone have the experiment been quite fairly tried, for here there has been excellent culture, no manipulation, & natural intercrossing allowed in whole beds of the same hybrid kind.

With animals, though first crosses can in some cases be so easily effected, yet it cannot be said that the perfect fertility of the successive generations of any one hybrid has been fully established; though we have no reason to doubt it between certain pheasants. But how few experiments have been tried between closely allied species, both of which will breed perfectly under confinement. Nor must the ill effects of close interbreeding be overlooked in those few cases in which hybrids all descended from the same two parents have been bred for some generations inter se./

140/When we consider that the Fertility both of first crosses & of hybrids graduates in different cases from zero to the normal degree of perfection; that it is in all such cases eminently susceptible of favourable & unfavourable conditions; that it fails easier on the male than female side; that the degree, in the same hybrid in the first & successive generations, is innately highly variable; that the degree does not run closely parallel with the amount of systematic difference between the two parent-species, even within the limits of the same genus; that it often differs widely in reciprocal crosses from between the same two species; that there is no absolute relation between the facility of getting a first cross & the degree of fertility of the hybrid offspring; that there is no close relation between the likeness, whether abnormal & occasional, or normal & regular, of the hybrid to one of its parent-species; when we consider all these & other such singular facts, I cannot believe that the lessened fertility of first crosses & of hybrids has been a specially endowed quality to prevent those forms which coexist in the same country from becoming blended together. The complexity/141/& singularity of the rules seem to me to indicate that they are incidental on differences in the organs & functions of Reproduction in different species, in some degree analogous to the differences in the organs & functions of vegetation, on which the capacity for grafting depends, & which, I presume, no one would suppose were specially endowed to facilitate or prevent one tree being grafted on another. No doubt, differences in different species in the organs & functions of Reproduction & of Vegetation will follow pretty closely as a general rule, systematic affinity, which means the sum of all resemblances of all kinds, & not of any one particular organ or function.

In first crosses, the sterility must depend on different causes: in plants widely different the pollen-tube does not penetrate the stigmatic tissue; in more closely related plants, though reaching the ovule it does not cause an embryo to germinate: in other cases, a large part of the sterility depends on the early death of the fertilised embryo: in these cases we can no more offer any explanation than we can why some trees belonging to the same genus cannot be grafted on each other./

142/In hybrids, the cause of sterility is widely different from that of the first cross between two pure species; for in hybrids the male & female sexual products are manifestly deteriorated, whereas in the parent species they were of course perfect. When we bear in mind the numerous facts given in our third Chapter, showing how eminently susceptible the sexual functions are to any change in the conditions of life, I think we need feel little surprised that hybrids with their double constitution & laws of growth confounded together should have their fertility affected in a somewhat analogous manner, whatever may be the conditions of life to which they may be exposed. Nor is it, I think, any more strange, that the sterility of hybrids & of first crosses though depending on very different causes, should run in some degree parallel, than that both should run in some, but far less close, degree parallel with the capacity of grafting; for all these depend though in different ways, on the amount of/143/resemblance & dissemblance in the species experimentised on.

The most surprising circumstance in our whole present subject is the almost universal fertility of the most distinct varieties when crossed, as in the case of the several breeds of fowls & pigeons. But in plants we have as good evidence, as we can ever get on the slighter degrees of sterility between closely related species; that varieties are in some few cases slightly infertile together. In the case of animals, it may be as Pallas hypothetically concluded, that domestication eliminates the tendency to sterility in crosses; if this be so, we could not expect that sterility should appear between the most distinct varieties if produced under domestication. We have seen that slight changes in the conditions of life are favourable to fertility, though greater changes or changes of another kind affect in so decided a manner the reproductive functions; hence it is not/144/surprising that crosses within certain limits should be favourable to fertility, though beyond such limits they should cause sterility. But it is surprising & could never have been a priori anticipated, that crosses between, for instance, such extremely distinct breeds, as those of the pigeons, should not have been in the least degree unfavourable to their fertility: yet we should bear in mind, that man by his selection,—the great agency in the production of domestic breeds—has no power or wish to modify either directly, or indirectly by selecting constitutional differences in the reproductive system; & it is on differences in the reproductive system, on which the sterility of species when crossed, seems incidentally to depend.

Finally in all other respects, besides fertility, the offspring of species & varieties seem to follow, often absolutely & always very closely, the same laws namely in their resemblance to their parents, their variability, equal or unequal reciprocity prepotency, reduction &c:—but it is needless to sum up the conclusions, just arrived at, on these heads. I will conclude by remarking, that/145/if the difference in fertility between species & varieties when crossed, has been rightly urged as so very important a distinction; then on the other hand, so close a resemblance in their progeny in all other respects, ought to weigh with us as an argument of not a little weight that Species & Varieties have not had an essentially different origin.

CHAPTER X

MENTAL POWERS
AND INSTINCTS OF ANIMALS

INTRODUCTION

Having completed chapter IX at the end of December 1857, Darwin wrote Asa Gray, the Harvard botanist, on the fourth of the following April that: 'I have just finished a chapter on instinct, and here I found grappling with such a subject as bees' cells, and comparing all my notes made during twenty years, took up a despairing length of time.'[1] The first entry for 1858 in his Pocket Diary corroborates this: 'March 9th Finished Instinct Chapter.'

In comparison with the manuscripts of some of the other chapters, the one for this chapter shows relatively few signs of later use by Darwin. On folio 13 he scrawled 'Used Man Book' alongside his mention of Lonsdale's anecdote about snails, 'an example of personal attachment in *Helix pomatia*', which he published in 1871 on page 325 of the first volume of the first edition of *The Descent of Man*. The only other direct evidence in the manuscript of Darwin's later use is simply that folios 11 and 12 were cut up evidently also for use in *The Descent of Man* (I, pp. 30–1; II, pp. 334–5).

In the portfolio of notes marked Instinct, as for those used in earlier chapters, Darwin continued to add dated notes marked 'Ch x' for some time after the first publication of the *Origin*. One such note is dated 'Jan 13, 61'.

Much of the text of this chapter was published with Darwin's approval by his younger friend George J. Romanes in two of his books which came out shortly after Darwin's death. One or two minor mysteries about these publications seem best approached by a review of their background in the warm friendship which grew up between Darwin and his devoted admirer. A common interest in animal instincts was to form one of the major bonds in this relationship. Before the end of 1874 Darwin had invited Romanes to call on him, and the beginning scientist, who had only passed his final examinations at Cambridge in 1870, the same year as Darwin's son Francis, was always to remember that visit. 'Mr. Darwin met him, as he often used to tell, with outstretched hands, a bright smile, and a "How glad I am that you are so young!"'[2] By 1878 Darwin was happy to offer some of his own notes and manuscript for Romanes' use in preparing his forthcoming evening lecture on Animal Intelligence at the Dublin meeting of the British Association for the Advancement of Science in August of 1878.

---

[1] L & L, II, 155; NY, I, 510. Here in a 'Note to the Fifth Thousand' Francis Darwin wrote: 'This letter should be omitted, the date [he had given 1859] being certainly incorrect.' He did omit the letter in the 'seventh thousand revised' of 1888. The correct year must be 1858, for this fits Darwin's statement that he had 'just finished a chapter on instinct.' An April letter would be unlikely to thus refer to the completion of the instinct chapter of the *Origin*, which was finished on November 13, 1858, according to the Pocket Diary.

[2] Ethel Duncan Romanes, *The Life and Letters of George John Romanes*, written and edited by his wife. (London, 1896), p. 14.

In a letter of June 16 Darwin wrote him:

Do just what you like in both cases.—The notes on insects were made about 40 years ago,—and I have just recollected that I have used them in drawing up a long chapter on Instinct, written 4 or 5 [sic] years before the "Origin" was published. I send the two pages out of this chapter which please return—I wish it had occurred to me to offer you this chapter of 110 pages to read, for in skimming over parts of it I find abundant references to many curious facts. It is I presume now quite too late to be of any use to you.[1]

On June 18 Romanes replied:

Very many thanks for your permission to use your observations, as well as for the additional information which you have supplied. If all the manuscript chapter on instinct is of the same quality as the enclosed portion, it must be very valuable. Time will prevent me from treating very fully of instinct in my lecture, but when I come to write the book for the International Science Series on Comparative Psychology, I shall try to say all that I can on instinct. Your letter, therefore, induces me to say that I hope your notes will be published somewhere before my book comes out (*i.e.* within a year or so), or, if you have no intention of publishing the notes, that you would, as you say, let me read the manuscript, as the references, &c., would be much more important for the purposes of the book than for those of the lecture. But, of course, I should not ask to publish your work in my book, unless you have no intention of publishing it yourself. I do not know why you have kept it so long unpublished, and your having offered me the manuscript for preparing my lecture makes me think that you might not object to lending it me for preparing my book. But please understand that I only think this on the supposition that, from its unsuitable length, isolated character, or other reason, you do not see your way to publishing the chapter yourself.[3]

On June 19 Darwin wrote further:

You are quite welcome to have my longer chapter on instinct. It was abstracted for the Origin. I have never had time to work it up in a state fit for publication, and it is so much more interesting to observe than to write. It is very unlikely that I should ever find time to prepare my several long chapters for publication, as the material collected since the publication of the Origin has been

[1] C.D. MSS., box 147, copy no. 39.   [2] Romanes, *Life and Letters*, pp. 71–2.

so enormous. But I have sometimes thought that when incapacitated for observing, I would look over my manuscripts, and see whether any deserved publication. You are, therefore, heartily welcome to use it, and should you desire to do so at any time, inform me and it shall be sent.[1]

On December 14, 1880, having finished the writing for some of his prior commitments, Romanes wrote to Darwin from the Linnean Society rooms in London: 'I have begun to come here (Burlington House) to read up systematically all the literature I can find on animal intelligence.'[2] The following April, Romanes reported to Darwin:

I have at length decided on the arrangement of my material for the books on Animal Intelligence and Mental Evolution. I shall reserve all the heavier parts of theoretical discussion for the second book—making the first the chief repository of facts, with only a slender network of theory to bind them into mutual relation, and save the book as much as possible from the danger that you suggested of being too much matter-of-fact. It will be an advantage to have the facts in a form to admit of brief reference when discussing the heavier philosophy in the second book, which will be the more important, though the less popular, of the two.[3]

Finally Mrs. Romanes tells us that on the last week end stay at Down in January 1881, 'Mr. Darwin was most particularly kind, and gave Mr. Romanes some of his own MSS., including a paper on "Instinct", which is bound up with Mr. Romanes' own book, "Mental Evolution in Animals."'[4]

In mid December 1883 a significant part of chapter x, constituting 47 out of the 116 folios of the manuscript, was published at the end of this book.[5] Starting with the section on migration on folio 50, Romanes published all the following folios except for 80 to 97 which he omitted, explaining in a note on page 373: 'Here follows a section on the Instincts of Parasitism, Slave-making and Cell-making, which is published in the *Origin of Species*.' This note probably provides a good part of the answer to the seemingly obvious question, why did not Romanes simply publish the whole chapter as it is in the manuscript? Perhaps he just decided to omit the parts of the chapter which he felt were already common knowledge.

In the preface to his book, Romanes explained how Darwin had made all his notes, clippings, and manuscript on instinct available to him and had asked him to publish any parts he chose. He then went on to make several slightly mystifying statements. First, he said the parts of the chapter included in his appendix were 'as much of this material as could be published in a consecutive form' and in addition he referred to 'numerous disjointed paragraphs' which he had 'woven into the text of this book', rather than presenting them as 'a string of disconnected passages'. This can only describe

[1] Romanes, *Life and Letters*, p. 72.   [2] Romanes, *Life and Letters*, p. 104.
[3] Romanes, *Life and Letters*, pp. 116–17.   [4] Romanes, *Life and Letters*, p. 128.
[5] See *Nature*, vol. 29, number 737 (Dec. 13, 1883), p. lv, for publisher's advertisement saying: 'Now Ready...Mental Evolution in Animals, By George J. Romanes,... With a Posthumous Essay on Instinct by Charles Darwin.'

Darwin's text after Romanes had rejected parts at the beginning of the chapter rather than the manuscript in its present complete form. The passages Romanes wove into the text of *Mental Evolution* come from folios 7 v, 8, 18–20, 24–28, 31–35, 36 v, 38, 43–45, and 48–49 of the manuscript. In Romanes' earlier book, *Animal Intelligence* (London, 1882), he also quotes from the manuscript.[1] In addition to these quotations from Darwin's text, he also printed a number of Darwin's footnotes.

The second mystifying statement in Romanes' preface to *Mental Evolution* is 'I therefore published at the Linnean Society...the chapter which...I have added as an appendix.' Although he read those sections of chapter x at a well publicised meeting of the Linnean Society on December 6, 1883,[2] the society did not publish the printed text. For some time I searched vainly for it, and I was puzzled by the statement in the preface until I remembered the first law of history, the primacy of sequence in time; then I noted the dates involved. Romanes dated his preface 'November, 1868' while the reading of the selections from chapter ten to the Linnean Society came only on December 6. Actually the Darwin text was set up in type for the *Journal of the Linnean Society*, Zoology. The Library of Christ's College, Cambridge, has a pamphlet marked 'Private uncorrected proof' showing it as pages 347 to 378 of number 102 in volume 17. This number was only to be dated as of February 29, 1884. The decision to omit the Darwin text from that number and to place other material on those pages may have been made because the appearance of Romanes' book had already made the text readily available before the number could be published.[3]

MENTAL POWERS
AND THE INSTINCTS OF ANIMALS

1/Our present subject might have been discussed under several heads in the previous chapters, but I have thought it best for simplicity sake to keep it separate. Here we shall consider whether the more remarkable instincts are so manifestly inexplicable on any view, except that of the separate creation of each species, that they suffice by themselves to prove my whole theory false. It is only under this point of view that I venture to approach the subject; for undoubtedly it is most natural to believe that the transcendant perfection & complexity of many instincts can be accounted for only by the direct interposition of the Creator: thus the comb modelled by the instinctive powers of the hive-bee has practically solved a most recondite problem in geometry, in even

[1] See p. 25, which quotes from fol. 13, and p. 26 which quotes from fol. 12.
[2] See *Nature*, vol. 29, number 735 (Nov. 29, 1883), p. xxxvii, 'Diary of Societies,' and p. 110, 'Notes,' and number 736 (Dec. 6, 1883), p. xlv, and 'The late Mr. Darwin on Instinct', pp. 128–9.
[3] Further details about Darwin's loan of his notes and manuscript might well be found in the 87 letters from Darwin to Romanes in the American Philosophical Library collection, see *Amer. Phil. Soc., Proc.* 98 (1954), 449.

perhaps a more perfect manner than has the structure of the eye optical problems./

2/I hope that it is hardly necessary for me to premise that here we are no more concerned with the first origin of the senses & the various faculties of the mind, than we are with the first origin of life: we have only to consider the various modifications of the mental powers & instincts of the several species within the same great classes.

My belief is, that, like corporeal structures, the mental faculties & instincts of animals in a state of nature sometimes vary slightly; & that such slight modifications are often inherited. Furthermore I cannot doubt, that an action performed many times during the life of an individual & thus rendered habitual, tends to become hereditary; but I look at this fact as of quite subordinate importance. It will not be disputed that instincts are as important to the welfare of an animal, as its corporeal structure, indeed they are generally correlated. Consequently I believe, under the slowly/3/ changing conditions of nature, that occasionally some slight modifications of instinct could not fail to be profitable to individual animals; & that such individuals would have a better chance, in the great battle of life, of surviving & of leaving offspring with the same inherited slight modifications of instinct. By this process of the gradual addition, through natural selection, of each profitable modification of instinct to instinct, I believe that the most complex & perfect instinctive actions, wondrous though they be, have been slowly acquired & perfected.

Authors have not agreed on a definition of Instinct: nor is this at all surprising, as nearly every passion, & the most complex dispositions, as courage, timidity, suspicion &c are often said to be instinctive; & when directed towards a particular object are always thus called. So again all/4/natural tastes & appetites are called instinctive; as we see in Galen's well-known experiment of a kid cut out of its mother's womb at once preferring milk to the other fluids placed before it. Reflex actions, or those excited independently of the will by certain nerves being stimulated, have been called by some authors instinctive; as in the case of an infant a few hours old sneezing: this is a good instance as a child cannot until several years old voluntarily coordinate nearly the same muscles, in blowing its nose, as I have heard Sir H. Holland remark. The will indeed seems actually antagonistic to the act of sneezing, for a set of men, if for a wager wishing to sneeze cannot sneeze, as I have seen though all taking snuff to which they were unaccustomed. On the other hand, the will can aid &

modify the reflex action of breathing. The act of sucking in a young animal which has so often been advanced as an example of an instinct ought, perhaps to be called a reflex action; for a puppy with all the intellectual part/5/of the brain removed, the medula oblongata alone being left, sucked a finger moistened with milk & placed between its lips.[1]

Another class of reflex actions (the sensori-motor of Carpenter) can be excited through the mind, as well as by the stimulus of certain nerves, as in vomiting from the idea of some disgusting object. Again some reflex actions can hardly or not at all be distinguished from habitual movements acquired during life: thus I have seen a carefully nurtured infant, between 8 & 9 weeks old, wink at suddenly noticing an object and at hearing a noise; & I presume there can be no doubt that this is an reflex or instinctive action, unconsciously excited through the mind, to protect the eyes. Not one person out of a dozen can by an effort of will/6/prevent ⟨himself from⟩ guarding his stomach from a pretended blow; or prevent extending his arms when falling on a feather-bed; so that those actions have the appearance of being reflex or instinctive like the winking of an infant. But as infants or even young children have no tendency to extend their arms or guard their stomachs under the above circumstances, I presume these movements in the old are due to habit.

I have made these few & imperfect remarks to show how complex the subject is. Notwithstanding the impossibility of defining an Instinct; certain actions have unanimously been called instinctive. The performance of any action without the aid of experience, or instruction or sufficient reasoning powers, which actions it might have been thought would have required such aids; its performance by all the individuals of the species at all known times in an almost unvarying manner; & ignorance of the end for which the action is performed,/7/have generally been looked at as the chief characteristics of Instincts.[2]

[1] Grainger, quoted by Carpenter. Comp. Physiology 4 Edit. p. 690.
[2] Kirby & Spence Introduction to Entomology seems to me to contain the best discussion on instincts ever published.—See, also, Lord Brougham's Dissertation on Science connected with Natural Theology 1839. vol. I—Professor Alison has published an admirable resumé on the subject, under Instinct, in Todd's Cyclopaedia of Anatomy & Physiology. See, also, Müller's Physiology (Eng Translat.) vol 2. p. 928–950 for excellent remarks on this subject. Müller says "All those acts of animals are instinctive which, though performed voluntarily, do not nevertheless primarily depend on the mere will of the animal, which have an object according with the wants of the organism but unknown to the animal, & of which the hidden cause, acting in accordance with the design of the system, incite the animal to the necessary acts by presenting to its sensorium the "theme" of the voluntary movements to be executed in detail by the influence of the will." [p. 946.]

But none of these characteristics can be considered as quite absolute. Thus it cannot be doubted that reason sometimes comes into play in the performance of instinctive actions. Huber after his immense experience says that nature has certainly given to insects "une petite dose de jugement". I will give only two or three instances: a very irregular piece of comb, when placed on a smooth table tottered much, so that the Humble-bee could not work well on it: to prevent this, two or three bees held the comb by fixing their front feet on the table & their hind feet on the comb, & this they continued to do, relieving guard, for three days, until they built/8/supporting pillars of wax: now such an accident as this could hardly have occurred in nature. Some other humble-bees shut up, where they could not get moss with which to cover their nests, tore threads from a piece of cloth, & "carded them with their feet into a felted mass", which they used for moss. A slip of glass having been placed by Huber in a Hive in front of a comb, the bees before actually coming in contact with the glass, began building the comb at right angles to its former plane, so as to avoid the smooth & hard surface which could never have occurred to them in nature.[1] Again in one of Huber's glass-hives, one of the combs slipped down, & was then fixed by buttresses & pillars of wax to the combs on each side:[2] this is not very surprising as Bees often strengthen the edges of their combs by fixing pillars to the walls; but as it was winter—, when the Bees do no work of the Kind, it is a marvellous fact that the Bees clearly seemed to take warning from the accident, & consequently strengthened all the other combs which seemed to Huber to be quite firm.[3]/

8 v/Sir B. Brodie[4] gives on the authority of a friend the following case. Hive-bees invariably build from the top of the Hive downwards. But "on one occasion, when a large portion of the honey-comb had been broken off they pursued another course. The fragment had somehow become fixed in the middle of the hive, & the bees immediately began to erect a new structure of comb on the floor, so placed as to form a pillar supporting the fragment, & preventing its further descent. They then filled up the space above, joining the comb which had become detached to that from which it had been separated, & they concluded their labours by removing the newly constructed comb below; thus proving that they had intended it to answer a merely temporary purpose."/

[1] [François Huber, *Nouvelles observations sur les abeilles*, 2nd ed., vol. 2 (Paris, 1814), p. 218.]
[2] [F. Huber, *Abeilles*, vol. 2, pp. 286–8.]
[3] For these several cases, see Kirby & Spence Introduction to Entomology. vol. 1. p. 382. vol. 2. 477, 495, 487.
[4] Psyc[h]ological Inquiries 1854. p. 188. [Darwin presented all the rest of this paragraph as a footnote.]

9/Kirby & Spence, like all other good observers, admit that instincts are occassionally in some slight degree modified by intelligence; yet they doubt whether reason has been the modifying agent in some of the foregoing & other such cases. They argue that these modifications of instinct have always been limited in degree & uniform in kind; but this perhaps is rather begging the question. We must not forget that as Bees have no written or traditionary knowledge, their power of acting intelligently under new circumstances must depend wholly on their innate degree of intelligence; which no doubt would remain for enormous periods uniform. By the same line of argument we might almost prove that the canoe & weapons of the Fuegians, which have remained the same for nearly three centuries were not the product of reason but of instinct. These authors[1] urge that if Bees acted by reason, why do they not copy the Martin & sometimes use mud or mortar instead of a precious wax or propolis; "show us but one instance of their having substituted mud for propolis. . . . ., & there could be no doubt of their having been here guided by reason." And they have answered to this appeal; for Andrew Knight saw his Bees repeatedly removing a cement of wax & turpentine, with which he had covered barked trees, & using it as propolis.[2]/10/ Nevertheless in nearly all variations of instinctive actions through reason, instinct continues to play by far the more important part; thus when the hive-bees built their comb at right angles to avoid the slip of glass; they made their cells on the outside of the bend three or four times as wide as/10 add./those on the inside; and as the cells on both sides had their bases in common, each cell on the outside of the bend had to be made wider & wider towards its mouth, whilst those on the inner side had to be made in the same proportion narrower & narrower; & this had to be done by a multitude of workmen;—a marvellous piece of architecture; quite transcending the powers of reason.[3]/

10/I do not doubt that we often underrate the intellect of the lower animals especially those insignificant in size; therefore it is well to remember, what the most capable judge von Baer, has said, namely "that the Bee is in fact more highly organized than a fish, although upon another type".[4] Look at the power of communicating intelligence amongst ants;/11/when from two adjoining nests of the same species, countless hosts join in deadly strife,

[1] Introduction to Entomology, vol. 2. p. 497.
[2] Philosophical Transactions 1807. p. 242.
[3] Kirby & Spence Introduction vol. 2. p. 495.
[4] Philosoph. Fragments, translated by Huxley in Scientific Memoirs: Taylor May 1853. p. 196.

each ant knows all its own comrades![1] We are not much surprised when we hear from Rengger[2] that the wild monkeys of Paraguay gather oranges, which is not a native fruit, & have learnt to beat them against the branches so as to crack the rind, & thus to peel them. Nor are we surprised at a dog, which could not leap a gate with a ham in its mouth, pushing the ham under the lower bar, then leaping the gate & carrying it onwards.[3] But we are surprised when we hear from an excellent[4]//[source that: "The cobra is rather a sluggish snake; ...it feeds principally on toads, which it captures in holes. I once watched one which had thrust its head through a narrow aperture and swallowed one. With this encumbrance he could not withdraw himself: finding this, he reluctantly disgorged the precious morsel, which began to move off; this was too much for snake-philosophy to bear, and the toad was again seized, and again, after violent efforts to escape, was the snake compelled to part with it. This time however a lesson had been learnt, and the toad was seized by one leg, withdrawn, and then swallowed in triumph."]

12/Let us now glance lower in the scale of life. An esteemed naturalist[5] whilst watching a shore-crab (Gelasimus) making a burrow threw some shells towards//["its hole, in order to see whether it would bring it up again or not; of the four that were thus thrown, one only entered the hole, the others remaining within a few inches of it. It was about five minutes before the animal again made its appearance, bringing with it the shell which had gone down, and carrying it to the distance of about a foot from its burrow, it there deposited it. Seeing the others lying near the mouth of the hole, it immediately carried them, one by one, to the place where the first had been laid down, and then returned to its former labour of carrying up sand. It was impossible not to conclude that the actions of this little creature, which holds so low a station in the chain of beings, were the result of reason, rather than of blind instinct..."]12/How unintellectual does a snail appear, but hear Mr. W. White,[6] who fixed a land-shell mouth uppermost in a chink of rock, in a short time the animal protruded itself to its utmost length, & attaching its foot vertically

---

[1] Kirby & Spence Introduct. vol. 2. p. 74, 525. [See appendix for further comment of Darwin's on this point.]

[2] Saugethiere von Paraguay. 1830. p. 39.

[3] [Here Darwin added in pencil: 'Yarrell's case of Gull'.]

[4] Layard on Cobra. Annals of Nat. Hist. ix. 1852. p. 333. [The rest of the sheet is sheared off at this point. The interpolated quotation is from Layard, loc. cit.]

[5] G. Gardner's Travels in the Interior of Brazil 1846, p. 111. [The text is sheared off after the next line and the interpolation is quoted from Gardner, loc. cit.]

[6] A Londoner's Walk to the Lands End. 1836, p. 55.

above tried to pull the shell into a straight line; then resting for a few minutes, it stretched out its body on the right side & pulled its utmost but failed; resting again, it protruded its foot on the left side/13/pulled with its full force & freed the shell. This exertion of force in three directions, which seems so geometrically reasoned, might have been instinctive. Mr. Lonsdale, the geologist, kept two snails (Helix pomatia) in a small garden ill provided with vegetables; one of the snails was weak; in a short time the sound one disappeared, & was traced by its slime across a wall into another well stocked garden: after an absence of 24 hours it returned to its sick companion & both started together by the same track & disappeared.[1] This looks like a power of communication, affection & even reason in a snail. Even the headless oyster seems to profit by experience, for Dicquemare[2] asserts that oysters taken from a depth never uncovered by the sea, open their shells, lose the water within & perish; but oysters taken from the same depth & place, if kept in reservoirs, where they are occasionally left uncovered for a short time & are otherwise incommoded, learn to keep their shells shut, & thus live for a much longer time, when taken out of water./

13 v/Mr. W. Kidd who had had such immense experience[3] states his conviction that the varied dispositions in the Canary strongly tend to be hereditary. He, also, like Bechstein insists on the diversity of disposition in nestling Larks taken from the nest. Humboldt[4] says that the Indians who catch monkeys to sell "know very well that they can easily succeed in taming those which inhabit certain islands; while monkeys of the same species, caught on the neighbouring continent die of terror or rage when they find themselves in the power of man". So with the crocodiles. These diverse dispositions seem here to run in families./

14/Lord Brougham has insisted that ignorance of the final end is highly characteristic of all instincts; & so it undoubtedly is in the vast majority of cases. No one supposes that the White Cabbage Butterfly knows why it deposits its eggs on the cabbage. We see how blind an impulse instinct is, in such cases, as caged Birds, which are not mated & are not going to build, yet having a taste for carrying bits of sticks in their beaks;[5] /14 v/in Thrushes reared from the nest in a room, where they could never have seen another

---

[1] [Darwin bracketed this sentence and in the MS. margin pencilled: 'Used Man Book.']
[2] Journal de Physique vol 28. p. 244.
[3] Gardeners Chronicle 1851, p. 181
[4] Personal Narrative vol. 3. p. 383
[5] Rev. L. Jenyns. Observations in Nat. History 1846. p. 162

thrush, amusing themselves with hammering a silver thimble against any hard substance, in exactly the same manner as these birds do snail-shells—1/14/in Beavers,[2] when kept in a place without water accumulating pieces of wood: in Squirrels when having no materials to cover up spare nuts, yet quickly patting them when placed on a bare table in exactly the same manner as they do, when covered up with moss & straw.[3] But in all cases in which intelligence comes into play, the animal must to a certain extent know what it wants to do. When the Humble-bee carded the threads of the cloth, it must have known that it wanted moss. The/15/caterpillar of the cabbage Butterfly, before changing into a chrysalis, covers a small space with a web of silk, to which the suspensor girth of the chrysalis can be firmly attached; but Kirby[4] found that when this metamorphosis was effected in a box covered by a muslin lid, the caterpillar perceived that the preparatory web was useless & did not make one, but fixed the girth to the muslin. The Tailor Bird weaves threads of cotton, with which to sew up the edge of a leaf to form its wonderful nest; but it has been seen.[5] to pick up & use pieces of artificially made thread, which shows that it before hand knows for what purpose it spins the cotton; though it cannot know that it makes its suspended nest that its eggs may be hatched, & its young reared safe from snakes & other enemies.

Perhaps the most striking character of an instinct is that the young perform the action without any experience, as perfectly as do the old: thus Reamur & Swammerdam[6] positively assert that/16/a young Bee, as soon as its wings are dry, will collect honey & fabricate a cell as adroitly as the most hoary inhabitant of the Hive. Young cuckoos migrate two months after their parents have started & they must be able to perform their first journey safely. Innumerable insects can never have seen their parents & yet they perform instinctive actions perfectly though only once in their lives; for instance there is an Ichneumon, which deposits its eggs within the body of a larva hidden between the scales of a fir-cone, which it can never have seen.[7] The manner in which a Hawk makes a swoop & seizes its prey on the wing, must be

[1] Penny Magazine vol 3. 1834. p 12
[2] Flourens sur L'Instinct des Animaux 1845. p 110.
[3] E. Blyth. in Charlesworth Mag. of Nat. Hist. vol. 1. 1837. p 7: Archbishop Whately narrated to me an analogous case of a Fox, chained up in a bare paved court-yard pretending to cover up his superfluous food, & being then content.
[4] Introduct to Ent. vol 2. p. 476
[5] [T. Hutton] Journal Asiatic Soc. of Bengal vol II. p. 502 [504].
[6] Kirby & Spence Introduction vol. 2. p. 470
[7] Kirby & Spence Introduct to Entomology vol. 1. p. 357

considered as instinctive; but Dureau de la Malle[1] witnessed the curious manner in which the young birds were trained by their parents first dropping dead birds & then letting live ones escape. It is a surprising instinct which leads the ferret to bite the back part of the head of the rat where the medulla oblongata lies, & where death can be easiest inflicted; but Professor Buchanan[2] states that young ferrets "instead of having for their single object to put themselves into the proper position to inflict the death wound, engaged in a conflict with the rats"; yet they had the proper instinct, though not perfectly developed, for they dashed in the right manner on a dead rat. The singing of Birds is instinctive, yet it is notorious that their notes are improved by practice; & that young birds at first sing very badly.[3] But it may, perhaps, be said that in these cases the instinct from the first is perfect, but that the muscles do not thoroughily obey the will, until they have been practiced.[4]/

18/The actions of animals are sometimes influenced by imitations. Seeing how adroitly a young kitten licked the inner side of its feet & then washed its face with the moistened surface; I concluded that trifling as this action was, it was instinctive: but Dureau de la Malle[5] brings Audouin as a witness that three puppies brought up under a cat learnt this habit of washing their faces. At the Eccalobeion[6] in 1840 I saw chickens hatched without a mother, and when exactly four hours old, they ran, jumped, chirped, scratched the ground & crowded together, as if round the hen,— all actions beautifully instinctive. It might have been thought that the manner in which fowls drink, by filling their beaks, lifting up their heads & allowing the water to run down by its gravity, would have been especially taught by instinct; but this is not so, for I was most positively assured that the chickens of a brood reared by themselves, generally required their beaks to be pushed into a trough; but that if there were older chickens present, who had learnt to drink, the younger ones imitated their movements, & thus acquired the act. It has been stated that lambs turned out without their mothers are very liable to eat poisoning herbs; & it

[1] Annales des Sciences Nat. Tom 22. p. 406. This account is confirmed by Brehm see Charlesworth Mag. of Nat. History vol 2. 1838 p 402

[2] Annals & Magazine of Nat. History vol 18. 1846. p 378

[3] Bechstein Stübenvogel 4 Edit. p. 7

[4] LeRoy (Lettres Philosoph. sur les animaux. 1802 p. 104) who is esteemed a good observer, states that the nests of young birds are not so well made or placed as those of old; but I doubt much whether this is to be believed.

[5] Annal des Sc. Nat. Tom. 22. p. 397

[6] [See William Bucknell, *Eccaleobion: a treatise on artificial incubation.* London, 1839.]

seems to be certain that cattle, when first/19/introduced into [a] country are killed by eating poisonous herbs, which the cattle already naturalised there have learnt to avoid.[1]

Animals understand & profit by the cry of danger of other species, as every sportsman knows: thus, in the United States, the inhabitants like the Martins to build on their houses,[2] as their cry, when a hawk appears, alarms the chickens, though these latter are not aborigines of the country. In the summer of 1857 I observed a much more curious case of one insect apparently imitating a complex action from another of a different genus. From some experiments, which I was making, I had occasion very closely to watch some rows of the tall Kidney-bean, & I daily saw innumerable Hive-Bees alighting as usual on the left wing-petal & sucking at the mouth of the flower. One morning for the first time I saw several Humble-Bees, (which had been extraordinarily rare all summer) visiting these flowers & I saw them in the act of cutting with their mandibles holes through the under/20/side of the calyx, & thus sucking the nectar: all the flowers in the course of the day were perforated, & the Humble-bees in their repeated visits to each flower were thus saved much trouble in sucking. The very next day I found all the Hive-bees without exception sucking through the holes, which the Humble-bees had made. How did the Hive-bees find out that all the flowers were bored, & how did they so suddenly acquire the habit of using the holes?—I never saw, though I have long attended to the subject or heard of Hive Bees themselves boring holes. The minute holes made by the Humble Bees are not visible from the mouth of the flower, where the Hive-bees had hitherto invariably alighted: nor do I believe from some experiments which I have made that they were guided by the scent of the nectar escaping through these orifices more readily than through the mouth of the flower. The Kidney-bean is, also, an exotic. I must think that the Hive-bees either saw the Humble-bees cutting the holes, & understood what they were doing, & immediately profited by their labour; or that they merely imitated the Humble-bees after they had cut the holes & were sucking at them. Yet I feel sure, that if anyone who had not known this previous history, had seen every single Hive Bee, without a second's hesitation, flying with the utmost celerity & precision from the under side of one flower to another; &/21/

[1] [G. Clark, in] Annals & Mag of Nat. History 2 series vol. 2. p. 364 [Linné, 'On the Use of Natural History,' in] Amoenitates Acad. vol 7. p. 409 [Linné, 'Swedish Pan' in] Stillingfleets Tracts p. 350. In regard to Lambs see Youatt on Sheep. p. 404.
[2] Kalm's Travels vol 2. p. 148

thus rapidly sucking the nectar, he would have declared that it was a beautiful case of instinct.[1]/

21a/I have published an account of this case in the Gardeners Chronicle 1857 p 725. Whether the perforations of Flowers is an instinctive action in the Humble-bee I know not: they perform it with much skill either on the upper or lower side, either through the corolla alone, or through calyx & corolla, according to the position of the nectary: they make two holes, when the flower is rather broad, & there are two nectaries as in Pentstemon. Exotic flowers, I think, are more commonly perforated than endemic species;—as Pentstemon, Phaseolus, Vicia faba, Azalea, Salvia patens & Grahami, Stachys coccinea, Mirabilis, Antirrhinum majus &c. but I have seen the common British Melampyrum & Lonicera perforated. In most of these flowers, I have also seen the Humble-bees extracting the nectar without a hole having been cut; & it is evident that they cut the holes in order to save time: it seemed to me that they could visit nearly twice as many perforated, as non-perforated flowers in the same time. In large beds of the same flower, which are frequented by numerous bees, I found *every single* flower perforated; whereas a single plant of the same species in another part of the same garden, & the later flowers in the very same bed, when few in number, though visited by Bees, were not perforated; & this seemed to me to be due to the lesser number of Bees working at same time & consequent less eager rivalry to get the nectar. Most of the above facts were published by me in Gardeners Chronicle 1841. Aug. 21.—/

21a v/Many curious facts could be added on this subject. For nearly 20 years I have attended to the actions of Bees in flowers; & never during this time did I see a Bee visit a tall perennial Phlox, until the summer of 1857, when every single flower was perforated & visited by Humble-bees.—I never saw a Hive-Bee visit a Viola tricolor; nor has Mr. Grant (Gardener's Chronicle 1844. p. 374) seen his own Bees, but a body of strangers arriving one day, as known by battling with his own bees, industriously sucked the pansies in front of his Hives.—/

21/The only remaining character which has generally been attributed to instincts, is that they are unvarying in all places & during all time. To assume that they never change during the long lapse of geological time is to beg the question; & we shall presently see that instincts are not quite immutable.

Although from the facts above given, & very many more might have been added, I must believe that instincts are occasionally subjected in some very slight degree to the influence of reason, experience, instruction & imitation; & though I believe that such modifications may be of some importance from at least becoming habitual, & from habitual actions becoming hereditary for which reason I have discussed at some length the intelligence of animals —yet I must fully admit that all such modifications are of subordinate importance in an extreme degree, to that blind impulse strictly called an Instinct./

[1] [Darwin presented the following long paragraph as a footnote.]

22/Several of the elder metaphysical writers of England & France, & of late years F. Cuvier[1] have compared instinct with habitual actions; & it seems to me, laying quite on one side the question of their origin, that nothing can be juster or give us a more correct notion of the nature of an instinct, than this comparison. Look at a person playing a familiar tune on the pianoforte (this is Bishop Berkeley's illustration), at the same time conversing with his whole attention on some subject, see how perfectly, yet how unconsciously, he performs most complex actions in a given time & order. If Mozart[2] instead of almost naturally coming to play at three years old & compose at four years, had without instruction played some one simple tune, which tune alone his parents had played, everyone, I think, would have called this tune as completely instinctive as the song of a bird. It has indeed been admitted that reflex actions, which are often called instinctive, can hardly or not at all be distinguished from thoroughily habitual movements.[3] Many habits once/23/acquired do not continue improving or altering but remain, like an instinct, the same throughout life. Habits, indeed, are very often performed by a blind impulse in direct opposition to the will, as in the case of Sir W. Scott's clerk-like flourish at the bottom of the page./23 v/I have heard it remarked & noticed it very many times, that almost everyone from the habit of blowing out a spill carefully blows out the mere remnant of one before throwing it into the fire: I have found it quite difficult to cure myself of this mistake of habit; which may be compared to such mistakes of instinct, as a kitten[4] carefully covering with ashes a drop of clean water spilt on a hearth: I have seen a kitten shake its feet, as if wetted, when it merely touched its nose with water; & another kitten did the same on hearing water poured out./23/Yet the will & intellect may readily come into play, as when a woman immersed in thought is knitting & she meets with some little accident; this will arouse her attention just enough to get over the difficulty by some slight modification in the habitual knitting movement; almost in the same way, as we have seen that the Bees built their comb to avoid the slip of glass, but still continued instinctively to make their cells as nearly as possible of the proper shape. Habitual actions seem in some sense to stand in opposition to intellect: at least it has been noticed that persons of weak intellect are very apt to fall into

[1] Mem. du Mus. d'Hist. Nat. 1823. Tom x. p. 243. Flourens de l'Instinct des Animaux 1845 p. 57 Sir H. Holland Chapters on Mental Physiology ch. x.
[2] Holmes Life of Mozart p. 9.
[3] Alison on Instinct in the Cyclopaedia of Anatomy & Physiology. p. 4.
[4] Darwin's Zoonomia, vol i. p. 160

habits; & I may mention that I once knew a decidedly idiotic dog/24/which had the instinct of turning round before lying down on the carpet (a remnant it may be supposed of the instinct of forming a seat in long grass)[1] so strongly developed, that he has been counted thus to turn round twenty times. Habits again, just like instinct, readily become associated with particular states of the body or periods of time.

In repeating anything by heart or in playing a tune, every one feels that if interrupted, it is easy to go back a little, but very difficult suddenly to resume the train of thought or action a few steps in advance. Now P. Huber has described a caterpillar which makes by a succession of processes a very complicated hammock for its metamorphosis; & he found that if he took a caterpillar which had completed its hammock up to, say, the sixth stage of construction & put it into a hammock completed up only to the third stage, the caterpillar did not seem puzzled but reperformed the fourth, fifth, & sixth/25/stages of construction: if, however, a caterpillar was taken out of a hammock made up, for instance, to the third stage & put into one finished up to the sixth stage, so that much of its work was done for it, far from feeling the benefit of this, it was much embarrassed, & seemed forced to go over the already finished work, starting from the third stage where it had left off before it could complete its hammock. So again the Hive Bee in the construction of its comb seems compelled to follow an invariable order of work. M. Fabre gives another curious instance how one instinctive action invariably follows another: a Sphex makes a burrow, flies away & searches for prey, which it brings paralised by having been stung to the mouth of its burrow, but always enters to see that all is right within before dragging in its prey: whilst the Sphex was within the burrow, M. Fabre removed the prey to a short distance; when the Sphex came out it soon found the prey & brought it again to the mouth of the burrow; but then came the instinctive/26/ necessity of reconnoitering the just reconnoitered burrow; & as often as M. Fabre removed the prey so often was all this gone over again, so that the unfortunate Sphex reconnoitered its burrow forty times consecutively! When M. Fabre, altogether removed the prey, the Sphex instead of searching for fresh prey & thus making use of its completed burrow felt itself under the necessity of following the rhythm of its instinct, & before making a new burrow, carefully closed up the old one as if it were all

[1] E. Jesse Gleanings in Nat. History (3 series) p. 141. [Actually p. 23.]

right, though in fact utterly useless as containing no prey for its larva.[1]

In another way we perhaps see the relation of habit & instinct, namely in the latter acquiring greater force if practised only once or for a short time: thus it/27/is asserted that if a calf or infant has never sucked its mother, it is very much easier to bring it up by hand than if it has sucked only once.[2] So again Kirby[3] states that larvae after having "fed for a time on one plant will die rather than eat another, which would have been perfectly acceptable to them if accustomed to it from the first."

Although as I have here attempted to show there is a striking & close parallelism between habits & instinct; & although habitual actions & states of mind do become hereditary & may then, as far as I can see, most properly be called instinctive; yet, it would be, I believe, the gravest error to look at the great majority of instincts as acquired through habit & become hereditary. I believe that most instincts are the accumulated results through natural selection of slight & profitable modifications of other instincts; which modifications, I look at/28/as due to the same causes which produce variations in corporeal structures. Indeed I suppose that it will hardly be disputed that when an instinctive action is transmitted by inheritance in some slightly modified form, that this must be caused by some slight change in the organization of the Brain.[4] But in the case of the many instincts, which, as I believe, have not at all originated in hereditary habit, I do not doubt that they may have been strengthened & perfected by habit; just in the same manner as we may select corporeal structures for fleetness of pace, but likewise improve this quality by training in each generation./

29/After these preliminary remarks on the nature of Instinct, we have, in order to render our theory at all probable, to discuss, whether the mental faculties & instincts of animals in a state of nature do vary at all, & whether the variations are inheritable. We have, also, to consider how much to attribute to the effect of habit & how much to the natural selection of what may be called chance variations of preexistent instincts. In these latter questions, & indeed as rendering variations of instinct in a state of nature, the more probable, it will be desirable briefly to consider the

---

[1] M. Fabre in Annales des Sciences Nat. 4 Ser. Tome VI p. 148.—with respect to Hive Bee, see Kirby & Spence Entomology. vol I. p. 497. For the hammock Caterpillar see P. Huber in Mém. Soc. Phys. de Genève. Tom VII p. 154.

[2] Hippocrates & the celebrated Harvey have, also, made analogous remarks on this subject, see Darwin's Zoonomia vol I. p. 140.

[3] From Reaumur, Introduction to Entomology. vol. I. p. 391.

[4] This is expressly Sir B. Brodie's opinion, in his Psychological Inquiries 1854. p. 199.

changed instincts & inherited habits of domesticated animals. As by our theory, instincts can be modified only by the addition through selection of numerous slight variations or by slowly changing habits, so it follows that every complex instinct must have been preceded by a gradated chain of simpler instincts & we ought to be able to show that such a/30/series has existed, or at least that there is no instinct, which might not thus have originated through a series of slight changes. But it should be remembered that we here necessarily lie under great disadvantages, as no instinct can be fossilised, & instincts cannot be brought like specimens from foreign & little-known lands; therefore we have no right to expect nearly such perfect series as in the case of corporeal structures. Merely for brevity sake I will not enter separately on all these several heads, but will do so occasionally whilst discussing some of the more remarkable instincts.

That animals born in a state of nature & captured young show a great diversity of dispositions is the unanimous conviction of all who have attended to menageries. I could give numerous cases from the Elephant to the Humming Bird.[1] That the same diversity is common to our domestic animals, even to those of the same litter is notorious; & that these infinitely diversified dispositions strongly tend to be/31/inherited is the decided opinion of all those who have written on our domestic quadrupeds. Some veterinary authors maintain that disposition is more strongly inherited in the horse than corporeal structures.

In the early part of Chapter VIII, I had occasion to give several cases of the same animal having different habits of life in different places, & of changes in habit in naturalised animals & in native animals, where the country has long been occupied by man. These several facts blend into & can hardly be separated from cases of changed Instincts.

With respect to the inheritance of all sorts of mental tendencies, peculiarities, consensual movements etc., it is quite superfluous to give examples: they may be found in all Treatises on Inheritance.[2]

---

[1] See Corse on the Elephant in the Philosoph. Transactions 1799 p. 32.—For Humming Birds & Doves see Mr. Gosses curious account in his Birds of Jamaica 1847. p. 120. After speaking of the sulky, timid or confiding dispositions of Humming Birds, he states his belief that there is perhaps as much individuality in the character of Humming birds as in men.—Bechstein, seems to have particularly attended to this (Stubenvögel 1840 s. 142, 251, 267) & makes the same remark: he insists on the difference in the facility of learning to sing & in all sorts of peculiarities of disposition in Bull-finches & Larks taken from the nest, & in Canary Birds bred under domestication.

[2] Traité L'Hérédité Naturelle Dr. Prosper Lucas Tom. 1. p. 340 to 584. I may here just recall to mind the remarks made on the authority of Prof. Stokes on the

Many of these are extremely curious & well authenticated. John Hunter has remarked that tricks, that is any peculiar way of performing some action or some odd movement often associated with a certain state of mind or body, are certainly inherited. My Father, who/32/practised as a physician for sixty years, gave me many cases of children, whose parents had died during their infancy, who inherited all sorts of the slightest peculiarities, as a peculiar manner of placing the heads whilst reading, playing with things in their fingers, manner of entering a room &c. I will give one single case, which I have myself witnessed & can vouch for its perfect accuracy; namely that of a child who as early as between her fourth & fifth years, when her imagination was pleasantly excited & at no other time, had a most peculiar & irresistible trick of rapidly moving her fingers laterally, with her hands placed by the side of her face; & her father had precisely the same trick under the same frame of mind & which was not quite conquered even in old age; in this instance there could not possibly have been any imitation./

32 v/The inheritance of any peculiar habit or trick seems to me only one step more marvellous than those many cases on record of something learnt in infancy & entirely forgotten, but recovered during the delirium of a severe illness. I will give one illustration, communicated to me by my Father. Miss C. had been born in Lisbon but had left so early that she could remember nothing about the place or her Portuguese nurse: in her delirium after epileptic fits, she continually hummed a very peculiar tune & after a time added broken words to the air: these & the tune were written down: on her recovery her Uncle, a physician played & sung the words to her: she declared she had never heard them or the air; her Uncle played & sung it many times over & translated the words to her; at last suddenly like a flash of lightning her memory came & she exclaimed that it was a song her old & forgotten Portuguese nurse had sung to her. It is a curious fact that after

extreme improbability on the doctrine of chances of the same rare peculiarity appearing in father & child, without some genetic connexion.—Jesse in his Gleanings in Nat. Hist. 3 series p 149. says he has had a breed of Terriers, which all shew their teeth & put out their paws when caressed. The Rev. W. Darwin Fox tells me that he had a Skye Terrier bitch which when begging rapidly moved her paws in a way very different from that of any other dog which he had ever seen: her puppy, which never could have seen her mother beg, now when full-grown performs the same peculiar movement exactly in the same way. Somerville (in his Autobiography of a Working Man p 46) gives a curious case of a cow which would always lead the herd & had, during *a certain corporeal state during the summer*, a strange propensity to eat clothes hung up to dry; all her progeny inherited these propensities, & were so troublesome that they were at last all fattened & sold.

this conscious recovery of the air in her subsequent fits of delirium, she never once sung it again, although before hand it had been her constant habit to do so.—/

32/Look at the several breeds of Dogs & see what different tendencies are inherited, many of which cannot, from being utterly useless to the animal, be inherited from their one or several wild prototypes. I have talked with several intelligent Scotch Shepherds, & they were unanimous that occasionally/33/a young sheep-dog without any instruction will naturally take to run round the flock, & all thorough-bred dogs can be easily taught to do this: though they intensely enjoy this exercise of their innate propensity, yet of course they do not worry the sheep, as any wild canine of the same size would have done. Look at the Retriever, which so naturally takes to bringing back any object to his master. Every naturalist has read of Magendies experiments on a dog imported from England. The Rev. W. D. Fox informs me that he had taught in a single morning, a young Retriever six months old to fetch & carry well, & in a second morning to return on the path to search for an object left purposely behind & not seen by the dog; yet I know from experience how difficult it generally is to teach this habit at least to terriers.

Let us consider one other case, though so often quoted, that of the Pointer: I have myself gone out with a young dog for the first time & his innate tendency was shown in a ludicrous manner for he pointed fixedly not only at the scent of game, but at sheep, & large white stones; & when he found a lark's-nest, we were actually/34/compelled to carry him away: he backed the other dogs. Generally young Pointers require some little instruction, & occasionally they give much trouble. It has commonly been supposed that pointing is the pause before the spring of a beast of prey, carried to an excess, quite useless to a wild animal, & become hereditary./34 v/Two pointers of Col. Thornton's are said to have stood for one hour & a quarter.[1]/34/This habit & the silence of Pointers is the more remarkable, as all who have studied dogs agree in classing them as a sub-breed of the Hound, which gives tongue so freely & dashes after his prey. The tendency in the young Pointer to back other dogs, or to point without any scent of game, when they see other dogs point is, perhaps, the most singular part of his inborn propensities.[2]

[1] Col. Hamilton Smith Dogs. 1840 p. 196.
[2] With respect to the inherited tendency to back, see Ch. St. John's Wild Sports of the Highlands 1846 p. 116. Col. Hutchinson on Dog Breaking 1850 p. 144. Blaine, Encyclopedia of Rural Sports. p. 791.—Besides the tendency to point, Pointers inherit a peculiar manner of quartering their ground.—It is, I think, impossible

Now if we were to see one kind of wolf, in a state of nature running round a herd of deer & skillfully driving them whither he liked, & another species of wolf, instead of chasing its prey, standing silent & motionless on the scent even for more than an hour, with the/35/wolves of the pack all assuming the same statue-like attitude, & cautiously approaching, we should surely call their actions Instinctive. The chief characters of an Instinct seem to me fulfilled in the Pointer. A young dog cannot be supposed to know why he points, anymore than the white butterfly knows why it lays its eggs on the cabbage: yet an old dog will sometimes modify his inborn propensity & sagaciously show that he does understand the use of pointing; for instance in the case of Col. Hutchinson's dog[1] who in cover without having been taught would break his point, & run round the game to drive it towards his master./

35 v/Mr. Colquhoun (in his Moor & Loch 1841) states that his Retriever, like other well-bred & trained dogs would never ruffle a feather of the game which he retrieved; but one day after repeatedly trying to bring two wounded mallards across a river, & finding that when one was left it scrambled away, "he deliberately killed one, brought over the other & then returned for the killed bird". Almost the same thing happened to Col. Hutchinson (p 46), when his retriever whilst carrying a winged partridge, came across a dead bird; after some trials, finding that he could not carry the winged & dead bird, he "deliberately murdered" the winged one, & then was enabled to carry both. It must be remembered that it is abhorrent to a good Retriever to hurt his game./

35/It seems to me to make no essential difference that pointing is of no use to the dog, only to man; for the habit has been acquired through man's selection & training; whereas ordinary instincts are acquired through natural selection & training, exclusively for the animal's own good./36/The young Pointer often points without any instruction, imitation or experience; though no doubt, as we have also seen is sometimes to be the case with true instincts, he often profits by these aids./

36 v/It is difficult to determine how much dogs learn by experience & imitation. I apprehend there can be little doubt that the manner of attack of the English Bull-dog is instinctive. Rollin [Roulin] (Mem. presentés par divers Savans a l'Acad. Tom IV. [VI.] p 339) believes that certain dogs in S. America without education rush at the belly of the stags which they hunt, & that certain other dogs when first taken out run round the herds of

to read Sir John Sebright's admirable Pamphlet on Instinct, or Andrew Knight's paper on the Hereditary Propensities of Animals in Philosoph. Transactions 1837 p. 365, without being convinced to what an extraordinary degree qualities become hereditary in all our domestic animals, & more especially in Dogs.—

[1] Dog-Breaking p. 39

Peccari. We are led to believe that these actions are instinctive, when we hear from Sir T. Mitchell (Australia vol. 1. p. 292) that his dogs did not learn how safely to seize the Emu by the neck, until the close of his second expedition. On the other hand Mr. Couch (Illustrations of Instinct. p. 191) gives the case of a dog who learned after a single battle with a Badger, the spot where it could inflict a fatal bite, & it never forgot the lesson.—In the Falkland Islands it seems that the dogs learned from each other (Antarctic Voyage Sir J. Ross. vol 2. p 246) the best way of attacking the wild cattle.—/

36/Each breed of dog delights in following his peculiar inborn propensities. The most important distinction between pointing & a true instinct, is that pointing is less strictly inherited & varies greatly in the degree of its inborn perfection: this, however, is just what might have been expected; for both mental & corporeal characters are less true in domestic animals than in those in a state of nature; in as much as their conditions of life are less constant & man's selection & training are far less uniform & have been continued for an incomparably shorter period, than in the case of nature's productions. If, then, the inherited tendencies of the foregoing several breeds of dogs, do not essentially differ from Instincts, we see that at least the simpler instincts may be acquired, without an act of Creation.

With respect to the origin of the acquired instincts of our domestic animals, they have been spoken of by some authors as simply hereditary *habits*; but/37/this, I think, is incorrect. To take the case of pointing, though I fully believe from facts presently to be given that the compulsory training or habit continued during many generations will have had much influence on the breed, yet I doubt whether anyone would have thought of training a dog to point, had it not first shown some innate propensity in this line.— I have seen a terrier (& I have heard of other cases) which would point for a short time. If those puppies of the terrier which inherited any tendency to point had been picked out, trained, bred-from, & again selected, I do not doubt that a breed of pointing Terriers in time might have been formed. But in this case the selection of a self-formed propensity would have had as much to do with the formation of the breed, as habit. Even to the present day, with our Pointers, the principle of selection is steadily at work; for everyone breeds from the best dogs. What the first origin of the instinct of the sheep-dog may have been, I will not pretend to conjecture; but that dogs are born with numerous, slightly different, tendencies/38/independently of the habits of their parents, cannot be doubted. I am assured by Masters of Hounds that their young dogs show decided differences in their nature, some being best to find their fox in cover, others to make casts or patiently to recover the lost scent, or to recover it on roads; some tend to run straggling, others pack well &c; & that these different tendencies are in some degree hereditary. Had there been any object, there can be no doubt that packs could be formed with different innate capacities, as indeed we see with Fox hounds & Harriers, which "even when taken out for the first time have a very different mode of hunting."[1]

To take another class of facts: so many independent authors[2] have stated that horses in different parts of the world inherit artificial paces, that I think the fact cannot be doubted. Dureau de la Malle asserts that three different paces have [been] acquired since the time of the Roman classics; that from his own observations these are inherited. In these ancient times it seems that the amble was taught by curious & laborious processes; &/39/those horses which do not naturally inherit this pace are now taught it by the S. American Abipones, who fasten their front & hind legs together. But is it likely that these laborious practices would ever have been thought, had not some horses shown a natural turn for these several paces?

Tumbler Pigeons offer an excellent instance of an instinctive action, acquired under domestication, which could not have been taught, but must have appeared naturally, but probably has been vastly improved by the continued selection of those birds which showed the strongest propensity, more especially in ancient times in the East, when Flying Pigeons were much esteemed. Tumblers have the habit of flying in a close flock to a great height, & as they rise tumbling many times head over tail. I have bred & flown young birds, which could not possibly have ever seen a Tumbler tumble; after a few attempts over they went in the air. Imitation aids the instinct for all Fanciers are agreed that it is highly desirable to/40/fly young birds with first-rate old ones. Still more marvellous are the habits of the Indian sub-breed of Tumblers, called Lowtun, on which I have given details in a former chapter,[3] showing that during at least the last 250 years these birds have been known to tumble on the ground, after being slightly shaken, & to continue tumbling until taken up & blown on. As this breed has gone on so long, the habit can hardly be called a disease.

---

[1] Sir J. Sebright on Instinct p. 13

[2] Molina Hist. Nat. Chile vol I. p. 302. [Actually p. 368.] Dobrizhoffer Account of the Abipones vol I p. 225. See—Rollin, [Roulin] Mem. divers Savans &c Tom IV. [VI] p 337 for the horses of New Granada.—For the inherited amble in the Mongolian Horses see [Bergmann] Mem. du Mus. d'Hist. Nat. Tom XVI. p. 456.— Dureau de la Malle in Annal. des Scien. Nat. Tom 21. p. 58 & Tom 27. p. 24. I may add that I was formerly struck by no horse on the grassy plains of La Plata, having the natural high action of some English Horses.

[3] [See *Variation*, 1st ed. I, pp. 150–1.]

I need hardly remark that it would be as impossible to *teach* one Pigeon to tumble; as impossible as to *teach* another kind to inflate its crop to that enormous size, which the Pouter pigeon habitually does. I may add that the Pouter offers a good instance of how each animal enjoys performing its own instinctive action: when a Fancier wants to show off his Pouter, which at the moment will not pout, he takes the bird's beak in his mouth & blows him up like a bladder; & the Pouter, when let free, conscious of his magnificent dimensions, uses his best exertion, as I have seen, to retain his balloon-like appearance./

41/On the other hand some instinctive propensities in our domestic animals must have originated wholly in hereditary habit; sometimes perhaps aided by the selection of those individuals which have most strongly inherited the desired habit, or by the destruction of those which have failed in inheritance. "The wild rabbit", says Sir J. Sebright[1] is by far the most untameable animal that I know & I have had most of the British mammalia in my possession. I have taken the young ones from the nest & endeavoured to tame them, but could never succeed. The domestic rabbit, on the contrary is, perhaps, more easily tamed than any other animal, excepting the dog". We have an exactly parallel case in the young of the wild & tame Duck./

41 v/In Chapter 7. I have given some facts showing that when races or species are crossed there is a tendency from quite unknown causes in the crossed offspring to revert to ancestral characters;—as, for instance, in crossed pigeons the assumption of the wing-bars &c.—A ⟨strong⟩ suspicion has crossed me that a slight tendency to primeval wildness sometimes thus appears in crossed animals. Mr. Garnett of Clitheroe in a letter to me states that his hybrids from the musk & common Duck "evinced a *singular* tendency to wildness." Waterton (Essays of Nat. History [Series 1] p 197) says that in his Duck, a cross from the wild & common, "their wariness was quite remarkable". Mr. Hewitt who has bred more hybrids between pheasants & fowls than any other man in letters to me, speaks in the strongest terms of this wild, bad & troublesome disposition; & this was the case with some which I have seen. Capt. Hutton made nearly the same remark to me in regard to the crossed offspring from a tame Goat & a wild species from the western Himalaya. Lord Powis' agent, without my having asked him the question, remarked to me that the crossed animals from the domestic Indian Bull & common cow "were more wild than the thorough-bred breed".—I do not suppose that this increased wildness is invariable; it does not seem to be the case according to Mr. Eyton with the crossed offspring from the common & Chinese geese; nor according to Mr. Brent with crossed birds from the Canary./41/In Norway, the Ponies are trained to obey the voice & not the rein: Andrew Knight imported some of them & he states[2] that "the horse-breakers complain & certainly with very good reason, that it is

---

[1] On Instincts 1836 p. 10.   [2] Philosophical Transactions 1837. p. 369.

impossible to give them what is called a mouth; they are nevertheless exceedingly docile & more than ordinarily obedient when they understand the commands of their master". Sturm[1]/42/says that it is notorious that young cattle in the district where the old are habitually used for draft are much more easily broken in than elsewhere. Our cows readily yield milk, when their calves are removed; but this [is] very far from the case in many less civilised regions.[2]

In most of these cases, at least with the inherited tameness of the rabbit & duck & with our Ponies readily learning to obey the rein, habit alone can have come into play; for probably no one has selected rabbits or ponies for their qualities.

We daily witness, but overlook from familiarity, a remarkable case of instinct, changed from habit or training, aided probably by the destruction of all individuals which fail in the desired frame of mind. I refer to our dogs, which are with such difficulty prevented chasing rabbits & other game, so seldom requiring to be taught not to worry sheep or poultry; yet every wild canine animal would at once attack/43/them. This was the case with a native dog from Australia, whelped on board a ship, which Sir J. Sebright tried for a year to tame, but which "if led near sheep or poultry became quite furious"; so again Captain Fitz-Roy[3] says that not one of the many dogs, procured from the natives of Tierra del Fuego & Patagonia, "which were brought to England could easily be prevented from indulgence in the most indiscriminate attacks on poultry, young pigs &c." As the natives of these countries do not keep domestic animals, their dogs have not been trained to spare them. Not only have our dogs lost the desire to attack poultry, but our chickens have quite lost that fear, which no doubt is as natural to them as to young pheasants; & Waterton[4] found that some young pheasants hatched under a hen, though tame to any person whom they knew, could never be cured of being so terrified at the mere sight of a dog, that some rushed into a pond & were thus drowned. Nor are our chickens or young turkeys become insensible to fear of all kinds,: let the hen give the danger-chuckle, & they will run from under her body (to allow their mother, who has almost lost the power of flight, to fly away!) scatter themselves, squat & hide.[5]

44/Again look at the utter indifference with which our domestic

---

[1] Ueber Racen. &c 1825 s. 85.

[2] Le Vaillant (First Travels Vol. 2. p. 194). gives curious particulars on this head with respect to the *well-tamed* Caffre cattle. It is the same in La Plata.

[3] In Col. Hamilton Smith Treatise on Dogs. 1840 p 214. Sir J. Sebright on Instinct p. 12.

[4] Watertons Essay on Nat. History. p. 197 [See Series 1, p. 99.]

[5] A friend of mine can imitate the danger-cry of the Hen so well, that he can make his chickens squat. To show that young Turkeys have not lost all instinctive fear,

cats pass by young chickens, which assuredly would be a delicious morsel to any wild feline animals; as indeed all those who have tamed several wild species know full well. Yet each kitten, even when taken early from its mother & reared solitary has not to be taught to avoid chickens. Pigeons are not as commonly kept as poultry, & every Fancier knows how difficult it is to keep his favorites safe from their incorrigible enemy, the cat.

The many cases of inborn fear or ferocity in young animals directed towards particular objects, as well as the loss of these individualised passions, seems to me extremely curious. Let anyone who doubts their existence, give a mouse to a kitten, taken early from its mother, & which has never before seen one, & observe how soon the kitten growls with hair erect, in a manner wholly different from when at play or when fed with ordinary food. We cannot suppose that the kitten has an inborn picture of a mouse graven in its mind. But as when an old Hunter snorts with eagerness at the very first sound of the horn, /45/ we must suppose that old associations excite him almost as instantly, as when a sudden noise makes him instinctively start; so I imagine, with the difference that the association has become hereditary instead of being only fixed deeply by habit, the kitten without any definite anticipation, thrills with excitement at the smell of the mouse.

From the several facts now given, we may conclude that *inherited* propensities & actions may originate, without any training of the parents, as with the tumbler-pigeon; or that they may thus appear in a slight degree at first, as in case of pointing, & be increased by training; or that they may wholly arise from habit in the parent, as with the tameness of rabbits, & without the aid of selection. But in most cases, the selection of those individuals which have inherited most strongly any propensity or action, whether self-originating or due to habit, and the destruction of those individuals which have inherited it freely, probably has played a most important part in its /46/ increase or perfection. I have discussed this subject at some length, as our knowledge of what takes place under domestication is our best guide in speculating on the origin of the Instincts of animals in a state of nature.

Instincts are lost under domestication. In the just quoted instances of animals born tame, it is in fact only that they have lost that timidity, suspicion & restlessness so characteristic of wild animals: the tameness, however of dogs, is something more, for,

---

I may add that the Rev. W. D. Fox saw a brood of his Turkeys with their mother in agonies of horror at a frog peeping out of a hole; as Mr. Fox remarked their instinct probably misled them to mistake the bright eyes of the frog for those of a deadly N. American snake.

as Sir J. Sebright has remarked, they appear to inherit an instinctive love of man. I could give several instances of partially or wholly lost instincts. Two examples will suffice: Chinese & Polynesian dogs, though so strictly carnivorous animals, from having been for many generations fed on vegetable food, have lost their instinctive taste for flesh.[1] Considering the general habits of Birds, it must be, as Paley has remarked, a most strong impulse which leads a bird to sit so closely on her eggs: yet some of our breeds of fowls, as the Polish & Spanish have quite lost this strong instinct./

47/It is probable that instincts acquired under domestication are much more easily lost than natural instincts. High-bred Tumbler-pigeons, which have been confined for several generations often quite lose the habit of tumbling, as was the case with some of mine, which I allowed to fly: but a good observer assures me that they more fixedly retain the habit of flying high in a compact flock. The barking of the Dog is an acquired instinct, very different in degree in different breeds, & certainly often lost when dogs have become feral. Poeppig[2] says that puppies of the feral dogs of Cuba, brought up in the house, always remain fearful & treacherous & with an inborn propensity to steal; so that the many excellent qualities of our dogs here seem to have been lost. On the other hand, Capt. Sulivan R.N. took some young rabbits at the Falkland Islands, where this animal has run wild for several generations, & he is convinced that they were more easily tamed than really wild rabbits in England, which as a boy he had tried to tame. The facility of breaking in the feral horses in La Plata, can, I think, be/48/accounted for on the same principle of some little of the effects of domestication being long inherent in the breed.—

The acquired instincts of our domestic animals, resemble natural instincts, when tried by what may be considered the severe test of crossing the breed. It is well known that when two distinct species are crossed the instincts are curiously blended, & vary in the successive generations, just like corporeal structures. To give one example; a dog kept by Jenner[3] which was grandchild or had a quarter-blood of the Jackall in it, was easily startled, was inattentive to the whistle, & would steal into the fields & catch mice in a peculiar manner. I may add, as showing what trifling

[1] Whites Nat. History of Selbourne Letter 57. Sturm (Ueber Racen &c. s. 82) states positively that in parts of Germany, where for hundreds of generations, the calves have been taken, directly after birth from the cow, she has lost much of her instinctive maternal love. Again he says Merino sheep will allow strange lambs to suck them, as it is the practice to change the lambs, whereas the native German sheep will not allow this; so that *individualised* maternal love seems lessened in this breed.—

[2] Reise in Chile and Peru [vol. I] s. 290   [3] Hunter's Animal Economy p. 325

peculiarities are affected by inheritance that a dog with only 1/8 wolf's blood in its veins, would come, when called, "mais non pas en ligne droite, comme font ordinairement les chiens".[1] Now I could give numerous examples of crosses between breeds of dogs, both having artificial instincts, in which these instincts have been most curiously blended, as between the Scotch & English Sheep dog, between the setter & pointer: the effect, moreover/49/ of such crosses can sometimes be traced for very many generations, as in the courage acquired by Lord Orfords famous greyhound[2] from a single cross with the bulldog:/49 v/this cross in the first generation has so much courage, that crossed dogs of this kind tried, in Atholl forest in hunting deer were all killed by attacking the deer in front.—/49/On the other hand a dash of the grey-hound, will give a family of sheep-dogs a tendency to hunt hares for many generations, as I was assured by an intelligent shepherd.

We will now pass on to the instincts of animals in a state of nature; & we will consider some of the best-known & largest classes of instincts, & some of those which have been universally ranked as the most wonderful, under the points of view which mainly concern us,—namely their variability ⟨⟨whether self-originating or due to habit)⟩ which will allow natural selection to seize on any profitable modification, & the present or possible former existence of a graduated chain in each class of instincts, for according to our theory the most complex instinct can have been arrived at only by innumerable, slight, intermediate modifications. The instincts, which we will now discuss, may be arranged under the following heads. (1) Migration (2) Fear of danger and Feigning death (3) Parasitism (4) Nidification & habitation (5) ants making slaves (6) the comb of the Hive-Bee./

50/*Migration.*—The migration of young birds across broad tracts of the sea, & the migration of young salmon from fresh into salt-water, & the return of both to their birth-places, have often been justly advanced as surprising instincts. With respect to the two main points which concern us: we have, firstly, in different birds, a perfect series from those which occasionally or regularly

---

[1] LeRoy Lettres Philosoph. 1802. p 228.
[2] Youatt on the Dog p. 31.—Daniel (see Blaine Encyclop. of Rural Sports p. 863) asserts that a cross of the Beagle "generations back, will give to a spaniel a tendency to hunt hares over feathers." See p. 793 for account of cross of Setter & Pointer. See, also Andrew Knight in Philosoph. Transactions 1837. Part 2. for an account of crossed instinctive propensities in Dogs.—See W. Scropes Art of Deer Stalking, p. 316 on the crossed Dogs being killed by Deer.

shift their quarters within the same country, to those which periodically pass to far distant countries, traversing often by night the open sea over spaces of from 240 to 300 miles, as from the north-eastern shores of Britain to Southern Scandinavia. Secondly in regard to the variability of the migratory instinct: the very same species often migrates in one country & is stationary in another; or different individuals of the same species, in the same country are migratory or stationary; & these can sometimes be distinguished from each other by slight differences.[1] Dr. Andrew Smith has often remarked to me how inveterate is the instinct of migration in some of the quadrupeds of S. Africa, notwithstanding the persecution to which they are in consequence subjected: in N. America, however, persecution has driven the Buffalo within a late period[2] to/51/cross in its migrations, the Rocky mountains; & those "great highways, continuous for hundreds [of] miles, always several inches & sometimes several feet in depth", worn by the migrating buffaloes on the eastern plains, are never found westward of the Rocky mountains./51 v/In the United States, Swallows & other birds have largely extended, within quite a late period, the range of their migration.[3]/

51/The migratory instinct in birds is occasionally lost; as in the case of the Woodcock, some of which have lately, without any assignable cause,[4] taken to breed & become stationary in Ireland & Scotland. In Madeira the first arrival of the Woodcock is known[5] & it is not there migratory; nor is our common Swift, though belonging to a group of birds, almost emblematical of migration. A Brent goose, which had been wounded, lived for nineteen years in confinement; & for about the first twelve years, every spring at the migratory period, it became very uneasy, & would, like other confined individuals of this species, wander as far northward

---

[1] Mr. Gould has observed this fact in Malta & in Tasmania in the southern hemisphere. Bechstein (Stubenvögel 1840 s. 293) says that in Germany the migratory & non-migratory Thrushes can be distinguished by the yellow tinge of the soles of their feet. The Quail is migratory in S. Africa, but stationary in Robin Island, only two leagues from the continent. (Le Vaillant Travels vol. 1. p. 105: Dr. Andrew Smith confirms this). In Ireland the Quail has late taken to remain in numbers & breed there. (W. Thompson Nat. Hist. of Ireland, Birds vol. 2. p. 70.

[2] Fremont, Report of Exploring Expedition 1845 p. 144.

[3] See Dr. Bachman's excellent memoir on this subject in Silliman's Philosoph. Journal. vol. 30. p. 81.
Mr. W. Thompson has given an excellent & full account of this whole subject. See Natural History of Ireland. Birds vol. 2. p. 247–257. where he discusses the cause. There seems reason to believe (p. 254) that the migratory & non-migratory individuals can be distinguished. For Scotland see. Ch. St. John's Wild Sports of the Highlands 1846. p. 220.
Dr. Heineken in Zoological Journal vol v. p. 75. See also Mr. E. V. Harcourt's Sketch of Madeira 1851. p 120.

as possible; but after this period "it ceased to exhibit any particular feeling at this season."[1] So that we here see the migratory impulse at last worn out./

52/In the migration of animals, the instinct which impels them to proceed in a certain direction ought, I think, to be distinguished from the unknown means by which they can tell one direction from another & by which, after starting, they are enabled to keep their course in a dark night over the open sea; & likewise from the means, whether some instinctive association with changing temperature or with want of food &c, which leads them to start at the proper period. In this, & other cases, the several parts of the problem have often been confounded together under the word instinct.[2] With respect to the period of starting; it cannot of course be memory, as young Cuckoos start for the first time two months after their parents have departed: yet it deserves notice that animals somehow acquire a surprisingly accurate idea of time: A. d'Orbigny shows that a lame Caracara Hawk in S. America, knew the period of three weeks & used at this interval to visit monasteries where food was distributed to the poor. Difficult though it may be to conceive how animals either intelligently or instinctively come to know a given period; yet we shall immediately see that in some cases, our domestic animals/53/have acquired an annually recurring impulse to travel, extremely like, if not identical with, a true migratory instinct; & which can hardly be due to mere memory.

It is a true instinct which leads the pinioned Brent goose to try to escape northward; but how the bird distinguishes north & south we know not. Nor do we know how a bird which starts in the night as many do, to traverse the ocean, keeps its course, as if provided with a compass. But we should be very cautious in attributing to migratory animals any capacity in this respect, which we do not ourselves possess; though certainly in them carried to a wonderful perfection. To give one instance; the experienced navigator Wrangell[3] expatiates with astonishment at the "unerring instinct" of the natives of N. Siberia, by which they guided him through an intricate labyrinth of hummocks of

---

[1] D. W. Thompson. Nat. History of Ireland: Birds. Vol 3. p. 63. In Dr. Bachman's paper just referred to, cases of Canada geese in confinement periodically trying to escape northward are given.

[2] See E. P. Thompson on the Passions of Animals 1851 p. 9. & Alison's remarks on this head in the Cyclopaedia of Anatomy & Physiology. Article Instinct p. 23.

[3] Wrangells Travels Eng. Translat. p. 146. See, also, Sir G. Grey Expedition to Australia vol 2. p 72. for an interesting account of the powers of the Australians in this same respect.—The old French Missionaries used to believe that the N. American Indians were actually guided by instinct in finding their way.—

ice, with incessant changes of direction; whilst Wrangell "was watching the different turns compass in hand & trying to resume the true route, the Native had always a perfect knowledge of it/ 54/empirically". Moreover the power in migratory animals of keeping their course is not unerring, as may be inferred from the numbers of lost Swallows often met with by ships in the Atlantic:[1] the migratory salmon, also, often fails in returning to its own river, "many Tweed salmon being caught in the Forth". But how a small & tender bird coming from Africa or Spain, after traversing the sea, finds the very same hedge-row in the middle of England, where it made its nest last year, is indeed truly marvellous.

Let us now turn to our domesticated animals: many cases are on record of animals finding their way home in a mysterious way; & it is asserted that Highland sheep have actually swam over the Firth of the Forth to their home, a hundred miles distant;[2] when bred for three or four generations in the lowlands they retain their restless disposition. I know of no reason to doubt the minute account given by Hogge[3] of a family of sheep, which had a *hereditary propensity* to return at the lambing season to a place, called Crawmell but only ten miles off, whence the first of the lot was bought; & after their/55/lambs, were old enough, they returned by themselves to the place where they usually lived: so troublesome was this inherited propensity, associated with the period of parturition, that the owner was compelled to sell the lot.—Still more interesting is the account given by several authors of the "trashumantes" sheep in Spain, which from ancient times have annually migrated during May from Estremadura to old Castille, a distance of about 400 miles: all the authors[4] agree that "as soon as April comes the sheep express by various uneasy

---

[1] The number of birds, which by chance visit the Azores (C. Hunt in Journal of Geograph. Soc. vol 15 Part 2. p. 282) so distant from Europe, is probably in part due to lost directions during migration: W. Thompson (Nat. History of Ireland. Birds vol. 2. p. 172) shows that N. American birds which occasionally wander to Ireland generally arrive at the period when they are migrating in N. America. In regard to the Salmon, see Scropes Days of Salmon Fishing p. 47.

[2] Gardener's Chronicle. 1852 p. 748 [Letter by C.N.D.]: other cases given by Youatt on Sheep p. 377.

[3] [James Hogg, 'the Ettrick Shepherd'] Quoted by Youatt in Veterinary Journal [i.e. *The Veterinarian*] vol. 5. p. 282.

[4] Bourgoanne Travels in Spain (Eng. Translat) 1789 vol. i. p. 38–54. In Mills' Treatise on Cattle 1776. p. 342 there is an extract of a letter from a gentleman in Spain to Peter Collinson, from which I have made extract. Youatt on Sheep p. 153. gives references to three other publications with similar accounts.— I may add that Von Tschudi (Sketches of Nature in the Alps Eng. Trans. 1856 p. 160) states that annually in the spring the cattle are greatly excited, when they hear the great bell which is carried with them; well knowing that this is the signal for their "approaching migration" to the higher Alps.—

motions a strong desire to return to their summer habitation." "The unquietude", says another author "which they manifest might in case of need serve as an almanack". "The shepherds must then exert all their vigilance to prevent their escaping", "for it is a known truth that they would go to the very place where they had been born". Many cases have occurred of three or four sheep having started & performed the journey by themselves, though generally these wanderers are devoured/56/by the wolves. It is very doubtful whether these migratory sheep are aborigines of the country; & it is certain that within a comparatively recent period their migrations have been widely extended; this being the case I think there can hardly be a doubt that this "natural instinct" as one author calls it, to migrate at one particular season in one direction has been acquired during domestication, based no doubt on that passionate desire to return to their birth-place, which as we have seen is common to many breeds of sheep. The whole case seems to me strictly parallel to the migrations of wild animals.

Let us now consider how the more remarkable migrations could possibly have originated. Take the case of a bird being driven, each year, by cold or want of food, slowly to travel southward, as is the case with some birds; & in time we may well believe that this compulsory travelling would become an instinctive passion, as with the sheep of Spain. Now during the long course of ages, let valleys become converted into estuaries, & these into wider & wider arms of the sea, & still I can/57/well believe that the impulse which leads the pinioned goose to scramble northward, would lead our bird over the trackless waters; & that by the aid of the unknown power by which many animals (& savage men) can retain a true course, it would safely cross the sea now covering the submerged path of its ancient land journey./

57 v 2/I do not venture to suppose that the line of navigation of birds always marks the line of formerly continuous land. It is, possible, that a bird accidentally blown to a distant land or island, after staying some time & breeding there, might be induced by its innate instinct to fly away, & again to return there in the breeding season. But I know of no facts to countenance this idea; & I have been much struck in the case of oceanic islands, lying at no excessive distance from the main-land, but which for reasons to be given in a future chapter, I do not believe have ever been joined to the mainland, with the fact that they seem most rarely to have any migratory Birds. Mr. E. V. Harcourt who has written on the birds of Madeira informs me that there are none at Madeira: so I am informed by Mr. Carew Hunt it is in the Azores; though he thinks that perhaps the Quail which migrate from island to island may leave the archipelago.[1]/

[1] ['Canaries none C. de Verdes' added in pencil.]

57 v 1/In the Falkland Islands as far as I can find out no *land-bird* is migratory. From inquiries which I have made there is no migratory bird in Mauritius or Bourbon: Colenso asserts (Tasmanian Journal vol. 2. p 227) that a cuckoo, Cuculus lucidus is migratory, remaining only 3 or 4 months in New Zealand; but New Zealand is so large an island that it may easily migrate to the south & remain then quite unknown to the natives of the North. Faröe, situated about 180 miles from the North of Scotland, has several migratory Birds (Graba Tagebuch 1830. p. 205) & Iceland seem to be the strongest exception to the apparent rule; but it lies only [   ] miles from the line of 100 fathoms./

⟨I will give one case of Migration which seemed to me at first to offer especial difficulty. It is asserted that in the extreme North of America, Elk & Reindeer annually cross, as if they could smell the herbage at the distance of a hundred miles, a tract of *absolute* desert, to visit certain islands, where there is a better but still scanty supply of food.—How could this migration have been first established? If the climate formerly had been a little more favourable, the desert a 100 miles in width might then have been clothed with vegetation sufficient to have just tempted the quadrupeds to have roamed over it, & so to have found out the more fertile northern islets. But the intense Glacial preceded our present climate; & the idea of a former better climate seemed quite untenable; but if those American Geologists are right, who believe, from the range of certain shells, that subsequently to the Glacial period there was one slightly warmer than the present period, then, perhaps we have the key to this migration across the desert of the Elk & Rein-deer.⟩/

58/*Instinctive fear.* I have already discussed the hereditary tameness of our domesticated animals: from what follows I have no doubt that the fear of man has always first to be acquired in a state of nature, & that under domestication it only is lost again. In all the few archipelagoes & islands uninhabited by man, of which I have been able to find an early account, the native animals were entirely void of fear of man: I have ascertained this in six cases in the most distant parts of the world, & with birds & animals of the most different kinds.[1] ⟨Old Dom Pernety says that the Ducks & Geese at the Falkland Islands walked before them as if mad [sic; Pernety wrote 'privés' i.e. tame.]⟩[2] At the Galapagos Islands I pushed a hawk off a tree with the muzzle of my gun, &

[1] I have given in my Journal of Researches (1845) p. 398 details on the Falkland & Galapagos Islands. Cada Mosto (Kerr's collection of Voyages vol. 2. p. 246) says that at the C. de Verde Islands that the pigeons were so tame as readily to be caught. These then are the only large groups of islands, with the exception of the Azores, (of which I can find no early account) which were uninhabited when discovered. Thos. Herbert in 1626 in his Travels p. 349 describes the tameness of the birds at Mauritius, & Du Bois in 1669–72 enters into curious details on this head with respect to all the birds at Bourbon. Capt. Moresby lent me a M.S. account of his survey of St. Pierre & Providence islands. north of Madagascar, in which he describes the extreme tameness of the Pigeons.—Capt. Carmichael has described the tameness of the birds at Tristan d'Acunha in Linn. Transact. vol. XII. p. 496.

[2] [Journal historique, II, p. 438.]

the little birds drank water out of a vessel which I held in my hand. But I have in my Journal given details on this subject; & I will here only remark that the tameness is not general, but is special towards man: for at the Falklands, the Geese build on the outlying islets on account of the foxes. These wolf-like foxes were here, as fearless of man, as were the birds; & the sailors in Byron's voyage, mistaking their curiosity for fierceness ran into the water to avoid them:[1] in all old/59/civilised countries, the wariness & fear of even young foxes & wolves is well known.[2] At the Galapagos Islands the great land-lizards (Amblyrhynchus) were extremely tame so that I could pull them by the tail whereas in other parts of the world *large* lizards are wary enough. The aquatic lizard of this same genus, lives on the coast-rocks, is adapted to swim & dive perfectly, & feeds on submerged algae: no doubt it must be exposed to danger from the sharks; & consequently, though quite tame on the land, yet I could not drive them into the water & when I threw them in, they always swam directly back to the shore: see what a contrast with all amphibious animals in Europe, which, when disturbed by the more dangerous animal, man, instinctively & instantly take to the water.

The tameness of the birds at the Falklands is particularly interesting, because most of the very same species, more especially the larger birds, are excessively wild in Tierra del Fuego, where for generations they have been persecuted by the savages. Both at these islands & at the Galapagos, it is particularly note-worthy, as I have shown in my Journal by the comparison of the several accounts up to the time when we visited these islands, that the birds are gradually getting less & less tame; & it is surprising, considering/60/the degree of persecution which they have occasionally suffered during the last one or two centuries, that they have not become wilder; it shows that the fear of man is not soon acquired.

In old inhabited countries, where the animals have acquired much general & instinctive suspicion & fear, they seem very soon to learn from each other, & perhaps, even from other species, caution directed towards any particular object. It is notorious that rats & mice cannot long be caught by the same sort of trap,[3] however tempting the bait may be; yet as it is rare that one which has actually been caught escapes, the others must have learnt the danger from seeing others suffer. Even the most terrific object, if never causing danger & if not *instinctively* dreaded, is immediately viewed with indifference; as we see in our railway-

---

[1] [Cf. Kerr, *Voyages*, xii, (1814), 46–7.]   [2] LeRoy Lettres Philosoph. p. 86.
[3] E. P. Thompson. Passions of Animals 1851. p 79.

trains: what bird is so wary & difficult of approach as the Heron; & How many generations would it not require to make Herons fearless of man; yet W. Thompson says[1] that these birds after a few days experience would fearlessly allow a train to pass within half-gun-shot distance. Although it cannot be doubted that the fear of man in old inhabited countries is partly acquired; yet it, also, certainly is instinctive, for nestling birds are generally terrified/61/at the first sight of man; certainly far more so than most of the old birds at the Falklands & Galapagos archipelagoes after years of persecution.

We have in England excellent evidence of the fear of man being acquired & inherited in proportion to the danger incurred; for, as was long ago remarked by Daines Barrington[2] that all our *large* birds, young & old, are extremely wild, yet there can be no relation between size & fear; for on unfrequented islands when first visited the large birds were as tame as the small. How excessively wary is our magpie; yet it fears not horses or cows, & sometimes alights on their back, just like the Dove at the Galapagos did in 1684 on Cowley. In Norway, where the Magpie is not persecuted, it picks up food "close about the doors, sometimes walking inside the houses":[3] the Hooded Crow (C. cornix) again is one of our wildest birds, yet in Aegypt[4] is perfectly tame. Every single young magpie & crow cannot have been frightened in England, & yet all are fearful of man in the extreme; on the other hand, at the Falkland & Galapagos Islands/62/many old birds & their parents before them, must have been frightened & seen others killed; & yet they have not acquired a salutary dread of that most destructive animal, man.

Animals feigning, as it is said Death,—an unknown state to each living creature—seemed to me a remarkable instinct. I agree with those authors[5] who think that there has been much exaggeration on this subject: I do not doubt that fainting (I have had a Robin faint in my hands) & the paralyzing effects of excessive fear have sometimes been mistaken for the simulation of death.[6]

---

[1] Nat. History of Ireland: Birds vol. 2. p. 133.

[2] Philosoph. Transact.—1773. p. 264.

[3] W. C. Hewitson in Magazine of Zoology & Botany vol 2. 1838. p. 311.

[4] Geoffroy St. Hilaire Annales du Museum Tom. ix p 471.

[5] Couch, Illustrations of Instinct. p. 201.

[6] The most curious case of apparently true simulation of death amongst the higher animals, is that given by Wrangel (Travels in Siberia p. 312 Eng. Translat) of the geese which migrate to the Tundras to moult, & are then quite incapable of flight.—He says, they feign death so well "with their legs & necks stretched out quite stiff, that I passed them by, thinking them dead." But the natives were not thus taken in. This simulation would not save them from foxes or wolves

Insects are most notorious in this respect. We have amongst them a most perfect series, even within the same genus (as I have observed in Curculio & Chrysomela) from species which feign only for a second, & sometimes imperfectly still moving their antenna (as with some Histers) & which will not feign a second time however much irritated to other species which, according to De Geer, may be cruelly roasted at a slow fire without the slightest movement, to others which will long remain motionless, as much as 23 minutes, as I/63/found with Chrysomela Spartii. Some individuals of the same species of Ptinus assumed a different position from that of others. Now it will not be disputed that the manner & duration of the feint is useful to each species, according to the kind of danger which it has to escape: therefore there is no more real difficulty in the acquirement, through natural selection, of this hereditary attitude than of any other. Nevertheless it struck me as a strange coincidence that insects should thus have come to exactly simulate the state which they took when dead. Hence I carefully noted the simulated positions of seventeen different kinds of insects (including one Iulus, Spider & Oniscus) belonging to the most distinct genera, both poor & first rate-shammers; afterwards I procured naturally dead specimens of some of these insects, & others I killed with camphor by an easy & slow death; the result was, that in no one instance was the attitude exactly the same, & in several instances the attitudes of the feigners & of the really dead were as unlike as they possibly could be./

64/*Nidification & habitation.*—We come now to more complex instincts. The nests of Birds have been carefully attended to, at least in Europe & the United States; so that we have a good & rare opportunity of seeing whether there is any variation in an important instinct, & we shall find that this is the case. We shall further find that compulsion, favourable opportunities & intelligence sometimes slightly modify the constructive instinct. In the nests of birds, also, we have an unusually perfect series, from those which build none but lay on the bare ground, to others

&c, which I presume inhabit the Tundras; would it save them from Hawks? The case seems a strange one. A lizard in Patagonia (Journal of Researches [2nd ed.] p. 97) which lives on the sand near the coast & is speckled like it, when frightened feigned death with outstretched legs depressed body & closed eyes: if further disturbed it buried itself quickly in the sand. If the Hare had been a small & insignificant animal & if she had closed her eyes, when on her form, should we not, perhaps, have said that she was feigning death? In regard to Insects, see Kirby & Spence, Introduction to Entomology. vol. 2 p. 234.—

which make a most imperfect & simple nest, to others more perfect, & so on, till we arrive at marvellous structures, rivalling the weaver's art.

Even in so singular a nest, as that of the Hirundo (Collocalia) esculenta, eaten by the Chinese,/64A/we can, I think, trace the stages by which the necessary instinct has been acquired. The nest is composed of a brittle white translucent substance very like pure gum-arabic or even glass, lined with adherent feather-down. The nest of an allied species in the British museum consists of irregularly reticulated fibres, some as fine as [   ] of the same substance: in another species bits of sea-weed are agglutinated together with a similar substance. This dry mucilaginous matter soon absorbs water & softens: examined under the microscope it exhibits no structure, except traces of lamination & many generally conspicuous in small dry fragments, & some bits looked almost like vesicular larva. A small, pure bit put into flame, crackles, swells, does not readily burn & smells strongly of animal matter. The genus Collocalia, according to Mr. G. R. Gray, to whom I am much obliged for allowing me to examine all the specimens in the British Museum, ranks in the same sub-family with our common Swift. This latter bird generally seizes on the nest of a sparrow, but Mr. Macgillivray has carefully described two nests, in which the confusedly felted materials were agglutinated together by extremely thin shreds of a substance which crackles but does not readily burn when put into a flame. In N. America,[1] another species of Swift causes its nest to adhere against the vertical wall of a chimney; & builds it of small sticks, placed parallel & agglutinated together with cakes of a brittle mucilage, which like, that of the esculent swallow, swells & softens in water; in flame, it crackles, swells, does not/64 bis/readily burn, & emits a strong animal odour; it differs only in being yellowish-brown, in not having so many large air-bubbles, in being more plainly laminated, & in having even a striated appearance, caused by

[1] For our Cypselus murarius, see Macgillivray British Birds. vol. 3. 1840. p. 625. For Cypselus pelasgius, see Mr. Peabodys excellent paper on the Birds of Massachusetts in the Boston Journal of Nat. History vol. 3. p. 187 M. E. Robert (Comptes Rendus, quoted in Annals & Mag. of Nat. History. vol. 8. 1842. p. 476) found that the nest of the Hirundo riparia, made in the gravelly banks of the Volga, had their upper surfaces plaistered with a yellow animal substance, which he imagined to be fishes' spawn. Could he have mistaken the species; for there is no reason to suppose that our bank-martin has any such habit? This would be a very remarkable variation of instinct, if it could be proved; & the more remarkable that this bird belongs to a different sub-family from the Swifts & Collocalia. Yet I am inclined to believe it, for it has been argued with apparent truth, that the House-martin moistens the mud, with which it builds its nest, with adhesive saliva.

innumerable elliptical excessively minute points, which I believe to be drawn out minute air-bubbles.—

Most authors believe that the nest of the esculent swallow is formed of either a Fucus or of the Roe of fish; others, I believe, have suspected that it is formed of a secretion from the salivary glands of the bird. This latter view I cannot doubt from the preceding observations is the correct one. The inland habits of the Swifts, & the manner in which the substance behaves in flame almost disposes of the supposition of Fucus. Nor can I believe, after having examined the dryed roe of fishes, that we should find no trace of cellular matter in the nests, had they been thus formed. How could our Swifts, the habits of which are so well known, obtain roe, without being detected? Mr. Macgillivray has shown that the salivary crypts of the Swift are largely developed, & he believes that the substance with which the materials of its nest are felted together, is secreted by these glands. I cannot doubt that this is the origin of the similar & more copious & purer substance in the nest of the N. American Swifts, & in that of the Collocalia esculenta. We can thus understand its vesicular & laminated structure, & the curious reticulated structure of the Philippine island species. The only change required in the instinct of these several birds is that less & less foreign material should be used. Hence I conclude, that the Chinese make soup of dried saliva!/

65/In looking for a perfect series in the less common forms of Birds' nests, we should never forget that all existing birds must be almost infinitely few compared with those which have existed since foot-prints were impressed on the beach of the New Red Sandstone of N. America.

It is be admitted that the nest of each bird, wherever placed & however constructed be good for that species under its own conditions of life; and if the nesting-instinct varies ever so little, when a bird is placed under new conditions, & the variations can be inherited, of which there can be little doubt, then natural selection in the course of ages might modify & perfect almost to any degree the nest of a bird in comparison with that of its progenitors in long past ages. Let us take one of the most extraordinary cases on record, & see how selection may possibly have acted; I refer to Mr. Goulds observations[1] on the Australian Megapodidae. The Talegalla Lathami scrapes together a great pyramid, from two to four cart-loads in amount, of decaying vegetable/66/matter;

---

[1] Birds of Australia [vol. i, pp. lxxii–lxxvii and vol. v., text to pls. 77–9.] and Introduction to the Birds of Australia 1848 p. 82.

& in the middle it deposits its eggs. The eggs are hatched by the fermenting mass, the heat of which was estimated at about 90° Fahr. & the young birds scratch their way out of the mound./ 66 v/The accumulative propensity is so strong, that a single un-mated cock confined in Sydney, annually collected an immense mass of vegetable matter. The Leipoa ocellata makes a pile, sometimes 45 feet in circumference & four feet in height, of leaves thickly covered with sand, & in the same way leaves its eggs to be hatched by the heat from fermentation./66/The Megapodius tumulus in the more Northern parts of Australia makes even a much larger mound but apparently including less vegetable matter; & other species in the Malayan archipelago are said to place their eggs in holes in the ground, where they are hatched by the heat of the sun alone. It is not so surprising that these birds should have lost the instinct of incubation, when the proper temperature is gained either from fermentation or the sun as that they should have been led to pile up before-hand a great mass of vegetable matter in order that it might ferment; for, however the fact may be explained, it is known that other birds will leave their eggs, when the heat is sufficient as in the case of the Fly-catchers, which built its nest in Mr. Knights hot-house:[1] even the snake takes advantage of a hot-bed/66 v/in which to lay its eggs; & what concerns us more, is that a common hen, according to Prof. Fischer, "made use of the artificial heat of a hotbed to hatch her eggs":[2]/ 66/& Reamur, as well as Bonnet, observed[3] that ants ceased their laborious task of daily moving their eggs to & from the surface, according to the heat of the sun, when they had built their nest between the two cases of a Bee-hive, where a proper &/67/equable temperature prevailed.

Now let us suppose that the conditions of life favoured the extension of a bird of this Family whose eggs were hatched by the solar rays alone, into a colder, damper & more wooded country: then those individuals, which chanced to have the accumulative propensity or [sic] so far modified as to prefer more leaves & less sand, would be favoured in their extension, for they would ac-cumulate more vegetable matter, & its fermentation would com-pensate for the loss of the solar heat, & thus more young birds would be hatched, which might as readily inherit the peculiar accumulative propensity of their parents, as one breed of dogs inheriting a tendency to retrieve, another point, another to dash

[1] Yarrells British Birds vol. I p. 166.
[2] Alison, Article Instinct. in Todd's Cyclop. of Anat. & Phys. p. 21.
[3] Kirby & Spence Introduct. to Entomology. Vol. 2. p. 519.

& another to dash round its prey. And this process of natural selection might be continued, till the eggs came to be hatched exclusively by the heat of fermentation; the bird of course being as ignorant of the cause of the heat, as that of its own body.

In the case of corporeal structures, when two closely allied species, one for instance semi-aquatic & the other terrestrial, are modified for their different manner of life, their main & general agreement in structure is due, according to our theory, to descent from common parents; & their slight differences/68/to subsequent modification through natural selection. So when we hear that the thrush (Turdus Falklandicus) of South America, like our European species, lines her nest in the same peculiar way with mud, though from being surrounded by wholly different plants & animals, she must be placed under somewhat different conditions;—or when we hear that in N. America, the males of two Kitty wrens,[1] like the male of our species, have the strange & anomalous habit of making several "cock-nests," not lined with feathers, in which they shelter themselves;—when we hear of such cases, & they are infinitely numerous in all classes of animals, we must attribute the similarity of the instinct to inheritance from common progenitors, & the dissimilarity either to selected & profitable modification, or to inherited & acquired habit. In the same manner, as the northern & southern thrushes have largely inherited their instinctive nidification from a common parent, so no doubt the Thrush & blackbird have likewise inherited much from their common progenitor, but with somewhat more considerable modifications of instinct in one or both species, from that of their ancient & unknown ancestor./

69/We will now consider the variability of the nesting instinct. The cases no doubt would have been far more numerous, had the subject been attended to in other countries with the same care as in Great Britain & the United States. From the general uniformity of the nest of each species, we clearly see that even trifling details, such as the materials used & the situation chosen whether on a high or low branch, on a bank or on level ground, whether solitary or in communities, are not due to chance, or to intelligence but to true instinct. The Sylvia sylvicola, for instance, can be distinguished from two closely allied wrens more readily by its nest never being lined with feathers than by almost any other character (Yarrells. British Birds).

Necessity or compulsion often leads birds to change the situation of their nests: numerous instances could be given in various parts of the world, of birds breeding in trees, but in tree-less countries on the ground, or amongst rocks. Audubon (quoted by Dr. Cabot in Boston Journ. of Nat. History

[1] Troglodytes Aedon & brevirostris; Peabody in Boston Journal of Nat. Hist. vol. 3. p. 144.—For our British Troglodytes vulgaris, see Macgillivray British Birds. vol. 3. p. 23. and Gardener's Chronicle 18. [blank].

vol. IV. p 249) states that the Gulls on an islet off Labrador, "in consequence of the persecution which they have met with, now build in trees," instead of on the rocks.—Mr. Couch (Illustrations of Instinct p. 218) states that three or four successive layings of the sparrow (F. domestica) having been/70/ "destroyed the whole colony, as if by mutual agreement, quitted the places & settled themselves amongst some trees at a distance,—a situation which though common in some districts, neither they nor their ancestors had ever before occupied here, where their nests become objects of curiosity".—The sparrow builds in holes in walls, on high branches, in ivy, under rook's nests, in the holes made by the sand-martin & often seizes on the nest made by the house martin: "the nest also varies greatly according to the place". (Montagu Ornith. Dict: Rennie p. 482) The Heron (Ardea cinerea; Macgillivray British Birds. vol 4. p. 446: W. Thompson Nat. Hist of Ireland. vol 2. p. 146) builds in trees, on precipitous sea-cliffs, & amongst heath on the ground. In the United States, the Ardea herodias (Peabody in Boston Journ. of Nat. History vol. 3. p 209) likewise builds in tall or low trees or on the ground; & what is more remarkable sometimes in communities or heronries & sometimes solitaryly.

Convenience comes into play: we have seen that the Taylor-Bird in India uses artificial thread instead of weaving it. A wild goldfinch (Bolton's Harmonia ruralis vol. 1. p. 492) first took wool, then cotton, & then down which was placed near its nest. The common Robin (Sylvia rubecula) will often build under sheds, four cases having/71/been observed one season at one place (W. Thompson. Nat. History of Ireland. vol I p 14 [actually p. 162]).). In Wales the Martin (Hirundo urbica) builds against perpendicular cliffs, but all over the lowlands of England against houses; & this must have prodigiously increased its range & numbers. In Arctic America, in 1825, the Hirundo lunifrons (Richardsons Fauna Boreali-Amer. p 331) for the first time, built against houses; & the nest instead of being clustered & each having a tubular entrance like a retort, were built under the eaves in a single line & without the tubular entrance or with a mere ledge. The date of a similar change in the habits of H. fulva is also known.—

In all changes whether from persecution or convenience, intelligence must come into play in some degree. The Kitty wren (Troglodytes vulgaris) which builds in various situations usually makes its nest to match with surrounding objects (Macgillivray vol. 3. p 21); but this perhaps is instinct; yet when we hear from White (Letter 14) that a willow wren (& I have known a similar case) having been disturbed by being watched, concealed the orifice of her nest we might argue that the case of the Kitty wren was one of intelligence./ 71 v/Neither the Kitty-wren or Water owzel (Cinclus aquaticus. [W. Thompson] Magazine of Zoology. vol 2. 1838. p. 429) invariably build domes to their nests, when placed in sheltered situations./71/Jesse describes a Jackdaw (Corvus monedula) which built its nest on an inclined/72/surface in a turret, reared up a perpendicular stack of sticks ten feet in height,—a labour of seventeen days: families of this bird, I may add (White's Selbourne Letter 21) have been known regularly to build in rabbit-burrows. Numerous analogous facts could be given. The Water-hen (Gallinula chloropus) is said usually to cover her eggs when she leaves her nest, but in one protected place, W. Thompson (Nat. Hist of Ireland. vol 2. p. 328) says that this was never done./72 v/Water-hens, & Swans, which build in or near the water, will instinctively raise their nest, as soon as they perceive the water to begin to rise (Couch Illustrations of Instinct p. 223–6). But the following seems a more curious case;/72/Mr. Yarrell showed me a sketch made by Sir R. Heron of the nest of a Black Australian Swan, which had been built directly under the drip of the eaves of a building; & to avoid this, the male & female conjointly added semicircles to the nest, until it extended close to the wall, within the line of drip; & then they pushed the eggs into the newly added portion, so as to be quite dry. The magpies (Corvus pica) under ordinary circumstances build a remarkable, but very uniform nest; in Norway they build in churches, on spouts under the eaves of houses, as well as in trees. In a treeless part of Scotland, a pair built for several years in a gooseberry bush, which they barricaded all round in an extraordinary manner with briars & thorns, so that "it would have cost a fox some days labour to have got in". On the other /73/hand in a part of Ireland, where a reward had been offered for each egg & the magpies had been much persecuted, a pair built at the bottom of a low thick hedge "without any large collection of materials likely to attract notice". In Cornwall, Mr. Couch says he has seen near each other, two nests, one in a hedge, not a yard from the ground " & unusually fenced with a thick structure of thorns"; the other "on the top of a very slender & solitary elm,—the expectation clearly being that no creature would venture to climb so fragile a column". I have been struck by the slenderness of the trees sometimes chosen by the magpie; but intelligent, as this bird is, I cannot believe that it foresees that boys could not climb such trees, but rather having chosen such a tree it has found from *experience* that it is a safe place.[1]

Although I do not doubt that intelligence & experience often come into play in the nidification of Birds, yet both often fail: a Jackdaw has been seen trying in vain to get a stick through a turret window & had not sense to draw it in lengthways; White (Letter 16) describes some martins which year after year built their nests on an exposed/74/wall, & year after year they were washed down. The Furnarius cunicularius in S. America makes a deep burrow in mud-banks for its nest; & I saw (Journal of Researches p. 216 [sic, see 2nd ed. p. 95]) these little birds vainly burrowing numerous holes through mud-walls, over which they were constantly flitting, without thus perceiving that the walls were not nearly thick enough for their nests.

Many variations cannot in anyway be accounted for; the Totanus macularius. (Peabody in Boston Journ. of Nat. History. vol. 3 p. 219) lays her eggs sometimes on the bare ground, sometimes in nests slightly made of grass. Mr. Blackwall has recorded the curious case of a yellow Bunting (Emberiza citrinella; given in Yarrell's British Birds [I, p. 444]), which laid its eggs & hatched them on the bare ground: this bird generally builds on or very close to the ground, but a case is recorded of its having built at a height of seven feet. A nest of a Chaffinch (Fringilla coelebs: [W. Thompson] Annals & Mag. of Nat. History vol. 8 1842. p 281) has been described, which was bound by a piece of whip-cord passing once round a branch of a pine-tree & then firmly interwoven with the materials of the nest: the nest/75/of the Chaffinch can almost be recognised by the elegant manner with which it is coated with lichen; but Mr. Hewitson (British Oology p. 7) has described one in which

---

[1] For Norway, see Hewitson in Mag. of Zoology & Botany 1838. vol. 2 p. 311 For Scotland. Rev. J. Hall Travels in Scotland. see Art. Instinct in Cyclop. of Anat. & Phys. p. 22.—For Ireland, W. Thompson Nat. History of Ireland. vol 1 p. 329. For Cornwall see Couch. Illustrations of Instinct p. 213.

bits of paper were used for lichen. The Thrush. (Turdus musicus) builds in bushes; but sometimes where bushes abound, in holes of walls or under sheds; & two cases are known of its having built actually on the ground in long grass & under turnip-leaves (W. Thompson. Nat. Hist of Ireland. vol I. p. 136: Couch Illustrations of Instinct p. 219). The Rev. W. D. Fox informs me that one "eccentric pair of Blackbirds" (Turdus merula) for three consecutive years built in ivy against a wall, & always lined their nest with black horse-hair; though there was nothing to tempt them to use this material: the eggs, also, were not spotted./75 v/The same excellent observer (in Hewitson British Oology. Pl. cx) has described the nests of two Redstarts, of which one alone was lined with a profusion of white feathers./75/The Golden-crested wren (Sylvia regulus: Mr. Sheppard in Linn. Transact. vol. xv. p 14 [cf. p. 20]) usually builds an open nest attached to the under side of a fir-branch: but sometimes on the branch, and Mr. Sheppard has seen one "pendulous with a hole on one side". Of the wonderful nest of the Indian weaver-bird (Ploceus Philippensis. [Burgess] Proc. Zoolog. Soc July. 27 1852) about one or two in every/76/fifty have an upper chamber, in which the males rest, formed by the widening of the stem of the nest with a pent-house added to it. I will conclude by adding two general remarks on this head by two good observers (Sheppard in Linn. Transact. vol xv. p. 14. [sic, see p. 20] & Blackwall quoted by Yarrell British Birds. vol 1. p. 444) "There are few birds which do not occasionally vary from the general form in building their nests" "It is evident", says Mr. Blackwall "that birds of the same species possess the constructive powers in very different degrees of perfection, for the nests of some individuals are finished in a manner greatly superior to those of others."

Some of the cases above given, such as the Totanus either making a nest or building on the bare ground, or that of the Water-owzel making or not making a dome to its nest, ought, perhaps, to be called a double instinct, rather than a variation. But the most curious case of a double instinct which I have met with, is that of the Sylvia cisticola, given by Dr. P. Savi: (Annales des Sc. Nat. Tom. 2. p. 126)/77/this bird in Pisa annually makes two nests; the autumnal nest is formed by leaves being sewn together with spiders' webs & the down of plants & is placed in marshes: the vernal nest is placed in tufts of grass in corn-fields & the leaves are not sewed together; but the sides are thicker & very different materials are used. In such cases, as was formerly remarked with respect to corporeal structure, a great & *apparently* abrupt change might be effected in the instinct of a bird by one form alone of the nest being retained.

In some cases, when the same species ranges into a different climate, the nest differs: the Artamus sordidus in Tasmania builds a larger, more compact & neater nest, than in Australia (Gould, Birds of Australia). The Sterna minuta, according to Audubon (Annals of Nat. History. vol. 2 1839. p. 462) in the southern & middle U. States merely scoops a slight hollow in the sand; "but on the coast of Labrador it makes a very snug nest, formed of dry moss, well matted together & nearly as large as that of the Turdus migratorius". Those individuals of Icterus Baltimore (Peabody in Boston Journ. of Nat. Hist. vol. 3. p. 97)/78/"which build in the South make their nests of light moss, which allows the air to pass through, & complete it without lining; while in the cool climate of New England they make their nests of soft substances closely woven, with a warm lining."

I think sufficient facts have now been given to show that the nests of birds do sometimes vary.

Habitations of Mammals.—On this head I shall make but few remarks, having said so much on the nests of Birds. The buildings erected by the Beaver have long been celebrated; but we see one step, by which its wonderful instincts might have been perfected, in the simpler house of an allied animal, the Musk Rat, (Fiber Zibethicus) which house, Hearne,[1] says is something like that of the Beaver. The solitary Beavers of Europe do not practice or have lost the greater part of their constructive instincts. Certain species of Bats, now uniformly inhabit the roofs of houses[2] but other species keep to hollow trees,—a change analogous to that in Swallows./79/Dr. Andrew Smith informs me that in the un-inhabited parts [of] S. Africa the hyaenas do not live in burrows whilst in the inhabited & disturbed parts they do.[3]/79 v/Several animals & birds usually inhabit burrows made by other species but where such do not exist, they excavate their own habitations.[4]/

79/In the genus Osmia, one of the Bee Family, the several species not only offer the most remarkable differences, as described by Mr. F. Smith[5] in their instincts; but the individuals of the same species vary to an unusual degree in this respect; thus illustrating a rule, which certainly seems to hold good in corporeal structure, namely that the parts which differ most in allied species, are apt also to vary most in the same species. Another Bee, the Megachile maritima, as I am informed by Mr. Smith,[6] near the sea, makes its burrows in sand-banks, whilst in wooded districts, it bores holes in posts./

80 v/*Parasitism*. The incalculable host of parasites which pass their whole lives on or in the bodies of other animals do not here especially concern us. But ever since classical days, the instinct which leads the/80/Cuckoo to lay its eggs in other birds' nests, has excited much surprise. Some species of the group, always build their own nests & hatch their own eggs. This is generally the case with the *Cuculus Americanus*, but *sometimes* this species lays its

---

1 Hearnes Travels p. 380. Hearne has given the best description (p. 227–236) ever published of the habits of the Beaver.
2 Rev L. Jenyns in Linn. Transactions vol xvi p. 166
3 A case sometimes quoted of Hares having made burrows in an exposed situation ([Otway] Annals of Nat. History vol. 5. p 362) seems to me to require verification: were not the old rabbit burrows used?
4 Zoology of the Voyage of the Beagle, Mammalia p. 90.
5 Catalogue of British Hymenoptera 1855. p. 158.
6 [See also Smith's *Catalogue*, p. 173.]

eggs in other birds nests;[1] & even our own Cuckoo (C. canorus) has not absolutely lost its aboriginal (according to our theory) instinct of nidification & incubation, as it certainly has been known to rear its young. Hence we have a series within the same small group;[2] & the only difficulty on our theory of natural section is to understand how the instinct of *occasionally* laying in other birds' nests could have first arisen, & how it could profit a species to do so habitually. In this latter respect, I think the fact ascertained in the case of two American species of Cuckoo, namely that in their nests, there are at the same time eggs just laid, young just hatched, & others ready to fly, throws some light on the subject; for the parent Cuckoos in these cases must have incubated more than twice as long as other birds,—no less than eleven young birds having been *successively* hatched in one nest.—and the first-born, when leaving the nest could hardly receive from both parents as much care as the fledglings of other birds. Hence it might well/81/be a great advantage to a Cuckoo to lay her eggs in other birds nests.[3] Nor is the first commencement of the habit so surprising as it at first seems; for numerous cases are on record of birds seizing on the nests of other species,[4] & likewise of laying in the nest of other individuals of the same species, as with Guinea-fowls, Curassows, Partridges & Thrushes,[5] & in the nests of distinct species, as the Guinea-fowls, & Land-Rails in that of the Partridge. Now according to our theory if a greater number of young were reared in consequence of such aberrant habits, then, it being probable that the propensity would sometimes be inherited, the habit might be rendered through natural selection more & more common, till it became characteristic of the species.

When a Cuckoo has laid an egg in another birds nest, it is not

---

[1] [C.D. added in pencil: 'Yarrell, I believe gives a reference.' See Yarrell, *Birds*, II p. 190.]

[2] [C.D. pencilled in margin: 'Audubon quoted by Yarrell [II, p. 192] and Couch.']

[3] M. Prevost (L' Institut 1834. p. 418) adduces evidence from a marked bird for believing that our C. canorus pairs with the male after each laying of one or two eggs; & this probably stands in intimate connexion with the supposed fact that this species like the American cuckoos lays its eggs at much longer intervals than do other birds. Why the male & female cuckoo should differ in the above respect, is not in the least known; but it is perhaps the ultimate (as far as we can now see) cause of their parasitic habits.

[4] As the sparrow on that of the swallow, the Hobby-hawk on that of the crow,; Mr. Gould in his Birds of Australia [vol. II, text for pl. 33] says that the Artamus leucopygialis often takes possession of other birds' nests.

[5] Mr. Couch, Illustrations of Instinct p. 233, has collected several instances. See Yarrell's Birds, & Azaras Nat History of Paraguay. In the Annals & Mag of Nat. Hist. vol XI. 1843 p 290, [W. Thompson,] a case is given of a nest in common of the Guinea-fowl, containing between 200 & 300 eggs.—Also Poultry Chronicle vol I. p. 456.

surprising that it should be hatched by the foster-parent, even if the period of incubation were different; for it is experimentally known[1] that birds do not instinctively know the duration of their own incubation. Nor is it surprising that the young parasite when hatched,/82/should be tenderly reared, for numberless cases are known, of birds carefully nursing & feeding the young of other species. But how that singular instinct of the nestling cuckoo, which lasts only for about twelve days, of ejecting the eggs or young of the foster-parent[2] has arisen, I will not pretend to conjecture; even when two eggs of the Cuckoo have been laid in the same nest—a good instance of a mistaken instinct—the stronger of the two nestlings ejects the weaker. The young of the American Molothrus, a bird belonging to a quite different family, but having precisely the same parasitic habits with the Cuckoo does not eject its foster brothers.[3] Can this instinct be a modification of that which teaches the young to eject their excrement over the sides of the nest? However this instinct may have arisen, if it be highly important to the young Cuckoo, as it probably is in order that so large & quickly growing a bird might be sufficiently fed by its foster-parent, then it is not in truth more surprising than any other instinct confined to the young, as a chicken picking up its own food, the nestling finch gaping to receive that ⟨food⟩ brought to it, or the young & blind pigeon inserting its bill into that of its parent to receive the regurgitated food.[4]

The case of the American ostrich (Rhea americana) is somewhat analogous to that of the Cuckoo. I have/83/elsewhere[5] shown, that four or five hens unite & together lay from 20 or 30 up to even 70 eggs first in one nest & then in a second nest & so on; & that

---

[1] Montagu Ornith. Dict. Rennie Edit. p. 161.

[2] See Jenner's celebrated paper in Philosoph. Transactions 1788. p. 226.—Macgillivray in his British Birds vol. 3. p 115 gives the fullest account of the habits of Cuckoos, which I have met with.

[3] [See Richardson, *Fauna Boreali-Americana*, part 2, Birds, p. 277.]

[4] [With this folio of the manuscript there is a typical Darwin note slip, representative of the numerous notes he originally assembled in classified portfolios. On the recto is written in a hand other than Darwin's: 'Montagu Dict. p. 164. Extraordinary growth of young cuckoo is same with what happens with others produced from eggs of disproportionately small size, as in raven, whose eggs not half size of kites & yet comes to maturity same time.' Darwin added in pencil: 'Problem if eggs of the Amer. [?] cuckoo are very small. What good? Is this false reasoning vide back of this [slip.]'

On verso Darwin added: 'Small eggs lead to choosing small birds.—Size to be arrived at & rapid maturation. Another consequence apparently of small eggs leads to the probably stronger instinct of ejection.—

'Begin with it has been supposed that small eggs adaptation to habit—wrong view Has Molothus small eggs?']

[5] Journal of Researches [2nd ed.] p. 90.

the males sit on these several accumulations of eggs. The first origin of this habit is not surprising, as we have seen that a similar & occasional habit is not very rare in the allied order of Gallinaceae; its advantages must be great, for as each hen probably lays in the several nests the above great number of eggs, & as it appears she does not lay each day successively, either the first eggs would have to remain for a great length of time before being sat upon, or the young would not come out together; and this must be almost necessary, for like chickens they follow their parents & search for their own food as soon as hatched. But I have alluded to this case chiefly to show that the instincts of the Rhea can hardly be considered as quite perfect; as owing apparently to some difficulty in the association of the females, a surprising number of eggs, called Huachos by the Spaniards, are dropped about the country & are never hatched: thus in one day's hunting on horseback in N. Patagonia I found two nests, containing together 44 eggs, & no less than twenty of these huachos or wasted eggs./

84/The great class of Hymenopterous insects abounds with infinitely numerous cases of parasitism. A multitude of Ichneumon-idae lay their eggs within the bodies of other insects; & the parasitic larva have the marvellous instinct to avoid the vital parts of their living prey.[1] But I will not attempt to discuss this case, for I know of no facts showing how these instincts could have been acquired.

In the great Bee Family, also, there are many parasites,— strictly comparable in habits with the Cuckoo; but we must suppose the Cuckoo to be parasitic on another species of the same Family. Thus the several species of Apathus closely resemble in structure & appearance common Humble Bees; but they do not live in communities, do not make nests, have no neuters, but lay their eggs in the nests of other Humble-Bees, where their larvae are tended & reared: when carefully examined, their posterior tibiae are seen not to be hollowed out & they are destitute of the instruments for collecting & carrying bee-bread, so that their structure has been modified in accordance with their parasitic habits. The/85/natural disposition & habits of Bees to a certain extent favours the numerous cases of parasitism in this family. The Bees preyed on are so pacific, that they & their parasites "mingle together in perfect harmony":[2] Mr. Smith has also seen the nests of five distinct species, belonging to three genera in close

---

[1] Westwood Modern Class. of Insects vol 2. p. 147.
[2] Catalogue of British Hymenoptera in Brit. Mus. 1855 p. 16, 46, & 225. Mr. Smith informs me that he has seen workers of Bombus pratorum in the nest of B. muscorum; & the workers o[f] this latter species in the nest of B. sylvarum.

proximity. This author has, also, many times noticed in autumn a stray *worker* of one species of Humble-Bee domiciled in the nest of another species.

Another Hymenopterous group, the Sphegidae, ordinarily have the habit of half-killing or paralysing other insects with their stings, thus storing up fresh food for their larvae to feed on when hatched. I cannot forbear here just alluding to the marvellous instinct[1] which leads one species of Sphex to sting the abdominal surface of a cricket in two or three spots, where its nervous ganglia lie, & another species of Sphex to sting a Curculio in one spot where its single ganglion is situated,—thus far transcending the instinct of the ferret, which after a little practice leads it to bite the medulla oblongata of the rat, or the intelligence of the dog, which after a single battle learnt the mortal places where to seize the badger./86/Though such are the ordinary habits of the Sphegidae, yet some species[2] are parasitic, depositing their eggs in the cells already stored with prey by other species. And in this case we can see, as with the Cuckoo, how the instinct might have been perfected through natural selection; for Mr. Fabre[3] has given good reason for believing that though the Tachytes nigra is organised to make its own burrows & catch its own prey, yet that when it finds a nest made & stored by a Sphex, it takes advantage of it & so becomes for the occasion parasitic.

Instincts of neuter Social Insects. In the eighth chapter, I have stated that the fact of a neuter insect often having a widely different structure & instinct from both parents, & yet never breeding & so never transmitting its slowly acquired modifications to its offspring, seemed at first to me an actually fatal objection to my whole theory. But after considering what can be done by artificial selection, I concluded that natural selection might act on the parents, &/87/continually preserve those which produced more & more aberrant offspring, having any structure or instinct advantageous to the community. Having already amply discussed this difficulty, which I do not at all wish to underrate, I will not here allude to it,/87 v/excepting to remark that it is highly important, as it shows, if our theory be true, that the most wonderful & complicated instincts may be acquired through the continued selection of slight modifications of the parental instincts, without the smallest aid having been derived from inherited habit./87/But

---

[1] See Mr. Fabre's most interesting paper in Annales des Sci. Nat. 4 Series. Tome vi. p. 157.

[2] Westwood Modern Class. vol 2. p 209, 212.   [3] Annal. des Sc. id. p. 147.

quite independently of this special difficulty of the neuters not having offspring, social insects have almost innumerable instincts, marvellous & most difficult of explanation. Every attempt to explain a complex instinct must rest on our knowledge of a graduated chain of simpler instincts of the same group, & on our finding that these are occasionally in some slight degree variable. But in the case of social insects, any accurate knowledge is confined to *extremely* few species (the long line of extinct species being of course excluded) and to those alone inhabiting Europe. I will, therefore, make very few remarks under the present head, but will choose the two cases, which seem to have struck all observers with the utmost astonishment; namely the slave-making expeditions of certain ants, & the comb-making powers of the Hive-bee./

88/Slave-making ants have been found in Europe, S. America & India. But the habits of only two species, in Europe, have been carefully observed. The workers of Formica (Polyerges) rufescens are so constructed that they cannot make their own nests or feed their own young: so helpless are they, that Huber found that when thirty of them were shut up in a glazed box, together with their own larvae & pupae, & with honey for food, that they actually could not or would not feed themselves, & many perished of hunger; but when he introduced a single slave-worker (F. nigra), she instantly fed & saved the survivors, made a cell & tended the larvae: when this species migrates from one nest to another, the masters are carried away by the slaves. The other pretty well-known slave-making species, F. sanguinea, has workers of its own of two kinds: the smaller workers seem to have the ordinary habits of workers, &, as I am informed by Mr. F. Smith, they search for food & tend their Aphides: the larger workers are warriors & slave-makers, defending the nest & capturing the slaves, which consist of workers of Formica fusca, F. flava & Myrmica rubra. So that, in the formicaries of Formica sanguinea, we have males, females & workers of two kinds of that species, all differing in appearance, & slave-workers of three/89/distinct species;[1] & on the neighbouring trees, we may find their aphides, which may almost be called their domesticated cattle![2]/

[1] F. Smith in British Formicidae. Transact. Entomology Soc. vol. 3. Part III. p. 95. I am much indebted to this gentleman for much interesting information on this subject. All the other facts on Slave-making ants are taken from Kirby & Spence Introduction to Entomology. vol. 2. ch XVII. Mr. Smith has never seen stray workers of one species domiciled in the nest of another not-slaving species; so that the pupa which according to Gould are sometimes seized on & on the fate of which Kirby doubts [,] are probably devoured.

[2] [Darwin presented the following passage as a long foot-note.]

89 v/Our cattle, in comparison with the cattle of less civilised countries, certainly seem to yield their milk far more readily through inherited habit. When ants visit their Aphides to collect the sweet excretion from their bodies; it is known that they tap them in a peculiar manner with their antennae & that the Aphides then spout out their excretion: is it too fanciful to believe that this habit during myriads of generations, has become hereditary or instinctive in the Aphides? if so they may truly be said to have become domesticated. The Cocci, also, yield their excretion to the ants, which give a different signal to them, from what they do to the Aphides. Several species of Staphylinidae are likewise kept in various parts of the world, apparently for some secretion./89 v¹/In Brazil the ants derive the sweet juice secretion by the larvae of Cicadella, Cercopis & Membracis; (Lund, in Annal des Sc. Nat. Tom 23 (1831) p. 126); & what is very remarkable, when Aphides were introduced on foreign plants near Rio de Janeiro, the native ants found them out, & used them, as European ants do their Aphides.—/89 v/Ants guard their Aphides like we do our cattle; & they confine the root-feeding species within their formicaries; & still more wonderful, they take great care of the eggs of the Aphides, & in order that they may be hatched soon, carry them about as Kirby witnessed, so as to be in the warmest places. Hence we see that man is not the only animal which keeps domesticated animals. (See Kirby & Spence's Introduction to Entomology vol. 2. Ch. XVII.)—/

89/The slaves of F. sanguinea, as far as Mr. Smith has seen, work only within the nests of their masters: when these latter migrate instead of being carried, they carry their slaves. Here then in comparing the habits of the only two well known slave-making species, (F. rufescens & sanguinea) we see some gradation. If the larger workers of F. sanguinea were rendered, by the natural selection of the parents which produced them, more & more instinctively inclined to take slaves, & slaves of the right kind, all their wants might come to be supplied by their slaves, & they might by disuse be unfitted to perform any work for themselves: & then the smaller workers of F. sanguinea might cease to be produced; & the communities of F. sanguinea would be in the condition of those of F. rufescens, & as abjectly dependent on their slaves as in that species.

How the instinct first originated of taking slaves, we are left to conjecture: but some circumstances favour the commencement of such a habit. If the pupae of one species of ant are placed near the nest of another they are greedily carried away: & King asks, whether they are carried away to be devoured or educated./90/ Ants of different species are, also, sometimes extremely sociable; Mr. F. Smith has seen Myrmica nitidula, M. muscorum & Formica rufa living in harmony within the very same nest. Education has a remarkable effect on the instinct of ants: F. nigra reared in its own nest has a horror of the slave-taking species, but reared in their masters nest, they seem quite contented with their lot; &

Huber found that even the two slave-making species if brought up together from the pupa-state in an artificial formicary lived together in perfect amity.

It is a marvellous part of the instinct of the slave-making ants, that they take only those larvae & pupae which will turn into workers, & which alone would be useful to them; but what is far more wonderful, the slaves of F. rufescens instinctively prevent their masters going on their marauding expeditions, until the time arrives when there are worker-larvae & pupae ready in the nests of the species to be attacked. This, if really well ascertained, would seem to imply that the instincts of the slave-species have been adapted to serve the ends of the distinct & hostile, master-species. If this could be proved to be the case, it would assuredly be a fatal objection to my theory; for it could not profit/91/the parents of the species which are subjected to slavery that their neuter off-spring should well serve another species & therefore there could be no natural selection for this end; nor could long-continued habit however originating, have become hereditary for the breeding individuals of the slave-species are never made slaves. But may it not be, that some instinct, as the prevention of a too early migration &c, proper to the slave-species in their own nests, come to be modified under the peculiar conditions of slavery, & thus incidentally is rendered serviceable to the master-species? We should never forget that if the slaves did not check their masters going out too early on their marauding expeditions, & that in consequence the community suffered, the simple result would be that this particular slave-making species would add one more to the myriads of species which have disappeared from the face of the earth; or to speak more correctly, the slave-making species would never have come into existence./

92/*Bees comb.*—A mere casual inspection of a Bee's comb must strike everyone with the liveliest admiration; & when we hear that the bees have practically solved a most difficult geometrical problem, namely that of making their cells with the least possible expence of wax, by constructing hexagons, alternately opposed to each other, with bases formed of three rhomboidal plates with angles of 109° 28' and 70° 32' and inclined to each other at an angle of 120°,[1] which angles mathematicians have tried & proved

---

[1] Lord Brougham, Dissertations on Natural Theology 1839 vol i. p 244. M. Lalanne in Annal. des Scien. Nat. 2 series. Zoologie. Tome 13. 1840. p. 361. gives the angles of the rhomdoidal plates, as 109°. 28' 16", and 70° 31' 44".—I may here add that M. Lalanne believes (p. 372) that the whole structure & all the angles of the cells follow from the formation of the trapezium of the first row which

to be the correct ones;—when we hear all this, our admiration is silenced into bewilderment. Here then, it might be thought we assuredly have a case of a specially endowed instinct, which could not possibly have been arrived at by slow & successive modifications, such as are indispensably necessary on our theory. Let us see what the facts are.—

Of Social Bees not many genera are known, & the habits of extremely few species in them have been carefully examined. Bees existed during the Tertiary periods & perhaps as far as back to the Jurassic era;[1] & all analogy tells us that if we could see the combs of these extinct species, we should have many intermediate structures. But looking to the known cells & combs of living species, extremely few though/93/they be, we can clearly see stages in the scale of perfection. At one end of the short series, we have the marvellously perfect comb of the Hive-Bee; at the other end, the nests of the Bombi or Humble-bees. These latter consist of oval cocoons & of spheroidal pots of various sizes made of soft wax (secreted from the abdominal rings), placed together in irregular combs, connected by small pillars of wax: the empty cocoons & pots made of wax are filled with honey & pollen: the cocoons are sometimes elongated by a cylinder of wax:[2] the cells & Irregular combs are placed either in a hole or within a nest made of moss which is internally coated with wax.[3] In all these circumstances we clearly see that the irregular pots are the analogues of the hexagonal cells of the Hive-Bee. In the instinctive construction of the pots, there is no greater difficulty than in a bird making its nest,—in the case of the Hirundo esculenta nest from its dried saliva.

Now let us look at the cells & comb of the Mexican Melipona domestica described by so high an authority as Pierre Huber[4] The comb consists of cylinders, which are first used for the larvae & sometimes subsequently for honey, and moreover of larger subspherical pots, aggregated together & filled with honey or pollen.

are attached to the hive; & these he thinks the insects might without much difficulty, taking their own bodies as measures, construct; but this theory does not seem applicable to wasp's combs, which start from their bases, & are not laterally attached by any trapeziums to the walls of their nest.—

[1] Pictet Paléontologie (2 Edit.) Tom. 2. p. 384.

[2] Kirby & Spence Entomology. vol 1. p. 504. Huber in Transact of Linnean Soc. vol vi p. [blank.—See pp. 238–9.]

[3] As Birds sometimes vary according to convenience, but sometimes without any assignable cause the materials of their nest, so it is, as Mr. F. Smith (Catalogue of British Hymenoptera 1855. p 212) has shown with Humble-Bees. He records one case of a nest made entirely of short pieces of horse-hair, to collect which the Bees entered a stable. In another case some Humble-Bees took possession of a Robin's nest, & adopted it for their purpose.—

[4] Mémoires de Soc. Phys. de Geneve Tom viii. p. 1 et seq.

The cylinders & pots are/94/made of soft wax: they are arranged in nearly circular discs, which increasing in diameter from the point of suspension downwards, and united one to the other by irregular pillars, are suspended within a wooden hive lined with wax. These parallel discs evidently make an approach to the parallel combs of the Hive Bee. The small cylinders are of equal sizes, & being packed close together, they give to the comb, as Huber remarks, the appearance of a hexagonal reticulation. The irregularly spherical pots are of about the same size & are *closely* agglomerated together: hence arises the very important fact, that in those places where the pots would have intersected each other if completed, there is a single intervening wall common to the two adjoining pots & this is formed into *a perfectly plane surface*; these intervening walls are of the same thickness as the outer curved parts of the pots, when not in contact with other pots. The flat intervening walls are of various size & shapes, dependent on the number & size of the pots, which come irregularly into contact. Where the base of one pot is surrounded by other pots, it consists wholly of plane surfaces meeting at various angles, which clearly is a feeble approach to the angular but incomparably more regular outline of the cell of the Hive-Bee. In the Hive Bee the walls of those cells on one side of the comb exactly correspond with, & are built on, the three ridges formed by the union of the three basal rhomboidal plates of the cell on the opposite side of the comb: now as Huber/95/remarks & shows by his figures—and the remark seems to me a most important one—"mais ici comme dans les célèbres fonds pyramidaux des abeilles d'Europe, on pouvait observer que les angles formés par la rencontre des plans d'une même loge ou case, répondaient exactement aux cloisons élevées de l'autre côté sur les bords de ces plans". Again speaking of "ces fonds pyramidaux tant vantés" of the Hive Bee, he says in the cells of the Melipona, "partout on voit trois arrêtes exterieures se réunir en un point qui correspond à la sommité de la pyramide formée par les plans voisins les uns des autres".... "partout il y a fond pyramidaux où trois fonds se rapprochent". It is impossible not to see in these several facts an approach to the admirable structure of the cell of the Hive-Bee: we must conclude with Huber that "on retrouvait ainsi sous une forme grossière, l'esprit de la construction allveolaire des ruches Européennes."

The genus Melipona in structure & instinct is nearer to Humble-bees (Bombus) than to Hive-Bees (Apis); & I think there is no great difficulty in believing/96/that the instinct of Bombus could be modified through natural selection so as to arrive at that modest degree of perfection, required by the Melipona to make its irregular combs. If the honey-pots of the Melipones had been made of exactly the same size & had been arranged at equal distances (& the cells for the larvae are arranged at equal distances) in a single series in the same plane, there can be no doubt, as we shall immediately show that here would have been produced the peculiar comb of our Hive-Bee.—⟨that each would have been an hexagonal outline; for each would necessarily have been surrounded by six pots, & the surfaces of intersection would have been, as they now are, flat, & they would have been equal, for the pots are supposed to be equal & placed at equal distances. If the pots had been aggregated in a double layer or series, each on one surface corresponding to the interspace between those on the opposite surface, taken from what we see in the actual comb of the Melipones, each pot on the inside would have had its bases formed of a three-sided pyramid; & as the pots are supposed to be equal & placed at equal distances, the sides of the pyramid would have been equal. The ridges on the outer or projecting side of the pyramid would have carried, as we see in the actual comb of the Melipona, the flat intervening walls of the adjoining pots on the outer or opposite side of the comb. Hence on the above suppositions, we should have had cells closely similar to, perhaps identical (for I am not geometrician to calculate)/97/those of the Hive bee.⟩ Wax is a precious secretion, & we can clearly see that its economy in the comb of the Hive-Bee, & the admirable strength gained by the form of the cells & by the manner in which the bases interlock, must be a great advantage to this species. Therefore I conclude that the marvellous comb of the Hive-Bee—the most marvellous instinct known—does not present an insuperable difficulty on our theory of natural selection or the preservation of each profitable modification of instinct; for the instinct which leads the Melipona to build its honey-pots so near to each other that they intersect, a plane surface consequently serving, for a common wall to two or more adjoining pots, might have been so far perfected as to lead the insect to make all the pots of still more exactly equal size, & to arrange them in a double series on a level surface, at equal distances; in which case, as we have just seen, we should have had cells most closely resembling or rather identical with the admirable cells of the Hive-Bee.[1]/

[1] [Darwin later added in pencil: 'Theory of the construction of the cells of Hive Bees.—Mr. Waterton has given a theory, very like that above hinted at'
    In the Darwin MSS. there is much material about bees. In the section of vol. 46.2 entitled 'Habits of Bees', are many note scraps made by Darwin from July 1840 at Maer to Sept. 1862 at Bournemouth. At the end of vol. 48 is a collection of notes, diagrams and drafts regarding the construction of bees' cells.]

98/I have now discussed several of the most extraordinary classes of instincts; but I have still a few miscellaneous remarks which seem to me worth making. First for a few cases of variation which have struck me: a spider which had been crippled & could not spin its web, changed its habit from compulsion into a hunter,— which is the regular habit of one large group of spiders.[1] Some insects have two very different instincts under different circumstances or at different times of life; & one of the two might through natural selection be retained & so cause an apparently abrupt difference in instinct in relation to the insects nearest allies: thus the larvae of a beetle (the Cionus scrophulariae) when bred on the Scrophularia exude a viscid substance, which makes a transparent bladder within which it undergoes its metamorphosis; but their larva when naturally bred, or transposed by man, on to a Verbascum, becomes a burrower & undergoes its metamorphosis within a leaf.[2] In the caterpillars of certain moths, there are two great classes, those which burrow in the parenchyma of leaves, & those which roll up leaves with consummate skill; some few caterpillars in their early age are burrowers, & then become leaf-rollers; & this change was justly considered so great, that it was only lately discovered that the caterpillars belonged to the same species.[3] The Angoumois moth/99/annually has two broods; the first are hatched in the spring from eggs laid in the autumn on grains of corn stored in granaries, & these immediately take flight to the fields & lay their eggs on the standing corn, instead of on the naked grains stored all round them: the Moths of the second & autumnal brood, (produced from the eggs laid on the standing corn) are hatched in the granaries, & these do not leave the granaries, but deposit their eggs on the grains around them; & from these eggs proceed the vernal brood which has the different instinct of laying on the standing ears.[4]/99 v/Some hunting spiders, when they have eggs, & young, give up hunting & spin a web wherewith to catch prey; this is the case with a Sa[l]ticus, which lays its eggs within snail-shells, & at that time spins a large vertical web.[5]/

99/The pupae of two species of Formica are *sometimes*[6] uncovered or not enclosed within cocoons: this certainly is a highly remarkable variation; the same thing is said to occur with the common Pulex.

[1] Leach on authority of Sir J. Banks in Transact. Linn. Soc. vol. [. see vol. II (1815), 393.]
[2] P. Huber in Mém. Soc. Phy. de Genève Tom. x. p. 33.
[3] J. O. Westwood, in Gardener's Chronicle 1852. p. 261.
[4] Bonnet, quoted in Kirby & Spence's Entomology vol. 2. p. 480.
[5] Dugès in Annal. des Sci. Nat. 2 series. Tom VI. p. 196.
[6] F. Smith in Transact. Ent. Soc. vol 3. N.S. Part III. p 97. And De Geer quoted in Kirby & Spence's Entolomogy. vol. 3 p. 227.

Lord Brougham[1] gives as a remarkable case of instinct, the chicken within the shell picking a hole & then "chipping with its bill-scale, till it has cut off a segment from the shell. It always moves from right to left, & it always cuts off the segment from the big end." But the instinct is not quite so invariable; for I was assured at the Eccalobeion (May 1840) that cases have occurred of chickens having commenced so close to the broad end, that they could not escape from the hole thus made, & had consequently to commence chipping again/100/so as to remove ⟨another & larger⟩ circular rim of shell; moreover occasionally they have begun at the narrow end of the shell.—The fact of the occasional regurgitation of its food by the Kangaroo[2] ought, perhaps, to be considered as due to an intermediate or variable modification of structure rather than of instinct, but it is worth notice. It is notorious that the same species of Bird has slightly different vocal powers in different districts; & an excellent observer remarks that "an Irish covey of Partridges springs without uttering a call, whilst the Scotch covey on the opposite coast shrieks with all its might when sprung."[3] Bechstein says that from many years experience he is certain that in the nightingale a tendency to sing in the middle of the night or in the day runs in families & is strictly inherited.[4] It is remarkable that many birds have the capacity of piping long & difficult tunes, & others, as the Magpie of imitating all sorts of sounds, & yet that in a state of nature, they never display these powers.[5]

As there is often much difficulty in imagining how an instinct could first have arisen, it may be worth while to give a very few out of many cases, of occasional & curious habits, which/101/cannot be considered as regular instinct, but which might according to our views give rise to such. Thus several cases are on record,[6] of insects which naturally have very different habits having been hatched within the bodies of man,—a most remarkable fact considering the temperature to which they have been exposed, & which may explain the origin of the instinct of the gad-fly or Oestrus. We can see how the closest association might be developed

[1] Dissertations on Natural Theology. vol I. p 117.
[2] W. C. Martin. in Mag. of Nat. History New S. vol. 2. 1838. p. 323.—
[3] W. Thompson in Nat. History of Ireland vol. 2. p. 65. says that he has observed this, & that it is well known to sportsmen.
[4] Stubenvögel. 1840. s. 323. See on different powers of singing of several species in different places. s. 205, 265.
[5] Blackwalls Researches in Zoology 1834. p. 158. Cuvier long ago remarked that all the passeres have apparently a similar structure in their vocal organs; & yet only a few & those the males, sing; showing that fitting structure does not always give rise to corresponding habits.
[6] Rev. L. Jenyns. Observations in Nat. History. 1846 p. 280.

in Swallows, for Lamarck[1] saw a dozen of these birds aiding a pair, whose nest had been taken, so effectually that it was completed on the second day; & from the facts given by Macgillivray[2] it is impossible to doubt that the ancient accounts are true that martins sometimes associate & entomb alive sparrows which have taken possession of one of their nests. It is well known that Hive-Bees, which have been neglected "get a habit of pillaging from their more industrious neighbours"[3] & are then called corsairs; & Huber gives a far more remarkable case of some Hive Bees which took almost entire possession of the nest of a Humble-Bee, & for three weeks these latter went on collecting honey & then regorged it at the solicitation, without any violence, of the Hive-Bees. We are thus reminded of those Gulls (Lestris) which exclusively live by pursuing other gulls & compelling them to disgorge their food.[4]

In the Hive-Bee, actions are occasionally performed, which we must rank amongst the most wonderful of instincts; &/102/yet these instincts must often have lain dormant during very many generations: I refer to the death of the queen, when several worker-larvae are mercilessly destroyed & ["one or more queen-grubs selected out of the unhoused working grubs"][5] being placed in large cells & reared on royal food are thus rendered fertile: so again when a Hive has its queen, the males are all infallibly killed by the workers in autumn; but if there be no queen, not a single drone is ever destroyed.[6] Perhaps a feeble ray of light is thrown by our theory on these mysterious but well ascertained facts, by considering that the analogy of other members of the Bee family would lead us to believe that the Hive-Bee is descended from other Bees, which regularly had many females inhabiting the same nest during the whole season & which never destroyed their own males; so that not to destroy the males & to give the normal food to additional larvae, perhaps is only a reversion to an ancestral instinct; & in case of corporeal structure reversions are apt to occur after many generations.

I will now refer to a few cases of special difficulty on our theory, —most of them parallel to those which I adduced when discussing in Ch. VIII corporeal structures. Thus we occasionally meet with the same peculiar instinct in animals widely remote in the scale

[1] Quoted by Geoffroy St. Hilaire in Annales du Museum Tom IX. p 471.
[2] British Birds vol. 3. p. 591.
[3] Kirby & Spence's Entomology vol. 2. p. 207. The case given by Huber is at p. 119.
[4] There is reason to suspect (Macgillivray British Birds vol. 5. p. 500) that some of the species can only digest food, which has been partially digested by other birds.
[5] [The point inadvertently skipped by Darwin is quoted from Kirby & Spence, *Entomology*, II, 511.]   [6] Kirby & Spence Entomology vol. 2. p 510–513.

of nature, & which consequently cannot have derived the peculiarity from community of descent. The Molothrus (a bird something like a starling) of/103/N. & S. America has precisely the same habits with the Cuckoo; but parasitism is so common throughout nature that this coincidence is not very surprising. The parallelism in instinct between White-ants, belonging to the Neuroptera & ants, belonging to the Hymenoptera is a far more wonderful fact; but the parallelism seems to be very far from close. Perhaps as remarkable a case as any on record of the same instinct having been independently acquired in two animals very remote from each other in relationship is that of a Neuropterous & Dipterous larva digging a conical pit-fall in loose-sand, lying motionless at the bottom, & if the prey is about to escape casting jets of sand all round.[1]

It has been asserted that animals are endowed with instincts, not for their own individual good or for that of their own social bodies, but for the good of other species, though leading to their own destruction: it has been said that fishes migrate that birds & other animals may prey on them;[2] this is impossible on our theory of the natural selection of self-profitable modifications of instinct. But I have met with no facts, in support of this belief worthy of consideration. Mistakes of instinct, as we shall presently see, may in some rare cases do injury to a species & profit another; one species may be compelled or even apparently induced by persuasion to yield up its food or its secretions to another species; but that any animal has been specially endowed with an instinct/ 104/leading to its own destruction or harm, I cannot believe without far better evidence than has hitherto been advanced.[3]

An instinct performed only once during the whole life of an animal appears at first as a great difficulty on our theory; but if indispensable to the animal's existence, there is no valid reason why it should not have been acquired through natural selection, like corporeal structures used only on one occasion, like the hard tip to the chicken's beak ⟨for breaking out of its egg⟩; or like the temporary jaws of the pupae of the Caddis-fly or Phryganea, which are exclusively used for cutting open the silken doors of its curious case, & which are then thrown off for ever.[4] Nevertheless it is impossible not to feel unbounded astonishment, when one reads of such cases, as that of a caterpillar first suspending itself by its

[1] Kirby & Spences Entomology vol. 1. p. 429–435.
[2] Linnaeus [Oratio de memorabilibus in Insectis] in Amoenitates Acad. vol. 2 [3rd ed. p. 389] Prof. Alison on "Instinct" p. 7, 15. in Todd's Cyclopaedia of Anat. & Physiology.
[3] [Here Darwin added in pencil: 'case of ants & aphis.']
[4] Kirby & Spences Entomology vol. 3 p. 287.

tail to a little hillock of silk attached to some subject & then undergoing its metamorphosis; then after a period splitting open on one side & exposing the pupa, destitute of limbs or organs of sense & lying loose within the *lower* part of the old bag-like split skin of the caterpillar; this skin serves as a ladder, which the pupa ascends by seizing on portions between the creases of its abdominal segments, & then searching with its tail, which is furnished with little hooks, thus attaches itself & afterwards laboriously disengages & casts off the skin which had served it for a ladder.[1] I am tempted to give one other analogous case, that of the caterpillar of a Butterfly, (Thekla [sic]), which feeds within the pomegranite, but when full-fed/105/gnaw their way out (thus making the exit of the butterfly possible before its wings are fully expanded) & then attach with silk-threads the fruit to the branch of the tree, that it may not fall before the metamorphosis is completed. Here, as in so many other cases, the larva works on this one occasion for the safety of the pupa & of the mature insect. Our astonishment at this manoeuvre is lessened in a very slight degree when we hear that several caterpillars attach more or less perfectly with silken threads leaves to the stem for their own safety; & that another caterpillar, before changing into a pupa bends the edges of a leaf together, coats one surface with a silk web & attaches this web to the footstalk & branch of the tree; the leaf afterwards becomes brittle & separates, leaving the silken cocoon attached to the footstalk & branch; in this case, the process differs but little from the ordinary formation of a cocoon & its attachment to any object.[2]

A really far greater difficulty is offered by those cases, in which the instincts of a species differ greatly from those of its related forms. This is the case with the above mentioned Thekla of the pomegranite; & no doubt many instances could be collected. But we should never forget what a small proportion the living must bear to the extinct, amongst insects, the several orders of which have so long existed on this earth. Moreover, just in the same way as/106/with corporeal structures, I have been surprised how often when I thought I had got a case of a perfectly isolated instinct, I found on further enquiry at least some traces of a graduated series.

I have not rarely felt that small & trifling instincts were a greater difficulty on our theory, than those which have so justly excited the wonder of mankind; for an instinct if really of no considerable importance in the struggle for life could not be modified or formed

[1] Kirby & Spences Entomology vol. 3 p. 208–211.
[2] J. O. Westwood in Transact. of Entomolog. Soc. vol. 2. p. 1.

through natural selection. Perhaps as striking an instance as can be given, is that of the workers of the Hive-Bee, arranged in files, & ventilating by a peculiar movement of their wings the well-closed Hive: this ventilation has been artificially imitated,[1] & as it is carried on even during winter there can be no doubt that it is to bring in fresh air & displace the carbonic acid gas; therefore in truth it is indispensable, & one may imagine the stages—a few bees first going to the orifice to fan themselves—by which the instinct might have been arrived at. We admire the instinctive caution of the Hen-pheasant, which leads her as Waterton has remarked to fly from her nest, & so leave no track to be scented/ 107/out by beast of prey; but this again may well be of high importance to the species. It is more surprising that instinct should lead small nesting birds to remove the broken eggs & the early mutings whereas with partridges, the young of which immediately follow their parents, the broken eggs are left round the nest; but when we hear that the nests of those birds (Halcyonidae), in which the mutings are not enclosed by a film, & so can hardly be removed by the parent, are thus "rendered very conspicuous;"[2] & when we remember how many nests are destroyed by cats, we cannot any longer consider these instincts of trifling importance. But some instincts, one can hardly avoid looking at as mere tricks or sometimes as play; an Abyssian pigeon when fired at, plunges down so as almost to touch the sportsman & then mounts to an immoderate height;[3] the Biz[c]acha (Lagostomus) almost invariably collects all sorts of rubbish, bones, stones, dry dung, near its burrow; Guanacoes have the habit of returning (like Flies) to the same spot to drop their excrement & I saw one heap eight feet in diameter; & as this habit is common to all the species of the genus, it must be instinctive, but is hard to believe that it can be of any use to the animal, though it is to Peruvians, who use the dryed dung for fuel.[4] Many analogous facts could probably be collected./

[1] Kirby & Spences Entomology vol. 2 p. 193.
[2] Blyth in Mag. of Nat. History New Series vol. 2. 1838 p. 354. From a fact given in a note in Jenyns Edit of Whites Selbourne 1843. p. 214, it would appear that the envelope of the mutings is connected with the inactive life of the young; & has only incidentally been advantageous in favouring their removal. The Rev. W. Darwin Fox informs me that he has attended to the case of the parents removing the broken egg-shells, & that there can be no doubt of its truth. See on this head Macgillivray British Birds vol. 3. p. 730.
[3] Bruce's Travels vol 5. p. 187.
[4] See my Journal of Researches p. 167 for the Guanaco; for the Bizcacha p. 125. Many odd instincts are connected with the excrement of animals, as with the wild Horse of S. America (see Azara's Travels. vol. I. p. 373.), with the common House Fly, & with dogs; see on the urinary deposits of the Hyrax, Livingstone's Missionary Travels. p. 22.

108/Wonderful & admirable as most instincts are, yet they cannot be considered as absolutely perfect; there is a constant struggle going on throughout nature between the instinct of the one to escape & of the other to secure its prey. If the instinct of the Spider be admirable, that of the Fly which rushes into the toils is so far inferior. Rare & occasional sources of dangers are not avoided; if death inevitably ensues & caution cannot be learnt by seeing others suffer, it seems that no guardian instinct is acquired; thus the ground within a solfatara in Java is strewed with the carcasses of tigers, birds & masses of insects killed by the noxious exhalation, with their flesh, hairs & feathers preserved, but their bones entirely corroded.[1] The migratory instinct not rarely fails & animals, as we have seen, are lost. What ought we to think of that strong impulse which leads Lemmings, Squirrels, ermines[2] & many other animals, which are not regularly migratory, occasionally to congregate & pursue a headlong course, across great rivers, lakes & even into the sea, where vast numbers perish; ultimately it would appear that all/109/perish. The country being overstocked seems to cause the original impulse; but it is doubtful, whether in all cases scarcity actually prevails. The whole case is quite inexplicable. Does the same feeling act on these animals, which causes men to congregate under distress & fear; & are these occasional migrations or rather emigrations a forlorn hope to find a new & better land? The occasional emigrations of insects of many kinds, associated together, which, as I have witnessed, must perish by countless myriads in the sea, are still more remarkable, as they belong to families none of which are naturally social or ever migrate.[3]

The social instinct is indispensable to some animals, useful to still more for the readier notice of danger, & apparently only pleasant to some few animals. But one cannot avoid thinking that this instinct is carried in some cases to an injurious excess: the

---

[1] Von Buch Descript. Phys. des Iles Canaries 1836 p. 423—on the excellent authority of Mr. Reinwardts.

[2] L. Lloyd, Scandinavian Adventures 1854. vol. 2. p. 77 gives an excellent account of the migrations of Lemmings; when swimming across a lake if they meet a boat, they crawl up one side & down the opposite side. Great migrations took place in 1789, 1807, 1808, 1813, 1823. Ultimately all seem to perish. See Högströms account in Swedish Acts. vol 4. 1763 of ermines migrating & entering the sea.— See Bachman's, in Mag. of Nat. History. N. S. vol 3. 1839. p. 226, account of the emigrations of Squirrels; they are bad swimmers & yet cross great rivers.

[3] Mr. Spence in his Anniversary address to Entomologic Soc. 1848 has some excellent remarks on the occasional migrations of insects, & well shows how inexplicable the case is.—See, also, Kirby & Spences Entomology. vol. 2, p. 12. See, also, Weissenborn in Mag. of Nat. Hist. N.S. 1839. vol. 3. p. 516, for interesting details on a great migration of Libellulae, generally along the course of rivers.

antelopes in S. Africa & the Passenger Pigeons in N. America are followed by hosts of carnivorous beasts & birds, which could hardly be supported in such numbers if their prey was scattered. The Bison in North America migrates in such vast bodies, that when they come to/110/narrow passes in the river-cliffs, the foremost, according to Lewis & Clarke (?) [sic][1] are often pushed over the precipice & are dashed to pieces. Can we believe that when a wounded herbivorous animal returns to its own herd & is then attacked & gored, that this cruel, but very common instinct is of any service to the species? It has been[2] remarked that with Deer, only those which have been much chased by Dogs are led by a sense of self preservation to expel their pursued or wounded companion, who would bring danger on the herd. But the fearless wild elephants will ungenerously attack one which has escaped into the jungle with the bandages still on its legs."[3] And I have seen domestic pigeons attack & badly wound sick or young & fallen birds.

The cock-pheasant crows loudly, as everyone may hear, when going to roost, & is thus betrayed to the poacher.[4] The wild Hen of India, as I am informed by Mr. Blyth, chuckles, like her domesticated offspring, when she has laid an egg; & the natives thus discover her nest. In La Plata, the Furnarius rufus builds a large oven-like nest of mud, in as conspicuous a place as possible; on a bare rock, on the top of a post, or cactus-stem;[5] & in a thickly/ 111/peopled country, with mischievous boys, would soon be exterminated. The great Butcher-bird conceals its nest very badly & the male during incubation & the female after her eggs are hatched betray the nest by their repeated harsh crys.[6] So again

---

[1] [Cf. Lewis & Clark, *Travels* new ed. (London, 1817) vol. I, pp. 371–2, entry for June 17, 1800.]

[2] W. Scrope's Art of Deer Stalking p. 23.

[3] Corse, in Asiatic Researches vol 3. p 271. This fact is the more strange as an Elephant, which had escaped from a pit was seen by many witnesses ([Roughsedge,] Athenaeum 1840 p. 238) to stop & assist with his trunk his companion in getting out of the pit. Capt. Sulivan R.N. informs me that he watched for more than half an hour, at the Falkland islands a logger-headed Duck defending a wounded Upland Goose from the repeated attacks of a carrion Hawk. The upland goose first took to the water & the Duck swam close along side her, always defending her with its strong beak; when the goose crawled ashore the Duck followed, going round & round her, & when the goose again took to the sea, the Duck was still seen vigorously defending her; yet at other times this Duck *never* associates with this goose for their food & place of habitation are utterly different. I very much fear from what we see of little birds chasing hawks that it would be more philosophical to attribute this conduct in the Duck to hatred of the carrion hawk rather than to benevolence for the goose.—

[4] Rev. L. Jenyns, Observations in Natural History. 1846. p. 100.

[5] [Darwin] Journal of Researches p. 95.

[6] Lanius excubitor. Jesse [actually Knapp], Journal of a Naturalist p 188.

a kind of shrew-mouse at the Mauritius continually betrays itself by screaming out as soon as approached. Nor ought we to say that these failures of instinct are unimportant, as probably concerning man alone, for as we see instinctive wildness directed towards man, there seems no reasons why other instinct should not be related to him.

The number of eggs of the American ostrich, scattered over the country & so wasted has already been noticed. The Cuckoo sometimes lays two eggs in the same nest, leading to the sure ejection of one of the two young birds. Flies, it has often been asserted, frequently make mistakes, & lay their eggs in substances not fitted for the nourishment of their larvae. A spider[1] will eagerly seize a little ball of cotton, when deprived of her eggs enveloped as they are in a silken envelope; but if a choice be given her, she will prefer her own eggs, & will not always seize/ 112/the ball of cotton a second time: so that we see sense or reason here correcting a first mistake. Little birds often gratify their hatred by pursuing a Hawk & perhaps by so doing distract its attention; but they often mistake & persecute (as I have seen) any innocent & foreign species. Foxes & other carnivorous beasts often destroy far more prey than they can devour or carry away: the Bee Cuckoo kills a vast number more Bees than she can eat; & "unweariedly pursues without interuption this pastime all the day long."[2]/112 v/A queen Hive-bee confined by Huber, so that she could not lay her worker eggs in worker cells would not oviposit, but dropped them, upon which the workers devoured them/112/An unfertilised queen can lay only male eggs, but these she deposits in worker & royal cells,—an aberration of instinct not surprising under the circumstances; but "the workers themselves act as if they suffered in their instinct from the imperfect state of their queen, for they feed these male larvae with royal jelly & treat them, as they would a real queen".[3] But what is more surprising the workers of Humble-bees habitually endeavour to seize & devour the eggs of their own queens; & the almost incessant activity of the mothers is "scarcely adequate to prevent this violence."[4] Can this strange instinctive habit be of any service to the Bee? Seeing the innumerable/113/& admirable instincts all directed to rear & multiply their young, can we believe with Kirby & Spence, that this strange aberrant instinct is given them

[1] A Lycosa, these facts are given by Dugès in Annal: des Science. Nat. 2 series. Zoolog. Tom VI p 196.
[2] Bruce's Travels in Abyssinia vol. 5. p. 179.
[3] Kirby & Spences Entomology (3 Edit.) vol. 2. p. 161.   [4] Idem vol I—p 380.

"to keep the population within due bounds."? Can the instinct, which leads the female spider savagely to attack & devour the male after pairing with him[1] be of service to the species? The carcase of her husband no doubt nourishes her; & without some better explanation can be given, we are thus reduced to the grossest utilitarianism compatible, it must be confessed, with the theory of natural selection. I fear that to the foregoing cases, a long catalogue could be added.

Conclusion.—We have in this chapter chiefly considered the instincts of animals under the point of view whether it is possible that they could have been acquired through the means indicated on our theory, or whether, even if the simpler ones could thus have been acquired, others are so complex & wonderful that they must have been specially endowed, & thus overthrow the theory. Bearing in mind the facts given on the acquirement, through the selection of self-originating tricks or modifications of instinct, or through training & habit, aided in some slight degree by imitation, experience & intelligence, of hereditary/114/actions & dispositions in our domesticated animals; & their parallelism (subject to being less true) to the instincts of animals in a state of nature: bearing in mind that in a state of nature instincts do certainly vary in some slight degree: bearing in mind how very generally we find in allied but distinct animals a gradation in the more complex instincts, which shows that it is at least possible that a complex instinct might have been acquired by successive steps; & which moreover generally indicates, according to our theory, the actual steps by which the instinct has been acquired, in as much as we suppose allied animals to have branched off at different stages of descent from a common ancestor, & therefore to have retained, more or less unaltered, the instincts of the several lineal ancestral forms of any one species; bearing all this in mind, together with the certainty that instincts are as important to an animal as is their generally correlated structure, & that in the struggle for life/115/under changing conditions, slight modifications of instinct could hardly fail occasionally to be profitable to individuals, I can see no overwhelming difficulty on our theory. Even in the most marvellous instinct known, that of the cell of the Hive-bee, we have seen how a simple instinctive action may lead to results which fill the mind with astonishment.

Moreover it seems to me that the very general fact of the

---

[1] Idem Vol I—p 280, a long list of several insects which either in their larval or mature condition will devour each other is given.

gradation in complexity of Instincts within the limits of the same group of animals; and likewise the fact of two allied species, placed in two distant parts of the world & surrounded by wholly different conditions of life, still having very much in common in their instincts supports our theory of descent, for they are explained by it; whereas if we look at each instinct as specially endowed, we can only say that so it is. Imperfections & mistakes of instinct on our theory cease to be surprising: indeed it would be wonderful that far more numerous & flagrant cases/116/could not be detected, if it were not that a species which failed to become modified & so far perfected in its instincts that it could continue struggling with the coinhabitants of the same region, would simply add one more to the myriads which have become extinct. It may not be logical, but to my imagination, it is far more satisfactory to look at the young Cuckoo ejecting its foster-brothers,—the larvae of the Ichneumonidae feeding within the live bodies of their prey—cats playing with mice, otters & cormorants with living fish, not as instincts specially given by the Creator, but as very small parts of one general law leading to the advancement of all organic beings,—Multiply, Vary, let the strongest forms by their strength Live & the Weakest forms Die.—

CHAPTER XI

GEOGRAPHICAL DISTRIBUTION

HISTORICAL INTRODUCTION AND
SPECIAL EDITORIAL COMMENT

Early in 1856 while Darwin was reviewing and organizing his materials, before and at the start of his writing of *Natural Selection*, his concentration on an evolutionary explanation of the geographical distribution of plant and animal species and concurrent experiments on transport of viable seeds in dirt and of small young land molluscs, and his opposition to the extremes of continental extensions or land bridges suggested by Edward Forbes, Hooker, Wollaston, and Woodward are all reflected in his letters.[1] In answer to the letter of April 16, 1856, in which Charles Bunbury had written encouragingly that 'I trust you will not on any account give up the idea of publishing your views' (see Introduction p. 8), Darwin sent the following letter:

Down Bromley Kent
April 21 [1856]

My dear Bunbury

You are quite right, I do take a very great interest about the Cape Flora & Fauna, & I thank you much for your letter, which, as all yours do, has pleased & instructed me much.—I have lately been especially attending to Geograph. Distrib., & most splendid sport it is,—a grand game of chess with the world for a Board. The fact you allude to about the zoology (at least mammifers) of the Cape not being nearly so peculiar as the Botany has often struck me much: I think the most probable HYPOTHETICAL explanation is that it was long a group of islands, since united with the continent allowing the vertebrates to enter.—Thank you about the Colletia, I called on Lindley, but c^d extract nothing & wrote to the Gardener who raised the seed, (but have not, & shall not receive any answer) to ask whether he ever had seed from S. America of any kind; undoubtedly the common form was in the Garden.

I am very glad to hear you are still thinking of Madeira; there seems to me much to be done there yet; but I hear from Mr. Lowe, he is going to publish a Flora, & he has sent me a curious account of vegetation of P. Santo. A careful comparison of the Floras of Madeira, Azores, & Canary Is^d would, I cannot doubt, lead to

---

[1] L & L, ii, 68, 72–8, 80–2; NY, i, 427, 431–6, 438–41.

some very curious results. You speak in FAR too flattering a way about my work, in which I will persevere; & I will endeavour (eheu how difficult) "to be cautious & candid & avoid dogmatism". My determination to put difficulties, as far as I can see them, on both sides is a great aid towards candour; because I console myself, when finding some great difficulty, in endeavouring to put in as forcibly as I can.—I am trying many *little* experiments, but they are hardly worth telling, though some I am sure will bear on distribution & I think on *aquatic* plants. As you say you like scientific chat, & your kind letter makes me sure that you will not think me an egotistical bore, I will tell you of a theory I am maturing (by the way please do not mention it to anyone, for 2 directly opposite reasons: viz. whether valueless or valuable).

As glacial action extended over whole of Europe, & in Himalaya, on *both* sides of N. America & *both* sides of Southern S. America & I believe in N. Zealand, within very late times (existence of recent species); I cannot but think the *whole* world must have been rather colder during the Glacial Epoch: (I know I ought to be able to show that the glacial action was *actually* & *absolutely* coincident in North & South, & this I cannot do, nor can I here enter in details to show *how far* I can show them coincident) At this period I look at the intertropical plants as somewhat distressed, but not (or only a few) exterminated. ⟨because there were no other species fitted to a ⟨warmer⟩ colder climate, & therefore able to seize their place.⟩—Under these conditions I consider it probable that some of the warmer temperate plants would spread into the Tropics, whilst the arctic plants reached the foot of the Alps & Pyrenees. (according to poor Forbes' view; by the way I had this part of the theory *written out*,[1] 4 years before Forbes published!) Some, I consider it possible might cross the Tropics & survive at C. of Good Hope, T. del Fuego & S. Australia; but within the Tropics, when warmth returned, all would be exterminated, except such as crawled up ⟨the⟩ mountains, as in Ceylon, Neilghiries, Java, Organ Mountains in Brazil. This theory, I conceive, explains certain aquatic productions in S. hemisphere &c &c. (& European Fish at C. of Good Hope)—But on the view that species change, it throws, I think, far more light on the analogous, but not identical species, on the summits of the above named mountains. Of course I cannot enter in details (& you would not care to hear them) on the subject, which I am sure in some degree would render the view more probable than it will seem to you at first.—You will probably object, why have so many more northern

---

[1] In the 1842 Sketch and the 1844 Essay, see *Foundations*, pp. 30–1, 165–8.

species & forms gone to the south, than southern forms come to the north; I can explain this only on a pure ⟨simple⟩ hypothesis of cold having come on first from the north; but there has been *some* migration from south to north, as of Australian forms on *mountains* of Borneo. And I am sure I have notes of a few S. African forms, as wanderers across the Tropics, into N. Africa & Europe: is not this so with Gledichia, Stapelia (?) Can you help me in this, either identical species, or allied forms, of well marked S. African forms? By the way I look at Abyssinia, *during the cold period*, as the channel of communication; for some (as I know from Richard) very northern temperate species of plants are found there; & some S. African forms likewise.—There, I am sure, you will agree that I have prosed enough on my own doctrines; which I may have to give up, but I strongly suspect that the theory is a sound vessel & will hold water. I look at the vegetation of the Tropics, during the cold period, as having been somewhat like the vegetation described by Hooker at foot of Himalaya, as essentially Tropical, but with an odd mixture of Temperate forms & even identical species, before they became mostly modified.—What will you say to such a dose of speculation! You will exclaim, "he is a pretty fellow to talk of caution"!

Pray believe me
Yours very sincerely
CHARLES DARWIN

If at any time you are inclined to write, pray attack my doctrine. —With respect to diffusion of water plants in *very distant* regions, it seems, as far as my doctrine is concerned, sufficient answer that the same species of water plants in the same continent are very widely diffused, & whatever the means of diffusion may be, the same means w^d tend to carry them to the most distant parts during the cold period.—The same argument is applicable to the Glumaceae to some extent; but Decandolle thinks that certain lowly organised phanerogams, which are very widely diffused (I forget whether he includes Glumaceae, which I think some authors consider the highest of the monocots?) are diffused owing to such species having been *very anciently* created & therefore having had *more time* to become diffused. I doubt whether he has any grounds for his belief, without it be a feeble analogy of the greater duration of mammifers compared with molluscs.—[1]

[1] Original letter at Bury St. Edmunds and West Suffolk Record Office; the reference is E.18/750/14.

During the summer of 1856 letters to his long time friend, Joseph Dalton Hooker, botanist at Kew, were particularly concerned with geographical distribution. That of July 13, notifying Hooker that he planned to send him a fair copy of about forty pages for comment, suggests that Darwin may then already have nearly or completely finished his first draft of the present chapter; but on July 30 he wrote: 'My MS. will not, I fear, be copied before you go abroad.'[1] After reorganizing and expanding his first draft, he had a fair copy made which was probably completed before October 10th. This copy he submitted for review to Hooker with the following questions:

Dr Hooker

Please read this first

I want, especially to know whether Botanical facts are *fairly* accurate. 2[d]. any general or special criticisms: please observe if you will mark margin with pencil, if your criticisms run to any length, I would gladly & gratefully come to Kew, to save you writing.

I *really* hope no other chapter in my book ⟨is⟩ will be so bad; how *atrociously* bad it is, I know not; but I plainly see it is too long, & dull, & hypothetical.

Do not be *too* severe, yet not too indulgent: remember that it will be *extra* dull to you, for it will be a compilation with hardly anything new to you.—

It is only fragment of chapter, & assumes some points as true, which will require *much* explanation,—as to close relation of plants to plants rather than to conditions: again I am unfortunately forced not to admit continental extension as you know.—

Glance at the notes, at back of Pages.—

In truth you are doing me a very GREAT kindness in reading it, for I am sorely perplexed what to do & how much to strike out.—[2]

After an earlier encouraging note[3], Hooker wrote on November 9, 1856:

I have finished the reading of your MS., and have been very much delighted and instructed. Your case is a most strong one...I never felt so shaky about species before....I have a page or two of notes for discussion, many of which were answered, as I got further on with the MS., more or less fully. Your doctrine of the cooling of the Tropics is a startling one, when carried to the length of supporting plants of cold temperate regions....[4]

The four pages of notes mentioned by Hooker, which accompanied Darwin's manuscript, are to be found in the appendix.

[1] ML, no. 49; L & L, II, 81; NY, I, 439.
[2] C.D. MSS. vol. 50: glacial sequence, fol. 9.
[3] See C.D.'s reply, ML no. 332.   [4] ML no. 333.

In four letters during November 1856,[1] Darwin clarified and maintained those of his points which Hooker had questioned, and in a letter of April 26 [1858] to Lyell mentioning this section, Darwin reported that recently Hooker 'was inclined to come round pretty strongly to my views of distribution and change during the glacial period,'[2] but this carries us well beyond the writing period for this part of the manuscript.

COMMENT

The consultation with Hooker reveals Darwin's thoughts and questions while composing this section, but, together with Darwin's revisions, this presents us with a complex chronological sequence of manuscript material.

Two stages of Darwin's holograph draft are recognizable from an earlier folio numbering which he later cancelled and replaced by a new number sequence. Where both folio numbers are present, they are both shown, with the earlier cancelled numbers enclosed in angle brackets, so that the unmistakably earlier folios can be recognized by the reader. During his revision Darwin occasionally cut off parts of the earlier folios including the top parts bearing the original numbers but where these are reasonably clear from the continuity of text, the probable original numbers are supplied within angle brackets with a question mark added.

The facts that, of the earlier stage of the draft represented by the earlier number sequence, the folios originally numbered 43 to 45 were set aside in the course of revision to be put in the short separate unit on representative species, and that, of this earlier state, the final folio preserved and incorporated into the separate long unit was originally numbered 40, all suggest that the earlier stage may already have been completed and that Darwin may actually have counted its folios by July 13, 1856, when he wrote to Hooker asking him to do him a favour in 'five or six weeks': 'to read, but well copied out, my pages (about forty!!) on Alpine floras and faunas, Arctic and Antarctic floras and faunas, and the supposed cold mundane period.'[3] It is clearly evident that during his revision Darwin expanded his introductory section, because the earlier folio, originally numbered 8 became folio 16. Of the whole earlier sequence of folios originally numbered from 8 to 40 most were reused and renumbered. Some original sheets were partly cut up, and occasionally new sheets were interpolated but very few sheets of this original folio sequence appear to have been completely discarded in favour of expanded or completely rewritten passages.

A third clear stage of the draft is the fair copy. Like everyone else, the copyist who made it had difficulty reading Darwin's handwriting. Darwin did not correct all of the resulting mistakes, so that significant errors such as 'Geological Society' instead of 'Survey', 'found' instead of 'formed', 'Vennica' instead of 'Veronica', and 'Mr. Burk' instead of 'Busk' remain uncorrected in the fair copy. This makes it unsatisfactory as the basic text, nor would its folio numbers reveal Darwin's earlier revisions. Therefore, Darwin's holograph is used as the basic text; where necessary, as in the case of Darwin's table of contents, Hooker's notes, and the marginalia, folio numbers referring to the fair copy have been changed to fit the numbering of the folios of this basic text. Darwin wrote various changes and additions

---

[1] ML nos. 334–6, and L & L, II, 86–7; NY, I, 444–5.
[2] L & L, II, 113; NY, I, 470.   [3] ML no. 49.

on the fair copy. Of these, single words and short phrases have been incorporated into the text here published without special note. Longer such additions are indicated by folio numbers preceded by 'Fair Copy' or 'F C'. Of these, Darwin wrote some in ink and clearly intended them as formal additions to his text, and they are so treated. Others written in pencil are memoranda of questions Darwin asked himself or of points and references he noted as agenda for further work. For one of these latter scribbles on the slip of paper designated Fair Copy 6 v, Darwin supplied a specific date: October 10, 1856. Cancellations related to the content of this addendum and made on pages of the fair copy suggest that this copy had probably been completed before this date.

On October 9, Darwin had received a copy of the September issue of Silliman's *American Journal of Science*, containing the first part of Asa Gray's article on 'Statistics of the Flora of the Northern United States.'[1] This replaced and greatly extended the information available in Gray's 1843 article in the *London Journal of Botany*. It would seem that Darwin was just about to send the fair copy off for Hooker's perusal, for he merely added a few rough changes to the fair copy without again resorting to the copyist. He cancelled the passage at the end of holograph folio 5 and its now superseded note. He added brief memoranda to himself regarding the information available in Gray's new article. On the half sheet designated Fair Copy 6 v, he scribbled an addendum about Oswald Heer's bold assumption about a land bridge connecting Europe and America, and added a memorandum regarding Gray's botanical evidence against Heer's theory, alluding to Gray's new article. Then in the margin of the fair copy he scribbled 'dele' and drew a vague line alongside of the passage from holograph folio 8 which I have indicated as cancelled. Later, in Chapter 4, on Variation under Nature, and in the addendum to this chapter, entitled 'Wide ranging, common and much diffused species tend most to vary', he made very good use of the statistics Gray supplied in this article.

Another stage in the history of the draft shows us the exchange of ideas between Darwin and Hooker. These comments and queries, which both men added in pencil to the fair copy, are here presented as footnotes with asterisks and daggers relating them to the text as closely as is possible for scribblings which may run along the margins in a direction at right angles to the copyist's lines. Besides substantial comments, Hooker made stylistic interpolations and substitutions on the fair copy as he mentioned in his letter to Darwin of November 9, 1856.[2] These I have simply omitted, as being non-Darwinian, of minor significance regarding the manuscript, and as being awkward to present without distracting interference with easy reading of Darwin's text. Hooker sometimes signed his notes and comments with his initials. In the other cases the two authors of the notes could be distinguished by their hand-writing, and they are identified by their initials added within square brackets.

Page long footnotes did not dismay Victorian publishers, nor discourage thousands of Victorian book buyers, and Darwin used this style of exposition when it suited him.[3] Several long passages which are presented as notes in the fair copy have here been incorporated into the text, but set off in reduced type, to conform more closely with present-day publishing practice.

[1] ML no. 331. [2] ML no. 333.
[3] *Variation* (1868), II, 375–6.

During Darwin's revision of the earlier state of his holograph draft he set aside a few sheets towards the end, including those originally numbered 43, 44, and 45. These dealt with 'representative species', that is, closely allied species filling the same place in the economy of nature in separate habitats. In the course of his revising he cancelled parts of his earlier text, cut up sheets, and rewrote, and he added new pages as well as renumbering the earlier ones, all in a separate sequence. This all constitutes a separate discussion of this topic, which is also distinguished in the fair copy by a separate series of folio numbers from 1 to 5. On folio. 13 of chapter XI, Darwin mentions that representative species will be considered in the 'next chapter', and his pencil note to Hooker at the top of folio 1 of the fair copy of this section on representative species explains that 'This will come towards end of *another* chapter.' In the *Origin*, of course, Darwin did devote two chapters to geographical distribution. To place this brief isolated section at the conclusion of Darwin's formal text would seem to end with an anticlimax, and so it will be found in the appendix, together with some other draft sections on foolscap sheets found with Darwin's notes on geographical distribution.

GEOGRAPHICAL DISTRIBUTION

0/*As I believe that all organic beings are produced by the ordinary laws of reproduction which includes, according to the theory under discussion, modification of specific forms, & as it is exceedingly improbable that the *same* species should ever have been generated in one place from one set of parents, & in another place, (especially if under different conditions) from another set of parents specifically dissimilar, the first & most obvious question is whether we can account on the ordinary notion of propagation for the existence of the same identical species in all quarters of the world.—This is the question, which has long agitated naturalists, namely whether the same species has been created once & therefore at a single point, or more than once at different points.

After giving general reasons in favour of single production; consider the many & grave difficulties. The most prominent of them may be grouped into three following classes.—First, insular productions of the temperate & tropical latitudes.†—Here give reasons for doubting vast continental extensions of Forbes & Co:/ 0 v/Give condensed means of dispersal.—

Secondly, range of Fresh Water Productions.—

Thirdly, as follows.—

1/We now come to our third class of facts, namely the existence of the same species of plants & animals on mountains distant from each other, & likewise on the lowlands nearer the pole, where

* Copy this page on separate paper. [C.D.]
† Perhaps allude to A. Decandolle on large-scaled Plants. [C.D.]

the climate is nearly similar; as alpine plants & animals could not possibly migrate through the lowlands from one distant alpine point to another, until lately this, perhaps, was one of the strongest cases which could have been adduced by those who believe in the same species having been created at more than one point of the earth's surface. It is familiar to every one that several* plants grow, for instance, on the summits of our Scotch mountains, & on the lowlands of Northern Scandinavia & not in the intermediate low country. So the mountain plants of the Alps are separated by a space of [1000] miles[1] from the northern land clothed by the same species. Similar observations have been made on insects.[2]/ 2/At the height of 8000–9000 feet on the Alps, we meet with northern grouse. But the most striking cases are afforded by mammals, for here we avoid accidental dispersions, as seeds borne by the wind or birds.—It would be a prodigy to find such northern animals as the Steinbock, which lives on the Alps at a height of 8000–9000 ⟨7000–9000⟩ feet, or the field-mouse (Arvicola oeconomica) which lives at the height of 10,000 feet, or the Variable hare &c, in the low country at the northern base of these mountains: equally striking it is to observe in Johnston's Physical Atlas of Europe the small brown patches, marking the far separated & Alpine homes of the Chamois.[3] Some few northern plants, but far more generally representatives of northern genera, have been observed on the more southern mountains of Spain & Greece; but to this subject we shall return.—†

These facts have been explained with beautiful simplicity by the late Professor E. Forbes: from the presence of the innumerable, ice-borne, great fragments of northern rocks scattered over the temperate zones of Europe,—from the former far lower descent of glaciers,—& more especially from the several Arctic shells imbedded in the drift, we know to a certainty that/3/Europe, during a quite recent geological period, suffered under a severe climate; & Prof. Forbes believes that at this Glacial epoch, the seas & land were colonised by arctic forms, which when the climate

* many [J.D.H.]

† I have not yet noticed migration from N. to So. America [?] [C.D.]

[1] See A. Decandolle, Bot: Geograph: p. [1007–13] for an excellent summary on this subject.

[2] Thus the Elaphrus Lapponicus which inhabits Lapland & Kamtschatka was found by Sir Charles Lyell on the Grampians: according to Erichson Tachinus elongatus is common to Sweden, Unalashka & the Alps of Switzerland. So again Latreille (Memoires du Museum Tom. 3, p. 39) says that Prionus depsarius is found in Sweden & on the Swiss mountains; Lycus minutus in the boreal regions & in Cantal. [Hooker here added: 'Get from Adam White the alpine Himalayan & Tibetan insects of Thomson.']   [3] [Pl]. 28.

became warmer, *remigrated** northward, but where the land was high the plants, insects & mammals ascended the mountain-peaks & have lived there ever since. Hence it has come, that we now see the same forms at distant points, impenetrably separated by wide extents of surface fully occupied by the productions of the now temperate regions.

Prof. Forbes believed[1] that the mountains of temperate Europe existed during the Glacial epoch as islands, & the seeds of northern plants were brought to these islands by icebergs.† That the greater part of Europe (& of Northern America) was under sea during some part of the glacial epoch is certain; but it is equally certain that this epoch endured for an enormous period, & that there were great contemporaneous changes of level,—as indeed most geologists would infer from the rocks scratched by floating ice at such different levels. I do not believe that there is any evidence to show/4/whether the greater part of Europe existed as land at the commencement of the glacial epoch, so as freely to allow of the southward migration of northern forms; but that there was continuous land before the close of the cold period may be safely inferred from the distribution of the arcto-alpine mammals; & this is likewise confirmed by such cases as that remarked on by Mr. H. C. Watson,[2] namely that the alpine flora of Britain is much more nearly related to that of the country northward of it, than to the floras of the Alps of middle Europe; if the remigration northward had been by various accidental means, this relation could hardly have been preserved.‡

We may indeed infer that the land in some parts of Europe stood even higher than at present, before the close of the glacial epoch, as is indicated by the presence of the Alpine Hare on the Scotch mountains, showing that Great Britain & Europe were then united. Let no one think that these great migrations southward & remigrations northward of a whole body of species are improbable, for it is hardly too strong an expression to say that in the case of the sea-shells the migration has/5/been witnessed[3] by Prof. E. Forbes, so beautifully do the fossil shells on the shores

* for the most part migrated. ?perhaps created on the southern land & sea when cold is Forbes exact on this point? [J.D.H.]

† It is difficult to believe that during glacial epoch the northern land was warm enough for any plants at all. see Note A. [J.D.H.] [See p. 575.]

‡ very good—it is this confounded relation that obtains so much in more distant plants. [J.D.H.]

1 Memoirs of Geological Survey Vol I, p. 399 &c.
2 Cybele Britannica, Vol I. p. 37.
3 Memoirs of Geolog. Survey. Vol I. p. 385.

of England & in the Mediterranean, compared with those living before & since, show the course of the migrations.

In North America we meet with similar facts of distribution. Dr. Asa Gray has most kindly given me a list of 59 plants (only 33 strictly alpine) growing near the summit of the White Mountains in N. Hampshire at the height of 5000–6000 feet. He believes that every one of them, with only two exceptions, inhabit Labrador situated 400 or 500 miles to the North: the great majority of these plants, viz: about 46, inhabit the circumpolar regions of Europe or Asia. No less than 33, or above half, grow on the Alps of Europe! about 22 of these same plants have been found on the summits of the mountains of New York, (separated from the White Mountains by a plain 60 or 70 miles in width) which afford a fitting site of only a few acres in extent.[1] ⟨Even in the mountains of Carolina[2] "a list of the shrubs & [herbaceous] plants [of this mountain] would be found to include a large portion of the common productions of the extreme northern states & Canada." But I do not know whether the Alleghanies do not now afford a highway* by which the plants could have travelled thus far south.⟩/

6/We shall, perhaps, better realise what formerly took place, if we imagine a glacial epoch now to come on again. The extreme arctic productions of all kinds are at present mostly the same round the pole: as the cold came on, whether or not strictly contemporaneously in Europe & N. America, then similar forms would slowly travel south, along the shores & land; & when the warmth returned they would return to their native north; but where there were mountains, as the ice & snow thawed & left the rocks uncovered, the northern forms would ascend, & would become surrounded by the stream of living beings flowing up from the south. And as the parent circumpolar productions were mostly the same, so would the alpine productions of the Old & New Worlds left on the mountain-summits on the returning warmth be to a great extent the same in the two worlds. We can thus understand the truly wonderful case of more than half the plants of the White Mountains being the same with those on the Alps of Europe,

* Yes there is [C.D.]

[1] [In the fair copy, Darwin here wrote: '(Briefly contrast these with *general* proportion of whole U. States Flora compared to Europe.)' He probably wrote this and cancelled the rest of the paragraph, including the next note, in October, 1856, when he found new and better facts in the article by Gray discussed in note 3, p. 538. He added: '(Yes there is)' after 'highway' in the last sentence of the paragraph, probably also at this time.]

[2] ⟨Dr. Asa Gray in London Journal of Botany. 1843. p. 114.—see Bartram for the Occone Mountains; Travels in N. America p. 335.⟩

though separated by the whole Atlantic ocean, & on each side by a broad belt of low land, on which these Alpine productions could not possibly exist.—Had not the glacial epoch been brought to/6 v/light & generally recognised, in main part owing to Agassiz, this case of the identical species on the White Mountains & Alps, might have been advanced as a grand proof of double creation./

7/But in our recent imaginary change of climate we found the circumpolar productions mostly the same; & I should infer that this must have been the case before the real glacial epoch came on. Prof. Forbes believes[1] that N. America and Europe were connected by continuous land, situated far to the north, during the glacial epoch, or towards its close: but the identity of several plants on mountains so far south as the Alps of Switzerland with those of America, seems/7 v/to show that the connexion between the Old and New Worlds had been established before the close of the cold period, otherwise the species common to America could not have got so far south as the Alps. During the most intense part of the glacial epoch, I can hardly doubt that land lying "far to the north"/7/would have been covered by ice & snow, like the Antarctic islands at the present day under corresponding latitudes, & hence would not have allowed of the passage or of the existence of barely a single terrestrial production. Hence it is that I infer that the connexion was anterior to the Glacial epoch.*/

Fair Copy 6 v/Dr. Oswald Heer[2] boldly supposes that land stretched continuously from West Europe to East America, sending promontories to Iceland in the north & to the Canary Islands in the south & that this land endured till the end of the Tertiary period, which would include the glacial epoch. To those who can freely admit such enormous geographical changes on no other evidence besides distribution, this view will satisfactorily[3] account for many of the phenomena./

* Certainly J.D.H. Note B [See p. 576.]

[1] Memoirs of Geolog: Survey Vol I. p. 383 and 402.

[2] Band xv der neuen Denkschriften des allgem. Schweizerischen Gesellschaft. [This whole paragraph, which seems to fit best here, is on a slip, all written in Darwin's hand, now marked 6 v, in with the fair copy.]

[3] [Darwin later underlined 'satisfactorily' thrice in pencil, and added two question marks and an arrow pointing to the following memorandum:] Give argument against continuous land between Europe & N. America from the proportion of plants common to Europe & U. States, *not* greatly exceeding those in common between Asia & U. States Oct. 10. 56. See A. Gray in Silliman [Darwin received a copy of Gray's article (*Amer. J. Sci.* vol. 22, no. 65 (Sept., 1856), 204–32.) on Oct. 9, 1856 (see ML, Ltr. no. 331.) On p. 229 Gray, tabulating Phaenogamia, gives as the 'whole number of species in northern United States':—2091; number 'extending into Asia':—308; number 'extending into Europe':—321.]

7/At the present day it appears[1] that almost every one of the few mammals and birds, common to the two worlds range up to the shores of the Arctic Seas. In regard to plants, I find in the table to Mr. Watson's Remarks on the Geographical distribution of British plants, that/8/about 500[2] of these British plants (approximately 1400 in number) are found both in N. America & Europe; ⟨& of these 500, 110 do not either in the Old or New World range into Mr. Watson's arctic & polar regions, i.e.; not further north than nearly the line of the arctic circle: 60 other plants range only on one side of the Atlantic, (generally on the European side, warmed by the Gulf-stream) into the Arctic Zone. Some, but extremely few,* of the plants common to the two worlds, do not range northward of the latitude of the northern point of Great Britain. But Mr. Watson informs me that since his publication in 1835 our knowledge has been much increased, & that the above numbers can be considered only as approximate.⟩ With respect to the sea-shells, common to the shores of Europe & N. America, I am informed by Mr. Woodward, that about one-third of their number do not range into Forbes' Arctic Sea,[3] which washes the northern shores of Asia & America: but this one-third includes most of the doubtful cases, so that the number of species in common, which do not reach the arctic zone, is probably considerably less than one-third. As in the case of the plants, some 4 or 5 shells, which are common to the Old &/9/New Worlds do not range even into Forbes' boreal province, & are therefore separated by the whole width of the Atlantic. But these shells/ 9 v/& the few plants similarly circumstanced probably belong to a distinct category, & were common to the old & new worlds long before the Glacial epoch: this, at least might well have been the case with some of the land shells in common, for these as I am informed by Mr. Woodward, are known to occur as older pliocene fossils. With respect to this more ancient connexion between the two worlds we shall have briefly to return./

9 A/If we now look to a map of the circumpolar regions, we find near the Arctic circle almost continuous land & sea-coast

---

* Asa Gray thinks there are not a few plants common to U.S. & Europe, which do *not* range to Arctic regions. [C.D.] "Certainly" J.D.H.

[1] See Sir J. Richardson's admirable Report on N. American Zoology to British Assoc: 1836; and Fauna Boreali America.

[2] [Darwin later underlined 500, and in parentheses between the lines added: 'NO give Asa Gray's facts, far more accurate,' and also wrote 'dele' in the margin of the lines giving Watson's figures, evidently preferring those he had found in the article by Gray referred to in the preceding note.]

[3] See map, of Marine Life, in Johnston's Physical Atlas 2nd Edit. [pl. 31.]

from Lapland to Eastern America, & by going further north even almost to Eastern Greenland. Therefore if all the organisms, which are now common to Europe & America, could flourish under the present climate between the Arctic circle & 70° (& a great majority do now live there) I can see no insuperable difficulty to their having in the course of ages circulated round the polar regions by this course. No doubt the distance is very great, viz in the parallel of 70°, between 6000 & 7000 miles; but we know that most of the productions on this long line are now the same; & many species of fish & marine shells have even a wider range in the Indo-Pacific ocean.[1]/9/Shortly before the Glacial epoch came on, during an earlier part of the pliocene period, when most of the organic beings were the same as now, we may fairly infer that the temperature of the* northern hemisphere was *slightly* warmer, perhaps more equable, than now; & as there can be no doubt judging from their southern limits that almost all arctic productions can well withstand a slightly warmer climate, I can see no great difficulty in supposing that all the organic beings, now common to the/10/two worlds, inhabited the long line of shore-land from eastern America to Lapland. As the glacial epoch came on, the species in common, associated with some not in common, would have migrated southward, & subsequently as the warmth returned they would have remigrated northward to their present homes. During these two great migrations, & with the local changes of climate which we might expect to ensue, it would be strange if several of the species did not become locally extinct. Hence we might expect to find in favourable situations, both on the southern high lands & in their latest northern homes, nests of species occuring elsewhere but not in the country immediately adjoining on either hand. Such nests seem to occur on the more temperate promontory of N. W. America, as I am informed by Dr. Hooker; and again near Lake Baikal[2] in Lat: 52°, in the very middle of Siberia. In this same country, Gmelin[3] gives several strong cases of plants

* ? Atlantic shores of—(do we know what pliocene temp. of Asia was? [J.D.H.]

[1] For Fish see Report to Brit. Assoc. for 1845 on the Ichthyology of the Seas of China by Sir J. Richardson, p. 190, 191.

[2] See M. Turczaninow in Bull. de la Soc. Imp. de Moscow. 1842, p. 15 account of this Flora. He is much surprised at half the Baikal Flora being the same with that round St. Petersburgh: it has, also, 452 species in common with Sweden: He specifies also species in common with N. America. See Gmelin, Flora Siberica p. cxiv, for the relations of the vegetation of the mountains near Baikal with that of Kamtshatka. See Ledebour (in Hooker's Miscellany. Vol 2. p. 241) for the similarity of the vegetation on the lower part of the Altai with that of Europe; & Sir W. Hooker in Linnean Transactions Vol xiv. p. 360.

[3] Flora Siberica. p. cix, p. cx.

with interrupted ranges & was thus led, even in the year 1747 to infer that/11/the same species had been created on diverse points of the earth's surface.

I have remarked that during the most intense part of the Glacial epoch, when arctic plants lived at the foot of the Alps, the most northern land & islands, as Spitzbergen, Iceland, Feroe, & Greenland, must probably have been icy deserts, like the Antarctic islands in the same latitude, & could have supported hardly any or no terrestrial productions. Hence these islands must have been colonised at a comparatively late period;* & it/11A/would appear from M. Martins' remarks[1] on the decreasing number of European plants on the islands the further we go from Europe, that they were colonised by various accidental means (a view, however, which would be rejected by many of the most competent judges) from Europe, aided by the probable greater height & extension of parts of the European continent/11/during the latter end of the Glacial epoch. Considering the lateness of the colonisation of these islands (for as explained in an early chapter on the view of species being formed by selection we can distinctly understand how time comes in as a most important element)† it ceases to be remarkable that islands so isolated as Spitzbergen, or Feroe, and Iceland, ⟨& seas like the Baltic⟩ should not possess, as I believe, a single endemic or peculiar inhabitant.‡/

11 v/It is possible that some few plants may have survived at the southern extremity of Greenland, in Lat: 60°. Four or five plants in Dr. Asa Gray's M.S. list common to Greenland & the White Mountains of N. Hampshire, & not found in Europe or Asia, perhaps indicate this; without indeed they were brought subsequently by icebergs.§ It may be objected to the foregoing view that Greenland, Iceland & Spitzbergen are inhabited by some mammals which could not have been transported by accidental means.—I have not seen complete lists of the aboriginal quadrupeds of these lands. In regard to wolves, foxes & bears, I am assured by Dr. Sutherland (see also Appendix [i. pp 489, 494, by Dr. John Richardson] to Capt. Back's Journey) that they have all been seen alive on icebergs far out at sea, & their introduction therefore offers no serious difficulty; & indeed Mackenzie (Travels p. 341) asserts that the Black Fox is at present thus sometimes introduced into Iceland. Possibly the same explanation may suffice for the Mustela, (likewise carnivorous) said to have been seen on Spitzbergen (Richardson's Report 1836, on N. American Zoology p. 162) & for the Gulo in the Parry Islands. Whether the Rein-deer, & the Lepus glacialis, which is found in Greenland,

* Note C [J.D.H.] [See p. 576.]
† Greenland ought also to have alpine species. [C.D.]
‡ Do Greenland or Lapland or Scotland even? [J.D.H.]
§ Which way?—great changes of level required. [J.D.H.]

[1] [In the draft MS., Darwin left an unfilled space for a citation here. See *Edinb. New Phil. J.* 46 (1849), 50–1.]

& the Mus oeconomicus, said to be an inhabitant of Iceland, & the rodents on which the Mustela must feed in Spitzbergen, could have crossed on ice, I know not; but the skeleton of the Lemming, found by Capt: Parry on the ice in Lat: 81 3/4° N. should not be forgotten. Even at present Iceland is connected with Greenland in the spring by continuous pack-ice./

12/On the other hand, at the very first return of the warmth, long before the northern islands could have been colonized, the arctic productions at the base of the more Southern mountains, such as Pyrenees and Alps, would have been completely isolated, as on an island in the sea, as soon as they had ascended the lower slopes of the mountains. Previously to the glacial epoch, these mountains must have had their Alpine species, such as Gentians &c which do not inhabit the arctic regions, & these on the returning warmth, after having, as it would appear, spread over the surrounding country* would together with the arctic species have reascended the mountains./

12 v/It would appear from Ledebour's account (Hookers Miscellany, Vol: II p. 241, 249) that the Alpine vegetation of the Altai in about Lat: 50°, had been able to keep itself during the glacial changes of climate unmixed, owing perhaps to the peculiar character of the climate of the steppes to which the old vegetation would have been adapted. Ledebour says that at the height of 4500 feet the vegetation has a greater similarity to that of Europe than on the surrounding plains, though some of the peculiar steppe-forms are yet found: between 4500 and 6500 feet, the European species gradually diminish in number and give place to the proper flora of the Altai.†/

12/Here then we have, according to the principles laid down in our fourth chapter, all the elements present which tend to modify species, though not in the highest degree,—namely, considerable lapse of time, isolation, & especially association with somewhat different sets of organic beings. Hence we might have expected that there would have been many representative species & strongly marked varieties, on the several alpine summits of Europe,‡ when compared one with another & with the arctic regions.§ I infer that this is the case from various scattered notions; & to give one/13/example, namely the chamois, which some Zoologists think specifically distinct on the Alps & Pyrenees, & others merely a variety of one & the same species. The arctic forms, which during their migration southward and remigration

* Note D [J.D.H.] [See p. 576.]
† Please give me this reference. I should like to know if these *proper* species are varieties of Arctic. J.H.
‡ Dr. Hooker: I wish I knew whether this was so: Forbes thought so, but I do not know whether he is to be trusted. [C.D.]
Certainly true J.H.
§ Note E [J.D.H.] [See p. 576.]

northward, did not become isolated & so differently associated, but kept in a body together under nearly similar conditions, would have undergone according to our principles very little modification./

13 v/Ireland, according to Prof: Forbes, was insulated at an earlier period than Great Britain, & indeed it is scarcely possible to look at W. Thompson's tables of distribution & doubt this. Its climate would have become fitted for its present productions at a later period than the isolation of the more southern alpine forms, but at an earlier period than the colonisation of the northern islands. The fact that there are several doubtful cases of representative species, or very strongly marked varieties in the animals & plants of Ireland, seems to accord with this./

13/The views here given may perhaps be extended further, though in doing so we are trespassing on the next chapter & considering representative species. If we compare the temperate productions of the lowlands of Europe & N. America, we find in all classes, terrestrial & aquatic, a vast number of species of the same genera, many obviously representatives of each other on the two sides of the Atlantic, & filling exactly the same place in the economy of nature, & not a few so closely allied, that the most practiced naturalists doubt whether to consider them as varieties or as true species. Many examples could be given in every single natural class of these doubtful species, & of quite distinct but representative species. If we look further west across the northern Pacific to Japan, we find many/14/most striking[1] representative species of European, but more especially of American* genera of plants, of mammals, birds & other beings. To complete the circle of the temperate zones, I may just allude to the many closely allied & a few perhaps even identical species of Crustacea in the seas of Japan & in the Mediterranean, as remarked on by Prof: Dana in his admirable Report on Crustacea:[2] yet the Mediterranean & Japan, even if we submerge the isthmus of Suez, are separated by a hemisphere of equatorial ocean. I may mention that I was myself much struck by finding two very close & obviously representative species of a very rare genus of parasitic cirripedes on crabs, from Madeira & Japan. Some of the fish, also, from Madeira, as I am informed by the Rev: R. B. Lowe represent those of Japan.†—/

* & Himalayan! [J.D.H.]   † Note F [J.D.H.] [See p. 576.]

[1] See the account of Zuccarini's observations in Silliman's Journal Vol [— see *Amer. J. Sci.*, 39 (1840), 175–6; 52 (1846), 135–6.] Decandolle has, also, insisted strongly on the representative species of Europe, America & Asia in the Dict. des Sciences naturelles. Art. Geograph. Bot. p. 414 (1820) [C.D.'s pencilled addition:] Refer to Asa Gray's most striking tables.

[2] United States Exploring Expedition, p. 1552, 1567, 1586.

15/Now if the view before given is in some degree probable, namely, that just before the glacial epoch when the climate was very slightly more favourable, there was nearly continuous circumpolar land & coast, as at present, inhabited by a nearly uniform flora & fauna; then it is not so very improbable that still earlier, during the older pliocene or even Miocene period, when the climate in these high latitudes was temperate,[1] there was likewise land & shore to some extent continuous, whence the closely related and often identical organic terrestrial & acquatic productions might have migrated southward, as the temperature fell, but long before the glacial epoch. As soon as this southern movement had taken place, the several existing floras & faunas of the northern hemisphere would have been separated from each other, as at present, would have been differently associated together, & exposed to somewhat different conditions. And as we are now dealing with comparatively ancient times, we might expect, according to the principles which we are testing in this work, that only a few species of those originally in common would have remained absolutely identical, but still we might expect plainly to see in the productions of the land & seas of temperate Europe, N. America, & Eastern Asia, evidence of their descent from a common home & common parentage; and this, I believe we do see, in the many representative species of these now quite separated countries. And/ Fair copy 15 v/it would appear from the observations of Brongniart, Agassiz & Heer that this relation between Europe & N. America was plainer during the later tertiary periods than at present; several American forms having since become extinct in Europe./

16⟨8⟩/We have seen that Prof. Forbes' theory explains the distribution of the alpine productions of Europe, in a manner which I think must be satisfactory to every one; & can be extended with no very great difficulty, as it seems to me, to N. America. But I believe that the theory is capable of a much greater extension. In South America I was formerly[2] much struck with finding numerous boulders on the island of Chiloe in Lat: 42°, where the rankest forests are now intertwined with almost arborescent canes & these gigantic boulders had been carried on ice from the Cordillera across a wide arm of the sea: on the plains of Patagonia, on the opposite side of the Cordillera, in Lat: 50° and over the southern

---

[1] Whether the abundant vegetable remains found by Sir John Richardson (Geological Journal, Vol [   ]) in the extreme arctic regions of America, belonged to the older pliocene age, or to a later period but anterior to the glacial epoch, is not known.

[2] Geological Transactions. Vol VI. p. 424 (1841)

extremity of the continent, ⟨& on the Eastern end of the Falkland Islands,⟩ boulders are very numerous. In central Chile, on the road to the Portillo Pass, I examined a mound of detritus which at the time never having read of Moraines greatly perplexed me, but now I can hardly doubt that it was a terminal moraine situated thousands of feet below the line where a glacier could now descend.[1] On/17⟨9⟩/the Cordillera of equatorial America, the marks of the former lower descent of glaciers have been observed.[2] In N. America, on the eastern side we have the plainest geological evidence of glacial action, as far south as Lat: 36°–37°. On the high plains near the Rocky Mountains boulders have likewise been observed; & on the shores of the Pacific in Lat: 46°.[3]

In Europe erratic boulders extend to near the Western base of the Oural, & in parts, southward to Lat: 45°–46°. In Siberia[4]/ 17A/they do not appear to occur; the surface, perhaps, at this period not having been under the sea: but Prof. Forbes' theory, of the community of alpine & arctic species having been caused by a former cold climate bears so strong an impress of truth, that where there is such community, we may almost safely turn round & argue from it, that the climate has been colder: if so, there can be no doubt from the several cases enumerated in a former note that Siberia has suffered from a cold climate./

18⟨10⟩/Looking south we find in the Himalaya abundant evidence[5] of the former much lower descent of the Glaciers, which have left behind them enormous Moraines. In India, using Prof: Forbes' theory, we have some evidence of a cooler climate in several plants,[6] (& in some mammals according to Mr. Blyth) being the same on the Nilghiri with those on the Khasia mountains & on the Himalaya; & again some* ⟨Dr. Hooker believes abundance of plants⟩ on the Nilghiri[7] are common with those on the mountains

* Dr. Hooker believes [C.D.] abundance of plants [J.D.H.]

[1] This mound blocking up the valley at the lower end of the lake-like expanse of the "valle del yeso," seemed to be composed wholly of alluvium, & was apparently 800 feet in thickness; "its surface consisted" (to quote my original notes) "of a confused hilly mass of rounded & angular fragments of rock, many of the latter of very large size."

[2] Bull. Geolog. Soc. [Acosta, Bull. soc. geol. de France Sér. 2: 8 (1851) 493. 9 (1852) 398.]

[3] Dana: Geology of the United States Exploring Expedition. Vol x. p. 674.

[4] [See appendix for passage cancelled in draft here & replaced by fol. 17a.]

[5] Dr. Hooker—Himalayan Journal. Vol I. p. 248, 380.

[6] Dr. Hooker—Flora Indica, p. 87, 99.

[7] Gardner—in London Journal of Botany, no. 47. 1845. Also Gardner in Journal of Horticultural Soc.—Vol IV. p. 37. Southward of the Nilghiri on the Pulney mountains (Madras Journal of Literature and Science. Vol v. p. 283) according to Dr. Wight, several of the same northern genera are met with at the height of

of Ceylon & the Himalaya.* There is much affinity, more especially as shown by the many European genera of plants & some species in common, between these Indian mountains[1] & those of Sumatra & Java;[2] the case of the Mydaus here comes into play, a quadruped found at the height of several thousand feet on the isolated volcanic mountains of Java, and never in the hot and low intermediate country.[3]

I have never heard of any marks of glacial action in S. Africa or in Australia; but in New/19⟨11⟩/Zealand, Mr. W. Mantell has shown me sketches of great fragments of quartz, lying on tertiary strata, which probably are erratic boulders: I saw myself boulders near the Bay of Islands, which appeared to me at the time as possibly of glacial origin. Dr. Hooker, moreover, informs me that there are certainly many plants common to the mountains of New Zealand, & not inhabiting the intermediate plains: some likewise are common to the mountains of New Zealand & Tasmania, & likewise to the lesser heights on the islands lying south of New Zealand: here then at the Antipodes we have the same sort of evidence of a cooler climate[4] as in the northern hemisphere.

With respect to the period of the glacial action or of the cooler climate at these several & very distant parts of the world, we can at least say that it has been in a geological sense recent: the phenomena are superficial; the evidence from scored rocks & moraines shows no great changes of surface have taken place since the moving ice covered the rocks. In most of the cases the glacial period has certainly supervened during/20⟨12⟩/the existence of the majority of living plants, & in the case of Europe &

* Note G [J.D.H.] [See p. 577.]

8000 feet.—[Addenda:] Flora of Pulney Mts. is identical with Nilghiri [J.D.H.] Also peculiar land-shells nearly or quite the same in mountains of Ceylon & Nilghiri [C.D.]

[1] Flora Indica, p. 104 Dr. Hooker says "constantly, during our examination of the temperate as well as tropical plants of the Nilghiri, Khasia, Ceylon & the Himalaya we find them identical in species with Japanese mountain plants."

[2] F. Junghuhn in his Java, Seine Gestalt, Pflanzendecke (1852) Vol I. p. 417, gives a long list of the genera found at the height of about 7000 feet, the names of which are familiar to every European. Dr. C. Reinwardt (Journal Hort. Soc. Vol IV. p. 233.) says the vegetation of the mountains of Java brings strongly to mind "our native home", but he asserts that all the species are distinct. With respect to Sumatra Temminck speaking of one of the mountains (Coup d'Oeil gen. sur les Possessions Neerlandaises Vol 2. p. 82.) says "La végétation sur son sommet porte tous les characteres [sic] des plantes alpestres de Europe"

[3] Sir C. Lyell, Principles of Geology. p. [638–9] Temminck says (Coup d'Oeil sur la Faune des iles de la Sonde p. 13) that the Turdus varius is common to the mountains of Java & the lesser heights of Japan.

[4] See Dr. Hooker's Remarks on this subject in the Introduction to the Flora of New Zealand. p. xxiii.

N. America, of living sea-shells. In Europe & N. America there certainly seems to have been a close parallelism in the whole phenomena of glacial action & in the coincident changes of land, but I am well aware that this does not prove strict contemporaneity./ 20⟨12⟩ v/In N. America, according to Sir C. Lyell, & in S. America & in Europe some large mammals have become extinct since the glacial period, namely the Mastodon, Cervus & Megaceros in the north & Macrauchenia[1] in the south.—/20⟨12⟩/As the northward curvature of the lines of equal temperature at the present day in Europe, compared with N. America is due to the warmth of the Gulf-stream, Mr. Hopkins[2] has inferred that the glacial epoch of Europe was probably caused by the Gulf Stream having formerly flowed up the central parts of N. America; how far the greater extension southward of the glacial action in N. America than in Europe, & likewise,[3]/21⟨13⟩/the apparent parallelisms of the miocene isotherms compared with the existing isotherms agrees with this theory may be doubted: for it might be argued from these facts that probably the course of the Gulf-Stream had long been constant.

Seeing how similar the superficial glacial phenomena are on both sides of the Cordillera of Southern America with those on both sides of N. America & bearing in mind that there is some evidence of glacial action in central Chile & all along the Cordillera & in the Equatorial Andes, there seems to me a primâ facie probability that both Americas were cooler strictly at the same time. No one who has carefully examined the effects of glacial action/21 v/when the phenomena are well developed, can doubt, that the cold, though in a geological sense recent, endured for an enormous lapse of years./21⟨13⟩/The vast number of boulders borne by icebergs & widely scattered—the thick masses of drift,— the great coincident changes of level both up & down,—the enormous amount of denudation,[4]—all bring this conviction most

---

[1] Darwin's Geolog: Observations on S. America, p. 97.

[2] [Geol. Soc. Quart. J., 8 (1852), p. lxiii.]

[3] Sir C. Lyell Travels in N. America, 1845, Vol I. p. 139. This inference is drawn from the comparison of the tertiary fossil shells with those of Europe.—

[4] At the head of the S. Cruz river in Patagonia ([Darwin] Geolog: Transact: Vol VI. p. 417) the great plain 1200 feet in height is strewed with boulders; this plain has been worn into a vast bay-like depression, facing the distant Cordillera, & is fringed by an 800 ft plain, also strewed with boulders, & the bay-like depression itself 440 feet above the level of the sea, again has great angular blocks on its surface, but different in kind from those on the upper plains. Here we see that the hard basaltic rocks of the upper plains which form on each side exactly corresponding strata, have been cut away; the intermediate plain has been formed: the whole country has been elevated from the height of at least 440 feet to 1200; & all this has taken place, whilst ice was transporting in Lat: 50° great boulders from the Cordillera. How many hundreds of thousands of years must have been required!

forcibly home to the mind. When I state that there seems a primâ facie probability that both Americas were at the same time cooler, I am far from wishing to infer that the cold in the North & South either began or ended at the same date, but that a part of the/22⟨14⟩/long period in the North & South was strictly co-incident, so that the intermediate zones were at the same time somewhat cooler. It is immaterial for our present discussion, whether the whole world was at one geologically recent time slightly cooler than at present, (as I confess seems to me most probable) or whether the two Americas were at one time cooler; Europe & Africa at another time; Siberia, the Himalaya, the plains of India, the eastern Archipelago & Australia at another time, but all within the age of the pleistocene formations.

Finally to sum up,—it is far from proved that any part of the cold period at these several & distant regions, was strictly co-incident, ⟨contemporaneous⟩, either over the whole world or along a few great meridional belts, indeed this is almost beyond the scope of simple geology; but there seems sufficient probability in this view, so that if it will explain several phenomena of organic distribution, otherwise inexplicable, it may be accepted as a theory worthy of consideration.

Let us then assume that at the period when the northern & southern portions of the world were colder than now, that either the whole, or first one & then another meridional belt of the intertropical regions was rendered slightly/23⟨15⟩/cooler, & what would be the result? The inter-tropical productions would retreat into the hottest districts; their proportional numbers would probably be considerably altered; some would become extinct; some, according to the principles which we are testing, would become modified. But according to these same principles, it may be doubted whether there would be very great modification; in as much as the great mass of surrounding organisms would remain the same, & we have seen reason to believe, that although changed conditions will cause variability, the selection of new specific forms is far more intimately related to the surrounding organic beings, amongst which each has to struggle for existence, & to seize on & occupy by selected changes in its structure any vacant place in the economy of nature./23⟨15⟩ v/I have previously remarked that on the same principles I should not expect great modification in the arctic species during their migration southward & remigrations northward, for they must have migrated in a body. To explain further by a metaphor what I mean: if a whole nation migrated in a body, each might retain almost his usual habits & business, but if only a few settled in a foreign land each probably would have more or less to change his habits, & occupy a different position in society./23⟨15⟩/For the same reason I should not anticipate very much extinction during the cold period within the Tropics, for we have seen good reason to believe that extinction depends far more on other organic beings seizing on the place of the dying forms than on changed conditions; & indeed we know that most organic beings, plants for instance, will endure a considerable change of climate, if protected from competing forms.

On the frontiers of the Tropics, the whole body of the temperate productions would invade, from the north & south, the cooled land; & as/24⟨16⟩/all the intertropical productions would be in some slight degree distressed, I can see no great improbability in some few temperate forms penetrating even to the equator and holding their own./24⟨16⟩ v/We might expect to see a vegetation like that so strikingly described by Dr. Hooker[1] at the base of the Himalaya, where true Tropical forms are mingled with such northern forms as Birches, Maples, whortle-berries, strawberries &c. Chains of mountains and high land running north and south would obviously favour the invasion of the temperate forms. After the glacial epoch, & even during an early portion of it, they would/ 24⟨16⟩/be eminently liable to have every slight variation, by which they would become still better adapted to struggle with their new compatriots, selected, & their structure thus specifically altered./24⟨16⟩ v/Though thousands of years might be required for their passage through the Tropics, slowly advancing as the cold came on, I can believe that they would not so much tend to be specifically altered, as afterwards when permanently settled on some isolated mountain associated with new organic beings: for in our chapter on natural selection it has been shown how excessively slow this process must be, counteracted as it must be in many ways; & that under changing conditions it could effect comparatively little, just as a breeder would be infinitely delayed if he changed his object or standard of perfection./

24⟨16⟩/⟨It is obvious that chains of mountains & high land

---

[1] Himalayan Journal, Vol I. p. 109. Vol. II. p. 319. For similar remarks, see Royle's Illustrations of the Botany of the Himalaya, p. 14, The Rev. F. W. Hope (Entomology, in Royle's Illustrations) describes a similar mixture of insect forms: he says (p. 15) we find in the valleys of the Himalaya where tropical forms abound, "European types & species in numbers sufficient to excite our astonishment." I may add as further showing the possibility of the commingling of tropical and temperate forms. that Lichtenstein [: see *Akad. Wiss. Berlin, Abh.* 1838, p. 422] states that this is the case with the birds in parts of Mexico.

would greatly favour the invasion of the temperate forms. One of the most obvious objections to the theory, is the enormous migratory power, though over continuous land, thus attributed to the temperate forms.⟩

We will consider some of the most obvious objections to this theory, after we have seen its local applications. First, for America: no one doubts that during the glacial epoch the northern pox... was inhabited by many old-world forms, the introduction of which we have already discussed. These would have a broad and eminently favourable high-road for migration southward, during this colder period, as far/25⟨17⟩/as near the isthmus of Panama. As just stated I can see no great difficulty in some temperate forms passing this hot & low barrier;[1] but it may have been then higher; we know, at least, that the isthmus has existed since the creation of the two distinct marine faunas on its two sides; & off Yucatan the coral-reefs favour the idea of considerable subsidence. After passing the isthmus the temperate forms would find in the Cordillera a grand line of communication to the southern part of the continent, as suggested by Dr. Hooker,[2] who supposes that at the period of migration the Cordillera were loftier; & therefore more temperate: geological evidence, from the equator southward, as far as it goes, is opposed to this view, & I think all the facts are better explained by change of climate, of which we have much independent evidence./

26/As the climate became warmer, towards the close of the glacial epoch, we may readily believe that nearly all the northern temperate species would be destroyed on the mountains of southern Peru & northern Chile,[3] owing to the extreme aridity of their present climate; for in Chile even at the greatest heights, glaciers are now hardly formed. As soon as the ice & snow, with which Tierra del Fuego was probably covered during the intensity of the glacial epoch, disappeared, this southern point of the continent would have been clothed with plants, including the northern temperate forms, which had travelled down the Cordillera. Some of these plants would be left on the mountains, where the climate was fitted for them: of this I saw one instance in Chiloe, where

---

[1] In Mr. Seeman's Narrative of the Voyage of the Herald (Vol. I. p. 253.) it is said that on the mountains of Panama, at the height of 2000 feet the vegetation resembles that of Mexico "with forms of the torrid zone harmoniously blended with those of the temperate."

[2] Introduction to the Flora of New Zealand. p. xxv.

[3] The alpine vegetation seems now to be very peculiar on these great mountains, but has been only very imperfectly described: see Meyen's Reise Band [: see I, 348, 466.] & Poeppig's Reise Band [   ].

at the height of about a thousand feet, the well known antarctic beech of Fuegia lived in a dwarfed condition. Thus, I think,/ 27⟨18⟩/we can understand the presence of so many European forms in Tierra del Fuego, as is so forcibly shown to be the case by Dr. Hooker in the Flora Antarctica, some absolutely identical, some presenting strongly marked varieties & some quite distinct, but still plainly related to their northern congeners. According to the principles discussed in this volume, we might have expected considerable modification in these forms which have wandered so immensely far from their native home, and which have/27 A/lived with new associates. But those northern forms which found a suitable home on the lofty mountains of America[1] must have been associated for a still longer period with new beings, namely with the American alpine forms which we cannot doubt existed previously to the glacial epoch; they must, also, have been exposed to still more different conditions, & hence we might expect that they would have undergone greater modification than those of Tierra del Fuego, and this I suspect is the case.[2]/

28⟨19⟩/Those few temperate forms which were able to penetrate the lowlands of Tropical America during the Glacial epoch, would be most of all modified, & when the climate again became hot, could only survive on high land: thus, perhaps, we can understand the presence of species of such temperate genera as Vaccinium, Andromeda, Gaultheria, Hypericum, Drosera & Habenaria found by Mr. Gardner[3] between 6000–7000 feet on the Organ Mountains of Brazil. It would appear that some truly American alpine forms had descended & spread over the plains of S. America during the cooler period; for thus apparently can only be explained the presence of the Andian genus Bejaria, & even the same species of Thibaudia on the Silla of Caraccas[4] & mountains of New Granada, where they are associated with some of the same genera found on the Organ Mountains & on the heights of Jamaica.

Now let us turn to Africa & briefly consider the period whether or not strictly coincident with the cold period of America, when arctic forms were living at the foot of the Alps. At this period I believe those few northern temperate species, which are now

---

[1] In Johnston's Physical Atlas. Botan: Geography it is said that of the 327 genera of plants found on the declivities of the Andes, at the height of 7000 feet & upwards, 180 genera, or more than half, are common to the temperate zone.

[2] [Here C.D. scribbled a memorandum:] See some paper in Eding. New Phil. Journal?

[3] Journal of the Horticult: Soc: Vol I. (1846) p. 281. Mr. Purdie found [on] the mountains of Jamaica (London Journal of Botany Vol. III. p. 512.) Vaccinium, Andromeda, Myrica Mexicana, & Viburnum.

[4] Humboldt—Personal Narrative (Eng. Translat.) Vol III. p. 494, 500.

found on the highland of Abyssinia[1] penetrated to that latitude though so near the equator. In Drege's enormous collection of plants from the Cape of Good Hope/29⟨20⟩/as described by E. Meyer.[2] there are 96 European phanerogams & ferns enumerated. Mr. Bunbury, who has personally collected at the Cape, has kindly looked over the list for me & has added three species: he considers many of these plants as probably naturalized by man. Some are littoral plants which may possibly have travelled by the coast; about 14 are aquatic or marsh plants which seem to have, as we have seen, some special means of diffusion; but 30 plants apparently do not come under either of these categories & I should infer (if really not naturalized by man's agency) had migrated through the tropics during the cold period. Considering the ordeal they must have gone through in having been so long associated with the very distant Cape species, this number is too great for my theory. If there exists, as some have supposed, near the East African Coast nearly continuous high land from Abyssinia to the Cape, their migration at least, into this colony would be rendered more probable. The fact that on the very arid & somewhat isolated mountains of the Cape, at the height of from ⟨6000⟩ 7000 to 8000 feet, there are some distinct species of such northern genera[3] as Geum, Epilobium, Pimpinella, Galium, Tanacetum, Myosotis, Dianthus & Anemone, associated with many species of Cape genera, harmonises better with the theory.

In the East, at the time when the glaciers descended low on the Himalaya & the prodigious moraines described by Dr. Hooker were forming, & when probably the woolly-covered Rhinoceros tichorinus and Elephas primigenius* were ranging over the/ 30⟨21⟩/plains of Siberia, I must believe that those plants, already alluded to as common to the Himalaya, the Nilghiri, & to the mountains of Ceylon & of Java, ranged over the intermediate now torrid country; & that during the cool period they reached

---

* Note H [J.D.H.] [See p. 577.]

[1] A. Decandolle, Geograp. Bot. p. [   ] To the West of northern Africa on the heights of Teneriffe a very few northern species, & several northern genera have been found; & lately, as I am informed by Mr. Wollaston, Erica cinerea has been found near the summit of Madeira. To the East in Lycia at heights between 6000 & 10,000 feet (Lieut. Spratt and E. Forbes Travels. [II] p. 157) Draba aizoides, Anemone Appenina, Scilla bifolia &c are found.

[2] Flora 1843 Band II. Zwei Pflanzengeograph. Doc. p. 9.

[3] Flora 1843 B. II p. 53 Mr. Bunbury thinks that the genera Dianthus, Franklinia Statice are the most striking cases of northern genera having representative species at the Cape. The Heaths offer a well known case, abounding at the Cape, & not known to reappear in the north nearer the Equator than Teneriffe & Arabia; but the species from the north & south I believe, show no especial affinity.

& subsequently ascended their present isolated & elevated homes. On the Himalaya, Dr. Hooker has shown that many plants are representatives & many specifically the same, (though often presenting varieties) with those of the regions lying north of them & of the European mountains; & this migration might well have happened during the cool period considering the latitude of these great mountains, & more especially the high but broken land to the north and northwest.[1] The majority of the species of northern genera on the Nilghiri & on the heights of Ceylon, as I infer from the writings* of Mr. Gardner[2] are representatives, as would ensue from their having been differently associated as compared one with another, having been, as compared with the Himalaya, isolated for a longer period, owing to their more southern position. Considering that Java, Sumatra & Borneo lie near each other & arise from a shallow bank[3] & that they have/31⟨22⟩/some few mammals in common there is a strong probability that the whole area within recent times may have stood at a higher level & been continuous; & therefore there is little more difficulty in the heights of these great islands having been colonized by northern forms, since modified, than in Ceylon having been thus colonized.

We now come to a more difficult case. Long since Robert Brown showed that there were several northern plants in Australia, which could not be considered as naturalized by man's aid. Recently,/ 31⟨22⟩ v/Dr. F. Muller has found on the Australian Alps several European plants, as Lysimachia vulgaris, Turrutes glabra, Veronica serpyllifolia, which species together with some others mentioned to me by Dr. Hooker, are not common in Australia & are found no where else in the southern hemisphere & are not very widely distributed in the northern hemisphere. I should suppose that these plants had migrated into/31⟨22⟩/Australia during the cold

---

* No I think *majority* identical [J.D.H.]

[1] We find exactly the same class of facts in the insects of the Himalaya. Mr. Hope (Entomology, in Dr. Royle's Illustrations) seems continually in doubt whether certain insects are identically the same or most close representatives of those of Europe and Siberia. Amongst Birds we have both identical species & some beautiful representatives of those of Europe, as in the bull-finch, goldfinch, shrikes &c, as represented in Mr. Gould's Century of Birds from the Himalaya. So again it is with many mammals, & Mr. Ogleby [Ogilby] seems in doubt (Royle's Illustrations) in regard to some of the mustelae, badgers, hedgehogs &c whether to consider them identical or representatives.

[2] Journal of the Horticult. Soc. London Vol IV. p. 37. A short table is given showing that many more species are representative than identical. ['Gardner however could only guess & not compare enough.' (J.D.H.)]

[3] Windsor Earl, on the Physical structure of the Indian Archipelago. Geograph. Journal, 1845. Vol 15. p. 358. [Cite not precisely in text—could be for map facing this page.]

period (when the mountain plants in common to Tasmania, the Australian Alps & New Zealand inhabited the low grounds) by the islands of the Malay archipelago: perhaps through New Guinea; but the vegetation of the lofty mountains of this island is unfortunately quite/32⟨23⟩/unknown. Between New Guinea & Java, where northern forms are found, the sea in parts is deep, but it is studded with an extraordinary number of islands, so that by strides of 50 miles the interspace can be crossed on dry land; & there/32⟨23⟩ v/is some evidence in parts of subsidence. The identity of the above specified & several other Australian plants with those of Europe, is certainly most remarkable, considering their long sojourn amongst foreigners; & is a parallel case with the European plants at the Cape of Good Hope. It might have been anticipated in both instances that fully as many plants would have undergone specific modification as on the mountains of Java or India.

In New Zealand Dr. Hooker has found[1] 60 European plants and in addition several striking cases of representative species or as I should consider them modified forms of northern genera. With respect to the introduction of these species, I will only remark that we/32⟨23⟩/have good evidence of prodigious recent subsidence in New Caledonia[2] (of which the mountain vegetation is unknown) & at the S. E. extremity of New Guinea, & that between New Zealand & New Caledonia (about 800 miles apart) there are some islets. But the number of species & genera of plants common to New Zealand & Australia though nearly 900 miles apart at their nearest points perhaps should lead to the belief that New Zealand had in chief part derived its northern forms through Australia.[3] New Zealand undoubtedly offers another great difficulty to the views here advocated, & is a strong case in favour of those who believe in multiple creations./

33/This discussion has as yet been almost confined to plants, & I will now make a very few remarks on other organic beings in relation to their migration from north to south during the glacial epoch. In mammals & reptiles, I know of no cases of the same or representative species being found in the opposite hemispheres & not in the intermedial Tropics. In Australia, as I am informed by Mr. Gould, there are some striking cases of Birds, chiefly aquatic, as the Australian coot, moor-hen, & some ducks which represent northern forms, & are not known to occur within the Tropics. In

---

[1] New Zealand Flora: Introduct. p. xxx.
[2] The barrier coral-reefs show that the island formerly extended 150 geographical miles further at its northern end.
[3] [For long note inserted here and later cancelled, see appendix.]

land-shells I can hear of no northern & southern identical or representative species; and this could hardly be expected; for land-shells have either been so frequently created, or as I should infer so easily modified, that there do not/33A⟨24[?]⟩/appear to be many species in common even on mountain-summits as near to each other as the Alps & Pyrenees, or, as I am informed by Mr. Benson, on the Nilghiri & heights of Ceylon. Nor should we be surprised at this, when we hear from so competent a witness, as Prof. Adams that in Jamaica, the collector in the course of every ten miles finds new species. In regard to insects, I carefully collected the beetles of Tierra del Fuego, & Mr. Waterhouse has examined them; but none are identical with, or closely/34⟨25⟩/ representative of, northern forms; Carabus, however, must be excepted, as it seems to have travelled, like many Fuegian plants, along the Cordillera from the north. In southern Australia & New Zealand, there are only a few very doubtful cases of representatives of northern forms. But it should be observed that insects are not nearly such wide rangers as might have been anticipated./

34/Turning now to marine productions, we hear from Sir J. Richardson,[1] that Arctic forms of fishes disappear in the seas of Japan & of northern China, are replaced by other assemblages in the warmer latitudes & reappear on the coast of Tasmania, southern New Zealand & the antarctic islands. He further states that the southern cod-fish are "much like those of the north, & Notacanthus & Macrourus, two very remarkable Greenland genera, which inhabit deep water, have recently been discovered on the coasts of New Zealand & S. Australia." In regard to sea-shells, Dr. A. Gould[2] says proceeding from the north, across the equatorial seas, "there is not a return to the same species & rarely to the same genera". But he adds: "along our northern seas, some of the most characteristic shells are Buccinum, Tritonium, Fusus &c. Around Cape Horn are shells of the same types, so closely allied that they have not yet been separated/35/as distinct genera, though peculiar in many important respects." Whether this resemblance depends on migration during the glacial epoch & subsequent modification, I can form no opinion. Considering the wide ranges of many shells, I am surprised that there is not more identity or very close representation* between the north & south. In the Bryozoa/35⟨26[?]⟩/or Polyzoa, Mr. Busk gives several cases[3] of

* C. of Good Hope [C.D.]

[1] Report on Icthyology, Brit. Assoc. 1845, p. 189, 191.
[2] Introduction to the Conchological part of the U. States Exploring Expedition p. xii.
[3] Catalogue of Marine Polyzoa in British Museum. 1852. p. 39, 67, 70, 83, 84, 94.

European corallines now inhabiting Tierra del Fuego, New Zealand, & the Cape of Good Hope, not yet found in the Tropics, but it may be objected that the intertropical seas have hardly yet been sufficiently searched. In the Ascideae, the genus Boltenia has allied species in the arctic & antarctic seas, & Prof. Huxley thinks that the genus is not Tropical; but here again from our ignorance much caution is requisite./

36/In regard to Crustacea we can refer to Prof. Dana's full & admirable memoir on their Geographical Distribution. Many species, belonging to many genera have very wide ranges, compared with most marine animals;[1] & this is important for us in allowing extensive migration during the cool period. Prof. Dana states that the sub-torrid shores of Natal, Japan, & even the Sandwich islands have several identical species & several representative species not found in the intervening torrid seas; & Prof. Dana doubts, though granting the possibility of wide migration, whether these species could possibly have passed from the southern to the northern zones;[2] but under a cooler climate this difficulty would be greatly lessened./37⟨29⟩/On the west coast of America, ⟨Prof. Dana[3] states that⟩ the Californian subtemperate province has a close resemblance in some of its genera to the subtemperate province of Chile, though separated by 3700 miles of warmer seas; but it does not appear that any of the species are in common./ 37A/In Prof. Dana's work and in that of Milne Edwards[4] I observe that the genera Cancer, Atelecyclus, Lithodes, Jaera & Anonyx have species on the west coast of S. America in the temperate & colder zones, both to the north & south, but none in the intermediate hotter latitudes./37⟨29⟩/The case of New Zealand again is similar for ⟨Prof. Dana[5] shows that⟩ there is a clear relationship between its crustacea and those of the northern hemisphere. A Palemon[6] is almost identical with a British species: Cancer is not elsewhere known out of the temperate zones of N. & S. America & of Europe. The species of Portunus "are representatives of the most characteristic of European genera, & they belong rather to the cold temperate than sub-temperate regions of the Australian

---

[1] Report on Crustacea: United States Exploring Expedit. by James D. Dana. At p. 1551–54, a list of 42 species are given with very wide ranges. At p. 1574, another list of 33 species common to the African coasts, Indian ocean & Pacific. Some few species p. 1585 are common even to the East & West coasts of America.

[2] Report on Crustacea, p. 1584: at page 1574, a list of 12 species in common to Natal & Japan is given.

[3] Report on Crustacea, p. 1557, 1561.

[4] Histoire Naturelle des Crustaces. Tome III p. 588.

[5] Report on Crustacea, p. 1578, 1587.

[6] Histoire Naturelle des Crustaces Tome II. p. 391.

& New Zealand seas," Well does Prof. Dana remark that "it is certainly a wonderful fact that New Zealand should have a closer resemblance in its Crustacea to Great Britain, its antipode, than to any other part of the world," [p. 1587]./

38/Finally I may add a most striking case on the authority of Dr. Hooker namely that 25 of the same species of algae or seaweeds, belonging to 20 genera, inhabit the shores of New Zealand & Europe, & have not been found in the intermediate tropical ocean./

38⟨[30?]⟩/In the theory now propounded of the cold of the Glacial epoch having affected at the same time the whole world, or at least broad meridional belts, during which period northern species, both terrestrial & aquatic, crossed the Tropics, (the terrestrial stopping where higher land allowed of their permanent existence);—these species in many cases when thrown amongst foreign associates having become modified, we encounter some serious difficulties. Besides special difficulties, such as how the northern plants got into New Zealand,—why sea-shells do not offer better evidence of migration from north to south &c, we encounter some difficulties of a more general nature. The theory supposes that certain species have migrated over an immense space, during a period considered short by Geologists & sometimes falsely spoken of as mere intercalated fraction of time; but no/ 39⟨31⟩/geologist who has examined the glacial phenomena for himself will doubt that the period measured by years has been enormous. Nor should we forget that by the very theory all tropical productions would be in a somewhat distressed condition, & therefore would not oppose so bold a front, as before or subsequently, to the intrusion of strangers; & we know in the case of naturalised plants how widely some few have spread even in the course of a few hundred years.

Those naturalists who believe in the modification of species, but attribute much to the direct action of external conditions or who believe that there is some law determining all species to change cotemporaneously, will object that the whole body of Tropical productions ought to have become changed; but I believe that this view is erroneous, & that there would be but little tendency to change as long as the great body of tropical productions coexisted; whatever modification there may have been, would have chiefly resulted from the altered proportions of the old forms & the intrusion of strangers, new places being thus made in the polity of nature, which would be better occupied by slight selected changes of structure. So it would be with the northern & southern temperate productions during their advance & retreat in mass from the poles towards the equator. Very different would/40⟨32⟩/ it be with many or most of those forms which either crossed the Tropics & gained the temperate regions on the other side, or remained on the mountain heights within the Tropics, for they would have been associated from a more or less early part of the glacial epoch with new animal & vegetable productions. Undoubtedly it is surprising according to the theory we are here discussing, that any temperate forms should have slowly crossed the Tropics, associating all the time with productions of most different natures & exposed to very different conditions, & yet have retained the same identical character. But during these long journeys variability might have ensued, without any new permanent modification having been selected, adapting the wanderers to the not very permanent conditions which they must have encountered during these migrations. Immensely long as was the Glacial epoch, we know not in the least, whether the subsequent period during which the temperate forms have lived with their new associates may not have been far longer ⟨than the glacial epoch itself. And we can distinctly understand on the theory of selection how simple time plays a most important part in the modification of specific forms.⟩/40 A/Hence perhaps it is a greater difficulty that several of the northern species which have reached the southern zones, should still remain identically the same, than that they should have not been modified during their migration across the Tropics.

But I must here observe that in several of the cases/41⟨32⟩/in which we have representative forms in the north & south, it by no means follows that all ⟨towards either pole⟩ have been modified since the Glacial epoch. A genus may formerly have extended, as many now do, from north to south, & have had species at both extremes, & since have become extinct in the equatorial zones, from causes independent of climate. It is also possible that one or two species of a northern genus might during the glacial epoch have migrated southward across the Tropics, & have left from subsequent extinction no individuals of the same species in the north; & in both these cases, we should falsely be led to attribute to modification during or since the glacial epoch, that which was due to migration & extinction, or to modification at a period no ways connected with the glacial epoch.

There is one other & curious difficulty to/42⟨33⟩/the foregoing theory of migration during a late cooler period. Dr. Hooker[1] has remarked how singular it is that in America, whilst many northern

---

[1] Flora of New Zealand. Introduction p. xxv. note

forms have penetrated to the south, no southern forms can be said to have migrated northwards: M. A. DeCandolle[1] has made the same remark/42/in regard to Australia; & indeed the same species or any species of Eucalyptus or Banksia in the north would be a prodigy! But we have a most curious exception to this remark in the recent discovery on a mountain of Borneo, at the height of 8000 feet, of "three of the most peculiar Antarctic, New Zealand & Tasmanian genera",[2] associated with Indian, & with Australian forms, such as the heath-like Epacridae. On the mountains of Java,[3] two Australian temperate genera, have been found, namely Leucopogon & Thelymitra; & it would appear from Dr. Hooker's[4] observation that some few other Australian genera have travelled up the Malay peninsula, & two or three have even spread over India:* some of these genera, as the above named Leucopogon & Lagenophora, I believe are confined to the southern temperate zones. In Africa, also, there seems to be some faint indication of migration from the south to the north, as well as from the north, southwards: I allude to the two Mediterranean species of the great Cape genus of Mesembryanthemum & the one species/43/of Ixia, compared by the elder Decandolle[5] to soldiers driven from their regiments.†

Notwithstanding these partial exceptions, there seems to be no doubt, that many more species & forms have passed from north to south than in the opposite direction. In attempting to explain this singular fact, we should not forget that in the northern temperate hemisphere, there is much more land than in the south, & that the plants inhabiting it are wider rangers than the more isolated species inhabiting the smaller areas in the southern hemisphere;[6] & therefore that there would be a better chance for some

---

* Stylidium a capital case [J.D.H.]
† The Pelargonium & Stapelia in Levant & Algiers, & various other cases. [J.DH.]

[1] Geographie Botanique p. [   ].
[2] Flora of New Zealand p. xxxvi. I observe that one of the three genera, mentioned by Dr. Hooker, Drimys, was found by Mr. Gardner in the Organ Mountains of Brazil, where it, likewise seems to be a wanderer from the south or from the Cordillera.—[Hooker addendum: 'is found all the way to Mexico. See Fl. Antarct. ii. sub Drimys.']
[3] F. Junghuhn, Java seine Gestalt &c. 1852. Vol i. p. 417.
[4] Flora Indica Introduction p. 103, 253.
[5] Dictionaire des Sciences Nat. Art. Geograph. Bot. p. 413. In Abyssinia, also, Cape forms are found, but the intermediate country is very little known; (see C. J. F. Bunbury's Residence at the Cape of Good Hope. p. 218. also Hooker's Flora Antarctica p. 210) and the southern forms here mingle, as on the mountains of Borneo and Java, with northern forms.
[6] A. Decandolle (Geograph. Bot. p. [   ].) gives a curious comparison of the greater range of the species of the same Families in the Russian Empire & at the Cape of Good Hope.

of the northern species than for the southern species being great wanderers ⟨enabled to cross the Tropics when slightly cooled⟩. To hazard a conjecture unsupported by any facts, I may remark, that if the cold of the glacial epoch first came on north of the equator, the northern forms would first penetrate the Tropics, & any southern species subsequently intruding would be opposed by the great body of tropical productions with the gaps already occupied by northern forms./43 A/I believe that all the few southern temperate forms occurring on high land near the equator are specifically different from their southern congeners,* in this respect differing from the several species of northern genera found in the south; it may be & probably is accidental, but this fact harmonises with the view so often referred to that the more complete the association with foreigners the greater the probability of specific modifications through selection./

44/I have reserved to the last some cases of distribution, the most extraordinary under our present point of view, as yet known, & which have been fully given by Dr. Hooker in his admirable Flora of the southern ocean; & I am greatly indebted to my friend for having endeavoured to make me appreciate the full force of the several difficulties. Kerguelen's Land is inhabited by only 18 phanerogamic plants; of these three are fresh-water plants found almost everywhere, & have been alluded to as most wonderful cases in an early part of this chapter. Two plants are distinct genera, known no where else; they baffle all inquiry, but do not immediately concern our immediate inquiry, Of the remaining 13 plants, 7 are endemic or aborigines, but one of them is too close to a Fuegian species; & five out of the seven genera to which these seven plants belong are genera found in but not confined to Fuegia. The remaining 8 plants are common to Fuegia, but three of them are likewise found in the New Zealand group of islands. Therefore, as remarked by Dr. Hooker, Kerguelen's Land has a much stronger Botanical affinity to Tierra del Fuego than to any other region. But these two points, measured along the parallel of 50° S. Latitude, are separated by no less than about 5000 miles of open ocean; & Kerguelen/45/Land is situated very much nearer to the southern points of Africa & Australia, between which it lies intermediate.

The island of Tristan d'Acunha is[1]/45.1/situated between America & Africa in Lat: 37°, about 700 miles nearer the equator than

---

* not so with various Cordillera species found in Fuegia [J.D.H.]

[1] [See appendix for earlier cancelled version of this paragraph.]

Kerguelen Land: it is inhabited, as I am informed by Dr. Hooker, by about 33 plants; of these some are not perfectly known to Dr. Hooker; from 7 to 16 are endemic; 12 are common to S. America & of the twelve, six are, in the Southern hemisphere, not found elsewhere. Hence Tristan d'Acunha, like Kerguelen Land, is botanically more nearly related to Fuegia (from which it is almost 2300 miles distant) than to any other country; & this is the more remarkable as it is only about 1700 miles distant from the southern point of Africa, to which it is related by only one or perhaps two forms, & differs in the most striking manner. Lastly, seven of the 33 plants are common to several of the antarctic islands & to the mountains, as I am informed by Dr. Hooker, of New Zealand, Tasmania and South America; so that ⟨if derived from those countries⟩ their introduction into these several countries & islands probably dates from the glacial epoch./

45/Dr. Hooker accounts for the close connexion of the floras of these distant points by supposing that within the existence of living species, there was once nearly continuous land. For reasons already given,* it seems to me that those who are inclined to believe in multiple creations, might object to the admission of such enormous changes of land & ocean without the concurrence of the weightiest evidence, both geological, zoological & botanical./

46⟨38⟩/Sir Charles Lyell[1] has suggested that plants may have been widely disseminated in the antarctic ocean by the agency of icebergs. There seems to me much probability in this view, especially if we bear in mind the prodigious number of great blocks of rock, which have been transported from both sides of the Cordillera from its southern extremity up to Latitude 42°.[2] The most obvious objection to this view is that the icebergs must have travelled a vast distance in nearly the same latitude; & in the northern hemisphere we know from the scratches on the rocks that the course of the icebergs was formerly, as now, approximately north & south. But the great difference between the northern hemisphere & the quite open ocean of the southern hemisphere must not be overlooked. We have, also, the following fact as a guide: a bottle was thrown overboard by Sir James Ross[3] a little northward and eastward/47⟨39⟩/of Cape Horn, and was picked up at Cape Liptrap, the extreme southern point of Australia ⟨north of Tasmania⟩;/

* in previous chapter [C.D.]

[1] [Principles, 9th ed., p. 622.]
[2] ⟨Geological Transactions Vol VI. C. Darwin on the Distribution of the erratic Boulders of S. America.⟩
[3] As stated before the Geographical Society, June 22, 1846. [See *The Athenaeum* (1846), 656.]

47 v/this bottle, during its voyage Eastward of about 9000 miles had gained only about 900 ⟨miles⟩ northing. The extraordinary prevalence of violent westerly gales, comparable in regularity with the trade-winds, in these latitudes must not be forgotten. But the course of the ice-bergs would be very much determined by their depth, & their reaching the underlying stream of cold water flowing to the equator. At the present day icebergs have been observed within a degree of the Cape of Good Hope,[1] & could hardly have travelled, bearing in mind the westerly winds, less than 3000 miles./

47⟨39⟩/With respect to icebergs occasionally carrying seeds, I think it would be quite extraordinary if they did not do so, just in the same way as seeds are carried in the ballast of ships & plants thus naturalised: we should remember the innumerable great fragments of rock which certainly have thus been carried many hundred miles. I have had the particulars given me of two icebergs in the antarctic ocean with great fragments of rock, at least 1200 miles from the nearest known land. Besides stones, "loads of earth", brushwood, live animals of several kinds, the skull of the musk ox which was landed in Greenland, the bones of the Lemming, & even the nest of a bird with its eggs[2] have all been observed on icebergs. Can we doubt that seeds of plants, with their vitality well preserved, might likewise be thus carried? Dr. Rae has suggested to me that the gales of winter, which sweep the ground bare of snow, can hardly fail to blow seeds on the fissured glaciers near the coast: stray birds resting on icebergs might occasionally leave hard seeds of fruit in their droppings: where there are rivers the autumnal frosts would freeze mud & seeds together, & such river-ice,/48⟨40⟩/as I am informed by Dr. Rae, is sometimes 6 or 8 feet thick, & when in the sea gets packed and crushed together. When an iceberg is stranded, great masses of ice, by the unanimous testimony of Arctic travellers, are pushed up high and dry by the pressure of the pack outside; & Dr. Rae assures me he has seen hundreds of instances of ice driven so high on land, that when it thawed any enclosed seeds would have had a good chance of growing. Seeds in earth, even if discharged in the sea on a shallow coast, would have a chance of being thrown up, like shells from deep water. We must never forget during how

---

[1] Horsburgh. Philosophical Transactions 1830 p. 117
[2] For these latter facts see Crantz, History of Greenland. Vol I. p. 26. Supplement to Parry's voyage by Capt. Sabine p. cxc. Also Richardson's British Assoc. Report for 1836. p. 163. Scoresby estimated the weight of "*the beds of earth & rock*" on many of the icebergs near Spitzbergen at from 50,000 to 100,000 tons; (Lyell's Principles of Geology. 9th Edit. p. 227.)

many hundred-thousand years this action must have gone on during the glacial period; & that during this period the native plants of many southern coasts would have been distressed & could not have resisted the intrusion of more vigorous ⟨southern⟩ strangers. Hence I can see no insuperable difficulty in the seeds of Fuegian plants having been carried to Kerguelen Land[1] & to Tristan d'Acunha during some part of the glacial period./Fair copy 42/This would account for the species in common, & by modification for some of the representative species. But a large inexplicable residuum is left; in regard to part of which, some remarks will presently be made./

48A/*I should infer from Capt. Carmichael's account that there were proportionally fewer species in Tristan d'Acunha identically the same with those of Fuegia, than in Kerguelen Land. If this be so these islands present a parallel case to the mountains of Scotland & the Alps of Switzerland compared with the Arctic regions; the cause being, I should suppose the same, namely, the points nearer the equator having been colonized, and the colonists isolated, at an earlier part of the Glacial epoch./

49/New Zealand offers in some respects a still more difficult case. Dr. Hooker/49A/states[2] that 89 plants are common to New Zealand & S. America; several of these are very wide ranging species & offer no more special difficulty than in other parallel cases: some few may have travelled from the north during the glacial period, and so got into these two distant southern points of the world. But of the 89, Dr. Hooker has given me a list of about 25 species which are absolutely confined to the southern temperate zone, & yet are identical to these two widely separated points. The interspace of ocean measured along the parallel of 45°, is about 4500 miles, without one single island now existing as a resting place.†/49/These 25 species may be divided into two classes,/49A'/namely 13 common to Fuegia & New Zealand (of which 6 likewise occur in the small Aukland & Campbell islands lying between 200 & 300 miles south of New Zealand)/49/and

---

* This will not do I fear.—[C.D.]

† The Fuschias [sic] & Calceolarieas [sic] are as great difficulties under your view of modification. [J.D.H.]

[1] It perhaps deserves notice that the *Modiolarca trapezina* is common to the Falkland Islands & Kerguelen Land (see Woodward's excellent Supplement to Treatise on Shells p. 371, 378) & as this shell is often attached to the masses of the gigantic Macrocystis, this seaweed is probably drifted from the one place to the other.— [Hooker added: "We pulled up immense masses with shells & stones attached, in *all* parts of Ant. Sea beyond the very icy regions J.H."]

[2] Flora of New Zealand, Introduction. p. xxxi.

12 common to Chile/50/& New Zealand. Of these twelve, it can only be said that they belong (with the very remarkable exception of a Myosurus) to genera having species in many parts of the world; but Dr. Hooker informs me that the species in these genera are neither particularly wide rangers nor particularly restricted./ FC 44 v/In his Flora of New Zealand[1] Dr. Hooker infers from the species in common & more especially from the representative species that both Chile & Fuegia were/49/formerly connected with New Zealand by intermediate land, but not necessarily continuous at any one time over the whole distance./50/In the next chapter when remarking on these very interesting representative species,[2] I shall have occasion to allude to the possibility of those southern islands which are now wholly covered by ice, having been clothed with vegetation before the commencement of the glacial epoch; & the seeds of some plants now in common to New Zealand, S. America & the other Antarctic islands, as Kerguelen Land &c, may have been carried by icebergs at an early part of the glacial epoch, from a common southern home. ⟨& the species not have been subsequently modified.⟩ The advocates of multiple creations, may, in my opinion, bring forward the species more especially those found in Chile & New Zealand as a very strong case in favour of their view:/51/but it should not be overlooked that they would find it very difficult to give any rational explanation of the community of these few species, for the great mass of organic productions, & all the external conditions are widely different in Chile & New Zealand; & it might well be asked, why should these few plants be identical & so vast a number of other productions widely different: it seems to me safer to rely on our ignorance of the means of diffusion.—

In regard to our general conclusion on the great amount of migration during the glacial epoch, of which epoch we have in many of the areas in question independent & decisive geological evidence, I think it has much probability, notwithstanding the

---

[1] Introduction p. xxiii.

[2] There are, according to Dr. Hooker, several identical & representative species on the heights of New Zealand & Tasmania. The possibility of great icebergs having been formed during the glacial epoch on the Eastern side of Tasmania, & having thus carried seeds should not be forgotten, for in the very same latitude on the shores of S. America immense blocks of rock (one of granite 15 feet by 11, & 9 in height) have been carried about 40 miles from the Cordillera (not very lofty in this part) to the island of Chiloe.—The case of the bottle carried from near Cape Horn to the southern point of Australia should not be overlooked; for it seems just possible, though very improbable that seeds might have been formerly carried from Tierra de Fuego to New Zealand; though the voyage at the rate of 25 miles a day, would take a year! [Hooker wrote: 'Why strike this out? The berg would perhaps travel faster with winds J.H.']

many cases of difficulty enumerated, some special, some general & others probably overlooked. This same view, may I believe be extended to some cases, which have not been here noticed from want of space. It explains in my opinion many anomalies in distribution, & removes some few of the greatest difficulties in admitting, in accordance with/52/the strong presumption derived from general laws, that each organic form was created or produced in one area. Moreover it strengthens the theory, in as much as it explains to a certain extent several facts otherwise inexplicable, that species under certain given conditions undergo modification. There is much interest in looking at the alpine productions of mountains in the most distant quarters of the world as monuments of not very high antiquity, yet often written in a changed dialect, recording the nature of the organic beings which once, when the world was cooler, surrounded their bases, & there perished. We have on these monuments the evidence of a great tide of life which slowly flowed from either pole towards the equator,—the waters, it may be said, breaking more freely over from the north than from the south. The two great tidal waves then slowly ebbed towards the Poles, but have not yet reached, & perhaps will never reach, their first & native source./

53/To sum up this chapter, already much too long, we commenced with showing that many general facts or laws indicate that each species has appeared at one point or rather area of the earth's surface; each species not being necessarily derived from a single pair, but by the very slow modification, through selection, of many individuals of another species. The supposed creation of the same species at more than one point of the earth's surface is admitted, even by those who hold this belief, to be an exceptional ⟨& even paradoxical⟩ case; yet it must be owned that such exceptional cases are not rare, & often present inexplicable, but not in my opinion overwhelming difficulties./FC 46 v/The difficulties will appear less to those many eminent naturalists, who see no great improbability in almost every island, having been within recent times connected with one & often with two neighbouring continents./53/We see ⟨clear⟩ indications of a law of single creation, & we cannot ⟨honestly⟩ deny that we are profoundly ignorant of the many possible means of diffusion, past & present. Who denies that the weather is due to regular laws, yet who can go into detail & say why the sun shone yesterday, or the rain falls today? The cases of the greatest difficulty are mostly included in the three/54/classes discussed in this chapter, namely in the floras & faunas of oceanic islands,—of fresh-water lakes or rivers, —& of mountain-summits with the polar regions; & I have collected together such explanations as have been given by others or have occurred to myself. To have collected the several isolated cases, would have been less serviceable & most tedious: yet some such are very curious & quite inexplicable. ⟨For instance the presence of Myrsine africana at the Cape of Good Hope, Abyssinia & the Azores & not as far as is known in any intermediate point.⟩ Besides our ignorance of the means of dispersion, & the chances of naturalisation by man's agency at some unknown time, we should never forget as long since urged by Lyell & by Forbes that a species/54A/may formerly have had under different conditions a more continuous range & become extinct in the intermediate regions, & secondly that some species have retained the same identical forms since even the commencement of the Miocene period, & this allows time for prodigious geographical changes. Hence I conclude that it has not as yet been absolutely proved that the same species has ever appeared, independently of migration,/55/on two separate points of the earth's surface: if this were proved or rendered highly probable, the whole of this volume would be useless,* & we should be compelled to admit the truth of the common view of ⟨absolute⟩ actual creation; & that organic beings are not exclusively produced by ordinary generation, with or without modification.

* No No—whether or no do not say so—it is not to the purpose. J.H.

APPENDICES

[The following fragments of the manuscript, letters, and related materials are arranged in the sequence of their relation to the text as here published. The numerical identification given at the beginning of each piece refers to chapter and folio numbers in the manuscript. Where necessary, the location of the source is stated in square brackets at the end of the passage.]

[Stray sheet, sole survivor of Chapter i.]
1, 40/   Var.[iation] under domest[ication]
Variation of Multiple Parts
Law of Variation [in] Nature.

Whenever any part or organ is repeated many times over in the structure of a species, it is variable in number, the same part or organ becoming numerically constant, either in other parts of the body in the same individual, or in other species, whenever the number is few: what can be more inconstant than the number of the feathers on a birds body, yet in the wing & tail, the principal feathers are remarkably constant in whole genera & even Families; but in some of those genera which have an unusual number of caudal feathers, the number is found to vary in the same species. It might be thought that the greater importance of the wing and tail feathers would account for their constancy; but I doubt this, for we find the same rule in the vertebrae, which are generally constant in mammals & birds, but in snakes, according to Schlegel [p. 27], the number varies greatly in the same species. So I believe it is in the teeth of fish & reptiles compared with the teeth of mammals./1, 40 v/According to Mr Wollaston it has been asserted that in insects bearing multi-articulate antennae, the number of joints in the antennae vary: in cirripedes the number of joints in the second & third pair of limbs, is usually pretty constant, but in Tetraclita their limbs are greatly elongated, & have very many articulations, & in them I could hardly find two individuals with the same number./1, 40/In plants, in those species which have many petals stamens or pistils their number is far more variable than when there are only few: & Gaertner (good → Kentniss Der Befruchtung s. 220, 364) has remarked that the number of seeds is far more constant in those plants, which have few, than in the polyspermous kinds. Why this rule of the variability in number of any part or organ which is already numerous, should/[End of page.

567

i, 40 v/Isidore Geoffroy's Law
Hist. Anomalies Tom 3, p. 456
["En recherchant quels sont les organes les plus sujets aux
variations anomales, j'ai d'abord établi cette généralité importante:
les organes les plus variables de tous sont ceux qui ont plusieurs
homologues placés en série; et l'on peut dire même que la variabilité
anomale d'un organe est en raison du nombre de ses homologues."]

[C.U.L., C.D. MSS. vol. 47, fol. 95.]

iii, 49/I then gathered a flower & breathed hard into it several
times, soon several very minute Flies crawled out/iii, 50/⟨dusted
all over, even to their wings, with⟩ pollen, & ⟨flew away. Three
of them I distinctly saw⟩ fly to another arum about a yard off;
they alighted on the inner [?] surface [?] of the spathe & then
suddenly flew down into the flower which I opened & although
not a single anther had burst, several grains of arum-pollen were
lying at the bottom ⟨of the spathe, near to but not on the stigmas.⟩
These must have been brought by the above or some other
Diptera ⟨midges, minute though they were⟩ from another plant
⟨individual arum.⟩ ⟨I may mention that in some other arums⟩
In other flowers, which ⟨had their anthers burst⟩ I opened, these
diptera ⟨I saw these midges⟩ were crawling about ⟨over the
stigmas⟩ & I saw them leave pollen on the stigmas. [Darwin later
pencilled an almost illegible addendum which may read:] in other
flowers in which there was also no dust [??].

iv, A 20/⟨From the common and widely diffused species in a
country presenting so large a proportion of varieties, and from
the evidence, such as it is, of these same common species occurring
more frequently in the larger genera, it might have been anticipated
that the species in the larger genera would, also, tend oftenest to
present varieties. But I was led to this anticipation from quite
other and theoretical grounds,—namely from looking at species
as only strongly marked and well defined varieties: for it follows
from this that wherever many closely related species⟩ have been
formed [,] many varieties, or as I look at them incipient species
ought, as a general rule, to be now forming.

v, 3, 4/⟨Let the cause be what it may, organisms/in a state of
nature are in some degree variable; & no doubt external conditions
produce some direct effect on them, to which subject we shall
have briefly to return; but mere fluctuating variability or the
direct effects of external conditions are wholly inadequate to

explain the infinitude of exquisitely ⟨beautifully⟩ correlated structures which we see on all sides of us. Look at the woodpecker, ⟨nut hatch⟩ or anteater with its long tongue & great claws; or the giraffe with its long tongue & long neck & high fore quarters—or look at what we are pleased to consider as the humblest parasite & see how beautifully its limbs are formed to cling to the hairs or feathers of the animal on which it lives. There are insects with admirably adapted structures formed to lay their eggs in the bodies of other species of insects, & others are adapted to lay their eggs on special plants, together with a poison, which no chemist can understand or imitate, which will cause the tissues of the special plant/v, 5/in question to develop a gall or abnormal growth of fixed form./⟩

[The Struggle for Existence.
Noteworthy in regard to Darwin's choice of this key phrase is the original opening for this paragraph, with its words added above the lines as alternatives or afterthoughts:]

v, 9/War of Nature

Struggle of Nature. The elder De Candolle in an eloquent passage has declared that all nature is at war. [Here Darwin added in pencil above the line an illegible phrase beginning: "plants...."]
　　　　　　　　fine　　　　　　　appearance of a landscape
When one views on a ⟨spring⟩ day the contented face of nature, or a tropical forest glowing with life, one may well doubt this; & at
　　such　　　　　　most of the inhabitants
⟨most⟩ periods ⟨nearly all living things⟩ are probably ⟨contented
living　　　　　　　　great danger hanging over them
& happy⟩ with no ⟨danger imminent⟩ & often with a superabundance of food.

[Darwin had already encountered the phrase 'struggle for existence' in a number of the works he had read, and he had used it in the 1844 Essay: Lyell, *Principles of Geology*, 1st edition, II (1832), p. 56; Edward Blyth 'Attempt to Classify Varieties', *Mag. Nat. Hist.* 8 (1835), p. 46; F. v. Wrangel, *Expedition to the Polar Sea*, 2nd ed. (1844), p. 47; and A. R. Wallace, *Amazon and Rio Negro* (1853), p. 121. Malthus, *Essay on...Population*, 6th ed., I (1826), p. 95, has the wording 'struggles for existence', (but cf. 1st ed. (1798), pp. 47–8, where the phrase is 'struggle for existence'.) For other references to 'struggle' in works Darwin had read, see note 47 regarding folio 30B of chapter v. In his 1842 Sketch, Darwin had already written of 'De Candolle's war of nature' and the related 'struggle' (*Foundations*, pp. 7, 8 note 3.) In his 1844 Essay, he wrote of this 'struggle' (ibid. pp. 91, 92), 'a recurrent struggle for life' (p. 148), and 'a severe struggle for existence' (p. 241.)]

[Notes on heath at Farnham, Surrey. C.D. MSS. vol. 46.1.]

v, 14/After Maer Heath give Farnham case—Cattle & Sheep destroy seedlings—I saw very MANY young oaks on bare *enclosed* common springing up I sh^d think ½ mile from any oak tree—so that enclosure of Maer Heath might be cause of young Oaks.

v, 37/I ⟨can⟩ often gaze ⟨for a long⟩ at a square yard of turf & reflect with astonishment at the play of forces which determine the presence & relative number of the 30 or 40 [figure not clear, could even be 30040.] plants which may be counted in it.— Apr 24/57

v, 38/May 5 1857
In Surrey about Crooksbury Hill [near Moor Park, Farnham], (which is covered by old Fir woods) as soon as the barren Heaths are enclosed, tens of thousands of young Scotch-Firs spring up. One can tell almost year of enclosure by seeing how old the oldest of the *innumerable* young Trees are. Now on Farnham Common, there are several large clumps of old Trees, & one might walk or ride over/39/hundreds of acres of surrounding common & say, if attention not drawn to it, that not *one single* seedling Scot fir c^d be seen; but on closer inspection ground cover[ed?] with seedlings scotfir 1 or 2 years old & a few older ones, not rising above shortest Heath, from being broused, & in all states of of decay—one of these pigmy trees was 26 years old.—So thick are the seedlings that on one place I found in square yard/39 v/30 seedlings—Now these hundreds of thousands had been sown for 30- or 40 years & not one had succeeded in escaping the cattle which ⟨very⟩ only rarely wander over this wild & barren Heath. Think of effect of drought for few seasons reducing number of cattle. I doubt whether sheep will eat young Firs. (I think they will to certain degree) I judge from Moor Park, into which only sheep are turned. (N. B. I have been again all over Farnham common: *part enclosed & part unenclosed & the case is very curious.*—Enclosed part studded with trees.
/38 v/It is curious how for about 15 ⟨10–20⟩ yards from edge of great woods, & when not in the least shadowed on all sides or exposures young Scotch firs *appear* not to grow; they do really grow in few numbers but the soil is exhausted & they keep quite dwarf, so that old woods are bounded in the *enclosed* Heath by *bare* strip of Heath without young trees, beyond which young trees. Good illustration of exhaustion of Soil.

[Regarding Farnham Heath, Scotch firs, see also Darwin's letter to Hooker dated June 3rd [1857] L & L, II, 99–100.]

Nov 10/1857/
v, 32/Looking at Crooksbury hill case of *apparent* absence of Firs within about 20 yards of the tall trees, the Heath being apparently not less vigorous, a very striking instance of rotation of crops.— Under trees they do not even germinate.—

v, 11/Even on very worst Heath, as near Waverly Abbey, Scotch firs will, if protected from cattle, *most* thickly spring up—(like grass patches) showing that others can grow there.—

23/Oct 30 [1858?] It is wonderful the number of little Oak Trees on Farnham Heath S. W. of Waverly & between it & the clump of old Firs. In some places all w^d not have been enabled to have grown up—they would cover Heath like New forest.— Judging from size of little Firs (it is now a *pinery* instead of a *Heath*) cattle are kept out *now*.—It is clear that oaks depend not on *mould*, but on absence of cattle—There were *very* few much beyond the great clumps, & this looks as if jay or wood-pigeons were part of cause: but I saw rooks chasing each other & playing over this side of Heath—All the trees about same size, or within 2 or 3 years old—Acorns w^d not last long on surface of Heath.—Do they carry as many as they can to feed at leisure? but I sh^d doubt rooks going to the clumps to rest & feed—Judging from Moor Park, Birches might be added; & Moor Park confirms the Scotch case.—It really is marvellous case.— The young oaks extended for about a mile from oaks & only in one direction.—I think rooks playing & dropping acorns best explanation.—

v, 46/1859 I observed at Moor Park, that when the Calluna had been cut, the surface *quite covered* with purple Heath, which would certainly be replaced by the Calluna—Aug 20. 59 I have been looking at the square patch of *poorest Heath* beyond the Leith Hill Tower [Surrey] which had been cleared of all *Heath*. There are several plants of grass, Carex, Potentilla, & Rumex growing up, & I suspect in grass would certainly become covered with vegetation, not Heath; i.e. if all Heath were picked out. None/46 v/of these plants grow [on] road—But I do not think vegetation would be quite same as on path, because of trampling & possibly manuring from animals—But, I believe, a struggle for life even on Heath does take place—

v, 53/May 17, 1862   Leith Hill Place
I have looked at the 2 square pieces of cleared Heath-Land; &
though a few more plants grow there than on surrounding Heath,
they may be said to be almost bare. Heaths struggle against each
other & against conditions of life.—I am now convinced that
manure [?] makes the grass paths, also[?] partly [?] by Heath
being killed.

vi, 26/⟨Forms produced by natural selection if considerably
different will be called species, if still more different, genera & so
on. But in these cases, besides inheritance & modification, extinction
which will always play a part, will here have played a very important
part in the destruction at some period of intermediate forms. To win⟩

[Folio 27 is gone, presumably discarded when it was replaced by folios
26* ff.]

vii. 14 v/[Notebook C, p.] 253.

Acclimatisation.—Bachman tells me in Audubon there is most
curious history of first appearance of the S. American Pipra Fly-
catcher which is now becoming common—likewise of the *Hirundo
fulva* (added by Audubon in Appendix) showing WHAT CHANGES
are taking place & how birds are extending their ranges even
migratory birds like swallows.—Of migrations of birds he mentioned
many most curious cases. the birds seem to follow narrow bands,
certain kinds as gallinules taking the low country near coast &
others the mountains, & then/p. 254/appearing to remain about
a fortnight that is succession of birds. See Silliman's Journal 1837
Paper by Bachman.—in some species as Tanagra males come first
& then females in flocks as in English nightingales.—other birds
(& this seems common kind migration of America) migrate singly
flying few miles every day & generally by night—one bird which
is strictly diurnal, migrates singly by night.—others in flocks.
kind of migration quite different in species of same genus. these
birds seem clearly directed by kind of country; the Muscicapa
solitaria stay about a fortnight in one particular part of country,
like White of Selborne Rock Ouzels.—If the line or bands of
country (These facts show the normal condition of migration).

[/p. 255/"gradually separate the birds might yet remember which way to
fly—"
This leaf, pages 253–4, was selected and cut out of transmutation Notebook
C, (C.D. MSS., item 122), and was placed here. See also Sir Gavin de Beer

and M. J. Rowlands, *Bull. Brit. Museum (Nat. Hist.) Historical Series* vol. 2 (1961), p. 191.
The following three note slips were also included with the MS here:]

VII, 14 v/In Portfolio "Instinct" some excellent facts from Bachman on change of ranges in N. American Birds even Pelidna.

VII, 14 v/Kalm 1/292 The maize thieves (icterus) & several sorts of Squirrels have increased owing to the greater cultivation of maize. 1/294. Codfish were formerly never caught at Cape Hinlopen, but now they are numerous there

VII, 14 v/[Steel] Silliman's Jour. vol. 19 p. 357. Describes the first appearance at *Union* in Maine, of a new kind of swallow, in the first 5 years there were about 50. At Saratoga they arrived in 1828, they have since increased rapidly, to that at in 1831 they were computed at some hundreds. Is not this spreading North.

VII, 38/[The following memoranda and questions are later than the original text of the chapter, for they are written on a foolscap sheet with the watermark: 'E. Towgood 1858'.]
The Umbelliferae with lax heads *oftenest*? [sic] have ray florets. Is it conceivable that pressure in Hasselquistia & Coriander could make seeds orthospermous & coelospermous.
In Marygold seeds are convex externally looking like pressure. Seeds differ in ray & centre of some Compos. without differences of corolla.
Would ovary or corolla be first formed? as first formed most likely to affect last formed.—
Heads of flower rendered more conspicuous in Viburnum & in Mussaenda (by the exterior ray of *sepal* being white; their flower *not* sterile) by the exterior flowers being developed. In Feather Hyacinth by central.—In carrot central flower also affected, as in peloria.—
These facts seem to show that some connection with more or less nourishment of central or exterior parts./38 v/The phenomenon not more frequent in densest heads.

VII, 44/[The gray foolscap sheet written in ink Darwin labelled: 'note p. 44 to Ch 7', and later he pencilled the comment: 'Metamorphosis & Embryology difficult—all this page confirms.' The sheet is now grouped with other notes mostly on embryology in C. 40. f.]
⟨In those animals, which according to our theoretical notions must have undergone very great modifications, it would appear

probable that owing to successive modifications of every part of the structure becoming earlier and earlier developed in the embryo, at last all traces of a distinct embryonic form might be absorbed & lost.⟩ Prof. Owen (Lectures on Comparative Anatomy: invertebrate animals. 1855. p. 638) has remarked that in the Cephalopoda, the highest or most modified mollusca, & in the Arachnidae considered by him the highest articulata, there is no distinct metamorphosis. But in the Acaridae or ⟨lowest⟩ Arachnidae of extremely low development ⟨& in the Brachiopoda or low Mollusca⟩ there is no great am't of embryonic change, some other & quite distinct principle must come into play in accounting for the amount diversity & duration of embryonic changes. Many annelids offer a strong instance of great and little change in form or metamorphosis in the same class. So again in certain pupiparous Flies, which I presume cannot be considered as having undergone modification in any extraordinary degree in comparison with other Diptera, Leon Dufour (Annal. des Sciences 3 Ser. Zoolog. Tom. 3 p. 79) could detect no trace of a larval stage.

But in all these [?] there is probably connected[?]metamorphosis.

VIII, 65/A   British Museum
                                                     10 Nov. 1857
My dear Sir.

   Some time ago you asked me to furnish you with remarkable instances of disparity in form &c in workers of Insects living in community—As one is apt to forget these things at the moment they are asked for I send you one that is a truly remarkable instance—In my Monograph on the Genus Cryptocerus I figured & described a species as C. discocephalus—Some time subsequent I received a letter from Mr. H. W. Bates from Brazil—he said— "I have met with your curious Species C. discocephalus—the creature figured is only the large size of the worker of the species— I send you both the workers taken from several nests constructed in dead branches of shrubs—" I send you tracings of the creatures in relative proportions—Dont trouble to reply to this but tell me what you think of it when you are next time at the Museum and believe me

                                                     Yours very Truly
                                                     FRED H SMITH

x, 11/I felt at first a little sceptical on this head, but this was unreasonable, for how could the hosts fight, if those on the same side did not know each other? Nevertheless to try this, I took several times some hill-ants (*F. rufa*) from their own nest & placed them on another; they were always extremely much agitated & were instantaneously attacked by the inhabitants: whereas when I returned several of the same lots to their own nest, they seemed immediately to recognise their comrades & be recognised by them. In Moor Park, near Farnham, there is an enormous nest, asserted by the the country people to have existed on the same spot, during their whole lives, for at least forty years, & inhabited by I should think, some hundreds of thousands of ants; & yet these as in the case of smaller nests, immediately recognised & attacked a stranger of the same species. Some ants, which I kept for 19 hours in a bottle & then put back on their own nest were not attacked though some were threatened: the bottle used in this case smelt of physic, & the ants must have been thus scented, & as they were not withstanding this recognised by their comrades, it would appear that the recognition is not owing to all the ants of the same nest having a common odour.—

CHAPTER XI APPENDIX

[J. D. Hooker's Notes regarding Darwin's Geography Chapter. C.U.L. C.D. MSS. vol. 100, fols. 109–10.]

Note A

Would Forbes suppose that the presence of the South Shetland Anas antarctica on the Falklands was due to Iceberg transportation North? Is it not more natural to suppose that A. ant. [arctica] was produced by creation or variation on the American continent & thence either transported South to S. Shetland or that it inhabited an intermediate sunk area. I am against making arctic regions centres of creation either by variation or by specific creation.

I think it would facilitate our researches much not to look beyond the epoch of the existence of those continents having the required climates for the existence of the scattered productions whose migrations we seek to account for. It is enough to admit a glacial land & sea over central Europe & do not let us speculate on the origin of its species. Never wander further back into Geological time than is necessary—it bewilders.

On the whole then I would perhaps confine this part of the discussion to the migration North & vertical ascent of species inhabiting a cold country.

Note B

Might not much of this difficulty be got over by supposing the E. & W. parts of the glacial continent differently heated, & that currents flowed East & West or NE & NW.

Thus the connecting land of Europe & America might be much warmer than those parts of either continent in the same latitude where the mountains were.

Note C

I cannot see why the colonisation of Iceland, Ferroe & Spitzbergen should come under a different category from other lands—this is most unphilosophical since a theoretical inflexion of the isothermals should not be wholly lost sight of, during the glacial epoch, as it manifestly is after it. The gradual accession of the Gulf Stream's influence would warm all that part of the glacial sea coast or chain of Islands that included Iceland Ferroe &c before any other part of the glacial region & induce migration along that line, however cold the preexisting arctic desert in which they were situated may be assumed to be.

Note D

I cannot understand this. Why do the Gentians not go North? these not being more Alpine than the Arctic species—Why should they have spread over the intervening country?

Note E

Then why no peculiar species or varieties in Iceland, Spitzbergen, &c.

Note F

The same argument must hold for the Arctic & Antarctic representative crustacea—on which Ross was always insisting & swearing that some were identical with what he had described in Capt. Parry's Voy[age] &c.[1]

[1] [Ross, 'Zoology', pp. 91–120 of Appendix of Parry, *Journal of a Third Voyage for the Discovery of a North-West Passage 1824–5, in his Majesty's ships Hecla and Fury*, London 1826; and Ross, 'Zoology', pp. 189–206 Appendix of Parry, *Narrative of an Attempt to reach the North Pole...MDCCCXXVII*, London, 1828.]

Note G

The fact that Flora ⟨character⟩ of analogous elevations of Ceylon, Nilghiri, Khasia & Himal. is to great extent specifically the same.

Note H

After which why did not any ascend the Himalaya?

An argument in favor of alteration induced by isolation afforded by fact that so many well known species when found isolated have as much difference as to deceive botanists & then when dried lose all distinguishing characters.

Change of Tropical Climate demanded is far too great Where were many tropical genera & orders—Also migration not always N & S. but across continents obliquely—Also all this leaves longitudinal distribution unaccounted for as Abyssinia & India—W. Austral. & Carnatic.

Ordinary laws of reproduction include modif. of specific forms.

But it is improbable that similar forms be generated from specifically different parents in different places.

Hence will propagation account for presence of identical forms in all parts of globe.

Plants, insects—common to Alps & Scandinavia Steinbock, Variable Hare, Chamois.

Forbes glacial epoch accounts for this

Help may be got by introducing humidity as an element—quote very different levels on Himal. Khasia & Ceylon for same species.

[ON REPRESENTATIVE SPECIES]

1/*We will now consider some of the most striking cases of difficulty on geological grounds opposed to the theory that closely allied or representative species, are due to the modification of the same species./1 v/This theory implies a communication of some kind between the areas occupied by the representative species at some former period, generally not very remote in a geological sense, as distinctly as does the theory of single centres of creation in regard to the same species when found at distant & separated points./ 1/Dr. Hooker has given a most curious list[1] of representative species, found in New Zealand, Australia & S. America. With respect to Australia there is no greater difficulty (but great enough)

---

* This will come towards end of *another* chapter. [C.D.]

[1] New Zealand Flora. Introduction. p. xxxiv.

than in other analogous cases, already discussed when considering the identical species of insular floras & faunas.* But with respect to S. America the case is different, owing to the vast space of open ocean between that continent & New Zealand. Yet even here we find an accordance with the general rule that the productions of an island are more or less allied to those of the land nearest to it; & again we see an accordance with the rule that where there are representative species there are some identical species in common, proving to those who believe/2/in single creations, that there has been at some time some channel of communication between the two areas in question. Dr. Hooker believes that/ 2 v/there was formerly a communication by more or less continuous land from both Chile & Fuegia to New Zealand. I cannot persuade myself to admit (though far better judges see no difficulty in admitting) such great geographical changes within so recent a period; & I think that a slight modification of Dr. Hooker's view will remove some *little* of the difficulty./

2/Of the 50 genera which afford the best instances of representative species in New Zealand & extra-tropical S. America, 7, as I am informed by Dr. Hooker, are northern genera,/2A/& 18 of very general distribution; & the representative species in these 25 genera (bearing in mind the glacial epoch) present nothing more remarkable than representative species in other parts of the world, to which in former times we may imagine the descendants of the same species to have travelled & subsequently to have become modified: but the other 25 genera are strictly confined to the south with all their species extratropical,—a few on the mountains within the Tropics being excepted. This fact of half the genera being confined to the South seems to me remarkable considering that out of the 89 species belonging to 76 genera[1] absolutely identical in New Zealand & S. America, only two species belong to genera confined to the South, namely Colobanthus subulatus & Rostkovia Magellanica: Goodenia repens need hardly be added as this is an Australian genus with one littoral wandering species./ 3⟨43⟩/Again it may be noticed in Dr. Hooker's list (which I am aware is not given as perfect) that of the *southern* genera, which have representatives in New Zealand & S. America, there are five which have none in Australia or Tasmania, & this is what might have been expected considering the greater distance of Australia, than of New Zealand from S. America; but Australia has four

---

* Find out whether any old Rocks in New Zealand; also about soundings, so to give chance of former union. [C.D.]

[1] [Ibid.] Introduction p. xxxi.

southern genera (viz; Eucryphia, Pernettya, Lebetanthus, & Lomatia) with representatives of S. American species, which genera do not occur in New Zealand or the Auckland islands. These facts, together perhaps with the genera Colobanthus, Ac[a]ena & Lagenophora having representative species, both on these two lands & in several of the circumpolar islands, seem to me to indicate some common centre of radiation.[1]/

4⟨44⟩/Now taking the northern hemisphere as our guide, I should look to the circumpolar regions as the centre of radiation for the representative & for some of the species still remaining identical in the above named several lands. If we look to a chart we see in the little explored regions between 62° & 80° several islands, & large tracts of land, with surroundings in one place 100 miles from the shore, & with indications[2] of other rocks besides volcanic./ 4⟨44⟩ v/Here then I should infer that it was no ways improbable that these lands & islands may have recently been of greater extent & more continuous./4⟨44⟩/On these islands not one single land plant can now live, but bearing in mind that in the north the space between these same parallels is the home of the whole Arctic Flora, it seems to me a not very improbable supposition that before the glacial epoch came on, these islands might have been covered by a /5⟨45⟩/not scanty vegetation.[3] According to all analogy, this antarctic vegetation from its isolation would have been very peculiar, but would have been in some degree related to that of the two nearest continents, America and Australia; & this antarctic vegetation though perhaps not nearly so uniform, as that now growing on the almost continuous arctic land, would probably have been[4]/5A/tolerably uniform. From this source, I am

[1] [Preceding lines sheared off foot of fol. 3 ⟨43⟩, and following passage cancelled.]
⟨.../representative species of the southern genera which in S. America grow in low land are confined in New Zealand ⟨& in Tasmania⟩ to high land; (Flora of New Zealand Introduct. p. xxiv & note D) & as inferred by Dr Hooker (Id. p. xxiii) in regard to the identical species of the two countries similarly situated & associated with them, the climate of New Zealand must have grown cooler since their introduction.⟩

[2] Sir J. Ross Voyage to S. Seas. Vol. 2, p. 421.

[3] In the southern hemisphere we have no distinct evidence that the climate was warmer during the older pliocene periods. The existence of burnt & silicified trees & thin beds of coal under the streams of lava in Kerguelen's Land & in [New South] Shetland Isld. (V. Dana's letter on Mr. Eights [*Amer. J. Sci.* 2nd ser., 22 (1856), 391]) where the vegetation is now so scanty should not be forgotten. I may, also, state that in Tierra del Fuego, ([Darwin] Geological Observation, p. 118) I found in beds *underlying* the drift or glacial deposits, many leaves of trees; which belong, according to Dr. Hooker to three species of Beech, apparently differing from the two species, which now clothe that forest-clad land.
[At this point the following passage was cancelled before the MS. was given to the copyist.] ⟨internally related or already pretty closely related. As the glacial epoch came on, we may imagine that the seeds of these antarctic plants were

inclined to suppose that some plants, either identical or allied, migrated before the glacial epoch by various accidental means, aided probably by more continuous land & by the several inter- mediate islands which we see still existing South of New Zealand; & that when the glacial epoch did come on, the seeds of other plants were brought in a N. Easterly course by icebergs from their common home, soon to be converted into an icy desert. The plants which arrived at this latter period, would, on the returning warmth have ascended the mountains of New Zealand & Tasmania, —most of the species especially those brought first, having sub- sequently undergone modification & now existing as representative species,—a few having remained identical. But those who receive the common view that every species has been created as we now see it, & that the same species has sometimes been created at more than one point of the earth's surface, may truly say with derision/6/what complicated theories are required, such as the one just given, or the more simple one but requiring much greater geographical changes given by Dr. Hooker, to account on the theory of descent for the same, & with subsequent modification, for the representative species in these two distant areas. On the other hand those who believe in simple creation can, in my opinion, give no explanation in the least degree satisfactory of the shades of affinity & degree of identity in the cases which we have been discussing: they in fact simply state so it is./

xi, 17⟨9⟩/In Siberia ⟨they [erratic boulders] have not been observed by [?] peasants [?]; how far the fossil remains of Elephas primi- genius & Rhinoceros trichorbinus under about the same latitude as found in Europe is any evidence of the climate having been formerly colder, I am doubtful.⟩

xi, 32⟨23⟩/[Cancelled note on verso.]
⟨Dr Hooker believes that New Zealand & Australia were united within the period of existing species; I have already given my reasons for not being able to admit this view, which would remove many difficulties, but at the same time cause some others. But

carried in a N. Easterly course from their native home & landed on the southern shores of Australia, New Zealand, S. America & the several antarctic islands,— already chilled & ready to receive southern colonists. After the glacial epoch, as the climate improved, these antarctic plants would ascend the hills, where we now see them, some few remaining the same in these now widely separated colonies, some having undergone modification since their arrival, & some having arrived distinct, as they existed on their antarctic native islands.

It may be asked whether any other organic//...//seem to have radiated from this//....[Bottom of folio torn off.]

I am far from wishing to deny the possibility that there may have existed larger & more numerous islands in the intermediate sea; or that the main coasts may have formerly extended somewhat nearer to each other.⟩

[After the fair copy was made, Darwin cancelled the following passage thereon and replaced it by the passage on the new holograph folio (designated 45.1) which is new folio 40 of the fair copy.]

XI, 45/The island of Tristan d'Acunha ⟨*a is inhabited by 29⟨27⟩ phanerogamic plants, some of which are Fuegian species: it is situated between America and Africa, to which latter continent it shows in two of its plants some slight affinity: it lies in Lat. 37° about 700 miles nearer the equator than Kerguelen Land, & is distant from Tierra del Fuego about 2300 miles.⟩

[Among the large group of notes and papers on geographical distribution the following seem of particular interest. The number 18 on these items was marked by Darwin in ochre crayon as if it were that of the portfolio and pigeonhole in which he filed them. The longer pieces are drafts on special topics written on the usual foolscap size sheets which Darwin used for the Natural Selection manuscript; the shorter ones are on miscellaneous smaller scraps of paper. These items afford evidence for my opinion that Darwin's whole collection of notes and papers on distribution invites further study.]

Oct. 25/50/   18
Agassiz doctrine that a number of individ. of a spec. created at once implies a previous gap in economy of that place, which is very improbable—If he supposes *all* the species in any spot co-created, that implies a catastrophe to form large new untenanted locality, which is very improbable—Hence one or only few individuals created at first—In Bees Agassiz right.—

[Darwin MSS. C. 40 c.]

THEORETICAL GEOGRAPH. DISTRIB.

Nov./54/   18
A species is well suited to its conditions, sports & becomes modified or becomes parent of another species either remaining itself generally for a time, & then usually replaced & dying: but ⟨when⟩ fact of one new species having been formed is evidence that it is suited to conditions, & will probable give rise to other forms. If the region will support so many composita, while one genus, has

* [on verso] ⟨Some account of the Island of Tristan d'Acunha by Capt. D. Carmichael. Linnean Transactions vol XII p. 483.⟩

sported & shown its adaptation as the most likely to yield more forms. Or thus,—When a species breaks & gives rise to another species, the chances seem favourable (for it has given birth to one simply because its whole constitution is well adapted to the conditions &c) to its giving birth to others. (No doubt here comes in question of how far isolation is necessary, 'I sh^d 'have thought more necessary than facts seem to show it.—In fact there never can be isolation for the parent form must always be present & tend to cross & bring back to ancestral form; it will *always* be a struggle against crossing, & will require either vigorous selection or some isolation from habits, farness nature of country to separate) Hence genera will be local owing to their origin from common point; & small genera (2–8 species) certainly, from S[c]hoenherr, are local in proportion of 215: 52 (& these 52 are not such small genera as the others).

As to make species is slow work, if genera increase to considerable ⟨great⟩ size much time w^d be required, hence as Forbes says w^d be local in their origin in past time: the species w^d extend over continuous spaces in area & time. But if as generally ⟨often⟩ happens during the time necessary to make a large genus, that geographical mutations & chance accident w^d disperse genera & the very fact of the genus having become large in one area, we may suppose w^d give it some better chance in another & continuous area, & thus the genus w^d get bigger & bigger. And certainly most larger genera are widely extended. When a genus began to fail & die out, if large, it w^d leave probably a few species in distant quarters of the world: Hence this w^d be another cause of small genera: these w^d be aberrant. The mere fact of large genera, generally being wider can be accounted for by creationists showing that if a genus be created in different distant quarters it w^d probably form so many local species.

If inhabitants of S. America & Australia turned into each other by an isthmus I sh^d expect the larger increasing genera of the two w^d persist & spread, & exterminate many of the smaller genera; probably inhabitants of one continent w^d prevail? considerably? over the other; when one continent formed another, usually one w^d be sinking &   decreasing.

Unusual powers of dispersion (mem. all insects are biaxial [?]) might account for some small genera being widely distributed, but I am much surprised if *small* genera with widely distributed species are not remnants of large genera & so AS SPECIES aberrant, or very distinct from each other though they may not belong to ABERRANT GENERA. I think thus alone we can account for "wandering species" of plants being generally very distinct, they are remnants after extinction of connecting species.—

[verso:]
All existing continents show signs of former connections with other continents.

Strictly still in the large genera which have arrived at their numbers from conditions having been favourable. they may be now increasing or decreasing. The increasing genera are in very nature genera with close species.

Ought there not to be as many small genera due to forming genera as to dying genera.—Yes I think. (?)

The dying genera will often be widely spread & will contain more distant species.

[Slip pinned on verso:]
I do not yet quite see why dying genera sh^d & therefore small genera sh^d not be often widely distributed. Only a few genera can survive to a future period, for these few make families. Is it that when a genus once becomes widely extended it generally does live on

March, 1855

1856 Feb
It is clear there are two very distinct causes for small genera, just forming & becoming extinct: the latter in affinity will have species very different from each other, & will be more apt to have wide ranges—Babington I remember remarks that where a genus has its metropolis the species are apt to run into each other, when, I think, discussing in Annals of Nat. Hist. [p. 388] the Batrachian Ranunculi—But yet, as I have shown, the small genera with widest ranges are not the most aberrant, but they may be broken genera with their few species not very closely allied.—

[Darwin MSS. C. 40 i.]

March 28/55   18
I believe in single creations, because (1) as a general rule species have non interrupted or scarcely interrupted ranges, such as offer no difficulty whatever in carrier transportal.

(2) because when we know means of transportal in different localities, as quadrupeds.

(3) because same species (Lyell) do not appear & reappear in time.—

(4) because under apparently similar conditions same species do not appear as New Zealand & parts of Europe.

(5) because extension of species as general rule bears relation to obstacles, preventing immigration.—which w^d not be case if created independently.—look at shells of W. Amer. & Pacific with respect to (1) quote De Candolle—most flagrant exceptions explained by Forbes, extended to S. Hemisphere, extended by him to world, so explaining Crustacea of New Zealand.

Lastly we might expect exceptions to continuity of geographical range to chance introductions, & still more to extinction of species in intermediate points.

Then go on to whether created in single pairs or in crowds—Agassiz's argument of no force in some cases, as shown by introductions & as shown by Beavers?—Bees difficult case—My theory grants that whole body of individuals being slowly altered.—N.B. J. Lubbock tells me that some of the Alpine Beetles are thought to be only *vars*. of the lowland Forms.—

[Darwin MSS. C. 40. c]

[CHAPTER XI on GEOGRAPH DIST.] 18

My proposition is that when *close* or representative species occur, then there is or has been channel of communication by which species have entered & become modified—as inferred from presence of identical species—at least in eyes of those who believe in single creation—the channel having been possible under altered climate or with land at different levels.—or as inferred by simple geographical position; for under whatever contingency it is generally more probable that forms derived from nearest lands—opposed to last proposition Kerguelen Land, already discussed—Vaccinium in mountains of Sandwich Is^d—

The Representative species of New Zealand—yet accord with geographical position & some species in common though it cannot be accounted how these species came to be in common.—

Sum up the cases of representative species in last chapter.

The law of geographical position is too perfect.—case of 2 sides of Australia related to Africa & S. America—other such cases—descend into detail—Such delicate shading seems too beautiful to be accounted for by accidental migration & accords best with continental extension of land.

[C.D. MSS. C. 40. c]

ON BARRIERS 18

If we take a general view of distribution, I think we must conclude that barriers, whatever the nature, in regard to powers of passage of organisms is the chief, I sh^d say decidedly the most important element in their distribution. For marine productions, landing [?] stretching N. & S. is a perfect barrier, if it has long existed, so again a wide space of ocean; now compare the shells on each side of I. of Panama, only one the same; so with crustacea, so with Fish.—(Isthmus of Suez so low) Again there is profound ocean, fully as wide as Atlantic ocean, west of S. America, without an island & here there is not a shell in common—but westward in ocean strewn with isl^d (& with evidence of former isl^d) the shells & fish extend with very many in common, even to W. coast of Africa, almost exactly an hemisphere. Again land shells of America, correlate with water shells on opposite sides of Alleghanies. (some fish cases of Hooker)—so all productions on opposite & alike sides of Cordilleras.—Looking to land mammals & considering their feeble means of transport, except in certain ice cases ⟨quadrupeds we must agree that S. America Australia, Madagascar & tropical S. America in late geological period to a certain extent, & Africa & Asia together⟩ & taking soundings as source & the only indication of the chance of former connection: for no geologist who has studied the astonishing accumulation of evidence of oscillation of level, will think it in the least impossible that land separated by only 200 or // 300 fathoms sh^d have been joined, we shall find very fair evidence/2/[of] concordance between identity of species & continuous land. East as E. Indian Archipelago. Celebes case— Australia, Van Diemen's land, New Guinea.—Japan.—Madagascar —Asia & Africa & Europe ⟨isthmus of Suez⟩—S. Africa itself deserts lines of sounding in Mediterranean hence isl^d of Mediterranean—Grt Britain, S. America, Falklands soundings, (evidence of ancient subsidence at S. Cruz) but here possibly Ice action. S. America itself rather a difficulty, formerly whole of America more united with old world & N. & S. America more united & Isl^d of West Indies: Isthmus of Panama does not seem to have been wide enough road.—* There are a few in common with S. America & N. America. Look at S. Africa, Australia under same

---

* Note for folio 2 re Barriers.

I am here confounding 2 considerations identity of species & connection of forms, I must reconsider whole case: now I do not know whether any species the same in New Guinea & Australia.^{a}

2 v/I think begin with same conditions do not produce same mammalia.— tropical Africa & tropical America—sub-tropical Australia & S. Africa conditions differ as far as organic beings are concerned.

^{a} Europe some species in common with Africa. give connections oscillation of Gibraltar.—depth of Mediterranean—Mediterranean isl'^{d} Britain.— Africa Celebes East. Australia & Tasmania—(India with Asia) N. America soundings —glacial action—S. America with N. America. Madagascar. Climate of coast [?] comes into play thus none of S. America common with Europe.—Contrast S. America, C. of Good Hope & Australia.

[Darwin MSS. C. 40c.]

latitude & in part with very similar climate & how preposterous the idea of any species being the same. The r[el]ationship of forms another consideration to which I will have to recur.—No oceanic Isl^d has same species of mammifers.—I do not mean to say that when no barriers the same species occur, for here other considerations come into play.—/

3 *Plants.** A. Decandolle has remarked that in N. when lands most united (has one sounding in Behring St. & bearing in mind more intimate connection anterior to glacial period.—) that most species in common in north and get fewer & fewer in south—We shall have to consider some exceptions far south—Cape of Good Hope one of the most distinct areas in world, deserts & tropical band. Most plants have such small areas that it w^d not be fair to compare (& so it is not fair, except in relation to higher connections those nine[?] species of mammals) S. America, S. Africa & Australia. The evidence from plants not too good, as we shall see when we come to isl^ds. But yet we see that Barriers have much to do, more species of plants in common between the warmer tangent regions of B.[anda] Oriental & S. Brazil, more between tropics of Australia & S. Australia than between S. America & Australia, & C. Good Hope. In Plants we see well the great distinctions without Barriers. East & West Australia, formerly isl^d preoccupation—Look in S. Africa in A. Decandolle I believe very few species common to the 20 areas in which this has been divided.

Upon the whole Barriers are a most important element in distribution, & thus I can understand only that species created in one spot, & wanders as far as it can, considering conditions & preoccupation, till it meets Barrier.—Perhaps even those who believe in double creation w^d admit this & say that double creations are exceptions, & the case probably sh^d be considered, whether former means of passage are not the exceptions.—

* I think plants had better be discussed earlier.

BIBLIOGRAPHY

INTRODUCTION

Wherever feasible the citations given by Darwin and listed in the following bibliography have been checked with the original sources. In a few instances, I could not find the editions with pagination fitting Darwin's page references. These titles have been marked by the sign [ed.?] after the date of publication. I have added a few entries enclosed in square brackets to list works not cited directly by Darwin, but mentioned either by his sources or in the editorial commentary.

Titles by writers who were anonymous, pseudonymous, or identified only by initials have been listed at the beginning of the bibliography, grouped together by the journal in which they appeared and then arranged chronologically.

The chapter and folio references after each entry locate Darwin's use of the sources in the text.

ABBREVIATIONS FOR PERIODICALS CITED IN THESE VOLUMES

Acad. sci. Belg. Bull.: Bulletin de l'académie royale des sciences et belles-lettres de Bruxelles.

Acad. sci. Berlin, Mém. Mémoires de l'Académie Royale des sciences et Belles-Lettres de Berlin.

Acad. sci. Paris, Mém. prés.: Mémoires présentés par divers, savans à l'académie royale des sciences de l'Institut de France et imprimés par son ordre, Sciences mathématiques et physiques.

Acad. sci. Petro. Acta: Acta academiae scientiarum imperialis petropolitanae.

Acad. sci. Petro. Mem.: Mémoires de l'académie impériale des sciences de St. Pétersbourg.

Acad. sci. Petro. Nova Acta: Nova acta academiae scientarum imperialis petropolitanae.

Akad. Wiss. Berlin Abh.: Abhandlungen der Königlich Preussischen Akadamie der Wissenschaften.

Albany Inst. Trans.: Transactions of the Albany Institute.

Allg. schweiz. Ges. gesamm. Naturw. neue Denkschr.: Neue Denkschriften der allgemeinen Schweizerischen Gesellschaft für die gesammten Natur-wissenschaften.

Amer. J. Sci.: [Silliman's] American Journal of Science and Arts.

Ann. Bot.: Annals of Botany.

Ann. Mag. Nat. Hist.: Annals and Magazine of Natural History.

Ann. Nat. Hist.: Annals of Natural History; or, Magazine of Zoology, Botany, and Geology.

Ann. sci. nat.: Annales des sciences naturelles. (Paris.)

Arch. Naturgesch.: [Wiegmann's] Archiv für Naturgeschichte.

Asiat. Soc. Bengal Trans. (or *Asiat. Soc. Bengal J.*): Asiatick Researches; or,

Transactions of the Society, instituted in Bengal, for inquiring into the history and antiquities, the arts, sciences and literature of Asia.

Athen.: Athenaeum.

Bd. Agric. Lond. Commun.: Communications to the Board of Agriculture, on subjects relative to the husbandry and internal improvement of the country.

Bd. Agric. N.Y., Mem.: Memoirs of the Board of Agriculture of the State of New York.

Boston J. Nat. Hist.: Boston Journal of Natural History, containing papers and communications, read to the Boston Society of Natural History.

Bot. Gaz.: The Botanical Gazette, a journal of the progress of British Botany and the contemporary literature of the science.

Bot. Misc.: [Hooker's] Botanical Miscellany.

Bot. Soc. Edinb. Trans.: Transactions of the Botanical Society of Edinburgh.

Brit. Ass. Advanc. Sci. Rep.: Report of the British Association for the Advancement of Science.

Calcutta J. Nat. Hist.: Calcutta Journal of Natural History, and miscellany of the arts and sciences in India.

Calcutta Rev.: Calcutta Review.

Cottage Gdnr.: Cottage Gardener (and country gentleman's companion).

Dict. class. hist. nat.: Dictionnaire classique d'histoire naturelle. 17 vols., Paris, 1822–31.

Edinb. J. Nat. Sci.: The Edinburgh Journal of Natural and Geographical Science.

Edinb. New Phil. J.: The Edinburgh New Philosophical Journal, exhibiting a view of the progressive discoveries and improvements in the sciences and the arts.

Edinb. Rev.: Edinburgh Review, or critical Journal.

Ent. Soc. Lond., J.: Journal of the Entomological Society.

Flora: Flora oder Botanische Zeitung welche Recensionen, Abhandlungen, Aufsätze, Neuigkeiten und Nachrichten, die Botanik betreffend, enthält.

Gdnrs. Chron.: The Gardeners' Chronicle and Agricultural Gazette.

Gdnrs. Mag.: [Loudon's] Gardeners Magazine and Register of rural and domestic Improvement.

Geogr. Soc. Lond. J.: The Journal of the Royal Geographical Society of London.

Geol. Soc. Lond. Proc.: Proceedings of the Geological Society of London.

Geol. Soc. Lond.; Quart. J.: The Quarterly Journal of the Geological Society of London.

Geol. Soc. Lond. Trans.: Transactions of the Geological Society of London.

Geol. Survey G.B. Mem.: Memoirs of the Geological Survey of Great Britain, and of the Museum of Economic Geology in London.

Highl'd Soc. Scotl. Trans.: Prize Essays and Transactions of the Highland Society of Scotland.

Hort. Soc. Lond. J.: Journal of the Horticultural Society of London.

Hort. Soc. Lond. Trans.: Transactions of the Horticultural Society of London.

India Sport Rev.: The India Sporting Review, a Record of the Turf, the Chase, the Gun, the Rod, the Spear.

Institut.: L'Institut, Journal des académies et sociétés scientifiques de la France et de l'étranger.

Isis: Isis, oder encyclopädische Zeitung von Oken.

J. Bot.: [Hooker's] Journal of Botany, being a second series of the Botanical Miscellany.

J. Phys.: Journal de Physique, de chimie, d'histoire naturelle et des arts. Paris.

Linn. Soc. Lond. J.: Journal of the Proceedings of the Linnean Society.

Linn. Soc. Lond. J. Bot.: The Journal of the Linnean Society, Botany.

Linn Soc. Lond. Trans.: Transactions of the Linnean Society of London.

Lond. J. Bot.: The London Journal of Botany; containing figures and descriptions of such plants as recommend themselves by their novelty, variety, history, or uses; together with botanica notices and information, and occasional memoirs of eminent botanists.

Madras. J. Lit. Sci.: Madras Journal of Literature and Science.

Mag. Nat. Hist.: Magazine of Natural History, and Journal of Zoology, Botany, Mineralogy, Geology, and Meteorology. Series 1 (1828–36) ed. by J. C. Loudon. Series 2 (1837–40) ed. by E. Charlesworth.

Mag. Zool. Bot.: Magazine of Zoology and Botany.

Med. Times: Medical Times and Gazette.

Muséum Hist. Nat. Mem.: Mémoires du Muséum d'histoire naturelle, par les professeurs de cet établissement.

Nova Acta Leop. Carol.: Nova Acta Academia Caesareae Leopoldino-Carolinae Germanicae Naturae Curiosorum. Halle a.S.

Phil. Trans.: Philosophical Transactions of the Royal Society of London.

Phytologist: The Phytologist: A popular botanical miscellany.

Poult. Chron.: The Poultry Chronicle.

Rev. deux mondes: Revue des deux mondes.

R. Instn. Proc.: Notices of the Proceedings at the Meetings of the Members of the Royal Institution of Great Britain, with Abstracts of the Discourses delivered at the evening meetings.

R. phys. Soc. Edinb. Proc.: Proceedings of the Royal Physical Society of Edinburgh.

Soc. Edinb. Proc.: Proceedings of the Royal Society of Edinburgh.

Soc. Geol. Fr. Bull.: Bulletin de la Société Géologique de France.

Soc. Hort. J.: Journal de la Société impériale et centrale d'horticulture.

Soc. linn. Bordeaux Actes: Actes de la société linnéenne de Bordeaux.

Soc. linn. Bordeaux Bull.: Bulletin de la Société linnéenne de Bordeaux.

Soc. linn. Norm. Mem.: Mémoires de la société linnéenne de Normandie.

Soc. Nat. Moscou, Bull.: Bulletin de la société Imperiale des Naturalistes de Moscou.

Soc. Phys. Genève Mém.: Mémoires de la société de physique et d'histoire naturelle de Genève.

Statist. Soc. Lond. J.: Journal of the Statistical Society London.

Tasmanian J. Nat. Sci.: Tasmanian Journal of Natural Science, Agriculture, Statistics &c.

Veterinarian: The Veterinarian [London].

Wernerian Soc. Mem.: Memoirs of the Wernerian Natural History Society, Edinburgh.

Westminster Rev.: Westminster and Foreign Quarterly Review.

Württemb. Landw. Verein Corres. Bl.: Wuerttembergischer landwirthschaftlicher Verein, Stuttgart, Correspondenz-blatt.

Zool. Bot. Ges. Wien, Verh.: Verhandlungen der K.K. Zoologisch-botanischen Gesellschaft in Wien.

Zool. Soc. Lond. Proc.: Proceedings of the Zoological Society of London.

Zool. J.: The Zoological Journal.

BIBLIOGRAPHY

ANONYMOUS, PSEUDONYMOUS AND INITIALLED ARTICLES LISTED BY JOURNALS

Amer. J. Sci. 18 (1830), 278–85. Anon. review of Felix Pascale's *Practical Instructions on the Culture of Silk and the Mulberry Tree in the United States*, New York, 1830. v, 10.

20 (1831), 177, Anon., 'Destruction of Live Stock by Wolves in Russia.' v, 21 A.

42 (1842), 195–7. Anon., 'Yellow Showers of Pollen.' III, 31.

Ann. sci. nat. Bot., 2nde sér., 3 (1835), 370–9. Anon. review: 'G. C. Roehling's *Deutschland's Flora.* Flore d'Allemagne de Roehlings, publiée...par W. D. G. Koch, vol. IV, Francfort, 1833.' IV, 64.

Bot. Gaz. 1 (1849), 307–8. Anon. 'Bracts in the Cruciferae.' IV, 27.

3 (1851), 84. Anon. 'Bracts in the Cruciferae.' IV, 27

Cottage Gdnr. 17 (1856), 100–1. Chanticleer. 'To Secure Colour and Vigour.' IX, 130.

Gdnr's. Chron. (1852), 747–8. C.M.D. 'Do not House Black-faced Sheep.' x, 54.

(1856), 729–30. Anon. 'Turnips inoculated with Cabbage.' III, 27.

(1856), 823–4. Anon. 'Messrs. Sharp and Co.'s Nursery, Sleaford, Lincolnshire.' III, 27.

(1857), 235. G. 'Wearing out of Races.' VI, 26*.

Gdnrs. Mag. 1 (1826), 200. Anon. 'Miscellaneous Intelligence...Grafting Pears on Apples.' IX, 66.

3 (1828), 380. N. A. B. 'Pears on Quince Stocks.' IX, 67.

8 (1832), 50. C. C. 'Perfect Seeds and Culture of Lathyrus grandiflorus.' III, 60.

Mag. Nat. Hist. 5 (1832), 149–56. W. A. B. 'On the Claim of certain Lepidopterous insects taken in England to be considered as Indigenous.' VIII, 12.

5 (1832), 718–723. Zoophilus 'The Stoat, the changes in the colour of its fur, and those changes rather referable to atmospheric temperature than to periodical changes of season; and the Stoat and its congeners trace their prey by the faculty of scent.' VII, 92.

Phytologist 2 (1847), 843–51, 871–7. C. 'Notice of a "Manual of British Botany. By Charles Cardale Babington, M. A., &c." second edition.' IV, 77.

Poult. Chron. 1 (1854), 456–7. 'Notes by "Alector".' x, 81.

2 (1855), 443–6. Anon. 'Manchester Poultry Exhibition.' IX, 102.

3 (1855), 486–8. Zenas. 'Spangled Hamburgs Sitting. IX, 96.'

LISTED BY AUTHORS

Abu Al-Fazl ibn Mubārak, called 'Allāmī' *Ayeen Akbery: or the Institutes of the Emperor Akber*. Translated from the original Persian by Francis Gladwin. 3 vols., Calcutta, 1783, 1784, 1786. IV, 48.

Acosta, Joaquin. 'Sur la géologie de la Sierra Tairona [Sierra Nevada de Santa Marta] (Nouvelle Grenade).' *Soc. Géol. Fr. Bull.*, 2nde sér., 9 (1852), 396–9. XI, 17.

'Sur les montagnes de Ruiz et de Tolima (Nouvelle-Grenade) et les éruptions boueuses de la Magdalena.' *Soc. Geol. Fr. Bull.*, 2nde sér., 8 (1851), 489–96. XI, 17.

Acosta, José de. *Histoire naturelle et moralle des Indes, tant orientalles qu'occidentalles*...Paris, 1600. cited by Robertson. v, 12v.

'*Acts of Swedish Academy*' i.e. *K. Swenska Wetenskaps Academiens Handlingar*. At Maer, the house of Darwin's uncle, Josiah Wedgwood, Darwin read an English version of this journal. In the Cambridge MSS. Darwin Scientific Papers, item 119, a notebook labelled 'Books Read', Darwin wrote: 'Maer (June 10 to Nov. 14, 1840)...Swedish Philosoph. Acts vol. 1 to 7. M.S. translat. from 1740, 2^d vol 1741, 3^d 1742 & 50 on. Erasmus Darwin also cited 'Swedish Acts', e.g. in the *Loves of the Plants*, 2nd. ed., London, 1790, p. 149, note. The articles Darwin read were also published in German translation in *Abhandlungen aus der Naturlehre, Haushaltungskunst und Mechanik*.

Adams, – . Quoted by Amos Eaton, 'Fish of Hudson River.' *Amer. J. Sci.*, 20 (1831), 150–2. v, 55b.

Adams, Charles Baker. *Contributions to Conchology*. No. 10 'On the Nature and Origin of the Species of Mollusca in Jamaica? ([N.Y.?] 1851) IV, A11.

Agassiz, Louis and E. Desor. 'Catalogue raisonné des familles, des genres et des espèces de la classe des échinodermes,' *Ann. sci. nat.*, 3 ser., zool., 6 (1846), 305–74. IV, 34.

Alison, W. P. Art. 'Instinct', in *The Cyclopaedia of Anatomy and Physiology*, ed. Robert B. Todd. vol. 3, London, 1847, 1–29. x, 7, 22, 52, 66, 73, 103.

Andersson, Mic. Joh. 'On *Carex ampullacea*, Good., and *Carex vesicaria*, L., with Remarks on their Modifications of Form.' *Bot. Gaz.* 2, (1850), 253–62. VII, 91.

Apperley, Charles James. Quoted in Delabere P. Blaine, *Encyclopaedia of Rural Sports*. London, 1840, p. 280. III, 3.

Appleby, T. 'The Petunia.' *Cottage Gdnr.*, 16 (1856), 205–6. IX, 25.

Armstrong, J. 'A Treatise on Gardening.' *Bd Agric.. N.Y., Mem.*, 2 (1823) 89–136. III, 62.

Audubon, John James. Quoted from 'Bibliographical Notices—*Ornithological Biography*...by John James Audubon, vol. iv, 1838'. *Ann. Nat. Hist.*, 2 (1839), 458–63. x, 77.

Audouin, Victor. 'Recherches anatomiques sur le thorax des animaux articulés et celui des insectes hexapodes en particulier,' *Ann. sci. nat.*, 1 (1824), 97–135, 416–32. VII, 48.

[Ayeen Akbery. L.C. entry: Abū al-Fadhl ibn Mubārak, *al-Hindi*, 1551–1602. B.M. entry: Abu Al-FAZL ibn Mubārak, called 'Allami'.]

Azara, Felix d'. *Essais sur l'histoire naturelle des quadrupèdes de la Province de Paraguay*. 2 vols. Paris An IX (1801). v, 19. XI, 80.

Voyages dans l'Amérique méridionale, par Don Félix d'Azara, commissaire et commandant des limites espagnoles dans le Paraguay depuis 1781 jusqu'en 1801; contenant la description géographique, politique et civile du Paraguay et de la riviére de la Plate; l'histoire de la découverte et de la conquête de ces contrées; des détails nombreux sur leur histoire naturelle, et sur les peuples sauvages qui habitent; le récit des moyens employés par les Jésuites pour assujétir et civiliser les indigènes, etc. Publiés d'après les manuscrits de l'auteur, avec une notice sur sa vie et ses escrits, par C. A. Walckenaer; enrichis de notes par G. Cuvier, secrétaire perpétuelle de la classe des sciences physiques de l'Institut, etc. Suivis de l'histoire naturelle des oiseaux du Paraguay et de la Plata, par le même auteur, traduite, d'après l'original espagnol et augmentée d'un grand nombre de notes, par M. Sonnini; accompagnés d'un atlas de vingt-cinq planches. Paris, 1809. V, 19. X, 107.

Babington, Charles Cardale. *Manual of British Botany containing the Flowering Plants and Ferns arranged according to the Natural Order.* 3rd ed., London, 1851. IV, 44; supp. a.

Babington, C. 'On the Batrachian Ranunculi of Britain,' *Ann. Mag. Nat. Hist.*, ser. 2, 16 (1855) 385–404. XI appendix C. 40. 1.

Bachman, John. 'On the Migration of Birds of North America.' *Amer. J. Sci.*, 30 (1836), 81–100. V, 25. VII, 14. X, 51.

'Monograph of the Genus *Sciurus*, with Descriptions of New Species and their Varieties.' *Mag. Nat. Hist.*, N.S. 3 (1839), 220–7. X, 108.

Back, George. *Narratives of the Arctic Land. Expedition to the Mouth of the Great Fish River, and along the Shores of the Arctic Ocean, in the Years 1833, 1834, and 1835; by Captain Back, R.N. Commander of the Expedition.* London, 1836. IX, 11.

Baer, Karl E. von. 'Fragments relating to Philosophical Zoology. Selected from the Works of K. E. von Baer', in *Scientific Memoirs, selected from the Transactions of Foreign Academies of science and from Foreign Journals —Natural History*—ed. Arthur Henfrey and T. H. Huxley, London, 1853. pp. 176–238. X, 10.

Baly, William and William Senhouse Kirkes. *Recent Advances in the Physiology of Motion, the senses, Generation and Development:—being a supplement to the second volume of Professor Müller's 'Elements of Physiology'.* London, 1848. VIII, 50.

Barnéoud, F. Marius. Mémoire sur le développement de l'ovule, de l'embryon et des corolles anomales dans les renonculacées et les violariées.' *Ann. sci. nat.*, 3 ser. Bot. 6 (1846), 268–96. VII, 41.

Barrande, Joachim.] 'Sur le système silurien de la Bohême.' *Soc. Geol. Fr. Bull.*, 2nd ser., 10 (1853), 403–23. VI, 26 hhv.

Barrington, Daines. 'An essay on the periodical appearing and disappearing of certain birds, at different times of the year.' *Phil. Trans.*, 62 (1772), 265–320. III, 83.

'Experiments and Observations on the Singing of Birds.' *Phil. Trans.*, 63 (1774), 249–91. X, 61.

Bartlett, A. D. 'Description of a New Species of Fuligula.' *Zool. Soc. London, Proc.*, (Part) 15 (1847), 48–50. IX, 85.

Bartram, William. *Travels through North and South Carolina, Georgia, East and West Florida, the Cherokee Country, the extensive Territories of the Muscogulges or Creek Confederacy, and the Country of the Chactaws, containing an Account of the Soil and Natural Productions of those Regions; together with Observations on the Manners of the Indians.* Embellished with copper plates. Philadelphia: Printed for James and Johnson, 1791. London: Re-printed for J. & J. 1792. V, 55 e. XI, FC 6A.

Beaton, D. 'Bedding Geraniums'. *Cottage Gdnr.*, 16 (1856), 44–6, 61–2, 93–5. IX, 24.

'Lobelia ramosoides—Breeding from shy-seeding Geraniums'. *Cottage Gdnr.*, 16 (1856), 109–10. IX, 24. X, 116.

Bechstein, Johann Matthäus. *Gemeinnützige Naturgeschichte Deutschlands nach allen drey Reichen. Ein Handbuch zur deutlichern und vollständigern Selbstbelehrung besonders für Forstmänner, Jugend lehrer und Oekonomen.*

Erster Band, welcher die nöthigen Vor Kenntnisse und die Geschichte der Säugethiere enthält. (new and enlarged) 2nd. ed. Leipzig, 1801. Zweiter Band, welcher die Einleitung in die Naturgeschichte der Vögel uberhaupt, und die Geschichte der Raubvögel Spechtartigen und Krähenartigen Vögel Deutschlands enthält (enlarged and improved). 2nd ed. Leipzig 1805. Dritter Band, welcher die Sumpf = und Hausvögel nebst einer Untersuchung über die Frischischen Vogel enthält. 1st ed., Leipzig, 1793. Vierter Band, welcher die Singvögel, den Vögelkalender, einige Zusätze zu den vorhergehenden Bänden und das Register über die drey Bänd der Vögel Deutschlands enthält. 1st. ed., Leipzig, 1795. IV, 43, 50, 56. IX, 79, 84, 102.

Bechstein, Johann M. *Naturgeschichte der Stubenvögel: oder Anleitung Kenntniss, Wartung, Zähmung, Fortpflanzung zum Fang derjenigen in = und auslandischen Vögel, welche man in der Stube halten Kann.* 4th ed. Halle, 1840. IX, 82, 84, 93, 125. X, 17, 30, 50, 100.

Bell, Thomas. *History of British Quadrupeds, including the cetacea.* London, 1837. IV, 42.

Bellamy, J. 'On the Distribution etc. of the Mammals of Devonshire.' *Brit. Ass. Advanc. Sci., Rep. for 1841 (1842)* part 1 p. 68.

Bennett, Edward Turner. Notes in Gilbert White's *Natural History...of Selbourne.* new ed. (London, 1837.) VI, 45.

Berg, Ernst von. 'Beiträge zur genauereu Kenntniss der Irisarten mit Schwerdtförmigen Blättern und bartigen Blumen.' *Flora, Beiblatter* (1833 vol. 1), 1–42. IV, 66.

'Fernerweitiger Bericht über die durch Samenaussaat erhaltenen Irisarten.' *Flora* (1835) vol. 2, 561–73. IV, 66.

Bergmann, B. 'Voyage de B. Bergmann chez les Kalmouks.' Trans. by J.-B. F. S. Ajusson de Grandsage. *Muséum Hist. Nat., Mém.,* 16 (1828), 431–60. X, 38.

Berlepsch, August von. 'Die italische Biene.' *Bienen-Zeitung,* 12 (1856) 3–6. VIII, 73.

Bernhardi, Johann Jacob. *Ueber den Begriff der Pflanzenart und seine Anwendung.* Erfurt, 1834. IV, 7, 64, 67–8.

Berry, Henry. 'Hybrid Birds produced between the Throstle and Blackbird in a State of Nature.' *Mag. Nat. Hist.,* 7 (1834), 598–9. IX, 79.

Bevan, Edward. *The Honey-Bee; its natural history, physiology and management.* London, 1827. VIII, 89.

Biberg, Isaac J. See Linné.

Blackwall, John. 'Notice of several cases of defective and redundant organization, observed among the Araneidea.' *Ann. Mag. Nat. Hist.,* 11 (1843), 165–8. IV, 34.

Researches in Zoology, illustrative of the Manners and Economy of Animals; with Descriptions of Numerous Species New to Naturalists; accompanied by Plates. London, 1834. V, 64. X, 100.

Blaine, Delabere P. *An Encyclopedia of Rural Sports; or, a complete account, historical, practical and descriptive, of hunting, shooting, fishing, racing, and other field sports of the present day.* London: Longman etc., 1840. III, 78. IV, 50. X, 34, 49.

Blyth, Edward. 'An Attempt to Classify the "Varieties" of Animals, with Observations on the Marked Seasonal and other Changes which Naturally take Place in Various British Species, and which do Not constitute Varieties.' *Mag. Nat. Hist.,* 8 (1835), 40–53. VII, 92.

'On the Psychological Distinctions between Man and all other Animals; and the consequent Diversity of Human Influence over the inferior Ranks of Creation, from any mutual and reciprocal Influence exercised among the Latter,' *Mag. Nat. Hist.*, N.S. 1 (1837), x, 14.

'Analytic Descriptions of the Groups of Birds composing the Order, Insessores Heterogenes.' *Mag. Nat. Hist.*, N.S. 2 (1838), 351–61. x, 103.

'Notices of various Mammalia, with Descriptions of Many New Species.' *Ann. Mag. Nat. Hist.*, 15 (1845), 449–75. VIII, 8.

'Report for May Meeting, 1855.' *Asiat. Soc. Bengal*, vol. 24 for 1855 (1856), 359–63. III, 81.

[Blyth, E.] 'Scrap Collator'—'More about Donkeys.' *India Sport. Rev.*, 1 (1856), 316–24. VII, 105.

[Blyth, E.] 'Zoophilus'.—'Natural History Notices' *India Sport Rev.*, N.S. 2 (1856), 131–7; 239–61. IX, 100.

[?Blyth] 'Zoophilus' see under *Mag. Nat. Hist.*

[Blyth, E.] 'British Birds in India.' *Calcutta Rev.*, 28 (1857), 129–95. VII, 60 bis.

Boitard, Pierre and – Corbié. *Les Pigeons de volière et de colombier ou histoire naturelle et monographie des pigeons domestiques, renfermant la nomenclature et la description de toutes les races et variétés constantes connues jusqu'à ce jour; la manière d'établir des colombiers et volières; d'élever, soigner les pigeons*, etc., etc., Paris, 1824. IX, 120, 128.

Bolton, James. *Harmonia Ruralis: or, An Essay towards a Natural History of British Song Birds: illustrated with Figures, the Size of Life, of the Birds, Male and Female in their most natural Attitudes: their Nests and Eggs*, etc. [H. G. Bohn.] Rev. ed., 2 vols., London, 1845. x, 70.

Bonaparte, Prince Charles Lucien (Prince of Canino). *A Geographical and Comparative List of the Birds of Europe and North America*. London, 1838. IV, 39 a.

Boothby, George. 'Influencing Colour-age of eggs for sitting,' *Cottage Gdnr.*, 17 (1856), 137. IX, 130.

Boreau, Alexandre. *Flore du centre de la France, ou description des plantes qui croissent spontanément dans la région centrale de la France et de celles, qui y sont cultivées en grand, avec l'analyse des genres et des espèces*. 2 vols., Paris, 1840. IV, 16, 69, c, A1.

Bourgoing, Jean Francois de. *Travels in Spain: containing a new, accurate and comprehensive View of the State of that Country. By the Chevalier de Bourgoanne. To which are added, copious Extracts from the Essay on Spain of M. Peyron.* Trans. from the French. 3 vols., London, 1789. x, 55.

Brand, Fitzjohn – Trans. & ed. *Select Dissertations from the* Amoenitates Academicae *a Supplement to Mr. Stillingfleet's Tracts relating to Natural History*. Translated by the Rev. F. J. Brand, M.A., London, 1781. v, 10 bis, 25, 38.

Brauer, Friedrich. 'Ueber den Dimorphismus der Weibchen in der Libellulinen-Gattung Neurothemis.' *Zool. Bot. Ges. Wien, Verh.*, 17 (1867), 971–6. VII, 70.

Brehm, Christian Ludwig. *Handbuch der Naturgeschichte aller Vögel Deutschlands*. Ilmenau, 1831. IV, 37.

'Observations on some of the Domestic Instincts of Birds.' *Mag. Nat. Hist.*, N.S. 2 (1838), 399–406. x, 16.

Breton, William H. 'Excursion to the Western Range, Tasmania.' *Tasmanian J. Nat. Sci.*, 2 (1843), 121–41. v, 40.

Brewer, T. M. 'Statement made in meeting of Oct. 15, 1839—'Proceedings of the Boston Society of Natural History, compiled from the Records of the Society, by Jeffries Wyman, M.D., Recording Secy.'' *Amer. J. Sci.*, 38 (1840), 392–3. v, 55 b, 77.

Brisson, Mathurin Jaques. *Ornithologie ou méthode contenant la division des oiseaux en ordres, sections, genres, espèces et leurs variétés. A laquelle on a joint une description exacte de chaque espèce, avec les citations des auteurs qui en ont traité, les noms qu'ils leur ont donnés, ceux qui leur ont donnés les différentes nations, et les noms vulgaires*, 6 vols., Paris, 1760. IV, 53.

[Brodie, Benjamin C.] Psychological Inquiries: in a series of essays. London, 1854. x, 8, 28.

Bromfield, William Arnold. 'Notes and occasional observations on some of the rarer British Plants growing wild in Hampshire.' *Phytologist*, 3 (1848), 205–13, 269–91, 332–44, 363–83, 401–41, 490–504, 519–36, 555–64, 571–80, 593–609, 617–35, 653–69, 685–703, 741–4. III, 96. IV, 45, 46, 67, 69, 72. IX, 31.

Brongniart, Adolphe. 'Mémoire sur la famille des Rhamnées.' *Ann. sci. nat.*, 10 (1827), 320–86. VIII, 43.

Bronn, Heinrich Georg. *Handbuch einer Geschichte der Natur*. Erster Band. Einleitung—I Theil: Kosmisches Leben—II Theil: Tellurisches Leben, Stuttgart, 1841. Zweiter Band. III Theil: Organisches Leben. Ergebnisse hauptsächlich aus der lebenden Welt über Entwicklung, Verbreitung und Untergang der früheren Bevölkerungen der Erde. Stuttgart, 1843. III, 83, 84. IV, 60, 66. VII, 6, 6 A. VIII, 14. IX, 79, 88 c, 126, 136.

Brooke, James. Letter to Waterhouse. *Ann. Mag. Nat. Hist.*, 9 (1842), 54–9. IV, 34.

Brougham, Henry. *Dissertations on Subjects of Science connected with Natural Theology: being the concluding volumes of the new edition of Paley's Work*. 2 vols., London, 1839. x, 7, 92, 98 a.

Brown, P. J. Letter about the primrose in Switzerland. *Ann. Mag. Nat. Hist.*, 9 (1842), 155–6. IV, 69, 75.

Brown, Robert. 'General Remarks, geographical and systematical, on the Botany of Terra Australis,' Appendix III, pp. 533–613 in vol. 2 of Matthew Flinders, *A Voyage to Terra Australis*...London, 1814. III, 48.
'Additional Observations on the Mode of Fecundation in Orchideae.' *Linn. Soc. Lond. Trans.*, 16 (1833), 739–45. III, 56.
'On the Organs and Mode of Fecundation in Orchideae and Asclepiadae.' *Linn. Soc. Lond. Trans.*, 16 (1833), 685–738. III, 32, 33, 52, 56. VIII, 78; IX, 44.

Bruce, James. *Travels to Discover the Source of the Nile in the Years 1768, 1769, 1770, 1771, 1772, and 1773*. 5 vols., 1790. Edinburgh, v, 34. x, 101, 112.

Brullé, Auguste. 'Observations sur l'absence des tarses dans quelques insectes.' *Ann. sci. nat.*, 2 ser. zool. 8 (1837), 246–9. VII, 27.
'Recherches sur les transformations des appendices dans les articulés.' *Ann. sci. nat.*, 3 ser. zool. 2 (1844), 271–374. VII, 41.

Buchanan, Andrew. 'On the Wound of the Ferret, with Observations on the Instincts of Animals.' *Ann. Mag. Nat. Hist.*, 18 (1846), 376–83. x, 16.

Buckland, William. Communication reported in note on page 85 of William Clift, A Description of the Fossil Bones Found in the Caves of Oreston. *Phil. Trans.*, (1823), pp. 81–90. v, 63.
Geology and Mineralogy considered with Reference to Natural Theology. [Bridgewater Treatise] 2 vols., London, 1836. VIII, 7.

Bucknell, Wm. *Eccaleobion: a treatise on artificial incubation*. Lond., 1839. x, 18.

Buffon, Georges Louis Leclerc de. *Natural History, general and particular.* Translated into English with occasional notes and observations by William Smellie, 3rd ed., 9 vols., London, 1791. IX, 83.

Bunbury, Charles James Fox. *Journal of a Residence at the Cape of Good Hope; with excursions into the interior, and notes on the natural history, and the native tribes.* London, 1848. XI, 43.

'Notes on the Vegetation of Buenos Ayres and the neighbouring district.' *Linn. Soc. Lond. Trans.,* 21 (1855), 185–98. VI, 11.

'Remarks on the Botany of Madeira and Teneriffe.,' *Linn. Soc. Lond. J.,* 1 (1856), 1–35. VI, 11.

Burgess, Lt. 'Note on the Indian Weaver-bird (Ploceus Philippensis).' *Zool. Soc. Lond. Proc.,* Part 20 (1852), 88–9. X, 75.

Burnes, Alexander. *Travels into Bokhara; being the Account of a Journey from India to Cabool, Tartary, and Persia; also, Narrative of a Voyage on the Indus, from the Sea to Lahore, with Presents from the King of Great Britain; performed under the Orders of the Supreme Government of India, in the Years 1831, 1832, and 1833.* 3 vols., London, 1835 or 1839 new ed. IX, 90.

[*Cabool: a personal narrative of a journey to, and residence in that city, in the years 1836, 7 and 8,* 1st ed. London, 1842. 2nd ed. London, 1843.] III, 78.

[Busk, George.] *Catalogue of Marine Polyzoa in the collection of the British Museum,* ed. J. E. Gray. Part I, London, 1852. Part II, London, 1854. XI, 35.

[Byron, John. 'An Account of a Voyage round the World, in the years 1764, 1765, and 1766, by the Honourable Commodore Byron, in his Majesty's Ship, the Dolphin.' Abridged version in Robert Kerr, *Voyages,* 12 (1814), 9–119.] X, 58.

Cabot, Samuel. 'Observations on the Characters and Habits of the Ocellated Turkey (*Meleagris ocellata,* Cuv.).' *Boston J. Nat. Hist.,* 4 (1842), 246–51. X, 69.

Cada Mosto, Alvise da. 'Original Journals of the Voyages of Cada Mosto and Padro de Cintra to the Coast of Africa: the former in the years 1455 and 1456, and the latter soon afterwards', in Robert Kerr, *Voyages,* 2 (1811), 200–57 re. Cada Mosto. V, 13.

Cambessèdes, Jacques. 'Monographie des Globulaires.' *Ann. sci. nat.,* 9 (1826), 15–31. IV, 28.

Candolle, Alphonse Louis Pierre Pyramus de. *Géographie botanique raisonnée, ou exposition des faits principaux et des lois concernant la distribution géographique des plantes de l'époque actuelle.* 2 vols., Paris, 1855. IV, 1, 6, 17, A1, A13, A19, A30; V, 46, 47, 50, 56, 58; VI, 26iv, 35; VII, 17; XI, 1, 28, 37, 42, 43.

'Mémoire sur la famille des Apocynacées.' *Ann. sci. nat.,* 3 ser., Bot. 1 (1844), 235–63. VII, 96.

'Revue sommaire de la famille des Bignoniacées,' *Ann. sci. nat.,* 2 ser. Bot. 11 (1839), 279–98. VII, 38c.

Candolle, Augustin Pyramus de. 'Géographie Botanique.' article in *Dictionnaire des Sciences Naturelles,* vol. 18. Paris, 1820. pp. 359–422. XI, 14, 43.

Candolle, Auguste Pyr. de and Alph. de Candolle. 'Huitième notice sur les plantes rares cultivées dans le jardin de Genève.' *Soc. Phys. Genève, Mém.,* 9 (1841–2), pp. 75–105. VII, 38b.

Prodromus systematis naturalis Regni vegetabilis, sive enumeratio contracta ordinum generum specierumque plantarum huc usque cognitarum, juxta methodi naturalis normas digesta. Paris, 1824–57. Parts 1–14. IV, supp.e.

Candolle, Auguste de. 'Rapport sur les plantes rares ou nouvelles qui ont fleuri dans le Jardin de Botanique de Genève pendant les années 1819, 1820 et 1821,' *Soc. Phys. Genève Mém.*, 1 (1822), 431–63. VII, 96 bis.

'Rapport sur les plantes rares ou nouvelles qui ont fleuri dans le jardin de botanique de Genève pendant les années 1822 et 1823.' *Soc. Phys. Genève, Mém.*, 2 part 2 (1824), 125–43. IV, 25.

'Revue de la famille des Lythraires.' *Soc. Phys. Genève, Mém.*, 3 part 2 (1826), 65–96. VII, 95.

Carmichael, Dugald. 'Some account of the Island of Tristan da Cunha and of its Natural Productions.' *Linn. Soc. Lond. Trans.*, 12 (1817), 483–513. X, 58. XI, appendix, 45.

Carpenter, William B. Lecture: 'On the Character of Species', given at Glasgow meeting of Brit. Ass., 1855. See *Athen.* (1855), 1090. IV, 1.

Principles of Comparative Physiology. 4th ed., London 1854. III, 99; V, 10. VIII, 38, 54, 78. X, 5.

Cassini, [Henri]. Referred to in Smith, Thomas. 'On certain species of Carduus and Onicus which appear to be dioecious.' *Linn. Soc. Lond. Trans.*, 13 (1822), 592–603. III, 26 bis.

['Observations sur le style et le stigmate des Synanthérées.' *Phys.*, 76 (1813), 97–128, 181–201, 249–75. Cited by Thomas Smith, *Linn. Soc. Lond. Trans.*, 13 (1822) 592–603: see *Nat. Sel.* ch. III, fol. 26 bis.]

'Tableau synoptique des Synanthérées.' *Ann. sci. nat.*, 17 (1829), 387–423. VII, 38.

Cavolini, Philip. 'On the flowering of Zostera oceanica LINN.' *Ann. Bot.*, 2 (1806), 77–91. III, 47. 4.

Chevreul, Michel Eugène. 'Considérations générales sur les variations des individus qui composent les groupes appelés, en histoire naturelle, variétés, races, sous-espèces, et espèces.' *Ann. sci. nat.*, 3 ser. Bot 6 (1846), 142–214. IX, 78, 90.

Christison, Professor. [It is observations mentioned in report of paper by John Lowe, 'On the Properties of Lolium temulentum', at the Edinburgh, Botanical Society.] *Gdnrs. Chron.*, (1857), p. 518. VIII, 53.

Churchill, Awnsham and John Churchill. *A collection of Voyages and Travels, some now first printed from Original Manuscripts. Others Translated out of Foreign Languages, and now first published in English. To which are added some few that have formerly appeared in English, but do now for their Excellency and Scarceness deserve to be reprinted.* Vols. III & V London, 1704, 1732. cited by Robertson.

Clark George. 'Extracts from a Letter to Thomas Bell, Esq., F.R.S., from George Clark, Esq., of Mauritius.' *Ann. Mag. Nat. Hist.*, N.S. 2 (1848), 361–4. X, 19.

Clark, William. 'On the Terebrating Mollusca.' *Ann. Mag. Nat. Hist.*, 5 (1850), 6–14. VII, 27.

Colenso, William. 'Memoranda of an excursion, made in the Northern Island of New Zealand, in the summer of 1841–2; intended as a contribution towards the ascertaining of the Natural Productions of the New Zealand Groupe: with particular reference to their Botany.' *Tasmanian J. Nat. Sci.*, 2 (1846), 210–34 etc. X, 57 vi.

Colin, Gabriel Constant. *Traité de physiologie comparée des animaux domestiques*, 2 vols. Paris, 1854–6. IX, 12 bis, 127.

Colquhoun, John. *The Moor and the Loch: containing practical hints on highland-sports, and notices of the habits of the different creatures of game and prey in the mountainous districts of Scotland; with instructions in river, burn and loch-fishing.* 2nd ed. London, 1841. IV, 50. X, 35.

Corse, John. 'An Account of the Method of Catching Wild Elephants at Tipura.' *Asiat. Soc. Bengal, Trans.*, London octavo reprint 3 (1799), 266–91, X, 110.

'Observations on the Manners, Habits and Natural History, of the Elephant.' *Phil. Trans.*, (1799), 31–55. X, 30.

'Observations on the different Species of Asiatic Elephants, and their Mode of Dentition.' *Phil. Trans.*, 1799: 205–36. IV, 48.

Couch, Jonathan. *Illustrations of Instinct, Deduced from the Habits of British Animals.* London, 1847. X, 36, 62, 69, 72, 73, 75, 81.

'Some particulars of the natural history of fishes found in Cornwall.' *Linn. Soc. Lond. Trans.*, 14 (1825), 69–92. IV, 30. V, 63.

Cranz, David. *The History of Greenland, containing a description of that country, and its inhabitants, and particularly a relation of the mission, carried on for above these thirty years by the Unitas Fratrum, at New Herrnhuth and Lichtenfels, in that country.* Translated from the High-Dutch. Ed. and in part translated by John Gambold. German ed: Barby, 1765. XI, 47.

Crawfurd, John. *Journal of an Embassy from the Governor-General of India to the Court of Ava, in the year 1827. By John Crawfurd, Esq., F.R.S. F.L.S. F.G.S. &c. Late Envoy. with an Appendix, containing a description of fossil remains by Professor Buckland and Mr. Clift.* 2nd ed. 2 vols. London, 1834. III, 74.

Crawfurd, John. *A descriptive dictionary of the Indian islands and adjacent countries.* London, 1856, III, 85. IV, 42.

Creplin, Friedrich Heinrich Christian 'Appendix' in Johan J. S. Steenstrup, *Untersuchungen über das Vorkommen des Hermaphroditismus...* Greifswald, 1946. III, 23.

Culley, George. *Observations on Live Stock; containing Hints for Choosing and Improving the best Breeds of the most useful Kinds of domestic Animals.* London; 4th ed. 1807, III, 3–4.

Cumming, Rousleyn Gordon. *Five Years of a Hunter's Life in the Far Interior of South Africa,* 2 vols. London: Murray, 1850. IV, 41–2.

Cuvier, Frédéric. 'Examen de quelques observations de M. Dugald-Stewart, qui tendent a détruire l'analogie des phénomènes de l'Instinct avec ceux de l'habitude.' *Museum Hist. Nat. Mém.*, 10 (1823), 241–60. X, 25.

Cuvier, Frédéric and Etienne Geoffroy-Saint-Hilaire. *Histoire naturelle des mammifères.* Paris, 1824–42. VII, 111.

[Dalyell, Sir John Graham. *The Powers of the Creator Displayed in the Creation; or, Observations on Life Amidst the Various Forms of the Humbler Tribes of Animated Nature.* Vol. 1, London, 1851.] V, 10.

Dana, James D. *Crustacea Part II. United States Exploring Expedition during the Years 1838, 1839, 1840, 1841, 1842, under the command of Charles Wilkes, U.S.N.* Vol. XIII, Phila., 1852. XI, 14, 36, 37.

Geology, United States Exploring Expedition during the Years 1838, 1839, 1840, 1841, 1842. Vol. X, Phila., 1849. XI, 17.

[Introductory letter for paper by James Eights] *Amer. J. Sci.*, 2nd. ser., 22 (1856), 391. XI, appendix on representative species, fol. 4.

Darwin, Charles. 'Bees and Fertilisation of Kidney Beans.' *Gdnrs Chron.*, (1857), 725. x, 21 a.

'Double Flowers—their Origin.' *Gdnrs Chron.*, (1843), 628. iii, 93.

Geological Observations on South America. Being the Third Part of the Geology of the Voyage of the Beagle. London, 1846. xi, 20; appendix on representative species fol. 4.

'Humble-Bees.' *Gdnrs Chron.*, (1841), 550. x, 21 a.

Journal of Researches into the Natural History and Geology of the Countries visited during the Voyage of H.M.S. Beagle round the World under the Command of Capt. Fitzroy, R.N. 2nd ed., London, 1845. v, 19, 37, 40, 27. viii, 12. x, 58, 62, 74, 110 a, 185 a, 83.

A Monograph on the Sub-Class Cirripedia, with Figures of all the Species. The Balanidae (or sessile cirripedes), the Verrucidae, &c. London: The Ray Society, 1854. iii, 21. vii, 55, 104.

'Nectar-secreting Organs of Plants.' *Gdnrs. Chron.*, (1855), 487. iii, 35.

'On the Character and Hybrid-like Nature of the Offspring from the illegitimate Unions of Dimorphic and Trimorphic Plants.' *Linn. Soc. Lond. J. Bot.*, 10 (1868), 393–437. ix, 83.

'On the Distribution of the Erratic Boulders and on the Contemporaneous Unstratified Deposits of South America.' *Geol. Soc. Lond. Trans.*, 2nd. ser., 6 (1841), 415–31. xi, 16, 21, 46.

On the Formation of Mould. *Trans. of the Geological Soc. of London.* 2nd ser. 5 (1840), 505–9. v, 39.

['On the specific difference between Primula veris, Brit. Fl. (var. officinalis, Linn.), P. vulgaris, Brit. Fl. (var. acaulis, Linn.), and P. elatior Jacq.; and on the hybrid nature of the common Oxlip. With Supplementary remarks on naturally produced hybrids in the genus Verbascum.' *Linn. Soc. Lond. J.* 10, Bot. (1868), 437–54. IV, introd., note 1.]

ed. *The Zoology of the Voyage of H.M.S. Beagle under the command of Captain Fitzroy, R.N., during the years 1832 to 1836. Published with the approval of the Lords Commissioners of her Majesty's Treasury. Edited and superintended by Charles Darwin, Esq., M.A., F.R.S., Sec.G.S., Naturalist to the Expedition.* London: Smith, Elder & Co. Parts appeared 1838–43.

ed. *The Zoology of the Voyage of H.M.S. Beagle. Part II, Mammalia by George R. Waterhouse,* London, 1839. x, 79.

ed. *The Zoology of the Voyage of H.M.S. Beagle. Part III, Birds by John Gould.* London, 1841. iv, 31. vi, 43. vii, 14, 20. viii, 10, 19.

The Zoology of the Voyage of H.M.S. Beagle...Part V. Reptiles by Thomas Bell. London, 1843. viii, 21.

Darwin, Erasmus. *Zoonomia; or, the Laws of organic Life.* 2 vols., London, 1794, 1796. x, 23, 27.

Davy, Humphry. *Salmonia, or Days of Fly-Fishing; in a Series of Conversations: with some Account of the Habits of Fishes belonging to the Genus Salmo.* [with] *Consolation in Travel, or the Last Days of a Philosopher.* London, 1840. [Published with additional title page as vol. ix of *The Collected Works of Sir Humphry Davy, Bart. LL.D. F.R.S. Foreign Associate of the Institute of France, etc.* edited by his brother, John Davy, M.D. F.R.S. London, 1840.] iv, 10.

Decaisne, J. 'Sur les caractères spécifiques des espèces du genre Herniaria, de la Flore française.' *Ann. sci. nat.*, 22 (1831), 97–101. iv, 29.

De Candolle see Candolle, de.

Defay, – 'Lettre de M. Defay sur l'accouplement de quelques oiseaux sauvages dans l'état de domesticité.' *J. Phys.*, 25 (1784), 293–5. III, 79.

Delafield, John, jr. *An Inquiry into the Origin of the Antiquities of America.* Cinncinnati, 1839. v, 52.

Denny, – . [Letter read by Dr. Lankester at meeting of Section D., Zoology and Botany, British Association Meeting. *Athen.*, (1843), 829.] III, 80.

[Letter read by Dr. Lankester. *Brit. Ass. Advanc. Sci. Rep. for 1843*, part 2 (1844), p. 71.] III, 80.

Desborough, J. G. 'On the Duration of Life in the Queen, Drone and Worker of the Honey Bee: to which are added Observations on the Practical Importance of this Knowledge in deciding whether to preserve Stocks or Swarms: being the Prize Essay of the Entomological Society of London for 1852.' *Ent. Soc. Lond. Trans.*, N.S. 2 (1853), 145–71. VIII, 92.

Deshayes, G. P. 'Observations sur l'estimation de la température des périodes tertiaires en Europe, fondée sur la considération des coquilles fossiles.' *Ann. sci. nat.*, 2 ser., zool. 5 (1836), 289–98. VII, 5 bis.

Desmarest, Anselm Gaëtan. *Mammalogie ou Description des espèces de mammifères. Seconde partie, contenant les ordres des rongeurs, des édentés, des pachydermes, des ruminans et des cétacés.* Paris, 1822. Volume in *Encyclopédie méthodique.* IX, 90 bis.

Desmoulins, Antoine. Article 'Chameau', in *Dict. class. hist. nat.*, vol. 3, Paris, 1823. IX, 83.

Art: 'Cheval', in *Dict. class. hist. nat.*, vol. 3, Paris, 1823, 554–65. VII, 105 A.

Art. 'Galéopithèque *Galeopithecus*', in *Dict. class. hist. nat.*, vol. 7, Paris 1825, 119–23. VIII, 7.

Des Moulins, Charles. 'De la Flore du centre, de M. Boreau.' *Soc. linn. Bordeaux, Actes*, 16 (1849), 53–62. IV, 36.

Dicquemare, – abbé. 'Suite des extraits du porte-feuille de l'abbé Dicquemare. Huîtres, Faculté locomotive etc.' *J. phys.*, 28 (1786), 241–4. X, 13.

Dixon, Edmund Saul. *The Dovecote and the Aviary: Being Sketches of the Natural History of Pigeons and other domestic Birds in a captive state, with Hints for their Management.* London, 1851. VI, 44. IX, 88 a.

Ornamental and Domestic Poultry: their history and management. London, 1848. IX, 79.

Dobrizhoffer, Martin. *An Account of the Abipones, an Equestrian People of Paraguay.* 3 vols., London, 1822. III, 1. X, 38.

Don, David. 'An Account of the Indian Species of *Juncus* and *Luzula*.' *Linn. Soc. Lond. Trans.*, 18 (1841), 317–26. VII, 8.

Doubleday, Edward. Verbal Report on Bardfield Oxlip etc. to Botanical Society of London. *Ann. Mag. of Nat. Hist.*, 9 (1842), 515. IV, 69, 75.

Doubleday, Thomas. *The True Law of Population shewn to be connected with the Food of the People.* 1st ed., London, 1842. III, 99.

Drège, J. F. and E. Meyer. Zwei pflanzen-geographische Documente Besondere Beigabe zur Flora 1843 Band II *Flora oder allgemeine botanische Zeitung.* Regensburg, 1843 v, 71. XI, 29.

Du Bois, – . *Les voyages faits par le Sieur D[u] B[ois] aux Isles Dauphine ou Madagascar, et Bourbon ou Mascarene, années 1669, 70, 71 et 72. Dans laquelle il est curieusement traité du caps Vert de la ville de Surate des isles de Saint Hélène, ou de l'Ascension, Ensemble les moeurs, religions, forces, gouvernemens & coûtumes des habitans desdites isles, avec l'histoire naturelle du païs.* Paris: C. Bachin, 1674. Trans. Hakluyt Soc. supp. to vol. LXXXIII, 1891. X, 58.

Duby, J. E. 'Mémoire sur la famille des primulacées.' *Soc. Phys. Genève, Mém.*, 10 (1843), 395–438. iv, 25. vii, 95.

Dugès, A. 'Observations sur les Aranéides.' *Ann. sci. nat.*, 2. ser. zool. 6 (1836), 159–218. viii, 39. x, 98a, 111.

Dureau de la Malle, A. J. C. A. 'Considérations générales sur la domestication des animaux.' *Ann. sci. nat.*, 27 (1832), 5–33. x, 38.

'De l'influence de la domesticité sur les animaux depuis le commencement des temps historiques jusqu'à nos jours.' *Ann. sci. nat.*, 21 (1830), 50–67. ix, 99; x, 38.

'Mémoire sur le développement des facultés intellectuelles des animaux sauvages et domestiques.' *Ann. sci. nat.*, 22 (1831), 388–419. x, 16, 18.

Durieu de Maisonneuve. 'Observations requested concerning the Fructification of the Mosses.' Translated from the 1st vol. of the *Bulletin d'Histoire Naturelle de la Société Linnéenne de Bordeaux. Amer. J. Sci.*, 21 (1832), 171–2. iii, 20.

Dzierzon, – . 'Was ist die italienische Biene?' *Bienen-Zeitung*, 12 (1856), 61–3. viii, 73.

Earle, Windsor. 'On the Physical Structure and Arrangement of the Islands of the Indian Archipelago.' *Geogr. Soc. Lond. J.*, 15 (1845), 358–65. xi, 30.

Eaton, John Matthews. *A Treatise on Domesticated and Fancy Pigeons, carefully compiled from the best authors with observations containing all that is necessary to be known of tame, domesticated and fancy pigeons.* London, 1852. iii, 2.

A Treatise on the Art of Breeding and Managing the Almond Tumbler. London, 1851. vi, 7a.

Edwards, Henri Milne. See Milne-Edwards, Henri.

Eights, James. 'Description of an Isopod crustacean from the Antarctic Seas, with observations on the New South Shetlands.' *Amer. J. Sci.*, 2nd ser. 22 (1856), 391–7. xi, appendix on representative species, fol. 4.

Ekmarck, Carl D. See Linné.

[Ekström, C. U. 'Beobachtungen über die Formver-änderung beyder Karausche (Cyprinus carassius L.).' Trans. by Creplin. [Oken's] *Isis*. (1840), 145–53, cited by Bronn. See *Nat. Sel.* ch. 4 fol. 60, (note 95).]

Ellis, William. *Narrative of a Tour through Hawaii, or Owhyhee; with Remarks on the History, Traditions, Manners, Customs, and Language of the Inhabitants of the Sandwich Islands.* London, 1826. iii, 1.

Encyclopaedia of Rural Sports. See Blaine, Delabere P.

Erichson, W. P. 'Report on the Contributions to the Natural History of Insects, Arachnida, Crustacea, and Entomostraca, during the year 1842.' *Ray Society. Reports on the Progress of Ecology and Botany. 1841, 1842.* Edinburgh, 1845. vi, 46.

Eudes-Deslongchamps, E. 'Mémoire sur les pleurotomaires des terrains secondaires du Calvados.' *Soc. linn. Norm. Mém.*, 8 (1849), 1–157. iv, 21.

'Note sur une anguille retirée d'un puits, au mois de juillet 1831.' *Soc. linn. Norm. Mém.*, 5 (1835), 47–51. vii, 32.

Paragraph in Résumé des travaux. *Soc. linn. Norm. Mém.*, 7 (1842), xxix. vii, 32.

Eyton, Thomas C. 'Remarks on the Skeletons of the common tame Goose, the Chinese Goose, and the Hybrid between the Two.' *Mag. Nat. Hist.*, N.S. 4 (1840), 90–2. ix, 84, 96.

'Some Remarks upon the Theory of Hybridity.' *Mag. Nat. Hist.*, N.S. 1 (1837), 357–9. IX, 96.

Fabre, Jean Henri. 'Étude sur l'instinct et les métamorphoses des Sphégiens.' *Ann. sci. nat.*, 4 ser. Zool., 6 (1856), 137–83. VIII, 91. X, 26, 85, 86.

Farquharson, J. 'Botanical Remarks.' *Bot. Misc.*, 3 (1833), 126–8. IV, 45.

Fennell, James H. 'Do Animals hybridise by Choice, or without Compulsion?' *Mag. Nat. Hist.*, 9 (1836), 615–16. IX, 79, 86.

Ferguson, Robert. *The Northmen in Cumberland and Westmorland*, London, Carlisle, 1856. V, 55 b.

Fernandez de Oviedo y Valdes, Gonzalo. La Historia generale & naturale delle Indie occidentali, in Ramusio, Giovanni Batista, *Navigationi et Viaggi*. Vol. 3, Venice, 1588. Cited by Robertson. V, 12v.

Feuillée, Louis. *Journal des observations physiques, mathématiques et botaniques, faites par l'ordre du roy sur les côtes orientales de l'Amerique meridionale, et dans les Indes occidentales, depuis l'année 1707, jusques en 1712*. 2 vols., Paris, 1714. Cited by Robertson. V, 12v.

Fink, John Henry. 'Answers to the Queries proposed by Sir John Sinclair, concerning the Breeding of Sheep in Germany, particularly in Upper Saxony, and the neighbouring Provinces.' *Bd. Agric. Lond. Commun.*, 1 (2nd ed., 1804), 276–94. IX, 82.

Fleming, John. *The Philosophy of Zoology, or A general view of the Structure, Functions, and classification of Animals*. 2 vols, Edinburgh, 1822. V, 21, 65 A.

A History of British Animals, exhibiting the descriptive characters and systematical arrangement of the genera and species of quadrupeds, birds, reptiles, fishes, mollusca, and radiata of the United Kingdom; including the indigenous, extirpated, and extinct kinds, together with periodical and occasional visitants. Edinburgh, 1828. VII, 27.

Flourens, Marie Jean Pierre. *De l'instinct et de l'intelligence des animaux. Résumé des observations de Frédéric Cuvier sur ce sujet*. 2nd ed., Paris, 1845. X, 14, 22.

De la longevité humaine et de la quantité de vie sur le globe. Paris, 1854. IX, 83, 84, 89, 132.

Forbes, Edward. 'On the Connexion between the distribution of the existing Fauna and Flora of the British Isles, and the Geological Changes which have affected their Area especially during the Epoch of the Northern Drift. *Geol. Survey G.B. Mem.*, (1846), 336–432. XI, 3, 5, 6.

'Report on the Distribution of Pulmoniferous Mollusca in the British Isles.' *Brit. Ass. Advanc. Sci. Rep. for 1839 (1840)*, part 1, pp. 127–247. IV, 36.

'Report on the Investigation of British Marine Zoology by Means of the Dredge Part I...', *Brit. Ass. Advanc. Sci. Rep. for 1850 (1851)*, 192–263. VI, 5 bis, 32.

'Report on the Mollusca and Radiata of the Aegean Sea, and on their distribution, considered as bearing on Geology.' *Brit. Ass. Advanc. Sci. Rep. for 1843 (1844)*, 130–93. VI, 57.

Forsskåhl, Jonas Gustavus. See Linné.

Fothergill, Charles. *An Essay on the Philosophy, Study, and Use of Natural History*, London, 1813. V, 13.

[Franklin, Benjamin. *Observations concerning the Increase of Mankind, Peopling of Countries & C.*, Published anon. Boston, 1755.] V, 9.

Franklin, John. *Narrative of a Journey to the shores of the Polar Sea, in the Years 1819, 20, 21, and 22.* London, 1823. IX, 78.

Fremont, J[ohn] C[harles]. *Report of the Exploring Expedition to the Rocky Mountains in the Year 1842, and to Oregon and North California in the years 1843–44.* Washington, 1845. X, 50.

Fries, Elias Magnus. 'A Monograph of the Hieracia; being an Abstract of Prof. Fries's 'Symbolae ad Historiam Hieraciorum,'' Translated and abridged from the 'Flora'. *Bot. Gaz.*, 2 (1850), 85–92, 185–8, 203–19. IV, A22. VII, 91.

Fürnrohr, August Emanuel. *Flora Ratisbonensis, oder Uebersicht der um Regensburg wildwachsenden Gewächse.* Vol. 2 of *Naturhistorische Topographie von Regensburg.* ed. A. E. Furnrohr *et. al.*, Regensburg, 1839. IV, supp.e.

Gallesio, Georges. *Traité du Citrus*, Paris, 1811. III, 27.

Gardner, George. 'Contributions to a history of the Relation between Climate and Vegetation in various parts of the Globe. No. 2, The Vegetation of the Organ Mountains of Brazil.' *Hort. Soc. Lond. J.* (1846), 273–85. XI, 28.

'Contributions to a History of the Relations between Climate and Vegetation, No. 9, The Vegetation of Ceylon.' *Hort. Soc. Lond. J.*, 4 (1849), 31–40. XI, 18, 30.

'Notes of a Botanical Visit to Madras, Coimbatore, and the Neelgherry Mountains.' *J. Bot.*, 4 (1845), 393–409, 551–67. XI, 18.

Travels in the Interior of Brazil, principally through the Northern Provinces, and the gold and diamond districts, during the years 1836–1841. London, 1846. V, 19. X, 12.

Gärtner, Carl Friedrich von. *Beiträge zur Kenntniss der Befruchtung der voll Kommeneren Gewächse.* Erster theil. Versuche und Beobachtungen über die Befruchtungsorgane der vollkommeneren Gewächse, und über die natürliche und künstliche Befruchtung durch den eigenen Pollen. Stuttgart, 1844. III, 5, 7, 28, 29, 30, 32, 42, 43, 66, 67, 87, 87 bis, 88, 90, 91, 92. IX, 9, 33. Appendix I, 40.

Versuche und Beobachtungen über die Bastarderzeugung im Pflanzenreich. Mit Hinweisung auf die ähnlichen Erscheinungen in Thierreiche, ganz umgearbeitete und sehr vemehrte Ausgabe der von der Königlich holländischen Akademie Wissenschaften gekrönten Preisschrift. Mit einem Anhang. Stuttgart, 1849. III, 8, 18, 26, 27, 30, 42, 61, 87 bis, 88, 93, 95, 96, 99. IV, 68, 69. VII, 99. IX, 9, 10, 12, 12 bis, 13, 14, 15, 16, 18, 19, 20, 22, 23, 29, 30, 31, 32, 33, 35, 38, 44, 45, 46, 47, 48, 49, 50, 51, 52, 53, 54a, 55, 56, 57, 58, 59, 60, 66, 68, 73, 83, 99, 102, 111, 112, 113, 115, 116, 117, 118, 119, 120, 120A, 121, 129.

Gay, – . 'Extrait d'une lettre de M. Gay, à M. de Blainville.' *Ann. sci. nat.*, 5 Zool. (1836), 224. VII, 5.

Gay, Jacques. 'Histoire de l'*Arenaria tetraquetra*, L.' *Ann. sci. nat.*, 3 (1824), 27–46. IV, 28.

Gay, J. & J. P. Monnard. 'Observations sur quelques crucifères décrites par M. Decandolle, dans le second volume de son *Systema naturale regni vegetabilis*.' par J. P. Monnard, avec des notes de M. Gay. *Ann. sci. nat.*, 7 (1826), 389–419. IV, 27.

Geoffroy-Saint-Hilaire, Isidore. Article 'Polatouche Pteromys.' *Dict. class. hist. nat.*, vol. 14, Paris, 1828. VIII, 6.

Essais de zoologie générale, ou mémoires et notices sur la zoologie générale, l'anthropologie, et l'histoire de la science, Paris, 1841. III, 72. IX, 88a, 124.

Article: 'Mammifères, des métis, et du croisement des espèces et des races chez les mammifères.' *Dict. class. hist. nat.*, vol. 10, Paris, 1826. IX, 124.

Histoire générale et particulière des anomalies de l'organisation chez l'homme et les animaux, ouvrage comprenant des recherches sur les caractères, la classification, l'influence physiologique et pathologique, les rapports généraux, les lois et les causes des monstruosités, des variétés, et des vices de conformation ou traité de tératologie, 4 vols, Paris, 1832–7. Appendix I, 40. IV, 32. VII, 35, 35a, 37, 46, 79, 80, 84.

Geoffroy-Saint-Hilaire, Etienne. 'Observations sur l'affection mutuelle de quelques animaux, et particulièrement sur les services rendus au requin par le pilote.' *Muséum Hist., Nat. Ann.*, 9 (1807), 469–76. X, 61, 101.

Principes de philosophie zoologique, discutés en mars 1830, au sein de l'Académie Royale des Sciences, Paris, 1830. VII, 79. VIII, 97.

Gillibert, Jean Emanuel. 'Singularité de la plante appellée *Menyanthes trifoliata*,' *Acad. sci. Petro. Acta*, I Part 2 for 1777 (1780), Histoire p. 45. III, 47 bis.

[Girou de Buzareingues, Ch. 'Mémoire sur les rapports des sexes dans le règne végétal.' *Ann. sci. nat.*, 24 (1831), 156–76. Cited by Gärtner, see *Nat. Sel.* ch. III, fol. 99.]

Gloger, Constantin W. Lambert. *Das Abändern der Vögel durch Einfluss des Klima's Nach zoologischen, zunächstron den europäischen Landvögeln entnommen Beobachtungen dargestellt, mit den entsprechenden Erfahrung bei den europäischen Säugthieren verglichen unt durch Thatsachen aus dem Gebiete der Physiologie, der Physik und der physischen Geographie erläutert.* Breslau, 1833. IV, 57.

Gmelin, Johann Georg. *Flora Sibirica sive Historia Plantarum Sibiriae.* Vol. I, St. Petersburg, 1747. XI, 10.

Godley, William. 'Note on Raising Cowslips (Primula veris) from seed.' *Phytologist*, 3 (1848), 180. IV, 74.

Godsall, William. 'Lathyrus grandiflorus.' *Gdnrs. Mag.*, Vol. 8 (1832), 733. III, 60.

'Will the Gooseberry take on the Current?' *Gdnrs. Chron.*, (1857), 757. IX, 66.

Godwin, William. *Of Population. An Enquiry concerning the Power of increase in the Numbers of Mankind, being an answer to Mr. Malthus's Essay on that Subject.* London, 1820. III, 99.

Goeppert, H. R. 'Bemerkungen über das Vorkommen von Pflanzen in Heissen Quellen und in ungewöhnlich warmen Boden.' *Arch. Naturgesch.*, 3–1 (1837), 201–10. VII, 9.

[Gordon, George.] *Collectanea for a Flora of Moray; or, a list of the phaenogamous plants and ferns hitherto found within the province.* Elgin, 1839. IV, 66.

Gosse, Philip Henry, and Richard Hill. *The Birds of Jamaica.* London, 1847. VII, 60, 60 bis, 30.

Gosse, Philip Henry. *A Naturalist's Sojourn in Jamaica.* London, 1851. V, 21, 23. IX, 102.

Gouan, Antoine. *Herborisations des environs de Montpellier, ou guide botanique à l'usage des élèves de l'École de santé; ouvrage destiné à servir de supplément au Flora monspeliaca.* Montpellier, [1796] V, 42.

Gould, Augustus Addison. *Mollusca and Shells, Vol. XII in United States Exploring Expedition during the years 1838, 1839, 1840, 1841, 1842. Under the Command of Charles Wilkes, U.S.N.* Philadelphia, 1852–6. XI, 34.

Gould, John. *The Birds of Australia*, Vols. I & II. London, 1848. x, 65, 77, 81.
Century of Birds, hitherto unfigured, from the Himalayan Mountains. London, 1831. xi, 30.
'Exhibition of Portion of a Collection of Birds Formed by Mr. Hauxwell on the Eastern Side of the Peruvian Andes.' *Zool. Soc. Lond. Proc.*, (1855), 77–8. vii, 6.
An Introduction to the Birds of Australia, London, 1848. (Printed for the Author). iv, A19. x, 65.
Graba, Carl Julian. *Tagebuch geführt auf Reise nach Färö im Jahre, 1828*. Hamburg, 1830. iv, 17, 53, 59. x, 57vl.
Grant, John. 'Bees.' *Gdnrs. Chron.*, (1844), 374. x, 21a.
Gray, Asa. *Manual of the Botany of the Northern United States: second edition including Virginia, Kentucky, and all East of the Mississippi: arranged according to the Natural System*. N.Y., 1856. iv, supp.d. vi, 26 i. vii, 99. ix, 31.
'Notes of a Botanical Excursion to the Mountains of South Carolina; with some Remarks on the Botany of the Higher Allegheny Mountains, in a letter to Sir W. J. Hooker.' *Lond. J. Bot.*, 2 (1843), 113–25. xi, 5.
Review of Hooker's *Icones Plantarum*...*Amer. J. Sci.*, 45 (1843), 214–16. iv, 24.
'Statistics of the Flora of the Northern United States.' *Amer. J. Sci.*, 2nd ser. 22 (1856), 204–32; 23 (1857), 62–84, 369–403. iv, 3, 39b, A5A, A14, A17.
Gray, John Edward. *Gleanings from the Menagerie and Aviary at Knowsley Hall*. Knowsley, 1846. iii, 81.
Gleanings from the Menagerie and Aviary at Knowsley Hall: Hoofed Quadrupeds. Knowsley, 1850. vii, 58, 106. ix, 91.
'On the genus Bradypus of Linnaeus.' *Zool. Soc. Lond. Proc.*, 17 (1849), 65–73. iv, 32.
'On the variation in the teeth of the crested seal, Cystophora cristata, and on a new species of the West Indies (C. Antillarum).' *Zool. Soc. Lond. Proc.*, 17 (1849), 91–3. iv, 32.
Greville, Robert K. ['On the Botanical Characters of the British Oaks.' *Bot. Soc. Edinb. Trans.*, 1 (1844), 65–9. Cited by Hooker and Arnott, in *British Flora*: see *Nat. Sel.* ch. iv, fol. 44.]
Grey, Sir George. *Journals of Two Expeditions of Discovery in Northwest and Western Australia, during the Years 1837, 38, and 39*. 2 vols. London, 1841. iii, 1. v, 46. x, 53.
Grisebach, August Heinrich Rudolf. *Spicilegium Florae Rumelicae et Bithynicae exhibens synopsin plantarum quas aes 1839 legit auctor*. 2 vols. Braunschweig, 1843–4. iv, supp.d.
Gyllenhal, Leonhard. *Insecta Suecica descripta a Leonardo Gullenhal* (Coleoptera sive Eleutcrata), 1808–27. iv, supp.e.
Hakluyt, Richard. *The THIRD AND LAST volume of the Voyages, Navigations, Traffiques, and Discoveries of the English Nation, and of some few places where they have not been, of strangers performed within and before the time of these hundred yeeres, to all parts of the newfound world of America, or the West Indies, from 73 degrees of Northerly to 57 of Southerly latitude*. London, 1600. Cited by Robertson. v, 12v.
Haldeman, S. S. 'Enumeration of the Recent Freshwater Mollusca which are common to North America and Europe; with observations on Species and their Distribution.' *Boston J. Nat. Hist.*, 4 (1843–44), 468–84. iv, 39.

Hall, James. *Travels in Scotland, by an Unusual Route: with a Trip to the Orkneys and Hebrides. Containing hints for Improvements in Agriculture and Commerce. With Characters and Anecdotes. Embellished with Views of Striking Objects and a Map including the Caledonian Canal.* 2 vols. London, 1807. x, 73.

Harcourt, Edward Vernon. 'Notes on the Ornithology of Madeira.' *Ann. Mag. Nat. Hist.*, 2nd ser., 15 (1855), 430–8. IV, 60. VI, 41.

A sketch of Madeira; containing information for the traveller or invalid visitor. London, 1851. IV, 60. x, 51, 51c.

Harlan, Richard. *Medical and Physical Researches, or original memoires in medicine, surgery, physiology, geology, zoology and comparative anatomy.* Philadelphia, 1835. IV, 41.

Harmer, Thomas. 'Remarks on the very different Accounts that have been given of the Fecundity of Fishes with Fresh Observations on that subject.' *Phil. Trans.* 57 (1767), 280–92. v, 10 bis.

Harvey, William H. *The Seaside Book*, London, 1849. VII, 8.

'Hasselgren'. See Linné.

Haygarth, Henry William. *Recollections of Bush Life in Australia during a Residence of Eight Years in the Interior.* London, 1848. v, 21.

Hearne, Samuel. *A Journey from Prince of Wales's Fort in Hudson's Bay to the Northern Ocean in the Years 1769–72.* London, 1795. v, 45. VIII, 11, 14. x, 78.

Heber, Reginald. [Narrative of a Journey through the Upper Provinces of India, from Calcutta to Bombay, 1824–1825. (With Notes upon Ceylon,) An Account of a Journey to Madras and the Southern Provinces, 1826, and Letters written in India. [ed. by Amelia Heber.] 2 vols. London, 1828. etc.] III, 74.

Heer, Oswald. 'Ueber die fossilen Pflanzen von St. Jorge in Madeira,' *Allg. schweiz. Ges. gesamm. Naturw. neue Denkschr.*, 15 (1857), Nr. 2, pp. 1–40. XI, FC 6.

Heineken, C. 'Observations on the Fringilla Canaria, Sylvia Atracapilla, and other Birds of Madeira.' *Zool. J.*, 5 (1830), 70–9. IV, 60. x, 51.

[Henschel, August Wilhelm Eduard Theodor. Von der sexualität der Pflanzen. Studien von Dr. August Henschel. Nebst einem historischen Anhange von Dr. F. J. Schelver. Breslau, 1820. Cited by Gärtner. See *Nat. Sel.* ch. III, fol. 99.]

Henslow, John Stevens. *A catalogue of British Plants, arranged according to the Natural System, with the Synonyms of De Candolle, Smith, Lindley, and Hooker.* 2nd ed. Cambridge, 1835. IV, supp.a.

'Flora Keelingensis. An account of the native plants of the Keeling Islands.' *Ann. Nat. Hist.*, 1 (1838), 337–47. VI, 26.

'On the Specific Identity of Anagallis arvensis and caerulea.' *Mag. Nat. Hist.*, 3 (1830), 537–8. IV, 67.

'On the Specific Identity of the Primrose, Oxlip, Cowslip, and Polyanthus,' *Mag. Nat. Hist.*, 3 (1830), 406–9. IV, 71.

'On the Triticoidal Forms of *Aegilops* and on the Specific Identity of *Centaurea nigra* and *C. nigrescens*.' *Brit. Ass. Advanc. Sci. Rep. for 1856, (1857)*, part 2, pp. 87–8. VI, 47.

'Retrospective criticism on the specific Identity of Anagallis arvensis and caerulea.' *Mag. Nat. Hist.*, IV, 67. 5 (1832), 493–4.

Herbert, Thomas. *Some Yeares TRAVELS into Divers Parts of Asia and Afrique. Describing especially the two famous Empires, the Persian, and great Mogull: weaved with the History of these later Times. As also, many rich and spacious Kingdoms in the Oriental India and other parts of Asia; Together with the adjacent Isles. Severally relating the Religion, Language, Qualities, Customes, Habit, Descent, Fashions, and other Observations touching them, with a revivall of the first Discoverer of America.* Revised and enlarged by the Author. London, 1638. x, 58.

Herbert, William. *Amaryllidaceae; preceded by an Attempt to Arrange the Monocotyledonous Orders, and followed by a Treatise on Cross-Bred Vegetables, and Supplement.* London, 1837. III, 5, 8, 9, 27, 90. VII, 98, 113. IX, 19, 20, 21, 25, 46, 49, 50, 51, 90, 113 a, 115.

'A History of the Species of Crocus.' *Hort. Soc. Lond. J.,* 2 (1847), 249–93. IV, 27.

Local Habitations and Wants of Plants, *Hort. Soc. Lond. J.,* 1 (1846), 44–9. V, 47. VII, 8.

'On Hybridization amongst Vegetables.' *Hort. Soc. Lond. J.,* 2 (1847), 1–28. III, 9, 87 bis. VII, 99. IX, 20, 21, 24, 25, 45, 49, 50.

'On the Production of Hybrid Vegetables; with the Result of many Experiments Made in the Investigation of the Subject.' *Hort. Soc. Lond. Trans.,* 4 (1822), 15–50. IV, 71, 73, 79.

Footnote in Gilbert White, *The National History and Antiquities of Selbourne.* Edward T. Bennett, London, 1837. V, 23.

Heron, Sir Robert. *Notes.* Grantham, 1850, etc. IX, 131. X, 72.

Hesselgren, Nicolaus L. See Linné.

Hewitson, William Chapman. *British Oology,* 2 vols. Newcastle-upon-Tyne, [1831–44.] IV, 30. X, 75.

'Notes on the Ornithology of Norway,' *Mag. Zool. Bot.,* 2 (1838), 309–17. X, 61, 73.

Hewitt, Edward. 'On Hybrids.' *Poult. Chron.,* 3 (1855), 14–15. IX, 81.

Hickson, W. E. 'Art. VI.—1. *An Essay on the Principle of Population.* By the Rev. T. R. Malthus, A.M., F.R.S. Sixth Edition. Murray.

2. *Principles of Political Economy,* by John Stuart Mill. Parker.

3. *The True Law of Population.* By Thomas Doubleday, Esq., G. Pierce, 310, Strand.

4. *General Report of the Sanitary Condition of the Labouring Population, from the Poor-Law Commissioners,* 1842.

5. *Reports of the Board of Health,* 1849.

6. *Journal des Economistes,* Gillaumin, Paris, and Luxford, London.'' *Westminster Rev.,* 52 (1849), 133–201. III, 99.

Hincks, Thomas. 'On a Peculiar Organ which occurs on some of the Marine Bryozoa, and which appears to indicate a Difference of Sex.' *Brit. Ass. Advanc. Sci. Rep. of 22nd. Meeting at Belfast for 1852...* London (1853), 75–6. III, 24.

Hodgson, Bryan Houghton. 'On the mammalia of Nepal.' *Asiat. Soc. Bengal, J.,* 1 (1832), 335–48. IV, 49.

Hofacker, J. D. *Ueber die Eigenschaften welche sich bei Menschen und Thieren von den Eltern auf die Nachkommen vererben, mit besonderer Rücksicht auf die Pferdezucht.* Tubingen, 1828. VII, 107.

Högström, Pehr. ['Anmarkingar vid den år 1763 Västerbotten infallna märkvardiga Vinter.' *K. Swenska wetenskaps Academiens Handlingar för År, 1764.* 25 (1764), 19–24.] X, 108 a.

['Anmerkungen bey dem 1763 in Westbothnien eingefallenen merkwürdigen Winter,' *Der Köngl. Schwedischen Akademie der Wissenschaften Abhand-*

lungen aus der Naturlehre, Haushaltung, Kunst und Mechanik auf das Jahre, 1764. 26 (1767), 19–24.] x, 108 a.

Holland, Henry. *Chapters on Mental Physiology.* London, 1852. x, 22.

Holmes, Edward. *The Life of Mozart, including his correspondence.* London, 1845. x, 22.

Hooker, Joseph Dalton. *Flora Antarctica*, 2 vols., London, 1844–7. Part I of The Botany of the Antarctic Voyage of H.M. Discovery Ships Erebus and Terror in the years 1839–1843, under the command of Captain Sir James Clark Ross. IV, supp.e XI, 42, 43.

Hooker, Joseph Dalton and Thomson. *Flora Indica; being a systematic Account of the Plants of British India, together with observations on the structure and affinities of their Natural Orders and Genera.* Vol. 1, London, 1855. IV, supp.e, 11–12. v, 68. VII, 35 a, 99. XI, 18, 42.

Hooker, Joseph Dalton. *Flora Novae-Zelandiae*, Part II of The Botany of the Antarctic Voyage of H.M. Erebus and Terror...2 vols. London, 1853–5. III, 47 a. IV, supp.e, A12, A19 A. VII, 9, 12. XI, 19, 25, 32, 42, FC44, XI, 49 A; appendix on representative species, 1, 2 A.

Himalayan Journals; or, Notes of a Naturalist in Bengal, the Sikkim and Nepal Himalayas, the Khasia Mountains, etc. 2 vols. London, 1854. III, 47 tres. v, 71. VII, 9. XI, 18, 24.

'Notices of Books: Géographie Botanique raisonnée...par M. Alph. de Candolle.' *J. Bot.*, 8 (1856), 82–8, 151–7, etc. VI, 38.

'On the Climate and Vegetation of the Temperate and Cold Regions of East Nepal and the Sikkim-Himalayan Mountains.' *Hort. Soc. Lond. J.*, 7 (1852), 69–131. VII, 6.

Hooker, Joseph Dalton and T. Thomson. 'Praecursores ad Floram Indicam: being sketches of the natural families of Indian Plants, with remarks on their distribution, structure and affinities.' [1857], *Linn. Soc. Lond. J.*, 2 (1858) Bot. 1–29, 54–103, 163–80. VII, 100.

Hooker, William Jackson and Arnott, G. A. W. *The British Flora: comprising phaenogamous or flowering plants, and the ferns.* The 7th edition, with additions and corrections, and numerous figures illustrative of the umbelliferous plants, the composite plants, the grasses, the ferns London, 1855. IV, 24, 44, 69.

Hooker, Sir William Jackson. 'Some Account of a Collection of Arctic Plants formed by Edward Sabine...during a voyage in the Polar Seas in the Year 1823.' *Linn. Soc. Lond. Trans.*, 14 (1825), 360–94. XI, 10.

Hope, Frederic William. 'On the Entomology of the Himalayas and of India', in Royle, John Forbes, *Illustrations of the Botany and other Branches of the Natural History of the Himalayan Mountains and of the Flora of Cashmere.* Vol. 1, London, 1839, pp. xxxvii–lii XI, 24, 30.

Hopkins, William. 'Anniversary Address of the President.' *Geol. Soc. Lond. Quart. J.*, 8 (1852), xxi–lxxx. XI, 20.

Hopkirk, Thomas. *Flora Anomola; A General View of the Anomalies in the Vegetable Kingdom.* Glasgow, 1817. VIII, 43.

Horsburgh, James. 'Remarks on several Icebergs which have been Met with in unusually low Latitudes in the Southern hemisphere.' *Phil. Trans.*, (1830), 117–20. XI, 47.

Hornschuch, Friedrich. 'Nachschrift', to preceding article by E. von Berg. *Flora*, Beiblätter (1833) vol. 1, 42–7. IV, 67.

'Ueber Ausartung der Pflanzen.' *Flora*, 31 (1848), 17–28, 33–44, 50–64, 66–8. IV, 15, 66.

Hoy, J. D. ['Observations on the Iceland and Ger Falcons (Fálco Icelándicus), tending to show that these Birds are of two distinct Species.' *Mag. Nat. Hist.*, 6 (1833) 107–10.] III, 78.

Huber, François. *Nouvelles Observations sur les abeilles.* 2nd ed., vol. 2. Paris, (1814). x, 7, 8.

Huber, Pierre. 'Mémoire pour servir à l'histoire de la chenille du hamac, Tinea Harisella Linnaei; oecophore de Latreille.' *Soc. Phys. Genève, Mém.*, 7 (1836), 121–60. x, 26.

'Mémoire sur le Charançon lozange (Cionus Scrophulariae).' *Soc. Phys. Genève, Mém.*, 10 (1843), 15–34. x, 98a.

'Notice sur la Mélipone domestique, abeille domestique mexicaine.' *Soc. Phys. Genève, Mém.*, 8 (1839), 1–26. x, 93.

'Observations on several species of the Genus Apis, known by the Name of Humble-bees, and called Bombinatrices by Linnaeus.' *Linn. Soc. Lond. Trans.*, 6 (1802), 214–98. x, 93.

Huc, Évariste R. *The Chinese Empire: forming a sequel to the work entitled 'Recollections of a Journey through Tartary and Thibet'.* 2 vols. London, 1855. III, 44.

Humboldt, Alexander von. *Personal Narrative of Travels to the Equinoctial Regions of the New Continent during the years 1799–1804*, trans. Helen M. Williams, 2nd ed. vol. III, London, (1822). x, 13. XI, 28.

Hunt, Consul Carew. 'A Description of the Island of St. Michael (Azores).' *Geogr. Soc. Lond., J.*, 15 (1845), 268–96. x, 54.

Hunter, John. 'Observations on Bees,' *Phil. Trans.*, (1792), 128–95. VIII, 90.

Hunter, John. *Observations on Certain Parts of the Animal Oeconomy. Inclusive of several Papers from the Philosophical Transactions, etc.*, with notes by Richard Owen. London, 1837. III, 2. VII, 54. IX, 78. x, 48.

Hutchinson, W. N. *Dog Breaking. The Most Expeditious, Certain, and Easy Method; whether Great Excellency or only Mediocrity is Required.* 2nd ed. London, 1850. x, 34, 35.

Hutton, Thomas. 'On the nest of the Tailor Bird.' *Asiat. Soc. Bengal, J.*, 2 (1823) 502–5. x, 15.

The Chronology of Creation; or, Geology and Scripture Reconciled, Calcutta, 1850. VII, 9.

'"Borz" of the Afghans—Wild Goat of Afghanistan, "The Paseng" of Authors.' *Calcutta J. Nat. Hist.*, 2 (1842), 521–35. IX, 93.

Huxley, Thomas H. 'Lectures on General Natural History.' Lecture 10, *Med. Times*, N.S. 14 (1857), 133–5, 181–3, 255–7, 353–5. VIII, 32.

'On Natural History, as Knowledge, Discipline, and Power.' *Roy. Inst. Proc.*, 2 (1856?), 187–95. VIII, 86.

Jardine, Sir William. ['Observations on a Collection of Birds lately received from Madeira, with the Description of some New Species from that Island.' *Edinb. J. Nat. Sci.*, 1 (1830), 241–5.] IV, 60.

Jenner, Edward. 'Observations on the Natural History of the Cuckoo.' *Phil. Trans.*, 78 (1788), 219–37. x, 82.

'Some Observations on the Migration of Birds.' *Phil. Trans.*, (1824), 11–44. v, 17.

Jenyns, Leonard. 'Notes on some of the smaller British Mammalia, including the description of a new species of Arvicola, found in Scotland.' *Ann. Mag. Nat. Hist.*, 7 (1841), 261–76. IV, 33.

Observations in Natural History, London, 1846. x, 14, 101, 110.

'Some observations on the Common Bat of Pennant: with an Attempt to prove its Identity with the Pipistrelle of French Authors.' *Linn. Soc. Lond. Trans.*, 16 (1833), 159–68. x, 78.

Jesse, Edward. *Gleanings in Natural History.* Third and last series. To which are added notices of some of the royal parks and residences, London, 1835. x, 24, 31.

Johns, C. A. 'Observations on the Plants of Land's End.' *Phytologist*, 2 (1847), 906–8. vi, 26f.

Johnston, Alexander Keith. *The Physical Atlas of Natural Phenomena*, Edinburgh and London, 1850. xi, 2, 8, 27A.

Journal of a Naturalist. See [Knapp, John Leonard].

Junghuhn, Franz. *Java, seine Gestalt, Pflanzendecke und innere Bauart.* Trans. J. K. Hasskarl. 3 vols. Leipzig, 1852–54. xi, 18, 42.

Jurine, Louis. 'Histoire abrégée des poissons du lac Léman.' *Soc. Phys. Genève, Mém.*, 3 (1825). vii, 32.

Jussieu, Adrien de. 'Monographie de la famille des Malpighiacées.' *Muséum Hist. Nat. Arch.*, 3 (1843), 5–151. vii, 38c, 96 bis. viii, 43.

Kalm, Pehr. *Travels into North America; containing its Natural History and a circumstantial Account of its Plantations and Agriculture in general, with the civil, ecclesiastical and commercial state of the country, the manners of the inhabitants, and several curious and important remarks on various subjects.* Trans. J. R. Forster. 3 vols, Warrington, 1770–1. iii, 96. iv, 9. v, 46. vii, 14. x, 19.

Kane, Elisha Kent. *Arctic Explorations: the second Grinnell expedition in search of Sir John Franklin, 1853, 54, 55.* 2 vols. Philadelphia, 1856. vii, 58.

Karkeek, [William Floyd.] 'On Breeding.' *Veterinarian*, 4 (1831), 1–12, etc. iii, 3.

Ker, Henry Bellenden. 'Account of the Cultivation of the Water Cress, as practised by Mr. William Bradbery....' *Hort. Soc. Lond. Trans.*, 4 (1822), 537–42. iv, 44.

Kerr, Robert. *A general History and collection of Voyages and Travels arranged in systematic order: Forming a complete history of the origin and progress of navigation, discovery, and commerce, by sea and land, from the earliest ages to the present time.* 18 vols. Edinburgh, 1811–24. v, 12. x, 58.

Kidd, William. 'British Song Birds (Cage Birds, No. 6).' *Gdnrs. Chron.*, (1851), 181–2. x, 13v.

Kingsley, Charles. *Glaucus; or the Wonders of the Shore.* 1st ed. Cambridge, 1855, etc. viii, 86.

Kirby, William and William Spence. *An Introduction to Entomology: or Elements of the Natural History of Insects.* Vol. 1, 3rd ed. London, 1818. Vol. 2, 2nd. ed. London, 1818. Vol. 3, London, 1826. Vol. 4, London, 1826. v, 38. vii, 26. viii, 49, 58, 59, 62, 63, 73, 91. x, 7, 8, 9, 10, 11, 15, 16, 26, 27, 62, 93, 98a, 101, 102, 103, 104, 106, 108a, 112, 113.

[Knapp, John Leonard]. *The Journal of a Naturalist.* 4th ed. London, 1838. v, 27. x, 111.

Knight, Thomas Andrew. 'An Account of some Experiments on the Fecundation of Vegetables.' *Phil. Trans*, (1799), 195–204. iii, 6A, 15, 44, 62. ix, 119.

'On the Economy of Bees.' *Phil. Trans.*, (1807), part 1, 234–44. x, 9.

'On the Effects of different Kinds of Stocks in Grafting.' *Hort. Soc. Lond. Trans.*, 2 (1817), 199–204. ix, 67.

'On the Hereditary Instinctive Propensities of Animals.' *Phil. Trans.*, (1837), 365–9. x, 34, 41, 49.

A Treatise on the Culture of the Apple and Pear, and on the Manufacture of Cider & Perry with an Appendix and Postscript. 4th ed., London, 1813. IX, 119.

Koch, Wilhelm Daniel Joseph. 'Essais de culture démontrant l'identité des *Taraxacum officinale* et *palustre.' Ann. sci. nat.*, ser. 2, Bot. 2, (1834), 119–20. IV, 63.

Report on vol. 4 of G. O. Roehlings *Deutschlands Flora*, published by W. D. G. Koch, *Ann. sci. nat.*, 2 ser. Bot 3 (1835), 370–9. IV, 64.

Synopsis Florae Germanicae et Helveticae, exhibens stirpes phanerogamas vite cognitas quae in Germania Helvetia, Borussia, et Istria sponte crescunt atque in hominum usum copiosus coluntur, secundum systema Candolleanum digestas, praemissa generum dispositione secundum classes et ordines systematis Linnaeani conscripta. 2nd ed. (par. 1), Frankfurt, 1843. (par. 2), Frankfurt, 1844. (Pars 1) pp. LX, 1–452. (Pars 2), pp. (451)–1164. III, 47 bis. IV, supp. c.

Kölreuter, Joseph Gottlieb. 'Daturae novae hybridae.' *Acad. sci. Petro. Acta*, vol. 5, for 1781, part 2 (1785), 303–13. III, 88. IX, 31.

'Dianthi novi hibridi.' *Acad. sci. Petro. Nova Acta*, vol. 3 for 1785 (1788), 277–84. IX, 34.

'Digitales hybrides.' *J. Phys.*, 21 (1782), 285–306. IV, 66. IX, 114, 118, 122.

'Dissertationis de antherarum pulvere continuatio.' *Acad. sci. Petro. Mém.*, vol. 3 for 1809–10 (1811), 159–99. III, 15, 26 bis.

'Lina hybrida.' *Acad. sci. Petro. Nova Acta*, vol. 1 for 1783 (1787), 339–46. IX, 52.

'Lucia Hybrida.' *Acad. sci. Petro. Acta*, vol. 2 for 1778, Part 1 (1780), 219–24. IX, 52.

'Malvacei ordinis plantae novae hybridae.' *Acad. sci. Petro. Acta* (1786), 251–88. Vol. 6 for 1782, Part 2. III, 63.

'Mirab. jalapae hybridae.' *Acad. sci. Petro. Nova Acta*, Vol. 11 for 1793 (1798), 389–99. III, 89. IX, 16, 52.

'Mirabil jalaparum hybridarum continuata descriptio.' *Acad. sci. Petro. Nova Acta*, vol. 12 for 1794 (1801), 378–98. IX, 45, 115.

'Mirabil jalaparum hybridarum ulterius continuata descriptio.' *Acad. sci. Petro. Nova Acta*, vol. 13 for 1795–6 (1802), 305–35. III, 8. IX, 15, 45, 114, 118, 131.

'Mirab. jalaparum hybridarum spicilegium ultimum' (373–95). 'Additamenta ad descriptarum quarundum Jalap. hybridarum naturam, pluribus exemplis illustrandam, maxime conducentia.' (396–408). *Acad. sci. Petro. Nova Acta*, vol. 14 for 1797–8 (1805), 373–408. IX, 13, 15, 114, 116, 118.

'Nouvelles observations et expériences sur l'irritabilité des étamines de l'épine-vinette (Berberis vulgaris. L.).' *Acad. sci. Petro. Nova Acta*, vol. 6 for 1788 (1790), 207–16. IX, 35.

'Verbasca nova hybrida.' *Acad. sci. Petro. Acta*, vol. 5 for 1781, Part 1 (1784), 249–70. IX, 122.

1. *Vorläufige Nachricht von einigen das Geschlecht der Pflanzen betreffenden. Versuchen und Beobachtungen.* Leipzig, 1761. 2. *Fortsetzung der vorläufigen Nachricht von einigen das Geschlecht der Pfanzen betreffenden Versuchen und Beobachtungen,* Leipzig, 1763. 3. *Zweyte Fortsetzung der vorläufigen Nachricht von einigen das Geschlecht Pflanzen betreffenden Versuchen und Beobachtungen.* Leipzig, 1764. 4. *Dritte Fortsetzung der vorläufigen Nachricht*

von einigen das Geschlecht der Pflanzen betreffenden Versuchen und Beo-bachtungen, Leipzig, 1766. III, 27, 35, 40, 66, 90, 92. V, 64. IX, 9, 13, 23, 30, 31, 34, 47, 56, 114, 117, 118, 120 A, 121, 122.

Kröyer, Henrik. 'On the Danish Oyster Beds.' *Edinb. New Phil. J.*, 29 (1840), 22–6. V, 27.

Lacaze-Duthiers, Henri de. 'Recherches sur l'armure génitale des insectes.' *Ann. sci. nat.*, 3 ser. zool., 12 (1849), 353–74, and 14 (1850), 17–52. VIII, 50, 90.

Laing, Samuel. [*Journal of a residence in Norway, during the years 1834, 1835, and 1836; made with a view to enquire into the moral and political economy of that country, and the condition of its inhabitants.* 2nd ed. London, 1837.] V, 36.

[*Notes of a Traveller, on the social and political state of France, Prussia, Switzerland, Italy, and other parts of Europe during the present century.* First series, London, 1854.] V, 36.

Lalanne, Léon. 'Note sur l'architecture des abeilles.' *Ann. sci. nat.*, 2nd ser., vol. 13, zool. (1840), 358–74. X, 92.

Lallemand, – . 'Observation sur le rôle des zoospermes dans la génération.' *Ann. sci. nat.*, 2 ser. zool. 15 (1841), 262–307. IX, 12 bis.

'Observations sur l'origine et le mode de développement des zoospermes.' *Ann. sci. nat.*, 2nd ser. zool., 15 (1841), 30–101. IX, 12 bis.

Lamarck, Jean Baptiste Pierre Antoine de Monet, Chevalier de. *Philosophie Zoologique, ou exposition des considérations relatives à l'histoire naturelle des animaux à la diversité de leur organisation et des facultés qu'ils en obtiennent; aux causes physiques qui maintiennent en eux la vie et donnent lieu aux mouvements qu'ils exécutent; enfin, à celles qui produisent les unes, le sentiment, et les autres l'intelligence de ceux qui en sont doués.* New issue I, Paris, 1830.] IV, 38.

Landt, Jørgen. *A Description of the Faroe Islands, containing an Account of their Situation, Climate, and Productions; together with the Manners, and Customs, of the Inhabitants, their Trade, & C.* By the Rev. G. Landt, trans. from the Danish. London, 1810. IV, 53. V, 55 a. VII, 8.

Latham, John. *General Synopsis of Birds*, 3 vols., London, 1781–5. 1st and 2nd supplements. London, 1787–1801. IX, 169.

Latreille, P. A. 'Introduction à la géographie générale des arachnides et des insectes, ou des climats propres à ces animaux.' *Museum Hist. Nat. Mém.*, 3 (1817), 37–67. XI, 1.

Lay, George Tradescant. 'Observations on a Species of Pteropus from Bonin.' *Zool. J.*, 4 (1829), 457–9. VIII, 8.

Layard, Edgar Leopold. 'Notes on the Ornithology of Ceylon, collected during an eight years' residence in the Island.' *Ann. Mag. Nat. Hist.*, 2nd ser., 14 (1854), 57–64, etc. IX, 88 a.

'Rambles in Ceylon.' *Ann. Mag. Nat. Hist.*, 2nd ser. 9 (1852), 329–39. X, 11.

Leach, William Elford. 'A Tabular View of the External Characters of Four Classes of Animals, which Linné arranged under Insecta; with the distribution of the Genera composing three of these classes into Orders, &c., and descriptions of several new genera and species.' *Linn. Soc. Lond. Trans.*, [1814] 11 (1815), 306–400. X, 98 a.

Le Conte, John L. 'General Remarks upon the Coleoptera of Lake Superior,' pp. 201–42, in Agassiz, Louis, *Lake Superior its physical character, Vegetation and Animals, compared with those of other and similar Regions, with a Narrative of the Tour by J. Elliot Cabot*, and contributions by other scientific gentlemen. Boston, 1850. IV, 34, 39 a.

Le Couteur, John. *On the varieties, properties and classification of wheat.* Jersey, 1836. III, 45.

Lecoq, Henri. *De la Fécondation naturelle et artificielle des végétaux et de l'hybridation, considérée dans ses rapports avec l'horticulture, l'agriculture et la sylviculture, ou études sur les croisements des plantes des principaux genres cultivés dans les jardins d'ornements, fruitiers et maraîchers, sur les végétaux economiques et de grande culture, les arbres, forestiers, etc. Contenant les moyens pratiques d'opérer l'hybridation et de créer facilement des variétés nouvelles.* Paris, 1845. III, 30, 43. IX, 23.

Ledebour, Carl Friedrich von. *Flora Rossica, sive enumeratio plantarum in totius imperii Rossici provinciis Europaeis, Asiaticis et Americanis hujusque observatarum.* 4 vols. Stuttgart, 1842–53. IV, supp. d.

'General Observations on the Flora of the Altaic Mountains, and the Neighbouring Steppes.' *Bot. Misc.,* 2 (1831), 245–57. V, 45, A30. XI, 10, 12.

Lees, Edwin. 'On Certain Forms of Species of Fruticose Brambles Experimentally Proved to be Permanent.' *Phytologist,* 3 (1848), 53–5. VII, 91.

Leighton, W. A. 'On the Form of the Capsule and Seeds as affording a Specific Character in Primula Vulgaris (Huds.), P. veris (Linn.), and P. elatior (Jacq.).' *Ann. Mag. Nat. Hist.,* ser. 2, 2 (1848), 164–6. IV, 68.

Leroy, Charles Georges. *Lettres Philosophiques sur l'intelligence et la perfectibilité des animaux, avec quelques lettres sur l'homme.* Paris, 1802. X, 17, 48, 59.

Le Vaillant, François. *Travels into the Interior Parts of Africa, by the Way of the Cape of Good Hope; in the years 1780–85.* Trans. from the French... [by E. Helme], 2 vols. Perth, 1791 [?] X, 42, 50.

Lewis, Meriwether and William Clark. *Travels to the Source of the Missouri River, and across the American Continent to the Pacific Ocean. Performed by Order of the Government of the United States in the Years 1804, 1805, and 1806. By Captains Lewis and Clarke.* New ed. 3 vols. London, 1817. X, 110.

Lichtenstein, Martin Heinrich Karl. ['Beitrag zur ornithologischen Fauna von Californien, nebst Bemerkungen über die Art-Kenn zeichen der Pelicane und über einige Vögel von den Sandwich-Inseln.' *Acad. Wiss.,* Berlin, 1838, 417–51.] XI, 30.

Travels in Southern Africa, in the years 1803, 1804, 1805 and 1806. Trans. by Anne Plumptre, 2 vols. London, 1812–15. IV, 42.

Lindley, John. *Digitalium Monographia; sistens Historiam Botanicam Generis, Tabulis omniam specierum Hactenus cognitarum illustratum, ut plurimum confectis ad icones Fernandi Bauer penes Gulielmum Cailley.* London, 1821. IV, 66.

'The Gardener's Chronicle.' *Gdnrs. Chron.,* (1856), 191–2. IV, 44.

'Home Correspondence.' *Gdnrs. Chron.,* (1856), 405. IV, 44.

'Notices to correspondents.' *Gdnrs. Chron.,* (1855), 776. IV, 44.

The Theory of Horticulture; or, an attempt to explain the principal operations of gardening upon physiological principles. London, 1840. III, 27, 94. IX, 49.

The vegetable kingdom; or, the structure, classification, and uses of plants, illustrated upon the natural system, (ed. 3). London, 1853. III, 54–5. IX, 65.

Linné, Carl von. *Amoenitates Academicae Seu Dissertationes variae Physicae, Medicae Botanicae antehac seorsim editae, nunc collectae et auctae cum tabulis aeneis,* vol. II, 2nd ed. Stockholm, 1762. v, 10 bis. x, 103.

[Stillingfleet, Benjamin, *Miscellaneous Tracts relating to Natural History and Physick* [trans. from the *Amoenitates Academicae*], to which is added the *Calendar of Flora,* 2nd ed. London, 1762. (See Stillingfleet.)

Linné, Carl von, praeses; Jonas G. Forsskåhl, resp; 'The Flora of Insects', pp. 345–68 in *Select Dissertations from the Amoenitates Academicae.* Trans. by F. J. Brand. London, 1781. v, 38.

Linné, Carl von, praeses; Isaac J. Biberg, resp; 'The Oeconomy of Nature', pp. 37–129 in *Miscellaneous Tracts relating to Natural History...*[trans. from the *Amoenitates Academicae*] by Ben. Stillingfleet, 2nd ed. London, 1762. v, 42.

Linné, Carl von, On the Increase of the Habitable Earth, pp. 71–127 in *Select Dissertations from the Amoenitates Academicae.* trans. F. J. Brand. London, 1781. v, 10.

Linné, Carl von, Carl D. Ekmarck, resp; 'On the Migration of Birds', pp. 215–63 in *Select Dissertations from the Amoenitates Academicae,* Trans. F. J. Brand. London, 1781. v, 25.

Linné, Carl von, H. C. D. Wilcke, resp; 'On the Police of Nature', pp. 129–66 in *Select Dissertations from the Amoenitates Academicae,* trans. F. J. Brand. London, 1781. v, 18.

Linné, Carl von, Aphonin, Matheus resp; 'On the Use of Natural History', *Amoenitates Academicae* VII pp. 409–37. [Also in Brand, pp. 1–70] x, 19.

Linné, Carl von. 'Oratio de Memorabilibus in Insectis', *Amoenitates Academicae* II, 2nd ed. Stockholm, 1762, pp. 356–77. [Also in Brand, pp. 309–43. v, 10] x, 103.

Linné, Carl von. 'Rön om Waxters Plantering, grundat på Naturen.' *Kongl. Swenska Wetenskaps Academiens Handlingar,* vol. 1 for 1739, (1741), 5–24. III, 87. Translated as:

'Versuch von Pflanzung der Gewächse, wie solche auf die Natur gëgrundet', K. Svensk, Vetens. Acad. *Abhandlungen aus der Naturlehre,* 1 (1749), 3–26. III, 87. [At Maer, Darwin read an English translation in manuscript of this article see entry under *Acts of Swedish Academy.*]

Linné, Carl von praeses, Nicolaus L. Hesselgren resp; Swedish Pan, pp. 339–62 in *Miscellaneous Tracts relating to Natural History...*[trans. from the *Amoenitates Academicae*] by Benjamin Stillingfleet, 2nd. ed. London, 1762. v, 37. x, 19.

Livingstone, David. *Missionary Travels and Researches in South Africa; including a sketch of Sixteen Years Residence in the interior of Africa, and a Journey from the Cape of Good Hope to Loanda on the West Coast; thence across the Continent, down the River Zambesi, to the Eastern Ocean.* London, 1857. x, 107.

Lloyd, Llewellyn. *Field Sports of the North of Europe; comprised in a personal Narrative of a Residence in Sweden and Norway, in the Years 1827–28,* 2nd ed., 2 vols. London, 1831. v, 21, 55a. VII, 15. IX, 88b, 126.

Scandinavian Adventures, during a Residence of upwards of Twenty Years. Representing Sporting Incidents, and Subjects of Natural History, and Devices for entrapping wild Animals. With some Account of the Northern Fauna, 2 vols. London, 1854. IV, 51. x, 108.

Loiseleur-Deslongchamps, Jean Louis Auguste. 1. *Considérations sur les céréales, et principalement sur les froments, (Partie historique).* Paris,

1842, pp. 1–109. III, 11, 44, 45. 2. *Considérations sur les céréales, et principalement sur les froments, (Partie pratique et expérimentale)*. Paris, 1843, pp. 1–248. [In Darwin's copy, these two parts are bound together with one publisher's paper wrapper.]

London Catalogue of British Plants. See Watson, Hewett C.

Loudon, John Claudius. *Arboretum et Fruticetum Britannicum or, the Trees and Shrubs of Britain, Native and Foreign, Hardy and Half-Hardy, Pictorially and Botanically Delineated, and Scientifically and Popularly Described; with their Propagation, culture, management and uses in the Arts, in Useful and Ornamental Plantations, and in Landscape-Gardening; Preceded by a Historical and Geographical Outline of the Trees and Shrubs of Temperate Climates Throughout the World*. London, 1838. IV, 46, 66. VII, 99.

An Encyclopaedia of Gardening, Comprising the Theory and Practice of Horticulture, Floriculture, Arboriculture and Landscape Gardening, including all the latest Improvements; a general history of gardening in all countries; and a statistical view of its present state; with suggestions for its future progress in the British Isles, 2nd ed. London, 1835. IX, 66, 67.

An Encyclopaedia of Trees and Shrubs; being the Arboretum et Fruticetum Britannicum abridged; containing the hardy trees and shrubs of Britain native and foreign, scientifically and popularly described; with their propagation, culture, and uses in the arts; and with engravings of nearly all the species, abridged from the large edition in 8 volumes, and adapted for the use of Nurserymen, Gardeners, and Foresters. London, 1842. III, 47a.

Lowcock, James. 'On Hybrid Pheasants,' *Ann. Mag. Nat. Hist.*, 6 (1841), 73. IX, 79.

Lucas, Prosper, *Traité philosophique et physiologique de l'hérédité naturelle*, 2 vols. Paris, 1847–50. IX, 82, 88a, 124, 126, 129, 130, 136. X, 31.

Lund, – . 'Lettre sur les habitudes de quelques fourmis de Brésil.' *Ann. sci. nat.*, 23 (1831), 113–38. X, 89.

Lund, Carl Fredr. 'Rön om Fiske-Plantering uti Insjöar.' *K. Svenska Vetenskaps Academiens Handlinger*, 22 (1761), 184–97. v, 10 bis.

'Von Pflanzung der Fische, in innländischen seen.' *K. Schwedische Akademie der Wissenschaften Abhandlungen*, trans. by A. Kastner, 1761. v, 10 bis. [At Maer, Darwin read an English translation in manuscript of Lund's article. See entry under *Acts of Swedish Academy*.]

Lyell, Charles. *A Manual of Elementary Geology*, 5th ed. London, 1855. VIII, 69.

Principles of Geology; or, the Modern Changes of the Earth and its Inhabitants considered as illustrative of Geology, 9th ed. London, 1853. v, 12, 30B. VI, 11. XI, 18, 46, 47.

'Silurian and Cambrian Rocks, and Mr. Barrande's Theory of Colonies.', pp. 34–39 in his *Supplement to the 5th ed. of a Manual of Elementary Geology*, 2nd ed. London, 1857. VI, 26hh.

Supplement to the 5th ed. of a Manual of Elementary Geology, 2nd ed., revised. London, 1857. VI, 26hh.

Travels in North America; with Geological Observations on the United States, Canada, and Nova Scotia, 2 vols. London, 1845. XI, 20.

MacCulloch, John. *The Highlands and Western Isles of Scotland, containing descriptions of their scenery and antiquities, &c., in letters to Sir Walter Scott*. 4 vols. London, 1824. VIII, 14.

McCulloch, Thomas. 'On the Importance of Habit as a Guide to Accuracy in Systematical Arrangement, Illustrated in the Instance of the *Sylvia petechia* of Wilson and All Subsequent Writers.' *Boston J. Nat. Hist.*, 4 (1844), 406–10. VIII, 10.

MacGillivray, William. *A History of British Birds, INDIGENOUS and MIGRATORY: including their organization, Habits and Relations; Remarks on Classification and Nomenclature; An account of the Principal Organs of Birds, and Observations Relative to Practical Ornithology*, 5 vols. London, 1837–52. V, 17, 27, 53, 55, 56, 57. VII, 59, 60, 60 bis. IX, 79, 88 c. X, 64, 68, 70, 82, 101, 107.

Remarks on the varieties of the fox observed in Scotland, *Wernerian Nat. Hist. Soc. Mém.*, 7 (1838), 481–2. IV, 50.

MacKenzie, George Steuart. *Travels in the Island of Iceland, during the summer of the Year MDCCCX*, 1st. ed. Edinburgh, 1811. XI, 11.

Malthus, Thomas R. *An Essay on the Principle of Population*, 6th ed., 2 vols. London, 1826. V, 9.

Marshall, William. *Minutes of Agriculture made on a farm of 300 Acres of various soils near Croydon...published as a sketch of the actual business of a Farm*, etc. London, 1778. III, 11.

[*The Rural Economy of Norfolk*: comprising the management of landed estates, and the present practice of husbandry in that county. 2 vols. London, 1787.] VI, 44.

[*The Rural Economy of Yorkshire, comprising the management of landed estates, and the present practice of husbandry in the agricultural districts of that county*, 2 vols. London, 1788.] VI, 7 a.

Martin, William Charles Linnaeus. *The History of the Horse*: its origin, physical and moral characteristics: its principal varieties and domestic allies....with an appendix on the diseases of the horse, by W. Youatt. London, 1845. VII, 109, 111, 112.

'A Monograph of the Genus Semnopithecus.' *Mag. Nat. Hist.*, N.S., 2 (1838), 320–6, etc. X, 98 a.

Martins, Charles François. 'On the Vegetable Colonisation of the British Islands, Shetland, Feroe, and Iceland.' *Edinb. New Phil. J.*, 46 (1849), 40–52. XI, 11 A.

Mauduyt, L. *Du loup et de ses races, ou variétés*. Poitiers, 1851. IX, 78.

Mauz, E. F. ['Versuche und Beobachtungen über Bastard-Bildung.' *K. Würtembergischer landwirthschaftlicher Verein, Correspondenzblatt*, 6 (1824), 141–8. Cited by Gärtner. See *Nat. Sel.* ch. III, fol. 28.]

Meyen, Franz Julias Ferdinand. *Reise um die Erde ausgeführt auf dem Seewandlungs-Schiffe Prinzess Louise*, Berlin, 1834–43. XI, 26.

Miller, Hugh. *My Schools and Schoolmasters; or, the Story of my Education*. Edinburgh, 1854. III, 99.

'On the Late Severe Frost.' *Phys. Soc. Edinb., Proc.*, 1 (1855), 10–14. V, 27.

Mills, John. *A Treatise on Cattle, shewing the most approved Methods of Breeding, Rearing, and Fitting for Use, Horses, Asses, Mules, Horned Cattle, Sheep, Goats, and Swine; with Directions for the proper Treatment of Them in their Several Disorders: to which is added A Dissertation on their Contagious Diseases. Carefully collected from the best Authorities, and interspersed with Remarks*. London, 1776. X, 55.

Milne, John. 'A Pair of Siskin Finches.' *Mag. Nat. Hist.*, 3 (1830), 440. IX, 79.

Milne-Edwards, Henri. 'Considérations sur quelques principes relatifs à la classification naturelle des animaux, et plus particulièrement sur la distribution méthodique des mammifères.' *Ann. sci. nat.*, 3 ser., zool. 1 (1844), 65–99. VII, 42.

Histoire naturelle des crustacés comprenant l'anatomie, la physiologie et la classification de ces animaux, (1837), 3 vols. Paris, 1834–40. VIII, 27. XI, 37, 37 A.

Introduction à la Zoologie générale ou considérations sur les tendances de la nature dans la constitution du regne animal, première partie. Paris, 1851. III, 68. VI, 261. VIII, 35, 38, 39, 47, 76.

Article 'Organisation' in *Dict. class. hist. nat.*, XII (Paris, 1827), 332–44. VI, 261.

'Recherches zoologiques faites pendant un voyage sur les côtes de la Sicile.' *Ann. sci. nat.*, 3 ser. zool. 3 (1845), 129–80, 257–307. VII, 42. VIII, 38.

'Recherches zoologiques pour servir à l'histoire des Lézards, extraites d'une monographie de ce genre.' *Ann. sci. nat.*, 16 (1829), 50–89. IV, 30.

Miquel, Frederik Anton Willem. *Disquisitio geographico-botanica de Plantarum Regni Batavi distributione.* Leyden, 1837. IV, supp. c.

Mitchell, Thomas Livingstone. *Three Expeditions into the Interior of Eastern Australia, with descriptions of the recently explored region of Australia Felix, and the present colony of New South Wales,* 2 vols. London, 1838. x, 36.

Mitchill, Samuel L. 'Facts and Observations intended to illustrate the natural and economical History of the Eatable Clam of New York and its vicinity.' *Amer. J. Sci.*, 10 (1826), 287–93. IV, 43.

Molina, Juan Ignacio. *Compendio de la Historia Geograficas Natural y Civil del Reyno de Chile, escrito en Italiano por el Abate Don Juan Ignacio Molina. Primera Parte, que abraza la Historia Geografica y Natural, Traducida en Español por Don Domingo Joseph de Arcuellada Mendoza.* Madrid, 1788. IX, 90. x, 38.

Montagu, George. *Ornithological Dictionary of British Birds,* 2nd ed., by James Rennie, London, 1831. v, 25. IX, 95. x, 70, 81.

Moquin-Tandon, Alfred. *Éléments de tératologie végétale, ou histoire abrégée des anomalies de l'organisation dans les végétaux,* Paris, 1841. VI, 6, 7, 35, 39, 46, 49, 79, 82, 100.

'Mémoires sur la famille des Chénopodées.' *Ann. sci. nat.*, 23 (1831), 274–325. IV, 29.

Morren, Ch. 'Recherches sur le mouvement et l'anatomie du style du Goldfussia anisophylla.' *Acad. sci. Belg. Mém.* (1839), [Mém 2]. III, 35.

'Recherches sur le mouvement et l'anatomie de Stylidium graminifolium.' *Acad. sci. Belg. Nouv. Mém.*, II (1838) [Mém. 13, 22 pp.] III, 35, 48.

Morton, Earl of. 'A Communication of a Singular Fact in Natural History.' *Phil. Trans.*, (1821), 20–2. VII, 109. IX, 136.

Morton, Samuel George. See Nott, J. C. and Gliddon, G. R.

Morton, Samuel George. 'Description of Two Living Hybrid Fowls, between Gallus and Namida.' *Ann. Mag. Nat. Hist.*, 19 (1847), 210–12. IX, 88 a.

'Hybridity in Animals considered in Reference to the Question of the Unity of the Human Species.' *Amer. J. Sci.*, 2nd. ser. 3 (1847), 39–50, 203–12. IX, 126.

'Hybridity in Animals and Plants, considered in Reference to the Question of the Unity of the Human Species.' *Edinb. New Phil. J.*, 43 (1847), 262–88. IX, 83.

Mowbray, William. Letter on fruiting Passifloras, *Hort. Soc. Lond. Trans.*, 7 (1830), 95–6. IX, 23.

Müller, Johannes. *Elements of Physiology*, trans. by William Baly, 2 vols. London, 1838–42. VIII, 27, 34, 78. X, 7.

Murray, Andrew. 'Monograph of the Genus Catops.' *Ann. Mag. Nat. Hist.*, 2nd ser., 18 (1856), 1–24, 133–56, 302–18, 391–404, 457–67. VII, 67.
'On some Insects from the Rocky Mountains received from the Botanical Expedition to Oregon under Mr. Jeffrey.' *Zoologist*, 11 (1853), 3893–5. IV, 39.

Neison, F. G. P. 'Contributions to Vital Statistics especially designed to Elucidate the Rate of Mortality, the Laws of Sickness, and the Influences of Trade and Locality on Health.' *Statist. Soc., Lond. J.* (1845), 290–343. V, 36.

Newman, –. 'A Word on the Pseudogynous Lepidoptera.' *Zoologist*, 15 (1857), 5764–5, VIII, 60.

Newman, H. W. 'On the Habits of the Bombinatrices.' *Ent. Soc. Lond. Trans.*, N.S. 1 (1851), 86–92, 109–12, 116–18. V, 24.

Nilsson, S. 'On Changes in the Fauna of Sweden.' abstracted from the Swedish by N. Shaw, M.D., *Brit. Ass. Advanc. Sci.—Notices and Abstracts of communications at the Oxford meeting 1847*, p 79. V, 32.

Nordman, Alexander von. 'Polype nouveau de la Mer-Noire.' *L'Institut.*, 7 (1839), 95. III, 24.

Nott, Josiah C. and Gliddon, Geo. R. *Types of Mankind: or, Ethnological Researches based upon the Ancient Monuments, Paintings, Sculptures, and Crania of Races, and upon their Natural, Geographical, Philological, and Biblical History: illustrated by Selections from the inedited Papers of Samuel George Morton, M.D. (Late President of the Academy of Natural Sciences, at Philadelphia), and by Additional Contributions from Prof. L. Agassiz, LLD; W. Usher, M.D.; and Prof. H. S. Patterson, M.D.* [4th ed.] Philadelphia, 1854. IX, 90 bis, 90, 100.

Oberlin, Jean Frédéric. *Memoirs of John Frederic Oberlin, Pastor of Waldbach in the Ban de la Roche*, ? 8th ed. London, 1838. III, 12.

Odart, Alexandre Pierre. *Ampélographie universelle ou traité des cépages*, 2nd ed. Paris, 1849. VIII, 70.

Ogilby, William. 'Memoir on the Mammalogy of the Himalayas.' in Royle, John Forbes, *Illustrations*...I. London, 1839, pp.[lvi]–lxxiv. XI, 30.

Orton Reginald. *Physiology of Breeding*. Two lectures to Newcastle Farmers' Club, Sunderland, 1855, Darwin Reprint Collection, 144. cf. *Variation* I 404n & II, 66 n. 8.

Otway, S. G. 'Remarkable Change of Habit in the Hare.' *Ann. Nat. Hist.*, 5 (1840), 362–3. X, 79.

[Ovalle. *Historica Relación del Regno de Chile*, trans. in Churchill, cited by Robertson. V, 12.]

Oviedo. See Fernandez de Oviedo y Valdès, Gonzalo.

Owen, Richard. *A History of British Fossil Mammals and Birds*, London, 1846. VII, 92 bis.
Lectures on the Comparative Anatomy and Physiology of the Invertebrate Animals, 2nd. ed. London, 1855. VII, appendix 44. VIII, 39, 41, 51.
Lectures on the Comparative Anatomy and Physiology of the Vertebrate Animals, delivered at the Royal College of Surgeons of England in 1844 and 1846. Part I, Fishes, London, 1846. VIII, 5, 39, 40, 54, 77.
'Notes on the Anatomy of the Nubian Giraffe.' *Zool. Soc. Lond., Proc.*, 6 (1838), 6–15. IV, 33.

'Osteological Contributions to the Natural History of the Chimpanzees (Troglodytes) and Orangs (Pithecus) No. VI. . .' (Read Dec. 9th, 1856), *Zool. Soc. Lond. Trans.*, 4 (1862), 165–78. VII, 58.

Resumé of 'Report on the Missourium now exhibiting at the Egyptian Hall, with an Inquiry into the Claims of the Tetracaulodon to Generic Distinction.' *Geol. Soc. Lond., Proc.*, 3 (1842), 689–95. VII, 38.

Owen, William Fitzwilliam. *Narrative of Voyages to Explore the Shores of Africa, Arabia, and Madagascar, performed in H.M. ships, Leven and Barracouta, under the Direction of Captain W. F. W. Owen R.N., by command of the Lords Commissioners of the Admiralty,* 2 vols. London, 1833. V, 22.

Paley, William. *Natural Theology; or Evidences of the Existence and Attributes of the Deity, collected from the Appearances of Nature.* London, 1802 [ed.?]. VIII, 88.

Pallas, – . 'Mémoire sur la variation des animaux; première partie.' *Acad. sci. Petro. Acta, Histoire de 1780*, part 2 (1784), 69–102. IX, 97, 99.

Pallas, Peter Simon. *Travels through the Southern Provinces of the Russian Empire,* trans. from the German, 2 vols. London, 1802–3. III, 87. V, 55 b.

Peabody, William B. O. 'A Report on the Birds of Massachusetts made to the Legislature in the Session of 1838–9.' *Boston J. Nat. Hist.*, 3 (1840), 65–266. X, 64, 68, 70, 74.

Pennant, Thomas. *History of Quadrupeds.* 3rd. ed., 2 vols, London, 1793. IX, 81, 90.

Pernety, Antoine Joseph. *Journal historique d'un voyage fait aux îles Malouines en 1763 & 1764 pour les reconnoître, & y former un établissement; et de deux voyages au détroit de Magellan avec une relation sur les Patagons,* 2 vols. Berlin, 1769. X, 58.

Persoon, Christian H. 'An Account of a remarkable variety of the Beech, Fagus sylvatica.' *Linn. Soc. Lond. Trans.*, 5 (1800), 232–3. IV, 24.

Pictet, F. G. 'On the Writings of Goethe relative to Natural History.' *Ann. Nat. Hist.*, 2 (1839), 313–22. VII, 45.

Pictet, Francois Jules. *Traité de paléontologie ou histoire naturelle des animaux fossiles considérés dans leurs rapport zoologiques et géologiques,* 2nd ed., 4 vols. Paris, 1835–7. X, 92.

Traité élémentaire de Paléontologie, ou histoire naturelle des animaux fossiles considérés dans leur rapports zoologiques et géologiques. 4 tom. Genève, 1844–6. VIII, 69.

Pierce, James. ['A Memoir on the Catskill Mountains, with Notices of their Mineralogy and Geology.' *Amer. J. Sci.*, 6 (1823), 86–97.] VI, 14.

Poeppig, Eduard Friedrich. *Reise in Chile, Peru, und auf dem Amazonen strome, während der Jahre 1827–1832,* 2 vols. Leipzig, 1835–6. X, 47. XI, 26.

Prescott, William Hickling. [*History of the Conquest of Peru. With a preliminary view of the civilization of the Incas.* 1st ed., 2 vols. New York, 1847.] III, 1.

Prévost, – . 'Sur les moeurs du Coucou d'Europe.' *Institut.*, 2 (1834), 418–19. X, 81.

Purdie, William. 'Journal of a Botanical Mission to the West Indies in 1843–4.' *J. Bot.*, 3 (1844), 501–33, etc. XI, 28.

Puvis, Marc Antoine. *De la Dégénération et de l'Extinction des variétés de Végétaux propagés par les grettes, boutures, tubercules, etc., et de la création des variétés nouvelles par les croisemens et les semis.* Paris, 1837. III, 45. IV, A39.

Quatrefages de Bréau, Armand de. 'Les métamorphoses.' *Extrait de la Revue des deux mondes*, 2 (1855), pp. 90–116, 275–314, 3 (1856), pp. 496–519, 859–83, 4 (1856), pp. 55–82. III, 20. IV, 2.

Ramusio, Giovanni Battista. *Navigationi et Viaggi*...3 vols. Venice, 1550, 1559, 1556, etc., cited by Robertson. V, 12.

Raspail, François Vincent. 'Classification générale des Graminées, fondée sur l'étude physiologique des caractères de cette famille.' Seconde partie, *Ann. sci. nat.*, 5 (1825), 287–311; 433–60. IV, 28.

Regel, E. 'Nouvelles variétés d'Achimènes et quelques faits relatifs aux hybrides.' *Journal de la société impériale et centrale d'horticulture*, 1 (1855), 251–2. IX, 120 A.

Reinwardt, G. C. 'The Vegetation of the Indian Archipelago.' (Freely translated and adapted from the German), *Hort. Soc. Lond. J.*, 4 (1849) 227–36, 'Contributions to a History of the Relation between Climate and Vegetation in various Parts of the Globe. No. 10'. XI, 18.

Rengger, Johann Rudolf. *Naturgeschichte der Säugethiere von Paraguay*. Basel, 1830. III, 76, 83. V, 12, 19, 63. IX, 102. X, 11.

Richard, Achille. 'Observations sur les prétendus bulbilles qui se développent dans l'intérieur des capsules de quelques espèces de *Crinum*,' *Ann. sci. nat.*, 2 (1824); 12–16. VII, 49.

Richardson, Sir John. *Fauna Boreali-Americana; or the Zoology of the Northern Parts of British America*. 4 parts, London, 1829–37. IV, 31, 32, 39 a, 51. V, 29, 61. VI, 45. VII, 58, 90, 101. VIII, 6, 8. IX, 100. X, 71, 82.

'Ichthyology.' *Encyclopedia Britannica*, 8th ed., XII (Edinburgh, 1856), p. 219. VII, 32.

'Report on the Ichthyology of the Seas of China and Japan.' *Brit. Ass. Advanc. Sci. Rep. (for 1845 meeting)*, 15 (1846), 187–320. XI, 9 A, 34.1.

'Report on North American Zoology.' *Brit. Ass. Advanc. Sci. Rep. (for 1836 Meeting)*, vol. 5 (1837), 121–224. XI, 7, 11 v.

Richardson, John M. D. 'Zoological Remarks' Appendix I, pp. 475–522 in George Back's *Narrative of the Arctic Land*...London, 1836. XI, 11 v.

Ritter, Carl. *Die Erdkunde*...2nd ed., 19 vols., Berlin, 1822–59. IX, 90 bis.

Robert, Eugene. 'Nests of the Hirundo riparia.' *Ann. Mag. Nat. Hist.*, 8 (1842), 476. X, 64.

Robertson, William. *The History of America*, 2 vols. London, 1777, etc. V, 12.

Robson, J. 'The Advantages of Having Seeds and Bulbs from Distant Places.' *Cottage Gdnr.*, 16 (1856), 186–7. III, 11.

Ross, James Clark. Societies: geographical society, June 22nd, *Athen.*, (1846), p. 656. XI, 46.

A Voyage of Discovery and Research in the Southern and Antarctic Regions during the Years 1839–43. 2 vols. London: Murray, 1847. X, 36. XI, appendix, 4.

'Zoology', pp. 91–120 of Appendix of William Edw. Parry, *Journal of a Third Voyage for the Discovery of a North-West Passage*...*1824–5, in his Majesty's ships Hecla and Fury*. London, 1826. XI appendix, note F.

'Zoology', pp. 189–206 in Appendix of Wm. Edward Parry, *Narrative of an Attempt to reach the North Pole*...*MDCCCXXVII*. London, 1828. XI appendix, note F.

Rothof, Lorens Wolt. ['Försök Gjoide vid Mossais upodling.' *K. [Swenska] Wetenskaps Academiens Handlingar*, 22 (1761), 50–5. Translated as: 'Versuche sumpfichte Gegenden zu verbessern.' *Akad. Wiss. Berlin,*

Abh., 23 (1764), 51–5. Darwin read an English translation in manuscript of this article. See entry under *Acts of Swedish Academy.*] v, 43.

Roughsedge, E. 'Battle with Wild Elephants.' (from the *Bombay Courier*), *Athen.*, (1840), 238. x, 110 a.

Roulin, F. 'Recherches sur quelques changemens observés dans les animaux domestiques transportés de l'ancien dans le nouveau continent.' *Mémoires présentés par divers savans à l'académie royale des sciences de l'institut de France, et imprimés par son ordre. Sciences mathématiques et physiques.* 6 (1835), 319–52. vii, 109, 113. x, 36, 38.

Rousseau, Lucien and Louis Vilmorin. 'Experiments on Sowing a Mixture of Varieties of Wheat (From the *Journal d'Agriculture Pratique*).' *Gdnrs. Chron.*, (1856), 859. vi, 26, 261.

Royle, John Forbes. *Essay on the Productive Resources of India.* London, 1840. vii, 13.

Illustrations of the Botany and other Branches of the Natural History of the Himalayan Mountains, and of the Flora of Cashmere, 2 vols. London, 1833–40. xi, 24, 30.

Sabine, Edward. 'Mammalia.' pp. clxxxiii–cxcii in [Parry, William Edward], *A Supplement to the Appendix of Captain Parry's Voyage for the Discovery of a North-West Passage in the Years 1819–20, Containing an Account of the Subjects of Natural History.* London, 1824. ix, 78. xi, 47.

Sabine, Joseph. 'Account of a newly produced Hybrid Passiflora.' *Hort. Soc. Lond. Trans.*, 4 (1822), 258–68. ix, 113.

Sageret, Augustin. *Pomologie physiologique, ou traité du perfectionnement de la fructification...* Paris, 1830. ix, 67, 68.

Saint-Ange, Martin. 'Mémoire sur les vices de conformation du rein, et sur les variétés qu'il présente dans sa structure chez les mammifères, et dans ses formes chez les oiseaux.' *Ann. sci. nat.*, 19 (1830), 306–33. vii, 34 bis.

Saint-Hilaire, Auguste de. 'Comparaison des genres Buttneria et Commersonia.' *Ann. sci. nat.*, 6 (1825), 134–8. viii, 43.

(Prouvençal de Saint-Hilaire, Augustin François César) *Leçons de Botanique, comprenant principalement la morphologie végétale, la terminologie, la botanique comparée, etc.* Paris, 1841. iii, 41, 48. vii, 48, 79, 96. viii, 35, 79.

'Monographie des Genres Sauvagesia et Lavradia.' *Museum Hist. Nat. Mém.*, 11 (1824), 11–68. iv, 26. vii, 97.

'Observations sur plusieurs genres de la famille des Salicariées.' *Ann. sci. nat.*, 2 ser. Bot., 1 (1834), 332–6. vii, 48.

'Premier mémoire sur le gunobase. Du gynobase considéré dans les polypétales.' *Museum Hist. Nat. Mém.*, 10 (1823), 129–64. iv, 25. viii, 43.

'Suite de la Description, Des principales Espèces nouvelles de la Flore du Brèsil cites dans le premier mémoire sur le Gynobase, *Museum Hist. Nat. Mém.*, 10 (1823), 356–77. iv, 26.

Saint John, Charles. *Short Sketches of the Wild Sports and Natural History of the Highlands.* London, 1846. iv, 50. vi, 13. x, 34, 51a.

A Tour in Sutherlandshire, with extracts from the field-books of a sportsman and naturalist, 2 vols., London, Murray, 1849. vi, 4.

Salter, T. Bell. 'A Descriptive Table of British Brambles.' *Bot. Gaz.*, 2 (1850), 113–31. vii, 91.

'On the Fertility of certain Hybrids.' *Phytologist*, 4 (1852), 737–42. iv, 15.

Savage, Thomas S. 'On the Habits of the "Drivers" or Visiting Ants of West Africa.' *Ent. Soc. Lond. Trans.*, 5 (1847), 1–15. viii, 65.

Savi, P. 'Notice sur le nid du Becquemouche (Sylvia cisticola Temminck), et observations sur les habitudes naturelles de cet Oiseau.' *Ann. sci. nat.*, 2 (1824), 126–8. x, 76.

Schlagintweit, Hermann and Robert. 'Notes on Some of the Animals of Tibet and India.' *Brit. Ass. Advanc. Sci., Rep. for 1857 (1858), Misc. communications*, pp. 106–8. ix, 91.

Schlegel, Hermann. *Essay on the Physiognomy of Serpents*, trans. by Thomas S. Traill, Edinburgh, 1843. Appendix, i, 40. vii, 34 bis, 60 bis, 102. viii, 47, 48.

Schomburgk, Robert Hermann. 'Visit to the Sources of the Takuty in British Guiana, in the year 1842.' *Geogr. Soc. Lond. J.*, 13 (1843), 18–75. iii, 81.

Scoresby, William, jr. *An Account of the Arctic Regions, with a history and description of the northern whale fishery.* 2 vols. Edinburgh, 1820. vii, 58.

Scrope, William. *The Art of Deerstalking; Illustrated by a Narrative of a few Days' Sport in the Forest of Atholl, with some Account of the Nature and Habits of Red Deer, and a short Description of the Scottish Forests; Legends; Superstitions; Stories; of Poachers and Freebooters, &c. &c.* 2nd ed. London, 1839. iii, 4. iv, 42. x, 49, 110.

Days and Nights of Salmon Fishing in the Tweed; with a short account of the Natural History and Habits of the Salmon, instructions to sportsmen, anecdotes, etc. London: Murray, 1843. x, 54.

Sebright, Sir John Saunders. [*The Art of Improving the Breeds of Domestic Animals.* In a letter addressed to the Right Hon. Sir Joseph Banks, K.B. London, 1809.] iii, 2 bis.

Sebright, John. *Observations upon the Instinct of Animals.* London, 1836. x, 34, 38, 41, 43.

Seeman, Berthold. *Narrative of the Voyage of H.M.S. Herald during the Years 1845–51 under the command of Captain Henry Kellett, R.N., C.B., being a Circumnavigation of the Globe, and Three Cruizes to the Arctic Regions in Search of Sir John Franklin,* 2 vols. London, 1853. xi, 25.

Selys-Longchamps, Edm. de. 'Récapitulation des hybrides observés dans la famille des Anatidées.' *Acad. sci. Belg. Bull.*, 12 part 2 (1845), 335–55. ix, 78, 85.

Seringe, Nicolas Charles. *Flore des jardins et des grandes cultures ou description des plantes de jardins, d'orangeries et des grandes cultures, leur multiplication, l'époque de leur fleuraison et de leur fructification et leur emploi,* 3 vols. Paris, 1845–9. ix, 66.

Sheppard, Revett. 'Observations.' [Extracts from the Minute Book] *Linn. Soc. Lond. Trans.*, 14 (1825), 587–8. iv, 42.

Sheppard, Revett and William Whitear. 'A Catalogue of the Norfolk and Suffolk Birds; with Remarks.' *Linn. Soc. Lond. Trans.*, 15 (1827), 1–62. iv, 30. v, 64. x, 75, 76.

Sherwill, Major W. S. reported by [E. Blyth] in 'Natural History Notices.' *India Sport. Rev.*, N.S. 2 (1857), 241. ix, 88c.

Shuckard, William Edward. *Essay on the Indigenous Fossorial Hymenoptera, comprising a Description of all the British Species of Burrowing Sand Wasps contained in the metropolitan collections.* London, 1837. vii, 103.

S[huckard?]. 'Doubtful Identity of Miscus campestris and Ammophila Sabulosa.' *Ann. Mag. Nat. Hist.*, 7 (1841), 526. ix, 79.

Sidebotham, Joseph. 'Experiments on the Specific Identity of the Cowslip and Primrose.' *Phytologist*, 3 (1849), 703–5. iv, 72.

Siebold, Carl Theodor Ernst von. *On A True Parthenogenesis in Moths and Bees; a contribution to the history of reproduction in animals*, trans. William S. Dallas. London, 1857. III, 25. VIII, 59, 73.

'Ueber die Sexualität der Muschelthiere.' *Arch. Naturgesch.*, 3–1 (1837), 51. III, 22.

Siemuszowa-Pietruski, Stanisl. Const. v. 'Beobachtungen über den Dachs.' *Arch. Naturgesch.*, vol. 3, Part I, (1837), 160–2. III, 76.

Sinclair, George. 'On Cultivating a Collection of Grasses in Pleasure-grounds or Flower-Gardens, and on the Utility of Studying the Gramineae.' *Gdnrs. Mag.*, 1 (1826), 112–16. VI, 26 e.

Sleeman, Sir William Henry. *Rambles and Recollections of an Indian Official*, 2 vols. London, 1844. III, 77.

Smeathman, Henry. 'Some Account of the Termites, Which are Found in Africa and Other Hot Climates.' *Phil. Trans.*, 71 (1781), 139–92. V, 30.

Smee, Walter. 'On the Maneless Lion of Guzerat.' *Zool. Soc. Lond. Trans.*, 1 (1835), 165–74. IV, 41.

Smellie, William trans. & ed., *Natural History, general and particular* by the Count de Buffon. 3rd ed., 9 vols. London, 1791. IX, 83.

Smith, Andrew. *Illustrations of the Zoology of South Africa; in the Years 1834–6*, 4 vols. London, 1849. IV, 33.

Smith, Charles Hamilton. *The Natural History of Dogs. Canidae or Genus Canis of Authors*...2 vols. London, 1839, 1840. X, 34, 43.

The Natural History of Horses. The Equidae or Genus Equus of Authors. Edinburgh, 1841. VII, 105 A, 108.

Smith, Frederick. *Catalogue of British Hymenoptera in the Collection of the British Museum, Part I, Apidae—Bees*. London, 1855. VIII, 72. X, 79, 85, 93.

'Descriptions of the British Wasps.' *Zoologist*, 1 (1843), 161–71. VIII, 59.

'Descriptions of some Species of Brazilian Ants belonging to the Genera Pseudomyrma, Eciton and Myrmica (with Observations on their Economy by Mr. H. W. Bates).' *Ent. Soc. Lond. Trans.*, N.S. 3 (1855), 56–69. VIII, 64, 74.

'Essay on the Genera and Species of British Formicidae.' *Ent. Soc. Lond. Trans.*, N.S. 3 (1855), 95–135. VIII, 64, 71, 73. X, 89, 98 a.

'Monograph of the Genus Cryptocerus, belonging to the Group Cryptoceridae—Family Myrmicidae—Division Hymenoptera Heterogyna.' *Ent. Soc. Lond. Trans.*, 2 (1854), 213–28. VIII, 64.

'Notes and Observations on the Aculeate Hymenoptera.' *Entomologists Annual* for 1857, 27–38. VII, 68.

Smith, James Edward. *The English Flora*, 4 vols. London, 1824–8. IV, 45.

Smith, Sydney. '*Works*.' the edition Darwin used has not been identified, but his information comes from Smith's first essay on 'Botany Bay', *Edinb. Rev.*, 32 (1819), 24–48. V, 12.

Somerville, Alexander. *The Autobiography of a Working Man*. London, 1848. X, 31.

Sonnerat, Pierre. *Voyage aux Indes Orientales et à la Chine, fait depuis 1774 jusq'en 1781; dans lequel on traite des moeurs, de la religion, des sciences et des arts des Indiens, des Chinois, des Pégouins et des Madegasses; Suivi des observations sur le Cap de Bonne-Esperance, les isles de France et de Bourbon, les Maldives, Ceylon, Malacca, les Philippines et les Moluques*, 2 vols. Paris, 1782. V, 13.

Spallanzani, Lazaro. *Dissertations relative to natural history of Animals and*

Vegetables, from the Italian trans. by T. Beddoes, 2 vols. London, 1784. III, 24.

Spence, William. *Address delivered at the Anniversary Meeting of the Entomological Society of London.* London, 1848. x, 109.

Spicer, John W. G. 'Note on Hybrid Gallinaceous Birds.' *Zoologist*, 12 (1854), 4294–6. IX, 79.

Spratt, Thomas A. B. also Forbes, Edward. *Travels in Lycia, and the Cibyratis, in company with the late Rev. E. T. Daniell*, 2 vols. London: John van Voorst, 1847. XI, 28.

Sprengel, Christian Konrad. *Das entdeckte Geheimniss der Natur im Bau und in der Befruchtung der Blumen.* Berlin, 1793. III, 26, 26 bis, 35, 39*, 43, 49, 54. V, 62. VII, 38. IX, 34.

Standish and Noble. 'A Chapter in the History of Hybrid Rhododendrons.' *Hort. Soc. Lond. J.*, 5 (1850), 271–5. IX, 62.

Stanley, Edward Smith, 13th Earl of Derby. *A [sale] Catalogue of the Menagerie and Aviary at Knowsley formed by the late [i.e. 12th] Earl of Derby...* Liverpool, 1851. IX, 94.

Stark, James. 'On the Existence of an Electrical Apparatus in the Flapper Skate and other Rays.' *R. Soc. Edinb. Proc.*, 2 (1844–45), 1–3. VIII, 54.

Steenstrup, Johan Japetus Smith. *On the Alternation of Generations; or, the Propagation and Development of Animals through Alternate Generations: a peculiar form of fostering the young in the lower classes of animals.* trans. from the German version of C. H. Lorenzen by G. Busk, London, Ray Society, 1845. VI, 7 a.

*Untersuchungen über das Vorkommen des Hermaphroditismus in der Natur: ein naturhistorischer Versuch...*Greifswald, 1846. III, 23.

Stenhouse, John. 'Examination of the proximate Principles of some of the Lichens.' *Phil. Trans.* (1848), 63–89. VIII, 53.

Stephens, James Francis. *Illustrations of British Entomology; or, A Synopsis of indigenous insects...Mandibulata, 7 vols, Haustellata, 4 vols.* London, 1827–46. VII, 66.

Stillingfleet, Benjamin, See Linné. *Amoenitates Academicae.*

Sturm, K. C. G. *Ueber Raçen, Kreuzungen und Veredlung der landwirthschaftlichen Hausthiere.* Elberfeld, 1825. IX, 132. X, 41, 46.

Sulivan, Bartholomew James. MSS. Letters in the Darwin Papers at the University Library, Cambridge. V, 20. VI, 44–5.

Sundevall, Carl J. Trans. by H. E. Strickland, 'The Birds of Calcutta.' *Ann. Mag. Nat. Hist.*, 19 (1847), 164–73, etc. IX, 97.

Sutherland, Peter C. *Journal of a Voyage in Baffin's Bay and Barrow Straits, in the Years 1850–1851, performed by H.M. Ships 'Lady Franklin' and 'Sophia', under the command of Mr. William Penny, in search of the missing crews of H.M. ships 'Erebus' and 'Terror': with a narrative of sledge excursions on the ice of Wellington Channel: and observations on the Natural History and Physical Features of the Countries and Frozen Seas visited*, 2 vols. London, 1852. XI, 11v.

Swainson, William and John Richardson. *Fauna Boreali-Americana...Part 2, Birds*, London, 1831. VII, 101.

'Swedish Acts'. i.e. K. [Swenska] Academiens Handlingar. See entry under 'Acts of Swedish Academy'.

Sykes, William Henry. *Catalogue of the Birds observed in Dukhun, East Indies.* London, 1832. VII, 17.

Tankerville, Charles Augustus Bennet, 5th Earl of. Letter quoted in J.

Hindmarsh, 'On the Wild Cattle of Chillingham Par.' *Brit. Ass. Advanc. Sci. Rep.*, 7 (1839), 100–4. III, 4. [Also published in *Ann. Nat. Hist.*, 2 (1839), 277.]

Tausch, J F. 'Botanische Beobachtungen.' *Flora*. . .1833, vol. 1, pp. 225–30. IX, 34.

'Classification des Ombellifères.' *Ann. sci. nat.*, 2 ser. Bot. 4 (1835), 41–8. VII, 38 b.

Taylor, Jeremy. [Quoted from his *Holy Living*, ch. II, 5, 'Of Modesty', pp. 155–6 in Basil Montagu, ed. *Selections from the Works of Taylor, Hooker, Barrow, South, Latimer, Brown, Milton and Bacon.* ? 3rd ed. London, 1829.] v, 54.

Taylor, Samuel. 'The Thick and Thin Sowing Discussion.' *Gdnrs. Chron.*, 1857, 178–9. VI, 26 f.

Teesdale, Robert. 'A Supplement to the Plantae Eboracenses printed in the second volume of these Transactions.' *Linn. Soc. Lond. Trans.*, 5 (1800) 36–95. IV, 67.

Tegetmeier, William Bernard. *The Poultry Book: including Pigeons and Rabbits (by Harrison Weir)*, (London, 1856–7). IX, 80, 136.

Temminck, Coenraad Jacob. *Coup d'oeil général sur les possessions Néerlandaises dans l'Inde Archipélagique*, 3 vols. Leiden, 1846–9. XI, 18.

Coup d'Oeil sur la faune des îles de la Sonde et de l'empire du Japon. Discours préliminaire destiné à servir d'introduction à la faune du Japon, in Ph. Fr. Siebold, *Fauna Japonica*. Leyden, 1833. XI, 18.

Histoire naturelle générale des pigeons et des gallinacés, 3 vols. Amsterdam and Paris, 1813–15. IX, 81, 88, 88 a, 88 c, 95, 126.

Manuel d'ornithologie, 2nd ed., part 3. Paris, 1835. IV, 53.

Thompson, Edward. *The Passions of Animals*. London, 1851. x, 52 a, 60.

Thompson, William. 'The Birds of Ireland.' *Ann. Mag. Nat. Hist.*, 8 (1841), 273–88, 11 (1843), 283–90, etc. x, 74, 81.

'Contributions to the Natural History of Ireland.' *Mag. Zool. Bot.*, 2 (1838), 427–40. x, 71.

The Natural History of Ireland, 4 vols. London, 1849. IV, 56. VII, 102. x, 50, 51, 54, 60, 70, 71, 72, 73, 75, 100.

Thouin, A. 'Des exemples de toutes les sortes de greffes. . .' *Museum Hist. Nat. Ann.*, 16 (1810), 209–39. IX, 66.

Thuret, Gustave. 'Recherches sur la fécondation des fucacées, suivies d'observations sur les anthéridies des algues.' *Ann. sci. nat.* 4th ser. Bot 2 (1854), 197–214; 3 (1855), 5–28. IX, 52, 71.

Todd, Robert Bentley. *The Cyclopaedia of Anatomy and Physiology*, 5 vols. London, 1836–59. x, 7.

Torrey, John and Asa Gray. *Flora of North America*, vol. II, part I. New York, 1841. IV, 14.

Tschudi, Friedrich von. *Sketches of Nature in the Alps*. London, 1856. x, 55.

Turczaninow, Nicolaus. 'Flora Baicalensi-Dahurica seu descriptio plantarum in regionibus cis- et Transbaicalensibus atque in Dahuria sponte nascentium.' *Soc. Nat. Moscou, Bull.*, 15 (1842), 3–105, etc. XI, 10.

Valenciennes, A. 'Nouvelles Recherches sur l'organe électrique du Malaptérure électrique. Lacép (Silurus electricus Linn).' *Museum Hist. Nat. Arch.*, 2 (1841), 43–61. VIII, 54.

Vaucher, Jean P. E. 'Mémoire sur le sève d'Août et sur les divers modes de développement des arbres.' *Soc. Phys. Genève, Mém.*, 1 (1822), 289–308. IV, 24. VII, 95.

Vieillot, Louis Pierre and P. Oudart. *La Galerie des Oiseaux*, 4 vols. Paris, 1825–34. IV, 53.

Vigors, Nicholas Aylward. 'Observations on the Natural Affinities that connect the Orders and Families of Birds.' *Linn. Soc. Lond., Trans.*, 14 (1825), 395–517. VIII, 19.

Visiani, Roberto de. *Flora Dalmatica: sive enumeratio stirpium vascularium quas hactenus in Dalmatia lectas et sibi observatas descripsit digessit rariorumque iconibus illustravit.* Leipzig, 1842–52. IV, supp. c.

Wagler, Johann Georg. *Systema Avium, Pars prima.* Stuttgart and Tübingen, 1827. IV, 53.

Wagner, Andr. 'Birds.' pp. 68–90 in *Reports on the Progress of Zoology and Botany, 1841, 42.* Edinburgh: Ray. Soc., 1845. IX, 78.

'Progress of Zoology in 1843—"Ornithology".' pp. 54–84 in *Reports on Zoology for 1843, 1844.* London: Ray. Soc., 1847. IX, 88.

'Progress of Zoology in 1844—"Birds".' pp. 277–300 in *Reports on Zoology for 1843, 44.* London: Ray. Soc., 1847. IX, 88 c.

Walker, Alexander. *Intermarriage.* London, 1838. IX, 102.

Walker, [John]. 'On the Cattle and Corn of the Highlands.' *Highl'd soc. Scotl. Trans.*, 2 (1803), 164–203. III, 11.

Wallace, Alfred Russel. *A Narrative of Travels on the Amazon and Rio Negro.* London, 1853. VIII, 87, 88.

Walton, William. *The Alpaca, its Naturalization in the British Isles considered as a National Benefit, and as an Object of Immediate Utility to the Farmer and Manufacturer.* Edinburgh and London, 1844. IX, 103.

Waterhouse, George Robert. 'Contributions to the entomology of the southern portions of South America.' *Ann. Nat. Hist.*, 13 (1844), 41–5. IV, 39 a.

'Descriptions of Coleopterous Insects collected by Charles Darwin, Esq., in the Galapago Islands.' *Ann. Mag. Nat. Hist.*, 16 (1845), 19–41. VI, 26 h.

A Natural History of the Mammalia, 2 vols. London, 1846–8. VII, 52, 58.

The natural history of Marsupialia or pouched animals. Edinburgh, 1841. VII, 101.

Waterton, Charles. *Essays on Natural History, chiefly Ornithology.* 1st Series: 2nd ed., London, 1838. 2nd Series: London, 1844. V, 25, 26. IX, 79, 84. X, 41, 43.

Waterton, Charles. *Wanderings in South America, the North-West of the United States, and the Antilles in the Years 1812, 1816, 1820, 1824*, 4th ed., London, 1839. V, 19.

Watson, Hewett Cottrell. *Cybele Britannica; or British Plants and their Geographical Relations*, vols. 1–3. London, 1847–52. III, 87. IV, 69. XI, 4.

'Explanations of some Specimens for Distribution by the Botanical Society of London in 1848.' *Phytologist*, 3 (1848), 38–48. IV, 70, 72.

'Further Experiments Bearing upon the Specific Identity of the Cowslip and the Primrose.' *Phytologist*, 2 (1847?), 852–4. IV, 72.

'Notes on the Affinity between Lysimachia nemorum (Linn.) and Lysimachia azorica (Hornem.).' *Phytologist*, 2 (1847), 975–9. VII, 12.

'Note on the Bardfield and Claygate Oxlips.' *Phytologist*, 1 (1844), 1001–2. IV, 72.

'On the Theory of "Progressive Development", applied in Explanation of the Origin and Transmutation of Species.' *Phytologist*, 2 (1845), 108–13, 140–7, 161–8, 225–8. IV, 15, 35, 64, 65.

Remarks on the Geographical Distribution of British Plants; chiefly in connection with latitude, elevation and climate, London, 1835. VI, 63. XI, 7–8.

'Report of an Experiment which Bears upon the Specific Identity of the Cowslip and Primrose.' *Phytologist*, 2 (1845), 217–19. IV, 12, 73.

'Supplementary Notes on the Botany of the Azores.' *Lond. J. Bot.*, ser. 2, 6 (1847), 380–97. IV, 65.

Watson, Hewett Cottrell and Syme, J. T. *The London Catalogue of British Plants*, 4th ed. London, 1853. III, 47a. IV, supp. b.

Webb, Philip Barker and Berthelot, Sabin. *Histoire naturelle des Îles Canaries*, 3 tomes. Paris, 1836–50. IV, supp. e. VI, 59.

Weissenborn, W. 'Great Migration of Dragon-flies observed in Germany.' *Mag. Nat. Hist.*, N.S. 3 (1839), 516–18. X, 109.

Wesmael, –. 'Sur une nouvelle espèce de fourmi du Mexique.' *Acad. sci. Belg.*, *Bull.*, 5 (1838), 766–71. VIII, 65.

Westwood, John Obadiah. *Addresses to the Entomological Society, 1851–53.* London, 1851–3. VII, 54.

'Description of the "Driver" Ants, described in the preceding Article.' *Ent. Soc. Lond. Trans.*, 5 (1847), 16–18. VIII, 72.

'Description of a Hybrid Smerinthus, with Remarks on Hybridism in General.' *Ent. Soc. Lond. Trans.*, 3 (1842), 195–202. IX, 79.

'Further Notices of the British Parasitic Hymenopterous Insects; together with the Transactions of a Fly with a Long Tail, observed by Mr. E. W. Lewis, and Additional Observations.' *Mag. Nat. Hist.*, 6 (1833), 414–21. VIII, 50.

An introduction to the Modern Classification of Insects, 2 vols, London, 1839–40. IV, 34. VII, 26, 65, 68, 119n. VIII, 20, 49, 50, 64, 73, 74, 90. X, 84, 86.

'The Long-horned Tortrix.' *Gdnrs. Chron.*, (1852), 261. X, 98a.

'Observations on the Genus *Typhlopone*, with descriptions of several exotic species of Ants.' *Ann. Mag. Nat. Hist.*, 6 (1840), 81–9. VIII, 64.

'Some Account of the Habits of an East Indian Species of Butterfly, belonging to the Genus *Thecla*.' *Ent. Soc. Lond. Trans.*, 2 (1837), 1–8. X, 105.

White, Gilbert. *The Natural History of Selborne, by the late Rev. Gilbert White, M.A., Fellow of Oriel College, Oxford. To which are added, The Naturalist's Calendar, Miscellaneous Observations, and Poems. A New Edition with Engravings in two Volumes.* London, 1825.

[*The Natural History and Antiquities of Selborne.* By the Rev. Gilbert White, M.A. With the Naturalist's Calendar; and Miscellaneous observations. Extracted from his papers. A New Edition, with Notes, by Edward Turner Bennett, Esq., F.L.S., &c., Secretary of the Zoological society; and others. London, 1837. VI, 45.

The Natural History of Selborne, by the late Rev. Gilbert White, M.A., A new edition, with notes by the Rev. Leonard Jenyns, M.A., F.L.S. London, 1843. pp. xvi, 398. V, 17, 26, 30. X, 46, 71, 72, 73, 107.

White, Walter. *A Londoner's Walk to the Land's End, and a Trip to the Scilly Isles.* London, 1856. X, 12.

Whitear, William. 'A Catalogue of the Norfolk and Suffolk Birds; with Remarks.' *Linn. Soc. Lond. Trans.*, part I, 15 (1826), 1–62. V, 64.

Wiegmann, Arend, Friedrich August. 'Erzeugung mehrerer cryptogamischen Gewächs aus der Priestleyischen grünen Materie. Abarten von Veronica longifolia. Saamen von Polygonum viviparum. Hyacinthus. Dianthus, Anagallis arvensis und coerulea.' *Flora*, (1821 vol. 1), 8–15. IV, 67.

*Ueber die Bastarderzeugung im Pflanzenreiche...*Braunschweig, 1828. III, 26, 30, 60, 92–3. IX, 116, 118, 125.

Wight, Robert. 'Statistical Observations on the Vurragherries, or Pulney Mountains.' *Madras J. Lit. Sci.*, 5 (1837), 280–9. xi, 18.

Wilcke, H. C. D. See Linné.

Wilkinson, John. *Remarks on the Improvement of Cattle, etc., in a Letter to Sir John Saunders Sebright*...Nottingham, 1820. ix, 104. [Not seen. The only listing found for this is in the Rothamsted Experimental Station, Harpenden, *Library Catalogue of Printed Books and Pamphlets on Agriculture published between 1471 and 1840*, compiled by Mary S. Aslin. 2nd ed. 1940.]

Willdenow, C. L. 'Détermination d'un nouveau genre de plante aquatique, nommé Caulinia; et observations générales sur les plantes aquatiques.' *Acad. Sci., Berlin, Mém., for 1798 (1801), Classe de Philos. Expérimentale*, pp. 78–90. Berlin.
'Determination of a new aquatic vegetable Genus, called Caulinia, with general observations on Water-plants.' *Ann. Bot.*, 2 (1806), 39–51. (Ed. Charles Konig and John Sims). Trans. from *Acad. Sci., Berlin, Mém.*, iii, 47 (4).

Wilson, James. 'Notice of the Occurrence in Scotland of the *Tetrao medius*, shewing that supposed species to be a hybrid.' *Roy. Soc. Edinb. Proc.*, 1 (1832–1844), 395. ix, 88 a.
A Voyage Round the Coast of Scotland and the Isles. 2 vols, Edinburgh, 1842. iv, 43. v, 17, 28.

Wilson, William. 'Further Remarks on the Pollen-Collectors of Campanula, and on the mode of fecundation.' *Lond. J. Bot.*, 7 (1848), 92–7. iii, 42.
'Observations on some British plants particularly with reference to the "English Fauna" of Sir James E. Smith.' *Bot. Misc.* 3 (1833) 109–18 etc. vii, 100 A.

Wollaston, Thomas Vernon. *Catalogue of the Coleopterous Insects of Madeira in the collection of the British Museum*. London, 1857. iv, supp. e.
Insecta Maderensia; being an account of the Insects of the Madeiran Group. London, 1854. iv, 39 b, 61, 62, A 4. vi, 26. vii, 22, 24, 65.
On the Variation of Species, with especial reference to the Insecta; followed by an inquiry into the nature of genera. London, 1856. iv, 8, 16, 34, 40, 61. v, 55 e, 56, 57. vi, 35, 63. vii, 6, 9, 22, 25 bis, 26, 103.

Woods, J. 'On the Genus Atriplex.' *Bot. Gaz.*, 1 (1849), 327–8. viii, 43.

Woodward, Samuel Pickworth. *A Manual of the Mollusca; or a Rudimentary Treatise of Recent and Fossil Shells, 3 parts*. London, 1851–56. xi, 48.

Wrangel, Ferdinand Ludwig von. *Narrative of an Expedition to the Polar Sea in the years 1820, 1821, 1822, and 1823*. Ed. E. Sabine, trans. Mrs. Sabine, London, 1840. v, 21. x, 53.

Yarrell, William. *A History of British Birds*. 3 vols, London, 1839–43. x, 66, 74, 76, 80, 81.
A History of British Fishes. 2 vols, London, 1836. v, 65 A.
'On the structure of the Beak and its Muscles in the Crossbill (Loxia curvirostra).' *Zool. J.*, 4 (1829), 459–65. vii, 60.

Youatt, William. *Cattle; their breeds, management, and diseases*. London, 1834. iii, 3. vi, 26*. viii, 66, 67. ix, 104.
The Dog, London, 1845. x, 49.
'The Intellectual and Moral Faculties of Brutes.' *Veterinarian*, 5 (1832), 275–84. x, 54.
Sheep: their breeds, management, and diseases. London, 1837. (Library of Useful Knowledge). x, 19, 54, 55.

BIBLIOGRAPHY

Zoological Society of London, Council. 'Report of the Council...IV The Garden Establishment' *Reports of the Council and Auditors of the Zoological Society of London, read at the annual general meeting, April 30th, 1855.* London: Taylor & Francis, 1855. IX, 94.

Zuccarini, Joseph Gerhard. 'Analogy between the Flora of Japan and that of the United States.' *Amer. J. Sci.*, 2nd ser., 2 (1846), 135–6. [Report on Zuccarini's work, signed 'A.Gr' (= Asa Gray).] XI, 14.

Anon. Review of his *Monographie der Amerikanischen Oxalis-Arten*, München, 1825. *Flora oder Botanische Zeitung*, 9 (1826), 337–43. III, 87.

['Dr. Siebold, Flora Japonica; Sectio prima, Plantae ornatui vel usui inservientes; digessit Dr. J. G. Zuccarini: fasc. 1–10, Leyden, 1835–9.' *Amer. J. Sci.* 39 (1840), 175–6.] XI, 14.

GUIDES TO THE TEXTS
OF THE LONG
AND THE SHORT VERSIONS

The substantial part of the long version referred to as 'Natural Selection' here transcribed and edited may be read as a pleasure and an education complete in itself. For Darwin built his argument on a considered selection of studies in natural history, breeding, geographical distribution, comparative anatomy and even physiology and philosophy. The text illumines much that is obscure in the Origin, provides cited examples where the Origin has generalisations, and not uncommonly presents views absent in the later work.

Four guides to the text are given:

(1) Darwin's own table of contents (pp. 25–32) relates subject headings and manuscript folio pages serially numbered for each chapter. Though the surviving chapters are not always complete, these headings still provide a useful first approach. Many of the headings are used in the index.

(2) The Bibliography surveys the uses to which Darwin put his sources. Each entry includes the folio page and chapter where the work was referred to in the manuscript.

(3) The Collation represents an attempt to identify any parallels that exist between Natural Selection and the first edition of On the Origin of Species by means of natural selection, or the preservation of favoured races in the struggle for life.* Each page of the first edition is matched with those printed pages of Natural Selection judged to deal with the same material. For readers with other editions the sub-headings Darwin provided for each chapter of the Origin are also given. Some few paragraphs are nearly identical between the texts. More often, a point discussed once in the Origin is referred to repeatedly in Natural Selection. Even when the Origin text closely parallels Natural Selection some facts and hypotheses discussed in one text are not found in the other. A single page of the Origin can be referred to on up to eight pages of Natural Selection. The development of a single sentence in the Origin illustrates the complex relations subsumed by the collation. For example, Darwin on p. 552

---

* Facsimile edited and introduced by Ernst Mayr with a newly compiled index to replace Darwin's original; Harvard University Press, Cambridge, Massachusetts, 1964.

analyses the rich flora of the Cape of Good Hope as listed in the description of Drege's collection and finds 96 European phanerogams and ferns of which roughly one third may have migrated through the highland tropics 'during the cold period'. The Alpine flora of Australia includes several European plants pp. 553–554 and these must have migrated through the Tropics in a like manner to that proposed for the European plants of the Cape of Good Hope p. 554. What does Darwin do with this carefully argued parallel between the floras of the Cape and of Australia? The rudiment in the *Origin* (p. 377) reads:

The tropical plants probably suffered much extinction; how much no one can say; perhaps formerly the tropics supported as many species as we see crowded together at the Cape of Good Hope, and in temperate Australia.

This is more than severe condensation. Yet knowledge of the contexts shows the genesis of the sentence. Other pages cited in the Collation for p. 377 of the *Origin* refer to other parts of that page.

The gross, as well as the fine structures of the *Origin* can be elucidated with the aid of the collation. The first sentence on page 67 of the *Origin* reads:

The face of Nature may be compared to a yielding surface, with ten thousand sharp wedges packed close together and driven inwards by incessant blows, sometimes one wedge being struck, and then another with greater force.

This confusing and ineffective passage is a hurried précis of the splendid metaphor on page 208 of *Natural Selection*:

Nature may be compared to a surface covered with ten-thousand sharp wedges, many of the same shape & many of different shapes representing different species, all packed closely together & all driven in by incessant blows: the blows being far severer at one time than at another; sometimes a wedge of one form & sometimes another being struck; the one driven deeply in forcing out others; with the jar and shock often transmitted very far to other wedges in many lines of direction: beneath the surface we may suppose that there lies a hard layer, fluctuating in its level, & which may represent the minimum amount of food required by each living being, & which layer will be impenetrable by the sharpest wedge.

This metaphor appears on page 135 of Notebook 'D', which was written on 28 September 1838 about the time Darwin read Malthus for the first time, in the following words:

one may say there is a force like a hundred thousand wedges trying to force every kind of adapted structure into the gaps in the oeconomy of nature, or rather forming gaps by throwing out weaker ones.

Darwin later returned to this passage inserting in pencil between the ink writing of the first passage:

The final cause of all this wedging must be to sort out proper structure & adapt it to change, to do that for form which Malthus shows is the final effect (by means however of volition) of this populousness on the energy of man.

The clear importance to Darwin of the concept of wedging would not be suspected from its appearance in the *Origin* if the notebook version and the text of *Natural Selection* had not survived.

(4) *The Index*

Darwin changed the language of biology. But to refer to Darwin's subject matter in the language of 1974 would impose a false imprimatur upon his work. *Natural Selection* was not written as prophecy and it is likely to endure without our contemporary congratulation for the author whenever Darwin's insights seize our attention for so remarkably foreshadowing our own. Thus Darwin's term 'reduction' persists in the index and is chosen in preference to its current nearest equivalent term – 'back-cross'. Usages such as cladogenesis, pleiotropy, and genotype do not appear. It is however a difficult, if not impossible, task fully to re-enter the period when *Natural Selection* was being written and to confine the vocabulary of the index to the usages of 1856 to 1858. The compilers accordingly seek forgiveness should they have perpetuated old and generated new anachronistic usages.

Authors are indexed in full. References, when numerous, are grouped under topics to reflect Darwin's use of the material. For example: Kölreuter believed the offspring of interspecific crosses less fertile than their parents. This empirical law caused Darwin some inconvenience, and in discussing it he emphasises the ten exceptions he was able to glean from Kölreuter's work. See page 391 for such an example.

Animals and plants are indexed by genus and sometimes family. The nomenclature is Darwin's: all land snails are deemed Helix; tapirs have yet to amble from the Pachydermata. References to genera are extensive but not exhaustive, the numberless indignities 'humble-bees' inflict on flowers are recorded under the name of the victim, but not the assailant.

Topics are somewhat selectively indexed. Ironical invocation of the Creator's foresight receives no mention, nor topics so broad that they are the subject of an entire chapter. Smaller topics occur in the index such as: contabescence, colonies, homologies, and under species reference is made to: representative; allied; protean; and dependent.

Subjects associated with one author, or a few, are best approached through the subheading of the author. For instance, the subheadings on hybridisation under Gärtner afford the best approach to the topic; and nearly complete coverage may be achieved if further reference be made to the similar subheadings under Kölreuter and Herbert. Comparisons of the same subhead are often instructive. Thus, one reads under 'Gärtner, Reduction', 'Hybrid continually crossed with parent assumes its characters, 395' while 'Kölreuter, Reduction' reads 'Reduced with pollen of one parent, yet offspring resemble other, 449'. When Darwin quotes a summary of one worker's idea from a secondary source, the entry appears under both names, with the original source indicated in parenthesis under the secondary source.

Conventions used in the index

Because of the considerable detail in the index to be organised so as to be accessible but not to do violence to Darwin's own use of the facts we have tried to follow the conventions given below. These are not the normal rules of indexing but this index aims to deal fairly with a particularly important book.

Capitalisation is used to indicate relationships under subheadings. The uses are shown in the example:

Nesting behaviour: Cuckoo sometimes builds own, 300; Lark feeds young 500 times a day, 401; Pied raven builds in trees, 121; nest larger than in common, 122; Rhea with imperfect, 478.

In this hypothetical example, the pattern is N b: C; L; P; n; R. Lower case n refers that entry to the subject of the immediately preceding capitalised entry, 'Pied raven'. Darwin would have capitalised Pied Raven and Common Raven – we have not done this so that capitalisation can be used as a convention 'a subsumed by immediately preceding B'. Only proper names are capitalised in an entry, in addition to the first word. Generic, but not specific, names are capitalised. Capitalised entries are nearly always arranged in alphabetical order (C, L, P, R, in the example). There are a few exceptions, for instance when a general law is given and followed by lists of examples and exceptions. Thus, 'Variable forms in all species, 200: Exceptions – brachiopods, 110; mites, 55.' A subheading is indicated by a colon: while groupings within a subheading are shown by a dash (–). Semicolons separate different entries, whether or not they refer to the same subject. Commas are used to separate page references to the same fact, or to very nearly the same fact (that is, same subject and predicate):

Bees: Attracted to red flowers, 200; Comb made by young workers, 500; Geometry of hive, 300; Instinct to sting not learned, 455, 475, 466; Visit clover, 123, daisy, 233, 244, pea, 190

Parentheses indicate the common name of a Latin genus or species, or that Darwin has used X's quote of (Y):

Lyell, Sir Charles

    All nature at war (Candolle), 175 n 3

In conclusion, the compilers of the index hope that the reader will experience as much pleasure and profit in its use as they did in drawing it up.

David Kohn
Graham Pawelec
Stan Rachootin
Sydney Smith

COLLATION BETWEEN THE *ORIGIN*
AND *NATURAL SELECTION**

* When several pages are quoted for each single *Origin* page, the order is intended to suggest their importance. Use of single brackets conveys insecurity in the collation, double brackets even greater insecurity. An asterisk is an admission of failure.

INDEX

For notes on the arrangement of entries see pages 632–4

Abipones (South American Indians): Abhor incest, 36 n; Teach horses to pace, 485

Acanthaceae, varieties, 153 table

Acarus (mite), hermaphroditism, 43 n 2

Accentor modularis (hedge sparrow), 184

Acclimatisation, *see* Naturalisation

Acosta, Joaquin

    Glaciers on the Cordillera once lower than now, 545 n 2

Adams, Charles Baker

    Land-shells: Local varieties, 139, and species, 155

Adams, —

    Proportions of allied species of shad altered, 200 n 6

Aesculus pavia, doubled flowers, 86

Agassiz, Louis

    Adopts Morton's term 'primordial forms', 97

    Evidence for glacial epoch, 538

    Large-eyed eels from deep wells same as common eel, 297

    Special creation fills gaps in economy of nature, 581

    Tertiary fauna of Europe and America more alike than at present, 544

Agassiz, Louis and Desor, E.

    Anal orifice of sea urchins highly variable, 112

Ajuga, fertilised by a fly, 55

Alauda (larks): Do not breed in captivity, 79; Variable dispositions, 472, 480 n 1

Alison, W. P.

    Hen uses hot bed to incubate eggs, 501 n 2

    Instinct, 468 n 2; and migration, 492 n 2; Fish migrate so that they may be preyed upon, 520

    Reflex actions approximate habitual, 477 n 3

Allium (onion), Sprengel asserts probable dichogamy, 54

Althaea (hollyhocks): Crosses, 71; Kölreuter, Sprengel assert fertilisation by bees, 65; Varieties, 65

Anagallis (pimpernel): A. arvensis (scarlet) and A. coerulea (blue) considered varieties, 128, 393, as species, 127–8; Crosses, 127 n 6, 128; Production of red and blue varieties from A. collina, 128

Analogy, *see* discussion, pp. 374–85.

Anatidae (waterfowl): Fertility in captivity, 79; Hybrid, 432 n 4

Andersson, Mic. Joh.

    Varieties of Carex, 323–4

Andral, —

    Malformation of early embryo modifies parts subsequently developed, 302 n 2

Andromeda, on mountains of Brazil and Jamaica, 551

Anemone apennina, on Lycian mountains, 552 n 1

Anser (goose): Hybrids, 427 n, 439, 440; Variable tarsi of A. canadensis, 111

Anthemis nobilis, contabescence, 85 n 3

Anthus (pipit): A. pratensis (meadow pipit) is 12 species, according to Brehm, 114; Variable beak, 104 n 3

Anthyllis vulneraria, varies with conditions, 283

Antirrhinum (snapdragon), insects perforate calyx, 43, 57, 476

Apathus, resembles, and parasitises, humble-bees, 509

Apis, *see* Kirby and Spence, Sprengel

661

Hudson, —
  Classification of British flora, 113
Humboldt, Alexander von
  Grasses bordering hot springs show no variation, 285 n
  Monkeys of same species from certain islands more easily tamed, 472
  Thibaudia on the Silla of Caraccas and mountains of New Granada, 551 n 4
Hunt, Consul Carew
  Birds of the Azores: Due to faulty navigation, 493 n 1; None migratory, 494
Hunt, James
  Common × pintail duck, 439
Hunter, John
  Animals disinclined to hybridise, 427; She-wolf must be held during union with dog, 36 n 2
  Bee dies from instinctive use of sting, 380
  Inheritance of behavioural idiosyncrasies, 481
  Jackal-like characters in a dog with one-quarter jackal blood, 489 n 3
  Secondary sexual characters depart from a typical structure of the group, 307 n 2
Hutchinson, W. N.
  Propensities of young pointers, 482 n 3, 483
Hutton, Thomas
  Hybrids, Anser, 439; Fringilla, 438
  Nanina vesicula wide-ranging but unvarying, 285 n
  Tailor bird will use artificial thread, 473 n 5
  Wild × tame goat offspring tends to wildness, 486
Huxley, Thomas Henry
  Development: Criticises theory of Brulle, 276–9, 303; Discussion of Milne Edwards, 277, 279; Mysis, 276; Of vertebrates, 276, 277, 278
  Geographical Distribution: Ascidian Boltenia represented in both polar seas, but not in the tropics
  Homology, Aquiferous vessels of Annelida and tracheae of Insecta, 357; Acoustic nerve in Mysis not homologous to other crustaceans, 353
  Possibility of crossing in invertebrates: Not physically impossible, 45 n 2, 46
    Structural hermaphroditism, but functional separation in oyster and ascidians, 45
  Utilitarianism: Disputes the necessity for a function for every part, 379 n 1
Hyaena, recovery from severe injury, 205
Hybrids and hybridisation, see Gärtner, Herbert, and Kölreuter
Hydra: Separation of sexes, 362; Turned inside out, tissues take up function appropriate to new position, 355
Hyla (tree frog), toes enlarged into sucker discs, 348 n 3
Hymenoptera, see Kirby and Spence, F. Smith, Westwood
Hypericum, on Organ Mountains of Brazil, 551

Ibla (barnacle); Males lack carapace, 306; Proper classification requires knowledge of descent, 96
Ichneumonidae: Larvae avoid vital parts of the grubs on which they feed, 509; Ovipositor can be used for defence, 361
Iridaceae, require insects for fertilisation, 52
Iris, conversion and variability of seedlings, 127
Isatis, conversion, 126
Isolation, Allows formation of new species, 252 et seq.; Rejoined areas, 266
Ixia, Cape form with representative in Mediterranean area, 559
Ixodes (tick), checks cattle in Brazil, 181

Jardine, Sir William
  Describes Madeiran variety of Sylvia articapilla as species, 124
Jenner, Edward
  Dog with one-quarter jackal blood has characters of jackal, 489

Miller, —
  Sonnerat's cock × common pheasant, 435 n 7
Milne, John
  Single record of siskins breeding in captivity, 427 n 3
Milne-Edwards, Henri
  Division of labour: Tends to perfection of every function, 73, 233; Crustacean limbs for swimming and respiration, 355; Nereid worms respire through vascularised areas of limbs, 356; Lower animals use same tissue for nutrition and respiration, 355
  Embryology: More widely two animals differ, the earlier their embryonic similarity ceases, 303; Most important of functional systems appear earliest in the embryo, 275, 303; Huxley comments, 277, 279
  Geographical distribution: Crustacea of temperate S. American coast not found in intervening torrid zone, 556
  Homology: Nature prodigal in variety, niggard in innovation, 354; Branchiae of higher Crustacea not modified from pre-existing part, 360; Transparent cones of compound eyes homologous with crystalline lenses of simple eyes, 351
  Law of economy; law of diversity of products, 374
  Variability of Lacerta, 110
Mimulus, fertilisation by insects, 53, 56, 57
Mitchell, Thomas Lingstone
  Dogs learn attack pattern slowly, 484
Modiolarca trapezina, common to Falkland Islands and Kerguelen's land, 563 n 1
Molina, Juan Ignacio
  Goat × sheep commonly in Chile, 437
  Inheritance of pace in horse, 485 n 2
Molothrus, cuckoo-like nesting habits, but does not eject foster parents' young, 508, 520
Monstrosities: Arrested embryo resembles others of the class, 320; Compensation, 306; Correlation of growth, 299, 302; Equally frequent in hybrids and in pure species, 445; Hybrids not monsters, 445; Resemble normal structures in related form, 318 *et seq.*; Show struggle not incessant, 205; Variable parts most likely to show malformations, 318, 334
Montagu, George
  Goldfinch lived in confinement 23 years, 184 n 3
  Nesting: Birds do not know instinctively period required for incubation, 508 n 1; House-sparrow varies nest to suit conditions, 503
  Phasianus torquatus in Northumberland, 438
Moquin-Tandon, Alfred
  Correlation: Malformation of plant axis affects appended structures, 302 n 3; Malformed flowers more likely away from axis, 320 n 1
  Developmental compensation holds for malformation, 305
  Effects of conditions: Cold lessens colour in hyacinths, 283 n 3; Dryness tends to produce hairs on plants, 283 n 6; Plants more woolly on mountains, 283; Plants near sea acquire fleshy leaves, 283
  Homologous parts show abnormal tendency to cohere, 298 n 3
  Malformation in one plant may be normal in another, 319
  Multiple parts may vary in number and form, 298
  Variation: Position of seeds in Suaeda altissima, 110; Stamens of Solanum dulcamara, 327
Moresby, Captain
  Bees and wasps swarm on Chagos islets, 178
  Tame pigeons of St Pierre and Providence Islands, 495 n 1
Mormodes, flowers eject pollen masses with considerable force, 66
Morren, Ch.
  Excitability of pistil and stigma when brushed by insects: Goodenia, 64; Goldfussia, Stylidium, 54

Pennant, Thomas

Dog-fox × dog, 437; Dog-wolf × Pomeranian bitch, 429 n 5

Pentstemon: Bees perforate calyx, 476; Herbert finds perfect fertility in P. angusti-folium × P. pulchellum, 398

Peplis portula, with and without petals, 325

Peristera (pigeon), differences between Galapagos form and P. brasiliensis, 115 n 1

Pernety, Antoine Joseph

Tame water-fowl of Falklands, 495

Persoon, Christian H.

Variability of Fagus sylvatica, 108

Petrocallis pyrenaica, difficulty interpreting embryo, 398

Petunia, Herbert finds hybrids very fertile, 398

Phanaeus (scarab beetle), Anterior tarsi lost from males, 294

Phaseolus (kidney bean); Bees perforate calyx, 53, 69, 382, 476; Ease of crossing, 69, 70

Phasianus (pheasant): Blending of P. colchicus (common), P. torquatus (ringed) and P. versicolor (Diard's), 432; Fertility of hybrids, 434 n 3, 435 n 1, 438, 440; Diard's × common reduced to common, 432 n 2, 458

Phlox, visited by insects, 55, 476

Phoenicopterus, webbed feet, but seldom even wades, 348

Phyteuma, Does not self-fertilise in the bud (Sprengel), 59

Pictet, F. G.

Carboniferous termites, 370 n 1; Jurassic bees, 514 n 1

Developmental compensation, 304 n 2

Picus (woodpecker): P. campestris feeds exclusively on the ground, 344; P. erythro-cephalus with deformed mandibles, 205–6; P. varius catches flies on the wing, 344 n 2

Pierce, James

Varieties of wolf, 220–1

Pigeon, Breeds of: Breed true only under selection, 316; Correlation does not account for small beak and feet of Tumbler, 300; Divergence, 228; Each prefers own kind, 258; Fertility of crosses, 443; Instincts acquired after domestication, 485; Inter-breeding has ill effects, 36, 37; Intermediate forms, 262; Length of bill decreased and increased in same, 293; Mongrels are intermediate, 454; exception, 455; Normal characters in one are variant for another, 321; Reversion to ancestral form, 322; Sex characters attached to single, 317; Shortfaced birds cannot escape shell, 206, 217

Pimpinella, on Cape Mountains, 552

Pinus sylvestris, status of varieties, 118 n 6

Pisum: Bees suck nectar without moving pistil or stamens, 68; Crosses, Gärtner's, 69; in nature, 69, 72

Pleurotomaria (slit shell), variable, 107

Podostemaceae: Possibility of occasional cross, 72; Self-fertilised underwater, 63

Poeppig, Edward Friedrich

Alpine flora endemic to central Andes, 550 n 3

Puppies of Cuban feral dogs steal when raised domestically, 489

Polygala vulgaris (common milkwort), grows under many conditions, 196, but pre-sents same range of variation, 284

Polygleae, opposite and alternate leaves on same plant, 325

Polygonaceae; Varieties in large and small genera, 153 table, Watson on classification, 104, 126

Pompilus, smooth sting, 381 n 2

Ponera (ant), blind, 368

Prepotency: 284, 417, 450, 456–8; Confounded with sex, 455; "Decided types" usually sterile, 417

Primula: Conversion, 96, 133; Intermediate forms, 130–1; Similarities and differ-ences between P. veris (cowslip), P. vulgaris (primrose) and P. elatior (oxlip), 128

Schlegel, Hermann (*cont.*)

number of vertebrae, 567, and position of spleen, 327; Vertebral processes in Tropidonotus serve as teeth, 361

Variable nasal prominences in Chameleon, 311

Schomburgk, Robert Hermann

Tame ducks and parrots in Guyana do not breed, 79

Scilla bifolia, present on Lycian mountains, 552 n 1

Sciurus cinereus, mates but sterile, in captivity, 77

Scops paradisea (crane), breeds in captivity, 79

Scoresby, William

Transport by icebergs, 562 n 2

Variable tusks of narwhal, 309 n 2

Scrope, William

Crossed dogs with mixed instincts, 490 n 2

Deer: Expel wounded when pursued by dogs, 524 n 2; Size reduced by interbreeding, 37 n 5; Varieties of Red, 117 n 3

Migratory direction-finding in salmon, 493 n 1

Scrophulariaceae: Dichogamy, 47; Irregular flowers need longer development, 303; Varieties in large and small genera, 153 table

Sebright, Sir John

Dogs: Different mode of hunting in hounds and harriers, 485 n 1; Instinctive love of man, 489; Native cannot learn to leave sheep alone, 487

Interbreeding: Ill effects, 36; on the owl-pigeon, 37

Traits of domestic animals become hereditary, 483 n; Domestic compared to wild rabbit, 486

Secondary sexual characters: Lost by interbreeding, 37; Sexual selection, 317, 335; Sexual selection in animals endowed with will and choice, 376; Variable, 315 *et seq.*, 338; in stag beetles mandibles related to larval nourishment, 366; and to male virility, 366

Seeman, Berthold

Tropical and northern floras mingle on Panamanian mountains, 550 n 1

Selby, —

Viability and distribution of pheasant in Northumberland, 438

Selys-Longchamps, Edm. de

Hybrids: Anatidae, 432 n 4; Anser cygnoides × A. canadensis, 426 n 1

Senecio (ragwort, groundsel): Outer florets not always present, 326; Protean genus, 106, 170

Separation of sexes, *see* Hermaphroditism

Seringe, Nicholas Charles

Trees with different characters do not graft easily, 419 n 2

Serranus (sea bass), hermaphroditism, 44

Sharp, —

Difficult to produce pure seed, 51 n 1, 70

Sheppard, Revett

Multicoloured mole, 117 n 2

Variability of birds' eggs, 110 n 7, and nests, 505

Sheppard, Revett, and Whitear, William

Deformed mandibles of birds, 206 n 1

Sherwill, Major W. S.

Hybrids of Clamator, 436 n 1

Shuckard, William Edward

Insects of different genera found in union, 428 n 2

Variability of neuration in insect wing, 328, 335

Siemuszowa-Pietruski, Stanisl. Const.

Badgers breed in captivity, 76 n 5

Sidebotham, Joseph

Conversion in Primula, 130–3

Smith, J. E. (*cont.*)
  Lychnis diurna and L. vespertina are varieties, 404
  On classification of British flora, 113
  Quercus robur and Q. sessiflora have different timber, 118
  Treats some of Henslow's varieties as species, 159, 168
Smith, Sydney
  Increase of sheep and cattle in New South Wales, 177 n 5
  Social Plants: see Colonies
Solanaceae, Varieties in large and small genera, 153 table
Solanum (nightshade), anomalous development of stamens in one species is normal for
    another, 327
Somerville, Alexander
  Idiosyncrasy inherited in cow, 481n
Sonnerat, Pierre
  Increase of animals introduced to Mauritius, 178 n 4
Sorbus, grafting onto allied species increases fecundity, 420
Sorex (shrew), variable length of intestine, 111
Spach, —
  Allied species of Matthiola, 403
Spallanzani, Lazaro
  Calculation of spermatozoa needed to fertilise a frog's egg, 46
Species: Definition of naturalists, 95, Darwin's, 164; Dependent, 187, 199, 206;
    Highly organised have narrow ranges, 144; Protean, 102, 113, 160, 252, 284;
    Representative, 227, 577 *et seq.*, communication between, 584; Wide ranging and
    common most variable, 134, 140
Spence, William
  see Kirby and Spence
Sphegidae, *see* Fabre
Spicer, John W. G.
  Crosses of grouse, 428 n 1
Spirifer rostratus (brachiopod), variations, 106
Spratt, Thomas A. B. and Forbes, Edward
  European plants on Lycian mountain, 552 n 1
Sprengel, Christian Konrad
  Allied species of Lychnis, 404
  Cross fertilisation: in legumes, 68; possible in Campanulaceae and Phyteuma, 59
  Dichogamy: 47, 51, 205; In apparently hermaphrodite trees, 61; In Epilobium and
    Oenothera, 59; In Parnassia, 54; Tested by other hybridisers, 47
  Insect attractants: Exterior florets of Mussaenda, 301; Intoxicating nectar, 52;
    Nectaries in bracts of legumes, 55 n 1; Streaks on petals are guides to nectaries, 52
  Insect fertilisation, Bees for Epilobium, Oenothera, 59, and Althaea, 65; Flies for
    Aristolochia, 64, and Epipactis, 66; Hymenoptera for Listera, 66; Necessity in
    some plants, 52; Viola odorata an example, 56–7
  Orchis militaris shows poor fertility compared to other orchids, 67 n 3
  Pericarps may differ on same plant, 300 n 2
Squalus acanthias (picked dogfish), swarms yet lays few eggs, 207
Staphylea, doubling of flowers, 86
Stark, James
  Modified caudal muscle of ray similar to electric organ, 363 n 5
Steenstrup, Johan Japatus Smith
  Alternation of generations, 217
Stenhouse, John
  Chemical variation in a lichen, 363
Stephens, James Francis
  Mandibles of male stag-beetles precisely formed, yet variable, 313
  Variability of Necrodes littoralis, 314 n 2
Sterna minuta, variable nesting habits, 505

Wilson, William
  Andromeda polifolia sometimes has stamens attached to corolla, 327 n 1
  Campanulaceae fertilised after flower opens, 59
Wingfield, William and Johnson, G. W.
  Pheasant prepotent over common fowl, 457 n 1
Wollaston, Thomas Vernon *Note*: All entries refer to Madeira
  Coleoptera: Diversity on Madeira, 124–5, 231, 292; Flightless – families with weakest
    fliers have most endemic species, 253; winged mainland species apterous on
    Madeira, 291, apterous condition still more prevalent on smaller islets, 292;
    Lurid coloration near the coast, 282, 283; Mainland families absent from Madeira
    contain strongest fliers, 292
  Colonies of land snails found living and fossil confined to single hillock, 201–2
  Erica cinerea near summit of Madeira, 552 n 1
  Isolation, species and varieties chiefly created in, 252; Barriers to insects, 15 n 1;
    Distance apart not a useful criterion for judging species in itself, 116
  Species: Change ascending a mountain or approaching a pole, 282; Percentage with
    varieties in large and small genera, 151 table, 170; Wide ranging most variable,
    135 n 1
  Sub-species a valid category for doubtful forms, 99
  Variable insect structures: Coloration of Gymnaetron, 327; Connateness of
    elytra, 112, in Harpalus, 313; Head of female Eurygnathus, 313; Mandible size
    in Scarites, 313; Number of joints in antenna, 567
  Varieties: Arbitrarily designated "technical varieties", 104; Fleeting variations not
    varieties, 159, 160; Have smaller range than parent form, 138; Intermediate
    between species, 124–5; rarity of individuals, 268
Wood, Searles
  Variability not constant in fossil shells, 106, 107
Woods, J.
  Variable seeds from same plant of Atriplex, 358 n 2
Woodward, Samuel Pickworth
  Brachiopods present little variation, 106
  Geographical distribution: Modiolarca trapezina common to Falklands and
    Kerguelen's Land, 563 n 1; Sea-shells shared by Europe and N. America, 539
  Pulmonary sac combined with gills in Ampullaria, 356 n 3
Wrangel, Ferdinand Ludwig von
  Geese feigning death, 497 n 6
  Reindeer checked by mosquitoes, 183 n 2
  Siberia: Directional sense of natives, 492–3; Long life of horses, 181 n 7; Tundra
    plants with edible berries, 383

Yarrell, William
  Charadrinus minor and C. hiaticula have different tail feathers, 327
  Egg production in cod-fish, 207 n 1
  Hybrids of common × changeless swan sterile, 433 n 2
  Interbreeding nearly causes extinction of owl-pigeon, 37
  Nesting: Birds leave nests if surroundings warm enough, 501 n 1; Laying eggs in
    others' nests, 507 n 5; not invariable in cuckoos, 506–7; Lining varies according
    to species of wren, 502; Nests vary from general form of species, 505; examples,
    black swan, 504, yellow bunting nests on bare ground, 504
  Use of reason by gull, 471 n 3
  Variable bill of Loxia europaea, 310 n 5
Youatt, William
  Cattle: Bakewell's breeding procedure, 369 n 3; Deterioration of Bakewell's
    cattle through interbreeding, 37; Extinction of breeds, 226; Herefordshire and
    Devonshire breeds of oxen, 369 n 4; Short- × long-horned cattle, 443 n 1
  Greyhound line acquires strength from single cross with bulldog, 369 n 4

Youatt, W. (*cont.*)
    Sheep: Hereditary homing instinct during lambing season, 493; Lambs grazed without mothers liable to eat poisonous herbs, 475 n 1

Zannichellia palustris (horned pondweed), variability, 108
Zanthoxylum monogynum, variability, 109
Zarco, J. G.
    Rabbits increase on Porto Santo, 178 n 1
Zea (maize): Gärtner on infertility of crossed varieties, 405–6; Number of seeds produced, 176 n 2; Pollen production, 72 n 3
Zephyranthes candida, thrives under many conditions, 84
Ziziphora, conversion, 127 n 2
Zostera oceanica (eel-grass), possibility of occasional crossing, 63 n 2 and n 4
Zuccarini, Joseph Gerhard
    American and European forms represented in Japan, 543 n 1
    S. African Oxalis sterile in Europe, 82